Lecture Notes in Computer Science   11376

Commenced Publication in 1973
Founding and Former Series Editors:
Gerhard Goos, Juris Hartmanis, and Jan van Leeuwen

Advanced Research in Computing and Software Science

Subline of Lecture Notes in Computer Science

More information about this series at http://www.springer.com/series/7407

Barbara Catania · Rastislav Královič
Jerzy Nawrocki · Giovanni Pighizzini (Eds.)

SOFSEM 2019:
Theory and Practice
of Computer Science

45th International Conference on Current Trends
in Theory and Practice of Computer Science
Nový Smokovec, Slovakia, January 27–30, 2019
Proceedings

 Springer

Editors
Barbara Catania 🆔
University of Genoa
Genoa, Italy

Jerzy Nawrocki
Poznań University of Technology
Poznań, Poland

Rastislav Královič 🆔
Comenius University
Bratislava, Slovakia

Giovanni Pighizzini 🆔
Università degli Studi di Milano
Milan, Italy

ISSN 0302-9743 ISSN 1611-3349 (electronic)
Lecture Notes in Computer Science
ISBN 978-3-030-10800-7 ISBN 978-3-030-10801-4 (eBook)
https://doi.org/10.1007/978-3-030-10801-4

Library of Congress Control Number: 2018965781

LNCS Sublibrary: SL1 – Theoretical Computer Science and General Issues

This Springer imprint is published by the registered company Springer Nature Switzerland AG
The registered company address is: Gewerbestrasse 11, 6330 Cham, Switzerland

Preface

This volume contains the invited and contributed papers selected for presentation at SOFSEM 2019, the 45th International Conference on Current Trends in Theory and Practice of Computer Science, which was held during January 27–30, 2019, in Nový Smokovec, High Tatras, Slovakia.

SOFSEM (originally SOFtware SEMinar) is an annual international winter conference devoted to the theory and practice of computer science. Its aim is to present the latest developments in research for professionals from academia and industry working in leading areas of computer science. While being a well-established and fully international conference, SOFSEM also maintains the best of its original Winter School aspects, such as a high number of invited talks, in-depth coverage of selected research areas, and ample opportunities to discuss and exchange new ideas. SOFSEM 2019 was organized around the following three tracks:

- Foundations of Theoretical Computer Science (chair Giovanni Pighizzini)
- Foundations of Data Science and Engineering (chair Barbara Catania)
- Foundations of Software Engineering (chair Jerzy Nawrocki)

With these three tracks, SOFSEM 2019 covered the latest advances in both theoretical and applied research in leading areas of computer science.

An integral part of SOFSEM 2019 was the traditional Student Research Forum (chair Roman Špánek) organized with the aim of giving students feedback on both the originality of their scientific results and on their work in progress. The papers presented at the Student Research Forum are published in separate local proceedings.

The SOFSEM 2019 Program Committee (PC) consisted of 70 international experts, representing the track areas with outstanding expertise. The committee undertook the task of assembling a scientific program for the SOFSEM audience by selecting from the 92 submissions entered in the EasyChair system in response to the call for papers. The submissions were carefully reviewed with approximately three reviews per paper, and thoroughly discussed. Following strict criteria of quality and originality, 35 papers were accepted for presentation as regular research papers. Additionally, based on the recommendation of the chair of the Student Research Forum, seven papers were accepted for presentation in the Student Research Forum.

SOFSEM 2019 added a new page to the tradition of SOFSEM dating back to 1974, which was possible thanks to the effort of many people. As editors of these proceedings, we are grateful to everyone who contributed to the scientific program of the conference. We would like to thank the invited speakers Uwe Assmann, Francesco Bonchi, Flavio Chierichetti, Christos Kapoutsis, Miroslaw Staron, and Martin Theobald for presenting their work to the audience of SOFSEM 2019. We thank all authors who submitted their papers for consideration. Many thanks go to the PC, and to all external referees, for their precise and detailed reviewing of the submissions. The work of the PC was carried out using the EasyChair system, and we gratefully acknowledge this contribution. Special thanks are due to Roman Špánek for his expert preparation and handling of the Student Research Forum, and to the SOFSEM Steering Committee headed by Július Štuller, for its support throughout the preparation of the conference.

We are also indebted to the Organizing Committee led by Dana Pardubská.

Finally we want to thank the Slovak Society for Computer Science, and the Faculty of Mathematics, Physics and Informatics of the Comenius University in Bratislava for their invaluable support.

January 2019

Barbara Catania
Rastislav Královič
Jerzy Nawrocki
Giovanni Pighizzini

Organization

Steering Committee

Barbara Catania	University of Genoa, Italy
Miroslaw Kutylowski	Wroclaw University of Technology, Poland
Tiziana Margaria-Steffen	University of Limerick, Ireland
Branislav Rovan	Comenius University, Slovakia
Petr Šaloun	Technical University of Ostrava, Czech Republic
Július Štuller (Chair)	Academy of Sciences, Prague, Czech Republic
Jan van Leeuwen	Utrecht University, The Netherlands

Program Committee

Program Chair

Rastislav Královič	Comenius University, Slovakia

Track Chairs

Barbara Catania	University of Genoa, Italy
Giovanni Pighizzini	University of Milan, Italy
Jerzy Nawrocki	Poznań University of Technology, Poland

Student Research Forum Chair

Roman Špánek	Technical University of Liberec, Czech Republic

Program Committee

Fabio Anselmi	Italian Institute of Technology, Italy
Ladjel Bellatreche	ISAE-ENSMA, Poitiers, France
Mária Bieliková	Slovak University of Technology in Bratislava, Slovakia
Stefan Biffl	Vienna University of Technology, Austria
Miklós Biró	Software Competence Center Hagenberg, Austria
Joan Boyar	University of Southern Denmark, Denmark
Stephane Bressan	National University of Singapore, Singapore
Francesco Buccafurri	University of Reggio Calabria, Italy
Davide Buscaldi	LIPN, Université Paris 13, Sorbonne Paris Cité, France
Barbara Catania	University of Genoa, Italy
Alfredo Cuzzocrea	University of Trieste, Italy
Jurek Czyzowicz	Université du Québec en Outaouais, Canada
Adam Dabrowski	Poznań University of Technology, Poland
Johann Eder	Alpen Adria Universität Klagenfurt, Austria

Michele Flammini	Gran Sasso Science Institute and University of L'Aquila, Italy
Paola Flocchini	University of Ottawa, Canada
Pierre Fraigniaud	CNRS and University of Paris Diderot, France
Johann Gamper	Free University of Bozen-Bolzano, Italy
Leszek Gasieniec	University of Liverpool, UK
Pawel Gawrychowski	University of Wroclaw, Poland
Loukas Georgiadis	University of Ioannina, Greece
Giovanna Guerrini	University of Genoa, Italy
Inge Li Gørtz	Technical University of Denmark, Denmark
Yo-Sub Han	Yonsei University, South Korea
Theo Härder	TU Kaiserslautern, Germany
Gabriel Istrate	West University of Timisoara, Romania, and the eAustria Research Institute, Austria
Mirjana Ivanovic	University of Novi Sad, Serbia
Johan Jeuring	Open Universiteit Nederland and Universiteit Utrecht, The Netherlands
Jarkko Kari	University of Turku, Finland
Ralf Klasing	CNRS and University of Bordeaux, France
Dennis Komm	ETH Zurich, Switzerland
Georgia Koutrika	Athena Research Center, Greece
Rastislav Královič	Comenius University, Bratislava, Slovakia
Orna Kupferman	Hebrew University, Israel
Martin Kutrib	Institut für Informatik, Universität Giessen, Germany
Martin Lange	University of Kassel, Germany
Julia Lawall	Inria/LIP6, France
Andrzej Lingas	Lund University, Sweden
Lech Madeyski	Wroclaw University of Science and Technology, Poland
Yannis Manolopoulos	Open University of Cyprus, Cyprus
Tomáš Masopust	Palacký University, Olomouc, Czech Republic
Elvira Mayordomo	Universidad de Zaragoza, Spain
Paolo Missier	Newcastle University, UK
Nelma Moreira	University of Porto, Portugal
Xavier Muñoz	Universitat Politècnica de Catalunya, Spain
Jerzy Nawrocki	Poznań University of Technology, Poland
Boris Novikov	St.-Petersburg University, Russia
Mirosław Ochodek	Poznań University of Technology, Poland
Dana Pardubská	Comenius University, Slovakia
Andrea Pietracaprina	University of Padua, Italy
Giovanni Pighizzini	University of Milan, Italy
Andrei Popescu	Middlesex University London, UK
Rajeev Raman	University of Leicester, UK
Gunter Saake	University of Magdeburg, Germany
Philippe Schnoebelen	CNRS, France
Shinnosuke Seki	The University of Electro-Communications, Japan

Arseny Shur	Ural Federal University, Russia
Daniel Stefankovic	University of Rochester, USA
Krzysztof Stencel	University of Warsaw, Poland
Ernest Teniente	Universitat Politècnica de Catalunya, Spain
Martin Theobald	University of Luxembourg, Luxembourg
Panos Vassiliadis	University of Ioannina, Greece
Valentino Vranić	Slovak University of Technology in Bratislava, Slovakia
Dorothea Wagner	Karlsruhe Institute of Technology, Germany
Bruce Watson	Stellenbosch University, South Africa
Abuzer Yakaryilmaz	University of Latvia, Latvia
Tomoyuki Yamakami	University of Fukui, Japan
Christos Zaroliagis	Computer Technology Institute and Department of Computer Engineering & Informatics, University of Patras, Greece
Norbert Zeh	Dalhousie University, Canada
Wolf Zimmermann	Martin Luther University Halle-Wittenberg, Germany

Organization Chair

Dana Pardubská	Comenius University, Slovakia

Organizing Institutions

Slovak Society for Computer Science
Faculty of Mathematics, Physics, and Informatics, Comenius University in Bratislava, Slovakia

Additional Reviewers

Aceto, Luca
Aloisio, Alessandro
Badouel, Eric
Balkenius, Christian
Bampas, Evangelos
Barth, Lukas
Berg, Christian
Bilò, Davide
Böckenhauer, Hans-Joachim
Cho, Da-Jung
Czyzewski, Andrzej
Das, Shantanu
Demri, Stéphane
Dimokas, Nikos

Fabrega, Josep
Fiol, Miquel Angel
Forišek, Michal
Fribourg, Laurent
Gainutdinova, Aida
Georgiou, Konstantinos
Giannis, Konstantinos
Gualandi, Stefano
Guillon, Bruno
Hundeshagen, Norbert
Jajcayova, Tatiana
Kapoutsis, Christos
Kawachi, Akinori
Killick, Ryan

Kim, Hwee
Ko, Sang-Ki
Komusiewicz, Christian
Kowaluk, Mirosław
Krüger, Jacob
Kľuka, Ján
Li, Yang
Loff, Bruno
Luettgen, Gerald
Malcher, Andreas
Mchedlidze, Tamara
Meister, Andreas
Mráz, František
Noceti, Nicoletta
Pająk, Dominik
Papadopoulos, Charis
Plachetka, Tomáš
Prigioniero, Luca
Prūsis, Krišjānis
Rafailidis, Dimitrios

Raghavendra Rao, B. V.
Rástočný, Karol
Rogalewicz, Adam
Rovetta, Stefano
Salehi, Özlem
Salo, Ville
San Felice, Mário César
Schmidt, Paweł
Toft, Bjarne
Tomás, Ana Paula
Tsichlas, Kostas
Tzouramanis, Theodoros
Ueckerdt, Torsten
van Ee, Martijn
Villagra, Marcos
Vinci, Cosimo
Wehnert, Sabine
Wendlandt, Matthias
Wong, Tom
Zetzsche, Georg

Contents

Cross-Layer Adaptation in Multi-layer Autonomic Systems (Invited Talk)

Uwe Aßmann[1]([✉]), Dominik Grzelak[1], Johannes Mey[1], Dmytro Pukhkaiev[1], René Schöne[1], Christopher Werner[1], and Georg Püschel[2]

[1] Institut für Software- und Multimediatechnik, Technische Universität Dresden, Dresden, Germany
{uwe.assmann,dominik.grzelak,johannes.mey,dmytro.pukhkaiev,rene.schoene, christopher.werner}@tu-dresden.de
[2] Wandelbots, Dresden, Germany
georg.pueschel@wandelbots.de

Abstract. This work presents a new reference architecture for multi-layer autonomic systems called *context-controlled autonomic controllers (ConAC)*. Usually, the principle of multiple system layers contradicts the principle of a global adaptation strategy, because system layers are considered to be black boxes. The presented architecture relies on an explicit context model, so a simple change of contexts can consistently vary the adaptation strategies for all layers. This reveals that explicit context modeling enables consistent meta-adaptation in multi-layer autonomic systems. The paper presents two application areas for the ConAC architecture, robotic co-working and energy-adaptive servers, but many other multi-layered system designs should benefit from it.

1 Introduction

Self-adaptive software systems (SAS) are used in many application areas, from autonomous driving [15] over co-working robotics [2] to energy-proportional servers and clouds [3]. It seems to be commonly understood that all SAS are *context-aware*, i.e., feel their context and adapt to its changes [29]. For instance, autonomous cars have to adapt to changing environments; their sensors have to feel obstacles, and their control has to take alternate routes or stop the vehicle to avoid accidents. Similarly, in robotic co-working, when robots are acting free of cages and collaborate directly with human workers, they have to feel the movements in their environment and self-adapt to changes of human positions, so

This project has received funding from the ECSEL Joint Undertaking under grant agreement No. 692480 (IoSense). This Joint Undertaking receives support from the EU Horizon 2020 research and innovation programme and Germany, Spain, Austria, Belgium, Slovakia. Also supported by the German Research Foundation (DFG) in the CRC 912 "Highly Adaptive Energy-Efficient Computing", the project RISCOS, the Research Training Group "Role-based Software Infrastructures for continuous-context-sensitive Systems (RoSI)", as well as the BMBF project OpenLicht.

© Springer Nature Switzerland AG 2019
B. Catania et al. (Eds.): SOFSEM 2019, LNCS 11376, pp. 1–20, 2019.
https://doi.org/10.1007/978-3-030-10801-4_1

that dangerous encounters are avoided. Also energy-proportional servers should be self-adaptive systems, because they must adapt their energy consumption to changes in their load. Because their load usually varies a lot throughout the work day, their energy consumption should follow the changes of the load in a reasonable way. It seems that architectures for such context-aware, self-adaptive, and self-optimizing software systems will play an important role in the software of the future. However, SAS differ in the self-* properties they support [29]. For instance, autonomous car systems and co-working robots solely guarantee functional safety (*self-governing system*), while an energy-proportional server system is a *self-optimizing system*, improving on the quality of the system's results.

Since several years, reference architectures for SAS are under development in the community. One of the most important reference architectures for SAS are *autonomic software systems*, which use an *autonomic controller* implementing a MAPE-K loop[1] to control the adaptation [17,23]. In particular, we focus on *autonomic software product lines (ASPL)* [1] in which the autonomic controller dynamically manages the variation of several *code variants* of a dynamic software product line.

In this paper, we deal with self-adaptive or autonomic architectures for *multiple system layers* in *Multi-Layer Autonomic Systems (MuLAS)* [34], in which a *system layer* is considered to be a loosely coupled, potentially independently developed, black-box subsystem (or component) that talks to other layers only via exchange of messages, data flow or web-service calls.[2] In a MuLAS, every layer is self-adaptive on its own, i.e., runs an autonomic controller following a specific adaptation strategy. To avoid *strategic conflicts* between the layers, the local, layer-specific adaptation strategies must somehow be coordinated by a global *meta-adaptation*, typically the task of a *global adaptation controller* who manages the autonomic controllers of each layer systematically. If the meta-adaptation manages to coordinate all system layers *consistently*, we call this *cross-layer adaptation*.

[34] calls a design structure of a MAPE-K loop a MAPE *pattern*. We follow this approach and specify three MAPE patterns for cross-layer adaptation of MuLAS. The main MAPE pattern of the paper, *context-controlled autonomic controllers (ConAC[3])*, serves for consistent cross-layer adaptation in hierarchic and multi-layer autonomic systems (MuLAS); the other MAPE patterns serve for self-optimization. Our contribution is that we use *explicit context modeling* [16,30] to specify self-adaptation behavior systematically, so that, during meta-adaptation, the self-adaptation of all layers is consistently changed across all levels of a MuLAS. Also, we show, how to integrate a KNOWLEDGEBASE into the

---

[1] In the literature, *autonomic controllers* or *autonomic managers* run a MAPE-K-loop with MEASURE-ANALYZE-PLAN-EXECUTE functions communicating via a KNOWLEDGEBASE to self-adapt a software system.

[2] Usually, hierarchic autonomic systems can be considered as a subclass of multi-layered systems, because the latter may share components on lower levels, i.e., their use relationship is a directed acyclic graph instead of a tree.

[3] ConAC is intended to be spelled ['kɔnjak].

patterns, going beyond pure MAPE functionality [34]. Thirdly, we show that *role-oriented programming* [32] provides a quite natural abstraction concept for cross-layer adaptation strategies. Fourthly, we hint at several implementation patterns in classical object-oriented languages, if no context- or role-based programming language is at hand.

Using Context Modeling to Model Multiple Layers and Cross-Cutting. In all presented MAPE patterns, the layers of a MuLAS as well as their cross-cutting contexts are interpreted as entries in a *context model for cross-layer adaptation*:

(System) Layers are contexts grouping and hiding subsystems, components, objects, and functions. Layers are independent and stacked on other layers. The layer itself may contain sublayers, i.e., intra-layer contexts. Layers may hide or share sublayers (hierarchically or partially ordered). Usually, in a system, the number of layers will be fixed.

Cross-cutting contexts do not belong to one layer, but cross-cut several layers, relating components, objects, and functions that are distributed over several layers. They do not *encapsulate*, but only *correlate* functionality. The number of cross-cutting contexts in a MuLAS system may be unknown. Contexts need not be hierarchically organized nor partially ordered, but can crosscut each other.

Therefore, every object in a ConAC application is always related to at least two contexts: its layer and one or several cross-cutting contexts. This implies that autonomic controllers of different layers can share the same cross-cutting contexts. Because cross-layer contexts couple cross-layer behavior, the behavior of autonomic controllers can be coupled and consistently varied. This coupling is the key to consistent cross-layer adaptation.

Paper Roadmap. First, we discuss state of the art (Sect. 2.1) and present two case studies in Sect. 2.2, robotic co-working and energy-adaptive servers. To prepare the presentation of the ConAC MAPE pattern, we repeat some important definitions from the literature (Sect. 2.3). ConAC, is presented in Sect. 3. The variant for self-optimizing systems, qConAC, is discussed in Sect. 4, as well as the one for energy-proportional servers, eConAC, in Sect. 5.

2   State of the Art

2.1   Multi-layer Autonomic Systems

This section discusses the state of the art in describing and constructing autonomic systems with one or more subsystems.

Autonomic Systems and the MAPE Patterns. A *dynamic software product line (dynamic SPL)* consists of a set of products (variants) that are bound and varied dynamically [6]. Its variability mechanism goes beyond dynamic binding and variation of system parameters, and forms the basis of an SAS. An *autonomic software product line (ASPL)* [1] is a dynamic SPL which contains a MAPE control loop to change variants of a controlled subsystem. An ASPL can be *quality-adaptive* or *self-optimizing*, i.e., the planning of the new variant (configuration) is controlled by an *objective model* describing the desired quality of the applications that run on the system (*self-optimizing SPL*) [1]. Examples for such objectives are *utilities* the system should deliver, such as maximized speed, the *resources (cost)* the system should spent, such as minimized energy consumption, or the *cost-utility objective* of the system, such as to maximize its speed-energy efficiency.

ASPL may be designed with many different forms of MAPE loops, for which MAPE *patterns* can be employed [34]. A MAPE pattern is a static pattern on the type level that can be instantiated to a *pattern instance* on the object level. [34] presents several patterns that structure the MAPE control loops of subsystems in *regions*: (i) The *regional planning pattern* controls subsystems in regions with a combined PLAN function for each region. While the regions structure the systems, the interaction of different planners is possible but unstructured. (ii) In contrast, the *hierarchical control pattern* structures the MAPE control loops hierarchically, i.e., a MAPE loop controls either a subsystem or other loops. While the MAPE patterns of [34] are well-suited to model multi-layer systems, they solely treat MAPE functionality, but abstract from the notions of *knowledge*, *state* and *context*.

Multi-layer Autonomic Systems (MuLAS). In this paper, we rely on a simple MAPE pattern, the *autonomic controller (autonomic manager)* [7], a specific object with functions for the different phases of the MAPE-K loop of a controlled system:

Definition 1. *An* autonomic controller (object) *is an object that offers clearly demarcated functions for all activities of an autonomic* MAPE-K *(control) loop: the activities* MEASURE, ANALYZE, PLAN, *and* EXECUTE, *as well as* KNOWLEDGE *management based on a* KNOWLEDGEBASE.

An autonomic quality controller *is an autonomic controller equipped with a model for a quality objective of the system. We call a MuLAS with a quality objective a* multi-layer self-optimizing system (MuLOS).

An autonomic meta-controller *(object) is an autonomic controller that runs a* MAPE-K *loop over many autonomic controllers of different system layers or components.*

The first MAPE patterns for MuLAS have appeared [5,13]. In these patterns, the *meta-controller* coordinates the subordinated autonomic controllers of a multi-layer system, i.e., it adapts their reconfiguration strategies. [24] presents a three-layer controller structure with a provisioning controller, application controllers, and component controllers, and discusses several variants for MAPE control loops with different performance models, e.g., queue-based models, static and dynamic models, and policy-based models. In particular, the approach supports the definition of quality objectives in all layers.

Few approaches on cross-layer adaptation of ASPL exist. [11] defines the concept of *cross-layer autonomic manager (CLAM)* for a meta-controller. Beyond parameter adaptation, this approach also adapts workflows on each layer, changing the functional behavior of a multi-layer system considerably. [1] shows that also ASPL can be organized hierarchically, however, restricts itself mostly to the design-time porting of an ASPL to other application domains.

2.2 Examples of MuLAS

Self-adaptive Co-working Robots (Cobots). *Cobots* are robots that collaborate with humans, without being fenced by safety cages. Therefore, they must take care of human movements and self-adapt their functionality to them, for instance, by slowing down their speed or by taking alternate routes avoiding an encounter. To this end, cobots are equipped with sensor networks that recognize the movement of the humans (*smart environment*). [12] expounded a safety architecture with a safety automaton for cobots. Whenever a human approaches the cobot, the smart environment around the cobot detects the approach and slows down or stops the cobot, transitioning between the safety states *autonomous mode, human-friendly mode, collaborative mode, and failure mode* (Fig. 1).

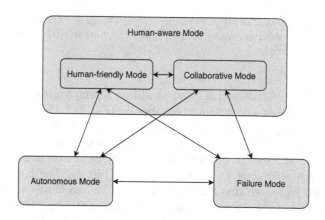

Fig. 1. Robotic co-working in Haddadin's safety automaton.

Looking at the Haddadin's safety architecture from the viewpoint of an autonomic system, it turns out to be a automaton-based MAPE-K loop for a single-layer autonomic system, as in [4]: (i) The afferent MEASURE function is realized by the sensors of the robot perceiving the environment. (ii) The input data streams are processed to detect complex events (ANALYZE function), e.g., to discover the presence of humans, and fed into the automaton's central event input stream. (iii) Based on previous and current status encoded in the states (KNOWLEDGEBASE), the cobot tries to make appropriate transition decisions

(PLAN function). (iv) Within the EXECUTE functions, called along the transitions, the actuators of the robot are triggered, for example, to decrease its velocity.

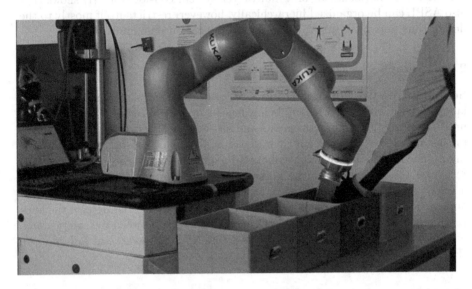

Fig. 2. Robotic co-working in the Cinderella case study. Every box has an associated autonomic controller watching safety of encounters of a human with robot. Box 1 is on the left, box 4 on the right.

In the case study Cinderella [27], we produced a prototypical pick-and-place system in which a cobot and a human collaborate (Fig. 2).[4] The four boxes store items for a pick-and-place task of the cobot, for example, to sort items according to their size. The cobot has two workflows: It can pick an item from boxes 2 or 3 (in the middle) and sort it, if surpassing a threshold in size, into another box, either on the left side (box 1) or the right side (box 4). The human wears a sensor-equipped jacket, so that the position of the human hand can be tracked. The cobot has to feel her and self-adapt its behavior according to her actions: (i) As soon as the human enters the room, the robot will switch from *autonomous* to *human-friendly mode* (driving slow). (ii) It will change from *human-friendly mode* into *collaborative mode* or into *failure mode*, if the human puts a hand into a box, e.g., to correct the cobot's decisions or to remove broken items. Collaborative mode means that if a human puts the hand into a box on the left side (box 1 or 2), the cobot can work on the right side (boxes 3 and 4), and vice versa. Thus, in the collaborative mode, the robot can switch its workflow to a second variant. The *failure mode* (stopping) is activated, if the cobot and the human's hand meet in the same box. Because such an encounter can harm the

---

[4] A video is found on https://www.youtube.com/watch?v=zk3ruVSTwCo.

human, the cobot must gracefully stop. The *failure mode* is switched again to the *collaborative mode*, when the human moves her arm outside of the current box.

In fact, cobots are *multi-layer self-adaptive systems* where the cobot's overall behavior can be modeled by a global MAPE-K loop. However, the Cinderella case study reveals that a single safety automaton is too simplistic for cobots. To be able to analyze the sensor data correctly, cobots need a *world model*, a model of the environment with further automata describing the states of the elements of the world [2]. In Cinderella, the four boxes form spaces, in which a human's hand and a cobot can meet. To model these spaces of encounter, the software architecture of Cinderella introduces a world model with a safety automata for every box. Cinderella discovers an encounter, because its smart environment continuously monitors the position of the human's hand as well as that of the cobot arm, and whether they meet in a box of the world model. In other words, Cinderella introduces an safety automaton for each of the four boxes, analyzing human movements and discovering the complex event of a near-encounter. Therefore, the architecture of the Cinderella software needs at least the following system layers:

Smart-Environment System. This layer comprises a world model with a safety automaton for every space of an encounter.

Robot System. This layer comprises the robot's operating system that drives the motors of the cobot and several robot subsystems.

Robot Application Workflow. In this layer, the overall workflow of the cobot, e.g., sorting items in boxes, is executed and adapted in order to avoid encounters between cobot and human.

All layers have to be adapted consistently: In case of an encounter of a human and a cobot, all layers have to immediately and *consistently* self-adapt their behavior. We will model more details of cobots in Sect. 3.

Energy-Proportional Servers as Cross-Layer Autonomic Systems. An energy-proportional server system consumes energy proportionally to its load [3]. An energy-proportional server is a large-scale full-fledged energy-adaptive system with multiple layers: Operating system, virtual machines (cloud level), database, application layer shall collaborate to tune a system either for performance, throughput, energy savings, or for a trade-off between all these qualities. Thus, an energy-proportional server is an example of a multi-layer self-optimizing system (MuLOS).

Energy-proportional servers can utilize several strategies for energy saving. In the research center "Highly-Adaptive Energy-Efficient Computing (HAEC)" in Dresden [8], several optimization techniques are investigated. The first technique is *pack-and-switch-off*, i.e., concentrate the load on a subset of the available hardware to switch off some of the computing components. To this end, the HAEC project develops special hardware that can be switched off or on with minimum delay. The second technique is to find an energy-optimal software configuration w.r.t. energy-utility objectives. This relates to all software levels of

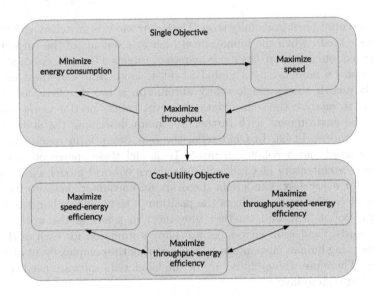

Fig. 3. An example of a meta-controller of an energy-proportional server with six quality objectives as states.

the server, operating system, database [20], compiler [9], and software application [26], which attempt to optimize towards the system's energy objectives, implementing a layer-local energy control loop in an autonomic controller. However, the configurations picked by the local controllers are heavily influenced by the global context of the server, its load, and the local adaptation strategies may contradict each other, so that a consistent cross-layer optimization is difficult. Also, whenever the energy objective of the server system changes, all layers of the system should switch to the new energy objective.

Example 1. A simple state-machine with global objectives for energy proportionality is shown in Fig. 3. The automaton has the states *minimize energy consumption, maximize speed,* and *maximize throughput* (single objectives on cost minimization or utility maximization), as well as three states for cost-utility objectives: *maximize speed-energy efficiency, maximize throughput-energy efficiency, maximize throughput-speed-energy efficiency.* Whenever a global objective of the system is changed, all autonomic controllers of its layers should be adapted accordingly. For example, if the system is in the energy-saving state and users start to require speed for their applications, the HAEC system should automatically: (i) switch on hardware units again and (ii) change the goal of the autonomic controllers at each layer to find an appropriate software configuration that runs its applications as fast as possible (see Sect. 5).

2.3   Background in Programming Technology

This subsection reviews programming technologies important for the rest of this paper.

Context-oriented Programming (COP). "Context-oriented Programming enables the expression of behavioral variation dependent on context." [16]. In a context-based object-oriented program, the execution does not only depend on the name of the method, its sender, and receiver, but also on the context (four-dimensional dispatch). [30] connects COP with autonomic software systems. To adapt, the self-adaptive system is built with application slices[5] and contexts, and the MAPE-K loop activates or deactivates slices depending on the currently active contexts. In Sect. 3, we generalize this approach to system layers.

Contexts, Roles and Teams. The notion of roles is an older research topic than COP originally stemming from linguistics. [32] collected 15 features that describe the relational and behavioral nature of roles. This publication is based on [22] which adds 11 new features to the classification of Steimann encompassing the context-dependent nature of roles. Roles are context-specific elements that extend objects (*players*) with new states and behavior, but do not form independent objects. They can always be traced back to their player and to their context. Roles interact with other roles (creating *collaboration slices*) and adapt the player's behavior according to their context. Roles are switched on or off depending on whether their context is active. Objects are able to take on, play, or drop roles at runtime and, thus, can step into, work in, and leave different contexts. These properties are useful for the implementation and modeling of runtime and self-adaptive systems. A practical role-based programming language is [14], which combines contexts and collaboration slices into *teams* of roles. Teams are a natural mechanism to implement dynamic software product lines and self-adaptive systems: Varying a team modifies the application by changing all involved roles and adapts it to new contexts. We use them in Sect. 3 to glue layers together.

Implementation Patterns for Role-Based Collaborations. Role- and team-based design simplifies the architecture of programs [33], however, the concepts are not available in classic programming languages. An implementation pattern for teams are *mixin layers*,[6] a technique to implement collaborative role-based designs based on the concept of *mixin classes* or *mixin objects*. On the type level, a mixin class describes a superclass that cannot be instantiated. Groups of mixin classes, *static mixin layers*, form static variants of a software product line [33], while *dynamic mixin layers* [31] can express dynamic variability in a dynamic SPL.

---

[5] We call the "layers" of a COP program its *application slices*, because, in this paper, layers are meant to be system layers.

[6] Again, mixin layers are application slices, while, in this paper, layers are system layers.

Dynamic mixin layers are similar to *delegation layers* [25]. In this approach, a complex object is composed of a set of collaborating sub-objects, and a mixin layer represents a variant of the set's implementation. The delegation-layer approach supports a precise dynamic semantics enabling on-the-fly extensibility. *Decorator layers* [28] are inspired by delegation layers, but can be introduced in any object-oriented language, because the decorators can be realized with the DECORATOR design pattern.

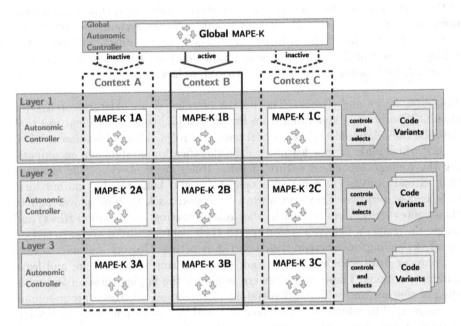

Fig. 4. Consistent multi-layer variation of autonomic controllers in a configuration space of three layers, with three variant MAPE-K role teams (or dynamic mixin layers) and their corresponding contexts.

Dynamic Aspect-oriented Programming. In, aspect-oriented programming (AOP) [19], functionality of *crosscutting concerns* is separated from the functionality of a *base program*. This separation enables a clear design, while an *aspect weaver* has to compose the aspects with the base program. Both AOP and COP need special language constructs for the modularization of crosscutting concerns. Compared to COP, which supports context-aware behavioral variation, AOP supports a general mechanism for the modularization of orthogonal functionalities. Furthermore, *dynamic AOP* deals with the activation of aspects at runtime and modifies the behavior of the base program during execution. Autonomic systems can be designed with dynamic AOP [10], because the autonomic behavior can be added as a crosscutting concern, with the execution of runtime aspects.

3 Consistent Multi-layer Variation of Autonomic Controllers with Cross-Cutting Contexts

Basically, the most simple MAPE pattern of our family, ConAC, follows the role-based team approach for dynamic software product lines, but transfers it to the autonomic controllers of a multi-layered system. This is summarized in the example of Fig. 4, which sketches an autonomic system with three layers and three autonomic controllers. It is assumed that the system runs in three global contexts, which exclude each other, but can be varied. The active context in Fig. 4 is Context B.

Definition 2. *A MuLAS structured according to the ConAC MAPE pattern has n system layers with autonomic controllers running a MAPE-K loop with MAPE-K functionality. All n MAPE-K functions of these autonomic controllers form a MAPE-K team (collaboration slice), and are related to a cross-cutting context. At most m variant contexts with m variant implementations of the MAPE-K team exist, of which only a subset is active at runtime of the MuLAS.*

In a ConAC architecture, whenever a cross-cutting context changes, all autonomic controllers are changed appropriately, by changing their MAPE-K functions of the context-related MAPE-K team. For instance, if Context B changes to Context C in Fig. 4, on each layer, the controllers MAPE-K-1B up to MAPE-K-3B are exchanged to their variants MAPE-K-1C to MAPE-K-3C. Cross-cutting contexts A-C are each related to a variant of the team of the MAPE-K functions of the layer-specific autonomic controllers. By changing a context, the MAPE-K behavior of all layers, the MAPE-K team, can be changed consistently. This leads to property 1 of the ConAC MAPE pattern:

Property 1. In a MuLAS structured according to ConAC, strategy conflicts in meta-adaptation can be avoided, because the adaptation strategies of the layers can be consistently varied by changing the global context together with the related MAPE-K team of the layer-local autonomic controllers.

The matrix-like product-line design of the ConAC pattern assures that every variant of the MAPE-K functionality of a layer-local autonomic controller is only related to *one* context, as in [14]. Under this assumption, the property follows from the principle of consistent variation of role-based teams in dynamic software product lines.

Example 2. As an example, consider the cobot scenario from Sect. 2.2. The cobot system should incorporate a global meta-controller with a MAPE-K loop implementing Haddadin's safety automaton [12], connected to autonomic controllers of the other system layers of a cobot. Figure 5 shows an appropriate meta-controller (upper layer), the Robot Application Workflow layer (middle layer), as well as the Smart-Environment System (lower layer). The global meta-controller has 7 states, belonging to contexts of the human hand with respect to the boxes and the cobot: (i) The human does not stay in the room; (ii) the human stays in the

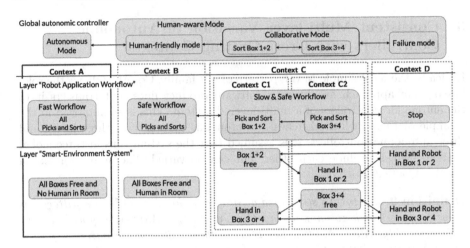

Fig. 5. Consistent self-adaptation of cobots by cross-cutting contexts.

room; (iii) the human stands near the cobot, but does not collaborate with it; (iv) the human collaborates with the cobot and all boxes for the pick-and-place task are free of the human's hands; (v) the human's hand is found in boxes 1 or 2; (vi) the human's hand is found in boxes 3 or 4; (vii) and finally, a critical state for safety: both human's hand and cobot are located in the same box.

The world model of the Smart-Environment System contains several autonomic controllers, each for a space of encounter (Fig. 5, lower layer right shows the safety automata of boxes 1+2, as well as boxes 3+4). In the Cinderella case study, the Smart-Environment System has mainly MEASURE and ANALYZE tasks, because the automata of the boxes monitor and analyse sensor signals from the 3D space of an afferent sensor network. Whenever such an analysis automaton in the Smart-Environment System signals to the meta-controller that a human's hand is detected in a box, it switches the global context accordingly. For instance, if the human grips into the box 1 or 2, the global context B changes to the subcontext C2. For the workflow layer of the cobot (Fig. 5, middle layer), this means that the global context changes the workflow so that the cobot works only with boxes 3 or 4 (middle layer, Context C2). On the other hand, if the human hand appears in the boxes 3 or 4, the global context changes to the context C1, and the workflow for pick-and-sort 1+2 is chosen. Layers are loosely coupled by signals: their MAPE-K adaptation strategy can be modified or extended independently. However, each modification of a layer has to be related to a global context - to keep the cross-layer adaptation intact.

In this architecture, a meta-adaptation step in the meta-controller's MAPE-K automaton changes the global context of all autonomic controllers of all layers of the MuLAS, as well as the involved MAPE-K team. The global context change can consistently steer other layers, e.g., in the robot's system or in other application layers.

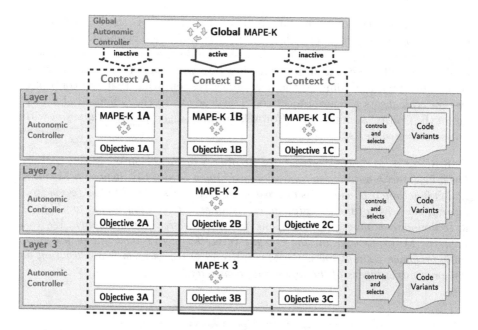

Fig. 6. Consistent quality-driven multi-layer variation of autonomic quality controllers in a configuration space of three layers. The upper layer has three variants of MAPE-K roles while Layer 2 and 3 share the MAPE-K loop. All layers have three variant objective models with management functions.

As stated above, a team can be implemented by a delegation or dynamic mixin layer. This leads us to the next property of ConAC:

Property 2. In a programming language without teams, collaborations, and roles, dynamic mixin layers or delegation layers can be used to implement the MAPE-K teams.

4 Cross-Layer Self-optimization in Multi-layer Autonomic Systems

As we have seen, the ConAC MAPE-K pattern is able to vary the adaptation behavior of a complex multi-layer autonomic system by consistently changing MEASURE, ANALYZE, PLAN, and EXECUTE functionality using crosscutting contexts. This section presents a variant of the ConAC pattern for self-optimization, in which the PLAN function optimizes the system w.r.t. an explicit *objective (goal)*, either represented by a symbolic objective or a numeric objective function in an objective model (see Fig. 6).

Definition 3. *Let the objective model of an autonomic quality controller be managed via an interface O. We call a MAPE-K loop with explicit quality objective model a MAPE-K-O loop (MAPE-K with objective model management interface).*

In a MAPE-K-O loop of an autonomic quality controller, the PLAN function depends on the objective model. In a MuLOS, an objective model of a layer expresses its quality goals. When an layer-local objective model of an autonomic quality controller is changed, its PLAN function will plan the local reconfigurations differently, according to the new objectives, ensuring that the controlled system layer meets the new objectives. This leads us to the definition of a variant of the ConAC architecture, qConAC, for *multi-layer quality adaptation* with *cross-cutting contexts*.

Definition 4. *A MuLOS structured according to the qConAC MAPE pattern has n layers with autonomic controllers running a MAPE-K-O loop with MAPE-K and O functionality, as well as m cross-cutting contexts related to two teams (collaboration slices) of the m autonomic controllers of each layer, one for MAPE-K (autonomic management) and one for O (objective management). We call them the MAPE-K team and the O team, respectively. In a modeling or programming language without roles, the MAPE-K and O teams form two delegation or dynamic mixin layers of the autonomic controllers.*

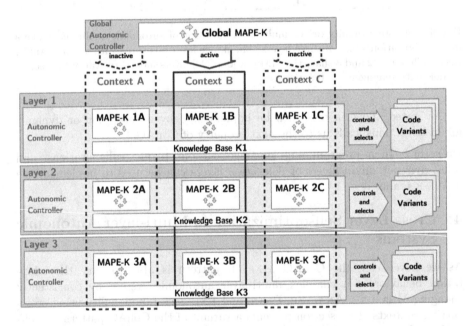

Fig. 7. Consistent multi-layer variation of autonomic controllers in a configuration space of three layers, with a KNOWLEDGEBASE (K) on all layers.

qConAC can consistently optimize the qualities of a multi-layer system w.r.t. a quality objective represented in an objective model:

Property 3. In a qConAC architecture, the behavior of the layers' self-optimization can be consistently varied by changing the O team, i.e., the objective models of all layer-local autonomic controllers, induced by the change of their contexts.

With qConAC, layers cannot self-optimize in an uncontrolled way; their objectives are correlated by a team of objective management functions and a cross-cutting context, which can be consistently changed by the meta-controller. To achieve this, qConAC realizes, via the cross-layer context, a mapping between the global and layer-local quality objectives. qConAC tracks the evolution of this mapping over time, by tracking context switches, leading to potentially complex, but consistent quality adaptations. Thus, qConAC is a MAPE pattern for consistent variation of self-optimization strategies in MuLOS (cross-layer self-optimization).

Property 4. Both teams layers of a qConAC system, MAPE-K and O team, can be varied independently.

The layers 2 and 3 depicted in Fig. 6 *share* the MAPE-K functions for all contexts, i.e., do not vary them during a context switch. On the other hand, in a multi-layer self-optimizing system with MAPE-K-O loops, the objective models should be context-specific, because self-optimization means to adapt to changing objectives. Therefore, a software architect can design a MuLOS with varying objective management functions, or additionally, also with varying MAPE-K functions.

The qConAC pattern is similar to the meta-layer architecture of [21] in that it relies on explicit objective models (goal models). However, instead of separating the objective level from the PLAN level, qConAC combines them in the concept of an autonomic quality controller with MAPE-K-O loop. This design is simpler, because compared to [21], it saves a metalevel (only 1 instead of 2), which enables qConAC to handle multi-layer systems.

5   Energy-Aware Self-optimization in Energy-Proportional Servers

To employ the qConAC pattern for *multi-layer energy adaptation (energy self-optimization)*, we have to introduce one further modification: An energy-adaptive system should have a shared KNOWLEDGEBASE so that all autonomic controllers can be informed on the global status of the system, e.g., on its current load. Thus, for an energy-adaptive MuLOS, we are interested in the case when the knowledge management task is divided into objective management (for managing the objectives of the controlled system) and in status data management (for managing the status data in the "digital twin" of the controlled system, Fig. 7). Of course, the objective function of a MuLOS has to express objectives on energy consumption and delivery of utility (see Fig. 3).

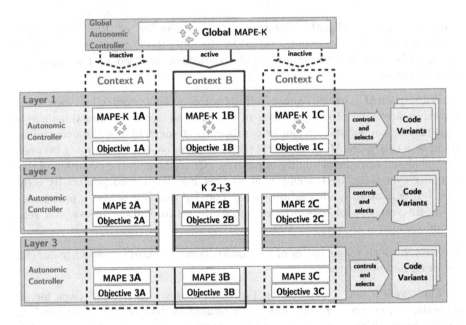

Fig. 8. Consistent multi-layer variation of autonomic energy controllers in a configuration space of three layers, with objectives on all layers, with a KNOWLEDGEBASE (K) shared between layers 2 and 3.

Definition 5. *An* autonomic energy controller *is an autonomic quality controller that offers an energy-utility objective function, as well as an explicitly layer-shared status model S (*MAPE-K-O-S, MAPE-K-O *with layer-shared status model).*

A MuLOS structured according to the eConAC pattern uses a qConAC architecture with MAPE-K-O-S *loops in all autonomic energy controllers.*

Extending a MAPE-K-O loop with explicit status information (system status model) has the advantage that this model can be shared within or between different layers of a MuLOS. This is important for the communication of global parameters such as load or resource pressure - on the one hand, for the local planning of resource allocation; on the other hand, for consistent global pack-and-switch-off decisions that do not only depend on energy objectives, but also on the system's status. Thus, for energy-adaptive MuLOS, not only a consistent reconfiguration of energy objectives on all layers is required, but also the precise, shared knowledge of resource pressure of all layers (Fig. 8).

Property 5. In a MuLOS structured by eConAC, layer-local MAPE-K-O-S loops as well as layer-local energy objectives can consistently be reconfigured by meta-adaptation.

Example 3. The eConAC pattern can serve as a reference architecture for servers that should support *energy proportionality*, i.e., to adapt energy consumption to the server's load, as well as other global energy objectives (Fig. 3). eConAC supports a meta-controller controlling an appropriate set of layer-local MAPE-K functions (correlated in the MAPE-K team), as well as a set of local energy objectives (correlated in the O team), related to a set of contexts. Whenever the global energy objective changes, the global context is changed, the related MAPE-K and O teams are changed, and all layer-local autonomic energy controllers are consistently varied by exchanging their MAPE-K behavior as well as their local objectives (Fig. 3).

Energy proportionality means energy consumption proportional to load. Therefore, in eConAC, the resource pressure generated by the overall load can be communicated to the layers by a shared status model, which all ANALYZE phases of the autonomic controllers take into account. To take consistent global place-and-switch-off decisions, the server system's meta-controller should evaluate global load and resource pressure, and then adapt the layer objectives consistently. Therefore, the eConAC architecture has the potential to consistently coordinate all layers of an energy-proportional server.

6   Conclusion

In this paper, we have used contexts as glue for the consistent variation of autonomic controllers. The MAPE pattern ConAC and its variants support the following design principles:

Loose Coupling of System Layers. In a ConAC architecture, layers are not aware of each other, except that in eConAC, they may share KNOWLEDGE. [18] shows how to deal with system layers that know of each other.

Model-based self-adaptation: All issues that may lead to a reconfiguration are contained in *explicit models*, not hidden in the code, e.g., contexts, quality objectives, system states.

Separation of concerns: The pattern family separates the concern of computation from the concern of self-adaptation, the concern of self-optimization, as well as from the concern of meta-adaptation.

While ConAC seems to be relatively simply structured into n layers and m cross-cutting contexts, its secret is that the contexts correlate the MAPE-K teams of the layer-specific autonomic controllers, enabling the consistent exchange of their implementations. With appropriate extensions of the basic design, quality objectives, as well as the access to shared knowledge bases, can be integrated. Thus, the ConAC MAPE pattern family enables consistent multi-layer reconfigurations for large self-adaptive and autonomic systems.

References

1. Abbas, N., Andersson, J.: Harnessing variability in product-lines of self-adaptive software systems. In: Schmidt, D.C. (ed.) Proceedings of the 19th International Conference on Software Product Line, SPLC 2015, 20–24 July 2015, Nashville, TN, USA, pp. 191–200. ACM (2015)
2. Aßmann, U., Piechnick, C., Püschel, G., Piechnick, M., Falkenberg, J., Werner, S.: Modelling the world of a smart room for robotic co-working. In: Pires, L.F., Hammoudi, S., Selic, B. (eds.) MODELSWARD 2017. CCIS, vol. 880, pp. 484–506. Springer, Cham (2018). https://doi.org/10.1007/978-3-319-94764-8_20
3. Barroso, L.A., Hölzle, U.: The case for energy-proportional computing. IEEE Comput. 40(12), 33–37 (2007)
4. Bencomo, N., Grace, P., Flores-Cortés, C., Hughes, D., Blair, G.: Genie: supporting the model driven development of reflective, component-based adaptive systems. In: Schäfer, W., Dwyer, M.B., Gruhn, V. (eds.) 30th International Conference on Software Engineering, ICSE 2008, 10–18 May 2008, Leipzig, Germany, pp. 811–814. ACM (2008)
5. Cámara, J., et al.: Self-aware computing systems: related concepts and research areas. In: Kounev, S., Kephart, J., Milenkoski, A., Zhu, X. (eds.) Self-Aware Computing Systems, pp. 17–49. Springer, Cham (2017). https://doi.org/10.1007/978-3-319-47474-8_2
6. Capilla, R., Bosch, J., Trinidad, P., Ruiz Cortés, A., Hinchey, M.: An overview of dynamic software product line architectures and techniques: observations from research and industry. J. Syst. Softw. 91, 3–23 (2014)
7. Ewing, J.M.: Autonomic performance optimization with application to self-architecting software systems. Ph.D. thesis, George Mason University, Fairfax, Virginia, USA (2015). http://hdl.handle.net/1920/9702
8. Fettweis, G., Nagel, W., Lehner, W.: Pathways to servers of the future: highly adaptive energy efficient computing (HAEC). In: Proceedings of the Conference on Design, Automation and Test in Europe, DATE 2012, San Jose, CA, USA, pp. 1161–1166. EDA Consortium (2012)
9. Goens, A., Khasanov, R., Castrillon, J., Hähnel, M., Smejkal, T., Härtig, H.: Tetris: a multi-application run-time system for predictable execution of static mappings. In: Proceedings of the 20th International Workshop on Software and Compilers for Embedded Systems, SCOPES 2017, New York, NY, USA, pp. 11–20. ACM (2017)
10. Greenwood, P., Blair, L.: A framework for policy driven auto-adaptive systems using dynamic framed aspects. In: Rashid, A., Aksit, M. (eds.) Transactions on Aspect-Oriented Software Development II. LNCS, vol. 4242, pp. 30–65. Springer, Heidelberg (2006). https://doi.org/10.1007/11922827_2
11. Guinea, S., Kecskemeti, G., Marconi, A., Wetzstein, B.: Multi-layered monitoring and adaptation. In: Kappel, G., Maamar, Z., Motahari-Nezhad, H.R. (eds.) ICSOC 2011. LNCS, vol. 7084, pp. 359–373. Springer, Heidelberg (2011). https://doi.org/10.1007/978-3-642-25535-9_24
12. Haddadin, S., Suppa, M., Fuchs, S., Bodenmüller, T., Albu-Schäffer, A., Hirzinger, G.: Towards the robotic co-worker. In: Pradalier, C., Siegwart, R., Hirzinger, G. (eds.) Robotics Research. STAR, vol. 70, pp. 261–282. Springer, Heidelberg (2011). https://doi.org/10.1007/978-3-642-19457-3_16

13. Heinis, T., Pautasso, C.: Automatic configuration of an autonomic controller: an experimental study with zero-configuration policies. In: Strassner, J., Dobson, S.A., Fortes, J.A.B., Goswami, K.K. (eds.) 2008 International Conference on Autonomic Computing, ICAC 2008, 2–6 June 2008, Chicago, Illinois, USA, pp. 67–76. IEEE Computer Society (2008)
14. Herrmann, S.: A precise model for contextual roles: the programming language ObjectTeams/Java. Appl. Ontol. **2**(2), 181–207 (2007)
15. Hillemacher, S., et al.: Model-based development of self-adaptive autonomous vehicles using the SMARDT methodology. In: Hammoudi, S., Pires, L.F., Selic, B. (eds.) Proceedings of the 6th International Conference on Model-Driven Engineering and Software Development, MODELSWARD 2018, 22–24 January 2018, Funchal, Madeira - Portugal, pp. 163–178. SciTePress (2018)
16. Hirschfeld, R., Costanza, P., Nierstrasz, O.M.: Context-oriented programming. J. Object Technol. **7**(3), 125–151 (2008)
17. Huebscher, M.C., McCann, J.A.: A survey of autonomic computing - degrees, models, and applications. ACM Comput. Surv. **40**(3), 7:1–7:28 (2008)
18. Kephart, J.O., et al.: Self-adaptation in collective self-aware computing systems. In: Kounev, S., Kephart, J., Milenkoski, A., Zhu, X. (eds.) Self-Aware Computing Systems. Springer, Cham (2017). https://doi.org/10.1007/978-3-319-47474-8_13
19. Kiczales, G., et al.: Aspect-oriented programming. In: Akşit, M., Matsuoka, S. (eds.) ECOOP 1997. LNCS, vol. 1241, pp. 220–242. Springer, Heidelberg (1997). https://doi.org/10.1007/BFb0053381
20. Kissinger, T., Hähnel, M., Smejkal, T., Habich, D., Härtig, H., Lehner, W.: Energy-utility function-based resource control for in-memory database systems live. In: Proceedings of the 2018 International Conference on Management of Data, SIGMOD 2018, New York, NY, USA, pp. 1717–1720. ACM (2018)
21. Kramer, J., Magee, J.: Towards robust self-managed systems. Prog. Inform. **5**, 1–4 (2008)
22. Kühn, T., Leuthäuser, M., Götz, S., Seidl, C., Aßmann, U.: A metamodel family for role-based modeling and programming languages. In: Combemale, B., Pearce, D.J., Barais, O., Vinju, J.J. (eds.) SLE 2014. LNCS, vol. 8706, pp. 141–160. Springer, Cham (2014). https://doi.org/10.1007/978-3-319-11245-9_8
23. Philippe, L., McCann, J.A., Diaconescu, A.: Autonomic Computing - Principles, Design and Implementation. Undergraduate Topics in Computer Science. Springer, London (2013). https://doi.org/10.1007/978-1-4471-5007-7
24. Litoiu, M., Woodside, M., Zheng, T.: Hierarchical model-based autonomic control of software systems. ACM SIGSOFT Softw. Eng. Notes **30**(4), 1–7 (2005)
25. Ostermann, K.: Dynamically composable collaborations with delegation layers. In: Magnusson, B. (ed.) ECOOP 2002. LNCS, vol. 2374, pp. 89–110. Springer, Heidelberg (2002). https://doi.org/10.1007/3-540-47993-7_4
26. Pukhkaiev D., Götz, S.: BRISE: energy-efficient benchmark reduction. In: 2018 IEEE/ACM 6th International Workshop on Green and Sustainable Software (GREENS), pp. 23–30, May 2018
27. Püschel, G.: Testing self-adaptive systems - a model-based approach to resilience. Ph.D. thesis, Technische Universität Dresden, June 2018. http://nbn-resolving.de/urn:nbn:de:bsz:14-qucosa-237791
28. Rosenmüller, M., Siegmund, N., Apel, S., Saake, G.: Flexible feature binding in software product lines. Autom. Softw. Eng. **18**(2), 163–197 (2011)
29. Salehie, M., Tahvildari, L.: Self-adaptive software: Landscape and research challenges. ACM Trans. Auton. Adapt. Syst. **4**(2), 14:1–14:42 (2009)

30. Salvaneschi, G., Ghezzi, C., Pradella, M.: Context-oriented programming: aprogramming paradigm for autonomic systems. The Computing Research Repository (CoRR). abs/1105.0069 (2011)
31. Smaragdakis, Y., Batory, D.: Implementing layered designs with mixin layers. In: Jul, E. (ed.) ECOOP 1998. LNCS, vol. 1445, pp. 550–570. Springer, Heidelberg (1998). https://doi.org/10.1007/BFb0054107
32. Steimann, F.: On the representation of roles in object-oriented and conceptual modelling. Data Knowl. Eng. **35**(1), 83–106 (2000)
33. VanHilst, M., Notkin, D.: Using role components in implement collaboration-based designs. SIGPLAN Notes **31**(10), 359–369 (1996)
34. Weyns, D., et al.: On patterns for decentralized control in self-adaptive systems. In: de Lemos, R., Giese, H., Müller, H.A., Shaw, M. (eds.) Software Engineering for Self-Adaptive Systems II. LNCS, vol. 7475, pp. 76–107. Springer, Heidelberg (2013). https://doi.org/10.1007/978-3-642-35813-5_4

Distance-Based Community Search (Invited Talk Extended Abstract)

Francesco Bonchi[1,2]([envelope])

[1] ISI Foundation, Turin, Italy
francesco.bonchi@isi.it
[2] Eurecat, Barcelona, Spain
http://francescobonchi.com/

1 Community Search

Suppose we have identified a set of subjects in a terrorist network suspected of organizing an attack. Which other subjects, likely to be involved, should we keep under control? Similarly, given a set of patients infected with a viral disease, which other people should we monitor? Given a set of companies trading anomalously on the stock market: is there any connection among them that could explain the anomaly? Given a set of proteins of interest, which other proteins participate in pathways with them? Given a set of users in a social network that clicked an ad, to which other users (by the principle of "homophily") should the same ad be shown?

Each of these questions can be modeled as a graph-query problem: given a graph $G = (V, E)$ where (V is a set of vertices representing entities and E is a set of edges modeling the relations among the entities) and given a set of query vertices $Q \subseteq V$, find a subgraph H of G which *"explains"* the connections existing among the vertices in Q, that is to say that H must be connected and contain all query vertices in Q.

Several problems of this type have been studied under different names, e.g., *community search* [3,6,17], *seed set expansion* [2,10], *connectivity subgraphs* [1,7, 15,18], just to mention a few. While optimizing for different objective functions, the bulk of this literature aims at finding a "community" around the set of query vertices Q: the (more or less) implicit assumption is that *the vertices in Q belong to the same community*, and a good solution will contain other vertices belonging to the same community of Q. As we showed in our work in [15], when such an assumption is satisfied, these methods return reasonable subgraphs, but when the query vertices belong to different modules of the input graph, these methods tend to return too large a subgraph, often so large as to be meaningless and unusable in real applications. Moreover, the assumption is not so realistic in practice. In fact, we have a set of vertices that we *believe* are of interest for the application at hand and we want to further investigate them: why should we assume they belong to the same community? Moreover, if we have already knowledge of the communities, then why do we need to *"reconstruct"* the community around Q?

© Springer Nature Switzerland AG 2019
B. Catania et al. (Eds.): SOFSEM 2019, LNCS 11376, pp. 21–27, 2019.
https://doi.org/10.1007/978-3-030-10801-4_2

2 The Minimum Wiener Connector

In our work in [15] we take a different approach: instead of trying to "reconstruct" the community around Q we seek a *small* connector, i.e., a connected subgraph of the input graph which contains Q and a small set of *important additional vertices*. These additional vertices could explain the relation among the vertices in Q, or could participate in some function by acting as important links among the vertices in Q. We achieve this by defining a new, *parameter-free* problem where, although the size of the solution connector is left unconstrained, the objective function itself takes care of keeping it small.

Specifically, given a graph $G = (V, E)$ and a set of query vertices $Q \subseteq V$, our problem asks for the connector H^* minimizing the sum of shortest-path distances among all pairs of vertices (i.e., the *Wiener index* [19]) in the solution H^*:

$$H^* = \underset{G[S]:Q \subseteq S \subseteq V}{\arg\min} \sum_{\{u,v\} \in S} d_{G[S]}(u, v)$$

where $G[S]$ denotes the subgraph induced by a set of nodes S, and $d_{G[S]}(u, v)$ denotes the shortest-path distance between nodes u and v in $G[S]$. We call H^* the *minimum Wiener connector* for query Q.

This is a very natural problem to study: shortest paths define fundamental structural properties of graphs, playing a role in all the basic mechanisms of networks such as their evolution [11] and the formation of communities [8]. The fraction of shortest paths that a vertex takes part in is called its *betweenness centrality* [4], and is a well established measure of the importance of a vertex, i.e., the extent to which an actor has control over information flow. A consequence of our definition of minimum Wiener connector is that our solutions tend to include vertices which hold an important position in the network, i.e., vertices with high betweenness centrality.

Consider social or biological networks with their modular structure [8] (i.e., the existence of communities of vertices densely connected inside, and sparsely connected with the outside). When the query vertices Q belong to the same community, the additional nodes added to Q to form the minimum Wiener connector will tend to belong to the same community. In particular, these will typically be vertices with higher "centrality" than those in Q: these are likely to be influential vertices playing leadership roles in the community. These might be good users for spreading information, or to target for a *viral marketing* campaign [9].

Instead, when the query vertices in Q belong to different communities, the additional vertices added to Q to form the minimum Wiener connector will contain vertices adjacent to edges that "bridge" the different communities. These also have strategic importance: information has to go over these bridges to propagate from a community to others, thus the vertices incident to bridges enjoy a strategically favorable position because they can block information, or access it before other individuals in their community. These vertices are said to span a *"structural hole"* [5]: they are the best candidates to target for blocking the spread of rumors or viral diseases in a social network, or the spread of malware in a network of computers. In a protein-protein interaction network these vertices can represent proteins that play a key role in linking modules and whose removal can have different phenotypic effects.

In [15] we show that, when the number of query vertices is small, the *minimum Wiener connector* can be found in polynomial time. However, in the general case our problem is **NP**-hard and it has no PTAS unless **P = NP**: note that, while the inapproximability result says that the problem cannot be approximated within *every* constant, it leaves open the possibility of approximating it within *some* constant. In fact, our central result is an efficient constant-factor approximation algorithm, which runs in $\widetilde{O}(|Q||E|)$ time. We also devise integer-programming formulations of our problem. We use them to compare our solutions for small graphs with those found using state-of-the art solvers, and show empirically that our solutions are indeed close to optimal. Our experiments confirm that our method produces solutions which are smaller in size, denser, and which include more central nodes than the methods in the literature, regardless of whether the query vertices belong to the same community or not.

3   The Minimum Inefficiency Subgraph

A common aspect of almost all the literature on community search is to require the solution to be a *connected* subgraph. The *requirement of connectedness* is a strongly restrictive one. Consider, for example, a biologist inspecting a set of proteins that she *suspects* could be cooperating in some biomedical setting. It may very well be the case that one of the proteins is not related to the others: in this case, forcing the sought subgraph to connect them all might produce poor quality solutions, while at the same time hiding an otherwise good solution. By relaxing the connectedness condition, the outlier protein can be kept disconnected, thus returning a much better solution to the biologist. Another consequence of the connectedness requirement is that by trying to connect possibly unrelated vertices, the resulting solutions end up being very large.

In our work in [16], we study the *selective connector problem*: given a graph $G = (V, E)$ and a set of query vertices $Q \subseteq V$, find a superset $S \supseteq Q$ of vertices such that its induced subgraph, denoted $G[S]$, has some good "cohesiveness" properties, but is not necessarily connected. Abstractly, we would like our selective connector $G[S]$ to have the following desirable properties:

- **Parsimonious vertex addition.** Vertices should be added to Q to form the solution S, if and only if they help form more "cohesive" subgraphs by better connecting the vertices in Q. Roughly speaking, this ensures that the only vertices added are those which serve to better explain the connection between the elements of Q (or a subset thereof).
- **Outlier tolerance.** If Q contains vertices which are "far" from the rest of Q, those should remain disconnected in the solution S and be considered as outliers. The necessity for this stems from the fact that real-world query-sets are likely to contain some vertices that are erroneously suspected of being related.

- **Multi-community awareness.** If the query vertices Q belong to two or more communities, then the connector should be able to recognize this situation, detect the communities, and refrain from imposing connectedness between them.

A natural way to define the cohesiveness of a subgraph $G[S]$ is to consider the shortest-path distance $d_{G[S]}(u, v)$ between every pair of vertices $u, v \in S$, as done in the previous section. One issue with shortest-path distance is that, when the connectedness requirement is dropped, pairs of vertices can be disconnected, thus yielding an infinite distance. A simple yet elegant workaround to this issue is to use the reciprocal of the shortest-path distance [13]; this has the useful property of handling ∞ neatly (assuming by convention that $\infty^{-1} = 0$). This is the idea at the heart of *network efficiency*, a graph-theoretic notion that was introduced by Latora and Marchiori [12] as a measure of how efficiently a network $G = (V, E)$ can exchange information:

$$\mathcal{E}(G) = \frac{1}{|V|(|V| - 1)} \sum_{\substack{u,v \in V \\ u \neq v}} \frac{1}{d_G(u, v)}.$$

Unfortunately, defining the selective connector problem as finding the subgraph $G[S]$ with $S \supseteq Q$ that *maximizes network efficiency* would be meaningless. In fact, the normalization factor $|V|(|V| - 1)$ allows vertices totally unrelated to Q to be added to improve the efficiency; clearly violating our driving principle of parsimonious vertex addition. Based on the above arguments, we introduce the measure of the *inefficiency* of a graph $G = (V, E)$, defined as follows:

$$\mathcal{I}(G) = \sum_{\substack{u,v \in V \\ u \neq v}} 1 - \frac{1}{d_G(u, v)}.$$

Hence, we define the selective connector problem as the *parameter-free* problem which requires extracting the subgraph $G[S]$, with $S \supseteq Q$, that *minimizes network inefficiency*. With this definition, each pair of vertices in the subgraph $G[S]$ produces a cost between 0 and 1, which is minimum when the two vertices are neighbors, grows with their distance, and is maximum when the two vertices are not reachable from one another. Parsimony in adding vertices is handled by the sum of costs over all pairs of vertices in the connector; adding one vertex v to a partial solution S incurs $|S|$ more terms in the summation. The inclusion of v is worth the additional cost only if these costs are small and if v helps reduce the distances between vertices in S. Moreover, note that by allowing disconnections in the solution, the second and third design principles above (i.e., outliers and multiple communities) naturally follow from the parsimonious vertex addition.

The *Minimum Inefficiency Subgraph* (**mis**) problem is **NP**-hard, and we prove that it remains hard even if we constrain the input graph G to have a diameter of at most 3. Therefore, we devise an algorithm that is based on first building a complete connector for the query vertices and then *relaxing* the connectedness

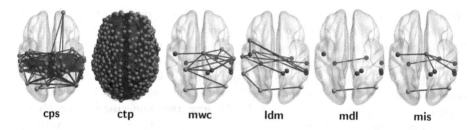

Fig. 1. Comparison between Minimum Inefficiency Subgraph (**mis**) and other notions in the literature on a cortical connectivity network. Query vertices are colored w.r.t. their known functionalities: memory and motor function (blue vertices), emotions (yellow vertices), visual processing (red vertices). The green vertices are the ones added to produce the solution. More details on the case study can be found in [16]. (Color figure online)

requirement by *greedily* removing non-query vertices. Our experiments show that in 99% of problem settings, our greedy relaxing algorithm produces solutions no worse than those produced by an exhaustive search, while at the same time being orders of magnitude more efficient. We empirically confirm that the **mis** is a selective connector: i.e., tolerant to outliers and able to detect multiple communities. Besides, the selective connectors produced by our method are smaller, denser, and include vertices that have higher centrality than the ones produced by the state-of-the-art methods. We also show interesting case studies in a variety of application domains (such as human brain, cancer, food networks, and social networks), confirming the quality of our proposal (Fig. 1).

4   Adaptive Community Search in Dynamic Networks

Although community search has received a great deal of attention in the last few years, most of the literature so far has focused on static networks. However, many of the networks of interest carry time information which can be very important for understanding the dynamics of interactions between the vertices. For instance, interactome, which is the set of molecular interactions in a cell, can be modeled as a network, in which the vertices are proteins and through their connections can perform biological functions. However, these connections are not constantly active, and therefore a dynamic analysis is more appropriate for understanding properly this complex network [14]. In communication networks, for example, the edges represent correspondence between two actors of the network. If a user A communicates with a user B at some time t_0 and later in time, the user B communicates with a user C the flow of information can pass from user A to user C, but not in the opposite direction.

In our ongoing work we are studying the problem of community search in dynamic networks with adaptive query updates. Our objective is to find a temporal connector that includes all the vertices of interest, connecting them with "temporal paths" that should be seen as paths both in space (i.e., network structure) and in time (i.e., network evolution). Since the network changes constantly in time, we expect that the connectors evolve as well. Therefore, it is natural that the query set is enriched during the evolution, with new vertices, that formed part of the solution of the previous time instances. As long as the added vertices remain related to the initial query set, they are maintained to it. Otherwise, they are removed from the query set. In this way, the connector keeps evolving in time and keeps monitoring the evolution of the interactions among the vertices of interest. We call this problem *temporal adaptive community search*.

Acknowledgements. I wish to thank all the co-authors of the various papers on which this invited talk is built: Natali Ruchansky, Ioanna Tsalouchidou, David García-Soriano, Francesco Gullo, Nicolas Kourtellis, Ricardo Baeza-Yates.

References

1. Akoglu, L., et al.: Mining connection pathways for marked nodes in large graphs. In: SDM (2013)
2. Andersen, R., Lang, K.J.: Communities from seed sets. In: WWW (2006)
3. Barbieri, N., Bonchi, F., Galimberti, E., Gullo, F.: Efficient and effective community search. DAMI **29**(5), 1406–1433 (2015)
4. Bavelas, A.: A mathematical model of group structure. Hum. Organ. **7**, 16–30 (1948)
5. Burt, R.: Structural Holes: The Social Structure of Competition. Harvard University Press (1992)
6. Cui, W., Xiao, Y., Wang, H., Wang, W.: Local search of communities in large graphs. In: SIGMOD (2014)
7. Faloutsos, C., McCurley, K.S., Tomkins, A.: Fast discovery of connection subgraphs. In: KDD (2004)
8. Girvan, M., Newman, M.E.J.: Community structure in social and biological networks. PNAS **99**(12), 7821–7826 (2002)
9. Kempe, D., Kleinberg, J.M., Tardos, É.: Maximizing the spread of influence through a social network. In: KDD (2003)
10. Kloumann, I.M., Kleinberg, J.M.: Community membership identification from small seed sets. In: KDD (2014)
11. Kossinets, G., Watts, D.J.: Empirical analysis of an evolving social network. Science **311**(5757), 88–90 (2006)
12. Latora, V., Marchiori, M.: Efficient behavior of small-world networks. Phys. Rev. Lett. **87**(19), 198701 (2001)
13. Marchiori, M., Latora, V.: Harmony in the small-world. Phys. A: Stat. Mech. Appl. **285**(3–4), 539–546 (2000)
14. Przytycka, T., Singh, M., Slonim, D.: Toward the dynamic interactome: it's about time. Brief. Bioinform. **11**(1), 15–29 (2010). https://doi.org/10.1093/bib/bbp057
15. Ruchansky, N., Bonchi, F., García-Soriano, D., Gullo, F., Kourtellis, N.: The minimum wiener connector problem. In: SIGMOD (2015)
16. Ruchansky, N., Bonchi, F., García-Soriano, D., Gullo, F., Kourtellis, N.: To be connected, or not to be connected: that is the minimum inefficiency subgraph problem. In: CIKM (2017)

17. Sozio, M., Gionis, A.: The community-search problem and how to plan a successful cocktail party. In: KDD (2010)
18. Tong, H., Faloutsos, C.: Center-piece subgraphs: problem definition and fast solutions. In: KDD, pp. 404–413 (2006)
19. Wiener, H.: Structural determination of paraffin boiling points. J. Am. Chem. Soc. **69**(1), 17–20 (1947)

Minicomplexity

Some Motivation, Some History, and Some Structure
(Invited Talk Extended Abstract)

Christos A. Kapoutsis[✉]

Carnegie Mellon University in Qatar, Doha, Qatar
cak@cmu.edu

Abstract. The term *minicomplexity* was first suggested in [2], as a name for the field of theory of computation which studies the *size complexity of two-way finite automata*, as outlined in [1]. In this talk, we discuss the motivation behind this field and enumerate some of its prominent results in their historical context. By reformulating these results, we then attempt to reveal additional structure which often passes unnoticed. The present report records the start of this attempt.

1 Machines vs. Machines

Central in [1] is the invitation to start viewing the results in this field similarly to how results are being viewed in standard complexity theory: not as statements about the relative power of various computational devices, but as statements about the relative difficulty of various computational problems.

To describe the difference between the two viewpoints and stress the benefits of such a shift, we go back to the seminal paper of Meyer and Fischer [5], which initiated the field, and to the three very first propositions in it. The first one[(1)]

> **Proposition 1.** For every n > 0, there is a regular set $R_n \subseteq \{0,1\}^*$ such that the reduced finite automaton accepting R_n has exactly 2^n states (and size 2^{n+1}), but there is an n-state nondeterministic finite automaton of size 3n-2 accepting R_n.

says that one-way nondeterministic finite automata (1NFAs) are strictly more powerful than deterministic ones (1DFAs), as some binary witness language R_n needs $\leq n$ states on 1NFAs but $\geq 2^n$ states on 1DFAs. In the proof, R_n is described only through its deciding 1NFA. After it, one more witness R'_n is given:

> An even simpler example suggested by Paterson is close to optimal: let $R'_n \subseteq \{0,1\}^*$ be the set of strings whose nth from the last digit is 1. Then R'_n is recognized by an n+1 state nondeterministic machine, the reduced deterministic machine has 2^n states, and the reduced machine for the reversal of R'_n has n+2 states. Moreover, $R'_n \cap \{0,1\}^n\{0,1,\lambda,\}^n$ is a <u>finite</u> event with similar properties.

B. Catania et al. (Eds.): SOFSEM 2019, LNCS 11376, pp. 28–38, 2019.
https://doi.org/10.1007/978-3-030-10801-4_3

which is (suboptimal, but) simpler, together with its restriction R_n'' (our name) to strings of length $<2n$, which shares the same properties.[2] The next proposition

> **Proposition 2.** For every n > 1, there is a regular set $F_n \subseteq \{0,1,2\}^*$ which can be recognized by a two-way finite automaton with at most 5n+5 states, but the reduced finite automaton accepting F_n has at least n^n states. Moreover F_n is a finite set.
>
> **Proof.** Let $F_n = \{0^{i_1} 1 0^{i_2} 1 \cdots 1 0^{i_n} 2^k 0^k \mid$
> $1 \leq k \leq n$ and $1 \leq i_j \leq n$ for $j = 1,\ldots,n\}$.

establishes a similar relation for two-way deterministic finite automata (2DFAs) and 1DFAs: now a witness language F_n needs $O(n)$ states on 2DFAs but $\geq n^n$ states on 1DFAs. Finally, the third proposition

> **Proposition 3.** For each n > 0 there is a regular event $P_n \subseteq \{0,1,2\}^*$ which is recognizable by a deterministic one pebble automaton with at most 3n+5 states. The reduced finite automaton accepting P_n has at least 2^{2^n} states. Moreover, P_n is finite.
>
> **Proof.** $P_n = \{x_1 2 x_2 2 \ldots 2 x_k 2 2 x_1 \mid x_j \in \{0,1\}^n$ for $1 \leq j \leq k$, $i \leq k$, and regarded as binary integer $x_j < x_{j+1}$ for $1 \leq j < k\}$. □

proves that, augmented with a pebble, 2DFAs are even more powerful: $O(n)$ states are now enough to decide a language P_n for which 1DFAs need $\geq 2^{2^n}$ states. In short, and in the standard parlance of the field, Propositions 1–3 tell us that the trade-off in the conversion from 1NFAs, 2DFAs, or single-pebble 2DFAs to 1DFAs is respectively $2^{\Omega(n)}, 2^{\Omega(n \lg n)}$, and $2^{2^{\Omega(n)}}$.

Overall, this is entirely a *"machines vs. machines"* discussion: computational devices compete against each other, and we want to know which is more powerful. In this competition, problems play only the secondary role of a witness task on which a stronger machine beats a weaker one by solving it with less resources.

2   Problems vs. Problems

The alternate viewpoint is *"problems vs. problems"*: computational tasks compete against each other, and we want to know which is more difficult. This time it is machines that play the secondary role, of a witness device on which a harder problem beats an easier one by requiring more resources.

This is the viewpoint of standard complexity theory, and it was proposed for minicomplexity, as well, by Sakoda and Sipser in their own seminal paper [6]. By switching to this viewpoint, we bring problems at the center of attention: we clearly describe them as computational tasks (as opposed to sets of strings); give them distinctive names; and collect them in complexity classes relative to the various machines and the polynomiality or not (as opposed to the asymptotics) of the used resources. For practical reasons, we also use h as the important parameter (instead of n, which is often needed as input length); and describe the instances over a large alphabet and with an associated promise for the format (if this simplifies the description without affecting the difficulty).

2.1 RETROCOUNT

For example, R'_h (Proposition 1) is the problem: "Given a bitstring, check that its h-th from the last digit is 1." Note that the reference "h-th from the last digit" is void on strings of length $<h$. We could get into a discussion of whether such strings should be accepted or not, but this would be a distraction from the main point of the task. A better description of the essence of this computational problem is one where the intended format of the input is taken for granted:

$$\text{"Given a bitstring of length } \geq h, \text{ check that its } h\text{-th last bit is 1."} \qquad (1)$$

so that a solving machine need not check that the length is $\geq h$; if it is not, then the machine is free to decide arbitrarily. We call this problem RETROCOUNT$_h$. Similarly, the restriction R''_h is the problem SHORT RETROCOUNT$_h$:

$$\text{"Given a bitstring of length } h \leq n < 2h, \text{ check that its } h\text{-th last bit is 1."} \qquad (2)$$

Again, it is not the job of a solving machine to check that the length of the input is appropriate; this is promised. The job is only to check the h-th last bit.

Formally, a *(promise) problem* $\mathfrak{L}$ over an alphabet Σ is a pair $(L, \tilde{L})$ of disjoint subsets of Σ^*. A string w is an *instance* of $\mathfrak{L}$ if $w \in L \cup \tilde{L}$, and is *positive*, if $w \in L$, or *negative*, if $w \in \tilde{L}$. To *solve* $\mathfrak{L}$ is to accept all $w \in L$ but no $w \in \tilde{L}$ (behaving arbitrarily on $w \notin L \cup \tilde{L}$). So, (1) and (2) are the promise problems

$$\text{RETROCOUNT}_h := \left(\quad \{0,1\}^* 1 \{0,1\}^{h-1}, \quad \{0,1\}^* 0 \{0,1\}^{h-1} \quad \right)$$
$$\text{SHORT RETROCOUNT}_h := \left(\quad \{0,1\}^{<h} 1 \{0,1\}^{h-1}, \quad \{0,1\}^{<h} 0 \{0,1\}^{h-1} \quad \right).$$

That both problems witness the exponential difference in number of states between 1NFAs and 1DFAs is expressed by the fact that both belong to the class

$$\text{1N} := \left\{ (\mathfrak{L}_h)_{h \geq 1} \; \middle| \; \begin{array}{l} \text{there exist 1NFAs } (N_h)_{h \geq 1} \text{ and polynomial } s \\ \text{such that every } N_h \text{ solves } \mathfrak{L}_h \text{ with } s(h) \text{ states} \end{array} \right\}, \qquad (3)$$

of "problems solved by small 1NFAs", but not in the respective class 1D for 1DFAs:

$$\text{RETROCOUNT, SHORT RETROCOUNT} \in \text{1N} \setminus \text{1D}, \qquad (4)$$

where RETROCOUNT $= (\text{RETROCOUNT}_h)_{h \geq 1}$ is the induced family, and similarly for SHORT RETROCOUNT. Note that, since the latter problem is a restriction of the former, all the information of (4) follows from the next two facts and lemma.

Fact 1. RETROCOUNT $\in$ 1N.

Fact 2. SHORT RETROCOUNT $\notin$ 1D.

Lemma 1. *If $\mathcal{L} \subseteq \mathcal{L}'$ and $\mathcal{L} \notin$ C, then $\mathcal{L}' \notin$ C.*

Here, $\mathcal{L} \subseteq \mathcal{L}'$ means that the family $\mathcal{L} = (\mathfrak{L}_h)_{h \geq 1}$ is a *restriction* of $\mathcal{L}' = (\mathfrak{L}'_h)_{h \geq 1}$ (so that SHORT RETROCOUNT $\subseteq$ RETROCOUNT); equivalently, we also say that $\mathcal{L}'$ is a *generalization* of $\mathcal{L}$. Formally, for $\mathfrak{L} = (L, \tilde{L})$ and $\mathfrak{L}' = (L', \tilde{L}')$, we write $\mathfrak{L} \subseteq \mathfrak{L}'$ if both $L \subseteq L'$ and $\tilde{L} \subseteq \tilde{L}'$; then we write $\mathcal{L} \subseteq \mathcal{L}'$ if $\mathfrak{L}_h \subseteq \mathfrak{L}'_h$ for all h.

2.2   PROJECTION

For another example, F_h (Proposition 2) is the problem: "Given a string, check that it consists of a tuple of h numbers from 1 to h (in unary, by 0s; delimited by 1s), an index k from 1 to h (in unary, by 2s), and the k-th number in the tuple (in unary, by 0s)." Clearly, the essence of this task is to check that the number after k equals the k-th number in the tuple. So, a better description is:

"Given a tuple of h numbers from 1 to h (in unary, by 0s; delimited by 1s), an index k from 1 to h (in unary, by 2s), and a number i from 1 to h (5) (in unary, by 0s), check that i equals the k-th number in the tuple."

so that, as above, checking that the input is correctly formatted is not important.

Also unimportant is the fact that the numbers and index are in unary. The problem preserves its essence, if we assume that the input cells are large enough to host any number from $[h] := \{1, \ldots, h\}$. So, an even better description of F_h is

"Given a tuple $i_1, i_2, \ldots, i_h$ of numbers in $[h]$, and two numbers k, i in $[h]$, (6) check that i equals i_k."

where the input alphabet is $[h]$. Now it is more clear what the problem is: to check that the projection of the given tuple to its k-th component returns i. So, we refer to (6) as PROJECTION$_h$;[3] and use UNARY PROJECTION$_h$ for (5).

With these clarifications, Proposition 2 says that UNARY PROJECTION $\in$ 2D $\setminus$ 1D, where the class 2D of "problems solved by small 2DFAs" is defined similarly to (3). Intuitively, the reasons for this fact are clear: When a solving 1DFA crosses the boundary between the tuple and the index, it must be able to answer any query of the form "does the k-th component equal i?", so it must store the full tuple, which needs $\geq h^h$ states. In contrast, a 2DFA can read $2^k 0^i$ and store k and i; rewind; then countdown to the k-th block of 0s to check that it contains exactly i of them, all doable with $O(h^2)$ states. Similarly, PROJECTION $\in$ 2D $\setminus$ 1D.

Now that we see why these problems witness 2D $\setminus$ 1D, we may further ask: Do we really need the tuple numbers in separate cells? Or k and i in separate cells? No. Over the alphabet $[h]^h \cup [h]^2$, where cells are large enough to host an entire h-tuple $\bar{i}$ or query $u = (k, i)$, we may define the problem COMPACT PROJECTION$_h$:

"Given a tuple $\bar{i} \in [h]^h$ and a query $u \in [h]^2$, check that $u_2 = i_{u_1}$." (7)

Intuitively, this is the best description of the essence of F_h, as it contains exactly the structure that is sufficient and necessary to place it in 2D $\setminus$ 1D.

Of course, problems (5), (6), and (7) are "essentially the same". To describe this intuition formally, we first define them as promise problems. E.g., (7) is:

$$\text{COMPACT PROJECTION}_h := \big(\{\bar{i}u \mid u_2 = i_{u_1}\},\ \{\bar{i}u \mid u_2 \neq i_{u_1}\}\big);$$

and similarly for (5) and (6), over alphabets $\{0,1,2\}$ and $[h]$. We then introduce reductions, as follows. For arbitrary problems $\mathfrak{L} = (L, \tilde{L})$ and $\mathfrak{L}' = (L', \tilde{L}')$ over alphabets Σ and Σ', we say that $\mathfrak{L}$ 1D-*reduces* to $\mathfrak{L}'$ ($\mathfrak{L} \leq_{1D} \mathfrak{L}'$) if there exists a one-way deterministic finite transducer (1DFT) T such that

$$w \in L \implies T(w) \in L' \quad \text{and} \quad w \in \tilde{L} \implies T(w) \in \tilde{L}' \qquad (8)$$

where $T(w)$ is the output of T on input w, if T accepts w, or undefined, otherwise. An alternative and more concise way to write (8) is:

$$T(\mathfrak{L}) \subseteq \mathfrak{L}' \qquad (9)$$

where $T(\mathfrak{L}) = T(L, \tilde{L}) = (T(L), T(\tilde{L})) = \big(\{T(w) \mid w \in L\}, \{T(w) \mid w \in \tilde{L}\}\big)$ is the pair of the images under T of all positive and all negative instances of $\mathfrak{L}$ (which is itself a problem iff $T(L) \cap T(\tilde{L}) \neq \emptyset$). As further alternative,

$$1\text{D}(\mathfrak{L}) \subseteq \mathfrak{L}' \qquad (10)$$

says the same, without identifying T (i.e., it is equivalent to $\mathfrak{L} \leq_{1D} \mathfrak{L}'$).

In the special case where the inclusion (9) is an equality, $\mathfrak{L}'$ is a 1D-*image* of $\mathfrak{L}$ under T, and we also write (10) as equality. E.g., (5) is a 1D-image of (6):

$$1\text{D}(\text{PROJECTION}_h) = \text{UNARY PROJECTION}_h \qquad (11)$$

via the $O(h)$-state 1DFT T which scans an instance $i_1 i_2 \ldots i_h k i$ and, for each symbol in $[h]$, prints the appropriate unary representation and delimiters. Note that T prints on its output tape only $h + 2 = \text{poly}(h)$ times; and, in each of these times, the printed string has length $\leq h + 1 = \text{poly}(h)$.

In another special case, where T has only 1 state and always accepts, T defines a homomorphism $H : \Sigma \cup \{\vdash, \dashv\} \to (\Sigma')^*$ such that $T(w) = H(\vdash w \dashv)$. We then say that $\mathfrak{L}$ *homomorphically reduces* to $\mathfrak{L}'$ ($\mathfrak{L} \leq_H \mathfrak{L}'$ or $H(\mathfrak{L}) \subseteq \mathfrak{L}'$), if $H(\mathfrak{L}) \subseteq \mathfrak{L}'$; or that $\mathfrak{L}'$ is a *homomorphic image* of $\mathfrak{L}$ ($H(\mathfrak{L}') = \mathfrak{L}$), if $H(\mathfrak{L}) = \mathfrak{L}'$. Hence,

$$H(\text{COMPACT PROJECTION}_h) = \text{PROJECTION}_h \qquad (12)$$

via the homomorphism H which maps every h-tuple $\bar{i} = (i_1, i_2, \ldots, i_h)$ to the string $i_1 i_2 \cdots i_h$, every query $u = (k, i)$ to the string ki, and each of $\vdash, \dashv$ to ϵ. Note that H maps every symbol to a string of length $\leq h = \text{poly}(h)$.

These definitions extend to problem families $\mathcal{L} = (\mathfrak{L}_h)_{h \geq 1}$ and $\mathcal{L}' = (\mathfrak{L}'_h)_{h \geq 1}$, if every $\mathfrak{L}_h$ reduces to some $\mathfrak{L}'_{h'}$ via a 1DFT T_h or a homomorphism H_h. However, to say that $\mathcal{L} \leq_{1D} \mathcal{L}'$, $1\text{D}(\mathcal{L}) = \mathcal{L}'$, $\mathcal{L} \leq_H \mathcal{L}'$, or $H(\mathcal{L}) = \mathcal{L}'$, we also need h' and the size of T_h to be small relative to h: namely, that $h' = \text{poly}(h)$ and T_h has $\text{poly}(h)$ states. If, in addition, every T_h prints only $\text{poly}(h)$ times, then we write

$\mathcal{L} \leq_{1D}^{lac} \mathcal{L}'$ and call the T_h *laconic*; if every printed string has length only poly(h), then we write $\mathcal{L} \leq_{1D}^{t} \mathcal{L}'$ (or $\mathcal{L} \leq_{H}^{t} \mathcal{L}'$) and call the T_h (or H_h) *tight*. Hence,

$$1D(\text{PROJECTION}) = \text{UNARY PROJECTION}$$
$$H(\text{COMPACT PROJECTION}) = \text{PROJECTION} \tag{13}$$

by the tight laconic transducers of (11) and the tight homomorphisms of (12).

In conclusion, (13) expresses the intuition that problems (5), (6), and (7) are "essentially the same". Now, the fact that they all witness 2D \ 1D follows from only two easy facts and standard lemmas [6, Sect. 3], [4, Corollary 3]:

Fact 3. UNARY PROJECTION $\in$ 2D.

Fact 4. COMPACT PROJECTION $\notin$ 1D.

Lemma 2. 2D *is closed under* $\leq_H$ *and* $\leq_{1D}^{lac}$.

Lemma 3. 1D *is closed under* $\leq_{1D}$ *(and thus also under* $\leq_H$ *and* $\leq_{1D}^{lac}$*).*

2.3 MEMBERSHIP

Let us now return to Proposition 1 and see how large alphabets can help us better understand the essense of its problems, too.

In SHORT RETROCOUNT$_h$, every instance w is of the form uv, where $|u| = h$ and $0 \leq |v| < h$. Note that the actual bits of v are unimportant; only $l := |v|$ matters: w is positive iff the $(l+1)$-st bit of u is 1. Namely, if $\alpha \subseteq [h]$ is the set of the indices of all 1's in u, then w is asking whether $l+1 \in \alpha$. So, the question is really whether a set α contains an element i; it's just that α is given in binary (by its characteristic vector u) and i is given in unary (by the length $i-1$ of v).

Let us also recall why SHORT RETROCOUNT $\in$ 1N \ 1D. On crossing the u-v boundary, a solving 1DFA must be able to handle any l, i.e., any query of the form "does the i-th bit of u equal 1?"; so it must store the full u, which needs $\geq 2^h$ states. In contrast, a 1NFA can scan u; guess the crucial 1; countdown from h on the next bits (entering v at count i, for i the index of the guessed 1); and accept iff the count is 1 on $\dashv$ (so $|v| = i - 1$), all doable with $O(h)$ states.

Now that we better understand what the problem is asking and why it is a witness, we may ask: Do we really need α in binary and i in unary? No. Over the alphabet $\{\alpha \mid \alpha \subseteq [h]\} \cup [h]$, we define the problem: "Given a set $\alpha \subseteq [h]$ and an element $i \in [h]$, check that $i \in \alpha$", or formally:[4]

$$\text{MEMBERSHIP}_h := (\{\alpha i \mid i \in \alpha\}, \{\alpha i \mid i \notin \alpha\}); \tag{14}$$

and claim that this best captures the essence of SHORT RETROCOUNT$_h$. That the two problems are "essentially the same" is formally expressed by the fact that the former homomorphically reduces to the latter via the obvious homomorphism which maps every α to its characteristic vector and every i to 0^{i-1}:

$$H(\text{MEMBERSHIP}) \subseteq \text{SHORT RETROCOUNT}. \tag{15}$$

What about RETROCOUNT$_h$? Its instances are derived by left-padding those of SHORT RETROCOUNT$_h$ by arbitrary bitstrings. Formally, let LPAD be the operator which maps $\mathfrak{L} = (L, \tilde{L})$ to the pair LPAD$(\mathfrak{L}) := (\{0,1\}^* L, \{0,1\}^* \tilde{L})$. This pair is not necessarily a promise problem: if there exist instances $w \in L$ and $\tilde{w} \in \tilde{L}$ and pad-strings $x, \tilde{x} \in \{0,1\}^*$ such that $xw = \tilde{x}\tilde{w}$, then the two sets in the pair are not disjoint. So, call $\mathfrak{L}$ *left-paddable*, if this does not happen. Then a family $\mathcal{L} = (\mathfrak{L}_h)_{h \geq 1}$ is *left-paddable* if every $\mathfrak{L}_h$ is; and LPAD$(\mathcal{L}) := (\text{LPAD}(\mathfrak{L}_h))_{h \geq 1}$.

Easily, SHORT RETROCOUNT$_h$ is left-paddable, since the sign of each instance is determined by its last h bits and these are unaffected by the padding; and

$$\text{LPAD(SHORT RETROCOUNT)} = \text{RETROCOUNT}. \qquad (16)$$

Overall, (15) and (16) formally relate (1), (2), and (14) to each other. Now, the fact that all three witness 1N \ 1D follows from only two easy facts and from suitable lemmas (Lemma 1, as $\mathcal{L} \subseteq \text{LPAD}(\mathcal{L})$; Lemma 3; and Lemmas 4–5):

Fact 5. SHORT RETROCOUNT $\in$ 1N.

Fact 6. MEMBERSHIP $\notin$ 1D.

Lemma 4. 1N *is closed under* $\leq_\mathsf{H}$.

Lemma 5. *If* $\mathcal{L}$ *is left-paddable and* $\mathcal{L} \in$ 1N, *then* LPAD$(\mathcal{L}) \in$ 1N.

Note how our earlier Facts 1 and 2 now become corollaries of Facts 5 and 6.

Retrocount vs. Projection. There is great similarity between our intuition why the projection problems are not in 1D and why the same holds for the retrocount problems. This suggests that the projection problems also have MEMBERSHIP at their core. Indeed:

$$\mathsf{H}(\text{MEMBERSHIP}) \subseteq \text{COMPACT PROJECTION} \qquad (17)$$

by the homomorphism which maps every set $\alpha \subseteq [h]$ to its "characteristic tuple" $\overline{i} \in [h+1]^{h+1}$ where $i_j = 1$ or $h+1$, based on whether $j \in \alpha$ or not, respectively; and each $i \in [h]$ to the query $(i, 1)$.[5] So, our earlier Fact 4 is now a corollary of Fact 6 (via (17) and Lemma 3). At the same time, MEMBERSHIP is also a witness of 2D \ 1D, because Fact 3 implies it is in 2D (via (13), (17), and Lemma 2).

2.4   LIST MEMBERSHIP

We now continue to problem P_h (Proposition 3): "Given a string, check that it is a strictly increasing list of h-long binary numbers (delimited by 2s), followed by a copy of one of them (separated by 22)." Clearly, the essence of this task is to check that the number after 22 appears in the preceding list. The condition that the list is strictly increasing is there to ensure that P_h is finite. Ignoring it (also dropping the finiteness of P_h from Proposition 3), we arrive at this better description:

"Given a list of h-long binary numbers (delimited by 2s) and an h-long binary number i (separated by 22), check that i is in the list." $\qquad (18)$

As previously, presenting the numbers in binary is unimportant; all that matters is that each block of h bits can host 2^h different strings. It is important, however, to know when we have arrived at i. So, to zoom into the essence of P_h, we switch to alphabet $[2^h] \cup \{\breve{\imath} \mid i \in [2^h]\}$, where each cell hosts a full number x (possibly ticked, as $\breve{x}$) in $[2^h]$ (as opposed to $\{0, \ldots, 2^h - 1\}$), and define the problem:

> "Given a list of numbers from $[2^h]$ and a ticked number $i \in [2^h]$, check that i is in the list." $\qquad(19)$

In it, one easily sees a variant of MEMBERSHIP$_h$, where elements are drawn (not from $[h]$, but) from $[2^h]$; and the set is given (not in a single cell, but) as a list over many cells, possibly with repetitions. To represent this problem, we first introduce its variant over the smaller alphabet $[h]$:[(6)]

LIST MEMBERSHIP$_h$:=

$$\left(\{i_1 i_2 \cdots i_t \breve{\imath} \mid t \geq 0 \ \& \ i_1, i_2, \ldots, i_t, i \in [h] \ \& \ (\exists j)(i_j = i)\}, \right. \qquad(20)$$
$$\left. \{i_1 i_2 \cdots i_t \breve{\imath} \mid t \geq 0 \ \& \ i_1, i_2, \ldots, i_t, i \in [h] \ \& \ (\forall j)(i_j \neq i)\} \right),$$

and refer to (19) itself, over $[2^h]$, as TALL LIST MEMBERSHIP$_h$. Then, for (18) we use the name BINARY TALL LIST MEMBERSHIP$_h$.

So, Proposition 3 says that BINARY TALL LIST MEMBERSHIP $\in \mathsf{P_1D} \setminus 2^{1D}$, where $\mathsf{P_1D}$ and 2^{1D} are the classes for small single-pebble 2DFAs and large 1DFAs, where "large" means "with $2^{\mathrm{poly}(h)}$ states". Once again, the intuitive reasons are clear: on crossing 22, a solving 1DFA must have stored the set of numbers occurring in the list, which needs $\geq 2^{2^h}$ states. In contrast, a single-pebble 2DFA can compare i against every i_j bit-by-bit, using the pebble to mark the current i_j, all doable with $O(h)$ states. By the same reasons, TALL LIST MEMBERSHIP $\in \mathsf{P_1D} \setminus 2^{1D}$. (Note that it is important for the pebble 2DFA to have the list spread across cells.)

Note that our intuition for the lower bound is the same as for MEMBERSHIP, except now there are exponentially more sets to remember. To represent this formally, let us first note that

$$\mathsf{H}(\text{MEMBERSHIP}) \subseteq \text{LIST MEMBERSHIP} \qquad(21)$$

via the homomorphism which maps every set $\alpha \subseteq [h]$ to a string $i_1 i_2 \cdots i_t$ of its members; and every $i \in [h]$ to its ticked variant $\breve{\imath}$. We then also note that (19) can be obtained from LIST MEMBERSHIP by applying an operator TALL,

$$\mathsf{TALL}(\text{LIST MEMBERSHIP}) = \text{TALL LIST MEMBERSHIP}, \qquad(22)$$

which maps a family $\mathcal{L} = (\mathfrak{L}_h)_{h \geq 1}$ to its sub-family $\mathsf{TALL}(\mathcal{L}) = (\mathfrak{L}_{2^h})_{h \geq 1}$ at indices which are powers of 2; and (18) can be obtained from (19) homomorphically

$$\mathsf{H}(\text{TALL LIST MEMBERSHIP}) = \text{BINARY TALL LIST MEMBERSHIP} \qquad(23)$$

by mapping every $i \in [2^h]$ to the h-long binary representation of $i - 1$, preceded or followed by 2, depending on whether i is ticked or not. Now, the lower bound of Proposition 3 follows from (21), (22), and (23) and a strengthening of Fact 6.

To see how, we start with some definitions and facts. The class quasi-1D corresponds to 1DFAs with quasi-polynomially many states (i.e., $2^{\text{poly}(\log n)}$ states). We can easily show the next strengthening of Fact 6 (by the standard reasoning, that MEMBERSHIP$_h$ needs $\geq 2^h$ states on a 1DFA) and variation of Lemma 3:

Fact 7. MEMBERSHIP $\notin$ quasi-1D.

Lemma 6. quasi-1D and 2^{1D} are closed under $\leq_H$.

A family $\mathcal{L} = (\mathfrak{L}_h)_{h \geq 1}$ is *self-homomorphic* if $\mathfrak{L}_h \leq_H \mathfrak{L}_{h'}$ for all $h \leq h'$; intuitively, if the instances of every $\mathfrak{L}_h$ can be seen as instances of every higher $\mathfrak{L}_{h'}$. Easily, LIST MEMBERSHIP is self-homomorphic, and we can prove that:

Lemma 7. If $\mathcal{L}$ is self-homomorphic and $\mathcal{L} \notin$ quasi-1D, then TALL$(\mathcal{L}) \notin 2^{1D}$.

Now, we can apply the following reasoning:

$$\begin{array}{ll}
\text{MEMBERSHIP} \notin \text{quasi-1D} & \text{(Fact 7)} \\
\implies \text{LIST MEMBERSHIP} \notin \text{quasi-1D} & \text{(Lemma 6 and (21))} \\
\implies \text{TALL LIST MEMBERSHIP} \notin 2^{1D} & \text{(Lemma 7 and (22))} \\
\implies \text{BINARY TALL LIST MEMBERSHIP} \notin 2^{1D} & \text{(Lemma 6 and (23))}
\end{array}$$

Overall, we see that the lower bound of Proposition 3 follows from the hardness of the core problem of Proposition 1 (Fact 7) and from properties of the classes.

For the upper bound of Proposition 3, we see that one of the two witnesses satisfies it because the other one does: TALL LIST MEMBERSHIP $\in$ P$_1$D follows from the next fact and lemma, and since the homomorphisms of (23) are tight:

Fact 8. BINARY TALL LIST MEMBERSHIP $\in$ P$_1$D.

Lemma 8. P$_1$D is closed under $\leq_H^t$.

3 Modular Witnesses

The three propositions of [5] offered witness languages for the differences $1N \setminus 1D$, $2D \setminus 1D$, and P$_1$D $\setminus 2^{1D}$. By analyzing these languages, we arrived at eight promise problems witnessing these differences. In the end, all bounds that we needed for these problems followed from Facts 3, 5, 8 (for the upper bounds) and Fact 7 (for the lower bounds), via the established relations between these problems and using a collection of lemmas of three distinct types:

– *preservation of hardness*: These are lemmas of the form

$$\mathcal{L} \notin C \implies \mathcal{L}' \notin C$$

where $\mathcal{L}'$ is derived from $\mathcal{L}$. E.g., Lemma 1 is such a lemma, with $\mathcal{L}'$ any generalization of $\mathcal{L}$. Similarly for every lemma for the closure of a class under $\leq_H$ or $\leq_{1D}$, with $\mathcal{L}'$ any generalization of H$(\mathcal{L})$ or 1D$(\mathcal{L})$.

– *propagation of hardness*: These are lemmas of the form

$$\mathcal{L} \notin \mathsf{C} \implies \mathcal{L}' \notin \mathsf{C}'$$

where $\mathcal{L}', \mathsf{C}'$ are derived from $\mathcal{L}, \mathsf{C}$ [3]. E.g., Lemma 7 is such a lemma, where $\mathcal{L}' = \mathsf{TALL}(\mathcal{L})$, and $\mathsf{C}' = 2^{\mathsf{1D}}$ is derived from $\mathsf{C} = \mathsf{quasi\text{-}1D}$ by an application of the general operator which raises the size bound $f(h)$ of a class to $f(2^h)$.

– *preservation of easiness*: These are lemmas of the form

$$\mathcal{L} \in \mathsf{C} \implies \mathcal{L}' \in \mathsf{C}$$

where $\mathcal{L}'$ is derived from $\mathcal{L}$. E.g., Lemma 5 is such, with $\mathcal{L}' = \mathsf{LPAD}(\mathcal{L})$ and $\mathsf{C} = \mathsf{1N}$. Same for any lemma for the closure of a class under an operation.

Hence, the propositions of [5] are connected via structural relations between their witnesses, which are easy to miss if we do not adopt the right point of view.

We now further observe that the relations between witnesses allow us to express each of them as a generalization of a problem that can be obtained from MEMBERSHIP by applying a sequence of operators. Specifically, in the following list, every witness on the left generalizes the problem on the right:

SHORT RETROCOUNT:	$\mathcal{H}_{15}(\text{MEMBERSHIP})$
RETROCOUNT:	$\mathsf{LPAD}(\mathcal{H}_{15}(\text{MEMBERSHIP}))$
COMPACT PROJECTION:	$\mathcal{H}_{17}(\text{MEMBERSHIP})$
PROJECTION:	$\mathcal{H}_{13}(\mathcal{H}_{17}(\text{MEMBERSHIP}))$
UNARY PROJECTION:	$\mathcal{T}_{13}(\mathcal{H}_{13}(\mathcal{H}_{17}(\text{MEMBERSHIP})))$
TALL LIST MEMBERSHIP:	$\mathsf{TALL}(\mathcal{H}_{21}(\text{MEMBERSHIP}))$
BINARY TALL LIST MEMBERSHIP:	$\mathcal{H}_{23}(\mathsf{TALL}(\mathcal{H}_{21}(\text{MEMBERSHIP})))$

where $\mathcal{H}_{15}$ is the family of the homomorphic reductions that justifies (15), and similarly for all other $\mathcal{H}_i$ and $\mathcal{T}_i$. Every lower bound for a witness on the left was established by proving a lower bound for the corresponding problem on the right and then using Lemma 1. Again by (the contrapositive of) Lemma 1, every upper bound for a witness on the left is also an upper bound for the respective problem on the right. Overall, the problems on the right witness the same differences as the problems on the left.

Let a *modular witness* for a difference $\mathsf{C}' \setminus \mathsf{C}$ be any problem which belongs to the difference and is derived from MEMBERSHIP by applying a sequence of operators. Our discussion above shows that every one of the differences in Propositions 1–3 of [5] admits a modular witness.

We conjecture that the same is true for all differences in minicomplexity. Namely that, for every two minicomplexity classes C and C':

If $\mathsf{C}' \setminus \mathsf{C}$ is not empty, then it contains a modular witness.

In the talk, we will examine as evidence supporting this conjecture several examples of known separations where the offered witnesses were indeed (generalizations of) modular ones or can be replaced by modular witnesses.

If this conjecture is true, then designing a witness for a separation reduces to (i) deciding which sequence of operators to apply to MEMBERSHIP, and then (ii) proving the corresponding necessary lemmas of hardness propagation and of hardness or easiness preservation. Gradually, this could lead to a library of operators and corresponding lemmas, available for reuse in (i) and (ii). It would, of course, also be interesting to see a proof of the conjecture, that explains why MEMBERSHIP is sufficient as the only "seed of hardness" in this domain.

If the conjecture is false, it would be interesting to see examples where it fails. Understanding these examples, one could then suggest conditions under which the conjecture remains valid, and work with this updated, restricted variant.

Notes

[1] Proposition 1 also includes a last sentence, that the reverse of R_n needs only $O(n)$ states on a 1DFA. We omitted that sentence, as it is redundant for our purposes.

[2] We write "$<2n$", although the definition of R_n'' also allows strings of length exactly $2n$. Excluding such strings does not change the desired properties of R_n'' in the context of Proposition 1; and is convenient for our purposes.

[3] In [2], the name PROJECTION was used for the reverse of this problem.

[4] In [2], the name MEMBERSHIP was used for the reverse of this problem.

[5] Note the redundant component i_{h+1}, which is always set of $h + 1$. We need $h + 1$ components, because we need $h + 1$ values; and we need $h + 1$ values, because we need to ensure that the max and min values are distinct even when $h = 1$.

[6] In [2], the reverse of this problem was called ∃EQUALITY.

References

1. Kapoutsis, C.A.: Size complexity of two-way finite automata. In: Diekert, V., Nowotka, D. (eds.) DLT 2009. LNCS, vol. 5583, pp. 47–66. Springer, Heidelberg (2009). https://doi.org/10.1007/978-3-642-02737-6_4
2. Kapoutsis, C.A.: Minicomplexity. In: Kutrib, M., Moreira, N., Reis, R. (eds.) DCFS 2012. LNCS, vol. 7386, pp. 20–42. Springer, Heidelberg (2012). https://doi.org/10.1007/978-3-642-31623-4_2
3. Kapoutsis, C., Královič, R., Mömke, T.: Size complexity of rotating and sweeping automata. J. Comput. Syst. Sci. 78(2), 537–558 (2012)
4. Kapoutsis, C., Pighizzini, G.: Two-way automata characterizations of L/poly versus NL. Theory Comput. Syst. 56, 662–685 (2015)
5. Meyer, A.R., Fischer, M.J.: Economy of description by automata, grammars, and formal systems. In: Proceedings of FOCS, pp. 188–191 (1971)
6. Sakoda, W.J., Sipser, M.: Nondeterminism and the size of two-way finite automata. In: Proceedings of STOC, pp. 275–286 (1978)

Action Research in Software Engineering: Metrics' Research Perspective (Invited Talk)

Miroslaw Staron[✉]

Chalmers | University of Gothenburg, Gothenburg, Sweden
miroslaw.staron@gu.se

Abstract. Software engineering is an applied discipline of science. Its focus on software development processes, technologies and organizations makes it broad and exciting to work with. However, it makes it also challenging to find the right research method to get the results that have impact on industrial practices and that help software organizations to provide more value to their customers. In this paper, I argue that the traditional empirical software engineering methods must be complemented with action research. In the paper, I provide an overview of the methodology of action research, briefly explain the phases and present my experiences from over a decade long applications of action research in industrial contexts. I focus on how action research helps to provide the most value to the collaborating companies, how it helps to build more robust software engineering theories and how it helps individual researchers to develop their careers. I conclude with a short description of how action research can evolve in the future.

1 Introduction – What Action Research Is

Empirical methods in software engineering have a long tradition. The advances in the last two decades, started by the book of Wohlin et al. [15]. Experiments, case studies and surveys became very popular as methods for collecting data about software engineering practices. They were conducted in increasingly orderly fashion and software engineering journals became increasingly vary about the quality of the empirical research.

Action research is one of the research methodologies that gained popularity in the second part of to 20th century [4]. The reason for its popularity is that action research focuses on organizational learning as part of the process of research. Today, action research is believed to swing the balance in software engineering towards industrial practices [5], mainly because it focuses on improvement of the practice, learning and focus on what practitioners do rather than what they say they do [2]. Compared to other research methodologies, where the focus is either on the observation and learning or evaluating.

Action research is defined by Sagor as *is a disciplined process of inquiry conducted by and for those taking the action. The primary reason for engaging*

© Springer Nature Switzerland AG 2019
B. Catania et al. (Eds.): SOFSEM 2019, LNCS 11376, pp. 39–49, 2019.
https://doi.org/10.1007/978-3-030-10801-4_4

in action research is to assist the "actor" in improving and/or refining his or her actions, [10]. As the definition indicates, it is focused on improving the work of the actors taking the action.

Another definition is presented by Reason and Bradbury as *a participatory, democratic process concerned with developing practical knowing in the pursuit of worthwhile human purposes, grounded in a participatory world view which we believe is emerging at this historical moment. It seeks to bring together action and reflection, theory and practice, in participation with others, in the pursuit of practical solutions to issues of pressing concern to people, and more generally the flourishing of individual persons and their communities* [8]. In this definition, Reason and Bradbury emphasize the aspect of participation and the aspect of practical solutions – creating new practices and new products.

Baskerville describes action research as an important example of modern research method in the area of information systems: "It is empirical, yet interpretive. It is experimental, yet multivariate. It is observational, yet interventionist". These characteristics make it perfect for software engineering research.

2   The Phases of Action Research Projects

There are a number of different ways of describing action research, which often differ in the number of phases of an action research project. The basic one contains five phases and is often referred to as canonical action research, as presented in Fig. 1 after [3] and [5].

Figure 2 shows the context of action research in terms of inputs, outputs, and stakeholders. This model of action research has been developed by my research team and we have worked according to it since 2006.

The important aspect of the figure is the input to the action research projects. Practitioners often bring the needs to improve their organizations, products or operations. The researchers often bring in the needs to evaluate or validate methods and tools in the new context of the collaborating organization. The different colors of these two inputs indicate that the two parties – industry and academia – often come from different directions and bring these inputs independently of each other. The mixed colors of the phases of action research and the outputs show that the rest of action research is done collaboratively and that it impacts both industry and academia.

The research team, which I call the action team, often has one or two leading researchers and one or two leading practitioners – stakeholders. Their goal is to develop their respective organizations as well as their own competence.

2.1   Diagnosing

Every action research cycle starts with addressing the question of *What is the problem that we need to solve?* Although the question is often partially answered when initiating the project, it's important to specify which part of the problem should be addressed in each cycle.

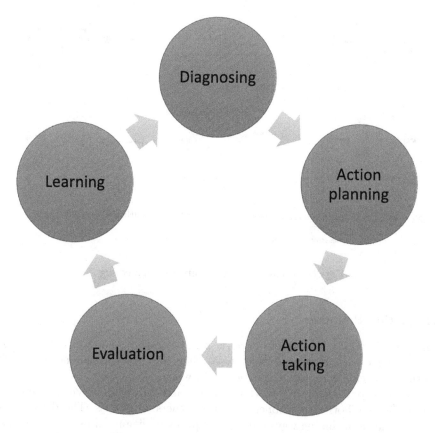

Fig. 1. Canonical action research cycle.

The first phase of each action research cycle – diagnosing – is unique for action research. Instead of starting a project with a detailed problem formulation, action research recognizes the fact that one needs to be embedded in the context in order to elicit the problem correctly. Therefore, every action research cycle starts with a precise diagnose of which problem should be solved.

Action researchers should start by collecting opinions and symptoms which they need to explore in order to decide which challenge to address during the action research cycle. It's important that the researchers focus on discussions with the practitioners when exploring the context and deciding what to do. The problem to be solved in each cycle should be limited in scope and its effects should be measurable.

One could see the diagnosing phase as similar to requirements elicitation phase in agile projects of the first part of the market analysis phase of the Build-Measure-Learn cycle of continuous deployment projects [9].

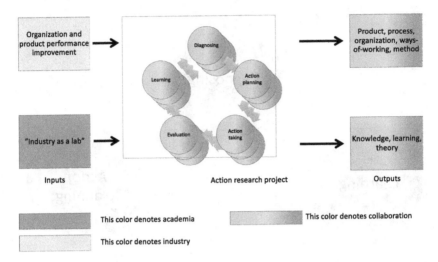

Fig. 2. Action research in software engineering

2.2   Action Planning

Planning of actions in a single cycle is always done in a collaborative manner. Academic researchers, industrial researchers and practitioners need to work together to decide who does the actions and when.

The collaborative nature of the action planning phase provides a unique opportunity for both practitioners and researchers to discuss. The discussions are often aimed at finding ways to solve the problem diagnosed in the first phase and identify resources, products and processes to be investigated and adjusted.

In the action planning activity, the research team discusses their plans with the reference groups and needs to get approval for the required resources from the management team. The plans need to be aligned with theoretical foundations of the work, i.e. the research team needs to identify theoretical or empirical work relevant for the diagnosed problem and plan the actions accordingly.

In this phase, the research team, together with the reference team, makes the plans for which data should be collected, from which objects, using which tools. The team also plans for which analysis methods should be employed to assess whether their actions lead to solving the diagnosed problem.

Often, although far from always, the research team plans their actions using standard project planning tools, like Gantt charts and work breakdown structures. However, these are often lightweight and documented only internally for the research team to follow and use as a communication tool to management.

2.3   Action Taking

The action taking phase is dedicated to making changes in the context – interventions. The phase is executed according to the plans laid out in the previous phase and is conducted by the action team. The reference group is involved on regular basis to provide feedback and to help the action team to solve the challenges that they encounter [1].

The action taking phase is specific for action research as it is one of the research methodologies where making changes are allowed midst operations of its context. For example, the research team is allowed to change the ways of working for software development teams midst sprint and observe these changes.

It is important to note that the action taking phase is both about making the change and observing its effect. As action research is a quantitative methodology, the data collection activities provide the possibility to reduce the bias of subjective observations and provide quantitative evidence. This quantitative evidence is used in the next phase – action evaluation – to assess the effects of the actions and is used as the input to the next's cycle diagnosing phase.

2.4 Evaluation

In the evaluation phase, the action team analyzes the data collected from the previous phase. The team uses statistical methods to make the analyses and presents the results to the reference team and the management.

In case when the data shows that the diagnosed problem is indeed solved using the actions taken, the outcome is straightforward. If the data is inconclusive, the action team needs to plan for either additional analyses, additional data to be collected or needs to pivot – i.e. finalize the current cycle, specify learning and find a new diagnose of the problem given the new data collected.

In the evaluation section, the team usually uses the same statistical methods as experimentation, i.e. descriptive and inferential statistics.

The team also needs to assure that the analysis of their data is aligned with the theories used in the cycle. This is important in order to make the contribution to the theory-building in the next phase.

2.5 Learning

The final part of the action research cycle is the specification of learning. It is done both as practical guidelines for the involved organizations and contexts, and as theory-building for the research community.

The practical guidelines are often specified in terms of guidebooks, white papers and instructions at the company's web. For example software development teams often use wiki-s to specify good practices and document good examples. That's often when the results of action research cycles can be found.

The contribution to the theory-building are often specified as scientific papers, with the scientific rigour and relevance. It is often the case that these are documented as experience reports from industrial studies, e.g. [6].

3   Maximizing Impact – Who Should be a Part of the Action Team

The main point of successful action research collaboration is the fact that industry and academia work together and the knowledge, theories, methods and tools impact both. It is important to note that there are no double cycles – one for academia and one for industry – although one could interpret it like that.

In action research projects, there usually are a number of actors. There is the action team, who is responsible for planning, executing and evaluating the research. There is a reference group, who is responsible for the advice and feedback for the research team. Finally, there is a management team, who is responsible for the management and governance of the project and providing important decisions for the institutionalization of change.

The action team consists of both practitioners and researchers. The practitioners are software engineers involved in planning and executing actions, e.g. architects, testers, designers, project managers and quality managers. It is important that they are involved as part of the research team, because they provide the context of the actions and it is their work that is changed as part of the research work.

So far we discussed the actors and practitioners in action research projects. However, the role of management is equally important, so we should also discuss the ways in which action research projects are managed.

Figure 3 shows an example organizational chart of an action research project. The figure contains three parts: (i) the research team, (ii) the reference group, and (iii) management.

The role of the action team is to plan and execute the research project. The role of the reference group is to provide the possibility to get feedback on the progress of the project. The reference group also helps the team to diagnose the problems and therefore steer the project in the right direction.

Finally, the management of the company is important as they decide upon the resources needed for the project. The resources, in turn, determine the scope of the project. The product and process management are important as they help to support the project in making the right impact of the results of their actions.

The researchers provide an external perspective on the organizational change and their role is to bring in theories and state-of-the-art research results to the collaboration. The researchers often ask critical questions and provide the possibility to bring in expertise from other projects.

In several countries, the legislation is not suited for the companies to directly engage in collaborations. Due to intellectual property rights management, resource allocation or anti-competitor regulations, companies are often discouraged by the amount of legal work required to formalize a collaboration.

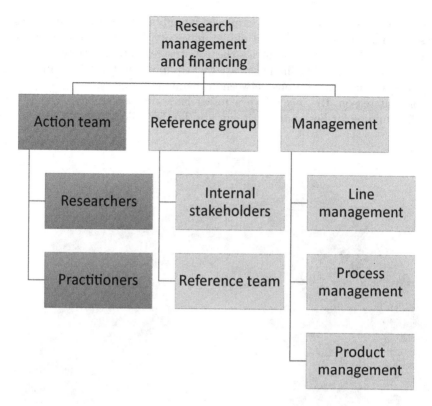

Fig. 3. "Organizational chart" of action research projects.

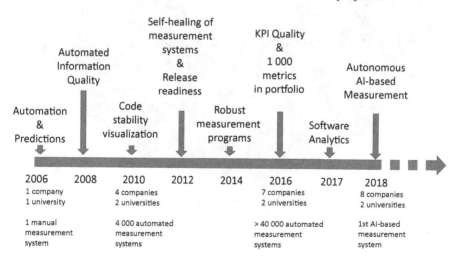

Fig. 4. Metrics action research timeline

However, researchers are often employed by public universities and it's easier to establish a collaboration with a public university, because of the established legal framework. In many countries the country regulators specify who owns research results and public financiers are well equipped with legal documents on how to establish collaborations between academia in industry.

For that reason, the mix between researchers and practitioners can be the most fruitful one for both parties; it can also be the easiest one.

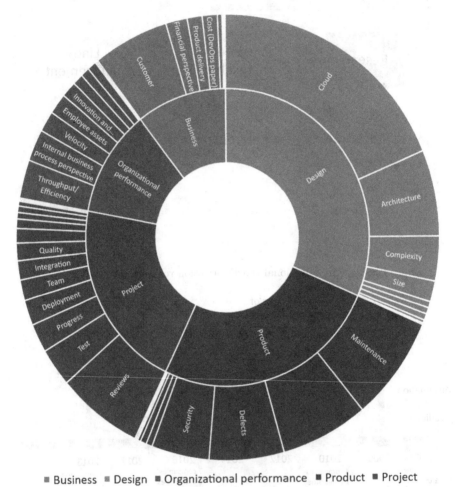

Fig. 5. Metrics portfolio – a catalogue of over 2000 metrics definition and experience of their use

4   Metrics Action Research

My research team has applied and developed action research methodology since 2006 in close collaboration with Ericsson and Volvo Cars. A timeline of the collaboration is shown in Fig. 4. One of the reasons why we have managed to sustain the collaboration is just action research. In particular, it is the fact that we could build an understanding and the flexibility to adjust to each others', constantly changing, needs.

Our first projects, and thus action research cycles, started by solving well-known, still unsolved, problems of predicting defects in software projects [11]. We started with standard, statistic-based approaches, which evolved into dynamic models based on bottleneck predictions and feature/defect flows [13]. We replicated these studies at other companies, yet with a timeline that was order of magnitude shorter (weeks instead of months) [7].

The project evolved from predictions, to infrastructure of measurement systems – understanding quality of information [12] and self-healing of measurement systems [14]. It has also constantly changed, just as software engineering changed, the company evolved and the action team matured.

Today, the action team has a portfolio of over 2000 metrics, documented from the experience and literature, shown in Fig. 5.

The team has written two books and over 200 papers; it has grown from one company and one university to eight companies and two universities. There are over 40,000 measurement systems that are developed based on the team's research.

5   Conclusions: The Future of Action Research

In this talk I describe the methodology of action research in the context of software engineering. The specifics of software development organizations, which is the focus on the development of new technologies, new processes and new ways-of-working, makes it unique and interesting for action researchers.

Action research consists of different phases, which are executed iteratively and which contribute to the development of the host organizations as well as the development of software engineering theories. The close collaboration in action research projects closes the gap between the researchers and practitioners – they all are part of the action team.

Almost all action teams, which I've observed, remove barriers between academia and industry. They focus on the task-at-hand and after a while, they often find themselves to build long-term collaborations build on team-feeling, trust, transparency, and altruistic goals.

It's the above close collaboration that will shape the future of action research and software engineering. The number of theories that are developed in closed university rooms will decrease. Instead, the action teams will increase, the collaboration will increase and so will the maturity of software engineering researchers.

We will see more software science done by action teams and the boundary between academic research and industrial practice will blur. When it does, the university professors will be more active in industry, they will help the students to get into industrial contexts faster. The industry practitioners will ask for academic more often and therefore constantly develop their university relations. The students will get more opportunities to work with companies who understand the scientific requirements on rigour of studies and rigour of reporting.

References

1. Antinyan, V., Staron, M., Sandberg, A., Hansson, J.: Validating software measures using action research a method and industrial experiences. In: Proceedings of the 20th International Conference on Evaluation and Assessment in Software Engineering, p. 23. ACM (2016)
2. Avison, D.E., Lau, F., Myers, M.D., Nielsen, P.A.: Action research. Commun. ACM **42**(1), 94–97 (1999). https://doi.org/10.1145/291469.291479
3. Baskerville, R.: Educing theory from practice. In: Kock, N. (ed.) Information Systems Action Research. Integrated Series in Information Systems, vol. 13, pp. 313–326. Springer, Boston (2007). https://doi.org/10.1007/978-0-387-36060-7_13
4. Brydon-Miller, M., Greenwood, D., Maguire, P.: Why action research? (2003)
5. Dos Santos, P.S.M., Travassos, G.H.: Action research can swing the balance in experimental software engineering. In: Advances in Computers, vol. 83, pp. 205–276. Elsevier (2011)
6. Meding, W.: Effective monitoring of progress of agile software development teams in modern software companies: an industrial case study. In: Proceedings of the 27th International Workshop on Software Measurement and 12th International Conference on Software Process and Product Measurement, pp. 23–32. ACM (2017)
7. Rana, R., et al.: Evaluation of standard reliability growth models in the context of automotive software systems. In: Heidrich, J., Oivo, M., Jedlitschka, A., Baldassarre, M.T. (eds.) PROFES 2013. LNCS, vol. 7983, pp. 324–329. Springer, Heidelberg (2013). https://doi.org/10.1007/978-3-642-39259-7_26
8. Reason, P., Bradbury, H.: Handbook of Action Research: Participative Inquiry and Practice. Sage, Thousand Oaks (2001)
9. Ries, E.: The Lean Startup: How Today's Entrepreneurs Use Continuous Innovation to Create Radically Successful Businesses. Crown Books, New York (2011)
10. Sagor, R.: Guiding School Improvement with Action Research. ASCD, Alexandria (2000)
11. Staron, M., Meding, W.: Predicting weekly defect inflow in large software projects based on project planning and test status. Inf. Softw. Technol. **50**(7–8), 782–796 (2008)
12. Staron, M., Meding, W.: Ensuring reliability of information provided by measurement systems. In: Abran, A., Braungarten, R., Dumke, R.R., Cuadrado-Gallego, J.J., Brunekreef, J. (eds.) IWSM 2009. LNCS, vol. 5891, pp. 1–16. Springer, Heidelberg (2009). https://doi.org/10.1007/978-3-642-05415-0_1
13. Staron, M., Meding, W.: Monitoring bottlenecks in agile and lean software development projects – a method and its industrial use. In: Caivano, D., Oivo, M., Baldassarre, M.T., Visaggio, G. (eds.) PROFES 2011. LNCS, vol. 6759, pp. 3–16. Springer, Heidelberg (2011). https://doi.org/10.1007/978-3-642-21843-9_3

14. Staron, M., Meding, W., Tichy, M., Bjurhede, J., Giese, H., Söder, O.: Industrial experiences from evolving measurement systems into self-healing systems for improved availability. Softw. Pract. Exp. **48**(3), 719–739 (2018)
15. Wohlin, C., Runeson, P., Höst, M., Ohlsson, M.C., Regnell, B., Wesslén, A.: Experimentation in Software Engineering. Springer, Heidelberg (2012). https://doi.org/10.1007/978-3-642-29044-2

From Big Data to Big Knowledge

Large-Scale Information Extraction Based on Statistical Methods (Invited Talk)

Martin Theobald[✉]

University of Luxembourg, Esch-sur-Alzette, Luxembourg
martin.theobald@uni.lu

Abstract. Today's knowledge bases (KBs) capture facts about the world's entities, their properties, and their semantic relationships in the form of subject-predicate-object (SPO) triples. Domain-oriented KBs, such as DBpedia, Yago, Wikidata or Freebase, capture billions of facts that have been (semi-)automatically extracted from Wikipedia articles. Their commercial counterparts at Google, Bing or Baidu provide back-end support for search engines, online recommendations, and various knowledge-centric services.

This invited talk provides an overview of our recent contributions—and also highlights a number of open research challenges—in the context of extracting, managing, and reasoning with large semantic KBs. Compared to domain-oriented extraction techniques, we aim to acquire facts for a much broader set of predicates. Compared to open-domain extraction methods, the SPO arguments of our facts are canonicalized, thus referring to unique entities with semantically typed predicates. A core part of our work focuses on developing scalable inference techniques for querying an uncertain KB in the form of a probabilistic database. A further, very recent research focus lies also in scaling out these techniques to a distributed setting. Here, we aim to process declarative queries, posed in either SQL or logical query languages such as Datalog, via a proprietary, asynchronous communication protocol based on the Message Passing Interface.

Keywords: Information extraction
Probabilistic databases · Distributed graph databases

1 Information Extraction

The World Wide Web is the most comprehensive—but likely also the most complex—source of information that we have access to today. A vast majority of all information in the Surface Web, i.e., the part of the Web that is publicly accessible either as static pages or in the form of dynamically created contents, in fact consists of unstructured text. This textual information just happens to occasionally be interspersed with semi-structured components such as form fields, lists,

© Springer Nature Switzerland AG 2019
B. Catania et al. (Eds.): SOFSEM 2019, LNCS 11376, pp. 50–53, 2019.
https://doi.org/10.1007/978-3-030-10801-4_5

and tables—or so-called "infoboxes" in Wikipedia. These infoboxes, plus perhaps some more metadata, however, still constitute the main source of information for all of the currently available, Wikipedia-centric knowledge bases such as DBpedia (Auer et al. 2007), YAGO (Suchanek et al. 2007), Freebase (Bollacker et al. 2008), and Wikidata (Vrandečić and Krötzsch, 2014). This means that we currently exploit only a very small fraction of the information that is published on the Web for the purpose of *information extraction* (IE) and *knowledge base* (KB) construction.

In a recent series of publications on systems like KORE (Hoffart et al. 2012), AIDA-light (Nguyen et al. 2014), J-NERD (Nguyen et al. 2016), J-REED (Nguyen et al. 2017b) and QKBfly (Nguyen et al. 2017a), we stepwisely investigated the transition from basic, domain-oriented IE tasks like *named-entity recognition* (NER) and *disambiguation* (NED) toward a more extensible, open-domain IE setting, which combines NER and NED with a flexible form of pattern matching of relational paraphrases (Nakashole et al. 2011; Nakashole et al. 2012) into a comprehensive framework for *relation extraction* (RE). A focal point of our efforts thereby lies in combining these IE steps via various forms of joint-inference, rather than by following the more traditional, pipelined architectures for IE and NLP: J-NERD, for example, merges the tasks of NER and NED by performing a form of joint-inference over an underlying probabilistic-graphical model (in the form of a conditional random field), while J-REED and QKBfly further integrate NER and NED with RE and other NLP tasks like pronoun and co-reference resolution (via a semantic graph representation and corresponding graph-densification algorithm). Compared to domain-oriented extraction techniques, we thereby aim to acquire facts for a much broader set of predicates. Compared to open-domain extraction methods, the SPO arguments of our facts are canonicalized, thus referring to unique entities with semantically typed predicates.

2   Probabilistic Databases

Probabilistic databases (PDBs) (Suciu et al. 2011) encompass a plethora of applications for managing uncertain data, ranging from scientific data management, sensor networks, data integration, to information extraction and reasoning with semantic knowledge bases. While classical database approaches benefit from a mature and scalable infrastructure for the management of relational data, PDBs aim to further combine these well-established data management strategies with efficient algorithms for probabilistic inference by exploiting given independence assumptions among database tuples whenever possible. Moreover, PDBs adopt powerful query languages from relational databases, including Relational Algebra, the Structured Query Language (SQL), and logical query languages such as Datalog. The Trio PDB system (Mutsuzaki et al. 2007), which we developed at Stanford University back in 2006, was the first such system that explicitly addressed the integration of data management (using SQL as query language), lineage (aka. "provenance") management via Boolean formulas, and probabilistic inference based on the lineage of query answers. The Trio data model, coined

"*Uncertainty and Lineage Databases*" (ULDBs) (Benjelloun et al. 2008), provides a closed and complete probabilistic extension to the relational data model under all of the common relational (i.e., SQL-based) operations. Our recent research contributions in the domain of PDBs comprise a lifted form of evaluating top-k queries over non-materialized database views (Dylla et al. 2013b), learning of tuple probabilities from user feedback (Dylla et al. 2014), as well as further temporal-probabilistic database extensions (Dylla et al. 2013a; Dylla et al. 2011; Papaioannou et al. 2018).

3   Distributed Graph Databases

The third part of the talk finally takes an in-depth look at the architecture of our TriAD (for "*Triple-Asynchronous-Distributed*") (Gurajada et al. 2014a; Gurajada et al. 2014b) engine, which provides an end-to-end system for the distributed indexing of large RDF collections and the processing of queries formulated in the SPARQL 1.0 standard. TriAD combines a novel form of sharded, main-memory-based index structures with an asynchronous communication protocol based on the Message Passing Interface (MPI). It thus aims to bridge the gap between shared-nothing MapReduce-based RDF engines, on the one hand, and shared-everything native graph engines, on the other hand (see (Abdelaziz et al. 2017) for a recent overview). TriAD is designed to achieve higher parallelism and less synchronization overhead during query executions than MapReduce engines by adding an additional layer of multi-threading for entire execution paths within a query plan that can be executed in parallel. TriAD is the first RDF engine that employs asynchronous join executions, which are coupled with a lightweight join-ahead pruning technique based on graph summarization. Our current work also considers the processing of multi-source, multi-target graph-reachability queries (coined "*Distributed Set Reachability*" (DSR)) (Gurajada and Theobald, 2016b), as they may occur, for example, in the recent "Property Paths" extension of SPARQL 1.1 (Gurajada and Theobald, 2016a).

References

Abdelaziz, I., Harbi, R., Khayyat, Z., Kalnis, P.: A survey and experimental comparison of distributed SPARQL engines for very large RDF data. PVLDB **10**(13), 2049–2060 (2017)

Auer, S., Bizer, C., Kobilarov, G., Lehmann, J., Ives, Z.: DBpedia: a nucleus for a web of open data. In: ISWC, pp. 11–15 (2007)

Benjelloun, O., Sarma, A.D., Halevy, A.Y., Theobald, M., Widom, J.: Databases with uncertainty and lineage. VLDB J. **17**(2), 243–264 (2008)

Bollacker, K., Evans, C., Paritosh, P., Sturge, T., Taylor, J.: Freebase: a collaboratively created graph database for structuring human knowledge. In: SIGMOD, pp. 1247–1250 (2008)

Dylla, M., Miliaraki, I., Theobald, M.: A temporal-probabilistic database model for information extraction. PVLDB **6**(14), 1810–1821 (2013a)

Dylla, M., Miliaraki, I., Theobald, M.: Top-k query processing in probabilistic databases with non-materialized views. In: ICDE, pp. 122–133 (2013b)

Dylla, M., Sozio, M., Theobald, M.: Resolving temporal conflicts in inconsistent RDF knowledge bases. In: BTW, pp. 474–493 (2011)

Dylla, M., Theobald, M., Miliaraki, I.: Querying and learning in probabilistic databases. In: Reasoning Web, pp. 313–368 (2014)

Gurajada, S., Seufert, S., Miliaraki, I., Theobald, M.: TriAD: a distributed shared-nothing RDF engine based on asynchronous message passing. In: SIGMOD, pp. 289–300 (2014a)

Gurajada, S., Seufert, S., Miliaraki, I., Theobald, M.: Using graph summarization for join-ahead pruning in a distributed RDF engine. In: SWIM, pp. 41:1–41:4 (2014b)

Gurajada, S., Theobald, M.: Distributed processing of generalized graph-pattern queries in SPARQL 1.1. CoRR, abs/1609.05293 (2016a)

Gurajada, S., Theobald, M.: Distributed set reachability. In: SIGMOD, pp. 1247–1261 (2016b)

Hoffart, J., Seufert, S., Nguyen, D.B., Theobald, M., Weikum, G.: KORE: keyphrase overlap relatedness for entity disambiguation. In: CIKM, pp. 545–554 (2012)

Mutsuzaki, M., et al.: Trio-one: layering uncertainty and lineage on a conventional DBMS. In: CIDR, pp. 269–274 (2007)

Nakashole, N., Theobald, M., Weikum, G.: Scalable knowledge harvesting with high precision and high recall. In: WSDM, pp. 227–236 (2011)

Nakashole, N., Weikum, G., Suchanek, F.M.: PATTY: a taxonomy of relational patterns with semantic types. In: EMNLP-CoNLL, pp. 1135–1145 (2012)

Nguyen, D.B., Abujabal, A., Tran, K., Theobald, M., Weikum, G.: Query-driven on-the-fly knowledge base construction. PVLDB 11(1), 66–79 (2017a)

Nguyen, D.B., Hoffart, J., Theobald, M., Weikum, G.: AIDA-light: high-throughput named-entity disambiguation. In: LDOW (2014)

Nguyen, D.B., Theobald, M., Weikum, G.: J-NERD: joint named entity recognition and disambiguation with rich linguistic features. TACL 4, 215–229 (2016)

Nguyen, D.B., Theobald, M., Weikum, G.: J-REED: joint relation extraction and entity disambiguation. In: CIKM, pp. 2227–2230 (2017b)

Papaioannou, K., Theobald, M., Böhlen, M.H.: Supporting set operations in temporal-probabilistic databases. In: ICDE, pp. 1180–1191 (2018)

Suchanek, F.M., Kasneci, G., Weikum, G.: YAGO: a core of semantic knowledge. In: WWW, pp. 697–706 (2007)

Suciu, D., Olteanu, D., Ré, C., Koch, C.: Probabilistic databases. Synth. Lect. Data Manag. 3(2), 1–180 (2011)

Vrandečić, D., Krötzsch, M.: Wikidata: a free collaborative knowledgebase. Comm. of the ACM 57(10), 78–85 (2014)

Sorting Networks on Restricted Topologies

Indranil Banerjee[1]($\boxtimes$) (iD), Dana Richards[2] (iD), and Igor Shinkar[3] (iD)

[1] Louisiana State University, Baton Rouge, LA 70803, USA
ibanerjee@lsu.edu
[2] George Mason University, Fairfax, VA 22030, USA
richards@gmu.edu
[3] Simon Fraser University, Burnaby, BC V5A 1S6, Canada
ishinkar@cs.sfu.ca

Abstract. The *sorting number* of a graph with n vertices is the minimum depth of a sorting network with n inputs and n outputs that uses only the edges of the graph to perform comparisons. Many known results on sorting networks can be stated in terms of sorting numbers of different classes of graphs. In this paper we show the following general results about the sorting number of graphs.

1. Any n-vertex graph that contains a simple path of length d has a sorting network of depth $O(n \log(n/d))$.
2. Any n-vertex graph with maximal degree Δ has a sorting network of depth $O(\Delta n)$.

We also provide several results relating the sorting number of a graph with its routing number, size of its maximum matching, and other well known graph properties. Additionally, we give some new bounds on the sorting number for some typical graphs.

Keywords: Sorting networks · Matchings in graphs
Routing via matchings

1 Introduction

In this paper we study oblivious sorting algorithms. These are sorting algorithms whose sequence of comparisons is made in advance, before seeing the input, such that for any input of n numbers the value of the i'th output is smaller or equal to the value of the j'th output for all $i < j$. That is, for any of the $n!$ possible permutations of the input, the output of the algorithm must be sorted. A sorting network, which typically arises in the context of parallel algorithms, is an oblivious algorithm, where the comparisons are grouped into *stages*, and in each stage the compared pairs are disjoint. In this paper we explore the situation where a given graph specifies which keys are allowed to be compared. We regard

---

The first author was partially supported by DTIC Contract FA8075-14-D-0002/0007.

B. Catania et al. (Eds.): SOFSEM 2019, LNCS 11376, pp. 54–66, 2019.
https://doi.org/10.1007/978-3-030-10801-4_6

a sorting network as a sequence of stages, where each stage corresponds to a matching in the graph and a comparator is assigned to each matched pair. There are fixed locations, identified with the vertices, each containing a key, and a comparator looks at the keys at the endpoints of the edges of the matching, and swaps them if they are not in the order desired by the underlying oblivious algorithm. Therefore, we say that the underlying algorithm induces a *directed matching*. The locations are ordered, and the goal is to have the order of the keys match the order of the locations after the execution of the algorithm. The *depth* of a sorting network is the number of stages, and the *size* is the total number of edges in all the matchings. Note that for an input of length n at most $\lfloor n/2 \rfloor$ comparisons can be performed in each step, and hence the well-known lower bound of $\Omega(n \log(n))$ comparisons in the sequential setting implies a $\Omega(\log(n))$ lower bound on the depth of the network, that is, the number of stages in the network.

A large variety of sorting networks have been studied in the literature. In their seminal paper, Ajtai, Komlós, and Szemerédi [1] presented a construction of a sorting network of depth $O(\log n)$. We will refer to it as the *AKS sorting network*. In this work we explore the question of constructing a sorting network where we are given a graph specifying which locations are allowed to be compared. We define the *sorting number* of a graph G, denoted by $st(G)$, to be the minimal depth of a sorting network that uses only the edges of G. The AKS sorting network can be interpreted as a sorting network on the complete graph, i.e., $st(K_n) = O(\log(n))$. More precisely, the AKS construction specifies *some* graph G_{AKS} whose maximal degree is $O(\log(n))$ and $st(G_{AKS}) = O(\log(n))$.

The setting where the comparisons are restricted to the n-vertex path graph, denoted by P_n, is perhaps the easiest case. It is well known that $st(P_n) \leq n$, which follows from the fact that the odd-even transposition sort takes n matching steps (see, e.g., [7]). For the hypercube graph Q_d on $n = 2^d$ vertices we can use the Batcher's bitonic sorting network, which has a depth of $O((\log n)^2)$ [5]. This was later improved to $2^{O(\sqrt{\log \log n})} \log n$ by Leighton and Plaxton [9]. We also have a lower bound of $\Omega(\frac{\log n \log \log n}{\log \log \log n})$ for a certain natural class of sorting networks on the hypercube due to Plaxton and Suel [10]. For the square mesh $P_n \times P_n$ it is known that $st(P_n \times P_n) = 3n + o(n)$, which is tight with respect to the constant factor of the largest term. This follows from results of Schnorr and Shamir [11], where they introduced the $3n$-sorter for the square mesh. We also have a tight result for the general d-dimensional mesh of $\Theta(dn)$ due to Kunde [8]. These results are, in fact, more general, as they apply to meshes with non-uniform aspect ratios.

2   Definitions

Formally, we study the following restricted variant of sorting networks. We begin by taking an undirected graph $G = (V, E)$, where the vertices correspond to the locations of an oblivious sorting algorithm, $V = \{1, 2, \ldots, |V|\}$. The keys will be modeled by weighted pebbles, one per vertex. Let a *sorted order* of G be given by a permutation π that assigns the rank $\pi(i)$ to the vertex $i \in V$. The edges of G represent pairs of vertices where the pebbles can be compared and/or swapped. Given a graph G the goal is to design a sorting network that uses only the edges of G. We formally define such a sorting network (Table 1).

Table 1. Known bounds on the sorting numbers of various graphs

Graph	Lower bound	Upper bound	Remark
Complete graph (K_n)	$\log n$	$O(\log n)$	AKS Network [1]
Hypercube (Q_n)		$2^{O(\sqrt{\log \log n})} \log n$	Plaxton et al. [9,10]
Path (P_n)	$n-1$	n	Odd-Even Trans. [7]
Mesh ($P_n \times P_n$)	$3n - 2\sqrt{n} - 3$	$3n + O(n^{3/4})$	Schnorr and Shamir [11]
d-dimensional mesh	$\Omega(dn)$	$O(dn)$	Kunde [8]
Tree		$O(\min(\Delta n, n \log (n/d)))$	This paper
d-regular expander	$\Omega(\log n)$	$O(d \log^3 (n))$	This paper
Complete p-partite	$\Omega(\log n)$	$O(\log n)$	This paper
Pyramid (d, N)		$O(dN^{1/d})$	This paper

Definition 1 (Sorting Network on a Graph). *A sorting network is a triple* $\mathcal{S}(G, \mathcal{M}, \pi)$ *such that:*

1. *$G = (V, E)$ is a connected graph with a bijection $\pi : V \to \{1, \ldots, |V|\}$ specifying the sorted order on the vertices. Initially, each vertex of G contains a pebble having some value.*
2. *$\mathcal{M} = (M_1, \ldots, M_t)$ is a sequence of matchings in G, for which some edges in the matching are assigned a direction. Sorting occurs in stages. At stage i we use the matching M_i to exchange the pebbles between matched vertices according to their orientation. For an edge $\overrightarrow{uv}$, when swapped the smaller of the two pebbles goes to u. If an edge is undirected then the pebbles swap regardless of their order.*
3. *After t stages the vertex labeled i contains the pebble whose rank is $\pi(i)$ in the sorted order. We stress that this must hold for all $(n!)$ initial arrangement of the pebbles. $|\mathcal{M}|$ is called the* depth *of the network.*

Definition 2 (Sorting Number). *Let G be a graph, and let π be an order on the vertices of G. Define $st(G, \pi)$ to be the minimum depth of a sorting network $\mathcal{S}(G, \mathcal{M}, \pi)$. The* sorting number *of G, denoted by $st(G)$, is defined as the minimum depth of any sorting network on G, i.e., $st(G) = \min_\pi st(G, \pi)$.*

In order to prove some of the results in this paper we need to define the model of routing via matchings, originally introduced by Alon et al. [2].

Definition 3 (Routing Number). *Given a connected labeled graph $G = (V, E)$, where each vertex $i \in V$ is initially occupied by a labeled pebble that has a unique destination $\pi(i)$, the routing time $rt(G, \pi)$ is defined as the minimum number of matchings required to move each pebble from i to its destination* *vertex labeled $\pi(i)$, where pebbles are swapped along matched edges. The routing number of G denoted by $rt(G)$ is defined as the maximum of $rt(G, \pi)$ over all such permutations $\pi : V \to V$.*

Note that the main difference between routing and sorting is that in the case of routing the destination of each vertex is specified by the permutation π. On the other hand, in sorting networks π specifies the target location of the pebbles according to their rank, which should hold for any initial arrangement of the pebbles.

3   Our Results

The AKS sorting network can be trivially converted into a network of depth $O(n \log(n))$ by making a single comparison in each round. However, it is not clear a priori whether for *any* graph there is a sorting network of depth $O(n \log(n))$. We show that this bound indeed holds for all graphs.

Theorem 1. *Let G be an n-vertex graph, and suppose that G contains a simple path of length d. Then $st(G) = O(n \log(n/d))$. In particular, for every n-vertex graph G it holds that $st(G) = O(n \log(n))$.*

If the maximal degree of G is small, it is possible to show a better upper bound on $st(G)$.

Theorem 2. *Let G be an n-vertex graph with maximal degree Δ. Then $st(G) = O(\Delta n)$.*

Next, we relate the sorting number of a graph to its routing number and the size of its maximum matching.

Theorem 3. *Let G be an n-vertex graph with routing number $rt(G)$ and matching of size $\nu(G)$. Then $st(G) = O\left(n \log(n) \cdot \frac{rt(G)}{\nu(G)}\right)$.*

In the following theorem we upper bound $st(G)$ for graphs G that contain a large subgraph H whose $st(H)$ is small.

Theorem 4. *Let G be an n-vertex graph, and let H be a vertex-induced subgraph of G on p vertices. Then $st(G) = O\left(\frac{n}{p} \log(\frac{n}{p}) \cdot (rt(G) + st(H))\right)$.*

Theorems 1, 3, and 4 above will be proven in Sect. 5. The proof of Theorem 2 appears in the Appendix. In Sect. 6 we prove bounds on some concrete families of graphs, including the complete p-partite graph, expander graphs, vertex transitive graphs, Cayley graphs, and the pyramid graph. We note here that there are instances of graphs for which all the above results are asymptotically tight. For example in the case of the star graph $K_{1,n-1}$ at most 1 comparison can be made per round ($d = 2$) and hence $st(K_{1,n-1}) = \Theta(n \log(n))$. Theorem 2 is optimal for balanced binary trees among other bounded degree graphs. Theorem 3 is tight for any graph with $rt(G)/\nu(G) = O(1/n)$ such as cliques and complete bipartite graphs etc. Same graphs also give tight lower bounds in Theorem 4.

We pause to emphasize that the deep question is how does the topology of the constraint graph G affect computation. This paper is a case study using the problem of sorting. A related problem that has been studied in the past is the problem of routing [2]; in the next section we revisit this problem and add some new results we will need. Since the graph must be connected perhaps the only required topological property is that the graph possesses a spanning tree. Therefore we begin by concentrating on trees; these results are subtle but nevertheless reflect bottlenecks on parallelism. From the examples given in the previous paragraph they are easily seen to be best possible bounds without making further assumptions. By working with graphs with fewer bottlenecks we get better bounds. In Sect. 6 we prove bounds on some concrete families of graphs, including the complete p-partite graph, expander graphs, vertex transitive graphs, Cayley graphs, and the pyramid graph. It is open question whether other graph properties (such as treewidth, connectedness, etc.) can lead to improved bounds on the sorting number.

4   Routing via Matchings

The following lemma is useful for constructing a sorting network for a given sorted order if a sorting network is already known for that graph for some arbitrary ordering. Additionally it implies that we incur at most a penalty of $rt(G)$ on the depth of the network with respect to an optimal ordering.

Lemma 1. *For any graph G and any order π of the vertices of G it holds that*

$$rt(G) \leq st(G, \pi) \leq st(G) + rt(G).$$

Proof. We show first that $rt(G, \sigma) \leq st(G, \pi)$ for any two permutations π, σ of the vertices. Indeed, suppose that the keys of the pebbles are $\{1, \ldots, |V|\}$. For all $i \in V$ place the pebble ranked i in the vertex $\sigma^{-1}(\pi^{-1}(i))$. Then, there exists a sorting network of depth $st(G, \pi)$ that sends the pebble ranked i to the vertex $\pi^{-1}(i)$ for all $i \in V$. That is, the pebble from the vertex $j = \sigma^{-1}(\pi^{-1}(i))$ is sent to the vertex $\pi^{-1}(i) = \sigma(j)$. Therefore, $rt(G, \sigma) \leq st(G, \pi)$ for all permutations σ, and thus $rt(G) \leq st(G, \pi)$.

For the upper bound let $\mathcal{S}(G, \mathcal{M}, \tau)$ be a sorting network on G of depth $st(G) = st(G, \tau)$. We use τ to create another sorting network $\mathcal{S}(G, \mathcal{M}', \pi)$ of depth at most $st(G) + rt(G)$. This is done in two stages. First we apply the sorting network $\mathcal{S}(G, \mathcal{M}, \tau)$. After this stage we know that the pebble at vertex i has a rank $\tau(i)$. Next, we apply a routing strategy with at most $rt(G)$ steps that routes to the permutation $\pi^{-1} \circ \tau$, i.e., sending a pebble in the vertex i to $\pi^{-1}(\tau(i))$ for all $i \in V$. After this step the vertex i contains the pebble of rank $\pi(i)$. This proves that $st(G, \pi) \leq st(G) + rt(G)$. □

4.1   Routing on Subgraphs of G

Next we introduce the notion of routing a subset of the pebbles to a specific subgraph. These results are later used to bound sorting number with typical graph parameters. First we discuss partial routing, where only a small number of pebbles are required to reach their destination.

Definition 4. *Given a graph $G = (V, E)$ let $A, B \subset V$ be two subset of vertices with $|A| = |B|$, not necessarily distinct. Let π_{AB} be a bijection between A and B. Routing of the pebbles from A to their respective destinations on B given by π_{AB} is a partial routing in G, where each pebble in $a \in A$ is required to reach $\pi_{AB}(a) \in B$ using the edges of G (and there are no requirements on the pebbles outside of A). Further,*

1. *Let $rt(G, A, B, \pi_{AB})$ be the minimum number of matchings needed to route every pebble $a \in A$ to $\pi_{AB}(a) \in B$ using the edges of G.*
2. *Let $rt(G, A, B) = \max_{\pi_{AB}} rt(G, A, B, \pi_{AB})$.*
3. *For $U \subseteq V$ let $rt_U(G) = \max_{A \subseteq V} rt(G, A, U)$.*
4. *For $p \in [1 \ldots |V|]$ let,*

$$rt_p(G) = \max_{A, B \subset V, |A| = |B| \leq p} rt(G, A, B)$$

Clearly, for any connected n-vertex graph G we have $rt(G) = rt_n(G)$. Some of the bounds for $rt(G)$ also hold for $rt_p(G)$. For example, $rt_p(G) \geq d$, where d is the diameter of G. Furthermore, $rt_p(G) = \Theta(rt(G))$ for any p if and only if $rt(G) = \Theta(d)$. We illustrate $rt_p(G)$ by computing it explicitly for some typical graphs. Recall from [2] that $rt(K_n) = 2$. It is easy to see that $rt_p(K_n) = 2$ for all $p \geq 3$, and $rt_2(K_n) = 1$. For the complete bipartite graph we have $rt_{n/2}(K_{n/2,n/2}) = 2$ and is $rt_p(K_{n/2,n/2}) = 4$ for $p > n/2$.

Theorem 5. *For any tree T with diameter d, $rt_p(G) = O((d+p) \min(d, \log \frac{n}{d}))$.*

Proof. The proof is similar to the proof used in [2] for determining the routing number of trees. Let r be the centroid whose removal disconnects the tree into a forest of trees each of which is of size at most $n/2$. Let $(T_1, \ldots, T_\ell)$ be the set of trees in the forest, with $r \in T_1$. For a tree T_i let S_i be the set of "improper" pebbles that need to be moved out of T_i. All other pebbles in T_i are "proper". In the first round we move all the pebbles in S_i as close to the root of T_i as possible, for all i. Using the argument used in [2] it can be shown that for a tree with diameter d this first phase can be accomplished in $c_1 d$ steps for some constant c_1. First we label each node in T_i as odd or even based on their distance from r_i, the root of T_i. In each odd round we match nodes in odd layers with proper pebbles to one of its children containing an improper pebble if one exists. Similarly, in even rounds we match nodes in even layers with proper pebbles to one of its children containing an improper pebble if one exists. Since T has diameter d any path from r_i to some leaf must be of length at most $d - 1$. Now consider an improper pebble u initially at distance k from the root. During a pair of odd-even matchings either the pebble moves one step closer to the root or one of the following must be true: (1) another pebble from one of its sibling nodes jumps in front of it or (2) there is some improper pebble already in front of it. It can then be argued (we omit the details here) that after $c_1 d$ matchings for some constant c_1 if u ends up in position j from r_i then all pebbles between u and r_i must be improper. Next we exchange a pair of pebbles between subtrees using the root vertex r, since at most $p/2$ pairs needs to be exchanged, the arguments used in [2] can be modified to show that this phase also takes $c_2 p$ steps for some constant c_2. After each pebble is moved to their corresponding destination subtrees we can route them in parallel. Noting that each tree T_i has diameter at most $d - 1$. Hence we have the recurrence

$$T(n, d, p) \leq T(n/2, d - 1, p) + c_1 d + c_2 p, \tag{1}$$

where $T(n, d, p)$ is the time it takes to route p pebbles in a tree of diameter d with n vertices. Taking $T(O(1), d, p) = O(d)$, and solving (1) gives the bound stated in the theorem. $\square$

The following lemma will be useful when proving an upper bound for the sorting number for trees. A more generic version of this result first appeared in [3].

Lemma 2. *Let T be a tree with diameter d, and let P be a path of length d in T. We can route any set of d pebbles to P in $3d - 2$ steps.*

5   General Upper Bounds on $st(G)$

Next we prove Theorem 4. Specifically, we will prove that if H is a vertex-induced subgraph of G on p vertices then $st(G) = O\left(\frac{n}{p} \log(\frac{n}{p}) \cdot (rt_H(G) + st(H))\right)$, where $rt_H(G)$ bounds the number of matchings required to route any set of p vertices to H. (Here we slightly abuse the notation from Definition 4, by identifying the subscript in $rt_H(G)$ with the vertex-set of H.) Later we will use this result to prove Theorem 1.

Proof (Theorem 4). Let us partition the vertex set V of G into $q = \lceil n/\lfloor p/2 \rfloor \rceil$ parts $V = A_1 \cup \cdots \cup A_q$ in a balanced manner (i.e., the size of each A_i is either $\lfloor n/q \rfloor$ or $\lfloor n/q \rfloor + 1$), so that $|A_i| + |A_j| \leq p$ for all i and j. Let K_q be a complete graph whose vertices are identified with $\{A_1, \ldots, A_q\}$, and let S be an oblivious sorting algorithm with $O(q \log q)$ comparisons on the complete graph K_q. (Here the sequence of comparisons is performed sequentially, and not in parallel.) In an ordinary sorting network in each step we perform a compare-exchange or a swap between two matched vertices (i, j) so that if $i < j$, then the pebble in the vertex i will be smaller than the pebble in j We will simulate S on G using a sorting network on H by sorting in each stage the elements in $A_i \cup A_j$. That is, for $i < j$ we are going to sort the elements in $A_i \cup A_j$ so that all the elements of A_i are smaller than every element of A_j, and the elements within each subset are internally sorted. This is done using an optimal sorting network in H, which we will denote by $\mathcal{S}_H$.

We can simulate any such compare-exchange in G between pairs of sets in A in $O(rt(G) + st(H))$ steps. Indeed, suppose the kth round in S compares the vertices $i < j$. In order to simulate this comparison we first route all the pebbles in $A_i \cup A_j$ to the subgraph H and relabel the vertices. This relabeling is done so that we can keep track of the vertices when sorting H. Then, we use $\mathcal{S}_H$ to sort $A_i \cup A_j$ which takes $st(H)$ steps. Once the sorting is done we split up the sets again and appropriately relabel the vertices so that the first $|A_i|$ vertices in the sorted order on H will now belong to A_i and the next $|A_j|$ vertices will belong to A_j. If instead the kth comparison is actually a swap then we simply swap the labels of the multisets (A_i is labeled A_j and vice versa). Hence performing the above simulation takes $O(rt_H(G) + st(H))$ steps per compare exchange or swap operation, which gives the result of the theorem. $\qquad\square$

In the proof of Theorem 4 above we only used an oblivious sorting algorithm with $O(q \log q)$ comparisons on the complete graph K_q, and did not use the fact that the comparisons can be done in parallel, e.g., using the AKS sorting network. This is because Theorem 4 only assumes that there is one subgraph H with small $st(H)$. If instead we assumed that there are many such subgraphs, then we could sort the A_i's in different subgraphs in parallel. This is described in the corollary below.

Corollary 1. *Let $G = (V, E)$ be an n-vertex graph. Let $V = V_1 \cup \cdots \cup V_q$ be a partition of the vertices, with $|V_i| = n/q$ for all $i \in \{1, \ldots, q\}$, such that H_i, the subgraph induced by V_i, is connected for each $i \in \{1, \ldots, q\}$. Then*

$$st(G) = O\left(\log(q) \cdot \left(rt(G) + \max_{k \in \{1,\ldots,q\}}\{st(H_k)\}\right)\right).$$

Proof (sketch). The proof uses the same idea that Theorem 4. We start by partitioning the vertex set V of G into $2q$ parts $V = A_1 \cup \cdots \cup A_{2q}$ of equal sizes. Then, we simulate oblivious sorting algorithm on K_{2q} with the sets A_i. The only difference is that instead of an oblivious sorting algorithm with $O(q \log(q))$ comparisons on the complete graph K_{2q} we use the AKS sorting network on $2q$ vertices of depth $O(\log(q))$. In each round of the sorting network there are at most q comparisons, and the corresponding sorting of $A_i \cup A_j$ can be performed in parallel, one in each H_k in time $st(H_k)$. $\qquad\square$

Below we prove Theorem 1 stating that if G contains a simple path of length d, then $st(G) = O(n \log (n/d))$.

Proof (Theorem 1). It is easy to see that if G contains a simple path of length d, then G has a spanning tree T with diameter at least d. The proof follows easily from Theorem 4 and Lemma 2. Indeed, in the setting of Theorem 4, let H be a path of length d in T. Then $st(H) \leq d$. By Theorem 4 $st(T) = O(\frac{n}{d} \log(\frac{n}{d}) \cdot (rt_H(T) + st(H)))$. The pebbles can be routed to H along the edges of the spanning tree T, and hence by Lemma 2 $rt_H(T) = O(d)$. Therefore, $st(G) \leq st(T) = O(n \log(n/d))$, as required. $\qquad\square$

Recall that Theorem 2 bounds the sorting number of a tree based on its maximum degree. The proof is essentially from [3], who proved that the acquaintance time of a G, defined in [6], is upper bounded by $20\Delta n$. The basic idea is to use an n round sorting network for P_n, and simulate this network in T with an overhead that depends only on Δ. The proof is given in the appendix.

Next, we prove Theorem 3 that states $st(G) = O(n\log(n) \cdot \frac{rt(G)}{\nu(G)})$, where $rt(G)$ is the routing number of G, and $\nu(G)$ is the size of the maximum matching in G.

Proof (Theorem 3). We prove the theorem by using G to simulate the AKS sorting network on the complete graph K_n of depth $O(\log(n))$. Specifically, we show that each stage (a matching) of the sorting network on K_n can be simulated by at most $O(\frac{n}{\nu(G)} rt(G))$ stages (matchings) in G. Let M be a matching at some stage of the AKS sorting network on the complete graph. We simulate the compare-exchanges and swaps in M by a sequence of matchings in G as follows. First we partition the edges in M into $t = \lceil n/\nu(G) \rceil$ disjoint subsets $M = M_1 \cup \cdots \cup M_t$, where $|M_i| = \nu(G)$ for all except maybe the last set M_t, which can be smaller. Let M_G be a maximum matching in G. Corresponding to each pair $(u,v) \in M_i$ we pick a distinct pair $(u',v') \in M_G$, this can always be done since the sets M_i and M_G are of the same size. Note that the pair (u,v) may not be adjacent in G, and so, we route each pair $(u,v) \in M_i$ to its destination in $(u',v') \in M_G$. This can be done in $rt(G)$ steps, where each step consists of only undirected matchings. Once the pairs have been placed into their corresponding positions we relabel the vertices such that the pair labeled (u',v') is now (u,v). Unmatched vertices keep their label. Since the pairs in M_i are now adjacent in G we can perform the compare-exchange or swap operation according to M_i. Therefore, the total number of matchings to execute the ith set of compare-exchanges and swaps in M_i is $rt(G) + 1$ in G. We remark that the routing maintains the oblivious nature of the network, and the swaps made while routing, are data independent. We can then reverse the oblivious routing that set up the exchanges for M_i. The set up for M_{i+1} invokes routing a different permutation. Note that the two phases of oblivious routing can be combined by using the composition of the two permutation. This implies that we can simulate M using at most $(rt(G)+1) \cdot t = O(\frac{n}{\nu(G)} \cdot rt(G))$ matchings in G. Therefore, since the depth of the AKS sorting network on the complete graph K_n is $O(\log(n))$, we conclude that $st(G) = O(n\log(n) \cdot \frac{rt(G)}{\nu(G)})$, as required.   □

6   Bounds on Concrete Graph Families

Below we state several results concerning the sorting time of some concrete families of graphs.

Proposition 1 (Complete p-partite graph). *Let G be the complete p-partite graph $K_{n/p,\ldots,n/p}$ on n vertices. Then $st(G) = \Theta(\log n)$.*

Proof. The lower bound is trivial. For the upper bound note that $K_{n/p,...,n/p}$ contains the bipartite graph $K_{\lfloor \frac{p}{2} \rfloor \frac{n}{p}, \lceil \frac{p}{2} \rceil \frac{n}{p}}$. In particular, it contains a matching of size $\nu(G) = \lfloor \frac{p}{2} \rfloor \cdot \frac{n}{p}$. Therefore, by Theorem 3 in [2] and the remark after the proof, we have $rt(G) \leq rt(K_{\lfloor \frac{p}{2} \rfloor \frac{n}{p}, \lceil \frac{p}{2} \rceil \frac{n}{p}}) \leq 2\lceil \frac{\lceil \frac{p}{2} \rceil}{\lfloor \frac{p}{2} \rfloor} \rceil + 2 \leq 6$, and hence by Theorem 3 it follows that $st(G) \leq O(\log(n))$.

Recall that a graph G is said to be a (n, d, λ)-expander if it is a d-regular graph on n vertices and the absolute value of every eigenvalue of its adjacency matrix other than the trivial one is at most λ.

Proposition 2 (Expander graphs). *Let G be an (n, d, λ)-expander. Then $st(G) \leq O(\frac{d^3}{(d-\lambda)^2} \log^3(n))$. In particular, if $\lambda < (1 - \frac{1}{\log^c(n)})d$, then $st(G) \leq O(d \cdot \log^{2c+3}(n))$.*

Proof. Recall from [2] that if G is an (n, d, λ)-expander, then $rt(G) = O\left(\frac{d^2}{(d-\lambda)^2} \log^2(n)\right)$. Therefore, since any d-regular graph contains a matching of size $n/2d$ it follows from Theorem 3 that $st(G) \leq O(\frac{d^3}{(d-\lambda)^2} \log^3(n))$.

Recall that a graph $G = (V, E)$ is said to be *vertex transitive* if for any two vertices $u, v \in V$, there is some automorphism[1] $f \colon V \to V$ of the graph such that $f(u) = v$.

Proposition 3 (Vertex transitive graphs). *Let G be a vertex transitive graph with n vertices of degree* polylog(n). *Then $diam(G) = O(\text{polylog}(n))$ if and only if $st(G) = O(\text{polylog}(n))$.*

Proof. It is trivial that $diam(G) \leq st(G)$. For the other direction, Babai and Szegedy [4] showed that for vertex-transitive graphs if the diameter of G is $O(\text{polylog}(n))$ then its vertex expansion is $\Omega(1/\text{polylog}(n))$. Therefore, $\lambda \leq d(1 - 1/\text{polylog}(n))$, where $d = O(\text{polylog}(n))$ is the degree of the graph. Therefore, by Proposition 2 we have $st(G) = O(\text{polylog}(n))$.

Since all Cayley graphs are also vertex transitive, the above bound is applicable to them as well.

Next we bound the sorting number of cartesian product of two given graphs. Recall that for two graphs $G_1(V_1, E_1)$ and $G_2(V_2, E_2)$ their Cartesian product $G_1 \square G_2$ is the graph whose set of vertices is $V_1 \times V_2$ and $((u_1, u_2), (v_1, v_2))$ is an edge in $G_1 \square G_2$ if either $u_1 = v_1$, $(u_2, v_2) \in E_2$ or $(u_1, v_1) \in E_1$, $u_2 = v_2$. Our next result bounds the sorting number of a product graph in terms of sorting numbers of its components (Fig. 1).

Proposition 4. *Let $G_1 = (V_1, E_1)$, $G_2 = (V_2, E_2)$ be two graphs and let $G = G_1 \square G_2$. Then*

$$st(G) \leq O(\min(\log|V_1|(rt(G) + st(G_2)), \log|V_2|(rt(G) + st(G_1)))).$$

---

[1] A mapping $f \colon V \to V$ is an automorphism of $G = (V, E)$ if for all $v_1, v_2 \in V$ it holds that $(v_1, v_2) \in E \Leftrightarrow (f(v_1), f(v_2)) \in E$.

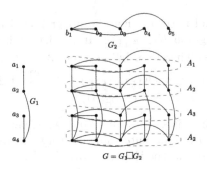

Fig. 1. The product graph $G = G_1 \square G_2$. The rows highlighted by blue regions represents copies of G_2. (Color figure online)

Proof. Since G has $|V_1|$ vertex disjoint subgraphs that are copies of G_2 we can apply Corollary 1 with these $q = |V_1|$ subgraphs, and all H_i being isomorphic to G_2. Therefore, we get $st(G) \leq O(\log(|V_1|) \cdot (rt(G) + st(G_2)))$. The bound $st(G) \leq O(\log(|V_2|) \cdot (rt(G) + st(G_1)))$ follows using the same argument by changing the roles of G_1 and G_2.

As an example of an application of the above corollary consider the d-dimensional mesh $M_{n,d}$ with n^d vertices. We know that $rt(M_{n,d}) \leq 2dn$ since $M_{n,d} = M_{n,d-1} \times P_n$. Therefore, $st(M_{n,d}) \leq O(\log(n^{d-1}) \cdot (rt(M_{n,d}) + st(P_n))) = O(dn \log(n))$. Although this bound is not optimal (it is known [8] that $st(M_{n,d}) = O(dn)$), we still find this example interesting.

6.1 The Pyramid Graph

A 1-dimensional pyramid with m-levels is defined as the complete binary tree of $2^m - 1$ nodes, where the nodes in each level are connected by a path (i.e., a one-dimensional mesh). We treat the apex (root) to be at level 0, and subsequent levels are numbered in ascending order. A 2-dimensional pyramid is shown in Fig. 2. In this case each level l is a $2^l \times 2^l$ square mesh. Similarly a d-dimensional pyramid having m levels, denoted by $\triangle_{m,d}$, is the graph with vertices partitioned into levels $\{0, \ldots, m\}$, where the vertices in level $l \in \{0, \ldots, m-1\}$ form a d-dimensional regular mesh of length 2^l in each dimension. Clearly, the size of layer l is $|M_l| = n_l = 2^{ld}$ and the number of vertices in the graph is $N = |\triangle_{m,d}| = \sum_{l=0}^{m-1} 2^{ld} = \frac{2^{md}-1}{2^d-1}$. We treat a vertex $x \in M_l$ as a vector in $[1, 2^l]^d$ which denotes its position on the mesh.

In this section we prove an upper bound on $st(\triangle_{m,d})$. In order to derive this bound we make use of the following bound on the routing number of pyramid whose proof is given in the appendix.

Lemma 3. *Let $\triangle_{m,d}$ be the d-dimensional pyramid graph with m-levels. Then $rt(\triangle_{m,d}) = O(dN^{1/d})$.*

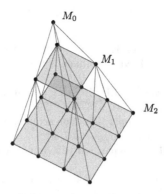

Fig. 2. $\triangle_{3,2}$ - a 2-dimensional pyramid.

Using the above theorem we give an upper bound on the sorting number of the pyramid.

Theorem 6 (pyramid). *The sorting number for a pyramid $\triangle_{m,d}$ is $O(d\, N^{1/d})$.*

We describe a sorting network on the pyramid with the above claimed depth. The proof of correctness is omitted due to space constraint.

Let $\triangle_{i,d}$ denote the sub-pyramid from level 0 to i and let M_i be the d-dimensional mesh at level i. Let π_i be some ordering of the mesh M_i. Note that $\pi_i : [1, 2^i]^d \to [n_i]$ is a bijection and π_0 is the identity permutation of order 1. Next we define a sorted order π for the pyramid $\triangle_{m,d}$ based on the π_i's. In π we assume the layers are sorted among themselves in ascending order starting from the apex. So the vertex labeled (with respect to M_i) i on layer j has a global rank $\pi(i) = \pi_j(i) + |\triangle_{j-1,d}|$. Recall that $st(M_i) = O(dn_i^{1/d})$ which is due to Kunde [8] where he used the general snake-like ordering. From Lemma 1 we see that this bound still holds if we replace the snake-like ordering with some arbitrary permutation. Obviously in this case $rt(M_i) = \Theta(st(M_i))$. Next we describe the matchings $\mathcal{M}$ of sorting the network $\mathcal{S}(\triangle_{m,d}, \mathcal{M}, \pi)$ in terms of an oblivious sorting algorithm described below.
$\mathcal{S}(\triangle_{m,d}, \mathcal{M}, \pi)$

1. Route all pebbles of $\triangle_{m-1,d}$ to M_{m-1} and sort them using the mesh.
2. Route these pebbles back to $\triangle_{m-1,d}$ such that they are in sorted order (according to π).
3. Sort the mesh M_{m-1} according to π_{m-1}.
4. Route a pebble of rank $i \le n_{m-2}$ at position $x_i \in M_{m-1}$ to $y_i \in M_{m-1}$ where

$$y_i[j] = 2\pi_{m-2}^{-1}(n_{m-2} + 1 - i)[j] - 1$$

   Let $Y = (y_1, \ldots, y_{n_{m-2}})$.
5. Merge Y with M_{m-2} using pair-wise compare-exchanges, where y_i is compared with $z \in M_{m-2}$ such that $\pi_{m-2}(z) = i$.
6. Repeat 1–5 once.
7. Repeat 1–3 once.

References

1. Ajtai, M., Komlós, J., Szemerédi, E.: An $O(n\log n)$ sorting network. In: Proceedings of the Fifteenth Annual ACM Symposium on Theory of Computing, pp. 1–9. ACM (1983)
2. Alon, N., Chung, F.R., Graham, R.L.: Routing permutations on graphs via matchings. SIAM J. Discrete Math. **7**(3), 513–530 (1994)
3. Angel, O., Shinkar, I.: A tight upper bound on acquaintance time of graphs. Graphs Comb. **32**(5), 1667–1673 (2016). arXiv:1307.6029
4. Babai, L., Szegedy, M.: Local expansion of symmetrical graphs. Comb. Probab. Comput. **1**(01), 1–11 (1992)
5. Batcher, K.E.: Sorting networks and their applications. In: Proceedings of the 30 April–2 May 1968, Spring Joint Computer Conference, pp. 307–314. ACM (1968)
6. Benjamini, I., Shinkar, I., Tsur, G.: Acquaintance time of a graph. SIAM J. Discrete Math. **28**(2), 767–785 (2014)
7. Knuth, D.E.: The Art of Computer Programming: Sorting and Searching, vol. 3. Addison Wesley Longman Publishing Co., Inc., Redwood City (1998)
8. Kunde, M.: Optimal sorting on multi-dimensionally mesh-connected computers. In: Brandenburg, F.J., Vidal-Naquet, G., Wirsing, M. (eds.) STACS 1987. LNCS, vol. 247, pp. 408–419. Springer, Heidelberg (1987). https://doi.org/10.1007/BFb0039623
9. Leighton, T., Plaxton, C.G.: Hypercubic sorting networks. SIAM J. Comput. **27**(1), 1–47 (1998)
10. Plaxton, C.G., Suel, T.: A super-logarithmic lower bound for hypercubic sorting networks. In: Abiteboul, S., Shamir, E. (eds.) ICALP 1994. LNCS, vol. 820, pp. 618–629. Springer, Heidelberg (1994). https://doi.org/10.1007/3-540-58201-0_103
11. Schnorr, C.P., Shamir, A.: An optimal sorting algorithm for mesh connected computers. In: Proceedings of the Eighteenth Annual ACM Symposium on Theory of Computing, pp. 255–263. ACM (1986)

Minimum Reload Cost Graph Factors

Julien Baste[1] , Didem Gözüpek[2] , Mordechai Shalom[3]([✉]) ,
and Dimitrios M. Thilikos[4,5]

[1] Institut für Optimierung und Operations Research, Universität Ulm, Ulm, Germany
julien.baste@uni-ulm.de
[2] Department of Computer Engineering, Gebze Technical University,
Gebze, Kocaeli, Turkey
didem.gozupek@gtu.edu.tr
[3] TelHai College, 12210 Upper Galilee, Israel
cmshalom@telhai.ac.il
[4] AlGCo project-team, LIRMM, CNRS, Université de Montpellier,
Montpellier, France
sedthilk@thilikos.info
[5] Department of Mathematics, National and Kapodistrian University of Athens,
Athens, Greece

Abstract. The concept of *Reload cost* in a graph refers to the cost that
occurs while traversing a vertex via two of its incident edges. This cost is
uniquely determined by the colors of the two edges. This concept has various applications in transportation networks, communication networks,
and energy distribution networks. Various problems using this model are
defined and studied in the literature. The problem of finding a spanning
tree whose diameter with respect to the reload costs is the smallest possible, the problems of finding a path, trail or walk with minimum total
reload cost between two given vertices, problems about finding a proper
edge coloring of a graph such that the total reload cost is minimized, the
problem of finding a spanning tree such that the sum of the reload costs
of all paths between all pairs of vertices is minimized, and the problem of
finding a set of cycles of minimum reload cost, that cover all the vertices
of a graph, are examples of such problems. In this work we focus on the
last problem. Noting that a cycle cover of a graph is a 2-factor of it, we
generalize the problem to that of finding an r-factor of minimum reload
cost of an edge colored graph. We prove several NP-hardness results for
special cases of the problem. Namely, bounded degree graphs, planar
graphs, bounded total cost, and bounded number of distinct costs. For
the special case of $r = 2$, our results imply an improved NP-hardness
result. On the positive side, we present a polynomial-time solvable special case which provides a tight boundary between the polynomial and
hard cases in terms of r and the maximum degree of the graph. We then
investigate the parameterized complexity of the problem, prove W[1]-
hardness results and present an FPT-algorithm.

Work supported by the bilateral research program CNRS/TUBITAK grant no.
114E731, TUBITAK 2221 programme. The last author was supported by projects
"DEMOGRAPH" (ANR-16-CE40-0028) and "ESIGMA" (ANR-17-CE23-0010).

B. Catania et al. (Eds.): SOFSEM 2019, LNCS 11376, pp. 67–80, 2019.
https://doi.org/10.1007/978-3-030-10801-4_7

Keywords: Parameterized complexity · Graph factors · Reload costs

1   Introduction

Edge-colored graphs can be used to model optimization problems in diverse fields such as bioinformatics, communication networks, and transportation networks. *Reload cost* in an edge-colored graph refers to the cost that occurs while traversing a vertex via two of its incident edges. This cost is uniquely determined by the colors of the two edges.

The reload cost concept has various applications in transportation networks, communication networks, and energy distribution networks. For instance, a multi-modal cargo transportation network involves different means of transportation, where the (un)loading of cargo at transfer points is costly [1]. In energy distribution networks, transferring energy between different carriers cause energy losses and reload cost concept can be used to model this situation [1]. In communication networks, routing often requires switching between different technologies such as cable and fiber, where data conversion incurs high costs. Switching between different service providers in a communication network also causes switching costs [1]. Recently, *dynamic spectrum access networks*, a.k.a. cognitive radio networks, received a lot of attention in the communication networks research community. Unlike other wireless networks, cognitive radio networks are envisioned to operate in a wide range of spectrum; therefore, frequency switching has adverse effects in delay and energy consumption [2–4]. This frequency switching cost depends on the frequency separation distance; hence, it corresponds to reload costs.

The reload cost concept was first introduced by Wirth and Steffan [5] who focused on the problem of finding a spanning tree whose diameter with respect to the reload costs is the smallest possible. Other works also focused on numerous optimization problems regarding reload costs: the problems of finding a path, trail or walk with minimum total reload cost between two given vertices [6], numerous path, tour, and flow problems [7], the minimum changeover cost arborescence problem [8–11], problems about finding a proper edge coloring of a graph such that the total reload cost is minimized [12], and the problem of finding a spanning tree such that the sum of the reload costs of all paths between all pairs of vertices is minimized [13].

An r-factor of a graph is an r-regular spanning subgraph. A 2-factor is also called a *cycle cover* and has many applications in areas such as computer graphics and computational geometry [14], for instance for fast rendering of 3D scenes. In an edge-weighted graph, the problem of finding a cycle cover with minimum cost can be solved in polynomial-time [15]. Its reload cost counterpart was studied by Galbiati et al. in [16]. In particular, they proved that the minimum reload cost cycle cover problem is NP-hard even when the number of colors is 2, the reload costs are symmetric and satisfy the triangle inequality. In this work, we build on this work by studying the minimum reload cost r-factor problem, which is a more generalized version of the minimum reload cost cycle cover problem.

In particular, we prove several NP-hardness results for the special cases of this problem. Namely, bounded degree graphs, planar graphs, bounded total cost, and bounded number of distinct costs. For the special case of $r = 2$, we prove an NP-hardness result stronger than the one in [16]. On the positive side, we present a polynomial-time solvable special case. We then investigate the parameterized complexity of this problem, prove W[1]-hardness results and present a fixed parameter tractable algorithm. Some of the proofs are omitted from this Extended Abstract and shown in the full paper [17].

2   Preliminaries

Sets, Vectors: Given a non-negative integer n, we denote by $\mathbb{N}_{\geq n}$ the set of all integers x such that $x \geq n$. If $n_1, n_2 \in \mathbb{N}_{\geq 0}$, we denote by $[n_1, n_2]$ the set of integers x such that $n_1 \leq x \leq n_2$. We also use $[n]$ instead of $[1, n]$. Given a finite set X and an integer $s \in \mathbb{N}_{\geq 0}$, we denote by $\binom{X}{s}$ the set of all subsets of A with exactly s elements. For a set X and an element x we use $X + x$ and $X - x$ as shorthands for $X \cup \{x\}$ and $X \setminus \{x\}$, respectively. For two vectors $\mathbf{u} = (u_1, \ldots, u_d)$ and $\mathbf{v} = (v_1, \ldots, v_d)$ over the reals, we write $\mathbf{u} \leq \mathbf{v}$ if $u_i \leq v_i$ for every $i \in [d]$.

Graphs: All graphs we consider in this paper are undirected, finite, and without self loops or multiple edges. Given a graph G, we denote by $V(G)$ the set of vertices of G and by $E(G)$ the set of edges of G. We say that a vertex $v \in V(G)$ and an edge $e \in E(G)$ are *incident* if $v \in e$, that is, v is an endpoint of e. Given a vertex $v \in V(G)$, we denote by $E_G(v)$ the set of edges of G that are incident to v. The *degree* of v in G, denoted by $\deg_G(v)$ is $|E_G(v)|$. We also define the *maximum degree* of G as $\Delta(G) = \max\{\deg_G(v) \mid v \in V(G)\}$, the *minimum degree* of G as $\delta(G) = \min\{\deg_G(v) \mid v \in V(G)\}$, and the average degree of G as $\overline{\deg}(G)$. For a subset X of $V(G)$, we denote by $G[X]$ the subgraph of G induced by X. A graph is *r-regular* if all its vertices have degree r.

   We say that a graph H is a *factor* of a graph G when $V(H) = V(G)$ and $E(H) \subseteq E(G)$. An r-regular factor of G is termed an r-*factor* of G.

Parameterized Complexity: We refer the reader to [18,19] for basic background on parameterized complexity, and we recall here only some basic definitions. A *parameterized problem* is a language $L \subseteq \Sigma^* \times \mathbb{N}$. For an instance $I = (x, k) \in \Sigma^* \times \mathbb{N}$, k is called the *parameter*.

   A parameterized problem is *fixed-parameter tractable* (FPT) if there exists an algorithm $\mathcal{A}$, a computable function f, and a constant c such that given an instance $I = (x, k)$, $\mathcal{A}$ (called an FPT *algorithm*) correctly decides whether $I \in L$ in time bounded by $f(k) \cdot |I|^c$. We use the $\mathcal{O}^*$ notation whenever we ignore polynomial factors. A parameterized problem is in XP if there exists an algorithm $\mathcal{A}$ and two computable functions f and g such that given an instance $I = (x, k)$, $\mathcal{A}$ (called an XP *algorithm*) correctly decides whether $I \in L$ in time bounded by $f(k) \cdot |I|^{g(k)}$. A parameterized problem with instances of the form $I = (x, k)$ is para-NP-*hard* if it is NP-hard for some fixed *constant* value of the parameter k. Note that, unless $\mathsf{P} = \mathsf{NP}$, a para-NP-hard problem cannot be in XP, hence it cannot be FPT either.

Within parameterized problems, the class W[1] may be seen as the parameterized equivalent to the class NP of classical optimization problems. Without entering into details (see [18,19] for the formal definitions), a parameterized problem being W[1]-hard can be seen as a strong evidence that this problem is *not* FPT. The following problem is a W[1]-hard problem that we will use in our reductions.

MULTICOLORED CLIQUE
Input: A graph G, an integer k, a coloring function $\chi : V(G) \to [k]$.
Parameter: k.
Question: Does G contain a clique on k vertices with one vertex from each color class?

MULTICOLORED CLIQUE is known to be W[1]-hard on general graphs, even in the special case where all color classes have the same number of vertices [20]. Clearly, we can also assume that every color class is an independent set since the problem is indifferent to edges within the same color class.

Tree Decompositions: A *tree decomposition* of a graph $G = (V, E)$ is a pair $\mathcal{D} = (T, \mathcal{X})$, where T is a tree and $\mathcal{X} = \{X_t \mid t \in V(T)\}$ is a collection of subsets of $V(G)$ such that:

- $\bigcup_{t \in V(T)} X_t = V$,
- for every edge $uv \in E$, there is a $t \in V(T)$ such that $\{u, v\} \subseteq X_t$, and
- for every $\{x, y, z\} \subseteq V(T)$ such that z lies on the unique path between x and y in T, $X_x \cap X_y \subseteq X_z$.

We call the vertices of T *nodes* of $\mathcal{D}$ and the sets in $\mathcal{X}$ *bags* of $\mathcal{D}$. The width of $\mathcal{D}$ is $\max_{t \in V(T)} |X_t| - 1$. The *treewidth* of G, denoted by $\mathsf{tw}(G)$, is the smallest integer w such that there exists a tree decomposition of G of width w. A tree decomposition in which the tree T is restricted to be a path is called a *path decomposition*. The *pathwidth* of G, denoted by $\mathsf{pw}(G)$, is the smallest integer w such that there exists a path decomposition of G of width w.

A tree decomposition is *rooted* if we distinguish in T some specific vertex r, and consider T as a rooted (on r) tree. We denote such a tree decomposition by a triple $\mathcal{D} = (T, \mathcal{X}, r)$.

Nice Tree Decompositions: Let $\mathcal{D} = (T, \mathcal{X}, r)$ be a rooted tree decomposition of G, and $\mathcal{G} = \{G_t \mid t \in V(T)\}$ be a collection of subgraphs of G. We say that the ordered pair $(\mathcal{D}, \mathcal{G})$ is *nice* if the following conditions hold:

- $X_r = \emptyset$ and $G_r = G$,
- every node of $\mathcal{D}$ has at most two children in T,
- for each leaf $t \in V(T)$, $X_t = \emptyset$ and $G_t = (\emptyset, \emptyset)$. Such a node t is called a *leaf node*,

– if $t \in V(T)$ has exactly two children t' and t'', then $X_t = X_{t'} = X_{t''}$, $G_t = G_{t'} \cup G_{t''}$, and $E(G_{t'}) \cap E(G_{t''}) = \emptyset$. The node t is called a *join node*.
– if $t \in V(T)$ has exactly one child t', then exactly one of the following holds.
  - $X_t = X_{t'} + v$ for some $v \notin X_{t'}$ and $G_t = (V(G_{t'}) + v, E(G_{t'}))$. The node t is called *vertex-introduce node* and the vertex v is the *introduced vertex* of X_t.
  - $X_t = X_{t'}$ and $G_t = (G_{t'}, E(G_{t'}) + e)$ where e is an edge of G with endpoints in X_t. The node t is called *edge-introduce node* and the edge e is the *introduced edge* of X_t.
  - $X_t = X_{t'} - v$ for some $v \in X_{t'}$ and $G_t = G_{t'}$. The node t is called *forget node* and v is the *forget vertex* of X_t.

The notion of a nice pair defined above is essentially the same as the one of nice tree decomposition in [21] (which in turn is an enhancement of the original one, introduced in [22]). As already argued in [21,22], given a tree decomposition, it is possible to transform it in polynomial time to a tree decomposition $\mathcal{D}$ of the same width and construct a collection $\mathcal{G}$ such that $(\mathcal{D}, \mathcal{G})$ is nice.

Reload Cost Model: For reload costs, we follow the notation and terminology defined by [5]. We consider an edge-colored graph G where edge colors are taken from a finite set X and the coloring function is $\chi : E(G) \to X$. The reload costs are given by a function $c : X^2 \to \mathbb{N}_{\geq 0}$ where $c(x_1, x_2) = c(x_2, x_1)$ for each $(x_1, x_2) \in X^2$. The cost of *traversing* two incident edges e_1, e_2 of G is $\mathsf{tc}(e_1, e_2) = c(\chi(e_1), \chi(e_2))$. Given a subgraph H of G and a vertex $v \in V(H)$, we define the *reload cost* of v in H as $\mathsf{rc}_{\chi,c}(H, v) = \sum_{\{e_1, e_2\} \in \binom{E_H(v)}{2}} \mathsf{tc}(e_1, e_2)$ and the *reload cost* of H as $\mathsf{rc}_{\chi,c}(H) = \sum_{v \in V(H)} \mathsf{rc}_{\chi,c}(H, v)$. When χ and c are clear from the context, we write $\mathsf{rc}(v)$ and $\mathsf{rc}(H)$ instead.

Problem Statement: The problem we study in this paper can be formally defined as follows for every $r \in \mathbb{N}_{\geq 2}$:

MINIMUM RELOAD COST r-FACTOR (r-MRCF)
Input: A graph G, an edge-coloring χ, a reload cost function c, and a non-negative integer k.
Output: Is there an r-factor H of G with reload cost at most k, i.e., $\mathsf{rc}(H) \leq k$?

Given an instance (G, χ, c, k) of r-MRCF we consider the following parameters:

– the maximum degree $\Delta(G)$ of G,
– the treewidth $\mathsf{tw}(G)$ of G,
– the pathwidth $\mathsf{pw}(G)$ of G,
– the number of colors $q = |X|$,
– the number of distinct costs: $d = |\{c(x_1, x_2) \mid (x_1, x_2) \in X^2\}|$,
– the minimum traversal cost $c_{min} = \min\{c(x_1, x_2) \mid (x_1, x_2) \in X^2\}$, and
– the total cost k.

Table 1. Summary of classical complexity results for the r-MRCF problem.

G	$\Delta(G)$	d	k	q	
	$\leq r+2$	2	k_{min}	$\min\{r,3\}$	NP-hard (Theorem 1)
	$r+1$				Polynomial (Theorem 2)
Planar	$\leq r+4$	2	k_{min}	7	NP-hard (Theorem 3)

Table 2. Summary of parameterized complexity results for the r-MRCF problem.

Parameter	d	k	Average degree	
$\mathsf{pw}(G)$	2	k_{min}	$< r + \begin{cases} 4 & \text{if } r = 2 \\ 4/3 & \text{otherwise.} \end{cases}$	W[1]-hard (Theorem 4)
$\mathsf{tw}(G) + \min\{q, \Delta(G)\}$				FPT (Theorem 5)

Clearly, we can assume that $\Delta(G) \geq r + 1$, and also $\delta(G) \geq r$ since otherwise the instance is trivial. Let $k_{min} = c_{min} \cdot |V(G)| \cdot \binom{r}{2}$. Note that the reload cost of every r-factor is at least k_{min}. Therefore, we can also assume that $k \geq k_{min}$.

A summary of the results regarding the classical and parameterized complexity of the r-MRCF problem is shown in Tables 1 and 2, respectively.

3   Classical Complexity of r-MRCF

The following construct will be used in our reductions. A *diamond* is a graph on four vertices and five edges, that is obtained by adding one chord to a cycle on four vertices. Clearly, a diamond contains two vertices of degree two and two vertices of degree three. The degree two vertices are termed as the *tips* of the diamond. A *joker* is a monochromatic diamond, that is, a diamond whose edges have the same color. In our reductions, every joker J will have exactly one vertex adjacent to other vertices of the graph. This vertex will always be a tip of J, and we will term it as the *connecting tip* of J. Given a joker J and a 2-factor F, it is easy to see that exactly one of the following happens:

– $F \cap E(J)$ is the 4-cycle of J, and $F \setminus E(J)$ does not contain any edges incident to J.
– $F \cap E(J)$ is a triangle of J, $F \setminus E(J)$ contains exactly two edges incident to J both of which are incident its connecting tip.

Furthermore, since our cost functions satisfy $c(\lambda, \lambda) = 0$ for every color λ, and $F \cap E(J)$ is always a cycle, the joker edges do not affect the cost of F. When we describe a 2-factor F, these properties allow us to leave the edges of $F \cap E(J)$ unspecified since they are implied by the edges of $F \setminus E(J)$. Such a partial description is valid if and only if the connecting tip have degree zero or two. Finally, a *5-joker* is a cycle on five vertices with an added chord. This graph has one triangle with one degree 2 vertex that we will refer to as *the tip* of the 5-joker. Note that a 5-joker has all the properties of a joker.

Another construct that we use in our reductions is a graph Q_ℓ that is obtained from the clique on $\ell+1$ vertices by subdividing $\ell-2$ arbitrary edges twice (into three edges) and removing the middle edge of each one. Clearly, Q_ℓ contains $\ell+1$ vertices of degree ℓ and $2(\ell-2)$ vertices of degree one. In total, Q_ℓ has $3\ell-3$ vertices.

Galbiati et al. proved in [16] that 2-MRCF, a.k.a. the minimum reload cost cycle cover problem, is NP-hard even when the number of colors is 2, the reload costs are symmetric and satisfy the triangle inequality. The following theorem whose proof appears in [17] states a hardness result for r-MRCF, which in particular implies a stronger hardness result for the special case of $r=2$.

Theorem 1. r-*MRCF is* NP-*hard for every* $r \geq 2$ *even when* $\Delta(G) \leq r+2$, $d=2$, $k=k_{\min}=0$ *and* $q=\min\{r,3\}$.

The value of the parameter d in Theorem 1 is clearly tight. When $d=1$, that is, when there is only one traversal cost, all the r-factors have the same reload cost. In this case the problem reduces to determining the existence of an r-factor. This problem is known to be polynomial-time solvable [23]. The following theorem states that the parameter $\Delta(G)$ of Theorem 1 is also tight.

Theorem 2. *For every* $r \in \mathbb{N}_{\geq 2}$, *if* $\Delta(G) = r+1$, *then* r-*MRCF can be solved in polynomial time.*

Proof. Recall that we assume $\delta(G) \geq r$. Let R be the set of vertices of degree r of G and $R^+ = V(G) \setminus R$ the vertices of degree $r+1$. Let also H be an r-factor of G. We observe that H is obtained by removing a perfect matching M of $G[R^+]$ from G. Every edge $e \in M$ reduces the reload cost by the sum of the traversal costs with all its incident edges. Therefore, $\mathsf{rc}(H) = \mathsf{rc}(G) - \sum_{e \in M} w(e)$ where

$$w(e) \stackrel{def}{=} \sum_{e' \in E_G(u)-e} c(e,e') + \sum_{e' \in E_G(v)-e} c(e,e')$$

for every edge $e = uv$ of G. Then, minimizing $\mathsf{rc}(H)$ boils down to the problem of finding a maximum weight perfect matching M of $G[R^+]$ with the edge weight function w, which can clearly be solved in polynomial time. $\square$

Both the result in [16] and Theorem 1 are for general graphs and hence leave the complexity of the problem open for special graph classes. In the following (in Theorem 3), we prove hardness of r-MRCF in planar graphs even under restricted cases. We need the following Lemma that is proven in [17] for our reduction.

Lemma 1. *Let* G *be a planar graph with an even number of vertices and* $\delta(G) \geq 2$. *There is a partition* M *of* $V(G)$ *into pairs such that the multigraph* G' *obtained by adding to* G *an edge between every pair of* M *is planar. Moreover, such a partition can be found in polynomial time.*

Theorem 3. *For every* $r \in [2,5]$, r-*MRCF is* NP-*hard even when the input graph* G *is planar,* $\Delta(G) \leq r+4$, $d=2$, $k=k_{\min}$, *and* $q = \begin{cases} 6 \text{ if } r = 2 \\ 7 \text{ otherwise.} \end{cases}$

Proof. We start by proving the theorem for $r = 2$ by reducing an instance (G, χ, c, k) of 2-MRCF with $\Delta(G) \leq 4$, $d = 2$, $k = k_{\min}$, $q = 2$ to an instance (G', χ', c', k') of 2-MRCF, with $\Delta(G') \leq 6$, $d = 2$, $k = k_{\min}$, $q = 6$, and G' is planar.

We rename the colors of the original graph as 1 and 2, and we add four colors $X_C = \{red, blue, green, yellow\}$. Therefore, $q = 6$. As for the cost function we use the following function c' that is an extension of c and uses only costs from $\{0, 1\}$, i.e. $d = 2$. $c'(\lambda, \lambda') = \begin{cases} 0 \text{ if } \lambda = \lambda' \\ 0 \text{ if } yellow \in \{\lambda, \lambda'\} \text{ and } \{1, 2\} \cap \{\lambda, \lambda'\} \neq \emptyset \\ 0 \text{ if } 1 \in \{\lambda, \lambda'\} \text{ and } 2 \notin \{\lambda, \lambda'\} \\ 1 \text{ otherwise.} \end{cases}$

We consider a planar embedding of G where all crossings are polynomial on $|V(G)|$ and no three edges cross the same point. G' is obtained by replacing every crossing point of two edges e, e' of G by a copy of the gadget depicted in Fig. 1, and 9 jokers such that one tip of each joker is identified with a distinct vertex of the gadget. We observe that $\Delta(G') \leq 6$, as required. For a gadget under consideration, the set of vertices that contain the index L (resp. R) are its *left* (resp. *right*) vertices, and the remaining vertices are its *middle* vertices. The *left part* (resp. *right part*) of the gadget is the subgraph induced by the left (resp. right) and middle vertices together. We refer to a four cycle $m_P a_{P,T} t a_{P,B}$ for $P \in \{L, R\}$ as a *green cycle*, to the triangle $t_T t_L t_R$ as the *middle triangle*, to a triangle that consists of an m-vertex, a t-vertex and an a-vertex as a *blue* or *red triangle*, and to the 8-cycle induced by the four c-vertices together with the four m-vertices as the *yellow rectangle*. Note that all these cycles have zero reload cost, though (despite their names) they are not monochromatic.

The two parts of the edge e (resp. e') inherit their color from e (resp. e'). Clearly, $V(G) \subseteq V(G')$. Finally, $k' = 0$.

In [17] we prove that G has a 2-factor with zero reload cost if and only if G' has one.

Now for any $r \in [3, 5]$, we have to reduce an instance (G, χ, c, k) of 2-MRCF where $\Delta(G) \leq 6$, $d = 2$, $k = k_{\min}$, $q = 6$, to an instance (G', χ', c', k') of r-MRCF, where $\Delta(G') \leq r + 4$, $d = 2$, $k = k_{\min}$, $q = 7$ and G' is planar. Note that, since whenever G is planar, we have $\delta(G) \leq 5$, it does not make sense to consider other values of r.

If G is odd, we subdivide an edge and add a 5-joker whose tip is identified with the newly created vertex. We color all the edges with the color of the original edge. This creates an equivalent instance having all the properties assumed for G. By Lemma 1, there is a pairing M of the vertices of G such that edges can be added between each pair of vertices while still preserving planarity. For every pair $\{u, v\} \in M$ we will add a gadget that will have the desired properties of Q_r, and in addition will preserve planarity. Namely, the gadget will have $2(r - 2)$ vertices of degree one, all the remaining vertices of degree r, and will have a planar embedding in which all the degree one vertices are in the outer face. Half (i.e., $(r-2)$) of the degree one vertices of the gadget are identified with u and the remaining half are identified with v. Now every new vertex has degree r and the degree of every original vertex increased by $r-2$. The new edges are colored with

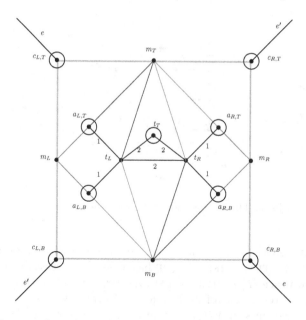

Fig. 1. The planar gadget that replaces two crossing edges of G. The vertices marked with a circle are identified with a tip of a joker. (Color figure online)

a new color λ' such that for every color λ we have $c'(\lambda', \lambda) = 0$. (G', χ', c', k') has an r-factor F' such that $\mathsf{rc}(F') = 0$ if and only if (G, χ, c, k) has a 2-factor F with $\mathsf{rc}(F) = 0$. We omit the proof of this fact since it is identical to the non-planar case.

We now describe the gadget depending on the value of r. If $r = 3$ the gadget is a Q_3 which has the desired planarity properties. If $r = 4$ the gadget is a Q_4. However, in order to get the planarity properties, the subdivided edges of the K_5 should constitute a matching. Finally, if $r = 5$ we use the gadget depicted in Fig. 2.

4   Parameterized Complexity of r-MRCF

Lemma 2. *For every $\alpha > 1$, 2-MRCF, parameterized by $\mathsf{pw}(G)$, is* W[1]*-hard even when $d = 2$, $k = k_{\min}$, and $\overline{\deg}(G) < 6 \cdot \alpha$.*

Proof. The proof is by an FPT reduction from MULTICOLORED CLIQUE that borrows ideas from Theorem 1 of [11]. We first note that MULTICOLORED CLIQUE is W[1]-hard even when the number of colors k is restricted to be odd. Indeed, the problem can be reduced to its special case as follows. Let (H, c, k) be an instance of MULTICOLORED CLIQUE where k is even. Let $(H', c', k + 1)$ be an instance of MULTICOLORED CLIQUE, where (H', c') is obtained by adding a universal vertex v colored $k + 1$ to H. It is easy to see that H contains a clique on k vertices with one vertex from each color class if and only if H' contains one. Also, by

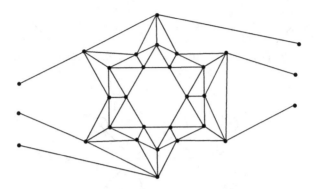

Fig. 2. The planar gadget used for the case $r = 5$.

adding dummy vertices, we can also assume that each chromatic class in the input graph of MULTICOLORED CLIQUE has at least t vertices, where $t = \frac{\alpha}{\alpha-1}$.

Given an instance (H, c, k) of MULTICOLORED CLIQUE with k odd, we construct an instance (G, χ, c, k) of 2-MRCF. Let $V(H) = V_1 \uplus V_2 \uplus \cdots \uplus V_k$, where V_i is the set of vertices of $V(G)$ that are colored i, for $i \in [k]$. Notice that $|V(G)| \geq t \cdot k$. Let W be an Eulerian circuit of the complete graph K_k, which exists since k is odd. We assume that $V(K_k) = [k]$. We also assume, for ease of exposition, that W starts with the sequence of vertices $1, 2, \ldots, k$, where every vertex i of this sequence is considered as the first occurrence of i in W. We also assume that the last vertex of W (i.e. the vertex before the first occurence of 1) is 3. Clearly, every $i \in [k]$ appears in W exactly $k' \stackrel{def}{=} (k-1)/2$ times. The vertex set of G is the disjoint union of three sets U, S, and T where

- U consists of k' copies $U_{i,1}, \ldots, U_{i,k'}$ of V_i for every $i \in [k]$, for a total of $\binom{k}{2}$ sets. For every $i \in [k]$, we number the vertices of V_i from 1 to $|V_i|$, and we number the vertices of every copy $U_{i,j}$ accordingly, as $u_{i,j,1}, \ldots u_{i,j,|V_i|}$.
- S consists of $\binom{k}{2}$ vertices $s_{i,j}$, one for every arc ij of W,
- $T = T_1 \cup T_2 \cup \cdots \cup T_k$, where every set T_i consists of $|V_i| - 1$ vertices $t_{i,1}, t_{i,2}, \ldots, t_{i,|V_i|-1}$.

We proceed with the description of the edge set of G, which contains three types of edges depending on their endpoints. Every edge has one endpoint in U and the other endpoint is in one of U, S or T.

- $S - U$ edges: Let $e = ij$ be an arc of W such that e is incident to the i'-th (resp. j'-th) occurence of i (resp. j) in W. Then $s_{i,j}$ is adjacent to every vertex of $U_{i,i'}$ and to every vertex of $U_{j,j'}$.
- $U - U$ edges: The $U - U$ edges form $|V(H)|$ vertex-disjoint paths, one path on k' vertices for every $v \in V(H)$. The path corresponding to the ℓ-th vertex of V_i is $u_{i,1,\ell}u_{i,2,\ell}\ldots u_{i,k',\ell}$ (see Fig. 3).
- $T - U$ edges: For every $i \in [k]$ and every $\ell \in [|V_i| - 1]$, the vertex $t_{i,\ell}$ is adjacent to $u_{i,1,\ell}$, $u_{i,1,\ell+1}$, $u_{i,k',\ell}$ and $u_{i,k',\ell+1}$ (see Fig. 3).

In [17], we prove that $\overline{\deg}(G) < 6\alpha$, and $\mathsf{pw}(G) \leq \binom{k}{2} + 3$.

We proceed with the description of the coloring function χ and the traversal cost function c. The color set is $V(H) \cup \{white\}$. In other words, the vertices of H corresponds to colors in the constructed instance H'; that is, there is a color in H' corresponding to each vertex in H. All the $U - U$ and $T - U$ edges are colored white, and the $S - U$ edges are colored upon their endpoint in U as follows. For every $S - U$ edge e with endpoint $u_{i,j,\ell}$, we define $\chi(e) = v_{i,\ell}$. The traversal cost $c(v, white)$ is 1 for every $v \in V(H)$ and $c(white, white) = 0$. For every $v, v' \in V(H)$ we set $c(v, v')$ to zero if $v = v'$ or vv' is an edge of H and to one otherwise. Finally, we set $k = 0$. This completes the construction of (G, χ, c, k), which can be clearly performed in polynomial time.

We now prove that H has a k-clique with one vertex from every color class if and only if G has a 2-factor F with $\mathsf{rc}(F) = 0$. Assume that H has a clique K with exactly one vertex from every color, and suppose without loss of generality that K consists of the first vertices of each color class. Let F be the 2-factor of G consisting of the following cycles.

- The clique cycle $G[C_K]$, where $C_K = S \cup \{u_{i,j,1} \mid i \in [k], j \in [k']\}$.
- The vertex cycles, one per every vertex $v \in V(H) \setminus K$. For every $i \in [k]$ and every $\ell \in [2, |V_i|]$, the vertex cycle corresponding to the vertex $v_{i,\ell} \in V(H) \setminus K$ is $G[C_{i,\ell}]$, where $C_{i,\ell} = \{t_{i,\ell-1}, u_{i,j,\ell} \mid j \in [k']\}$.

It is easy to see that every vertex is in exactly one of these cycles. Furthermore, the edges of the vertex cycles are all white, incurring a reload cost of zero. It remains to show that that the edges of the clique cycle also incur a zero cost. Indeed, the two edges incident to a vertex $u_{i,j,\ell} \in U$ incur a cost $c(v_{i,\ell}, v_{i,\ell}) = 0$, and the two edges incident to a vertex $s_{i,j}$ incur a zero cost since the adjacent vertices are $u_{i,i',1}$ and $u_{j,j',1}$ that correspond to the vertices $v_{i,1}$ and $v_{j,1}$ of K, which are adjacent in H.

Conversely, suppose that H contains a 2-factor F with $\mathsf{rc}(F) = 0$. Every vertex $s_{i,j}$ is adjacent to two distinct vertices of U that correspond to two distinct vertices v, v' of $V(H)$. Furthermore, since the cost at vertex $s_{i,j}$ is zero, we conclude that v and v' are adjacent in H. In particular, v and v' are in different color classes. Therefore, $s_{i,j}$ is adjacent to one vertex from each of $U_{i,i'}$ and $U_{j,j'}$. The second edge incident to these vertices in F is not white, since otherwise it incurs a positive traversal cost. Thus, the other edge is also a $U - S$ edge. We conclude that the $\binom{k}{2}$ vertices S are all in one cycle C_K that also contains $\binom{k}{2}$ vertices from U, one from every set $U_{i,i'}$. Furthermore, two consecutive vertices of U in C_K correspond to two adjacent vertices of H.

All the remaining cycles must be in $G \setminus S$, which consists of k connected components $T_i \cup \bigcup_{i'=1}^{k'} U_{i,i'}$, for every $i \in [k]$. The result is now apparent from Fig. 3 that depicts this connected component. If a vertex $u_{i,i',\ell}$ is in one of the remaining cycles, then all the vertices $u_{i,i'',\ell}$ for $i'' \in [k']$ are in the same cycle as $u_{i,i',\ell}$. Therefore, if a vertex $u_{i,i',\ell}$ is in C_K, then all the vertices $u_{i,i'',\ell}$ for $i'' \in [k']$ are in C_K. Recall that C_K contains one vertex $u_{i,i',\ell}$ from every set $U_{i,i'}$. We conclude that these vertices correspond to the same vertex $v_{i,\ell}$ of V_i.

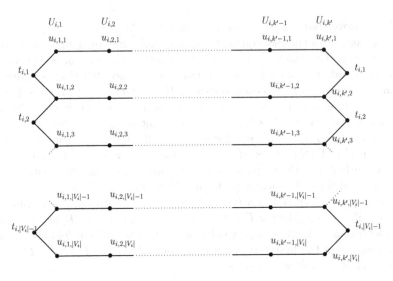

Fig. 3. The $U - U$ and $T - U$ edges of G.

We recall that consecutive U-vertices of C_K correspond to two adjacent vertices of H. Therefore, the vertices $U \cap C_K$ correspond to the vertices of a clique of H.

Given a graph G on n-vertices where n is even, a *pairing collection* of G is a collection M of pairs of vertices in G that forms a partition of $V(G)$. We denote $G + M = (V(G), E(G) \cup M)$. Notice that $G + M$ can be a multi-graph as the pairings in M might already be edges of G. In [17] we prove the following lemma that will prove useful in our proof.

Lemma 3. *Let G be a n-vertex graph where n is even and $\mathsf{pw}(G) \le k$ for some $k \ge 1$. Then G has a pairing collection M such that $\mathsf{pw}(G + M) \le k + 1$.*

Theorem 4. *For every $r \in \mathbb{N}_{\ge 2}$, r-MRCF, parameterized by $\mathsf{pw}(G)$, is $\mathsf{W}[1]$-hard even when $d = 2$, $k = k_{\min}$ and $\overline{\deg}(G)$ is less than $r + \begin{cases} 4 & \text{if } r = 2 \\ 4/3 & \text{otherwise.} \end{cases}$*

Proof. For $r = 2$ the Theorem is equivalent to Lemma 2. For $r > 2$ we present an FPT reduction from the base case, i.e. $r = 2$. using the same technique as in the proof of Theorem 1. Namely, given an instance (G, χ, c, k) of 2-MRCF with $d = 2$ and $k = k_{\min} = 0$ where $\overline{\deg}(G) < 6$, we construct an instance (G', χ', c', k') of r-MRCF where $d = 2$ and the average degree of G' is less than $r + 4/3$ by adding to G a pair collection, and replacing every new edge by the gadget Q_r. By the last part of the proof of Theorem 1, G' has an r-factor of zero cost if and only if G has a 2-factor of zero cost. In [17] we prove that the average degree of G' is at most $r + 4/3$, and $\mathsf{pw}(G') \le (\mathsf{pw}(G) + 1) \cdot \frac{r+1}{r-1} - 1$. Finally, we note that r is a constant of the problem. Therefore, the function the latter is a function of $\mathsf{pw}(G)$ as required.

In [17], we present a dynamic programming algorithm proving

Theorem 5. *For every* $r \in \mathbb{N}_{\geq 2}$, r-MRCF, *parameterized by* $\min\{q, \Delta(G)\}$ *and* $\mathrm{tw}(G)$ *is in* FPT.

References

1. Galbiati, G.: The complexity of a minimum reload cost diameter problem. Discrete Appl. Math. **156**(18), 3494–3497 (2008)
2. Arkoulis, S., Anifantis, E., Karyotis, V., Papavassiliou, S., Mitrou, N.: On the optimal, fair and channel-aware cognitive radio network reconfiguration. Comput. Netw. **57**(8), 1739–1757 (2013)
3. Gözüpek, D., Buhari, S., Alagöz, F.: A spectrum switching delay-aware scheduling algorithm for centralized cognitive radio networks. IEEE Trans. Mob. Comput. **12**(7), 1270–1280 (2013)
4. Celik, A., Kamal, A.E.: Green cooperative spectrum sensing and scheduling in heterogeneous cognitive radio networks. IEEE Trans. Cogn. Commun. Netw. **2**(3), 238–248 (2016)
5. Wirth, H.C., Steffan, J.: Reload cost problems: minimum diameter spanning tree. Discrete Appl. Math. **113**(1), 73–85 (2001)
6. Gourvès, L., Lyra, A., Martinhon, C., Monnot, J.: The minimum reload s-t path, trail and walk problems. Discrete Appl. Math. **158**(13), 1404–1417 (2010)
7. Amaldi, E., Galbiati, G., Maffioli, F.: On minimum reload cost paths, tours, and flows. Networks **57**(3), 254–260 (2011)
8. Galbiati, G., Gualandi, S., Maffioli, F.: On minimum changeover cost arborescences. In: Pardalos, P.M., Rebennack, S. (eds.) SEA 2011. LNCS, vol. 6630, pp. 112–123. Springer, Heidelberg (2011). https://doi.org/10.1007/978-3-642-20662-7_10
9. Gözüpek, D., Shalom, M., Voloshin, A., Zaks, S.: On the complexity of constructing minimum changeover cost arborescences. Theor. Comput. Sci. **540**, 40–52 (2014)
10. Gözüpek, D., Shachnai, H., Shalom, M., Zaks, S.: Constructing minimum changeover cost arborescenses in bounded treewidth graphs. Theor. Comput. Sci. **621**, 22–36 (2016)
11. Gözüpek, D., Özkan, S., Paul, C., Sau, I., Shalom, M.: Parameterized complexity of the mincca problem on graphs of bounded decomposability. Theor. Comput. Sci. **690**, 91–103 (2017)
12. Gözüpek, D., Shalom, M.: Edge coloring with minimum reload/changeover costs. arXiv preprint arXiv:1607.06751 (2016)
13. Gamvros, I., Gouveia, L., Raghavan, S.: Reload cost trees and network design. Networks **59**(4), 365–379 (2012)
14. Meijer, H., Núñez-Rodríguez, Y., Rappaport, D.: An algorithm for computing simple k-factors. Inf. Process. Lett. **109**(12), 620–625 (2009)
15. Schrijver, A.: Combinatorial Optimization: Polyhedra and Efficiency, vol. 24. Springer, Heidelberg (2003)
16. Galbiati, G., Gualandi, S., Maffioli, F.: On minimum reload cost cycle cover. Discrete Appl. Math. **164**, 112–120 (2014)
17. Baste, J., Gözüpek, D., Shalom, M., Thilikos, D.M.: Minimum reload cost graph factors. CoRR abs/1810.11700 (2018). http://arxiv.org/abs/1810.11700
18. Cygan, M., et al.: Parameterized Algorithms. Springer, Cham (2015). https://doi.org/10.1007/978-3-319-21275-3

19. Downey, R.G., Fellows, M.R.: Fundamentals of Parameterized Complexity. TCS. Springer, London (2013). https://doi.org/10.1007/978-1-4471-5559-1
20. Pietrzak, K.: On the parameterized complexity of the fixed alphabet shortest common supersequence and longest common subsequence problems. J. Comput. Syst. Sci. **67**(4), 757–771 (2003)
21. Cygan, M., Nederlof, J., Pilipczuk, M., Pilipczuk, M., van Rooij, J.M.M., Wojtaszczyk, J.O.: Solving connectivity problems parameterized by treewidth in single exponential time. In: Proceedings of the 52nd Annual Symposium on Foundations of Computer Science (FOCS), pp. 150–159. IEEE Computer Society (2011)
22. Kloks, T. (ed.): Treewidth. LNCS, vol. 842. Springer, Heidelberg (1994). https://doi.org/10.1007/BFb0045375
23. Pulleyblank, W.R.: Faces of matching polyhedra. Ph.D. thesis, University of Waterloo (1973)

Stable Divisorial Gonality is in NP

Hans L. Bodlaender[1,2], Marieke van der Wegen[1]([⊠]),
and Tom C. van der Zanden[1]

[1] Department of Information and Computing Sciences, Universiteit Utrecht,
Princetonplein 5, 3584 CC Utrecht, The Netherlands
M.vanderWegen@uu.nl
[2] Department of Mathematics and Computer Science,
Eindhoven University of Technology,
PO Box 513, 5600 MB Eindhoven, The Netherlands

Abstract. Divisorial gonality and stable divisorial gonality are graph parameters, which have an origin in algebraic geometry. Divisorial gonality of a connected graph G can be defined with help of a chip firing game on G. The stable divisorial gonality of G is the minimum divisorial gonality over all subdivisions of edges of G.

In this paper we prove that deciding whether a given connected graph has stable divisorial gonality at most a given integer k belongs to the class NP. Combined with the result that (stable) divisorial gonality is NP-hard by Gijswijt, we obtain that stable divisorial gonality is NP-complete. The proof consists of a partial certificate that can be verified by solving an Integer Linear Programming instance. As a corollary, we have that the number of subdivisions needed for minimum stable divisorial gonality of a graph with n vertices is bounded by $2^{p(n)}$ for a polynomial p.

1 Introduction

The notions of the divisorial gonality and stable divisorial gonality of a graph find their origin in algebraic geometry and are related to the abelian sandpile model (cf. [8]). The notion of divisorial gonality was introduced by Baker and Norine [1,2], under the name gonality. As there are several different notions of gonality in use (cf. [1,6,7]), we add the term *divisorial*, following [6]. See [7, Appendix A] for an overview of the different notions.

Divisorial gonality and stable divisorial gonality have definitions in terms of a chip firing game. In this chip firing game, played on a connected multigraph $G = (V, E)$, each vertex has a non-negative number of chips. When we *fire* a set of vertices $S \subseteq V$, we move from each vertex $v \in S$ one chip over each edge with v as endpoint. Each vertex v in S has its number of chips decreased by the number of edges from v to a neighbour not in S, and each vertex v not in S has

---

H. L. Bodlaender—This work was supported by the NETWORKS project, funded by the Netherlands Organization for Scientific Research NWO under project no. 024.002.003.

B. Catania et al. (Eds.): SOFSEM 2019, LNCS 11376, pp. 81–93, 2019.
https://doi.org/10.1007/978-3-030-10801-4_8

its number of chips increased by the number of edges from v to a neighbour in S. Such a firing move is only allowed when after the move, each vertex still has a nonnegative number of chips. The *divisorial gonality* of a connected graph G can be defined as the minimum number of chips in an initial assignment of chips (called *divisor*) such that for each vertex $v \in V$, there is a sequence of allowed firing moves resulting in at least one chip on v. Interestingly, this number equals the number for a *monotone* variant, where we require that each set that is fired has the previously fired set as a subset. See Sect. 2 for precise definitions.

A variant of divisorial gonality is *stable divisorial gonality*. The stable divisorial gonality of a graph is the minimum of the divisorial gonality over all subdivisions of a graph; we can subdivide the edges of the graph any nonnegative number of times. (In the application in algebraic geometry, the notion of *refinement* is used. Here, we can subdivide edges but also add new degree-one vertices to the graph in a refinement, but as this never decreases the number of chips needed, we can ignore the possibility of adding leaves. Thus, we use subdivisions instead of refinements).

It is known that treewidth is a lower bound for stable divisorial gonality [10]. The stable divisorial gonality of a graph is at most the divisorial gonality, but this inequality can be strict, see for example [4, Fig. 1].

In this paper, we study the complexity of computing the stable divisorial gonality of graphs: i.e., we look at the complexity of the STABLE DIVISORIAL GONALITY problem: given an undirected graph $G = (V, E)$ and an integer k, decide whether the stable divisorial gonality of G is at most k. It was shown by Gijswijt [11] that divisorial gonality is NP-complete. The same reduction gives that stable divisorial gonality is NP-hard. However, membership of stable divisorial gonality in NP is not trivial: it is unknown how many subdivisions are needed to obtain a subdivision with minimum divisorial gonality. In particular, it is open whether a polynomial number of edge subdivisions are sufficient.

In this paper, we show that stable divisorial gonality belongs to the class NP. We use the following proof technique, which we think is interesting in its own right: we give partial certificates that describe only some aspects of a firing sequence. Checking if a partial certificate indeed corresponds to a solution is non-trivial, but can be done by solving an integer linear program. Membership in NP follows by adding to the partial certificate, that describes aspects of the firing sequence, a certificate for the derived ILP instance. As a corollary, we have that the number of subdivisions needed for minimum stable divisorial gonality of a graph with n vertices is at most $2^{p(n)}$ for a polynomial p.

We finish this introduction by giving an overview of the few previously known results on the algorithmic complexity of (stable) divisorial gonality. Bodewes et al. [4] showed that deciding whether a graph has stable divisorial gonality at most 2, and whether it has divisorial gonality at most 2 can be done in $O(n \log n + m)$ time. From [9] and [3], it follows that divisorial gonality belongs to the class XP, i.e. there is an algorithm that decides in time $O(n^{f(k)})$ whether $\mathrm{dgon}(G) \le k$. It is open whether stable divisorial gonality is in XP. NP-hardness of the notions was shown by Gijswijt [11].

2   Preliminaries

In this paper, we assume that each graph is a connected undirected multigraph, i.e., we allow parallel edges. In the algebraic number theoretic application of (stable) divisorial gonality, graphs can also have selfloops (edges with both endpoints at the same vertex), but as the (stable) divisorial gonality of graph does not change when we remove selfloops, we assume that there are no selfloops.

A *divisor* D is a function $D\colon V(G) \to \mathbb{Z}$. We can think of a divisor as an assignment of chips, each vertex v has $D(v)$ chips. The *degree* of a divisor is the total number of chips on the graph: $\deg(D) = \sum_{v \in V} D(v)$. We call a divisor *effective* if $D(v) \geq 0$ for all vertices v. Let D be an effective divisor and A a set of vertices. We call A *valid*, if for all vertices $v \in A$ it holds that $D(v)$ is at least the number of edges from v to a vertex outside A. When we *fire* a set A, we obtain a new divisor: for every vertex $v \in A$, the value of $D(v)$ is decreased by the number of edges from v to vertices outside A and for every vertex $v \notin A$, the value $D(v)$ is increased by the number of edges from v to A. We are only allowed to fire valid sets, so that the divisor obtained is again effective.

Two divisors D and D' are called *equivalent*, if there is an increasing sequence of sets $A_1 \subseteq A_2 \subseteq \ldots \subseteq A_k \subseteq V$ such that for every i the set A_i is valid after we fired $A_1, A_2, \ldots, A_{i-1}$ starting from D, and firing $A_1, A_2, \ldots, A_k$ yields D'. We write $D \sim D'$ to denote that two divisors are equivalent. For two equivalent divisors D and D', the difference $D' - D$ is called *transformation* and the sequence $A_1, A_2, \ldots, A_k$ is called a *level set decomposition* of this transformation. A divisor D *reaches* a vertex v if it is equivalent to a divisor D' with $D'(v) \geq 1$.

A subdivision of a graph G is a graph H obtained from G by applying a nonnegative number of times the following operation: take an edge between two vertices v and w and replace this edge by two edges to a new vertex x.

The *stable divisorial gonality* $\mathrm{sdgon}(G)$ of a graph G is the minimum number k such that there exists a subdivision H of G and a divisor on H with degree k that reaches all vertices of H.

There are several equivalent definitions, which we omit here. If we do not require that the sequence of firing sets is increasing, i.e., we omit the requirements $A_i \subseteq A_{i+1}$, then we still have the same graph parameter (see [9]). The notion of a firing set can be replaced by an algebraic operation (see [2]); instead of subdivisions, we can use *refinements* where we allow that we add subdivisions and trees, i.e., we can repeatedly add new vertices of degree one. The definition we use here is most intuitive and useful for our proofs.

3   A (Partial) Certificate

Assume that we are given a yes-instance (G, k) of the problem. Without loss of generality, we assume that $k \leq n$. There exists a subdivision H and a divisor D on H with k chips that reaches all vertices. We do not know whether the number of subdivisions in H is polynomial in the size of the graph, i.e. in the number of vertices and edges of the graph, so we cannot include H in a polynomial certificate for this instance. But the chips in D can be placed on added vertices of H, so we cannot include D in our certificate either. We will prove that when we subdivide every edge once, we can assume that there is a divisor D' that reaches all vertices and has all chips on vertices of this new graph, and hence we can include D' in a polynomial certificate.

Definition 1. *Let G be a graph. Let G_1 denote the graph obtained by subdividing every edge of G once.*

Lemma 2. *Let G be a graph. The stable divisorial gonality of G is at most k if and only if there is a subdivision H of G_1 and a divisor D on H such that*

- *D has at most k chips, i.e. has degree at most k,*
- *D reaches all vertices of H,*
- *D has only chips on vertices of G_1.*

Proof. Suppose that there exists a subdivision H of G_1 and a divisor with the desired properties. Then it is clear that the stable divisorial gonality of G is at most k, since H is a subdivision of G as well.

Suppose that G has stable divisorial gonality at most k. Then there is a subdivision H of G and a divisor D on H with degree at most k that reaches all vertices. If not every edge of G is subdivided in H, then subdivide every edge of H to obtain H_1. Consider the divisor D on H_1. By [12, Corollary 3.4] D reaches all vertices of H_1.

Let $e = uv$ be an edge of G, and let $a_1, a_2, \ldots, a_r$ be the vertices that are added to e in H_1. Suppose that D assigns more than one chip to those added vertices, say it assigns one chip to a_i and one to a_j with $i \leq j$. Then we can fire sets $\{a_h \mid i \leq h \leq j\}$, $\{a_h \mid i - 1 \leq h \leq j + 1\}$, ... until at least one of the chips lies on u or v. Hence, D is equivalent to a divisor which has one chip less on added vertices. Repeat this procedure until there is for every edge of G at most one chip assigned to the vertices added to that edge. The divisor obtained in this way is equivalent to D, so it reaches all vertices of H_1 and has at most k chips. Thus we have obtained a divisor with the desired properties. $\square$

Now a certificate can contain the graph G_1 and the divisor D as in Lemma 2. From now on we assume D to have chips on vertices of G_1 only. A divisor D as in Lemma 2 reaches all vertices, so for every vertex $w \in V(G_1)$ there is a divisor $D_w \sim D$ with a chip on w and a level set decomposition $A_1, A_2, \ldots, A_r$ of the transformation $D_w - D$. Again we do not know whether r is polynomial in the size of G, so we cannot include this level set decomposition in the certificate. However, we can define some of the sets to be 'relevant', and include all relevant sets in the certificate.

Definition 3. *Let G be a graph and H a subdivision of G. Let D be a divisor on H and $A_1, A_2, \ldots, A_r$ a level set decomposition of a transformation $D' - D$. Let $D_0, D_1, \ldots, D_r$ be the associated sequence of divisors. We call A_i relevant if any of the following holds:*

- A_i moves a chip from a vertex of G, i.e. there is a vertex v of G such that $D_i(v) - D_{i-1}(v) < 0$, or
- A_i moves a chip to a vertex of G, i.e. there is a vertex v of G such that $D_i(v) - D_{i-1}(v) > 0$, or
- there is a vertex of G such that A_i is the first level set that contains this element, i.e. $(A_i \backslash A_{i-1}) \cap V(G)$ is not empty.

Lemma 4. *Let G be a graph and H a subdivision of G. Let D be a divisor on H with k chips and $A_1, A_2, \ldots, A_r$ a level set decomposition of a transformation $D' - D$. Let $D_0, D_1, \ldots, D_r$ be the associated sequence of divisors. Then there are at most $2kn + n$ relevant level sets.*

Proof. Each chip can reach each vertex at most once and can depart at most once from each vertex. So, there are at most kn sets A_i that fulfil the first condition of Definition 3 and at most kn sets that fulfil the second condition. Clearly, the number of sets A_i that fulfil the third condition is upper bounded by the number of vertices of G. □

This lemma shows that the number of relevant sets in a level set decomposition is polynomial, since $k \leq n$. However, the number of elements of each of these sets can still be exponential, so we cannot include those sets in a polynomial certificate. Instead, for a relevant set A_i, we will include $A_i \cap V(G_1)$ in our certificate. Moreover, for each relevant set, we will describe which chips move from/to a vertex of G_1 by firing A_i. When chip j is moved from a vertex v along edge e, we include a tuple $(v, j, -1, e)$, and when a chip j is moved towards a vertex v along edge e, we include a tuple $(v, j, +1, e)$.

Now, a *partial certificate* $\mathcal{C}$ consists of

- a divisor D on G_1 with k chips, where the chips are labelled $1, 2, \ldots, k$,
- for every vertex $w \in V(G_1)$, a series of pairs $(A_{w,1}, M_{w,1})$, $(A_{w,2}, M_{w,2})$, ..., (A_{w,a_w}, M_{w,a_w}) for some integer a_w, such that
  - $A_{w,1} \subseteq A_{w,2} \subseteq \ldots \subseteq A_{w,a_w} \subseteq V(G_1)$,
  - $M_{w,i} = \{(v, j, \sigma, e) \mid v \in V(G_1), 1 \leq j \leq k, \sigma \in \{-1, +1\}, e \in E(G_1)\}$.

This partial certificate should satisfy a lot of conditions, which are implicit in the intuitive explanation of this partial certificate. We list the intuition behind these conditions below and give the formal definition between brackets.

Incidence requirement. The edge along which a chip is fired is incident to the vertex from/to which it is fired. (For every $M_{w,i}$ and every tuple $(v, j, \sigma, e) \in M_{w,i}$, it holds that e is incident to v.)

Departure requirement. If a chip leaves a vertex, then this vertex is fired and its neighbour is not. (For every $M_{w,i}$ and $(v, j, -1, uv) \in M_{w,i}$, it holds that $v \in A_{w,i}$ and $u \notin A_{w,i}$.)

Arrival requirement. If a chip arrives at a vertex, then this vertex is not fired and its neighbour is. (For every $M_{w,i}$ and $(v, j, +1, uv) \in M_{w,i}$, it holds that $v \notin A_{w,i}$ and $u \in A_{w,i}$.)

Unique departure per edge requirement. For every vertex at most one chip leaves along each edge. (For every $M_{w,i}$ and $(v, j_1, -1, e), (v, j_2, -1, e) \in M_{w,i}$, it holds that $j_1 = j_2$.)

Unique arrival per edge requirement. For every vertex at most one chip arrives along each edge. (For every $M_{w,i}$ and $(v, j_1, +1, e), (v, j_2, +1, e) \in M_{w,i}$, it holds that $j_1 = j_2$.)

Unique departure per chip requirement. A chip can leave a vertex along at most one edge. (For every $M_{w,i}$ and $(v_1, j, -1, e_1), (v_2, j, -1, e_2) \in M_{w,i}$, it holds that $v_1 = v_2$ and $e_1 = e_2$.)

Unique arrival per chip requirement. A chip can arrive at a vertex along at most one edge. (For every $M_{w,i}$ and $(v_1, j, +1, e_1), (v_2, j, +1, e_2) \in M_{w,i}$, it holds that $v_1 = v_2$ and $e_1 = e_2$.)

Immediate arrival requirement. If a chip leaves a vertex v and arrives at another vertex u at the same time, then the chip is fired along the edge uv. (For every $M_{w,i}$ and $(v_1, j, -1, e_1), (v_2, j, +1, e_2) \in M_{w,i}$, it holds that $e_1 = e_2 = v_1 v_2$.)

Departure location requirement. If a chip leaves a vertex, then this chip was on this vertex, that is, either the last movement of this chip was to this vertex, or it was assigned to this vertex by D and did not move. (For every $M_{w,i}$ and $(v, j, -1, e) \in M_{w,i}$, the following holds. Let $i' < i$ be the greatest index such that there is a tuple $(u, j, \sigma, e') \in M_{w,i'}$, if it exists. Then there is a tuple $(v, j, +1, e') \in M_{w,i'}$ for some e'. If no such index i' exists, then D assigns j to v.)

Arrival location requirement. If a chip arrives at a vertex, then this chip was moving along an edge to this vertex, that is, either this chip just left the other end of the edge, or it left before and did not yet arrive. (For every $M_{w,i}$ and $(v, j, +1, e) \in M_{w,i}$, either $(u, j, -1, e) \in M_{w,i}$ where $u \neq v$, or the following holds. Let $i' < i$ be the greatest index such that there is a tuple $(u, j, \sigma, e') \in M_{w,i'}$. There is a tuple $(u, j, -1, e) \in M_{w,i'}$ with $u \neq v$ and $(v, j, +1, e) \notin M_{w,i'}$.)

Outgoing edges requirement. A chip is fired along each outgoing edge, that is, for each outgoing edge uv either a new chip leaves u or there is a chip that left u already and did not yet arrive at v. (For every $A_{w,i}$ and for every edge uv such that $u \in A_{w,i}$, $v \notin A_{w,i}$, the following holds. Either $(u, j, -1, uv) \in M_{w,i}$ for some j, or there is a $1 \leq j \leq k$ and an $i' < i$ such that $(u, j, -1, uv) \in M_{w,i'}$ and $(v, j, +1, uv) \notin M_{w,i''}$ for all $i' \leq i'' < i$.)

Previous departure requirement. If a chip leaves a vertex v along some edge e, and v was in the previous firing set as well, then a chip left v along e when the previous set was fired. (For every $A_{w,i}$ and $M_{w,i}$, the following holds. If $v \in A_{w,i}$, $v \in A_{w,i+1}$ and $(v, j, -1, e) \in M_{w,i+1}$ for some j and e, then $(v, j', -1, e) \in M_{w,i}$ for some $j' \neq j$.)

Next arrival requirement. If a chip arrives at a vertex v along some edge e, and v is not in the next firing set as well, then a chip will arrive at v along e when the next set is fired. (For every $A_{w,i}$ and $M_{w,i}$, the following holds. If $v \notin A_{w,i}$, $v \notin A_{w,i+1}$ and $(v, j, +1, e) \in M_{w,i}$ for some j and e, then $(v, j', +1, e) \in M_{w,i+1}$ for some $j' \neq j$.)

Reach all vertices requirement. For all vertices w, at the end of the sequence $A_{w,1}, \ldots, A_{w,a_w}$, there is a chip on w. (For every vertex w, either there is a $1 \leq j \leq k$ and an i such that $(w, j, +1, e) \in M_{w,i}$ for some e and $(w, j, -1, e') \notin M_{w,i'}$ for all $i' \geq i$, or there is a $1 \leq j \leq k$ that D assigns to w and $(w, j, -1, e) \notin M_{w,i}$ for all i.)

Now for a given graph G, and such a partial certificate $\mathcal{C}$, we want to decide whether there is a subdivision of G_1 such that for every vertex $w \in V(G_1)$ there is a divisor $D_w \sim D$ with a chip on w such that the sets $A_{w,1}, \ldots, A_{w,a_w}$ are the relevant sets of the level set decomposition of the transformation $D_w - D$. To decide this, we will construct an integer linear program $\mathcal{I}_\mathcal{C}$, such that this program has a solution if and only if there is such a subdivision of G_1. Since integer linear programming is in NP, we know that if there is a solution to $\mathcal{I}_\mathcal{C}$, then there is a polynomial certificate $\mathcal{D}$ for the ILP instance. In order to obtain a certificate for the STABLE DIVISORIAL GONALITY problem, we add the certificate for the ILP instance to the partial certificate, as defined above. Thus, a certificate for the STABLE DIVISORIAL GONALITY problem is then of the form $(\mathcal{C}, \mathcal{D})$.

For the integer linear program $\mathcal{I}_\mathcal{C}$, we introduce some variables. For every vertex $w \in V(G_1)$ and every $1 \leq i < a_w$, we define a variable $t_{w,i}$. This variable represents the number of sets that is fired between $A_{w,i}$ and $A_{w,i+1}$, including $A_{w,i}$ and excluding $A_{w,i+1}$. For every edge e of G_1, we define a variable l_e, which represents the length of e, i.e. the number of edges that e is subdivided into. Now we construct $\mathcal{I}_\mathcal{C}$:

- For every edge $e \in E(G_1)$, include the inequality $l_e \geq 1$. (Every edge has length at least one.)
- For every vertex $w \in V(G_1)$ and $1 \leq i < a_w$, include the inequality $t_{w,i} \geq 1$. (The set $A_{w,i}$ is fired, so $t_{w,i} \geq 1$.)
- For every edge $e = uv$ of G_1 such that there is a set $M_{w,i}$ with $(v, j, -1, e)$, $(u, j, +1, e) \in M_{w,i}$ for some j, include $l_e = 1$ in $\mathcal{I}_\mathcal{C}$. (If a chip arrives immediately after it is fired, then the edge has length one.)
- For every vertex $w \in V(G_1)$ and $1 \leq i < a_w$ such that there are v, j_1, j_2, e such that $(v, j_1, -1, e) \in M_{w,i}$ and $(v, j_2, -1, e) \in M_{w,i+1}$, include $t_{w,i} = 1$ in $\mathcal{I}_\mathcal{C}$. (If there is a set A that is fired between $A_{w,i}$ and $A_{w,i+1}$, then $A_{w,i} \subseteq A \subseteq A_{w,i+1}$. It follows that A fires a chip from v along e as well. But then A is a relevant set. We conclude that $t_{w,i} = 1$.)
- For every vertex $w \in V(G_1)$ and $1 \leq i \leq a_w$ such that there are v, j, e such that $(v, j, +1, e) \in M_{w,i}$, include $t_{w,i} = 1$ in $\mathcal{I}_\mathcal{C}$. (Notice that the set fired after $A_{w,i}$ either contains v or causes a chip to arrive at v, so this set is relevant.)
- For every vertex w and edge $e = uv$ of G_1, let i_0 be the smallest index such that $(v, j, -1, e) \in M_{w,i_0}$ for some j, i_1 the greatest index such that $(v, j, -1, e) \in M_{w,i_1}$ for some j, i_2 the smallest index such that $(u, j, +1, e) \in M_{w,i_2}$ for some j, and i_3 the greatest index such that $(u, j, +1, e) \in M_{w,i_3}$ for some j. Include the following inequalities in $\mathcal{I}_\mathcal{C}$:

$$(i_1 - i_0 + 1)l_e - (i_1 - i_0) + (i_3 - i_2) \geq \sum_{i=i_0}^{i_3} t_{w,i} \tag{1}$$

$$(i_3 - i_2 + 1)l_e + (i_1 - i_0) - (i_3 - i_2) \leq \sum_{i=i_0}^{i_3} t_{w,i}. \tag{2}$$

(There are $i_1 - i_0 + 1$ chips that left v along edge e, and $i_3 - i_2 + 1$ chips that arrived at u along e. There are $\sum_{i=i_0}^{i_3} t_{w,i}$ sets fired since the first chip left until the last chip arrives, and every of these sets causes one chip to move one step. The chips that arrived at u took l_e steps, the chips that did not arrive took at least one and at most $l_e - 1$ steps. This yields the inequalities.)

Now a certificate for the stable divisorial gonality problem is a pair $(\mathcal{C}, \mathcal{D})$. Here, the partial certificate $\mathcal{C}$ contains a divisor D on G_1 with labelled chips and for every vertex $w \in V(G_1)$ a series of pairs $(A_{w,1}, M_{w,1}), (A_{w,2}, M_{w,2}), \ldots,$ (A_{w,a_w}, M_{w,a_w}), and satisfies all requirements above. And $\mathcal{D}$ is a certificate of the integer linear program $\mathcal{I}_\mathcal{C}$.

4   Correctness

It remains to prove that there exists a certificate $(\mathcal{C}, \mathcal{D})$ if and only if $\text{sdgon}(G) \leq k$.

Lemma 5. *Let G be a graph with $\text{sdgon}(G) \leq k$. There exists a certificate $(\mathcal{C}, \mathcal{D})$.*

Proof. By Lemma 2 we know that there is a subdivision H of G_1 and a divisor D with k chips, all on vertices of G_1, that reaches all vertices. Choose a labeling of the chips and let D be the divisor in $\mathcal{C}$.

For every vertex $w \in V(G_1)$, there is a divisor $D_w \sim D$ with a chip on w and a level set decomposition $A_{w,1}, \ldots, A_{w,a_w}$. Let $A_{w,i_1}, \ldots, A_{w,i_{b_w}}$ be the subsequence consisting of all relevant sets. Let $B_{w,1} = A_{w,i_1} \cap V(G_1), \ldots,$ $B_{w,b_w} = A_{w,i_{b_w}} \cap V(G_1)$.

Fire the sets $A_{w,1}, \ldots, A_{w,a_w}$ in order. For every i_j, set $M_{w,j} = \emptyset$. When firing the set A_{w,i_j}, check for every chip h whether it arrives at a vertex v of G_1 or leaves a vertex v of G_1. If so, add the tuple (v, h, σ, e) to $M_{w,j}$, where $\sigma = +1$ if h arrives at v and $\sigma = -1$ if h leaves v, and e is the edge of G_1 along which h moves.

The divisor D together with the sequences $(B_{w,i}, M_{w,i})$, for every vertex $w \in V(G_1)$, is the partial certificate $\mathcal{C}$. Notice that by definition $\mathcal{C}$ satisfies all conditions: Incidence requirement, Departure requirement, Arrival requirement, Unique departure per edge, Unique arrival per edge, Unique departure per chip, Unique arrival per chip, Immediate arrival, Departure location, Arrival location, Outgoing edges requirement, Previous departure, Next arrival and Reach all vertices.

For every edge e of G_1, define l_e as the number of edges that e is subdivided into in H. For every vertex w of G_1 and $1 \leq j \leq b_w - 1$, define $t_{w,i}$ as the number of sets between $A_{w,i+1}$ and $A_{w,i}$, including $A_{w,i}$ and excluding $A_{w,i+1}$. Notice that this is a solution to the integer linear program $\mathcal{I}_\mathcal{C}$. So this is a certificate for this program, write $\mathcal{D}$ for this certificate. Now $(\mathcal{C}, \mathcal{D})$ is a certificate for (G, k).

□

We illustrate our proof with an example.

(a)

(b)

Fig. 1. (a) A graph G (b) A subdivision of G and divisor

Example 6. Consider the graph in Fig. 1a. Consider the subdivision in Fig. 1b and the divisor D with 7 chips on u. This divisor reaches v, for example by firing the following sets:

$$\{u\}, \{u\}, \{u\}, \{u, y_1\}, \{u, x_1, y_1\}, \{u, x_1, y_1\},$$
$$\{u, x_1, y_1, y_2\}, \{u, x_1, y_1, y_2\}, \{u, x_1, x_2, y_1, y_2\},$$
$$\{u, x_1, x_2, y_1, y_2, y_3\}, \{u, x_1, x_2, y_1, y_2, y_3\}, \{u, x_1, x_2, y_1, y_2, y_3\}.$$

We describe the corresponding partial certificate $(\mathcal{C}, \mathcal{D})$. The divisor D will be included in $\mathcal{C}$. Notice that there are 8 relevant sets. We obtain the following series of pairs, after labelling the chips $1, 2, \ldots, 7$:

$$
\begin{aligned}
A_{v,1} &= \{u\}, & M_{v,1} &= \{(u, 1, -1, e_1), (u, 2, -1, e_2)\} \\
A_{v,2} &= \{u\}, & M_{v,2} &= \{(u, 3, -1, e_1), (u, 4, -1, e_2)\} \\
A_{v,3} &= \{u\}, & M_{v,3} &= \{(u, 5, -1, e_1), (u, 6, -1, e_2)\} \\
A_{v,4} &= \{u\}, & M_{v,4} &= \{(u, 7, -1, e_1)\} \\
A_{v,5} &= \{u\}, & M_{v,5} &= \{(v, 1, 1, e_1)\} \\
A_{v,6} &= \{u\}, & M_{v,6} &= \{(v, 3, 1, e_1), (v, 2, 1, e_2)\} \\
A_{v,7} &= \{u\}, & M_{v,7} &= \{(v, 5, 1, e_1), (v, 4, 1, e_2)\} \\
A_{v,8} &= \{u\}, & M_{v,8} &= \{(v, 7, 1, e_1), (v, 6, 1, e_2)\}
\end{aligned}
$$

This gives the partial certificate $\mathcal{C}$. The partial certificate $\mathcal{D}$ consists of a solution to the integer linear program $\mathcal{I}_\mathcal{C}$. Here, the corresponding program is:

$$l_{e_1} \geq 1$$
$$l_{e_2} \geq 1$$
$$t_{v,i} \geq 1 \qquad \text{for } i \in \{1, 2, \ldots, 8\}$$
$$t_{v,i} = 1 \qquad \text{for } i \in \{1, 2, 3\}$$
$$t_{v,i} = 1 \qquad \text{for } i \in \{5, 6, 7, 8\}$$
$$4l_{e_1} \geq \sum_{i=0}^{8} t_{v,i}$$

$$4l_{e_1} \leq \sum_{i=0}^{8} t_{v,i}$$
$$3l_{e_2} - 1 \geq \sum_{i=0}^{8} t_{v,i}$$
$$3l_{e_2} + 1 \leq \sum_{i=0}^{8} t_{v,i}$$

We can simplify this to:

$$l_{e_1} \geq 1$$
$$l_{e_2} \geq 1$$
$$t_{v,i} = 1 \qquad \text{for } i \in \{1, 2, 3, 5, 6, 7, 8\}$$
$$t_{v,4} \geq 1$$

$$4l_{e_1} \geq t_{v,4} + 7$$
$$4l_{e_1} \leq t_{v,4} + 7$$
$$3l_{e_2} \geq t_{v,4} + 7$$
$$3l_{e_2} \leq t_{v,4} + 7$$

We see that $l_{e_1} = 3$, $l_{e_2} = 4$, $t_{v,4} = 5$ and $t_{v,i} = 1$ for $i \neq 4$ is a solution to this program, let this solution be the certificate $\mathcal{D}$.

Lemma 7. *Let G be a graph and k a natural number. If there exists a certificate $(\mathcal{C}, \mathcal{D})$, then $\mathrm{sdgon}(G) \leq k$.*

The idea of the proof of this lemma is as follows. Suppose we are given a certificate. Subdivide every edge of G_1 in l_e edges. Make $t_{w,i}$ copies of set $A_{w,i}$. For every edge $e = uv$ we distribute the added vertices over the copies of $A_{w,i}$ such that as many chips depart from u along e as described by the tuples and as many chips arrive at v along e as described by the tuples. Using the conditions that our certificate satisfies, we can prove that all chips are moved as described by the tuples in the sets $M_{w,i}$. We illustrate this idea in the following example. For all details see [5, Lemma 4.3].

Example 8. Again consider the graph in Fig. 1a and the certificate in Example 6. Since $l_{e_1} = 3$, we subdivide e_1 with two vertices x_1 and x_2 and since $l_{e_2} = 4$, we subdivide e_2 with three vertices y_1, y_2 and y_3.

We make 5 copies of set $A_{w,4}$, since $t_{w,4} = 5$. The first set that fires a chip along e_1 is $A_{v,1}$ and the last such set is $A_{v,8}$, in total there are 12 sets that fire a chip along e_1. When we fire the first four sets, a chip departs from u along e_1, so we will not add x_1 and x_2 to the first four sets. When we fire the last four sets, a chip arrives at v along e_1, so we add x_1 and x_2 to the last four sets. We add x_1 to the middle four sets, so that the chips move from x_1 to x_2. This yields:

$$A_{v,1} = \{u\}, A_{v,2} = \{u\}, A_{v,3} = \{u\}, A_{v,4,1} = \{u\}, A_{v,4,2} = \{u, x_1\},$$
$$A_{v,4,3} = \{u, x_1\}, A_{v,4,4} = \{u, x_1\}, A_{v,4,5} = \{u, x_1\}, A_{v,5} = \{u, x_1, x_2\},$$
$$A_{v,6} = \{u, x_1, x_2\}, A_{v,7} = \{u, x_1, x_2\}, A_{v,8} = \{u, x_1, x_2\}.$$

Analogously for e_2, we add the vertices y_1, y_2 and y_3:

$$\{u\}, \{u\}, \{u\}, \{u, y_1\}, \{u, x_1, y_1\}, \{u, x_1, y_1\},$$
$$\{u, x_1, y_1, y_2\}, \{u, x_1, y_1, y_2\}, \{u, x_1, x_2, y_1, y_2\},$$
$$\{u, x_1, x_2, y_1, y_2, y_3\}, \{u, x_1, x_2, y_1, y_2, y_3\}, \{u, x_1, x_2, y_1, y_2, y_3\}.$$

We see that we obtained the same subdivision and firing sets as we started with in Example 6.

As ILP's have certificates with polynomially many bits (see e.g., [13]), and the partial certificate is of polynomial size (see also Lemma 4), we have that, using Lemmas 5 and 7, the problem whether a given graph has divisorial gonality at most a given integer k has a polynomial certificate, which gives our main result.

Theorem 9. STABLE DIVISORIAL GONALITY *belongs to the class NP.*

Combined with the NP-hardness of STABLE DIVISORIAL GONALITY by Gijswijt [11], this yields the following result.

Theorem 10. STABLE DIVISORIAL GONALITY *is NP-complete.*

5   A Bound on Subdivisions

In this section, we give as corollary of our main result a bound on the number of subdivisions needed. We use the following result by Papadimitriou [13].

Theorem 11 (Papadimitriou [13]). *Let A be an $m \times n$ matrix, and b be a vector of length m, such that each value in A and b is an integer in the interval $[-a, +a]$. If $Ax = b$ has a solution with all values being positive integers, then $Ax = b$ has a solution with all values positive integers that are at most $n(ma)^{2m+1}$.*

Corollary 12. *Let G be a graph with stable divisorial gonality k. There is a graph H, that is a subdivision of G, with the divisorial gonality of H equal to the stable divisorial gonality of G, and each edge in H is obtained by subdividing an edge from G at most $m^{O(km^2)}$ times.*

Proof. By Lemma 5, we know that there is a certificate whose corresponding ILP has a solution. The values l_e in this solution give the number of subdivisions of edges in G_1. If we have an upper bound on the number of subdivisions per edge needed to obtain H from G_1, say α, then $2\alpha+1$ is an upper bound on the number of subdivisions per edge to obtain H from G. Applying Theorem 11 to the ILP gives such a bound, as described below.

The ILP has at most $n' \cdot (2kn' + n')$ variables of the form $t_{w,i}$, by Lemma 4, and m' variables of the form l_e, with n' the number of vertices in G_1 and m' the number of edges in G_1. We have $n' = n + m$, and $m' = 2m$, with n' the number of vertices of G and m the number of edges of G.

The number of equations and inequalities in the ILP is linear in the number of variables. An inequality can be replaced by an equation by adding one variable. This gives a total of $O(kn'^2 + m')$ variables and $O(kn'^2 + m')$ equations. Note that $O(kn'^2 + m') = O(km^2)$; as G is connected, $n \leq m - 1$. Also, note that all values in matrix A and vector b are -1, 0, or 1, i.e., we can set $a = 1$ in the application of Theorem 11. So, by Theorem 11, we obtain that if there is a solution to the ILP, then there is one where all variables are set to values at most

$$O(kn'^2 + m') \cdot O(kn'^2 + m')^{O(kn'^2 + m')} = O(km^2) \cdot O(km^2)^{O(km^2)} = m^{O(km^2)}.$$

Denoting by k the stable divisorial gonality of G, we know there is at least one certificate with a solution, so we can bound the number of subdivisions in G_1 by $m^{O(km^2)}$, which gives our result.   □

6   Conclusion

In this paper, we showed that the problem to decide whether the stable divisorial gonality of a given graph is at most a given number k belongs to the class NP. Together with the NP-hardness result of Gijswijt [11], this shows that the problem is NP-complete. We think our proof technique is interesting: we give a certificate that describes some of the essential aspects of the firing sequences; whether there is a subdivision of the graph for which this certificate describes the firing sequences and thus gives the subdivision that reaches the optimal divisorial gonality can be expressed in an integer linear program. Membership in NP then follows by adding the certificate of the ILP to the certificate for the essential aspects.

As a byproduct of our work, we obtained an upper bound on the number of subdivisions needed to reach a subdivision of G whose divisorial gonality gives the stable divisorial gonality of G. Our upper bound still is very high, namely exponential in a polynomial of the size of the graph. An interesting open problem is whether this bound on the number of needed subdivisions can be replaced by a polynomial in the size of the graph. Such a result would give an alternative (and probably easier) proof of membership in NP: first guess a subdivision, and then guess the firing sequences.

There are several open problems related to the complexity of computing the (stable) divisorial gonality of graphs. Are these problems *fixed parameter tractable*, i.e., can they be solved in $O(f(k)n^c)$ time for constant c and some function f that depends only on k? Or can they be proven to be $W[1]$-hard, or even, is there a constant c, such that deciding if (stable) divisorial gonality of a given graph G is at most c is already NP-complete? Also, how well can we approximate the divisorial gonality or stable divisorial gonality of a graph?

Acknowledgements. We thank Gunther Cornelissen and Nils Donselaar for helpful discussions.

References

1. Baker, M.: Specialization of linear systems from curves to graphs. Algebra Number Theory **2**(6), 613–653 (2008). https://doi.org/10.2140/ant.2008.2.613
2. Baker, M., Norine, S.: Riemann-Roch and Abel-Jacobi theory on a finite graph. Adv. Math. **215**(2), 766–788 (2007). https://doi.org/10.1016/j.aim.2007.04.012
3. Baker, M., Shokrieh, F.: Chip-firing games, potential theory on graphs, and spanning trees. J. Comb. Theory Ser. A **120**(1), 164–182 (2013). https://doi.org/10.1016/j.jcta.2012.07.011
4. Bodewes, J.M., Bodlaender, H.L., Cornelissen, G., van der Wegen, M.: Recognizing hyperelliptic graphs in polynomial time. In: Brandstädt, A., Köhler, E., Meer, K. (eds.) Graph-Theoretic Concepts in Computer Science, pp. 52–64 (2018). (extended abstract of http://arxiv.org/abs/1706.05670)
5. Bodlaender, H.L., van der Wegen, M., van der Zanden, T.C.: Stable divisorial gonality is in NP (2018). http://arxiv.org/abs/1808.06921
6. Caporaso, L.: Gonality of algebraic curves and graphs. In: Frühbis-Krüger, A., Kloosterman, R., Schütt, M. (eds.) Algebraic and Complex Geometry, vol. 71, pp. 77–108. Springer, Cham (2014). https://doi.org/10.1007/978-3-319-05404-9_4
7. Cornelissen, G., Kato, F., Kool, J.: A combinatorial Li-Yau inequality and rational points on curves. Math. Ann. **361**(1–2), 211–258 (2015). https://doi.org/10.1007/s00208-014-1067-x
8. Corry, S., Perkinson, D.: Divisors and Sandpiles: An Introduction to Chip-Firing. American Mathematical Society, Providence (2018)
9. van Dobben de Bruyn, J.: Reduced divisors and gonality in finite graphs. Bachelor thesis, Leiden University (2012). https://www.universiteitleiden.nl/binaries/content/assets/science/mi/scripties/bachvandobbendebruyn.pdf
10. van Dobben de Bruyn, J., Gijswijt, D.: Treewidth is a lower bound on graph gonality (2014). http://arxiv.org/abs/1407.7055v2
11. Gijswijt, D.: Computing divisorial gonality is hard (2015). http://arxiv.org/abs/1504.06713
12. Hladký, J., Král', D., Norine, S.: Rank of divisors on tropical curves. J. Comb. Theory Ser. A **120**(7), 1521–1538 (2013). https://doi.org/10.1016/j.jcta.2013.05.002
13. Papadimitriou, C.H.: On the complexity of integer programming. J. ACM **28**(4), 765–768 (1981). https://doi.org/10.1145/322276.322287

Coalition Resilient Outcomes in Max
k-Cut Games

Raffaello Carosi[1] , Simone Fioravanti[2] , Luciano Gualà[2] ,
and Gianpiero Monaco[3(✉)]

[1] Gran Sasso Science Institute, L'Aquila, Italy
raffaello.carosi@gssi.it
[2] University of Rome "Tor Vergata", Rome, Italy
simonefi92@gmail.com, guala@mat.uniroma2.it
[3] DISIM, University of L'Aquila, L'Aquila, Italy
gianpiero.monaco@univaq.it

Abstract. We investigate strong Nash equilibria in the *max k-cut game*,
where we are given an undirected edge-weighted graph together with a
set $\{1, \ldots, k\}$ of k colors. Nodes represent players and edges capture their
mutual interests. The strategy set of each player v consists of the k colors.
When players select a color they induce a k-coloring or simply a coloring.
Given a coloring, the *utility* (or *payoff*) of a player u is the sum of the
weights of the edges $\{u, v\}$ incident to u, such that the color chosen by u
is different from the one chosen by v. Such games form some of the basic
payoff structures in game theory, model lots of real-world scenarios with
selfish agents and extend or are related to several fundamental classes of
games.

Very little is known about the existence of strong equilibria in max
k-cut games. In this paper we make some steps forward in the comprehension of it. We first show that improving deviations performed by
minimal coalitions can cycle, and thus answering negatively the open
problem proposed in [13]. Next, we turn our attention to unweighted
graphs. We first show that any optimal coloring is a 5-SE in this case.
Then, we introduce x-local strong equilibria, namely colorings that are
resilient to deviations by coalitions such that the maximum distance
between every pair of nodes in the coalition is at most x. We prove that
1-local strong equilibria always exist. Finally, we show the existence of
strong Nash equilibria in several interesting specific scenarios.

1 Introduction

We consider the *max k-cut game*. This is played on an undirected edge-weighted
graph where the n nodes correspond to the players and the edges capture their
mutual interests. The strategy space of each player is a set $\{1, \ldots, k\}$ of k available colors (we assume that the colors are the same for each player). When
players select a color they induce a k-coloring or simply a coloring. Given a coloring, the *utility* (or *payoff*) of a player u is the sum of the weights of edges

© Springer Nature Switzerland AG 2019
B. Catania et al. (Eds.): SOFSEM 2019, LNCS 11376, pp. 94–107, 2019.
https://doi.org/10.1007/978-3-030-10801-4_9

$\{u, v\}$ incident to u, such that the color chosen by u is different from the one chosen by v. The objective of every player is to maximize its own utility.

This class of games forms some of the basic payoff structures in game theory, and can model lots of real-life scenarios. Consider, for example, a set of companies that have to decide which product to produce in order to maximize their revenue. Each company has its own competitors (for example the ones that are in the same region), and it is reasonable to assume that each company wants to minimize the number of competitors that produce the same product. Another possible scenario is in a radio setting; radio towers are players and their goal is selecting a frequency such that neighboring radio-towers have a different one in order to minimize the interference.

In such games on graphs it is beneficial for each player to anti-coordinate its choices with the ones of its neighbors (i.e., selecting a different color). As a consequence, the players may attempt to increase their utility by coordinating their choices in groups (also called coalitions). Therefore, in our studies we focus on equilibrium concepts that are resilient to deviations of groups. Along this direction, a very classic notion of equilibrium is the strong Nash equilibrium (SE) [2] that is a coloring in which no coalition, taking the actions of its complements as given, can cooperatively deviate in a way that benefits all of its members, in the sense that every player of the coalition strictly improves its utility. The notion of SE is a very strong equilibrium concept. A weaker one is the notion of q-Strong Equilibrium (q-SE), for some $q \leq n$, where only coalitions of at most q players are allowed to cooperatively change their strategies. Notice that the 1-SE is equivalent to the Nash equilibrium (NE), while the n-SE is equivalent to the SE.

When it exists, an SE is a very robust state of the game and it is also more sustainable than an NE. However, while NE always exists in these games [9, 15, 18], little is known about the existence of strong equilibria in Max k-cut games. Indeed, to the best of our knowledge, there are basically two papers of the literature dealing with such issue. In [12] the authors show that an optimal strategy profile (or optimal coloring), i.e., a coloring that maximizes the sum of the players' utilities or equivalently, a coloring that maximizes the k-cut, is an SE for the max 2-cut game, and it is a 3-SE, for the max k-cut game, for any $k \geq 2$. Moreover, they further show that an optimal strategy profile is not necessarily a 4-SE, for any $k \geq 3$. In [13] they show that, if the number of colors is at least the number of players minus two, then an optimal strategy profile is an SE. Finally, they show that the dynamics, where at each step a coalition can deviate so that all of its members strictly improve their utility by changing strategy, can cycle. The main consequence of this latter fact is that no *strong potential function*[1] can exist for the game, and hence the existence of an SE cannot be proved by simply exhibiting it. It is worth noticing that strong potential functions are one of the main tools used to prove the existence of an SE.

All the above results suggest that it is hard to understand whether SE always exist for max k-cut games. In this paper, although we do not prove or disprove

---

[1] See Sect. 2 for the definition of strong potential function.

that every instance of the max k-cut game possesses a strong equilibrium, we make some step forward in the comprehension of it.

Our Results. As pointed out in [13], sometimes the existence of an SE is proved by means of a potential function in which the set of deviating coalitions is restricted to minimal coalitions only, where a coalition is minimal if none of its proper subsets can perform an improvement themselves (see for example [14]). Understanding whether this approach can be used in the max k-cut game is mentioned as an open problem in [13]. We answer this question negatively (see Proposition 2) by showing an instance in which there is a cycle of improving deviations performed by minimal coalitions only.

We then focus on the unweighted case, where the utility of a player in a coloring is simply the number of neighbors with different color from its own, and we provide some non-trivial existential results for it. In particular, in Sect. 4 we show that 5-SE always exist for the max k-cut game. This is an improvement with respect to the existence of 3-SE [12].

Besides q-SE, we also consider another equilibrium concept that is weaker than the notion of SE. Observe that in a q-SE two players can form a coalition even if they are far from each other in the graph. This is unrealistic in many practical scenarios. In oder to encompass this aspect, in Sect. 5, we introduce the concept of x-*Local* SE (x-LSE). A coloring is an x-LSE if it is resilient to deviations by coalitions such that the shortest path between every pair of nodes in the coalition is long at most x. Therefore, the notion of x-LSE also takes into account that certain players may not have the possibility of communicating to each other and thus to form a coalition. This seems an important point to consider when modeling a situation of strategic interaction between agents. Here we suppose that the input graph also represents knowledge between players, that is two nodes know each other if they are connected by an edge. In this paper we focus on the case $x = 1$, that is each player in the coalition must have a social connection (namely an edge) towards every deviating player. We show that, for any k, a 1-LSE always exists. Interestingly enough, our analysis also provides a characterization of the set of local strong equilibria which relates 1-LSE to q-SE.

Finally, in Sect. 6, we show that an SE always exists for some special classes of unweighted graphs. More precisely, in Corollary 12, we prove that in graphs with large girth, any optimal strategy profile is an SE, for any $k \geq 2$. Moreover, in Proposition 13, we prove that whenever the number of colors k is large enough with respect to the maximum degree of the graph, then any optimal strategy profile is an SE.

Due to space constraints some proofs have been omitted. All the details can be found in the full version of the paper [7].

Further Related Work. The max k-cut game has been first investigated in [15,18], where the authors show that, when the graph is unweighted and undirected, it is possible to compute a Nash Equilibrium in polynomial time by exploiting the potential function method. When the graph is weighted undirected, even if the potential function ensures the existence of NE the problem of computing an equilibrium is PLS-complete even for $k = 2$ [22]. In fact, for such a value of k, it coincides with the classical max cut game. In [9] the authors show the existence of NE in generalized max k-cut games where players also have an extra profit depending on the chosen color. When the graph is directed, the max k-cut game in general does not admit a potential function. Indeed, in this case, even the problem of understanding whether they admit a Nash equilibrium is NP-complete for any fixed $k \geq 2$ [18]. In [8], the authors present a randomized polynomial time algorithm that computes a constant approximate Nash equilibrium for a large class of directed unweighted graphs.

Studies on the performance of Nash equilibria and strong Nash equilibria can be found in [9,15,18] and in [11–13], respectively.

A related stream of research considers coordination games. The idea is that agents are rewarded for choosing common strategies in order to capture the influences. Apt et al. [1] propose a coordination game modeled as an undirected graph where nodes are players and each player has a list of allowed colors. Given a coloring, an agent has a payoff equal to the number of adjacent nodes with its same color. The authors show that NE and 2-SE always exists, and give an example in which no 3-SE exists. Moreover, they prove that strong equilibria exist for various special cases.

Panagopoulou and Spirakis [21] study games where Nash equilibria are proper node coloring in undirected unweighted graphs setting. In particular, they consider the game where each agent v has to choose a color among k available ones and its payoff is equal to the number of nodes in the graph that have chosen its same color, unless some neighbor of v has chosen the same color, and in this case the payoff of v is 0. They prove that this is a potential game and that a Nash equilibrium can be found in polynomial time.

Max k-cut games are related to many other fundamental games considered in the scientific literature. One example is given by the graphical games introduced in [17]. In these games the payoff of each agent depends only on the strategies of its neighbors in a given social knowledge graph defined over the set of the agents, where an arc (i, j) means that j influences i's payoff. Max k-cut games can also be seen as a particular hedonic game (see [3] for a nice introduction to hedonic games) with an upper bound (i.e., k) to the number of coalitions. Specifically, given a k-coloring, the agents with the same color can be seen as members of the same coalition of the hedonic game. In order to get the equivalence among the two games, the hedonic utility of an agent v can be defined as the overall number of its neighbors minus the number of agents of its neighborhood that are in the same coalition. Nash equilibria issues in hedonic games have been largely investigated under several different assumptions [5,6,20] (just to cite a few).

Concerning local coalitions, a notion of equilibrium close in spirit to our LSE has been studied in the context of network design games in [19]. Moreover, locality aspects have been also considered when restricting the strategy space in single-player deviations (see for example [4,10]).

Finally, it is worth mentioning the classical optimization max cut problem, a very famous problem in graph theory that was proven to be NP-Hard by Karp [16].

2 Preliminaries

Let $G = (V, E, w)$ be an undirected weighted graph, where $|V| = n$, $|E| = m$, and $w : E \rightarrow \mathbb{R}_+$. Let $\delta^v(G) = \sum_{u \in V:\{v,u\} \in E} w(\{v, u\})$ denote the *degree* of v, that is the sum of the weights of all the edges incident to v. Let $\delta^M(G) = max_{v \in V} \delta^v(G)$ denotes the maximum degree in G. Given a set of nodes $V' \subseteq V$, let $G(V') = (V', E', w)$ be the subgraph induced by V', where $E' = \{\{v, u\} \in E \mid v \in V' \wedge u \in V'\}$. For any pair of nodes $v, u \in V$, the *distance* $dist_G(v, u)$ between v and u in G is equal to the length of the shortest path from v to u^2.

Given G and a set of colors $K = \{1, \ldots k\}$, the *max k-cut* problem is to partition the vertices into k subsets $V_1, \ldots, V_k$ such that the sum of the weights of the edges having the endpoints in different sets is maximized. A strategic version of the max k-cut problem is the *max k-cut game*, and it is defined as follows. There are $|V|$ players, and each node of G is controlled by exactly one rational player. Players have the same strategy set, and it is equal to the set of colors $\{1, \ldots k\}$. A *strategy profile*, or *coloring* $\sigma : V \rightarrow K$, is a labeling of nodes of G in which each player v is colored $\sigma(v)$. Given a coloring σ, let $E(\sigma) = \{\{u, v\} : \sigma(u) \neq \sigma(v)\}$ be the edges that are proper with respect to σ, and let $\delta_u^i(\sigma) = \sum_{v \in V} w(\{u, v\})_{\sigma(v)=i}$ be the sum of the weights of the edges incident to u and towards nodes colored i in σ. The utility (or payoff) of player u is defined as $\mu_u(\sigma) = \sum_{v \in V:\{u,v\} \in E \wedge \sigma(u) \neq \sigma(v)} w(\{u, v\})$. The *cut-value*, or *size of the cut*, of a coloring $S(\sigma)$ is defined as follows: $S(\sigma) = \sum_{\{u,v\} \in E \wedge \sigma(u) \neq \sigma(v)} w(\{u, v\})$. The *social welfare* of a coloring σ is defined as the sum of players' utilities, that is $SW(\sigma) = \sum_{v \in V} \mu_v(\sigma) = 2S(\sigma)$. Moreover, an optimal strategy profile (or optimal coloring) is defined to be a strategy profile which maximizes the sum of the players' utilities and thus the cut-value.

Given a coalition $C \subseteq V$ and a coloring σ, let $C_K(\sigma) = \{i \in K \mid \exists v \in C \text{ s.t. } \sigma(v) = i\}$ be the set of colors used by the coalition C in σ. Moreover, for each color i, let $C_i(\sigma) = \{v \in C \mid \sigma(v) = i\}$ be the set of players in C that are colored i in σ.

Given a strategy profile σ, a player v and a coalition C, we denote by σ_{-v} and σ_{-C} the strategy profile σ besides the strategy played by v and by C, respectively. Moreover, we denote by σ_C the coloring σ restricted only to players in C, and we use $(\sigma_{-v}, \sigma(v))$ and (σ_{-C}, σ_C) to denote σ.

A profile σ is a *Nash Equilibrium* (NE) if no player can improve its payoff by deviating unilaterally from σ, that is, $\mu_v(\sigma_{-v}, i) \leq \mu_v(\sigma)$ for each player $v \in V$ and for each color $i \in K$. For each $1 \leq q \leq n$, σ is a *q-Strong Equilibrium* (q-SE) if there exists no coalition C with $|C| \leq q$ that can cooperatively deviate from σ_C to σ_C' in such a way that every player in C strictly improves its utility in (σ_{-C}, σ_C'). The 1-strong equilibrium is equivalent to the Nash equilibrium, while for $q = n$ an n-strong equilibrium is called *strong equilibrium (SE)*. When a coalition C deviates so that all of its members strictly improve their utility, then we say it performs a *strong improvement*. A strong improvement is said

---

^{2} Even if the graph is weighted, we consider here the *hop-distance*, where the length of a path is defined as the number of its edges.

to be *minimal* if no proper subsets of the deviating coalition can perform an improvement themselves, and the coalition itself is said to be minimal. A strong improving dynamics (shortly dynamics) is a sequence of strong improving moves. A game is said to be convergent if, given any initial state, any sequence of improving moves leads to a strong Nash equilibrium. Given a coloring σ, if a coalition C induces a new coloring σ' after deviating, then we say that the set of edges $E(\sigma')\backslash E(\sigma)$ *enters* the cut, and that the set of edges $E(\sigma)\backslash E(\sigma')$ *leaves* the cut.

A *potential function* Φ is a function mapping strategy profiles into real values in such a way that, for each coloring σ and each player v, whenever v can profitably deviate from σ yielding a new coloring σ', it holds that $\Phi(\sigma') > \Phi(\sigma)$. When this is true also for profitably deviations performed by coalitions, the function is called *strong potential function*.

We conclude this section by stating some properties about minimal coalitions that will be useful later.

Proposition 1. *Let σ be a coloring, and let C be a minimal coalition that can perform a strong improvement from σ. Let σ' be the resulting coloring. Then, the following properties hold: (i) $C_K(\sigma) = C_K(\sigma')$; and (ii) if $G(C)$ is acyclic, then changing from σ to σ' strictly increases the size of the cut.*

3   Non-existence of a Minimal Strong Potential Function

In this section we focus on weighted graphs and, as discussed in the introduction, we close an open problem stated by Gourvès and Monnot in [13] by providing an instance in which there is a cycle of improving deviations performed by minimal coalitions only. More specifically, the loop is composed by the deviation of a clique, followed by four improvements performed by single players.

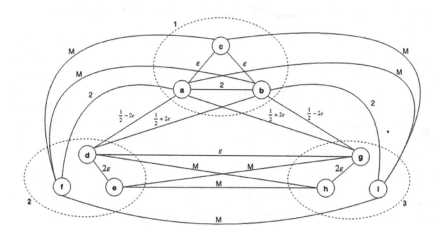

Fig. 1. Instance for which the strong improvement dynamics cycles.

Proposition 2. *No strong potential function exists for the max k-cut game, even if only minimal coalitions are allowed to deviate.*

Proof. Consider the graph G and the coloring σ depicted in Fig. 1, where, if a node v is contained in the dashed ellipse labeled i, then v is colored i in σ, and where M and ε denote a very large and small positive value, respectively.

Consider coalition $C = \{a, b, d, g\}$ and consider the deviation σ' where $\sigma'(a) = 2$, $\sigma'(b) = 3$, $\sigma'(d) = 1$, $\sigma'(g) = 1$. It is easy to check that this deviation is profitable for players in C. In fact, they all improve their utility by ε.

Player a has utility $3 + M$ in σ, and since it has an edge of weight M towards node i having color 3, a can only deviate to color 2. Hence, a strictly improves its utility from $3 + M$ to $3 + M + \varepsilon$ only if node d changes color. Analogously, in σ, d has one edge of weight M towards node h having color 3. Thus, d can only switch to color 1 and this is convenient for it only if both a and b leave color 1. If this happens, d's utility increases by at least ε. Similarly to a, player b deviates to color 3 only if player g switches to color 1, and this happens only if both a and b deviates too. To sum up, both d and g deviate if and only if a and b deviate too. Thus, C is minimal. Note that edge $\{d, g\}$ becomes monochromatic, but d and g's new payoffs make the deviation worth it anyway, since they both increase their utility by ε.

After the players in C jointly deviate from σ, player a, who is now colored 2, can go back to color 1, improving its utility from $3 + \varepsilon + M$ to $4 + M$. Because of a's deviation, d's utility goes down to $1/2 + 4\varepsilon + M$. Thus, it goes back to color 2, achieving $1 + \varepsilon + M$. Also g, whose utility is now $1/2 + \varepsilon + M$, deviates to its old color in σ, that is color 3, and it gets $1/2 + 3\varepsilon + M$. In this configuration b' utility is $5/2 + 3\varepsilon + M$. Thus going back to color 1 its utility improves to $3 + M$ and we are now back to the initial configuration σ. $\square$

4   The Existence of a 5-SE in Unweighted Graphs

From now on we will focus on unweighted graphs. In [12] it is shown that in the weighted case, any optimal strategy profile is always a 3-SE, and there are weighted graphs in which every optimal coloring is not a 4-SE. In this section we improve this result for unweighted graphs, by showing that a 5-SE always exists. This also establishes a separation between the weighted and unweighted case. In particular, we show that by performing minimal strong improvements with coalitions of size at most five, the cut value increases. It implies that the cut value is a potential function and thus the dynamics converges to 5-SE. We start by showing a simple lemma that is used in the rest of the section.

Lemma 3. *Let σ be an NE and let C be a minimal coalition which would profit by deviating from σ to σ'. If there exists two players $u, x \in C$ such that:*

(i) $\sigma(u) \neq \sigma(x)$
(ii) $\sigma'(u) = \sigma(x)$
(iii) $\{y \in C | \{u, y\} \in E, \sigma(y) = \sigma(x)\} = \{x\}$

then $\delta_u^{\sigma(u)}(\sigma) = \delta_u^{\sigma'(u)}(\sigma)$. *Moreover, if there exists a third player* $v \in C$ *such that* $\sigma(v) \neq \sigma(x)$ *and* $\sigma'(v) = \sigma(x)$, *then* $\{u, v\} \notin E$.

Proof. Since σ is an NE we know that u cannot improve its utility by deviating alone to $\sigma'(u)$ which implies:

$$\delta_u^{\sigma(u)}(\sigma) \leq \delta_u^{\sigma'(u)}(\sigma).$$

By (iii) we know that, moving from σ to σ', the only neighbor of u which leaves $\sigma'(u)$ is x which means that its new neighbors colored $\sigma'(u)$ are at least $\delta_u^{\sigma'(u)}(\sigma) - 1$, because, a priori, other players could move to the same strategy in σ', hence $\delta_u^{\sigma'(u)}(\sigma') \geq \delta_u^{\sigma'(u)}(\sigma) - 1$. Moreover, player u strictly improves its utility, which means $\delta_u^{\sigma'(u)}(\sigma') < \delta_u^{\sigma(u)}(\sigma)$. Using both inequalities we obtain:

$$\delta_u^{\sigma(u)}(\sigma) > \delta_u^{\sigma'(u)}(\sigma) - 1.$$

As a consequence, we have $\delta_u^{\sigma(u)}(\sigma) \geq \delta_u^{\sigma'(u)}(\sigma)$, and hence $\delta_u^{\sigma(u)}(\sigma) = \delta_u^{\sigma'(u)}(\sigma)$, which in turn implies in particular that $\delta_u^{\sigma'(u)}(\sigma') = \delta_u^{\sigma'(u)}(\sigma) - 1$, i.e. u's utility improves exactly by one. Thus, given a player v like in the hypothesis, if $\{u, v\} \in E$ then u's utility would not increase after the deviation. $\qquad\square$

Proposition 1 and Lemma 3 can be used to prove the following proposition, which shows that when the size of a deviating coalition C is related in a certain way to the number of colors used by the players in C, then the improving deviation always increases the size of the cut.

Proposition 4. *Let* σ *be an NE and let* C *be a minimal coalition which would profit by deviating from* σ *to* σ'. *If* $|C_K(\sigma)| \in \{2, |C| - 1, |C|\}$, *then the deviation strictly improves the size of the cut.*

Gourvès and Monnot [12] show that in weighted graphs an optimal solution is always a 3-strong equilibrium, that is, it is resilient to any joint deviation by at most three players. Proposition 4 already extends this result since it implies that unweighted graphs admit a potential function when minimal coalitions of at most four players are allowed to deviate, implying that 4-SE always exists. We now prove that the cut value is a potential function even when the deviation is extended to coalitions of size at most five. This implies that a 5-SE always exists in unweighted graphs.

Theorem 5. *Any optimal strategy profile is a 5-SE.*

5   Local Strong Equilibria

In this section we introduce and discuss local strong equilibria. As our main result, we show that, for any k, such an equilibrium always exists. Interestingly enough, our analysis also provides a characterization of the set of local strong equilibria which relates them to q-SE.

Let $C \subseteq V$ be a set of players. We say that C is an *x-local coalition* if the distance in G between any two players in C is at most x. Moreover, we define an *x-Local Strong Equilibrium* (*x-LSE*) to be a coloring in which no x-local coalition can profitably deviate. In this section, we will consider only the case $x = 1$, that is, the coalition C induces a clique. We will use LSE in place of 1-LSE.

Let us introduce some additional notation. Given a node u and a strategy profile σ, we denote by $c_u(\sigma)$ the *cost* of u in σ, namely the number of neighbors of u that have the same color of u in σ, i.e. $\delta_u^{\sigma(u)}(\sigma)$. Notice that $c_u(\sigma) = \delta_u - \mu_u(\sigma)$. Given a coalition C, we also define $c_{u,C}(\sigma) = |\{(u, v) \in E | v \in C, \sigma(v) = \sigma(u)\}|$.

We now prove a technical lemma which gives some necessary conditions for a clique to deviate profitably from an NE.

Lemma 6. *Let σ be an NE. Suppose there exists a deviation σ' such that all the members of C can lower their cost changing from σ to σ'. The following conditions must hold:*

(i) $|C_i(\sigma)| = |C_i(\sigma')|$ for all $i = 1, \ldots, k$;
(ii) $c_u(\sigma) - c_u(\sigma') = 1$ for all $u \in C$;
(iii) $c_u(\sigma) = c_u(\sigma_{-u}, \sigma'(u))$ for all $u \in C$.

The following lemma underlines a very interesting property of deviating cliques, which allows us to study only cliques formed by at most k players.

Lemma 7. *Let C be a clique which profits by deviating from an NE σ to σ'. Then there exists $j \leq |C_K(\sigma)|$ and a subcoalition $C' = \{u_i, \ldots, u_j\} \subseteq C$ whose players can improve their payoffs by deviating alone to the strategy they use in σ'. Moreover it holds:*

(i) $\sigma'(u_i) = \sigma(u_{i+1})$ for all $i = 1, \ldots, j - 1$;
(ii) $\sigma'(u_j) = \sigma(u_1)$.

Proof. Consider a player $v_1 \in C$ and assume without loss of generality that $\sigma(v_1) = 1$ and $\sigma'(v_1) = 2$. By Lemma 6, there is at least one player, say v_2, in C such that $\sigma(v_2) = 2$. If $\sigma'(v_2) = 1$, then $C' = \{v_1, v_2\}$. Otherwise, call without loss of generality $\sigma'(v_2) = 3$. Then, once again by Lemma 6 there is a node in C, say v_3, with $\sigma(v_2) = 2$. We can iterate this argument until we get a node v_h with $\ell := \sigma'(v_h) \in \{1, 2, \ldots, h - 1\}$. We set $C' = \{v_\ell, v_{\ell+1}, \ldots, v_h\}$ and set $j = |C'|$. Notice that C' already satisfies the properties 1 and 2 of the statement of the lemma (observe also that it could be $C' = C$).

Let σ^* be the strategy profile in which the players in C' play as in σ' while the others play as in σ. It remains to show that all players in C' is improving its utility by changing from σ to σ^*. Let $\nu = \sigma^*(v_i)$. By definition of σ^*, we claim that $c_{v_i}(\sigma^*) = c_{v_i}(\sigma_{-v_i}, \nu) - 1$. This is true because there is exactly one player in C' that leaves color ν and thus the number of v_i's neighbors with such a color decreases by 1. Hence,

$$c_{v_i}(\sigma^*) = c_{v_i}(\sigma_{-v_i}, \nu) - 1$$
$$= c_{v_i}(\sigma_{-v_i}, \sigma'(v_i)) - 1$$
$$= c_{v_i}(\sigma) - 1,$$

where in the last equality we used property (iii) of Lemma 6 on σ'. ◻

Lemma 7 allows us to prove the main result of this section, which is the following:

Theorem 8. *Any optimal strategy profile is an LSE.*

Proof. Let σ be an optimal strategy profile and assume σ is not an LSE. Clearly, σ is an NE. Then there exists a coalition $C = \{u_1, \ldots, u_j\}$ of $j \leq k$ players and a strategy profile σ' which satisfy the conditions of Lemma 7. We will show that the size of the cut increases by exactly j from σ to σ', which is a contradiction.

First of all, observe that all the edges between players of C are in the cut both in σ and σ'. Moreover, from property (ii) of Lemma 6, we have that the utility of each $u_i \in C$ increases exactly by one. Let $E_i = \{\{u_i, v\} | v \notin C\}$. As a consequence, we have that, for each $u_i \in C$, the number of edges in E_i crossing the cut increases by exactly one. Since E_i and E_j are disjoint for $i \neq j$, the size of the cut increases exactly by j from σ to σ'. ◻

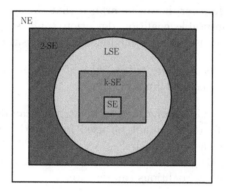

Fig. 2. Equilibria in the unweighted max-k-cut game

We conclude this section by discussing some consequences of our analysis about how LSE is related to q-SE: these results are depicted in Fig. 2. Some inclusions are straightforward from the definition of q-SE. Here we show that an LSE is always a 2-SE and a k-SE is always an LSE. Concerning the former fact, note that a coalition of 2 players can profitably deviate from an NE σ if and only if there exists an edge between them. In fact, otherwise, they could profitably deviate alone from σ. This means that such a coalition is a clique of two players, and hence a local coalition. As far as the latter relation is concerned, we prove the following:

Proposition 9. *A k-SE is always an LSE.*

Proof. Let σ be a k-SE and, by contradiction, let C be a clique which would profit deviating to σ'. By Lemma 7 there exists a minimal subcoalition C' of at most k players which can profit deviating alone, which is a contradiction.   □

It is worth noticing that, as a consequence, when $k = 2$ the set of LSE coincides exactly with the set of 2-SE. On the other hand, for $k \geq 3$, it is possible to show that all inclusions are proper (see the full version of the paper [7]).

6   Existence of SE for Special Cases

In this section we show that an SE always exists for some special classes of unweighted graphs. More precisely, we prove that in graphs with large girth or large degree, any optimal strategy profile is an SE. It is worth noticing that, for general graphs, we have already proved that any optimal coloring is both a 5-SE and an LSE. We conjecture that it is indeed always an SE, even if this seems to be challenging to prove in general. A natural approach could be that of using the size of the cut as a strong potential function, that is $\Phi_S(\sigma) = S(\sigma)$, as it has already been done for proving that max k-cut games admit a Nash equilibrium [15,18]. However, it can be argued that this approach cannot work in general, since a profitable coalition deviation could sometimes result in a cut-value decrease. This is stated in the following proposition.

Proposition 10. *The size of the cut is not a strong potential function for the max k-cut game on unweighted graphs.*

Even though there exist strong improvements that can decrease Φ_S, it does not mean that such function cannot be used in some interesting special setting. Indeed, there are cases in which Φ_S's value always increases after a strong improvement, that is, they admit a strong potential function. From now on we assume that only minimal coalitions can deviate.

Bounded Girth. Given a graph G, let $\rho(G)$ be its *girth*, that is the size of the minimum cycle. We show that a graph with girth $\rho(G)$ always admits a q-SE, for $q \leq 2\rho(G) - 3$. This implies that when $\rho(G) \geq (|V| + 3)/2$ then there always exists a strong equilibrium.

Proposition 11. *Given an unweighted graph G with girth $\rho(G)$ and any number of colors k, an optimal coloring is a $(2\rho(G) - 3)$-SE.*

Corollary 12. *If $\rho(G) \geq (|V| + 3)/2$, then an optimal coloring is always an SE.*

Bounded Degree. Here we show that whenever the number of colors k is large enough with respect to the maximum degree of the graph, then any optimal strategy profile is an SE. More precisely, we prove the following:

Proposition 13. *Any optimal strategy profile is an SE when $k \geq \lceil (\delta^M + 1)/2 \rceil$.*

Proof. Let σ^* be an optimal strategy profile and assume σ^* is not an SE. Then a coalition C and a strategy profile σ' exist such that all players in C strictly improves their utility by deviating to σ'. We will show that in this case the size of the cut will strictly increase in σ', which contradicts the optimality of σ^*.

As we already pointed out, σ^* is NE. Moreover, consider any node u. Since its degree δ^u is at most $\delta^M \leq 2k - 1$, we have that in any coloring, by the pigeonhole principle, there must exist a color that appears at most once in u's neighborhood. As a consequence, since σ^* is an NE, it holds that $\mu_u(\sigma^*) \geq \delta^u - 1$. On the other hand, since all nodes in C must strictly improve their utility, we have that, for every $u \in C$, $\mu_u(\sigma^*) = \delta^u - 1$ and $\mu_u(\sigma') = \delta^u$. This implies that the size of the cut must strictly increase. Indeed, consider the edge set $F = \{\{u, v\} | u \in C \text{ or } v \in C\}$. Clearly, only edges in F can enter or leave the cut when the strategy profile changes from σ^* to σ'. Moreover, all edges in F belong to the cut $E(\sigma')$ while there is at least an edge that is not in $E(\sigma^*)$. $\square$

7 Conclusions and Future Work

We investigated coalition resilient equilibria in the max k-cut game. We solved an open problem proposed in [13] on weighted graphs by showing that improving deviations performed by minimal coalitions can cycle. We then provided some positive results on unweighted graphs. More precisely, we proved that any optimal coloring is both a 5-SE and a 1-LSE. We also showed that SE exist for some special cases, namely, when the graph has a large girth or the number of colors is large enough with respect to the maximum degree.

Even though we made a progress on the topic, the problem of understanding whether any instance of the max k-cut game admits strong equilibria is still open on both weighted and unweighted graphs. We conjecture that an optimal strategy profile is always an SE in the unweighted case. However, proving that seems to be really challenging. Another possible way to prove the existence of an SE would be that of providing a strong potential function. We proved in Proposition 2 that such function cannot exist on weighted graphs even when only minimal coalitions can deviate but it is still unknown whether a strong potential function exists or not on unweighted graphs. Along this direction, an interesting intermediate step could be that of proving the existence of q-SE for possibly non-constant values of $q > 5$.

Regarding x-local coalitions, our results are only about the case $x = 1$ on unweighted graphs. Some other research questions could be the study of the existence of x-local strong equilibrium for $x > 1$, and how to extend our results to weighted graphs. For instance, it would be interesting to investigate whether any instance of the max k-cut game on weighted graphs admits local strong equilibria.

References

1. Apt, K.R., de Keijzer, B., Rahn, M., Schäfer, G., Simon, S.: Coordination games on graphs. Int. J. Game Theory **46**(3), 851–877 (2017). https://doi.org/10.1007/s00182-016-0560-8
2. Aumann, R.J.: Acceptable points in games of perfect information. Pac. J. Math. **10**, 381–417 (1960)
3. Aziz, H., Savani, R.: Hedonic games. In: Handbook of Computational Social Choice, chapter 15. Cambridge University Press (2016)
4. Bilò, D., Gualà, L., Leucci, S., Proietti, G.: Locality-based network creation games. In: 26th ACM Symposium on Parallelism in Algorithms and Architectures, SPAA, pp. 277–286 (2014)
5. Bilò, V., Fanelli, A., Flammini, M., Monaco, G., Moscardelli, L.: Nash stable outcomes in fractional hedonic games: existence, efficiency and computation. J. Artif. Intell. **62**, 315–371 (2018)
6. Bogomolnaia, A., Jackson, M.O.: The stability of hedonic coalition structures. Games Econ. Behav. **38**, 201–230 (2002). https://doi.org/10.1006/game.2001.0877
7. Carosi, R., Fioravanti, S., Gualà, L., Monaco, G.: Coalition resilient outcomes in max k-cut games. CoRR, abs/1810.09278 (2019)
8. Carosi, R., Flammini, M., Monaco, G.: Computing approximate pure nash equilibria in digraph k-coloring games. In: Proceedings of the 16th Conference on Autonomous Agents and MultiAgent Systems, AAMAS, pp. 911–919 (2017)
9. Carosi, R., Monaco, G.: Generalized graph k-coloring games. In: Proceedings of the 24th International Conference on Computing and Combinatorics, COCOON, pp. 268–279 (2018). https://doi.org/10.1007/978-3-319-94776-1_23
10. Cord-Landwehr, A., Lenzner, P.: Network creation games: think global – act local. In: Italiano, G.F., Pighizzini, G., Sannella, D.T. (eds.) MFCS 2015. LNCS, vol. 9235, pp. 248–260. Springer, Heidelberg (2015). https://doi.org/10.1007/978-3-662-48054-0_21
11. Feldman, M., Friedler, O.: A unified framework for strong price of anarchy in clustering games. In: Halldórsson, M.M., Iwama, K., Kobayashi, N., Speckmann, B. (eds.) ICALP 2015. LNCS, vol. 9135, pp. 601–613. Springer, Heidelberg (2015). https://doi.org/10.1007/978-3-662-47666-6_48
12. Gourvès, L., Monnot, J.: On strong equilibria in the max cut game. In: Leonardi, S. (ed.) WINE 2009. LNCS, vol. 5929, pp. 608–615. Springer, Heidelberg (2009). https://doi.org/10.1007/978-3-642-10841-9_62
13. Gourvès, L., Monnot, J.: The max k-cut game and its strong equilibria. In: Kratochvíl, J., Li, A., Fiala, J., Kolman, P. (eds.) TAMC 2010. LNCS, vol. 6108, pp. 234–246. Springer, Heidelberg (2010). https://doi.org/10.1007/978-3-642-13562-0_22
14. Harks, T., Klimm, M., Möhring, R.H.: Strong nash equilibria in games with the lexicographical improvement property. Int. J. Game Theory **42**(2), 461–482 (2013). https://doi.org/10.1007/s00182-012-0322-1
15. Hoefer, M.: Cost sharing and clustering under distributed competition. Ph.D. thesis, University of Konstanz (2007)
16. Karp, R.M.: Reducibility among combinatorial problems. In: Miller, R.E., Thatcher, J.W., Bohlinger, J.D. (eds.) Complexity of Computer Computations. IRSS, pp. 85–103. Springer, Boston (1972). https://doi.org/10.1007/978-1-4684-2001-2_9

17. Kearns, M.J., Littman, M.L., Singh, S.P.: Graphical models for game theory. In: Proceedings of the 17th Conference in Uncertainty in Artificial Intelligence, UAI, pp. 253–260 (2001)
18. Kun, J., Powers, B., Reyzin, L.: Anti-coordination games and stable graph colorings. In: Vöcking, B. (ed.) SAGT 2013. LNCS, vol. 8146, pp. 122–133. Springer, Heidelberg (2013). https://doi.org/10.1007/978-3-642-41392-6_11
19. Leonardi, S., Sankowski, P.: Network formation games with local coalitions. In: Proceedings of the Twenty-Sixth Annual ACM Symposium on Principles of Distributed Computing, PODC, pp. 299–305 (2007)
20. Monaco, G., Moscardelli, L., Velaj, Y.: Stable outcomes in modified fractional hedonic games. In: Proceedings of the 17th International Conference on Autonomous Agents and MultiAgent Systems, AAMAS, pp. 937–945 (2018)
21. Panagopoulou, P.N., Spirakis, P.G.: A game theoretic approach for efficient graph coloring. In: Hong, S.-H., Nagamochi, H., Fukunaga, T. (eds.) ISAAC 2008. LNCS, vol. 5369, pp. 183–195. Springer, Heidelberg (2008). https://doi.org/10.1007/978-3-540-92182-0_19
22. Schäffer, A.A., Yannakakis, M.: Simple local search problems that are hard to solve. SIAM J. Comput. **20**(1), 56–87 (1991). https://doi.org/10.1137/0220004

Phase Transition in Matched Formulas and a Heuristic for Biclique Satisfiability

Miloš Chromý[ID] and Petr Kučera[✉][ID]

Faculty of Mathematics and Physics, Department of Theoretical Computer Science
and Mathematical Logic, Charles University,
Malostranské nám. 25, 118 00 Praha 1, Czech Republic
{chromy,kucerap}@ktiml.mff.cuni.cz

Abstract. A matched formula is a CNF formula whose incidence graph admits a matching which matches a distinct variable to every clause. We study phase transition in a context of matched formulas and their generalization of biclique satisfiable formulas. We have performed experiments to find a phase transition of property "being matched" with respect to the ratio m/n where m is the number of clauses and n is the number of variables of the input formula φ. We compare the results of experiments to a theoretical lower bound which was shown by Franco and Van Gelder [11]. Any matched formula is satisfiable, and it remains satisfiable even if we change polarities of any literal occurrences. Szeider [17] generalized matched formulas into two classes having the same property—varsatisfiable and biclique satisfiable formulas. A formula is biclique satisfiable if its incidence graph admits covering by pairwise disjoint bounded bicliques. Recognizing if a formula is biclique satisfiable is NP-complete. In this paper we describe a heuristic algorithm for recognizing whether a formula is biclique satisfiable and we evaluate it by experiments on random formulas. We also describe an encoding of the problem of checking whether a formula is biclique satisfiable into SAT and we use it to evaluate the performance of our heuristic.

Keywords: SAT · Matched formulas · Biclique SAT · var-SAT
Phase transition · Biclique cover

1 Introduction

In this paper we are interested in the *problem of satisfiability* (SAT) which is central to many areas of theoretical computer science. In this problem we are given a formula φ in propositional logic and we ask if this formula is satisfiable,

This research was supported by Charles University project UNCE/SCI/004 and SVV project number 260 453. Access to computing and storage facilities owned by parties and projects contributing to the National Grid Infrastructure MetaCentrum provided under the programme "Projects of Large Research, Development, and Innovations Infrastructures" (CESNET LM2015042), is greatly appreciated.

ⓒ Springer Nature Switzerland AG 2019
B. Catania et al. (Eds.): SOFSEM 2019, LNCS 11376, pp. 108–121, 2019.
https://doi.org/10.1007/978-3-030-10801-4_10

i.e. if there is an assignment of values to variables which satisfies φ. This is one of the best known NP-complete problems [8]. In this paper we study special classes of formulas whose definition is based on the notion of incidence graph.

Given a formula φ in *conjunctive normal form* (CNF) we consider its *incidence graph* $I(\varphi)$ defined as follows. $I(\varphi)$ is a bipartite graph with one part consisting of the variables of φ and the other part consisting of the clauses of φ. An edge $\{x, C\}$ for a variable x and a clause C is in $I(\varphi)$ if x or $\overline{x}$ appears in C. It was observed by the authors of [2] and [18] that if $I(\varphi)$ admits a matching of size m (where m is the number of clauses in φ), then φ is satisfiable. Later the formulas satisfying this condition were called *matched formulas* in [11]. Since a matching of maximum size in a bipartite graph can be found in polynomial time (see e.g. [14, 15]), one can check efficiently whether a given formula is matched.

It is clear that if φ is a formula on n variables and m clauses then φ can be matched only if $m \leq n$. The authors of [11] asked an interesting question: What is the probability that a formula φ is matched depending on the ratio $\frac{m}{n}$? We can also ask if the property "being matched" exhibits a phase transition.

A phase transition was studied in context of satisfiability [5, 7, 9, 12, 16]. The so-called satisfiability threshold for a given k is a value r_k satisfying the following property: A random formula φ in k-CNF (each clause has exactly k literals) on n variables and m clauses is almost surely satisfiable if $\frac{m}{n} < r_k$ and it is almost surely unsatisfiable if $\frac{m}{n} > r_k$. For instance the value r_3 is approximately 4.3 [7, 9].

In the same sense we can study threshold for property "being matched". It was shown in [11] that a 3-CNF φ on n variables and m clauses is almost surely matched if $\frac{m}{n} < 0.64$. This is merely a theoretical lower bound, and in this paper we perform experimental check of this value. It turns out that the experimentally observed threshold is much higher than the theoretical lower bound. Moreover, we observe that the property "being matched" has a sharp threshold or phase transition as a function of ratio $\frac{m}{n}$.

Matched formulas have an interesting property: If a formula φ is matched then we pick any occurrence of any literal and switch its polarity (i.e. change a positive literal x into a negative literal $\overline{x}$ or vice versa). The formula produced by this operation will be matched and thus satisfiable as well. This is because the definition of incidence graph completely ignores the polarities of variables. The formulas with this property were called *var-satisfiable* in [17] and they form a much bigger class than matched formulas. Unfortunately, it was shown in [17] that the problem of checking whether a given formula φ is var-satisfiable is complete for the second level of polynomial hierarchy (Π_2^P-complete).

Szeider in [17] defined a subclass of var-satisfiable formulas called *biclique satisfiable formulas* which extends matched formulas. It was shown in [17] that checking if φ is biclique satisfiable is an NP-complete problem. In this paper we describe a heuristic algorithm to test whether a formula is biclique satisfiable. Our heuristic algorithm is based on an heuristic for covering a bipartite graph with bicliques described in [13]. We test our heuristic algorithm experimentally on random formulas. Our heuristic algorithm is incomplete, in the sense that, whenever it finds that a formula is biclique satisfiable, then it is so, but it may happen that a formula is biclique satisfiable even though our algorithm is unable to detect it. In order to check the quality of our heuristic, we propose a SAT based approach to checking biclique satisfiability of a formula. We compare both approaches on random formulas.

In Sect. 2 we recall some basic definitions and related results used in the rest of the paper. In Sect. 3 we give the results of experiments on matched formulas. In Sect. 4 we describe our heuristic algorithm for determining if a formula is biclique satisfiable and we give the results of its experimental evaluation. In Sect. 5 we describe a SAT based approach to checking biclique satisfiability and compare it experimentally with the heuristic approach. We close the paper with concluding remarks in Sect. 6.

2 Definitions and Related Results

In this section we shall introduce necessary notions and results used in the paper.

2.1 Graph Theory

We use the standard graph terminology (see e.g. [4]). A *bipartite graph* $G = (V_v, V_c, E)$ is a triple with vertices split into two parts V_v and V_c and the set of edges E satisfying that $E \subseteq V_v \times V_c$. Given a bipartite graph G we shall also use the notation $V_v(G)$ and $V_c(G)$ to denote the vertices in the first and in the second part respectively. For two natural numbers n, m we denote by $K_{n,m}$ the *complete bipartite graph* (or a *biclique*) that is the graph $K_{n,m} = (V_v, V_c, E)$ with $|V_v| = n$, $|V_c| = m$ and $E = V_v \times V_c$.

Given a bipartite graph $G = (V_v, V_c, E)$ the *degree* of a vertex $v \in V_v \cup V_c$ is the number of incident edges. A pairwise disjoint subset of edges $M \subseteq E$ is called a *matching* of G. A vertex v is *matched* by matching M if v is incident to some edge from M. M is a *maximum matching* if for every other matching M' of G we have that $|M| \geq |M'|$. A maximum matching of a bipartite graph $G = (V_v, V_c, E)$ can be found in time $O(|E|\sqrt{|V_v| + |V_c|})$ [14,15].

2.2 Boolean Formulas

A *literal* is a variable x or its negation $\overline{x}$. A *clause* is a finite disjunction of distinct literals $C = (l_1 \vee l_2 \vee \ldots \vee l_k)$, where k is the *width of clause C*. A formula in *conjunctive normal form* (*CNF*) is a finite conjunction of clauses $\varphi = C_1 \wedge C_2 \wedge \ldots \wedge C_n$. Formula φ is in k-CNF if all clauses in φ have width at most k. We shall also often write (k-)CNF φ instead of φ being in (k-)CNF.

Let us now recall the definition of probability space $\mathcal{M}_{m,n}^k$ from [11].

Definition 1 (Franco and Van Gelder [11]). *Let $V_n = \{v_1, \ldots, v_n\}$ be a set of Boolean variables and let $L_n = \{v_1, \overline{v_1}, \ldots, v_n, \overline{v_n}\}$ be the set of literals over variables in V_n. Let C_n^k be the set of all clauses with exactly k variable-distinct literals from L_n. A random formula in probability space $\mathcal{M}_{m,n}^k$ is a sequence of m clauses from C_n^k selected uniformly, independently, and with replacement.*

2.3   Matched Formulas

A CNF formula φ with m clauses is *matched* if its incidence graph $I(\varphi)$ has a matching of size m, i.e. each clause is matched with a unique variable. The following result on density of matched formulas in the probability space $\mathcal{M}_{m,n}^k$ was shown in [11].

Theorem 1 (Franco and Van Gelder [11]). *Under $\mathcal{M}_{m,n}^k$, $k \geq 3$, the probability that a random formula φ is matched tends to 1 if $\frac{m}{n} < 0.64$ as $n \to \infty$.*

One of the goals of this paper is to check experimentally how good estimate of the real threshold the theoretical value 0.64 is.

2.4   Biclique Satisfiable Formulas

One of the biggest limitations of matched formulas is that if φ is a matched formula on n variables and m clauses, then $m \leq n$. To overcome this limitation while keeping many nice properties of matched formulas, Stefan Szeider introduced biclique satisfiable formulas in [17].

A biclique $K_{n,m}$ is *bounded* if $m < 2^n$. Let φ be a CNF on n variables and m clauses and let us assume that $I(\varphi) = K_{n,m}$ where $m < 2^n$. Then φ is satisfiable [17]. This is because we have $m < 2^n$ clauses each of which contains all n variables. Each of these clauses determines exactly one unsatisfying assignment of φ, but there is 2^n assignments in total. Thus one of them must be satisfying.

A bipartite graph $G = (V_v, V_c, E)$ has a *bounded biclique cover* if there is a set of bounded bicliques $\mathcal{B} = \{B_1, \dots, B_k\}$ satisfying the following conditions.

- every $B_i, i = 1, \dots, k$ is a subgraph of G,
- for any pair of indices $1 \leq i < j \leq k$ we have that $V_v(B_i) \cap V_v(B_j) = \emptyset$, and
- for every $v \in V_c(G)$ there is a biclique $B_i, i = 1, \dots, k$ such that $v \in V_c(B_i)$.

If every biclique $B_i \in \mathcal{B}$ in the cover satisfies that $|V_v(B_i)| \leq k$, then we say the graph G has a bounded k-biclique cover. A formula φ is *(k-)biclique satisfiable* if its incidence graph $I(\varphi)$ has a bounded $(k$-)biclique cover.

Any biclique satisfiable formula is satisfiable, however, it is an NP-complete problem to decide if a formula is biclique satisfiable even if we only restrict to 2-biclique satisfiable formulas (see [17]). We can observe that 1-satisfiable formulas are matched formulas, because a single edge is a bounded biclique.

2.5   Generating Experimental Data

Whether a formula φ in CNF is matched or not depends only on its incidence graph $I(\varphi)$. Instead of random formulas from probabilistic space $\mathcal{M}_{m,n}^k$ we thus consider random bipartite graphs $G = (V_v, V_c, E)$ from the probabilistic space $\mathcal{G}_{m,n}^k$.

Definition 2. *Probability space $\mathcal{G}_{m,n}^k$ is defined as follows. A random bipartite graph $G \in \mathcal{G}_{m,n}^k$ is a bipartite graph with parts V_v, V_c where $|V_v| = n$, $|V_c| = m$. Each vertex $v \in V_c$ has k randomly uniformly selected neighbours from V_v.*

In our experiments we generated bipartite graphs $G \in \mathcal{G}_{m,n}^k$. Since we consider choosing clauses in formula $\varphi \in \mathcal{M}_{m,n}^k$ with replacement, we can have several copies of the same clause in φ. It follows that given a bipartite graph $G \in \mathcal{G}_{m,n}^k$, we have exactly $2^k m$ formulas $\varphi \in \mathcal{M}_{m,n}^k$ which have $I(\varphi) = G$—each vertex $c \in V_c$ can be replaced with 2^k different clauses by setting polarities to variables $x \in V_v$ adjacent to v in G. In particular, the probability that a random formula $\varphi \in \mathcal{M}_{m,n}^k$ is matched is the same as the probability that a random bipartite graph $G \in \mathcal{G}_{m,n}^k$ admits a matching of size m. The same holds for the biclique satisfiability.

3   Phase Transition on Matched Formulas

In this section we shall describe the results of experiments we have performed on matched formulas. In particular we were interested in phase transition of k-CNF formulas with respect to the property "being matched" depending on the ratio of the number of clauses to the number of variables. We will also compare the results with the theoretical bound proved in [11] (see Theorem 1).

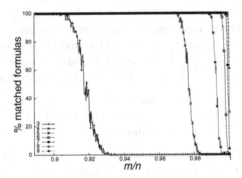

Fig. 1. Results of experiments on random graph $\mathcal{G}_{m,n}^k$ with $n = 4000$ and $k = 3, \dots, 9$.

In our experiments we considered values of number of variables $n = 100, 200, 500, 1000, 2000, 4000$ and $k = 3, 4, \dots, 10$. For each such pair n, k we have generated 1000 random graphs $G \in \mathcal{G}_{m,n}^k$ for ratios $\frac{m}{n} = 0.64, 0.65, \dots, 1$. Figure 1 shows the graph with the results of experiments for value $n = 4000$. The graph contains a different line for each value of $k = 3, \dots, 9$ which shows the percentage of graphs which admit matching of size m among the generated random graphs depending on the ratio $\frac{m}{n} = 0.64, 0.65, \dots, 1$. The complete results of the experiments are shown in Table 1. For each value of k we distinguish two values *high* and *low* where only 1% of the graphs generated in $\mathcal{G}_{m,n}^k$ with $\frac{m}{n} \geq high$ admit matching of size m, and on the other hand 99% of the graphs generated in $\mathcal{G}_{m,n}^k$ with $\frac{m}{n} < low$ admit matching of size m.

Table 1. Phase transition intervals as two values *high* and *low*. We provide only *low* value for $k \geq 7$, because the *high* value was 1 for all such configurations.

k	3		4		5		6		7	8	9	10
n	Low	High	Low	High	Low	High	Low	High	Low	Low	Low	Low
100	0.85	0.98	0.95	1	0.97	1	0.98	1	1	1	1	1
200	0.88	0.96	0.96	0.99	0.98	1	0.99	1	1	1	1	1
500	0.89	0.95	0.96	0.99	0.99	1	1	1	1	1	1	1
1000	0.895	0.939	0.97	0.989	0.986	0.999	0.99	0.995	0.997	0.998	0.999	0.999
2000	0.903	0.9325	0.97	0.985	0.988	0.9965	0.995	0.9995	0.998	0.9985	0.9995	0.9995
4000	0.909	0.929	0.9715	0.982	0.99	0.995	0.995	0.992	0.998	0.999	0.9995	0.9995

We can see that for higher values of n the interval $[low, high]$ gets narrower and we can thus claim that the property "being matched" indeed exhibits a phase transition phenomenon. Moreover, we can say that the average of values *low* and *high* limits to the threshold of this phase transition. We can see that the threshold ratio for $k = 3$ is around 0.92 which is much higher than the theoretical bound 0.64 from [11] (see Theorem 1). In all configurations with $k \geq 7$ the *high* value was 1 while the *low* value was close to 1 as well. Thus in the experiments we made with $k \geq 7$ even in the case $m = n$ almost all of the randomly generated graphs admitted matching of size m.

4 Bounded Biclique Cover Heuristic

The class of biclique satisfiable formulas forms a natural extension to the class of matched formulas. In this section we shall describe a polytime heuristic algorithm for finding a bounded biclique cover of a graph.

4.1 Description of Heuristic Algorithm

Our heuristic approach is described in Algorithm 1. It is based on a heuristic algorithm for finding a smallest biclique cover of a bipartite graph described in [13]. The algorithm expects three parameters. The first two parameters are a bipartite graph G and an integer t which restricts the size of the first part of bounded bicliques used in the cover, in other words only bicliques S satisfying that $|V_v(S)| \leq t$ are included in the cover which is output by the algorithm. The last parameter used in the algorithm is the strategy for selecting a seed.

Let G be a bipartite graph $G = (V_v, V_c, E)$. A *seed* in G is a biclique S which is a nonempty subgraph of G with $|V_v(S)| = 2$ and $V_c(S) \neq \emptyset$. We say that S is a *maximal seed* if there is no seed S' so that $V_v(S) = V_v(S')$ and $V_c(S) \subsetneq V_c(S')$.

After initializing an empty cover $\mathcal{C}$, the algorithm starts with a pruning step (unitGPropagation) which is used also in the main loop. In this step a simple reduction rule is repeatedly applied to the graph G: If a vertex $C \in V_c$ is present in a single edge $\{v, C\}$, then this edge is added into cover $\mathcal{C}$ as a biclique in order to cover C. In this case vertices v and C with all edges incident to v are removed from the graph G. If a vertex $C \in V_c$ with no incident edges is encountered during this process, the heuristic algorithm fails and returns an empty cover.

The algorithm continues with generating a list S of all maximal seeds induced by all pairs $\{v_i, v_j\} \subseteq V_v, i < j$. The input graph is modified during the algorithm by removing edges and vertices. In the following description $G = (V_v, V_c, E)$ always denotes the current version of the graph.

The main loop of the algorithm repeats while there are some seeds available and G does not admit a matching of size $|V_c|$. This is checked by calling function testMatched which also adds the matching to C if it is found.

The body of the main loop starts with selecting one seed S by function chooseSeed. This choice is based on a given strategy. We consider three strategies for selecting a seed: Strategy S_{min} chooses a seed with the smallest second part. Strategy S_{max} chooses a seed with the largest second part. And strategy S_{rand} chooses a random seed. Seed S is then expanded by repeatedly calling function expandSeed. This function selects a vertex $v \in V_v \backslash V_v(S)$ which maximizes the size of the second part of the biclique induced in G with left part being $V_v(S) \cup \{v\}$ (the second part is induced to be all the vertices incident to all vertices in $V_v(S) \cup \{v\}$). The expansion process continues while the size of the first part $V_v(S)$ satisfies the restriction imposed by the parameter t and while S is not a bounded biclique (that is while $2^{|V_v(S)|} \leq |V_c(S)|$).

If the expansion process ends due to the restriction on the size $|V_v(S)|$ given by t, S is not necessarily a bounded biclique. Function restrictSeed is then used to remove randomly chosen vertices from $V_c(S)$ to make S a bounded biclique.

Once a bounded biclique S is found, it is removed from the graph and it is added to the cover C. This is realized by a function removeBiclique which simply sets $V_v \leftarrow V_v \backslash V_v(S)$, $V_c \leftarrow V_c \backslash V_c(S)$, and $E \leftarrow E \cap (V_v \times V_c)$. Then we call unitGPropagation to prune the graph. After that the function removeInvalidSeeds removes from S all seeds S' with $V_v(S') \cap V_v(S) \neq \emptyset$. For remaining seeds $S' \in S$ the function sets $V_c(S') \leftarrow V_c(S') \cap V_c$.

After the loop finishes, the current cover C is returned.

The complexity of each step is noted in comments. Altogether the complexity of Algorithm 1 is $\mathcal{O}(n^2 \ell)$. If a nonempty set of bicliques C is returned by the algorithm, then it is a bounded biclique cover of G. It should be noted that the opposite implication does not necessarily hold; if the seeds are chosen badly then the algorithm may fail even if there is some bounded biclique cover in G. In the next section we aim at evaluating our heuristic algorithm experimentally.

4.2   Experimental Evaluation of Heuristic Algorithm

In this section we shall describe the experiments performed with our heuristic Algorithm 1 described in Sect. 4.1 for bounded 2-biclique cover. In the extended version [6] we describe the results of experiments with general bounded biclique cover as well.

Algorithm 1 works with bipartite graphs. We have tested proposed heuristic on random bipartite graphs $G \in \mathcal{G}_{m,n}^k$ with $n = 100, 200$ and with the degrees of vertices in the second part being $k = 3, \ldots, 100$. This corresponds to formulas in k-CNF for these values. We have considered different sizes of the second part given by ratios $\frac{m}{n} = 1, 1.01, \ldots, 1.5$. Note that there is no bounded 2-biclique cover for graphs with $\frac{m}{n} > 1.5$. We have tried the three strategies S_{min}, S_{rand} and S_{max} for selecting a seed. Due to time complexity of Algorithm 1 we have only generated a hundred random graphs in $\mathcal{G}_{m,n}^{k}$ for each configuration (given by a strategy and ratio $\frac{m}{n}$).

Data: Bipartite graph $G(V_v, V_c, E)$, $t \in \{2, \ldots, |V_v|\}$ — maximal size of $|V_v(S)|$
for a biclique S which we put into the cover and a seeds selection
strategy $\mathtt{st} \in \{S_{min}, S_{rand}, S_{max}\}$.
Result: biclique cover $\mathcal{C}$ of graph G if a heuristic found one, $\emptyset$ otherwise
$\mathcal{C} \leftarrow \emptyset$
if $\mathtt{unitGPropagation}(G, \mathcal{C})$ *fails* then return $\emptyset$ // $\mathcal{O}(nm)$
$S \leftarrow \mathtt{generateSeeds}(G)$ // $\mathcal{O}(n\ell)$
while $|S| > 0$ and not $\mathtt{testMatched}(G, \mathcal{C})$ do   // $\mathcal{O}(\ell\sqrt{n})$
$\quad$ $S \leftarrow \mathtt{chooseSeed}(S, \mathtt{st})$ // $\mathcal{O}(n^2)$
$\quad$ while $|V_v(S)| < t \wedge 2^{|V_v(S)|} \leq |V_c(S)|$ do
$\quad\quad$ $S \leftarrow \mathtt{expandSeed}(S)$ // $\mathcal{O}(\ell + m)$
$\quad$ end
$\quad$ if $2^{|V_v(S)|} \leq |V_c(S)|$ then $S \leftarrow \mathtt{restrictSeed}(S)$ // $\mathcal{O}(|V_c(S)|)$
$\quad$ $G \leftarrow \mathtt{removeBiclique}(G, S)$ // $\mathcal{O}(\ell)$
$\quad$ $\mathcal{C} \leftarrow \mathcal{C} \cup \{S\}$
$\quad$ if $\mathtt{unitGPropagation}(G, \mathcal{C})$ *fails* then return $\emptyset$ // $\mathcal{O}(nm)$
$\quad$ $S \leftarrow \mathtt{removeInvalidSeeds}(S)$ // $\mathcal{O}(n\ell)$
end
return $\mathcal{C}$

Algorithm 1. An heuristic for checking if there is a bounded biclique cover of a bipartite graph $G = (V_v, V_c, E)$. The complexity of each step is noted in comments where we consider $n = |V_v|$, $m = |V_c|$, and $\ell = |E|$.

Table 2. Results of experiments with our heuristic algorithm on graphs with size of second part $|V_v| = 100$. A more detailed explanation can be found in the main text.

	1		1.1		1.2		1.3		1.4		1.5	
	Low	High	Low	High	Low	High	Low	High	Low	High	Low	High
S_{min}	4	5	5	6	7	8	9	15	13	24	33	47
S_{rand}	4	5	5	6	7	8	9	15	13	24	33	47
S_{max}	4	5	5	6	7	8	9	15	14	24	41	87

Table 2 summarizes the results of our experiments. Each row corresponds to different strategy in bounded 2-biclique cover. Each column corresponds to a ratio $\frac{m}{n}$, we have included only ratios 1, 1.1, 1.2, 1.3, 1.4, and 1.5 in the table.

For each configuration we have two bounds *low* and *high* on degree k of vertices in the second part V_c of graph G. Our heuristic algorithm succeeded only on 1% of graphs with degree $k \leq low$ and on the other hand it succeeded on 99% of graphs with degree $k \geq high$.

For bounded 2-biclique cover, the strategies S_{min} and S_{rand} are never worse than S_{max} and they even get better for higher ratios. This makes S_{rand} the best strategy for bounded 2-biclique covers—it is easiest to implement and randomness means that repeated calls of our heuristic algorithm may eventually lead to finding a biclique cover. As we can expect, heuristic performs quite well on lower values of ratio $\frac{m}{n}$ and it gets worse on higher values of this ratio.

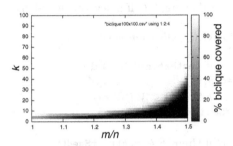

Fig. 2. Results of experiments with our heuristic algorithm with strategy S_{rand} and $n = 100$. The more white pixel is, the more random graphs were covered by a bounded biclique cover found by Algorithm 1.

We can observe a phase transition behaviour in the results of experiments. As we can see on Fig. 2 there is a phase transition $r_{\frac{m}{n}}$ for a fixed ratio $\frac{m}{n}$. Most of random graphs $G \in \mathcal{G}_{m,n}^k$ with $k \geq r_{\frac{m}{n}}$ have a biclique cover and our heuristic algorithm will find it. However, since our heuristic is incomplete, it is not clear how many random graphs $G \in \mathcal{G}_{m,n}^k$ with $k \leq r_{\frac{m}{n}}$ have biclique cover. We will take a closer look on this completeness gap in Sect. 5.

The most interesting case is when $\frac{m}{n} \leq 1.4$. As the ratio $\frac{m}{n}$ gets close to 1.5 we can expect smaller percentage of graphs $G \in \mathcal{G}_{m,n}^k$ having a 2-biclique cover, hence our heuristic algorithm fails to find one in most cases.

5   Bounded Biclique SAT Encoding

We shall first describe the encoding of the problem of checking if a bipartite graph has a bounded biclique cover into SAT, then we will describe and evaluate the experiments we have performed to compare this approach with Algorithm 1. We will also describe the environment we have used to run the experiments.

5.1   Description of SAT Encoding

We shall describe the encoding for the case of bounded 2-biclique cover, since this is the version we used in experiments for comparison with Algorithm 1.

The encoding can be easily generalized to bounded k-biclique cover or for using all kinds of bicliques, this is described in the extended version of the paper [6].

Let us consider a bipartite graph $G = (V_v, V_c, E)$, let us define $\mathcal{B}_2 = \{B \mid B$ is a subgraph of G which is either $K_{1,1}$ or $K_{2,3}\}$. We shall associate the following formula ψ with graph G. With each biclique $B \in \mathcal{B}_2$ we associate a new variable x_B. Every assignment of boolean values to variables x_B, $B \in \mathcal{B}_2$ then specifies a set of bicliques. We will encode the fact that the satisfying assignments of ψ exactly correspond to bounded 2-biclique covers of G. To this end we use the following constraints.

For each vertex $v \in V_v$ we add to ψ an at-most-one constraint on variables x_B, $v \in V_v(B)$. This encodes the fact that the first parts of bicliques in the cover have to be pairwise disjoint. We use a representation of the at-most-one constraint with a quadratic number of negative clauses of size 2.

For each clause $C \in V_c$ we add to ψ a clause representing an at-least-one constraint on variables x_b, $C \in V_c(B)$. This encodes the fact that each vertex of the second part belongs to a biclique in the cover.

5.2 Experimental Evaluation of Heuristic Algorithm

We used the encoding described in Sect. 5.1 to check the success rate of Algorithm 1 on random bipartite graphs and to check the phase transition for an existence of bounded 2-biclique cover. We ran the experiments on 100 random bipartite graphs $G \in \mathcal{G}_{m,n}^k$ with $n = 40$ for combinations of $k = 1, \ldots, 8$ and the size of the second part $m = rn$ for $r = 1.00, \ldots, 1.25$ with step 0.05. For each k we tested random graphs only for the ratios around the expected phase transition as observed in the Table 2.

The results of experiments are contained in Tables 3 and 4. Both tables have a similar structure. Each cell represents a single configuration (row corresponding to a value of k and column corresponding to the ratio $\frac{m}{n}$ where m denotes the number of clauses and $n = 40$ denotes the number of variables). In Table 3 each cell contains three numbers separated with slashes. The first is the number of instances (out of 100) on which Algorithm 1 successfully found a bounded 2-biclique cover. The second is the number of instances on which the SAT solver successfully solved the encoding and answered positively. The third is the number of instances on which the SAT solver finished within time limit which was set to 4 h for each instance. In some cells the values are missing, for these configurations we did not run any experiments, because they are far from the observed phase transition (see Table 2). We expect that the results would be 100/100/100 in case of black colored cells and 0/0/100 in case of white colored cells. The gray colored cells mark the borders of observed phase transition intervals of existence of bounded biclique cover, light gray corresponds to the results given by the SAT solver, dark gray to the results given by Algorithm 1 which form an upper bound on the correct values. We can see that in most cases the number of positive answers given by Algorithm 1 is close to the number of positive answers given by the SAT solver. However, there are some cases where the SAT based approach was more successful. Namely in cases of $k = 5, \frac{m}{n} = 1.15$ and $k = 5, \frac{m}{n} = 1.2$. However, in the latter case the SAT solver ran over time limit (4 h) in $\frac{2}{3}$ cases.

We compared the running time of our Algorithm 1 and the SAT solver. As we can see in the Table 4 our heuristic algorithm is much faster in average case. Standard deviation of runtime of our heuristic algorithm is around 10^{-2} and the standard deviation of running times of the SAT solver is up to 10^4 (where we have evaluated the average value and the standard deviation only on instances in which the SAT solver finished within the time limit). These values are quite high compared to the running times. One of the reasons is perhaps the fact that the experiments were not run on a single computer, but on several comparable computers (see Sect. 5.3 for more details). Although in all cases on a single instance, the SAT solver and Algorithm 1 were run on a single computer and it makes thus sense to look at the ratio between the running times of these two. We also computed these ratios and they are close to ratios implied by Table 4. We refer readers to extended version of the paper [6] for details.

Table 3. Number of bipartite graphs with bounded biclique cover (found by Algorithm 1/SAT finished with true/SAT finished within time limit). See the description within the text for more details.

	1	1.05	1.1	1.15	1.2	1.25
3	18/ 18/100	0/ 0/100				
4	95/ 95/100	50/ 56/100	7/ 10/100	0/ 1/100	0/ 0/100	
5	100/100/100	100/100/100	90/ 98/ 98	52/ 87/ 87	10/ 34/ 34	1/ 1/ 1
6		100/100/100	100/100/100	100/100/100	85/ 99/ 99	42/ 53/ 53
7				100/100/100	99/100/100	89/ 90/ 90
8						99/100/100

Table 4. Average runtime of Algorithm 1/average runtime of SAT on encoding in seconds.

	1	1.05	1.1	1.15	1.2	1.25
3	0.005/0.014	0.005/0.01				
4	0.005/1.7	0.006/25	0.006/21	0.005/13	0.006/7	
5	0.005/0.2	0.005/0.98	0.006/11	0.006/216	0.006/993	0.005/4589
6		0.006/0.4	0.006/2.4	0.006/50	0.006/838	0.006/3646
7				0.006/73	0.006/166	0.007/1666
8						0.006/272

We can see from the results that on random graphs Algorithm 1 has a success rate close to the one of the SAT based approach and it is much faster.

5.3   Experimental Environment

Let us say more on the environment in which the experiments were run. We used Glucose parallel SAT solver [3,10]. Our experiments were executed on grid computing service MetaCentrum NGI [1]. All experiments were run on a single processor machine (Intel Xenon, AMD Opteron) with 4 cores and frequency 2.20 GHz–3.30 GHz. On each random bipartite graph $G \in \mathcal{G}_{m,n}^k$, Algorithm 1 and the SAT solver were always run on the same computer. However, for the same configuration and different formulas, the experiments may have run on different computers. As we have noted in Sect. 5.2, this could be a reason of significantly high values of standard deviation of runtimes. The fact that the computer speed varied while the time limit for the SAT solver was still the same (4 h) could have led to situations where the SAT solver would not finished, because it was run on a slower computer, and could potentially finish had it been run on a faster computer. However, 1912 out of the total 2200 instances finished within an hour, then only 26 finished between an hour and 2 h, only 15 finished between 2 h and 3 h and only 9 finished between 3 h and 4 h. We can thus expect that the number of the border cases is similarly small. We can conclude that the variance in computer speeds had only minor influence on the number of SAT calls which finished within the time limit.

6   Conclusion

The first result of our paper is that the experimental threshold of phase transition of property "being matched" of 3-CNFs is around 0.92, which is much higher than the theoretical lower bound 0.64 proved for 3-CNF in [11]. This can be seen in Fig. 1. Our experiments also suggest that for $k \geq 6$ almost all formulas in k-CNF are matched (if they have at most as many clauses as variables).

We have also proposed a heuristic algorithm for finding a bounded biclique cover of an incidence graph $I(\varphi)$ of a given formula φ. In other words the algorithm tries to decide if φ is biclique satisfiable. We suggested three different strategies for selecting a seed in our heuristic and compared them. We can deduce from Fig. 2 that the success rate of our heuristic algorithm exhibits a phase transition phenomenon similar to the case of matched formulas. The exact values are shown in Table 2. Our results suggest it is better to use a strategy S_{rand} to find a 2-biclique cover using our heuristic algorithm.

The success rate of Algorithm 1 exhibits a very similar phase transition to matched formulas. In case of 3-CNFs, the phase transition is almost the same with *low* bound 0.9 and *high* bound 0.93. In case of 5-CNFs the situation is different. The observed *low* bound of the phase transition interval for property "being matches" is 0.985 and in case of our algorithm the *low* bound of the phase transition interval is 1.02. A formula can be matched only if the ratio $\frac{m}{n}$ of the number of clauses m to a number of variables n is at most 1. According to the results of our experiments a random k-CNF with $k > 5$ is matched with high probability even in case the ratio $\frac{m}{n}$ is 1. However, for 7-CNFs the *low* value of phase transition of our algorithm equals 1.14 and for $k \geq 10$ it is even more than 1.3, which means that if φ is a formula in 10-CNF with n variables and at most $1.3n$ clauses, Algorithm 1 will most likely find a bounded biclique cover of the incidence graph of φ. These results are summarized in Table 2.

Our heuristic algorithm is not complete; in particular, it can happen that a formula is biclique satisfiable, but Algorithm 1 is unable to detect it. It means that we can only trust a positive answer of the algorithm. We have compared our heuristic with a SAT based approach which can also check that a formula is not biclique satisfiable. We can see in Table 3 that formulas on which Algorithm 1 fails to answer correctly, are concentrated around the observed phase transition, and that the algorithm answers correctly in most cases for other configurations. We can say that the success rate of Algorithm 1 is not far from the complete SAT based method. Moreover, as we can see in Table 4, our heuristic is significantly faster than a SAT solver on the encoding we have described.

References

1. Metacentrum grid computing, June 2017. https://metavo.metacentrum.cz/en/
2. Aharoni, R., Linial, N.: Minimal non-two-colorable hypergraphs and minimal unsatisfiable formulas. J. Comb. Theory Ser. A **43**(2), 196–204 (1986). https://doi.org/10.1016/0097-3165(86)90060-9. http://www.sciencedirect.com/science/article/pii/0097316586900609
3. Audemard, G., Simon, L.: The glucose sat solver, June 2017. http://www.labri.fr/perso/lsimon/glucose/
4. Bollobás, B.: Modern Graph Theory. GTM, vol. 184. Springer, New York (1998). https://doi.org/10.1007/978-1-4612-0619-4
5. Cheeseman, P., Kanefsky, B., Taylor, W.M.: Where the really hard problems are. In: Proceedings of the 12th International Joint Conference on Artificial Intelligence, IJCAI 1991, vol. 1, pp. 331–337. Morgan Kaufmann Publishers Inc., San Francisco (1991). http://dl.acm.org/citation.cfm?id=1631171.1631221
6. Chromý, M., Kučera, P.: Phase transition in matched formulas and a heuristic for biclique satisfiability. ArXiv e-prints, August 2018. https://arxiv.org/abs/1808.01774
7. Connamacher, H.S., Molloy, M.: The satisfiability threshold for a seemingly intractable random constraint satisfaction problem. CoRR abs/1202.0042 (2012). http://arxiv.org/abs/1202.0042
8. Cook, S.A.: The complexity of theorem-proving procedures. In: Proceedings of the Third Annual ACM Symposium on Theory of Computing, STOC 1971, pp. 151–158. ACM, New York (1971). https://doi.org/10.1145/800157.805047
9. Dubois, O., Boufkhad, Y., Mandler, J.: Typical random 3-SAT formulae and the satisfiability threshold. In: Proceedings of the Eleventh Annual ACM-SIAM Symposium on Discrete Algorithms, SODA 2000, pp. 126–127. Society for Industrial and Applied Mathematics, Philadelphia (2000). http://dl.acm.org/citation.cfm?id=338219.338243
10. Eén, N., Sörensson, N.: The minisat, June 2017. http://minisat.se/
11. Franco, J., Van Gelder, A.: A perspective on certain polynomial-time solvable classes of satisfiability. Discrete Appl. Math. **125**(2–3), 177–214 (2003). https://doi.org/10.1016/S0166-218X(01)00358-4. http://www.sciencedirect.com/science/article/pii/S0166218X01003584

12. Gent, I.P., Walsh, T.: The SAT phase transition. In: Proceedings of the 11th European Conference on Artificial Intelligence, pp. 105–109. Wiley (1994)
13. Heydari, M.H., Morales, L., Shields Jr., C.O., Sudborough, I.H.: Computing cross associations for attack graphs and other applications. In: 2007 40th Annual Hawaii International Conference on System Sciences, HICSS 2007, p. 270b, January 2007. https://doi.org/10.1109/HICSS.2007.141
14. Hopcroft, J.E., Karp, R.M.: An $n^{5/2}$ algorithm for maximum matchings in bipartite graphs. SIAM J. Comput. **2**(4), 225–231 (1973)
15. Lovász, L., Plummer, M.D.: Matching Theory. North-Holland, Amsterdam (1986)
16. Mitchell, D., Selman, B., Levesque, H.: Hard and easy distributions of SAT problems. In: Proceedings of the 10th National Conference on Artificial Intelligence (AAAI 1992), San Jose, CA, USA, pp. 459–465 (1992)
17. Szeider, S.: Generalizations of matched CNF formulas. Ann. Math. Artif. Intell. **43**(1), 223–238 (2005). https://doi.org/10.1007/s10472-005-0432-6
18. Tovey, C.A.: A simplified NP-complete satisfiability problem. Discrete Appl. Math. **8**(1), 85–89 (1984). https://doi.org/10.1016/0166-218X(84)90081-7. http://www.sciencedirect.com/science/article/pii/0166218X84900817

On Infinite Prefix Normal Words

Ferdinando Cicalese, Zsuzsanna Lipták, and Massimiliano Rossi[✉]

Dipartimento di Informatica, University of Verona,
Strada le Grazie, 15, 37134 Verona, Italy
{ferdinando.cicalese,zsuzsanna.liptak,massimiliano.rossi_01}@univr.it

Abstract. Prefix normal words are binary words that have no factor with more 1s than the prefix of the same length. Finite prefix normal words were introduced in [Fici and Lipták, DLT 2011]. In this paper, we study infinite prefix normal words and explore their relationship to some known classes of infinite binary words. In particular, we establish a connection between prefix normal words and Sturmian words, between prefix normal words and abelian complexity, and between prefix normality and lexicographic order.

Keywords: Combinatorics on words · Prefix normal words
Infinite words · Sturmian words · Abelian complexity
Paperfolding word · Thue-Morse sequence · Lexicographic order

1 Introduction

Prefix normal words are binary words where no factor has more 1s than the prefix of the same length. As an example, the word 11100110101 is prefix normal, while 11100110110 is not, since it has a factor of length 5 with 4 1s, while the prefix of length 5 has only 3 1s. Finite prefix normal words were introduced in [12] and further studied in [5,6,9,22].

The original motivation for studying prefix normal words comes from the problem of *Indexed Binary Jumbled Pattern Matching* [1,4,8,10,13]. Given a finite word s of length n, construct an index in such a way that the following type of queries can be answered efficiently: For two integers $x, y \geq 0$, does s have a factor with x 1s and y 0s? As shown in [12], prefix normal words can be used for constructing such an index, via so-called *prefix normal forms*.

Prefix normal words have been shown to form bubble languages [5,20,21], a family of binary languages with efficiently generable combinatorial Gray codes; they have connections to the Binary Reflected Gray Code [22]; and they have recently found application to a certain class of graphs [3]. Indeed, three sequences related to prefix normal words are present in the On-Line Encyclopedia of Integer Sequences (OEIS [23]): A194850 (the number of prefix normal words of length n), A238109 (a list of prefix normal words over the alphabet $\{1, 2\}$), and A238110 (equivalence class sizes of words with the same prefix normal form).

In [9], we introduced infinite prefix normal words and analyzed a particular procedure that, given a finite prefix normal word, extends it while preserving

B. Catania et al. (Eds.): SOFSEM 2019, LNCS 11376, pp. 122–135, 2019.
https://doi.org/10.1007/978-3-030-10801-4_11

the prefix normality property. We showed that the resulting infinite word is ultimately periodic. In this paper, we present a more comprehensive study of infinite prefix normal words, covering several classes of known and well studied infinite words. We now give a quick tour of the paper.

There exist periodic, ultimately periodic, and aperiodic infinite prefix normal words (for precise definitions, see Sect. 2): for example, the periodic words 0^ω, 1^ω, and $(10)^\omega$ are prefix normal; the ultimately periodic word $1(10)^\omega$ is prefix normal; and so is the aperiodic word $10100100010000\cdots = \lim_{n\to\infty} 1010^2 \cdots 10^n$. In Sect. 3, we fully characterize periodic and ultimately periodic words in terms of their minimum density, a parameter introduced in [9].

Regarding aperiodic words, we show that a Sturmian word w is prefix normal if and only if $w = 1c_\alpha$ for some α, where c_α is the characteristic word of slope α (Theorem 2). The Fibonacci word $f = 01001010010010100101001001001001\cdots$ is thus not prefix normal, but we can turn it into a prefix normal word by prepending a 1, i.e. the word $1f$ is prefix normal. We show in fact that every Sturmian word w can be turned into a prefix normal word by prepending a fixed number of 1s, which only depends on the slope of w. This follows from a more general result regarding c-balanced words (Lemma 7).

The Thue-Morse word $\mathsf{tm} = 0110100110010110100101100110 1001\cdots$ is not prefix normal, but $11\mathsf{tm}$ is. However, the binary Champernowne word, which is constructed by concatenating the binary expansions of the integers in ascending order, namely $\mathfrak{c} = 011011100101110111100010011010 1011\cdots$ is not prefix normal and cannot be turned into a prefix normal word by prepending a finite number of 1s, because $\mathfrak{c}$ has arbitrarily long runs of 1s.

One might be tempted to conclude that every word with bounded abelian complexity can be turned into a prefix normal word by prepending a fixed number of 1s, as is the case for the words above: f has abelian complexity constant 2, tm has abelian complexity bounded by 3, and $\mathfrak{c}$ has unbounded abelian complexity. This is not the case, as we will see in Sect. 5.

We further show in Sect. 5 that the notion of prefix normal *forms* from [12] can be extended to infinite words. As in the finite case, these can be used to encode the abelian complexity of the original word. The study of abelian complexity of infinite words was initiated in [18], and continued e.g. in [2,7,14,16,25]. We establish a close relationship between the abelian complexity and the prefix normal forms of w. We demonstrate how this close connection can be used to derive results about the prefix normal forms of a word w. In some cases, such as for Sturmian words and words which are morphic images under the Thue-Morse morphism, we are able to explicitly give the prefix normal forms of the word. Conversely, knowing its prefix normal forms allows us to compute the abelian complexity function of a word.

Another class of well-known binary words are Lyndon words. Notice that the prefix normal condition is different from the Lyndon condition[1]: For finite words, there are words which are both Lyndon and prefix normal (e.g. 110010),

---

[1] For ease of presentation, we use Lyndon to mean lexicographically *greatest* among its conjugates; this is equivalent to the usual definition up to renaming characters.

words which are Lyndon but not prefix normal (11100110110), words which are prefix normal but not Lyndon (110101), and words which are neither (101100). In the final part of the paper, we will put infinite prefix normal words and their prefix normal forms in the context of lexicographic orderings, and compare them to infinite Lyndon words [24] and the max- and min-words of [17].

The paper is organized as follows. In Sect. 2, we introduce our terminology and give some simple facts about prefix normal words. In Sect. 3, we introduce the notion of *minimum density* and show its utility in dealing with certain prefix normal words. In Sect. 4, we study the relationship of Sturmian and prefix normal words. Section 5 treats prefix normal forms and their close connection to abelian complexity, and in Sect. 6 we study the relationship with lexicographic order. Due to space restrictions, all proofs were omitted, and will be included in the full version of the paper.

2   Basics

In our definitions and notations, we follow mostly [15], wherever possible. A finite (resp. infinite) binary word w is a finite (resp. infinite) sequence of elements from $\{0, 1\}$. Thus an infinite word is a mapping $w : \mathbb{N} \to \{0, 1\}$, where $\mathbb{N}$ denotes the set of positive integers. We denote the i'th character of w by w_i. Note that we index words starting from 1. If w is finite, then its length is denoted by $|w|$. The empty word, denoted ε, is the unique word of length 0. The set of binary words of length n is denoted by $\{0, 1\}^n$, the set of all finite words by $\{0, 1\}^* = \cup_{n \geq 0} \{0, 1\}^n$, and the set of infinite binary words by $\{0, 1\}^\omega$. For a finite word $u = u_1 \cdots u_n$, we write $u^{\mathrm{rev}} = u_n \cdots u_1$ for the reverse of u, and $\overline{u} = \overline{u}_1 \cdots \overline{u}_n$ for the complement of u, where $\overline{a} = 1 - a$.

For two words u, v, where u is finite and v is finite or infinite, we write uv for their concatenation. If $w = uxv$, then u is called a prefix, x a factor (or substring), and v a suffix of w. We denote the set of factors of w by $Fct(w)$ and its prefix of length i by $\mathrm{pref}_w(i)$, where $\mathrm{pref}_w(0) = \varepsilon$. For a finite word u, we write $|u|_1$ for the number of 1s, and $|u|_0$ for the number of 0s in u, and refer to $|u|_1$ as the *weight* of u. The *Parikh vector* of u is $pv(u) = (|u|_0, |u|_1)$. A word w is called *balanced* if for all $u, v \in Fct(w)$, $|u| = |v|$ implies $||u|_1 - |v|_1| \leq 1$, and *c-balanced* if $|u| = |v|$ implies $||u|_1 - |v|_1| \leq c$.

For an integer $k \geq 1$ and $u \in \{0, 1\}^n$, u^k denotes the kn-length word $uuu \cdots u$ (k-fold concatenation of u) and u^ω the infinite word $uuu \cdots$. An infinite word w is called *periodic* if $w = u^\omega$ for some non-empty word u, and *ultimately periodic* if it can be written as $w = vu^\omega$ for some v and non-empty u. A word that is neither periodic nor ultimately periodic is called *aperiodic*. We set $0 < 1$ and denote by $\leq_{\mathrm{lex}}$ the *lexicographic order* between words, i.e. $u \leq_{\mathrm{lex}} v$ if u is a prefix of v or there is an index $i \geq 1$ s.t. $u_i < v_i$ and $\mathrm{pref}_u(i - 1) = \mathrm{pref}_v(i - 1)$.

For an operation op : $\{0, 1\}^* \to \{0, 1\}^*$, we denote by $\mathrm{op}^{(i)}$ the ith iteration of op; $\mathrm{op}^*(w) = \{\mathrm{op}^{(i)}(w) \mid i \geq 1\}$; and $\mathrm{op}^\omega(w) = \lim_{i \to \infty} \mathrm{op}^{(i)}(w)$, if it exists.

Definition 1 (Prefix weight, prefix density, maximum and minimum 1s and 0s functions). *Let w be a (finite or infinite) binary word. We define the following functions:*

- $P_w(i) = |\operatorname{pref}_w(i)|_1$, *the* weight *of the prefix of length i,*
- $D_w(i) = P_w(i)/i$, *the* density *of the prefix of length i,*
- $F_w^1(i) = \max\{|u|_1 : u \in Fct(w), |u| = i\}$ *and* $f_w^1(i) = \min\{|u|_1 : u \in Fct(w), |u| = i\}$, *the* maximum *resp.* minimum number *of 1s in a factor of length i,*
- $F_w^0(i) = \max\{|u|_0 : u \in Fct(w), |u| = i\}$ *and* $f_w^0(i) = \min\{|u|_0 : u \in Fct(w), |u| = i\}$, *the* maximum *resp.* minimum number *of 0s in a factor of length i.*

Note that in the context of succinct indexing, the function $P_w(i)$ is often called $rank_1(w, i)$. We are now ready to define prefix normal words.

Definition 2 (Prefix normal words). *A (finite or infinite) binary word w is called 1-prefix normal, if $P_w(i) = F_w^1(i)$ for all $i \geq 1$ (for all $1 \leq i \leq |w|$ if w is finite). It is called 0-prefix normal if $i - P_w(i) = F_w^0(i)$ for all $i \geq 1$ (for all $1 \leq i \leq |w|$ if w is finite). We denote the set of all finite 1-prefix normal words by $\mathcal{L}_{\mathrm{fin}}$, the set of all infinite 1-prefix normal words by $\mathcal{L}_{\mathrm{inf}}$, and $\mathcal{L} = \mathcal{L}_{\mathrm{fin}} \cup \mathcal{L}_{\mathrm{inf}}$.*

In other words, a word is prefix normal (i.e. 1-prefix normal) if no factor has more 1s than the prefix of the same length. Note that unless further specified, by prefix normal we mean 1-prefix normal. Given a binary word w, we say that a factor u of w *satisfies the prefix normal condition* if $|u|_1 \leq P_w(|u|)$.

Example 1. The word 110100110110 is not prefix normal since the factor 11011 has four 1s, which is more than in the prefix 11010 of length 5. The word 110100110010, on the other hand, is prefix normal. The infinite word $(11001)^\omega$ is not prefix normal, because of the factor 111, but the word $(11010)^\omega$ is.

The following facts about infinite prefix normal words are immediate.

Lemma 1

1. *For all $u \in \mathcal{L}_{\mathrm{fin}}$, the word $w = u0^\omega \in \mathcal{L}_{\mathrm{inf}}$.*
2. *Let $w \in \{0, 1\}^\omega$. Then $w \in \mathcal{L}$ if and only if for all $i \geq 1$, $\operatorname{pref}_w(i) \in \mathcal{L}$.*

Definition 3 (Minimum density, minimum-density prefix, slope). *Let $w \in \{0, 1\}^* \cup \{0, 1\}^\omega$. Define the* minimum density *of w as $\delta(w) = \inf\{D_w(i) \mid 1 \leq i\}$. If this infimum is attained somewhere, then we also define $\iota(w) = \min\{j \geq 1 \mid \forall i : D_w(j) \leq D_w(i)\}$ and $\kappa(w) = P_w(\iota(w))$. We refer to $\operatorname{pref}_w(\iota(w))$ as the* minimum-density prefix, *the shortest prefix with density $\delta(w)$. For an infinite word w, we define the* slope *of w as $\lim_{i \to \infty} D_w(i)$, if this limit exists.*

Remark 1. Note that $\iota(w)$ is always defined for finite words, while for infinite words, a prefix which attains the infimum may or may not exist. We note further that density and slope are different properties of (infinite) binary words.

In particular, while $\delta(w)$ exists for every w, the limit $\lim_{i\to\infty} D_w(i)$ may not exist, i.e., w may or may not have a slope. As an example, consider $w = v_0 v_1 v_2 \cdots$, where for each i, $v_i = 1^{2^i} 0^{2^i}$; then $\delta(w) = 1/2$ and $\lim_{i\to\infty} D_w(i)$ does not exist, since $D_w(i)$ has an infinite subsequence constant $1/2$, and another which tends to $2/3$. But even for words w whose slope is defined, it can be different from $\delta(w)$. If w has slope α, then $\alpha = \delta(w)$ if and only if for all i, $D_w(i) \geq \alpha$. For instance, the infinite word 01^ω has slope 1 but its minimum density is 0. On the other hand, the infinite word $1(10)^\omega$ has both slope and minimum density $1/2$.

3   A Characterization of Periodic and Aperiodic Prefix Normal Words with Respect to Minimum Density

In [9], we introduced an operation which takes a finite prefix normal word w containing at least one 1 and *extends* it by a run of 0s followed by a new 1, in such a way that this 1 is placed in the first possible position without violating prefix normality. This operation, called flipext, leaves the minimum density invariant.

Definition 4 ([9] Operation flipext). *Let $w \in \mathcal{L}_{\mathrm{fin}} \setminus \{0\}^*$. Define* flipext$(w)$ *as the finite word $w0^k 1$, where $k = \min\{j \mid w0^j 1 \in \mathcal{L}\}$. We further define the infinite word $v = $* flipext$^\omega(w) = \lim_{i\to\infty}$ flipext$^{(i)}(w)$.

Proposition 1 ([9]). *Let $w \in \mathcal{L}_{\mathrm{fin}} \setminus \{0\}^*$ and $v \in $* flipext$^*(w) \cup \{$flipext$^\omega(w)\}$. *Then it holds that $\delta(v) = \delta(w)$, and as a consequence, $\iota(v) = \iota(w)$ and $\kappa(v) = \kappa(w)$. Moreover, $D_v(k \cdot \iota(w)) = \delta(w)$ for each $k \geq 1$.*

The following result shows that every ultimately periodic infinite prefix normal word has rational minimum density.

Lemma 2. *Let v be an infinite ultimately periodic binary word with minimum density $\delta(v) = \alpha$. Then $\alpha \in \mathbb{Q}$.*

Next we show that conversely, for every $\alpha \in (0,1)$, both rational and irrational, there is an aperiodic prefix normal word with minimum density α.

Lemma 3. *Fix $\alpha \in (0,1)$, and let $(a_n)_{n\in\mathbb{N}}$ be a strictly decreasing infinite sequence of rational numbers from $(0,1)$ converging to α. For each $i = 1, 2, \ldots$, let the binary word $v^{(i)}$ be defined by*

$$v^{(i)} = \begin{cases} 1^{\lceil 10a_1 \rceil} 0^{10 - \lceil 10a_1 \rceil} & i = 1 \\ \mathrm{pref}_{\mathrm{flipext}^\omega(v^{(i-1)})}(k_i |v^{(i-1)}|) 0^{\ell_i} & i > 1 \end{cases} \qquad \ell_i = \begin{cases} 10 - \lceil 10a_1 \rceil & i = 1 \\ \left\lceil k_i \left(\frac{|v^{(i-1)}|_1 - a_i |v^{(i-1)}|}{a_i} \right) \right\rceil & i > 1, \end{cases}$$

and k_i is the smallest integer greater than one such that $\ell_i > \ell_{i-1}$. Then $v = \lim_{i\to\infty} v^{(i)}$ is an aperiodic infinite prefix normal word such that $\delta(v) = \alpha$.

Summarizing, we have shown the following result.

Theorem 1. *For every $\alpha \in (0,1)$ (rational or irrational) there is an infinite aperiodic prefix normal word of minimum density α. On the other hand, for every ultimately periodic infinite prefix normal word w the minimum density $\delta(w)$ is a rational number.*

4 Sturmian Words and Prefix Normal Words

The results of the previous section show that there is a relationship between the rationality or irrationality of the minimum density of an infinite prefix normal word and its aperiodic or periodic behaviour. This is reminiscent of the characterization of Sturmian words in terms of the slope. Led by this analogy, in this section we provide a complete characterization of Sturmian words which are prefix normal. We refer the interested reader to [15], Chap. 2, for a comprehensive treatment of Sturmian words. Here we briefly recall some facts we will need, starting with two equivalent definitions of Sturmian words.

Definition 5 (Sturmian words). *Let $w \in \{0,1\}^\omega$. Then w is called* Sturmian *if it is balanced and aperiodic.*

Definition 6 (Mechanical words). *Given two real numbers $0 \leq \alpha \leq 1$ and $0 \leq \tau < 1$, the* lower mechanical word *$s_{\alpha,\tau} = s_{\alpha,\tau}(1)\, s_{\alpha,\tau}(2) \cdots$ and the* upper mechanical word *$s'_{\alpha,\tau} = s'_{\alpha,\tau}(1)\, s'_{\alpha,\tau}(2) \cdots$ are given by*

$$s_{\alpha,\tau}(n) = \lfloor \alpha n + \tau \rfloor - \lfloor \alpha(n-1) + \tau \rfloor$$
$$s'_{\alpha,\tau}(n) = \lceil \alpha n + \tau \rceil - \lceil \alpha(n-1) + \tau \rceil \qquad (n \geq 1).$$

Then α is called the slope *and τ the* intercept *of $s_{\alpha,\tau}, s'_{\alpha,\tau}$. A word w is called* mechanical *if $w = s_{\alpha,\tau}$ or $w = s'_{\alpha,\tau}$ for some α, τ. It is called* rational mechanical *(resp.* irrational mechanical*) if α is rational (resp. irrational).*

Fact 1 (Some facts about Sturmian words [15])

1. *An infinite binary word is Sturmian if and only if it is irrational mechanical.*
2. *For $\tau = 0$, and α irrational, there exists a word c_α, called the* characteristic *word with slope α, s.t. $s_{\alpha,0} = 0c_\alpha$ and $s'_{\alpha,0} = 1c_\alpha$. This word c_α is a Sturmian word itself, with both slope and intercept α.*
3. *For two Sturmian words w, v with the same slope, we have $Fct(w) = Fct(v)$.*

4.1 From Flipext to Lazy-α-Flipext

Recall the operation flipext(w) defined above (Definition 4). We now define a different operation that, given a prefix normal word w, extends it by adding 0s as long as the minimum density of the resulting word is not smaller than $\delta(w)$, and only then adding a 1. We show that this operation preserves the prefix normality of the word. The operation lazy-α-flipext is then applied to show that, by extending a prefix normal word w of minimum density at least α, in the same way as we compute the upper mechanical word of slope α, we obtain an infinite prefix normal word with prefix w.

Definition 7. *Let $\alpha \in (0,1]$ and $w \in \mathcal{L}_{\text{fin}}$ with $\delta(w) \geq \alpha$. Define* lazy-α-flipext(w) *as the finite word $w0^k1$ where $k = \max\{j \mid \delta(w0^j) \geq \alpha\}$. We further define the infinite word $v = $ lazy-α-flipext$^\omega(w) = \lim_{i \to \infty}$ lazy-α-flipext$^{(i)}(w)$.*

Example 2. Let $w = 111$ and let $\alpha = \sqrt{2} - 1$, then lazy-α-flipext$(w) = 11100001$, since $\delta(1110000) = 3/7 \geq \alpha$ and $\delta(11100000) = 3/8 < \alpha$; and lazy-$\alpha$-flipext$^{(2)}(w) = 1110000101$, since $\delta(111000010) = 4/9 \geq \alpha$ and $\delta(1110000100) = 2/5 < \alpha$.

Lemma 4. *Let $\alpha \in (0,1]$. For every $w \in \mathcal{L}_{\text{fin}}$ with $\delta(w) \geq \alpha$, the word $v = $ lazy-α-flipext(w) is also prefix normal, with $\delta(v) \geq \alpha$.*

Corollary 1. *Let $\alpha \in (0,1]$, $w \in \mathcal{L}_{\text{fin}}$ with $\delta(w) \geq \alpha$. Then $v = $ lazy-α-flipext$^{\omega}(w)$ is an infinite prefix normal word and $\delta(v) = \alpha$.*

We now show that the word lazy-α-flipext$^{\omega}(1)$ coincides with the upper mechanical word $s'_{\alpha,0}$, which also implies that $s'_{\alpha,0}$ is prefix normal.

Lemma 5. *Fix $\alpha \in (0,1]$ and let $v = $ lazy-α-flipext$^{\omega}(1)$. Let $s = s'_{\alpha,0}$ be the upper mechanical word of slope α and intercept 0. Then $v = s$.*

Corollary 2. *For $\alpha \in (0,1]$, the word $s'_{\alpha,0}$ is prefix normal and $\delta(s'_{\alpha,0}) = \alpha$.*

The following theorem fully characterizes prefix normal Sturmian words.

Theorem 2. *A Sturmian word s of slope α is prefix normal if and only if $s = 1c_\alpha$, where c_α is the characteristic Sturmian word with slope α.*

5   Prefix Normal Words, Prefix Normal Forms, and Abelian Complexity

Given an infinite word w, the *abelian complexity* function of w, denoted ψ_w, is given by $\psi_w(n) = |\{pv(u) \mid u \in Fct(w), |u| = n\}|$, the number of Parikh vectors of n-length factors of w. A word w is said to have bounded abelian complexity if there exists a c s.t. for all n, $\psi_w(n) \leq c$. Note that a binary word is c-balanced if and only if its abelian complexity is bounded by $c + 1$. We denote the set of Parikh vectors of factors of a word w by $\Pi(w) = \{pv(u) \mid u \in Fct(w)\}$. Thus, $\psi_w(n) = \Pi(w) \cap \{(x, y) \mid x + y = n\}$. In this section, we study the connection between prefix normal words and abelian complexity.

5.1   Balanced and c-Balanced Words

Based on the examples in the introduction, one could conclude that any word with bounded abelian complexity can be turned into a prefix normal word by prepending a fixed number of 1s. However, consider the word $w = 01^\omega$, which is balanced, i.e. its abelian complexity function is bounded by 2. It is easy to see that $1^k w \notin \mathcal{L}$ for every $k \in \mathbb{N}$.

Sturmian words are precisely the words which are aperiodic and whose abelian complexity is constant 2 [18]. For Sturmian words, it is always possible to prepend a finite number of 1s to get a prefix normal word, as we will see next. Recall that for a Sturmian word w, at least one of $0w$ and $1w$ is Sturmian, with both being Sturmian if and only if w is characteristic [15].

Lemma 6. *Let w be a Sturmian word. Then*

1. *$1w \in \mathcal{L}$ if and only if $0w$ is Sturmian,*
2. *if $0w$ is not Sturmian, then $1^n w \in \mathcal{L}$ for $n = \lceil 1/(1 - \alpha) \rceil$.*

Lemma 7. *Let w be a c-balanced word. If there exists a positive integer n s.t. $1^n \notin \mathrm{Fct}(w)$, then the word $z = 1^{nc} w$ is prefix normal.*

In particular, Lemma 7 implies that any c-balanced word with infinitely many 0s can be turned into a prefix normal word by prepending a finite number of 1s, since such a word cannot have arbitrarily long runs of 1s. Note, however, that the number of 1s to prepend from Lemma 7 is not tight, as can be seen e.g. from the Thue-Morse word tm: the longest run of 1s in tm is 2 and tm is 2-balanced, but $11\mathsf{tm}$ is prefix normal, as will be shown in the next section (Lemma 10).

5.2   Prefix Normal Forms and Abelian Complexity

Recall that for a word w, $F_w^a(i)$ is the maximum number of a's in a factor of w of length i, for $a \in \{0, 1\}$.

Definition 8 (Prefix normal forms). *Let $w \in \{0, 1\}^\omega$. Define the words w' and w'' by setting, for $n \geq 1$, $w_n' = F_w^1(n) - F_w^1(n - 1)$ and $w_n'' = F_w^0(n) - F_w^0(n - 1)$. We refer to w' as the prefix normal form of w w.r.t. 1 and to w'' as the prefix normal form of w w.r.t. 0, denoted $\mathrm{PNF}_1(w)$ resp. $\mathrm{PNF}_0(w)$.*

In other words, $\mathrm{PNF}_1(w)$ is the sequence of first differences of the maximum-1s function F_w^1 of w. Similarly, $\mathrm{PNF}_0(w)$ can be obtained by complementing the sequence of first differences of the maximum-0s function F_w^0 of w. Note that for all n and $a \in \{0, 1\}$, either $F_w^a(n + 1) = F_w^a(n)$ or $F_w^a(n + 1) = F_w^a(n) + 1$, and therefore w' and w'' are words over the alphabet $\{0, 1\}$. In particular, by construction, the two prefix normal words allow us to recover the maximum-1s and minimum-1s functions of w:

Observation 1. *Let w be an infinite binary word and $w' = \mathrm{PNF}_1(w), w'' = \mathrm{PNF}_0(w)$. Then $P_{w'}(n) = F_w^1(n)$ and $P_{w''}(n) = n - F_w^0(n) = f_w^1(n)$.*

Lemma 8. *Let $w \in \{0, 1\}^\omega$. Then $\mathrm{PNF}_1(w)$ is the unique 1-prefix normal word w' s.t. $F_{w'}^1 = F_w^1$. Similarly, $\mathrm{PNF}_0(w)$ is the unique 0-prefix normal word w'' s.t. $F_{w''}^0 = F_w^0$.*

Example 3. For the two prefix normal forms and the maximum-1s and maximum-0s functions of the Fibonacci word $\mathsf{f} = 01001010010010100101 \cdots$, see Table 1.

Now we can connect the prefix normal forms of w to the abelian complexity of w in the following way. Given $w' = \mathrm{PNF}_1(w)$ and $w'' = \mathrm{PNF}_0(w)$, the number of Parikh vectors of k-length factors is precisely the difference in 1s in the prefix of length k of w' and of w'' plus 1. For example, Fig. 1 shows the prefix normal forms of the Fibonacci word. The vertical line at 5 cuts through points $(5, -1)$

Table 1. The maximum number of 0s and 1s ($F_f^0(n)$ and $F_f^1(n)$ resp.) for all $n = 1, \ldots, 20$ of the Fibonacci word f, and the prefix normal forms of f.

n	1	2	3	4	5	6	7	8	9	10	11	12	13	14	15	16	17	18	19	20
$F_f^0(n)$	1	2	2	3	4	4	5	5	6	7	7	8	9	9	10	10	11	12	12	13
$F_f^1(n)$	1	1	2	2	2	3	3	4	4	4	5	5	5	6	6	7	7	7	8	8
$\mathrm{PNF}_0(f)$	0	0	1	0	0	1	0	1	0	0	1	0	0	1	0	1	0	0	1	0
$\mathrm{PNF}_1(f)$	1	0	1	0	0	1	0	1	0	0	1	0	0	1	0	1	0	0	1	0

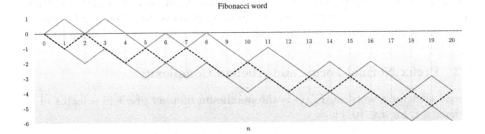

Fig. 1. The Fibonacci word (dashed) and its prefix normal forms (solid). A 1 corresponds to a diagonal segment in direction NE, a 0 in direction SE. On the x-axis the length of the prefix, on the y-axis, number of 1s minus number of 0s in the prefix.

and $(5, -3)$, meaning that there are two Parikh vectors of factors of length 5, namely $(2, 3)$ and $(1, 4)$. The Fibonacci word, being a Sturmian word, has constant abelian complexity 2. An example with unbounded abelian complexity is the Champernowne word, whose prefix normal forms are 1^ω resp. 0^ω.

Theorem 3. *Let $w, v \in \{0, 1\}^\omega$.*

1. *$\psi_w(n) = P_{w'}(n) - P_{w''}(n) + 1$, where $w' = \mathrm{PNF}_1(w)$ and $w'' = \mathrm{PNF}_0(w)$.*
2. *$\Pi(w) = \Pi(v)$ if and only if $\mathrm{PNF}_0(w) = \mathrm{PNF}_0(v)$ and $\mathrm{PNF}_1(w) = \mathrm{PNF}_1(v)$.*

Theorem 3 means that if we know the prefix normal forms of a word, then we can compute its abelian complexity. Conversely, the abelian complexity is the *width* of the area enclosed by the two words $\mathrm{PNF}_1(w)$ and $\mathrm{PNF}_0(w)$. In general, this fact alone does not give us the PNFs; but if we know more about the word itself, then we may be able to compute the prefix normal forms, as we will see in the case of the paperfolding word. We will now give two examples of the close connection between abelian complexity and prefix normal forms, using some recent results about the abelian complexity of infinite words.

1. The paperfolding word. The first few characters of the ordinary paperfolding word are given by

$$\mathfrak{p} = 0010011000110110001001110011011 \cdots$$

The paperfolding word was originally introduced in [11]. One definition is given by: $\mathfrak{p}_n = 0$ if $n' \equiv 1 \bmod 4$ and $\mathfrak{p}_n = 1$ if $n' \equiv 3 \bmod 4$, where n' is the unique odd integer such that $n = n'2^k$ for some k [16]. The abelian complexity function of the paperfolding word was fully determined in [16], giving the following initial values of $\psi_{\mathfrak{p}}(n)$, for $n \geq 1$: 2, 3, 4, 3, 4, 5, 4, 3, 4, 5, 6, 5, 4, 5, 4, 3, 4, 5, 6, 5, and a recursive formula for all values. The authors note that for the paperfolding word, it holds that if $u \in Fct(\mathfrak{p})$, then also $\overline{u^{\mathrm{rev}}} \in Fct(\mathfrak{p})$. This implies

$$F_{\mathfrak{p}}^1(n) = F_{\mathfrak{p}}^0(n) \text{ for all } n, \text{ and thus } \mathrm{PNF}_0(\mathfrak{p}) = \overline{\mathrm{PNF}_1(\mathfrak{p})}.$$

Moreover, from Theorem 3 we get that $F_{\mathfrak{p}}^1(n) = P_{\mathrm{PNF}_1(\mathfrak{p})}(n) = (\psi_{\mathfrak{p}}(n) + n - 1)/2$, and thus we can determine the prefix normal forms of $\mathfrak{p}$ as shown in Fig. 2. This same argument holds for all words with a symmetric property similar to the paperfolding word:

Lemma 9. *Let $w \in \{0,1\}^{\omega}$. If for all $u \in Fct(w)$, it holds that $\overline{u} \in Fct(w)$ or $\overline{u^{\mathrm{rev}}} \in Fct(w)$, then $F_w^1(n) = F_w^0(n)$ for all $n, \mathrm{PNF}_0(w) = \overline{\mathrm{PNF}_1(w)}$, and $F_w^1(n) = (\psi_w(n) + n - 1)/2$.*

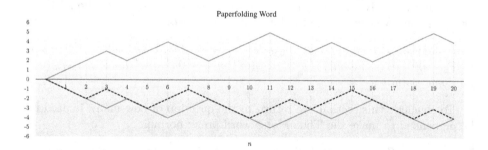

Fig. 2. The paperfolding word (dashed) and its prefix normal forms (solid).

2. Morphic images under the Thue-Morse morphism. The Thue-Morse word beginning with 0, which we denote by **tm**, is one of the two fixpoints of the Thue-Morse morphism μ_{TM}, where $\mu_{\mathrm{TM}}(0) = 01$ and $\mu_{\mathrm{TM}}(1) = 10$:

$$\mathbf{tm} = \mu_{\mathrm{TM}}^{(\omega)}(0) = 0110100110010110100101100110100 1 \cdots$$

The word **tm** has abelian complexity function $\psi_{\mathbf{tm}}(n) = 2$ for n odd and $\psi_{\mathbf{tm}}(n) = 3$ for $n > 1$ even [18]. Since **tm** fulfils the condition that $u \in Fct(\mathbf{tm})$ implies $\overline{u} \in Fct(\mathbf{tm})$, we can apply Lemma 9, and compute the prefix normal forms of **tm** as $\mathrm{PNF}_1(\mathbf{tm}) = 1(10)^{\omega}$ and $\mathrm{PNF}_0(\mathbf{tm}) = 0(01)^{\omega}$, see Fig. 3.

For the proof of the abelian complexity of **tm** in [18], the Parikh vectors were computed for each length, so we could have got the prefix normal forms directly (without Lemma 9). Moreover, a much more general result was given in [18]:

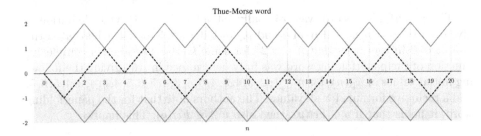

Fig. 3. The Thue-Morse word (dashed) and its prefix normal forms (solid).

Theorem 4 ([18]). *For an aperiodic infinite binary word w, $\psi_w = \psi_{\mathbf{tm}}$ if and only if $w = \mu_{TM}(w')$ or $w = 0\mu_{TM}(w')$ or $w = 1\mu_{TM}(w')$ for some word w'.*

The abelian complexity function does not in general determine the prefix normal forms, as can be seen on the example of Sturmian words, which all have the same abelian complexity function but different prefix normal forms. However, $\psi_{\mathbf{tm}}$ does, due to its values $\psi_{\mathbf{tm}}(n) = 2$ for n odd and $= 3$ for n even, and to the fact that both $F^1_{\mathbf{tm}}$ and $F^0_{\mathbf{tm}}$ have difference function with values from $\{0,1\}$: notice that the only pair of such functions with width 2 resp. 3 are the PNFs of $\mathbf{tm}$. Therefore, we can deduce the following from Theorem 4:

Corollary 3. *For an aperiodic infinite binary word w, $\mathrm{PNF}_1(w) = 1(10)^\omega$ and $\mathrm{PNF}_0 = 0(01)^\omega$ if and only if $w = \mu_{TM}(w')$ or $w = 0\mu_{TM}(w')$ or $w = 1\mu_{TM}(w')$ for some word w'.*

To conclude this section, we return to the question of how many 1s need to be prepended to make the Thue-Morse word prefix normal.

Lemma 10. *We have $11\mathbf{tm} \in \mathcal{L}$. This is minimal since $1\mathbf{tm}$ is not prefix normal.*

5.3 Prefix Normal Forms of Sturmian Words

Let w be a Sturmian word. As we saw in Sect. 4, the only 1-prefix normal word in the class of Sturmian words with the same slope α is the upper mechanical word $s'_{\alpha,0} = 1c_\alpha$.

Theorem 5. *Let w be an irrational mechanical word with slope α, i.e. a Sturmian word. Then $\mathrm{PNF}_1(w) = 1c_\alpha$ and $\mathrm{PNF}_0(w) = 0c_\alpha$, where c_α is the characteristic word of slope α.*

6 Prefix Normal Words and Lexicographic Order

In this section, we study the relationship between lexicographic order and prefix normality. Note that for coherence with the rest of the paper, in the definition of Lyndon words, necklaces, and prenecklaces, we use lexicographically *greater* rather than *smaller*. Clearly, this is equivalent to the usual definitions up to renaming characters.

Thus a finite *Lyndon word* is one which is lexicographically strictly greater than all of its conjugates: w is Lyndon if and only if for all non-empty u, v s.t. $w = uv$, we have $w >_{\text{lex}} vu$. A *necklace* is a word which is greater than or equal to all its conjugates, and a *prenecklace* is one which can be extended to become a necklace, i.e. which is the prefix of some necklace [15, 19]. As we saw in the introduction, in the finite case, prefix normality and Lyndon property are orthogonal concepts. However, the set of finite prefix normal words is included in the set of prenecklaces [6].

An infinite word is *Lyndon* if an infinite number of its prefixes is Lyndon [24]. In the infinite case, we have a similar situation as in the finite case. There are words which are both Lyndon and prefix normal: $10^\omega, 110(10)^\omega$; Lyndon but not prefix normal: $11100(110)^\omega$; prefix normal but not Lyndon: $(10)^\omega$; and neither of the two: $(01)^\omega$. Next we show that a prefix normal word cannot be lexicographically smaller than any of its suffixes. Let $shift_i(w) = w_i w_{i+1} w_{i+2} \cdots$ denote the infinite word v s.t. $w = w_1 \cdots w_{i-1} v$, i.e. v is w starting at position i.

Lemma 11. *Let $w \in \mathcal{L}_{\text{inf}}$. Then $w \geq_{\text{lex}} shift_i(w)$ for all $i \geq 1$.*

In the finite case, it is easy to see that a word w is a prenecklace if and only if $w \geq_{\text{lex}} v$ for every suffix v of w. This motivates our definition of infinite prenecklaces. The situation is the same as in the finite case: prefix normal words form a proper subset of prenecklaces.

Definition 9. *Let $w \in \{0, 1\}^\omega$. Then w is an infinite prenecklace if for all $i \geq 1$, $w \geq_{\text{lex}} shift_i(w)$. We denote by $\mathcal{P}_{\text{inf}}$ the set of infinite prenecklaces.*

Proposition 2. *We have $\mathcal{L}_{\text{inf}} \subsetneq \mathcal{P}_{\text{inf}}$.*

Another interesting relationship is that between lexicographic order and the prefix normal forms of an infinite word. In [17], two words were associated to an infinite binary word w, called $\max(w)$ resp. $\min(w)$, defined as the word whose prefix of length n is the lexicographically greatest (resp. smallest) n-length factor of w. It is easy to see that these words always exists. It was shown in [17]:[2]

Theorem 6 ([17]). *Let w be an infinite binary word. Then*

1. *w is (rational or irrational) mechanical with its intercept equal to its slope if and only if $0w \leq_{\text{lex}} \min(w) \leq_{\text{lex}} \max(w) \leq_{\text{lex}} 1w$, and*
2. *w is characteristic Sturmian if and only if $\min(w) = 0w$ and $\max(w) = 1w$.*

Lemma 12. *For $w \in \{0, 1\}^\omega$, $\text{PNF}_1(w) \geq_{\text{lex}} \max(w)$ and $\text{PNF}_0(w) \leq_{\text{lex}} \min(w)$.*

From Theorems 5 and 6 we get the following corollary:

---

[2] Note the different terminology in [17]: characteristic word → proper standard Sturmian, Sturmian → proper Sturmian, rational mechanical word → periodic Sturmian.

Corollary 4. *Let w be an infinite binary word. Then w is characteristic Sturmian if and only if $0w = \mathrm{PNF}_0(w) = \min(w)$ and $1w = \mathrm{PNF}_1(w) = \max(w)$.*

Acknowledgements. We wish to thank the participants of the Workshop on Words and Complexity (Lyon, February 2018), for interesting discussions and pointers, and to Péter Burcsi, who first got us interested in Sturmian words.

References

1. Amir, A., Chan, T.M., Lewenstein, M., Lewenstein, N.: On hardness of jumbled indexing. In: Esparza, J., Fraigniaud, P., Husfeldt, T., Koutsoupias, E. (eds.) ICALP 2014. LNCS, vol. 8572, pp. 114–125. Springer, Heidelberg (2014). https://doi.org/10.1007/978-3-662-43948-7_10
2. Blanchet-Sadri, F., Fox, N., Rampersad, N.: On the asymptotic abelian complexity of morphic words. Adv. Appl. Math. **61**, 46–84 (2014)
3. Blondin Massé, A., de Carufel, J., Goupil, A., Lapointe, M., Nadeau, É., Vandomme, É.: Leaf realization problem, caterpillar graphs and prefix normal words. Theoret. Comput. Sci. **732**, 1–13 (2018)
4. Burcsi, P., Cicalese, F., Fici, G., Lipták, Zs.: Algorithms for jumbled pattern matching in strings. Int. J. Found. Comput. Sci. **23**, 357–374 (2012)
5. Burcsi, P., Fici, G., Lipták, Zs., Ruskey, F., Sawada, J.: On combinatorial generation of prefix normal words. In: Kulikov, A.S., Kuznetsov, S.O., Pevzner, P. (eds.) CPM 2014. LNCS, vol. 8486, pp. 60–69. Springer, Cham (2014). https://doi.org/10.1007/978-3-319-07566-2_7
6. Burcsi, P., Fici, G., Lipták, Zs., Ruskey, F., Sawada, J.: On prefix normal words and prefix normal forms. Theoret. Comput. Sci. **659**, 1–13 (2017)
7. Cassaigne, J., Kaboré, I.: Abelian complexity and frequencies of letters in infinite words. Int. J. Found. Comput. Sci. **27**(05), 631–649 (2016)
8. Chan, T.M., Lewenstein, M.: Clustered integer 3SUM via additive combinatorics. In: Proceedings of the 47th Annual ACM Symposium on Theory of Computing (STOC 2015), pp. 31–40 (2015)
9. Cicalese, F., Lipták, Zs., Rossi, M.: Bubble-Flip - a new generation algorithm for prefix normal words. Theoret. Comput. Sci. **743**, 38–52 (2018)
10. Cunha, L.F.I., Dantas, S., Gagie, T., Wittler, R., Kowada, L.A.B., Stoye, J.: Fast and simple jumbled indexing for binary run-length encoded strings. In: 28th Annual Symposium on Combinatorial Pattern Matching (CPM 2017). LIPIcs, vol. 78, pp. 19:1–19:9 (2017)
11. Davis, C., Knuth, D.: Number representations and dragon curves, I, II. J. Recreat. Math. **3**, 133–149 and 161–181 (1970)
12. Fici, G., Lipták, Zs.: On prefix normal words. In: Mauri, G., Leporati, A. (eds.) DLT 2011. LNCS, vol. 6795, pp. 228–238. Springer, Heidelberg (2011). https://doi.org/10.1007/978-3-642-22321-1_20
13. Gagie, T., Hermelin, D., Landau, G.M., Weimann, O.: Binary jumbled pattern matching on trees and tree-like structures. Algorithmica **73**(3), 571–588 (2015)
14. Kaboré, I., Kientéga, B.: Abelian complexity of Thue-Morse word over a ternary alphabet. In: Brlek, S., Dolce, F., Reutenauer, C., Vandomme, É. (eds.) WORDS 2017. LNCS, vol. 10432, pp. 132–143. Springer, Cham (2017). https://doi.org/10.1007/978-3-319-66396-8_13

15. Lothaire, M.: Algebraic Combinatorics on Words. Cambridge University Press, Cambridge (2002)
16. Madill, B., Rampersad, N.: The abelian complexity of the paperfolding word. Discrete Math. **313**(7), 831–838 (2013)
17. Pirillo, G.: Inequalities characterizing standard Sturmian and episturmian words. Theoret. Comput. Sci. **341**(1–3), 276–292 (2005)
18. Richomme, G., Saari, K., Zamboni, L.Q.: Abelian complexity of minimal subshifts. J. London Math. Soc. **83**(1), 79–95 (2011)
19. Ruskey, F., Savage, C., Wang, T.: Generating necklaces. J. Algorithms **13**(3), 414–430 (1992)
20. Ruskey, F., Sawada, J., Williams, A.: Binary bubble languages and cool-lex order. J. Comb. Theory Ser. A **119**(1), 155–169 (2012)
21. Sawada, J., Williams, A.: Efficient oracles for generating binary bubble languages. Electron. J. Comb. **19**(1), P42 (2012)
22. Sawada, J., Williams, A., Wong, D.: Inside the Binary Reflected Gray Code: Flip-Swap languages in 2-Gray code order (2017, unpublished manuscript)
23. Sloane, N.J.A.: The On-Line Encyclopedia of Integer Sequences. http://oeis.org
24. Siromoney, R., Mathew, L., Dare, V., Subramanian, K.: Infinite Lyndon words. Inf. Process. Lett. **50**, 101–104 (1994)
25. Turek, O.: Abelian complexity function of the Tribonacci word. J. Integer Seq. **18** (2015). Article 15.3.4

Priority Scheduling in the Bamboo Garden Trimming Problem

Mattia D'Emidio[1]([✉]), Gabriele Di Stefano[1], and Alfredo Navarra[2]

[1] Department of Information Engineering, Computer Science and Mathematics, University of L'Aquila, Via Vetoio, 67100 L'Aquila, Italy
{mattia.demidio,gabriele.distefano}@univaq.it
[2] Department of Mathematics and Computer Science, University of Perugia, Via Vanvitelli 1, 06123 Perugia, Italy
alfredo.navarra@unipg.it

Abstract. We consider the *Bamboo Garden Trimming* (BGT) problem introduced in [Gąsieniec et al., SOFSEM'17]. The problem is NP-hard due to its close relationship to *Pinwheel scheduling*. The garden with n bamboos is an analogue of a system of n machines which have to be attended (e.g., serviced) with different frequencies. During each day, bamboo b_i grows an extra height h_i, for $i = 1, \ldots, n$ and, on the conclusion of the day, at most one bamboo is cut all its current height. The goal is to design a perpetual schedule of cuts to keep the height of the tallest ever bamboo as low as possible.

Our contribution is twofold, and is both theoretical and experimental. In particular, we focus on understanding what we call *priority schedulings*, i.e. cutting strategies where priority is given to bamboos whose current height is above a threshold greater than or equal to $H = \sum_{i=1}^{n} h_i$. Value H represents the total daily growth of the system and it is known that one cannot keep bamboos in the garden below this threshold indefinitely.

We prove that for any distribution of integer growth rates $h_1, \ldots, h_n$ and any priority scheduling, the system stabilises in a fixed cycle of cuts. Then, we focus on the so-called ReduceMax strategy, a greedy priority scheduling which each day cuts the tallest bamboo, regardless of the growth rates distribution. ReduceMax is known to provide a $O(\log n)$-approximation, w.r.t. the lower bound H. We prove that, if ReduceMax stabilises in a round-robin type cycle, then it guarantees 2-approximation. We conjecture that ReduceMax is 2-approximating for the BGT problem, hence we conduct an extended experimental evaluation, on all bounded in size integer instances of BGT, to support our conjecture and to compare ReduceMax with other relevant scheduling algorithms. Our results show that ReduceMax provides 2-approximation

The work has been supported in part by the European project "Geospatial based Environment for Optimisation Systems Addressing Fire Emergencies" (GEO-SAFE), contract no. H2020-691161 and by the Italian National Group for Scientific Computation GNCS-INdAM.

B. Catania et al. (Eds.): SOFSEM 2019, LNCS 11376, pp. 136–149, 2019.
https://doi.org/10.1007/978-3-030-10801-4_12

in such instances, and it always outperforms other considered strategies, even those for which better worst case approximation guarantees have been proven.

1   Introduction

We consider a perpetual scheduling problem in which n machines, denoted later as *bamboos*, need to be attended with possibly *known* and likely different frequencies. In other words some machines may have to be attended more often than others. This problem was proposed and studied in [13] under the name of *Bamboo Garden Trimming* (BGT) problem. We are given a collection (garden) G of n bamboos b_1, b_2, ..., b_n along with respective daily growth rates h_1, h_2, ..., h_n greater than 0. The authors in [13] assume that initially the height of each bamboo is set to zero, whereas in this paper we allow bamboos to start with arbitrary heights.

The robotic gardener maintaining the garden trims one bamboo per day to height zero according to some predefined schedule. The main goal in BGT is to design perpetual schedules of cuts which keep the height of the tallest ever bamboo as low as possible. The gardener is allowed to cut exactly one (of their choice) bamboo at the end of each day which corresponds to one round in the schedule. The problem, while of independent combinatorial interest, originates from perpetual testing of virtual machines in cloud systems [1]. In such systems, the frequency in which virtual machines are tested for undesirable symptoms varies depending on the importance of dedicated cloud operational mechanisms.

The problem considered here is also a close relative of several classical algorithmic problems which focus on *monitoring* including *Perpetual Graph Exploration* [14,18], *Art Gallery Problem* [21] and its dynamic extension *k-Watchmen Problem* [24]. In the work on *Patrolling* [11,12], the studies focus on monitoring a set of points with the same frequency of attendance, whereas in [10] the frequency may vary.

Our paper, similarly to [13], focuses on the case where each bamboo has its own attendance factor, which makes it related to *periodic scheduling* [23], several variants of *Pinwheel* related problems [7,8,15,17] including *periodic Pinwheel* problem [16,19] and *Pinwheel scheduling* [22], as well as the concept of *P-fairness* in sharing multiple copies of some resource among various tasks [2,3]. We point out here that the NP-hardness of BGT results from the intractability of Pinwheel scheduling proved in [4,20].

In related research on minimising the maximum occupancy of a buffer in a system of n buffers, the usual setting is a game between the player and the adversary [5,6,9]. The adversary decides how a fixed total increase of data in each round is distributed among the buffers and tries to maximise the maximum occupancy of a buffer. The player decides which buffer (or buffers) should be emptied next and tries to minimise the maximum buffer size. The upper bounds developed in this more general context can be translated into upper bounds for our BGT problem. However, our aim is to derive tighter bounds for the case where the knowledge of growth rates is not exploited, hence adopting strategies able to deal with input data being either partially or entirely unknown.

Probably the most natural algorithm to keep the elevation of the bamboo garden low is the greedy approach of always cutting the currently tallest bamboo. This method, that is agnostic w.r.t. growth rates, was first coined in [1] under the name of ReduceMax and further studied in [13]. This strategy was also considered independently (and under a different name) in the adversarial setting of buffer minimisation problems [6]. Another method studied in [13] is the ReduceFastest$_2$ approach, in which the fastest growing bamboo is cut but only among those having height above the threshold $2H$, where $H = \sum_{i=1}^{n} h_i$.

Value H represents the total daily growth of the system of bamboos and it is known [13] that one cannot keep bamboos below this threshold level. From [13] we have that ReduceFastest$_2$ provides a constant approximation to BGT, i.e., none of the bamboos grows above height $4H$. However, its applicability depends on the knowledge of value H and at least an ordering of the bamboos w.r.t. their growth rates. More interestingly, ReduceMax does not require such a knowledge but surprisingly it is not so well understood. While there are insights that it should perform better than ReduceFastest$_2$, the only upper bound known for the maximum height of bamboos is $O(H \cdot \log n)$ [13].

In this paper, the main contribution is twofold, and it refers to both theoretical and experimental studies on the BGT problem. In particular, we focus on better understanding of what we call *priority schedulings* that operate on any 'reasonable' (involving all bamboos) strategy of cuts in which priority is given to bamboos with the current height above a threshold greater than or equal to H. Both ReduceMax and ReduceFastest$_2$ fall into the priority category. Moreover, we require our scheduling strategies to be fully deterministic. Hence, in case of ties, i.e., when two or more bamboos are eligible to be cut w.r.t. the considered strategy, we select the bamboo with the biggest index.

We first prove that for any distribution of integer growth rates $h_1, \ldots, h_n$ and any priority scheduling, the system stabilises in a fixed cycle of cuts, eventually. However, the time needed to converge depends on the initial heights of bamboos. Then, we show that, whenever ReduceMax stabilises in a round-robin type cycle, it guarantees 2-approximation for BGT. Finally, we conduct extended experiments to compare ReduceMax with other strategies, including ReduceFastest$_2$. Note that [1] contains some experiments on ReduceMax, focusing only on selected distributions of bamboo growth rates. In contrast, here the focus is on all possible distributions of bounded size. Our experiments show ReduceMax being 2 approximating in all considered instances, and provide evidence that ReduceMax outperforms all other tested strategies. We conclude by conjecturing the following.

Conjecture 1. Algorithm ReduceMax is 2-approximating for the BGT problem.

2   Notation

We are given a collection $\{b_1, b_2, \ldots, b_n\}$ of n bamboos (a.k.a. a *garden*) along with respective daily growth rates $h_1, h_2, \cdots, h_n$. We assume that each h_i is a positive integer, for any $1 \leq i \leq n$, and call a *configuration* C^t the sequence $(\ell_1^t, \ell_2^t, \ldots, \ell_n^t)$ of the bamboos' heights at a given day t. Any configuration $C^t = (\ell_1^t, \ell_2^t, \ldots, \ell_n^t)$ is determined by a *growth mechanism* applied on its predecessor $C^{t-1} = (\ell_1^{t-1}, \ell_2^{t-1}, \ldots, \ell_n^{t-1})$, i.e., for any $1 \leq i \leq n$, we have $\ell_i^t = \ell_i^{t-1} + h_i$. The only exception to this behavior is what we call a *trimming* operation. In particular, we say a bamboo b_i is *trimmed* (equivalently *cut*) at a given day t if the height ℓ_i^t is reduced to zero and hence, trivially, $\ell_i^{t+1} = h_i$. For the sake of simplicity, in what follows we omit t from all notations when the number of the day is clear from the context. Finally, given a configuration C, we denote by $V(C)$ the *volume* of configuration C which is the sum of all bamboos' heights in C, i.e., $V(C) = \sum_{i=1}^{n} \ell_i$.

An input instance I to the BGT problem is a set of growth rates $\{h_i\}_{1 \leq i \leq n}$, complemented by the initial configuration C^0. We are interested in designing perpetual schedules of cuts which allow to keep the tallest bamboo in the garden as low as possible. In particular, we assume that at most one trimming operation can take place every day. Therefore, we aim at designing an algorithm $\mathcal{A}$ that, for a given input instance I, computes a *perpetual schedule* $\mathcal{A}(I) = (i_1, i_2, \ldots)$, i.e. a sequence of indices $i_j \in \{0, 1, 2, \ldots, n\}$ that determines, for any day $t > 0$, the bamboo to be trimmed, unless $i_j = 0$ when no bamboo is cut. In other words, a schedule of this kind defines an ordered sequence of trimming operations on the bamboos. In what follows, we call an algorithm determining perpetual schedules a *perpetual scheduling*, or simply *scheduling* algorithm.

Given an input instance I and a perpetual scheduling $\mathcal{S}$, an *execution* $E(I, \mathcal{S})$ is the sequence $(C^0, C^1, \ldots)$ obtained by applying the schedule computed by $\mathcal{S}$ on C^0. Moreover, given an execution E and a configuration C, we denote by $M(E)$ ($M(C)$, resp.) the maximum height reached by a bamboo in E (C, resp.). Finally, we denote by $H = \sum_{i=1}^{n} h_i$ the sum of the growth rates. It is known that no algorithm can compute a schedule that keeps the heights of all the bamboos below H indefinitely (i.e. such that $M(E) < H$) [13].

3 Theoretical Results

In this section we first introduce a formal definition of priority schedulings and analyse their performance (see Sect. 3.1). Then, we focus on the strategy ReduceMax and show it provides 2-approximation for BGT under specific assumptions (see Sect. 3.2).

3.1 Priority Schedulings

Let $\mathcal{C}$ be the set of any configuration of n bamboos.

Definition 1. *An oblivious scheduling $\sigma : \mathcal{C} \to \{0, 1, \ldots, n\}$ is a function which for any configuration of heights in $\mathcal{C}$ returns an index i of the bamboo to be cut, and $i = 0$ means that none of the bamboos is scheduled to be cut.*

In other words, in oblivious schedulings the next cut is solely based on the current configuration, without exploiting any knowledge about past cuts.

Definition 2. *A configuration* $C = (\ell_1, \ell_2, \ldots, \ell_n)$ *is said to be* ordered *whenever* $i < j$ *implies* $h_i \geq h_j$.

The above implies the order of the sequence in C reflects a non-increasing ordering of the growth rates, i.e. ℓ_1 is the height of the bamboo with the biggest growth rate.

Definition 3. *An* ordered oblivious scheduling $\sigma : \mathcal{O} \to \{0, 1, \ldots, n\}$ *is an oblivious scheduling where* $\mathcal{O} \subset \mathcal{C}$ *is the set of ordered configurations in* $\mathcal{C}$.

Definition 4. *Given a threshold* $\tau \geq H$ *and any (ordered) configuration* $C \in \mathcal{C}$, *let* L *be the set of indices of all bamboos whose height is strictly greater than* τ. *An oblivious scheduling* σ_τ *is a (ordered)* τ-priority scheduling *if and only if* $L \neq \emptyset$ *implies* $\sigma_\tau(C) \in L$.

In the remainder of the paper, a (ordered) τ-*priority scheduling* will be simply referred to as a *priority scheduling* when the ordering and the value of τ are either clear from the context or irrelevant. Clearly, ReduceFastest$_2$ is an ordered $2H$-priority scheduling as it cuts only bamboos above threshold $2H$ on the basis of the ordering of the growth rates. For ReduceMax, instead, we can prove it is a priority scheduling regardless the ordering of the bamboos' growth rates. This means it can be applied to a wider range of input configurations, not only to ordered ones as required by ReduceFastest$_2$.

Fact 1. ReduceMax *is a priority scheduling.*

Proof. Given any threshold $\tau \geq H$ (as required by Definition 4) each day ReduceMax cuts the tallest bamboo. This includes also the case when there are bamboos taller than τ. This in turn means that ReduceMax gives priority to bamboos higher than τ, if any.   □

The next lemma gives an upper bound on the volume the garden may reach in a priority scheduling.

Lemma 1 (Upper Bound on Volume). *Given a* τ-priority scheduling σ_τ *and an input* I. *There exists a time* t, *s.t., for any* $t' > t$ *we have* $V(C^{t'}) \leq n\tau$, *where* $C^{t'} \in E(I, \sigma_\tau)$.

Proof. First assume that $M(E(I, \sigma_\tau)) > \tau$, as otherwise $V(C) \leq n\tau$, for each $C \in E(I, \sigma_\tau)$, and the lemma holds. If we also assume $M(C^0) > \tau$ we are able to prove that, within finite time, a configuration C with $M(C) \leq \tau$ is reached by applying σ_τ. In particular, note that for as long as there are bamboos having height greater than τ, the total volume decreases each day, being $\tau \geq H$, of at least 1. Thus there must exist a time t when eventually $M(C^t) \leq \tau$, since the volume cannot decrease indefinitely. At this time, we have $V(C^t) \leq n\tau$.

Therefore, let $t' > t$ be the first day after time t such that $C^{t'}$ has a bamboo having height greater than τ. Clearly, we have $V(C^{t'}) < V(C^{t'-1}) - \tau + H$, as

σ_τ cuts a bamboo with height greater than τ and, at the end of the day, all bamboos grow by H in total. Moreover, by hypothesis $V(C^{t'-1}) \leq n\tau$, since $\ell_i \leq \tau$, for each $\ell_i \in C^{t'-1}$. Hence, $V(C^{t'}) < (n-1)\tau + H \leq n\tau$, as $H \leq \tau$. Now, let us focus on $V(C^{t'+1})$. Since in $C^{t'}$ the scheduling algorithm cuts a bamboo having height bigger than τ, and since the sum of daily growths is exactly H, it follows that $V(C^{t'+1}) < V(C^{t'})$, and this is true for any following configuration until day t'' where $\ell_i \leq \tau$ for any $\ell_i \in C^{t''}$. Notice that in t'' we fall in the same set of hypotheses as of day t, then by repeating the reasoning, the claim follows. $\qquad\square$

By Lemma 1 we derive the next corollary, which guarantees that any priority scheduling stabilises into a cycle of finite length, i.e., that the sequence of cuts becomes periodic.

Corollary 1 (Existence of a Cycle). *Given a τ-priority scheduling σ_τ and an input I. There exist two days t and t', where $t < t'$, $C^t = C^{t'}$, and $C^t, C^{t'} \in E(I, \sigma_\tau)$.*

Proof. The claim follows from Lemma 1, since the number of configurations having volume at most $n\tau$ is finite. $\qquad\square$

By the above corollary, we know that any priority scheduling stabilises in a cycle, eventually. In fact, by Definition 1, a priority scheduling (which is an oblivious scheduling) computes the same trimming operation if the same configuration shows up again. We remind that, in case of ties, the bamboo having the biggest index is cut. In this paper, we study the properties of such cycles, as they represent the perpetual trimming process that have to be executed indefinitely. Hence, we tend to disregard configurations preceding L_E where, given an execution E, then L_E denotes its periodic part (i.e., the sequence of configurations in the cycle). In particular, to better characterise such cycles, we introduce the following notation:

- $L_E = (C_1 = C^{t'}, C_2 = C^{t'+1}, C_3, \ldots, C_{\lambda_E})$, where $C^{t'}$ is the first configuration belonging to L_E, reached from C^0, and $\lambda_E = |L_E|$ is the length of the cycle, i.e., the number of configurations in L_E;
- l^t is the height of the bamboo cut in $C_t \in L_E$. We assume $l^t = 0$ if no bamboo is cut;
- c_i is the number of times bamboo b_i is cut in L_E, for each $i = 1, \ldots, n$, which is equal to the number of relative maximum heights reached by b_i in the cycle;
- m_i^j is the relative maximum height reached by b_i in the cycle just before the j-th cut, for $i = 1, \ldots, n$ and $j = 1, \ldots, c_i$. Note that by definition, during day t, $l^t = m_i^j$ for some values of i and j;
- $M_i = \sum_{j=1}^{c_j} m_i^j$ is the sum of the relative maximum heights reached by b_i.

The next lemma provides a very useful property of the cuts that are performed within a cycle of a priority scheduling. In particular, we can show that the average value of heights reached by bamboos in L_E, just before a cut, is always H.

Lemma 2 (On Average Height of Cuts within a Cycle). *Given an execution E of a priority scheduling, then $\frac{1}{\lambda_E} \sum_{t=1}^{\lambda_E} l^t = H$.*

Proof. Let $\Delta_V^t = V(C_t) - V(C_{t-1})$ be the change of the volume from C_{t-1} to C_t, for any $t = 1, \ldots \lambda_E$. $C_0 \equiv C_{\lambda_E}$, since cycle L_E exists by Corollary 1. Thus we have $\Delta_V^t = H - l^t$, because at day t the bamboo of height l^t is cut and all bamboos grow by H in total. Now, since $\sum_{t=1}^{\lambda_E} \Delta_V^t$ must be equal to zero, as configurations in L_E come periodically, we get: $\sum_{t=1}^{\lambda_E} \Delta_V^t = \sum_{t=1}^{\lambda_E} (H - l^t) = \lambda_E \cdot H - \sum_{t=1}^{\lambda_E} l^t = 0$. Therefore, we have $\sum_{t=1}^{\lambda_E} l^t = \lambda_E \cdot H$, and the claim holds. □

In what follows, we characterise the length of L_E in terms of the maximum height reached by any bamboo having growth rate equal to h_i.

Lemma 3 (Characterisation of λ_E). *Given an execution E of a priority scheduling and an index $i \in \{1, 2, \ldots, n\}$, then $\lambda_E = \frac{M_i}{h_i}$.*

Proof. Let $p_i^j = m_i^j / h_i$. The value of p_i^j is an integer representing the number of days between the j-th cut of b_i and the previous one. Then $M_i = \sum_{j=1}^{c_j} m_i^j = \sum_{j=1}^{c_j} h_i \cdot p_i^j = h_i \cdot \sum_{j=1}^{c_j} p_i^j = h_i \cdot \lambda_E$. □

An immediate consequence is the following corollary.

Corollary 2 (Properties of Rates and Maximum Heights). *Given an execution E of a priority scheduling, and two indices $i, j \in \{1, 2, \ldots, n\}$ then $\frac{M_i}{M_j} = \frac{h_i}{h_j}$.*

3.2 ReduceMax Scheduling

In this section we focus on ReduceMax. By Fact 1, we have already shown that ReduceMax is a priority scheduling, regardless of the chosen threshold $\tau \geq H$ required by Definition 4. It follows that ReduceMax inherits all results obtained in the previous section for priority schedulings. In particular, given an execution E obtained by applying ReduceMax, let l^t be the height of the bamboo cut in $C_t \in L_E$, i.e., $l^t = M(C_t)$, and let $\overline{M_E} = \frac{1}{\lambda_E} \sum_{t=1}^{\lambda_E} l^t$ denote the average of the maximum heights reached by bamboos in L_E. Then, for ReduceMax, Lemma 2 can be reformulated as follows.

Corollary 3 (On Average Height of Cuts of ReduceMax). *Let E be an execution of ReduceMax. Then $\overline{M_E} = H$.*

A natural intuitive property is provided in the next lemma.

Lemma 4 (On the Amounts of Cuts within a Cycle). *Let E be an execution of* ReduceMax. *For any two indices $i, j \in \{1, 2, \ldots, n\}$ such that $h_i \geq h_j$ we have $c_i \geq c_j$.*

Proof. By contradiction, assume that $c_i < c_j$. Then, in cycle L_E there must be at least two cuts of b_j between two consecutive cuts of b_i. However, since $h_i \geq h_j$, then b_j after the first cut will grow less than b_i and it will never reach b_i before its second cut, which is a contradiction. □

A direct consequence of Lemma 4 is that, if two bamboos exhibit the same growth rate they also have the same number of cuts in the cycle.

Corollary 4 (Sufficient Condition for Same Number of Cuts). *If E is an execution of* ReduceMax, *$h_i = h_j$ implies $c_i = c_j$.*

Let $m_i = \max\{m_i^j : j = 1, 2, \ldots, c_i\}$ be the maximum height reached by bamboo b_i in a cycle L_E of ReduceMax. We show that $c_i = 1$ suffices to guarantee $m_i < 2H$. Basically, the next lemma guarantees that the approximation factor of ReduceMax reduces from $O(\log n)$ [13] to 2 if all bamboos are cut only once within the cycle, i.e., when the cycle of ReduceMax is equivalent to the round-robin strategy.

Lemma 5 (Sufficient Condition for Bounded Maximum). *Let E be an execution of* ReduceMax. *Then, $c_i = 1$ for some $i \in \{1, 2, \ldots, n\}$ implies $m_i < 2H$.*

Proof. First of all, notice that $c_i = 1$ suffices to immediately obtain $m_i = \lambda_E \cdot h_i$. Moreover, by Corollary 3 we have $\overline{M_E} = H$, i.e., that $\lambda_E \cdot H - \sum_{t=1}^{\lambda_E} l^t = 0$.

Now, we know that b_i is cut only once and, as L_E is periodic, and C_1 is equal to the configuration where the height of bamboo b_i is h_i. Thus each term l^t is trivially lower bounded by the height of bamboo b_i at day t, i.e., $l^t \geq t \cdot h_i$. Hence, $\lambda_E \cdot H - \sum_{t=1}^{\lambda_E} t \cdot h_i \geq 0$. Therefore $\lambda_E \cdot H \geq \sum_{t=1}^{\lambda_E} t \cdot h_i \geq \frac{\lambda_E \cdot (\lambda_E + 1)}{2} \cdot h_i$ which implies $\frac{\lambda_E^2}{2} \cdot h_i + \frac{\lambda_E}{2} \cdot h_i - \lambda_E \cdot H \leq 0$, i.e. $\lambda_E \cdot h_i + h_i - 2 \cdot H \leq 0$. This in turn implies $\lambda_E \cdot h_i < 2H$ and hence $m_i < 2H$. □

If $c_i = 1$, for some $i = 1, 2, \ldots n$, then similarly to the proof of Lemma 5, we have a limit on the value of λ_E, summarised by the following corollary.

Corollary 5 (Sufficient Condition for Bounded Length of Cycles). *Let E be an execution of* ReduceMax. *Then $c_i = 1$ for some $i \in \{1, 2, \ldots, n\}$ implies $\lambda_E \leq \frac{2 \cdot H}{h_i} - 1$.*

Reminding that, when dealing with ReduceMax, it is not required to have bamboos ordered according to their growth rates, let b_x be the bamboo having the biggest growth rate. In what follows, we refer to the executions having $c_x = 1$ as *minimum-cycle executions*, since $c_x = 1$ implies that $c_i = 1$ for each bamboo b_i, and then the execution exhibits a minimum-length cycle, that is $\lambda_E = n$. Note that λ_E cannot be smaller than n, as each bamboo must be cut at least once. These observations are summarised in the next corollary.

Corollary 6 (Characterisation of Minimum-cycle Executions). *Let E be an execution of* ReduceMax. *Then $c_x = 1 \iff \lambda_E = n$.*

Finally, we can characterise the minimum-cycle executions of ReduceMax.

Corollary 7 (On the Maximum Height in Minimum-cycle Executions). *If E is a minimum-cycle execution of* ReduceMax *then $M(L_E) < 2H$.*

Proof. In a minimum-cycle execution the maximum height reached by a bamboo b_i is exactly $m_i = nh_i$. Hence, the maximum height reached during an execution by any bamboo is due to h_x and, in particular, is given by $M(L_E) = m_x = n \cdot h_x$. Hence the thesis is an immediate consequence of Lemma 5, as $c_x = 1$.   □

Corollary 7 is of particular interest, since by Conjecture 1 we basically state that $M(E) < 2H$ for all the executions E obtained via ReduceMax. Hence, Corollary 7 represents a partial proof to our conjecture, holding for the case of minimum-cycle executions. In the next section, we experimentally show the validity of our conjecture on a large set of inputs.

4   Experimental Results

In what follows, we provide an extensive experimental evaluation of four priority scheduling strategies. As already pointed out, we require our scheduling strategies to be fully deterministic. Hence, in case of ties, i.e., when two or more bamboos are eligible to be cut w.r.t. the considered strategy, we select the bamboo having the biggest index. The considered strategies are:

- ReduceMax (RMax, for short): This is the heuristic which performance is the most relevant to our studies. In particular, in [13], based on [6], a $O(\log n)$-approximation guarantee has been established. However, we are interested in determining whether such a bound is tight in practice, i.e., whether the logarithmic factor is an accurate estimation. The strategy works in a greedy fashion by cutting each day the tallest bamboo.
- ReduceFastest$_2$ (RFast$_2$, for short): This is another greedy strategy introduced in [13]. It guarantees 4-approximation. However, this method requires to order the input configurations according to the non-increasing order of the bamboos' growth rates. In fact, each day it cuts the fastest growing bamboo (the one having the biggest h_i) among those whose height exceeds threshold $2H$. If none of the bamboos is taller than $2H$, no cuts are performed.
- ReduceFastest$_1$ (RFast$_1$, for short): This is a variant of RFast$_2$, introduced here for the first time, obtained by decreasing the threshold from $2H$ to H, and by allowing the cut of the fastest growing bamboo also below the threshold. Basically, if none of the bamboos has reached height H, the fastest growing bamboo is cut. This is a natural extension of RFast$_2$, and the aim of defining it is to check whether there are chances to obtain better performance w.r.t. RFast$_2$ and RMax. Note that RFast$_1$ is an ordered H-priority scheduling.

– ReduceMin (RMin, for short): This priority algorithm cuts each day the shortest bamboo, giving priorities to those above H. The aim of defining this
strategy is to evaluate performance of counter-intuitive methods, i.e. to see
whether even in an adversarial approach one may obtain acceptable perfor-
mances. RMin is a H-priority scheduling.

To evaluate the behaviour of the scheduling strategies w.r.t. different metrics,
we implemented them within SageMath (v.7.5) under GNU/Linux and performed different types of experiments. For a fair comparison w.r.t. [13], in the
experiments we assume the heights of all bamboos are initially null, that is
$C^0 = (0, 0, \ldots, 0)$. Moreover, as our experiments involve ordered priority schedulings (that is RFast$_1$ and RFast$_2$), we consider ordered configurations. In what
follows, we fix parameter H and consider all possible instances of n bamboos
whose growth rates sum up to H. Such instances are generated by considering,
for a given $H \in \mathbb{N}$, all integer partitions of H. Hence, clearly $n \in \{1, 2, \ldots, H\}$,
where for $n = 1$ there is only one bamboo having growth rate equal to H while for
$n = H$ all bamboos have unitary growth rate (e.g., for $H = 3$ we have instances
[3], [2, 1] and [1, 1, 1]). Note that, we selected values of H ranging from 5 to 25.
This choice was dictated by the fact that the number of integer partitions, and
therefore instances to consider, grows very fast as H increases and hence too
large values of H induce a computationally prohibitive number of simulations.
It is known that the number of integer partitions $p(k)$ of a natural number k
grows asymptotically as $p(k) \approx \frac{1}{4k\sqrt{3}} e^{\pi \sqrt{\frac{2k}{3}}}$ as k approaches infinity [25].
    The results of the application of all the considered scheduling algorithms
on instances induced by the integer partitions $H = 25$ are shown in Fig. 1,
resp. where panels on the left show how the maximum $M(E)$, obtained for all
configurations and strategies, varies as a function of n. Reference lines $2H$ and
$4H$ are plotted to emphasise performance of the strategies. On the right, instead,
we show how the achieved maximum λ_E changes as a function of n, given that
all strategies are guaranteed to stabilise into cycles. Note that we report the
maximum measured value as we can have many instances having the same n.
Notice also that the results for $H \in [5, 10, 15, 20]$ lead to similar considerations
w.r.t. those provided below, hence we omit them.
    An alternative view of the results of the experiments for $H = 25$ is given in
Fig. 2. It shows how the obtained values of $M(E)$ and λ_E are distributed over
the considered instances. In detail, each value on the x-axis simply represents
one instance, to which we associate the corresponding values of $M(E)$ and λ_E on
the y-axis. Instances are sorted on the x-axis in non-decreasing order according
to their values on the y-axis to highlight the amount of inputs providing a same
value of $M(E)$ and λ_E.

On the Maximum Height. The main and the most interesting outcome of this
experiment is that, notwithstanding the $O(\log n)$ approximation factor, RMax
exhibits properties of a 2-approximation, i.e. $M(E) \leq 2H$. Another surprising
evidence is that also RFast$_1$ is always below $2H$ in terms of $M(E)$. However,
RFast$_1$ seems to follow an asymptotic trend toward $2H$ as n increases. This

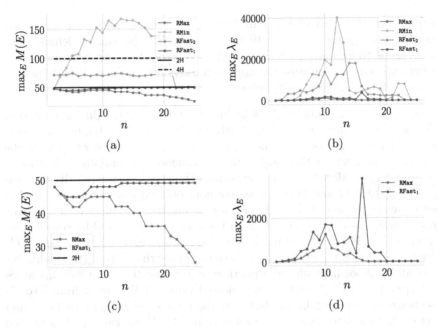

Fig. 1. Experiments conducted on all possible ordered instances obtained by setting $H = 25$ and hence considering n varying in $\{1, 2, \ldots, 25\}$. Panels (a) and (c) refer to maximum $M(E)$, whereas panels (b) and (d) refer to maximum λ_E. Panels (c) and (d) show strategies that, experimentally, exhibit 2-approximation.

suggests that, for these strategies, there could be a way to prove the worst case 2-approximation. Concerning RFast$_2$, from the literature, we know that $M(E)$ is guaranteed to be below $4H$ and this is confirmed by our tests. However, we observe also that $M(E)$ is always above $2H$, since no actions are performed by RFast$_2$ when there are no bamboos having height above $2H$. Still, the strategy exhibits a rather uniform behaviour, never overpassing $3H$. This suggests that perhaps also the bound of $4H$ guaranteed for RFast$_2$ is an overestimation of the true bound. Finally, regarding RMin, for low/high values of n (see Fig. 1), it behaves better than RFast$_2$, whereas for higher values of H its performance gets worse. Still, it seems to exhibit a constant approximation as well, with $M(E) \leq 7H$. As a final remark, in Fig. 2a we observe that, for all strategies, values of maximum $M(E)$ tend to have small variance among all instances having a same H, with curves assuming rather similar (flat) trends, and values always being very close to the average. The only exception is algorithm RMin, whose values of $M(E)$ are quite different across instances having a same H. Moreover, RMax achieves values of $M(E)$ that are far better than all other strategies, including RFast$_1$, being below $2H$ and, in some cases, below $\frac{3}{2}H$. This is even more evident in Fig. 1c where we focus on RMax and RFast$_1$ where $M(E)$ is experimentally always below $2H$.

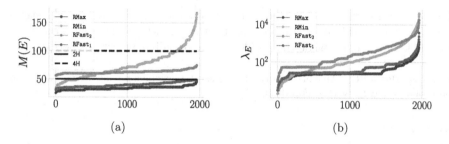

Fig. 2. Distribution of values of $M(E)$ (a) and λ_E (b) exhibited by all algorithms on instances induced by all partitions of $H = 25$. Instances are sorted by non-decreasing values of $M(E)$ in (a) and λ_E in (b). To magnify the differences, the y-axis in panel (b) is log-scaled.

On the Length of the Cycle. In Fig. 1b we show the maximum λ_E obtained by all strategies. Such values can be considered as proxy of the complexity of their cycle, as higher values of λ_E correspond to larger spaces of configurations that are explored by the strategies. This translates to higher variance in terms of height and volume, which can be seen as an undesired behaviour. Moreover, there might be also a relationship between such length and the quality of the provided factor of approximation, and it would be worth to study such relationship to define new bounds on this factor. Our data shows a very big gap between the results obtained for RMax, RFast$_1$, and those measured for RFast$_2$, RMin. In particular, as shown in Fig. 1b, when $n = 12$, RMin takes 40000 steps whereas the worst case for RFast$_2$ is obtained for $n = 14$, with around 20000 steps. It is worth mentioning that RMax and RFast$_1$ behave rather differently w.r.t. other strategies, exhibiting very low values of λ_E. This may lead to more accurate arguments about changes in the volume of the garden, to be exploited in proofs of constant approximation. Different considerations can be done by observing Fig. 2b. In particular, the largest values of λ_E are achieved in the great majority of the cases by RFast$_2$ while RMax results to be the best strategy also in this sense.

5   Conclusion

We have investigated the BGT problem to establish whether constant approximation deterministic algorithms can be designed. We have defined a new class of scheduling strategies called priority schedulings and provided theoretical results on such methods. In particular, we have proved that any priority scheduling eventually brings the system to perpetually repeated sequences of configurations. We have also analysed ReduceMax, a priority scheduling for which we conjecture 2-approximation. We have conducted extensive experimentation confirming our intuitions and showing that ReduceMax outperforms any other known strategy, including the 4-approximation ReduceFastest$_2$, which unlike ReduceMax relies on the knowledge of an ordering on the rates. In terms of knowledge required by the cutting strategies, a research direction that surely deserves further investigation is that of considering the more realistic scenario where the input data is not entirely known, i.e. to tackle the problem from an online algorithms perspective.

Acknowledgments. Authors deeply thank Leszek Gąsieniec for introducing them to the problem and for very useful discussions.

References

1. Alshamrani, S., Kowalski, D.R., Gąsieniec, L.: How reduce max algorithm behaves with symptoms appearance on virtual machines in clouds. In: Proceedings of the IEEE International Conference CIT/IUCC/DASC/PICOM, pp. 1703–1710 (2015)
2. Baruah, S.K., Cohen, N.K., Plaxton, C.G., Varvel, D.A.: Proportionate progress: a notion of fairness in resource allocation. Algorithmica **15**(6), 600–625 (1996)
3. Baruah, S.K., Lin, S.-S.: Pfair scheduling of generalized pinwheel task systems. IEEE Trans. Comput. **47**(7), 812–816 (1998)
4. Baruah, S., Rosier, L., Tulchinsky, I., Varvel, D.: The complexity of periodic maintenance. In: Proceedings of the 1990 International Computer Symposium, pp. 315–320 (1990)
5. Bender, M.A., et al.: The minimum backlog problem. Theoret. Comput. Sci. **605**, 51–61 (2015)
6. Bodlaender, M.H.L., Hurkens, C.A.J., Kusters, V.J.J., Staals, F., Woeginger, G.J., Zantema, H.: Cinderella versus the wicked Stepmother. In: Baeten, J.C.M., Ball, T., de Boer, F.S. (eds.) TCS 2012. LNCS, vol. 7604, pp. 57–71. Springer, Heidelberg (2012). https://doi.org/10.1007/978-3-642-33475-7_5
7. Chan, M.Y., Chin, F.Y.L.: General schedulers for the pinwheel problem based on double-integer reduction. IEEE Trans. Comput. **41**(6), 755–768 (1992)
8. Chan, M.Y., Chin, F.: Schedulers for larger classes of pinwheel instances. Algorithmica **9**(5), 425–462 (1993)
9. Chrobak, M., Csirik, J., Imreh, C., Noga, J., Sgall, J., Woeginger, G.J.: The buffer minimization problem for multiprocessor scheduling with conflicts. In: Orejas, F., Spirakis, P.G., van Leeuwen, J. (eds.) ICALP 2001. LNCS, vol. 2076, pp. 862–874. Springer, Heidelberg (2001). https://doi.org/10.1007/3-540-48224-5_70
10. Chuangpishit, H., Czyzowicz, J., Gąsieniec, L., Georgiou, K., Jurdziński, T., Kranakis, E.: Patrolling a path connecting a set of points with unbalanced frequencies of visits. In: Tjoa, A.M., Bellatreche, L., Biffl, S., van Leeuwen, J., Wiedermann, J. (eds.) SOFSEM 2018. LNCS, vol. 10706, pp. 367–380. Springer, Cham (2018). https://doi.org/10.1007/978-3-319-73117-9_26
11. Collins, A., et al.: Optimal patrolling of fragmented boundaries. In: Proceedings of the Twenty-Fifth Annual ACM Symposium on Parallelism in Algorithms and Architectures, SPAA, New York, USA, pp. 241–250 (2013)
12. Czyzowicz, J., Gąsieniec, L., Kosowski, A., Kranakis, E.: Boundary patrolling by mobile agents with distinct maximal speeds. In: Demetrescu, C., Halldórsson, M.M. (eds.) ESA 2011. LNCS, vol. 6942, pp. 701–712. Springer, Heidelberg (2011). https://doi.org/10.1007/978-3-642-23719-5_59
13. Gąsieniec, L., Klasing, R., Levcopoulos, C., Lingas, A., Min, J., Radzik, T.: Bamboo garden trimming problem (perpetual maintenance of machines with different attendance urgency factors). In: Steffen, B., Baier, C., van den Brand, M., Eder, J., Hinchey, M., Margaria, T. (eds.) SOFSEM 2017. LNCS, vol. 10139, pp. 229–240. Springer, Cham (2017). https://doi.org/10.1007/978-3-319-51963-0_18

14. Gąsieniec, L., Klasing, R., Martin, R., Navarra, A., Zhang, X.: Fast periodic graph exploration with constant memory. J. Comput. Syst. Sci. **74**(5), 802–822 (2008)
15. Holte, R., Mok, A., Rosier, L., Tulchinsky, I., Varvel, D.: The pinwheel: a real-time scheduling problem. In: II: Software Track, Proceedings of the Twenty-Second Annual Hawaii International Conference on System Sciences, vol. 2, pp. 693–702, January 1989
16. Holte, R., Rosier, L., Tulchinsky, I., Varvel, D.: Pinwheel scheduling with two distinct numbers. Theoret. Comput. Sci. **100**(1), 105–135 (1992)
17. Hsueh, C., Lin, K.: An optimal pinwheel scheduler using the single-number reduction technique. In: Proceedings of the 17th IEEE Real-Time Systems Symposium, RTSS 1996, pp. 196–205 (1996)
18. Kosowski, A., Navarra, A.: Graph decomposition for memoryless periodic exploration. Algorithmica **63**(1–2), 26–38 (2012)
19. Lin, S.-S., Lin, K.-J.: A pinwheel scheduler for three distinct numbers with a tight schedulability bound. Algorithmica **19**(4), 411–426 (1997)
20. Mok, A., Rosier, L., Tulchinski, I., Varvel, D.: Algorithms and complexity of the periodic maintenance problem. In: Proceedings of the 15th Symposium on Microprocessing and Microprogramming (EUROMICRO), pp. 657–664 (1989)
21. Ntafos, S.: On gallery watchmen in grids. Inf. Process. Lett. **23**(2), 99–102 (1986)
22. Romer, T.H., Rosier, L.E.: An algorithm reminiscent of euclidean-gcd for computing a function related to pinwheel scheduling. Algorithmica **17**(1), 1–10 (1997)
23. Serafini, P., Ukovich, W.: A mathematical model for periodic scheduling problems. SIAM J. Discret. Math. **2**(4), 550–581 (1989)
24. Urrutia, J.: Art gallery and illumination problems. In: Handbook of Computational Geometry, vol. 1, no. 1, pp. 973–1027 (2000)
25. Hardy, G.H., Ramanujan, S.: Asymptotic formulas in combinatorial analysis. Proc. Lond. Math. Soc. **17**, 75–115 (1918)

Patrolling on Dynamic Ring Networks

Shantanu Das[1(✉)], Giuseppe A. Di Luna[1], and Leszek A. Gasieniec[2]

[1] Aix-Marseille University, CNRS, LIS, Marseille, France
shantanu.das@lis-lab.fr, g.a.diluna@gmail.com
[2] University of Liverpool, Liverpool, UK
L.A.Gasieniec@liverpool.ac.uk

Abstract. We study the problem of patrolling the nodes of a network collaboratively by a team of mobile agents, such that each node of the network is visited by at least one agent once in every $I(n)$ time units, with the objective of minimizing the idle time $I(n)$. While patrolling has been studied previously for static networks, we investigate the problem on dynamic networks with a fixed set of nodes, but dynamic edges. In particular, we consider 1-interval-connected ring networks and provide various patrolling algorithms for such networks, for $k = 2$ or $k > 2$ agents. We also show almost matching lower bounds that hold even for the best starting configurations. Thus, our algorithms achieve close to optimal idle time. Further, we show a clear separation in terms of idle time, for agents that have prior knowledge of the dynamic networks compared to agents that do not have such knowledge. This paper provides the first known results for collaborative patrolling on dynamic graphs.

1 Introduction

In recent years patrolling is gaining on popularity in the area of algorithms and in particular algorithmics of mobile agents and applications. Patrolling naturally occurs in daily routines requiring regular visits to specific (possibly mobile) objects and areas. It can also refer to monitoring of complex network processes or systems behaviour. Typical applications of patrolling include safety or security related surveillance, regular updates, data gathering, and other perpetual tasks.

We consider the patrolling problem in networks (graphs) with the objective of visiting all nodes of the graph perpetually, optimizing the *idle time* - the maximum time period during which any node is left unvisited. Unlike all previous results on the patrolling problem, we study the problem on dynamic graphs where some links of the graph may be missing for certain duration of time. This complicates the problem and requires a strong coordination between the agents, in order to reduce the idle time, even in simple networks. We restrict our attention to dynamic ring networks, in this paper. In the case of a static ring network, the simple strategy of periodically cycling the nodes of the ring, is known to provide the optimal idle time. However, for patrolling dynamic rings, more involved strategies are required depending on the number of the agents,

© Springer Nature Switzerland AG 2019
B. Catania et al. (Eds.): SOFSEM 2019, LNCS 11376, pp. 150–163, 2019.
https://doi.org/10.1007/978-3-030-10801-4_13

the capabilities of the agents and whether or not the dynamic structure of the network is known to the agents. Among various known dynamic graph models, we consider interval connected dynamic networks which ensures that the network is connected at any time interval. We distinguish between the KNOWN setting, where the agents know in advance about the changes in the graph structure, from the UNKNOWN setting, where such information is not available to the agents. We show a clear separation between the two cases, in terms of the minimum idle time for patrolling. For both cases, we provide lower bounds and almost matching upper bounds on the idle time for patrolling, supported by deterministic algorithms for collaborative patrolling.

Related Work

Patrolling. The problem of patrolling is a close relative to several classical algorithmic challenges which focus on monitoring and mobility. These challenges include the Art Gallery Problem [32], where one is interested in determining the smallest number of inert guards and their location to constantly monitor all artefacts, and its dynamic alternative referred to as the k-Watchmen Problem [7,10]. In further work on fence patrolling [12,13,25] the authors focus on monitoring vital (possibly disconnected) parts of a linear environment where each point is expected to be visited with the same frequency. A similar approach is adopted in [14] where we find studies on monitoring of a linear environment by agents prone to faults. The problem of patrolling objects which require different frequencies of visits was first considered in [20], where the authors assume availability of a single mobile agent. They also showed a close relationship between these type of patrolling and the Pinwheel scheduling problem [9]. In a more recent work [20] the authors consider monitoring by two agents of n nodes located on a line and requiring different frequencies of visits. The authors provide several approximation algorithms concluding with the best currently known $\sqrt{3}$-approximation. Patrolling of segments and circles (equivalent to static ring) by many agents have been studied in [12].

Dynamic Networks and Mobile Agents. The field of dynamic networks is an hot and active research topic [8,21,29,30]. In the message passing model a lot of attention has been devoted to classic problems such as agreement [3,5,28], information dissemination [4,11,27,33], and counting [15,26]. Surprisingly, the investigation of mobile agents on dynamic networks started only recently. In the centralised setting (when agents know the dynamic of the graph apriori) the problem of exploring a graph in the fastest possible way has been studied in several papers [1,18,31]. The task is NP-hard on general graphs and it becomes polynomial on special topologies [2,22]. Notably, in the case of interval connected ring the exploration can be done in $\mathcal{O}(n)$ rounds [24].

The distributed setting (when agents do not know the dynamic of the graph) has been mostly overlooked, or limited to restrictive dynamic assumptions, such as periodic [19,23] or recurrent [24] graphs. The exploration with termination of

Table 1. Results for the idle time in dynamic rings of n nodes, with k uniformly placed agents having global visibility.

Adversary		Number of agents	
		$k = 2$	$k > 2$
KNOWN	Upper bound	$3\lceil\frac{n}{2}\rceil$	$3\lceil\frac{n}{k}\rceil$
	Lower bound	n	$\frac{2n}{k}$
UNKNOWN	Upper bound	$2n - 2$	$4\lceil\frac{n}{k}\rceil$
	Lower bound	$2n - 6$	$\frac{2n}{k}$

interval connected rings has been studied in [16]. For rings that are connected over time, a perpetual self-stabilizing exploration algorithm has been proposed in [6]. Finally, the gathering problem on interval connected rings has been studied in [17]. To the best of our knowledge there is no previous work studying the patrolling of a dynamic network.

Our Contributions

We show, first of all, in Sect. 4, that when the agents have local visibility, limited to the current node, then patrolling has an idle time of $n - \alpha \cdot k$ rounds, where $\alpha = 1$ when the agents may be arbitrarily placed by an adversary (and $\alpha = 2^b$ for uniform initial placement when the agents have b-bits of persistent memory). This means that using multiple agents reduces the idle time by only an additive factor. In contrast, for a *static* ring, the idle time for patrolling with k agents is $\frac{n}{k}$, achieving a multiplicative factor efficiency over single agent patrolling.

Thus, for the rest of paper, we consider agents having global visibility, allowing them to see the current configuration of the ring with the set of available links. We start with team size of $k = 2$ agents in Sect. 5 and then generalize these results to $k > 2$ agents in Sect. 6. The results of these two sections are summarized in Table 1. The bounds denoted here are for the stable idle time, after a stabilization time that is at most $O(n)$. These results show a clear distinction between the case of KNOWN adversary (where the dynamic structure of the network is known apriori) and the case of UNKNOWN adversary when the agents do not have prior knowledge of the dynamic network.

2   Model

A set of agents, $A : \{a_0, \ldots, a_{k-1}\}$, operates on a dynamic graph $\mathcal{G}$. Each agent follows the same algorithm (all agents are identical) executing a sequence of Look, Compute, Move cycles. In the Look phase of each cycle, the agent acquires a *snapshot* of the environment. In the Compute phase the agent uses the information from the snapshot and the contents of its local persistent memory to compute the next destination, which may be the current node or one of its neighbors. During the Move phase an agent traverses an edge to reach the destination node. The information contained in the persistent memory is the only thing that is preserved among cycles.

Synchronous System. The system is *synchronous*, that is the time is divided in discrete units called rounds. Rounds are univocally mapped to numbers in $\mathbb{N}$, starting from round 0. In each round, each agent in A executes exactly one entire Look, Compute, Move cycle.

Interval Connected Ring. A dynamic graph $\mathcal{G}$ is a function mapping a round $r \in \mathbb{N}$ to a graph $G_r : (V, E(r))$ where $V : \{v_0, \ldots, v_{n-1}\}$ is a set of nodes and $E : \mathbb{N} \to V \times V$ is a function mapping a round r to a set of undirected edges. We restrict ourselves to 1-interval-connected rings. A dynamic graph $\mathcal{G}$ is a 1-interval-connected ring when the union of the graph instances $G_\infty = (V, E_\infty) = (V, \cup_{r=0}^{+\infty} E(r))$ is a ring graph, and at each round r, the graph G_r is connected (in other words, at each round at most one edge is missing). The graph $\mathcal{G}$ is anonymous, i.e. all nodes are identical to the agents. The endpoints of each edge are labelled as either *clockwise* or *counter-clockwise*, in a consistent manner (i.e. the ring is oriented).

Local Versus Global Snapshot

- Local Snapshot: the snapshot obtained by an agent at a node v in round r contains only information about the node v, i.e. the number of agents in v and the set of available edges incident to node v at round r.
- Global Snapshot: the snapshot obtained by an agent contains the graph G_r (where the current location of the agent is marked), and for each node in V the number of agents present in that node at round r.

Knowledge of $\mathcal{G}$. We examine two different settings: the one with known $\mathcal{G}$ (KNOWN) and the one without such knowledge (UNKNOWN). In the KNOWN setting during the Compute phase agents have access to the dynamic graph $\mathcal{G}$. In this case, the decision taken by the agent depends on the snapshot, on the content of its local memory, and on the knowledge of the past history and future structure of dynamic graph $\mathcal{G}$. On the contrary in the UNKNOWN setting, during the Compute phase, an agent uses only the snapshot and its local memory (no other information is available). Another way to see the UNKNOWN setting is to imagine that $\mathcal{G}$ is adaptive to the strategy of algorithm $\mathcal{A}$: there exists an adversarial entity, namely the *scheduler*, that decides the graph $\mathcal{G}$ according to the strategy of algorithm $\mathcal{A}$.

Configurations and Initial Placement of Agents. Given a graph G_r and the set of agents A, a configuration at round r is a function $C_r : A \to V$ that maps agents in A to nodes of V where agents are located. We say that there is a *uniform initial placement*, if C_0 is such that the segments of consecutive rings nodes not occupied by agents have size $\lfloor \frac{n}{k} \rfloor$ or $\lceil \frac{n}{k} \rceil$. We say that there is an *arbitrary initial placement* if the configuration C_0 is injective (no two agents may start on the same node).

Idle Time. An algorithm $\mathcal{A}$ running on a graph $\mathcal{G}$, generates an execution $\mathcal{E}$, which is an infinite sequence of configurations $\{C_0, C_1, C_2 \ldots\}$, one for each round r. Given a node v and an execution $\mathcal{E}$, the set $S_{\mathcal{E},v} : \{r_1, r_2, r_3, \ldots\}$ of visits of v, is a set containing all rounds in which v has been visited by some agent in execution $\mathcal{E}$; more formally, $r_j \in S_{\mathcal{E},v}$ if and only if $C_{r_j}(a) = v$ for some $a \in A$. The idle set $I_{\mathcal{E},v}$ of node v is a set containing all the intervals of time between two consecutive visits of node v in execution $\mathcal{E}$; more formally, $x \in I_{\mathcal{E},v}$ if and only if there exists r_i, r_{i-1} in $S_{\mathcal{E},v}$ and $x = r_i - r_{i-1}$. We assume that each node has been visited at round -1.

We say that an algorithm solves patrolling on a graph $\mathcal{G}$, if each node of the graph is visited infinitely often. Given an algorithm $\mathcal{A}$ and an integer $n \geq 5$, we define as T_n the set of all executions of algorithm $\mathcal{A}$ over any (1-interval-connected) dynamic ring $\mathcal{G}$ with n nodes. The idle time of algorithm $\mathcal{A}$ is the function $I(n) = \max_{\forall \mathcal{E} \in T_n} (\cup_{\forall v \in V} I_{\mathcal{E},v})$.

Stable Idle Time. Given an execution $\mathcal{E}$ we define as $\mathcal{E}[r, \infty]$ the execution obtained by removing the first r configurations from $\mathcal{E}$. An algorithm $\mathcal{A}$ is said to have a stable idle time $I_{r_s}(n)$ with stabilisation time r_s, if for some round $r_s, I_{r_s}(n) = \max_{\forall \mathcal{E} \in T_n} (\cup_{\forall v \in V} I_{\mathcal{E}[r_s, \infty], v})$.

3   Preliminaries

We devote this section to some simple observations based on previous results on dynamic rings. Note that for a single agent moving in a dynamic ring, an adaptive adversary can keep the agent confined to the starting node and one of its neighbors.

Observation 1 ([27,33]). *In a dynamic ring $\mathcal{G}$ under the UNKNOWN model with global snapshot, a single agent can visit at most 2 nodes.*

Observation 2 ([27,33]). *In a dynamic ring $\mathcal{G}$ under the KNOWN model, a single agent can reach any given node v in at most $n - 1$ rounds.*

Due to the above observations, the only interesting cases for patrolling is for $k \geq 2$ which we investigate in this paper. For any k agents, we have the following observation derived from the proof of Proposition 1 in [24].

Observation 3 ([24]). *In a dynamic ring $\mathcal{G}$ under the KNOWN model, for any round r and any $1 \leq h \leq n-1$, there are $n - h$ distinct nodes, such that if $n - h$ agents are placed in these nodes and they all move in the same direction from round r until round $r + h - 1$, then they visit exactly $h + 1$ nodes.*

It is also possible to show an easy lower bound on the idle time of any algorithm under the strongest model considered in this paper (i.e. under global visibility and knowledge of $\mathcal{G}$).

Theorem 4. *Consider the* KNOWN *model with Global Snapshot. Let $\mathcal{A}$ be any patrolling algorithm for k agents with uniform initial placement. We have that $I_{r_s}(n) \geq \frac{2n}{k}$ for any stabilization time r_s.*

Proof. The scheduler removes the same edge forever. At this point the k agents have to patrol a line and the lower bound for idle time on a line with k agents is $\frac{2n}{k}$ (See [12] for a proof).

4   Patrolling with Local Visibility

In this section we analyse the Local Snapshot model, we first examine the case in which the placement of the agents is arbitrary and then we examine the case in which the placement is uniform.

Theorem 5. *Consider a dynamic ring under the* KNOWN *model with Local Snapshot and arbitrary initial placement. Then any patrolling algorithm $\mathcal{A}$ for k agents has stable idle time $I_{r_s}(n) \geq n - k$, for any stabilisation time r_s.*

Proof. Let us consider a static ring of n nodes $G = (V = \{v_0, \ldots, v_{n-1}\}, E = \{(v_0, v_1), (v_1, v_2), \ldots\})$ and a set of agents $\{a_0, \ldots, a_{k-1}\}$. Configuration C_0 is such that $C(a_j) = v_j$, that is agents are placed one for each node in $\{v_0, \ldots, v_{k-1}\}$. If the ring is oriented and the nodes are anonymous, each agent would have the same local view and they take the same action at each step. Thus, at any round r, the configuration C_r is a rotation of either one step counter-clockwise or one step clockwise of configuration C_{r-1}. This implies that the best idle time is obtained by having agents to perpetually move in the same direction. The idle time of this strategy is $I_{r_s}(n) = n - k$ for any possible stabilization time r_s. □

The above result assumes the agents to be placed on consecutive nodes, and its proof does not hold when there is an uniform initial placement of agents. However, even in the case of uniform placement, we show the following result for agents having constant amount of persistent memory (b bits), under the UNKNOWN model.

Theorem 6. *Consider a dynamic ring under the* UNKNOWN *model with local snapshots and uniform initial placement. Given any patrolling algorithm $\mathcal{A}$ for k agents, with $c = \mathcal{O}(1)$ bits of memory, the idle time for patrolling is $I(n) \geq n - 7 \cdot 2^c k$.*

5   Two Agents with Global Visibility

In this section we assume that the agents have access to a global snapshot of the configuration at each round during Look phase. We first consider the simpler case of $k = 2$ agents and show upper and lower bounds on patrolling for both the UNKNOWN and the KNOWN setting.

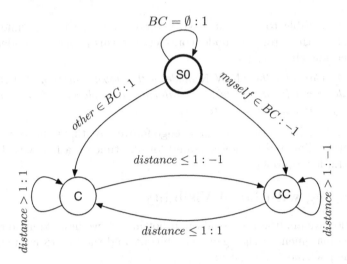

Fig. 1. Algorithm TICK-TOCK state diagram. The starting state is S0. Transition are of the form *Predicate : Movement* where values of $1, -1, 0$ denotes clockwise, counter-clockwise or no move, respectively.

5.1 UNKNOWN Setting

Given the graph G_r at round r, we define as BC_r (resp. BCC_r) the set of all agents that are attempting to move clockwise (resp. counter-clockwise) from a node v that has the clockwise (resp. counter-clockwise) edge missing at the round r. We will remove the subscript r when it is clear that we are referring to the current round.

We now describe a patrolling algorithm called TICK-TOCK for $k = 2$ agents in the UNKNOWN setting. Initially, both agents move in the clockwise direction in each round, until they reach a round r in which BC_r is not empty. At this point the symmetry between agents is broken, and we assign to the agent in BC_r the counter-clockwise direction while the other agent keeps the clockwise direction. Starting from round r, the agents continue to move according to the following rule: Move in the assigned direction until the minimum distance between the agents is less or equal to 1; When this happen, both agents reverse their direction (i.e, the agents bounce off each other). The state diagram of the algorithm is presented in Fig. 1.

Theorem 7. *For any dynamic ring in the* UNKNOWN *model with Global Snapshot and arbitrary initial placement, Algorithm* TICK-TOCK *allows two agents to patrol the ring with an idle time* $I(n) \leq 2(n-1)$.

Proof. The algorithm has two distinct phases. In the first phase, both agents move in the same direction, while in the second phase the agents always move in opposite directions. We need to show that for any node v, given two consecutive visits of v at round r_0 and r_1 it holds that $r_1 - r_0 \leq 2(n-1)$. First, let r_0 and

r_1 be both in the first phase of the algorithm. Observe that in this phase each agent loops around the ring visiting each node once in every n rounds. Since the agents on distinct nodes we have at most $n - 1$ rounds between two visits of node v; thus $r_1 - r_0 \leq n - 1$.

Now we examine the case when r_0 and r_1 are both in the second phase. It takes at most $n - 1$ rounds for the distance between the two agents to be 1 or less–the agents are moving on opposing direction and at most one of them can be blocked at any round. This means that during a period that is upper bounded by $n - 1$ all nodes are visited. Thus, there are at most $2(n - 1)$ rounds between consecutive visits of a node v.

Finally, we have to show that the bound still holds if r_0 is in the first phase and r_1 in the second. Let r be the round in which the algorithm switches phase. We necessarily have $r - r_0 = x \leq n - 1$, by the previous discussion regarding the first phase. At round r, one agent is at distance x from node v and thus, the distance between the agents on the segment not containing v, is at most $(n - x - 1)$. Now, if both agents are move towards v, then v would be visited in the next $(n - 1)$ rounds. Otherwise, the agents move away from v, therefore in at most $(n - x - 2)$ rounds, the two agents would be at distance one or less. In the subsequent $n - 1$ rounds all nodes would be visited (recall our previous discussion for the second phase). This implies that $r_1 - r_0 \leq 2(n - 1)$ in both cases. $\square$

Surprisingly, the algorithm TICK-TOCK is almost optimal.

Theorem 8. *Under the* UNKNOWN *model with global snapshot and uniform initial placement, any patrolling algorithm* $\mathcal{A}$ *for two agents has idle time* $I(n) \geq 2n - 6$.

We prove the above result by showing that the adversarial scheduler can (1) entrap one of the agents on two neighboring nodes of the ring, say, nodes v_{n-1}, v_{n-2}, and at the same (2) prevent the other agent from performing a full tour of the ring. Under the above two conditions, patrolling the ring by two agents reduces to patrolling a line of $l = n - 2$ nodes by a single agent, for which we have an idle time of $2(l - 1) = 2n - 6$.

5.2 KNOWN Setting

In this subsection we examine the KNOWN setting. We first present a solution algorithm, namely PLACE-&-SWIPE, that solves the problem with an idle time of $3\lceil \frac{n}{2} \rceil$ rounds, when there is an uniform initial placement of the agents. We then discuss how the algorithm can be adapted to work under arbitrary initial placement by having a stabilisation time of $\lfloor \frac{n}{2} \rfloor$ and a stable idle time of $3\lceil \frac{n}{2} \rceil$ rounds.

Patrolling Algorithm. The algorithm PLACE-&-SWIPE perpetually alternates between two phases of fixed length (each phase lasts $\lceil \frac{n}{2} \rceil$ rounds). During the first phase, called *Placement Phase*, the agents position themselves on a specially choosen pair of antipodal[1] nodes – the *swiping nodes*. In the second phase, called the Swipe Phase, the agents together visit all nodes of the ring by both moving clockwise for $\lfloor \frac{n}{2} \rfloor$ rounds without stop. A Placement Phase followed by Swipe Phase is an epoch of the algorithm, we use $i \geq 0$ to indicate the epoch number. Since every node is visited once in every Swipe phase, in the worst case, a node may be visited at the beginning of a Swipe phase and subsequently at the end of the next Swipe Phase, giving an idle time of at most $3\lceil \frac{n}{2} \rceil$ rounds.

We now show that for each epoch i, there exists a special pair P_i of antipodal nodes which allow the Swipe Phase to cover all nodes in $\lfloor \frac{n}{2} \rfloor$ rounds. Let $start_i = i \cdot n$, and $end_i = \lceil (\frac{1}{2} + i)n \rceil - 1$ be the starting and ending round of the i-th Placement Phase.

Lemma 1. *Given any dynamic ring $\mathcal{G}$ and any round $r = end_i + 1$, there exists a pair of antipodal nodes P_i, such that two agents placed on P_i and moving clockwise from round $end_i + 1$ to the end round $start_{i+1} - 1$, explore all nodes of the ring.*

Proof. The key idea to prove the existence of P_i is Observation 3. By plugging $t = \lceil \frac{n}{2} \rceil - 1$ in the statement of the observation. We have that there are $\lfloor \frac{n}{2} \rfloor + 1$ nodes, let E_i be this set, such that an agent being on one of these nodes at round $end_i + 1$ moving clockwise visits exactly $\lceil \frac{n}{2} \rceil$ nodes by the end of round $start_{i+1} - 1$. Now we have to prove that E_i contains a pair of antipodal nodes. But this is obvious since the ring contains at least $\lfloor \frac{n}{2} \rfloor$ antipodal pairs and the cardinality of E_i is $\lfloor \frac{n}{2} \rfloor + 1$. Being the pair P_i antipodals, when each agent visits $\lceil \frac{n}{2} \rceil$ nodes the ring has been explored.   □

To prove correctness of the algorithm, we need to show that agents starting from any uniform configuration, the two agents can reach the chosen nodes P_i in $\lceil \frac{n}{2} \rceil$ rounds. Note that, for computing P_i in each epoch, the algorithm needs only the knowledge of the future n rounds of $\mathcal{G}$.

Theorem 9. *Consider the* KNOWN *model with Global snapshot and uniform initial placement. The algorithm* PLACE-&-SWIPE *allows two agents to patrol a ring with an idle time $I(n) \leq 3\lceil \frac{n}{2} \rceil$.*

Arbitrary Initial Placement. Theorem 9 assumes that agents are starting at uniform distance. However, it is possible to easily adapt the algorithm to work under any initial placement sacrificing the stabilization time. Essentially, we need an initialization phase in which agents place themselves in antipodal positions. This can be done in $\lfloor \frac{n}{2} \rfloor$ rounds: in each round, agents move apart from each other increasing the distance by at least one unit per round. Thus, we obtain an algorithm with stabilization time $r_s = \lfloor \frac{n}{2} \rfloor$ and $I_{r_s}(n) \leq 3\lceil \frac{n}{2} \rceil$.

---

[1] A pair of nodes is antipodal if the distance between them in the ring is $\lfloor \frac{n}{2} \rfloor$.

Lower Bounds. A lower bound of n for the KNOWN setting is immediate from Theorem 4. However, when the initial placement of the agents is arbitrary we can show a slightly better bound.

Theorem 10. *Let $\mathcal{A}$ be a patrolling algorithm for two agents with arbitrary initial placement under the KNOWN model with Global snapshot. For any even $n \geq 10$, there exists a 1-interval connected ring where $\mathcal{A}$ has an idle time $I(n) \geq \lfloor (1 + \frac{1}{5})(n - 1) \rfloor$.*

6   Patrolling with $k > 2$ Agents Having Global Visibility

In this section we examine the case of $k > 2$ agents, showing how to generalize the algorithms of Sect. 5 for this case.

6.1   UNKNOWN Setting: Generalising TICK-TOCK for k Agents

We generalize TICK-TOCK for k agents assuming that: k divides n, k is even, and that there is uniform initial placement. At the end of the section we discuss how to remove such assumptions. The new algorithm, called K-TICK-TOCK is divided in two phases, *Single-Group-Swiping* and *Two-Groups-Swiping*, as described below.

The Single-Group-Swiping Phase starts at round $r = 0$ and all agents move clockwise in this phase, keeping uniform distribution. The phase ends at the first round r' when an agent is blocked. Starting from round r', the Two-Groups-Swiping phase starts. Recall that $BC_{r'}$ is the set of agents trying to move clockwise in round r' that encounter a missing edge. Since the agents are in distinct nodes, only one agent, say agent $a_j \in BC_{r'}$. This breaks the symmetry among the agents and they can partition themselves in two groups: group clockwise G_C and group counter-clockwise G_{CC}. The group G_C contains agent $a_{(j+2t) \bmod k}$ with $t \in \mathbb{N}$, and group G_{CC} contains all other agents. The partition into groups happens during the computation phase of round r'. From round r', the agents move according to the following rules: (See Fig. 2)

- Rule 1 (Group Movement): For $X \in C, CC$, an agent in G_X moves in direction X if no agent in G_X is blocked, i.e. $\nexists a \in BX_r \bigcap G_X$.
- Rule 2 (Membership Swapping): If at some round r'' agents in both groups are blocked, then the agents in $BC_{r''}$ and $BCC_{r''}$ swap their role, i.e. they exchange their states and thus their group membership in this round. Any other agent in G_X moves in direction X during this round.

Intuitively, for Rule 1 a group G_X moves when all the agents in the group would be able to move without trying to cross a missing edge. Rule 2 is applied only when two agents, one from group G_C and one from group G_{CC} are on two nodes that share the same missing edge, and this allows the groups to perform a *"virtual movement"*, i.e. the two blocked agents swap roles to simulate a move across the missing edge.

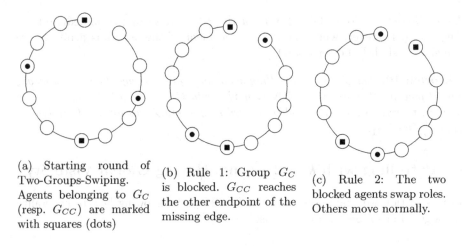

(a) Starting round of Two-Groups-Swiping. Agents belonging to G_C (resp. G_{CC}) are marked with squares (dots)

(b) Rule 1: Group G_C is blocked. G_{CC} reaches the other endpoint of the missing edge.

(c) Rule 2: The two blocked agents swap roles. Others move normally.

Fig. 2. Algorithm K-TICK-TOCK, depiction of salient cases.

Theorem 11. *The* K-TICK-TOCK *algorithm has an idle time of* $\frac{4n}{k}$.

The above result is based on the facts that: (1) in each round at least one group moves, and (2) After each visit of a node v, the distances between node v and the closest agents in G_C (or G_{CC}) that are moving towards v are at most $\frac{2n}{k} - 1$. So, in the successive $\frac{4n}{k}$ rounds, at least one group would reach v.

When k Is Not a Divisor of n. In the case k does not divide n, we have that in the initial placement the minimum distance between two agents is $\lfloor \frac{n}{k} \rfloor$ and the maximum distance is $\lfloor \frac{n}{k} \rfloor + 1$. We can use the same analysis of Theorem 11, taking into account the difference in the distance, which gives a bound of $\lfloor \frac{4n}{k} \rfloor + 2$.

When Agents Are Not Uniformly Placed. If agents are not uniformly placed initially, they can arrive at a uniform configuration in $O(n)$ steps.

Observation 12. *Consider a set of $k \geq 2$ agents arbitrarily placed in a dynamic ring under the* UNKNOWN *model with global snapshot, then the agents need at most $2n$ rounds to reach an uniform placement in the ring.*

When k is Odd. The problem for odd k is that once the algorithm switches to the Two-Group-Swiping phase, the groups G_C, G_{CC} do not have equal sizes. One group has size $\frac{k-1}{2}$ and the other $\frac{k+1}{2}$. Moreover, within each group the members are not uniformly placed. The last problem is easy fixable at the price of stabilization time using Observation 12. Once the groups are uniformly placed, we can bound the idle time to $\frac{4nk}{k^2-1} + 4$, as shown in the following lemma:

Lemma 2. *When one group has size $\frac{k-1}{2}$ and the other $\frac{k+1}{2}$, the Two-Groups-Swiping phase of* K-TICK-TOCK *has an idle time of at most $\frac{4nk}{k^2-1} + 4$ rounds.*

Proof. W.l.o.g. let G_C be the group of size $\frac{k-1}{2}$ and G_{CC} be the other group. Let r_0, r_1 be the times between two successive visits of some node v. In the worst case at round $r_0 + 1$, node v could be at distance at most $\frac{2n}{k-1} + 1$ from an agent in group G_C, and distance at most $\frac{2n}{k+1} + 1$ from an agent in G_{CC}. The sum of these distances is $\frac{4nk}{k^2-1} + 2$, and since only one group can be blocked at each round, this distance decreases by one at each round. This implies that $r_1 - r_0 \leq \frac{4nk}{k^2-1} + 2$, thus proving the bound. $\qquad\square$

From the previous Lemma and using the same proof strategy of Theorem 11 we have that $\frac{4nk}{k^2-1} + 4$ is the idle time of the algorithm. Unfortunately, it is not possible to bound the stabilization time of the algorithm. The adversary decides when, and if, the algorithm goes to the Two-Groups-Swiping phase, and when this happen a certain number of rounds has to be payed to position in an uniform way the members of each group. However, in any infinite execution of the algorithm, there are only finitely many times in which two consecutive visits of a node are spaced by more than $\frac{4nk}{k^2-1} + 4$ rounds.

6.2 KNOWN Setting: PLACE-&-SWIPE for k Agents

Generalising the algorithm Sect. 5.2, for k agents is immediate. The algorithm is essentially the same, the only variations are: each phase now lasts $\lfloor \frac{n}{k} \rfloor$ rounds and P_i is not a pair of nodes but k nodes uniformly placed. Also in this case we assume that agents start uniformly placed, such assumption can be dropped sacrificing the stabilization time (see Observation 12). Lemma 3 below is an equivalent of Lemma 1 for $k \geq 2$ agents. Further, we can show that starting from any uniform configuration, the agents can reach, using the knowledge of $\mathcal{G}$, any given target uniform configuration in at most $\lceil \frac{n}{k} \rceil$ steps.

Lemma 3. *Given any 1-interval connected dynamic ring $\mathcal{G}$, for any round r_i, there exists a set P_i of k uniformly spaced nodes, such that k agents placed on P_i and moving clockwise from round r_i to round $r_i + \lfloor \frac{n}{k} \rfloor$, together explore all nodes of the ring.*

Theorem 13. *Consider the KNOWN model with global snapshots. The algorithm PLACE-&-SWIPE allows k agents with uniform initial placement to patrol a ring with an idle time $I(n) \leq 3\lceil \frac{n}{k} \rceil$.*

7 Conclusion

We provided the first results on the patrolling problem in dynamic graphs. As patrolling is usually performed on boundaries of territories, it is natural to study the problem for ring networks. The results may be extended to other topologies e.g. by moving on any cycle containing all the nodes of a graph. Our results on the dynamic ring networks are almost complete, but there exists a small gap between the lower and upper bounds, specially for the case of $k > 2$ agents which can be reduced by future work. In particular, we believe the lower bound for $k > 2$ agents in the UNKNOWN setting can be improved.

References

1. Aaron, E., Krizanc, D., Meyerson, E.: DMVP: foremost waypoint coverage of time-varying graphs. In: Kratsch, D., Todinca, I. (eds.) WG 2014. LNCS, vol. 8747, pp. 29–41. Springer, Cham (2014). https://doi.org/10.1007/978-3-319-12340-0_3
2. Aaron, E., Krizanc, D., Meyerson, E.: Multi-robot foremost coverage of time-varying graphs. In: Gao, J., Efrat, A., Fekete, S.P., Zhang, Y. (eds.) ALGOSEN-SORS 2014. LNCS, vol. 8847, pp. 22–38. Springer, Heidelberg (2015). https://doi.org/10.1007/978-3-662-46018-4_2
3. Augustine, J., Pandurangan, G., Robinson, P.: Fast byzantine agreement in dynamic networks. In: Proceedings of the 32nd Symposium on Principles of Distributed Computing, PODC, pp. 74–83 (2013)
4. Awerbuch, B., Even, S.: Efficient and reliable broadcast is achievable in an eventually connected network. In: Proceedings of the 3rd Symposium on Principles of Distributed Computing, PODC, pp. 278–281 (1984)
5. Biely, M., Robinson, P., Schmid, U.: Agreement in directed dynamic networks. In: Even, G., Halldórsson, M.M. (eds.) SIROCCO 2012. LNCS, vol. 7355, pp. 73–84. Springer, Heidelberg (2012). https://doi.org/10.1007/978-3-642-31104-8_7
6. Bournat, M., Datta, A.K., Dubois, S.: Self-stabilizing robots in highly dynamic environments. In: Bonakdarpour, B., Petit, F. (eds.) SSS 2016. LNCS, vol. 10083, pp. 54–69. Springer, Cham (2016). https://doi.org/10.1007/978-3-319-49259-9_5
7. Carlsson, S., Jonsson, H., Nilsson, B.J.: Finding the shortest watchman route in a simple polygon. Discrete Comput. Geom. **22**(3), 377–402 (1999)
8. Casteigts, A., Flocchini, P., Quattrociocchi, W., Santoro, N.: Time-varying graphs and dynamic networks. Int. J. Parallel Emergent Distrib. Syst. **27**(5), 387–408 (2012)
9. Chan, M.Y., Chin, F.Y.L.: Schedulers for larger classes of pinwheel instances. Algorithmica **9**(5), 425–462 (1993)
10. Chin, W., Ntafos, S.C.: Optimum watchman routes. Inf. Process. Lett. **28**(1), 39–44 (1988)
11. Clementi, A., Monti, A., Pasquale, F., Silvestri, R.: Information spreading in stationary markovian evolving graphs. IEEE Trans. Parallel Distrib. Syst. **22**(9), 1425–1432 (2011)
12. Collins, A., et al.: Optimal patrolling of fragmented boundaries. In: 25th ACM Symposium on Parallelism in Algorithms and Architectures, SPAA 2013, Montreal, pp. 241–250 (2013)
13. Czyzowicz, J., Gąsieniec, L., Kosowski, A., Kranakis, E.: Boundary patrolling by mobile agents with distinct maximal speeds. In: Demetrescu, C., Halldórsson, M.M. (eds.) ESA 2011. LNCS, vol. 6942, pp. 701–712. Springer, Heidelberg (2011). https://doi.org/10.1007/978-3-642-23719-5_59
14. Czyzowicz, J., Gasieniec, L., Kosowski, A., Kranakis, E., Krizanc, D., Taleb, N.: When patrolmen become corrupted: monitoring a graph using faulty mobile robots. Algorithmica **79**(3), 925–940 (2017)
15. Di Luna, G.A., Baldoni, R.: Brief announcement: investigating the cost of anonymity on dynamic networks. In: Proceedings of the 34th Symposium on Principles of Distributed Computing, PODC, pp. 339–341 (2015)
16. Di Luna, G.A., Dobrev, S., Flocchini, P., Santoro, N.: Live exploration of dynamic rings. In: Proceedings of the 36th IEEE International Conference on Distributed Computing Systems, ICDCS, pp. 570–579 (2016)

17. Di Luna, G.A., Flocchini, P., Pagli, L., Prencipe, G., Santoro, N., Viglietta, G.: Gathering in dynamic rings. In: Das, S., Tixeuil, S. (eds.) SIROCCO 2017. LNCS, vol. 10641, pp. 339–355. Springer, Cham (2017). https://doi.org/10.1007/978-3-319-72050-0_20

18. Erlebach, T., Hoffmann, M., Kammer, F.: On temporal graph exploration. In: Halldórsson, M.M., Iwama, K., Kobayashi, N., Speckmann, B. (eds.) ICALP 2015. LNCS, vol. 9134, pp. 444–455. Springer, Heidelberg (2015). https://doi.org/10.1007/978-3-662-47672-7_36

19. Flocchini, P., Mans, B., Santoro, N.: On the exploration of time-varying networks. Theoret. Comput. Sci. **469**, 53–68 (2013)

20. Gąsieniec, L., Klasing, R., Levcopoulos, C., Lingas, A., Min, J., Radzik, T.: Bamboo garden trimming problem (perpetual maintenance of machines with different attendance urgency factors). In: Steffen, B., Baier, C., van den Brand, M., Eder, J., Hinchey, M., Margaria, T. (eds.) SOFSEM 2017. LNCS, vol. 10139, pp. 229–240. Springer, Cham (2017). https://doi.org/10.1007/978-3-319-51963-0_18

21. Harary, F., Gupta, G.: Dynamic graph models. Math. Comput. Model. **25**(7), 79–88 (1997)

22. Ilcinkas, D., Klasing, R., Wade, A.M.: Exploration of constantly connected dynamic graphs based on cactuses. In: Halldórsson, M.M. (ed.) SIROCCO 2014. LNCS, vol. 8576, pp. 250–262. Springer, Cham (2014). https://doi.org/10.1007/978-3-319-09620-9_20

23. Ilcinkas, D., Wade, A.M.: On the power of waiting when exploring public transportation systems. In: Fernàndez Anta, A., Lipari, G., Roy, M. (eds.) OPODIS 2011. LNCS, vol. 7109, pp. 451–464. Springer, Heidelberg (2011). https://doi.org/10.1007/978-3-642-25873-2_31

24. Ilcinkas, D., Wade, A.M.: Exploration of the T-interval-connected dynamic graphs: the case of the ring. Theory Comput. Syst. **62**(5), 1144–1160 (2018)

25. Kawamura, A., Kobayashi, Y.: Fence patrolling by mobile agents with distinct speeds. Distrib. Comput. **28**(2), 147–154 (2015)

26. Kowalski, D., Miguel, A.M.: Polynomial counting in anonymous dynamic networks with applications to anonymous dynamic algebraic computations. In: Proceedings of the 45th International Colloquium on Automata, Languages, and Programming, ICALP (2018, to appear)

27. Kuhn, F., Lynch, N., Oshman, R.: Distributed computation in dynamic networks. In: Proceedings of the 42nd Symposium on Theory of Computing, STOC, pp. 513–522 (2010)

28. Kuhn, F., Moses, Y., Oshman, R.: Coordinated consensus in dynamic networks. In: Proceedings of the 30th Symposium on Principles of Distributed Computing, PODC, pp. 1–10 (2011)

29. Kuhn, F., Oshman, R.: Dynamic networks: models and algorithms. SIGACT News **42**(1), 82–96 (2011)

30. Michail, O.: An introduction to temporal graphs: an algorithmic perspective. Internet Math. **12**(4), 239–280 (2016)

31. Michail, O., Spirakis, P.G.: Traveling salesman problems in temporal graphs. Theoret. Comput. Sci. **634**, 1–23 (2016)

32. Ntafos, S.C.: On gallery watchmen in grids. Inf. Process. Lett. **23**(2), 99–102 (1986)

33. O'Dell, R., Wattenhofer, R.: Information dissemination in highly dynamic graphs. In: Proceedings of the Joint Workshop on Foundations of Mobile Computing, DIALM-POMC, pp. 104–110 (2005)

Gathering of Robots in a Grid with Mobile Faults

Shantanu Das[1], Nikos Giachoudis[2], Flaminia L. Luccio[3],
and Euripides Markou[2]([✉])

[1] Aix-Marseille University, CNRS, LIS, Marseille, France
[2] DCSBI, University of Thessaly, Lamia, Greece
emarkou@dib.uth.gr
[3] DAIS, Università Ca' Foscari Venezia, Venezia, Italy

Abstract. The gathering of two or more agents in a graph is an important problem in the area of distributed computing and has been extensively studied especially for the fault free scenario. In this paper we consider the mobile agents gathering problem in the presence of an adversarial *malicious agent* which by occupying an empty node might prevent honest agents from entering this node. The honest agents move in synchronous rounds and at each round an agent can move to an adjacent node only if this node is not occupied by the malicious agent. We model the honest agents as identical finite state automata moving in an anonymous oriented grid topology and having no information about the size of the graph, while the malicious agent is assumed to be arbitrarily fast and to have full knowledge of the locations and the strategy of the honest agents at all times. The agents cannot leave messages at nodes or communicate with each-other unless they meet at a node. Previous studies consider the problem for ring networks and for asynchronous grids, where rendezvous was solved only for the special case of agents starting already in connected configurations. In this paper, we study the problem for synchronous agents in anonymous oriented grid networks for any number of agents starting in distinct locations. We first show that rendezvous is impossible for 2 agents even when the agents can see the locations of each-other at all times, while 3 agents can gather if they have global visibility. We then present a universal deterministic algorithm that solves the problem for 4 or more agents having only local visibility and constant memory, in any oriented grid with a malicious mobile adversary.

1 Introduction

Consider a set of mobile entities that are able to move in an environment and whose task is to meet at the same location. This is a fundamental problem in the area of distributed computing with mobile agents (or robots), e.g. they may need to meet to share information and to coordinate. The problem, called *rendezvous*, when there are two mobile agents, and *gathering* otherwise [16], has been widely studied when the environment is a graph and the agents can move along the edges of the graph. However, most of the studies are restricted to

© Springer Nature Switzerland AG 2019
B. Catania et al. (Eds.): SOFSEM 2019, LNCS 11376, pp. 164–178, 2019.
https://doi.org/10.1007/978-3-030-10801-4_14

fault-free environments and little is known about gathering in faulty or hostile environments. Possible faults can be a permanent failure of a node, like for example the so called *black hole* that destroys agents arriving at a node [12], or, transient faults that can appear anywhere in the graph and are controlled by a mobile hostile entity (an *intruder*) that behaves maliciously [2].

Most of the research that has been done in hostile environments is on the direction of how to locate a malicious node in a graph (see, e.g., [13,15]). Protecting the network against a malicious entity able to move along the edges of the graph, is generally a more difficult problem. Problems in this direction include the so-called *network decontamination* or *intruder capture* problem (see, e.g., [14,17]). Other types of faults or malicious behavior that have been considered in the context of the gathering problem are *Byzantine agents* [5,6,11], and *delay-faults* [4], and *edge evolving graphs* [18]. A Byzantine agent is indistinguishable from the legitimate or *honest* agents, but it may behave in an arbitrary manner and may provide false information to the good agents in order to induce them to make mistakes, thus preventing the rendezvous of the good agents. However, the Byzantine agents cannot actively harm the good agents or physically prevent the agents from moving. Delay-faults [4] can prevent an agent from moving for an arbitrary but finite time (i.e., they must eventually allow the agent to move), whereas probabilistic edge evolving graphs are dynamic networks where the set of communication links continuously changes thus preventing the use of standard gathering algorithms that work for static networks [18]. Gathering in dynamic ring networks has been studied in [10], under the 1-interval-connectivity model where the edges of the ring can disappear and reappear. Finally, the gathering problem has been also studied for robots moving on an unbounded plane under crash failures [1,3].

In this paper we consider a relatively new type of *malicious agent* that was first introduced in [9] and successively investigated in [7,8]. This malicious agent can move arbitrarily fast along the edges of a graph, it has full information about the graph and the location of all other honest agents, and it even has full knowledge of the actions that will be taken by the agents. The objective of the malicious agent is to prevent the good agents from gathering by blocking their path. More precisely, when the malicious agent occupies a node u of the graph, it can prevent (or *block*) the movement of any *honest* agent to node u, and at the same time is detected by those agents. In [7–9] it was shown how this malicious agent is a stronger adversary than the Byzantine agent or the Intruder agent, or the delay faults, as even one malicious agent can prevent rendezvous of honest agents in many cases.

This paper investigates the gathering of multiple honest agents scattered in a graph in the presence of such a malicious adversary. We assume that the honest agents are much weaker than the malicious agent; they are autonomous and identical anonymous processes with some constant internal memory, and two agents can communicate with each other only when they meet at a node. One or more honest agents present at a node v can prevent the malicious agent from entering node v. Thus, at any step the malicious agent can move along a path in the graph if it does not contain any honest agent.

The gathering problem has been investigated in [8,9] for *asynchronous* mobile agents moving in a ring or in a grid with one malicious agent. The asynchrony among the good agents combined with the ability of the malicious agent to move arbitrarily fast, gave the power to the malicious agent to prevent the next move of several agents at any stage of the algorithm. Thus, gathering of the good agents was shown to be impossible [9], in all cases except when the agents started from an initial configuration where the subgraph induced by all the occupied nodes was connected, and additionally, in the case of the grid graph, when the honest agents were able to "see" nodes at distance two in order to check if they were occupied or not.

To allow the possibility of gathering the good agents in more scenarios, we relax the constraint of asynchrony and assume that the honest agents move in synchronous steps. Note that, if two synchronous honest agents try to move to two distinct nodes at the same time, then the malicious agent can block the move of at most one agent, even though the malicious agent can execute moves of arbitrary distances between two consecutive steps of the good agents. This model with synchronous agents was studied in [7,8] only for the case of a ring, where it was already possible to solve gathering in more cases than with asynchronous agents. This paper explores the feasibility of gathering synchronous agents in anonymous oriented (2-dimensional) grid graphs of arbitrary size. Note that, such graphs can still be explored by agents having constant memory.

Contributions. We consider the gathering of a set of anonymous and synchronous honest mobile agents located at distinct nodes of an anonymous oriented grid graph in the presence of an arbitrarily fast malicious agent that can prevent any honest agent from visiting the node that it occupies. We assume that the agents have constant memory and our goal is to find the minimum number k of synchronous mobile agents that are able to gather in a grid despite the presence of the malicious agent and to design an algorithm for gathering of k or more honest agents in any grid network. First, in Sect. 3, we show an impossibility result for $k = 2$ agents, which holds even if the agents have unbounded memory and global visibility. We then present an algorithm that works for $k = 3$ agents with global visibility and finite memory. In Sect. 4 we show a general algorithm for any $k \geq 4$ when the agents have only local visibility. We conclude the paper in Sect. 5 and discuss open problems. Due to the space constraint, proofs and formal description of some results have been omitted.

2   Our Model

The Network: We consider an oriented grid graph $G = (V, E)$ consisting of n rows and m columns. We consider the non trivial case where $n, m > 1$. The nodes of G are anonymous, whereas the edges incident to a node are distinctly and consistently labeled with *North, South, East, West* labels. The edges of the network are FIFO links, i.e., all agents, including the malicious one, that move in the same link respect a FIFO ordering. Thus, an agent will pass another agent only if the latter stops at a node. When one or more *honest* agents are at a node v, we say that v is *occupied*. Otherwise we say that v is *free* or *unoccupied*.

Honest Agents: The agents are independent and identical, anonymous synchronous processes that move along the edges of the graph. They are initially scattered in the grid graph (i.e., there is at most one agent at a node), they have the same initial state and they start executing the same deterministic algorithm at the same time. An agent located at a node u can decide whether there are 1, 2, or at least 3 agents at u and can communicate with a co-located agent by reading its state. The agents cannot mark the nodes or edges of the graph. An agent arriving at a node v, learns the label of the incoming port and the labels of the outgoing ports of v. Two agents traveling on the same edge in different directions do not notice each other, and cannot meet on the edge. Their goal is to gather at a node. The agents do not know their number k and the size $n \times m$ of the grid network. We consider two visibility models. In the *local visibility* model at any time, an agent cannot see the state of any other node (i.e., whether it is occupied or not) apart from the node it occupies. In the *global visibility* model each agent, at any time, can see the locations of all the other agents in the grid. In both models an agent can communicate only with another agent at the same node, i.e., communication is face to face.

The agents have a constant number of states (independent of k, n, m), i.e. it has $O(1)$ bits of persistent memory. Since agents are synchronous, time is discretized into atomic time units. During each time unit, a (honest) agent arriving at a node u through a port q takes the following three actions: (1) It reads its own state, the outgoing port labels of u, decides whether there are 1, 2, or at least 3 agents at u, and reads the state of each agent at u[1]. (2) Based on the above information it performs some computation to decide its next destination and state. (3) The agent changes its state and either moves using the computed port label or waits at u. If the agent has decided to move on edge (u, v), and the node v is not occupied by the malicious agent, then the agent is located at node v in the next time unit. Otherwise, the agent is still located at node u in the next time unit, with a flag set in its memory notifying the agent that the move was unsuccessful.

Malicious Agent: The malicious agent M is a very powerful entity compared to honest agents: At any time, it has full knowledge of the graph and the positions and states of the honest agents; it has unlimited memory and knows the algorithm the honest agents follow. If M occupies a (free) node y at time t then M can either stay at y or appear at time $t + 1$ (i.e., move arbitrarily fast along the edges of the graph) at any other node w, as long as there is a path from y to w, such that no node on this path, including w, is occupied by any honest agent at time t. When M resides at a node u it prevents any honest agent A

---

[1] We notice that since an agent has only constant memory, it cannot simultaneously store in its memory the states of more than a constant number of agents.

from visiting u, i.e., it "blocks" A: If an agent A attempts to visit u, the agent receives a signal that M is in u.

3   Agents with Global Visibility

In this section we assume that the agents have global visibility, so they can see the entire grid and the locations of other agents in it, at all times during the algorithm. We will start by showing some basic properties for agents navigating in a grid.

When there is no malicious agent, any number of agents can easily gather in an oriented grid of size $n \times m$ within at most $n+m-2$ steps. Each agent simply moves to the North-East corner of the grid. We show below that when there is a malicious mobile agent in the grid, a similar strategy can gather at least $(k-1)$ out of the k agents: the malicious agent can block at most one agent at each step while the others make progress towards the North-East corner of the grid; if two or more agents are blocked at the same time, they could choose different paths.

Lemma 1. *In an oriented grid of size $n \times m$, with $n, m \geq 3$, k honest agents and one malicious agent, at least $(k-1)$ honest agents can gather within $k(n+m)/(k-1)$ steps under the global visibility model.*

Let us now recall the following property of a grid graph:

Property 1. An $n \times m$ grid graph $G = (V, E)$, with $n, m \geq 3$ has a minimum vertex cut of size 2, and every minimum vertex cut consists of a pair of nodes which are neighbours of a corner of the grid. This is depicted by the marked nodes in configuration C_2 of Fig. 1 (or any symmetric configuration).

3.1   Impossibility Result for Two Honest Agents

Lemma 2. *In an oriented grid of size $n \times m$, where $n, m > 3$, two agents cannot gather in the presence of one malicious agent, even if they have unlimited memory and global visibility.*

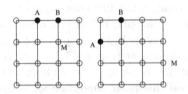

Fig. 1. Two agents gathering configurations C_1 and C_2.

Proof. Let $G = (V, E)$ be an oriented grid. Suppose that there is an algorithm $\mathcal{A}$ that gathers the two agents at a node of the grid at time t. Consider the last configuration of the agents before they meet (i.e., the configuration at time $t-1$). We first prove that the last configuration should either be the one in which: (1) the two agents are adjacent somewhere in the grid (let us call this configuration C_1) or, (2) they are both at distance one from the same corner of the grid (let us call this configuration C_2). Examples of those two configurations are shown in Fig. 1. Suppose for the sake of contradiction, that the last configuration C is not C_1 or C_2. In that case the agents in C have to be at distance two (but not as in C_2) at time $t - 1$, otherwise they cannot meet at time t. This means that both agents have to move to the same (free) node u in order to meet. From Property 1, if $n, m \geq 3$, we have that the only vertex cut of size less than 3 is a vertex cut of size 2 which is the one of configuration C_2, or the symmetric ones. Thus, in all other configurations, the subgraph induced by all free nodes (i.e., nodes which are not occupied by honest agents) is connected. Hence, in that case the malicious agent M can always reach the free node u before the agents, and thus prevents them from meeting each-other, even if the agents are in different states, have unlimited memory or they can see each-other locations. Thus, the last configuration before the gathering should be C_1 or C_2.

Let us now define configuration C_3 in which one of the agents occupies a corner node u of the grid and the other agent occupies the node v which is at distance 2 from u and not in the same row or column with u. We will show that unless the agents initially start in configuration C_1, C_2, or C_3, it is impossible to form any of those configurations, and hence it is impossible to gather. Suppose for the sake of contradiction that the agents are able to form a configuration of type $C \in \{C_1, C_2, C_3\}$ starting from a different configuration and let $C' \notin \{C_1, C_2, C_3\}$ be the last configuration before C is formed.

First observe that since C' is different than configuration C_2 and its symmetric ones, according to Property 1 configuration C' cannot have a vertex cut smaller than 3. Hence, if both agents at configuration C' try to move to the same node z, or only one agent tries to move to a node z in order to form configuration C, then M can reach node z before the agent(s) (since, in any configuration apart from C_2, M is able to reach any node which is not occupied). Therefore configuration C cannot be formed if at C' only one agent tries to move, or if both agents try to move at the same node. Let us study now the remaining case in which both agents at C' try to move to two distinct nodes.

Consider a configuration $C' \notin \{C_1, C_2, C_3\}$ composed of a node x containing agent A and a node y containing agent B. Suppose that the two agents A, B located at nodes x, y in configuration C' are trying to move to two distinct nodes z and w respectively in order to form configuration C. If z, w are not the occupied nodes of a configuration $C \in \{C_1, C_2, C_3\}$, then the malicious agent M does not block anyone and therefore C cannot be formed. If z, w are indeed the occupied nodes of a configuration $C \in \{C_1, C_2, C_3\}$ but either the pair (x, w) or the pair (z, y) are not the occupied nodes of a configuration $C \in \{C_1, C_2, C_3\}$, then the malicious agent M could block either node z (so that the new occupied nodes are (x, w)) or node w (so that the new occupied nodes are (z, y)) respectively. Hence, again the malicious agent M has a strategy to prevent the agents from forming configuration C.

The only remaining hypothetical scenario in which the malicious agent M would not be able to prevent the formation of configuration C from C' is when all pairs of nodes (x, w), (z, y), and (z, w) are the occupied nodes of configurations in $\{C_1, C_2, C_3\}$. We show below that this is impossible.

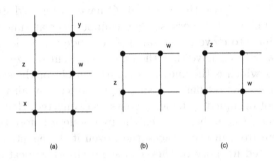

(a)   (b)   (c)

Fig. 2. Two agents in a $n \times m$ grid trying to move to two distinct nodes (z, w) from two distinct nodes (x, y) not in a configuration of type $\{C_1, C_2, C_3\}$ respectively. Nodes (z, w) are the occupied nodes of a configuration of type: (a) C_1 and $n, m > 3$, (b) C_2, (c) C_3.

- Suppose that the nodes (z, w) (in configuration C) are the occupied nodes of configuration type C_1. In other words, if both agents A, B move then the resulting configuration is connected (see an example in Fig. 2(a)). Node x could not be at w, since then configuration C' would be of type C_1. Node x can only be either North or South of w in Fig. 2(a), since otherwise (x, w) cannot be the occupied nodes of any configuration in $\{C_1, C_2, C_3\}$. Suppose without loss of generality that x is as shown in Fig. 2(a). Then y cannot be adjacent to x (otherwise configuration C' would be of type C_1) and can only be as shown in Fig. 2(a), since otherwise (z, y) cannot be the occupied nodes of any configuration in $\{C_1, C_2, C_3\}$. However, even in that case, the pairs of nodes (x, w) and (z, y) cannot be the occupied nodes of configuration types C_2 or C_3 if $n, m > 3$.
- Suppose that the nodes (z, w) (in configuration C) are the occupied nodes of configuration C_2 (see an example in Fig. 2(b)). Then nodes x, y have to be the other two nodes of the corner 2×2 subgrid in Fig. 2(b), otherwise (x, w) or (z, y) cannot be the occupied nodes of any configuration in $\{C_1, C_2, C_3\}$. However, this means that configuration C' was of type C_3, which is a contradiction.
- Suppose that the nodes (z, w) (in configuration C) are the occupied nodes of configuration C_3 (see an example in Fig. 2(c)). Then nodes x, y have to be the two nodes of the corner 2×2 subgrid in Fig. 2(c), otherwise (x, w) or (z, y)

cannot be the occupied nodes of any configuration in $\{C_1, C_2, C_3\}$. However, this means that configuration C' was of type C_2, which is a contradiction.

Hence, if the two agents initially start at a configuration of type different than C_1, C_2, or C_3, then they cannot form a configuration of type C_1, C_2, or C_3, and therefore they cannot gather. Notice that, this impossibility result holds even when the agents have unlimited memory and can see each-other's location at any time on the grid.   □

3.2   Gathering of Three Honest Agents

In this section, we show that under the global visibility model, even three honest agents (with constant memory) can gather in an oriented grid in presence of a malicious agent.

Some notations:

- Let C_0^3 be the set of all connected configurations with 3 agents (i.e., the nodes occupied by the agents form a connected subgraph of the grid).
- Let C_1^3 be the set of all configurations with 3 agents, where two agents are colocated and the third agent is at distance two from them on a straight line (i.e., either on the same row or on the same column, see an example in Fig. 3).
- Let C_2^3 be the set of all configurations with 3 agents, where two agents are colocated and the third agent is at distance two from them not on a straight line (i.e., the agents are not all on the same row or on the same column, see an example in Fig. 3).

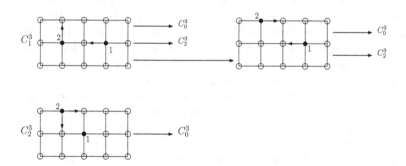

Fig. 3. Three agents in a grid: a tower of 2 and a single agent at distance two.

Theorem 1. *Three honest agents with global visibility can gather in an oriented grid in spite of one malicious agent.*

Proof. It is sufficient to show that the agents can form a connected configuration (it is straightforward to gather from a connected configuration if the agents can see each other). Due to Lemma 1, we know that 2 of the 3 agents can always gather at a node, if they have global visibility. So, let us assume that we start from a configuration where 2 agents are colocated (form a tower) and the third agent is in some distinct node of the grid. Due to the global visibility capability, the agents can approach each other, i.e., they can try to move to reduce the (vertical and then horizontal) distance between them. The two agents in the tower will move together during this process. Note that, if the distance between them is more than two then M can block either the tower or the third agent but not both at the same time. Thus, at each time step, the distance will be reduced until the distance is no more than two. If the distance is less than two then the agents already form a connected configuration and they can immediately gather. So, suppose that the agents reach a configuration where the distance between the tower and the third agent is exactly two.

Thus, this configuration can be either of type C_1^3 or of type C_2^3. We show that: (i) From a configuration of type C_1^3 we can reach a configuration of type C_2^3 or a connected configuration, and (ii) from a configuration of type C_2^3 we can always reach a connected configuration.

To prove (i), let us consider, w.l.o.g., the particular configuration $C \in C_1^3$ where the third agent is two steps to the EAST of the tower of two agents (see Fig. 3). Note that, for the other configurations in C_1^3, similar arguments hold, with rotation of directions etc. In configuration C, the algorithm will instruct the tower to perform GO(NORTH) and the solitary agent to perform GO(WEST). If both moves succeed, then the resulting configuration is in C_2^3 and we are done. If only the move of the tower is blocked then we have a connected configuration. So, we need to consider the only other case where the move of the solitary agent is blocked by M. The resulting configuration has the tower one step North and two steps West of the solitary agent (as shown in Fig. 3). From this configuration, in the next step, the algorithm will instruct the tower to perform GO(EAST) and the solitary agent to perform GO(WEST). If both moves succeed then we have a connected configuration and if either one of the moves is blocked then the resulting configuration is in C_2^3. Thus, we have proved (i).

To prove (ii), note that in any configuration of type C_2^3, there are two unoccupied nodes of the grid that are both at distance one from the tower and from the solitary agent. The algorithm will instruct the agents in the tower to split and move towards those two nodes respectively[2]. The malicious agent M cannot block the moves of both and if at least one of the moves succeeds then the resulting configuration is a connected configuration. Thus we have proved (ii). So, three agents with global visibility can always gather starting from any configuration.   □

Notice that, the above result can be extended to any number of k agents, by first forming a tower of $(k-1)$ agents (cf., Lemma 1) and repeating the technique of Theorem 1 with the $k-2$ agents of the tower acting as a single agent. However,

---

[2] We remind the reader, that as we have noticed in the beginning of the section, the agents can assign to themselves distinct identities and therefore they can perform distinct moves.

we will show in the next section that for $k > 3$ agents, gathering is still possible in the more challenging model with local visibility.

4 Agents with Local Visibility

In this section, we consider mobile agents that have local visibility and local communication capabilities. Thus, an agent has no knowledge of the location of any other agent unless it actually meets with an agent at a node. Although gathering is more difficult in this setting, we show that 4 or more agents can always gather starting from any starting configuration of scattered agents. We present a universal algorithm for gathering $k \geq 4$ agents, even without the prior knowledge of k. The high level description of the algorithm is the following: The algorithm first performs a partial gathering of at least 3 agents at a node. Such a group of colocated agents is called a *Tower*. As a next stage, the tower moves to the South-West corner of the grid, and starting from that corner, the tower moves towards the North-East corner, sweeping through all nodes and on the way collecting all the agents not belonging to any tower (note that multiple towers may be formed). Finally, all agents meet at the North-East corner of the grid (or at its adjacent node to the West).

The main algorithm, called GridWalk, is formally described in Algorithm 1. The algorithm uses a number of procedures. The first one is called Procedure FormTower and creates at least one tower of at least 3 colocated agents. We will show that a tower of 3 or more agents can always sweep through the grid, even if the malicious agent tries to block it. This is accomplished by two procedures. Procedure TowerWalk moves a tower to the next node in the intended direction, in spite of the malicious agent. Finally, Procedure ExploreLine is used to move a tower in a straight line in the chosen direction. We now describe each procedure in more detail.

Procedure FormTower: Each agent initially navigates towards the North-East corner of the grid, by first moving North up to the North border and then moving East up to the East border. During this navigation, if an agent is blocked by the malicious agent while moving North and it is alone, in the next move it will try to move East for one step (or West if it is located on the East border column). If at some point the agent (which still moves towards North) meets with exactly one more agent and both of them were blocked while trying to move North, then one of them moves East for one step (or West if it is located on the East border), while the other one moves North. If two agents which have already reached the North border are blocked on the North border while they try to move East, then one of the agents moves South for one step. The agents that reach the North-East corner wait until at least 3 agents are there.

Lemma 3. *If $k \geq 4$ agents execute Procedure FormTower in a $n \times m$ oriented grid, $n, m > 1$, with one malicious agent, then at least 3 agents will gather at a node within $O(n + m)$ time units.*

Algorithm 1. GridWalk (Gathering of $k > 3$ agents)

Perform FormTower;
if *there is an edge towards South* **then**
    | Perform ExploreLine(South);
if *there is an edge towards West* **then**
    | Perform ExploreLine(West);
/* The tower has reached the West border   */
while *there is an edge towards East* **do**
    | Perform ExploreLine(North);
    | **if** *ExploreLine returns* 1 **then**
    | | Perform TowerWalk(West, South);
    | Perform ExploreLine(South);
    | **if** *ExploreLine returns* 0 **then**
    | | Perform TowerWalk(East, North);
    | **if** *ExploreLine returns* 1 *and the ExploreLine was not performed at the West border* **then**
    | | Let u be the current node;
    | | Perform TowerWalk(East, North);
    | | **if** *tower ended up at distance* 2 *North-East of* u **then**
    | | | Let v be the current node;
    | | | Perform TowerWalk(South, East);
    | | | **if** *tower ended up South of* v **then**
    | | | | Perform TowerWalk(East, North);
    | **else**
    | | Perform TowerWalk(East, North);

/* The tower has reached the East border   */
Perform ExploreLine(North);
if *ExploreLine returns* 1 **then**
    | **while** *there is an edge towards East* **do**
    | | move East;

Consider $k \geq 3$ agents which gather at a node to form a tower. Even though the agents are all identical, since they initially start from distinct nodes, it is always possible to order the agents after they gather at a node (e.g., by comparing incoming directions, arrival times, etc.) Thus, we assume there exists a unique order on the agents constituting a tower, and we will denote the first three agents by $A1$, $A2$ and $A3$ in this order, while the fourth and subsequent agents (if any) would simply follow the actions of agent $A3$. We can thus explain the algorithm assuming that each tower contains only three agents $A1$, $A2$ and $A3$.

Procedure TowerWalk: This procedure moves all agents associated with the tower either to an adjacent node towards a given direction or to a node at distance 2 from the tower's current position. More precisely, given direction X

and Y (which is $90°$ clockwise or counterclockwise from X), the tower moves from position (x, y) to the node $(x + 1, y)$ which is adjacent to node (x, y) towards direction X; In case this node was blocked by the malicious agent, then TowerWalk(X, Y) moves the tower to node $(x + 1, y + 1)$ (the node adjacent to $(x + 1, y)$ in the direction Y). For example, an execution of Procedure TowerWalk(North, East) will move all agents associated with the tower either to node v adjacent to u in the North direction (if v is not blocked) or to node w at distance 2 from u in direction North-East (if v was blocked). The algorithm achieves this by splitting two agents of the tower in the direction X and Y, with the third agent staying in the initial location. At least one of the moves must be successful. If the intended node $(x + 1, y)$ is not reached, the agents are on two adjacent nodes (x, y) and $(x, y + 1)$, and one agent from each node tries to move in direction X. Thus, at least one agent reaches node $(x + 1, y)$ or $(x + 1, y + 1)$ as intended. The other agents now join this agent to reconstruct the tower.

Lemma 4. *Consider $k \geq 4$ agents in a $n \times m$ oriented grid, $n, m > 1$, with one malicious agent. If a tower of at least $l \geq 3$ agents located at node $u = (x, y)$ at time t, execute Procedure TowerWalk(X, Y), then after at most 6 time units, all agents in the tower reach either (1) node $v = (x + 1, y)$ which is adjacent to u towards direction X, only if v was not blocked at time $t + 1$ or $t + 3$, or, (2) node $w = (x + 1, y + 1)$ which is at a distance 2 from u towards direction $X - Y$, only if v was blocked at time $t + 1$ and $t + 3$.*

Procedure ExploreLine(Dir): This procedure moves a tower consisting of at least 3 agents from a node u, on a straight line direction Dir, reaching either a node v at the border of the grid on the same line, or a node v' adjacent to v on the border (in case v was blocked by the malicious agent). The procedure uses calls to Procedure TowerWalk(X, Y), where $X = Dir$ and Y is a direction $90°$ clockwise or counterclockwise from Dir. In simple words, the procedure moves the tower on a straight line sweeping a complete column (or a row) until it reaches the border. Suppose the tower was sweeping a column towards North: whenever the node on the North is blocked, the tower moves to the next column to the East; At each subsequent step, it tries to move back to the original column and otherwise continues towards North. The procedure ends when the tower reaches the border either on the original column or the adjacent column.

Lemma 5. *Consider a tower of $l \geq 3$ colocated agents located at a node u in a $n \times m$ oriented grid, $n, m > 1$, with one malicious agent. Let v be the node on the border of the grid towards direction Dir from u and let p be the distance (i.e., the length of the shortest path) between nodes u and v. If the tower-agents execute Procedure ExploreLine(Dir), then after $O(p)$ time units the agents will end up either at node v (in this case the procedure returns $Parity = 0$) or at a node on the border adjacent to v (in that case the procedure returns $Parity = 1$). During the execution of the procedure, the tower agents visit all nodes on the segment from u to v, except those that are blocked by the malicious agent.*

Any group of at least 3 agents (tower) tries to explore the whole grid from the South-West corner to the North-East corner by traversing each column back and forth and changing columns only through the South border. Any agent H which has not yet been associated with a tower is still executing Procedure FormTower. Let u be the first node in which the tower agents are blocked while they try to visit it at time t. Clearly, agent H could not be in u at time t. If node u is not on the North or East border nor adjacent to the North or East border then, according to Procedure FormTower agent H could only approach node u by moving either North or East (i.e., coming from nodes already 'cleared' by the tower agents). Hence agent H could not be in u at time t or later. For the remaining cases we show that if agent H moves again to node u at a later time then the tower agents are still there and meet agent H.

Theorem 2. *Consider $k \geq 4$ agents in a $n \times m$ oriented grid with a malicious agent, where $n, m > 1$. If the (honest) agents execute Algorithm GridWalk then they gather within at most $O(nm)$ time units.*

5   Conclusions

We studied the problem of gathering synchronous agents in oriented grids when there is a malicious mobile adversary. We showed that $k > 3$ synchronous agents with only local visibility capability can gather starting from any configuration without multiplicities, while for asynchronous agents it has been previously proved that the agents can gather only if they start from a connected configuration and have visibility at distance two. On the negative side we proved that two synchronous agents cannot gather even in the most powerful model with full visibility and unbounded memory.

Our studies on the oriented grids give an almost complete characterization of the solvable cases in these networks, leaving only one open question: Whether $k = 3$ agents can still gather in a grid starting from any configuration when they have only local visibility. As we proved in Lemma 1, $k - 1$ out of $k \geq 3$ agents with global visibility can easily gather. In fact, the proof of this lemma is only based on a two-distance visibility capability. However, the proof of Theorem 1 for 3 agents is heavily based on the global visibility property. In order for 3 agents to gather without the property of the global visibility, a different technique should be probably used.

A natural extension to this research would be the study of this problem in other topologies such as oriented multidimensional grids, and other well structured graphs that are easy to explore by constant memory agents. Another scenario is one with a less powerful malicious agent which has limited speed capabilities, or with multiple such mobile adversaries. For instance since the problem cannot be solved in a k-connected graph with k mobile adversaries an interesting question is to determine the maximum number of mobile adversaries for which the problem is solvable in a k-connected graph.

References

1. Agmon, N., Peleg, D.: Fault-tolerant gathering algorithms for autonomous mobile robots. SIAM J. Comput. **36**(1), 56–82 (2006)
2. Bampas, E., Leonardos, N., Markou, E., Pagourtzis, A., Petrolia, M.: Improved periodic data retrieval in asynchronous rings with a faulty host. Theoret. Comput. Sci. **608**, 231–254 (2015)
3. Bouzid, Z., Das, S., Tixeuil, S.: Gathering of mobile robots tolerating multiple crash faults. In: ICDCS 2013, pp. 337–346 (2013)
4. Chalopin, J., Dieudonne, Y., Labourel, A., Pelc, A.: Rendezvous in networks in spite of delay faults. Distrib. Comput. **29**, 187–205 (2016)
5. Chuangpishit, H., Czyzowicz, J., Kranakis, E., Krizanc, D.: Rendezvous on a line by location-aware robots despite the presence of byzantine faults. In: Fernández Anta, A., Jurdzinski, T., Mosteiro, M.A., Zhang, Y. (eds.) ALGOSENSORS 2017. LNCS, vol. 10718, pp. 70–83. Springer, Cham (2017). https://doi.org/10.1007/978-3-319-72751-6_6
6. Czyzowicz, J., Killick, R., Kranakis, E., Krizanc, D., Morale-Ponce, O.: Gathering in the plane of location-aware robots in the presence of spies. In: Lotker, Z., Patt-Shamir, B. (eds.) SIROCCO 2018, pp. 361–376. Springer, Cham (2018). https://doi.org/10.1007/978-3-030-01325-7_30
7. Das, S., Focardi, R., Luccio, F.L., Markou, E., Moro, D., Squarcina, M.: Gathering of robots in a ring with mobile faults. In: 17th Italian Conference on Theoretical Computer Science (ICTCS 2016), Lecce, Italy. CEUR, vol. 1720, pp. 122–135, 7–9 September 2016
8. Das, S., Focardi, R., Luccio, F.L., Markou, E., Squarcina, M.: Gathering of robots in a ring with mobile faults. Theor. Comput. Sci. (in press). https://doi.org/10.1016/j.tcs.2018.05.002
9. Das, S., Luccio, F.L., Markou, E.: Mobile agents rendezvous in spite of a malicious agent. In: Bose, P., Gąsieniec, L.A., Römer, K., Wattenhofer, R. (eds.) ALGOSENSORS 2015. LNCS, vol. 9536, pp. 211–224. Springer, Cham (2015). https://doi.org/10.1007/978-3-319-28472-9_16
10. Di Luna, G.A., Flocchini, P., Pagli, L., Prencipe, G., Santoro, N., Viglietta, G.: Gathering in dynamic rings. Theor. Comput. Sci. (in press). http://www.sciencedirect.com/science/article/pii/S030439751830639X
11. Dieudonne, Y., Pelc, A., Peleg, D.: Gathering despite mischief. ACM Trans. Algorithms **11**(1), 1 (2014)
12. Dobrev, S., Flocchini, P., Prencipe, G., Santoro, N.: Multiple agents RendezVous in a ring in spite of a black hole. In: Papatriantafilou, M., Hunel, P. (eds.) OPODIS 2003. LNCS, vol. 3144, pp. 34–46. Springer, Heidelberg (2004). https://doi.org/10.1007/978-3-540-27860-3_6
13. Dobrev, S., Flocchini, P., Prencipe, G., Santoro, N.: Mobile search for a black hole in an anonymous ring. Algorithmica **48**(1), 67–90 (2007)
14. Flocchini, P., Santoro, N.: Distributed security algorithms for mobile agents. In: Cao, J., Das, S.K. (eds.) Mobile Agents in Networking and Distributed Computing, pp. 41–70. Wiley, Hoboken (2012). Chap. 3
15. Klasing, R., Markou, E., Radzik, T., Sarracco, F.: Hardness and approximation results for black hole search in arbitrary graphs. TCS **384**(2–3), 201–221 (2007)
16. Lin, J., Morse, A.S., Anderson, B.D.O.: The multi-agent rendezvous problem. An extended summary. In: Kumar, V., Leonard, N., Morse, A.S. (eds.) Cooperative Control, vol. 309, pp. 257–289. Springer, Heidelberg (2004). https://doi.org/10.1007/978-3-540-31595-7_15

17. Luccio, F.L.: Contiguous search problem in Sierpinski graphs. Theory Comput. Syst. **44**, 186–204 (2009)
18. Yamauchi, Y., Izumi, T., Kamei, S.: Mobile agent rendezvous on a probabilistic edge evolving ring. In: ICNC, pp. 103–112 (2012)

Probabilistic Parameterized Polynomial Time

Nils Donselaar[✉]

Donders Institute for Brain, Cognition and Behaviour, Radboud University,
Montessorilaan 3, 6525 HR Nijmegen, The Netherlands
n.donselaar@donders.ru.nl

Abstract. We examine a parameterized complexity class for random-ized computation where only the error bound and not the full runtime is allowed to depend more than polynomially on the parameter, based on a proposal by Kwisthout in [15,16]. We prove that this class, for which we propose the shorthand name PPPT, has a robust definition and is in fact equal to the intersection of the classes paraBPP and PP. This result is accompanied by a Cook-style proof of completeness for the corresponding promise class (under a suitable notion of reduction) for parameterized approximation versions of the inference problem in Bayesian networks, which is known to be PP-complete. With these definitions and results in place, we proceed by showing how it follows from this that derandom-ization is equivalent to efficient deterministic approximation methods for the inference problem. Furthermore, we observe as a straightforward application of a result due to Drucker in [8] that these problems cannot have polynomial size randomized kernels unless the polynomial hierarchy collapses to the third level. We conclude by indicating potential avenues for further exploration and application of this framework.

Keywords: Parameterized complexity theory
Randomized computation · Bayesian networks

1 Preliminaries

The simple yet powerful idea which lies at the heart of the theory of parameter-ized complexity is that the hardness of computational problems may be better studied by analyzing the effects of particular aspects of its instances, treating these as a distinguished problem parameter and allowing the time (or other measures such as space) required to find a solution to depend on this parameter by an unbounded factor. This leads to an account of *fixed-parameter tractabil-ity* (FPT) and a hardness theory based on classes from the W-*hierarchy* which together mirror the parts played by P and NP in classical complexity theory.

In the two decades since the appearance of [6], the book by Downey and Fel-lows which largely formed the foundations of the field, research in parameterized complexity theory has gone far beyond this initial outlook and revealed a rich

Research for this paper has been funded through NWO EW TOP grant 612.001.601.

B. Catania et al. (Eds.): SOFSEM 2019, LNCS 11376, pp. 179–191, 2019.
https://doi.org/10.1007/978-3-030-10801-4_15

structure and many interesting questions to pursue, a lot of which covered in the updated [7]. Yet with a few notable exceptions, little attention has been paid to probabilistic computation in the parameterized setting. The most encompassing effort thus far has been made by Montaya and Müller in [18], where they show amongst other things that the natural analogue BPFPT relates to other complexity classes in much the same way as does BPP in the classical setting.[1]

Our aim with the present paper is to improve on this situation by demonstrating that studying parameterized probabilistic computation amounts to more than simply reconstructing results from the classical setting (which can already be a non-trivial task, as evidenced by the work done in [18]), and that results obtained in this way can have broader theoretical and practical significance. In particular we study a complexity class intended to capture *probabilistic parameterized polynomial time computability*, which we shall thus refer to as PPPT for this reason.[2]

This class PPPT is informally defined by considering probabilistic algorithms for parameterized problems, except allowing not the runtime but instead only the error bound to depend on the parameter by more than a polynomial factor. As such, PPPT can be thought of as containing those problems in PP which are nevertheless close to being in BPP and hence randomized tractable in a certain sense. This perspective, which we will explore more rigorously later on, formed much of the motivation of the class's original proposal in [15].

In what follows, we assume the reader to be familiar with the basics of classical and parameterized complexity theory. However, we repeat the definitions of the complexity classes used here, mostly to facilitate the comparison with the class PPPT which we formally define in the next section. First recall the probabilistic complexity classes BPP and PP:

Definition 1. BPP *is the class of decision problems computable in time $|x|^c$ for some constant c by a probabilistic Turing machine which gives the correct answer with probability more than $\frac{1}{2} + |x|^{-d}$ for some constant d.*

Definition 2. PP *is the class of decision problems computable in time $|x|^c$ for some constant c by a probabilistic Turing machine which gives the correct answer with probability more than $\frac{1}{2}$.*

We mostly follow the original definition presented in [10] in that we consider a probabilistic Turing machine $\mathcal{M}$ to be a Turing machine with access to random bits which it may query at every step of its execution, and whose transition function may depend on the values read off in this way. However, we include the generalization that a probabilistic Turing machine $\mathcal{M}$ may query not one but $r_{\mathcal{M}}$ many random bits at each step, where as usual we drop the subscript when it can be inferred from the context. Before continuing we note that $r_{\mathcal{M}} \leq \log |\mathcal{M}|$.

---

[1] We also mention [3] which studies PFPT, the parameterized counterpart to PP.

[2] In [15,16] this class was proposed under the name FERT for *fixed-error randomized tractability*, intended to be reminiscent of FPT. We believe that the name PPPT is more appropriate as it calls into mind the class PP as well as ppt-reductions.

Definition 3. FPT *is the class of parameterized decision problems computable in time* $f(k)|x|^c$, *where* f *is a computable function in* k *and* c *is a constant.*

Definition 4. paraBPP *is the class of parameterized decision problems computable in time* $f(k)|x|^c$ *by a probabilistic Turing machine which gives the correct answer with probability more than* $\frac{1}{2} + |x|^{-d}$, *where* f *is a computable function in* k *and* c *and* d *are constants.*

The method of converting classical complexity classes C to parameterized classes paraC illustrated above originates from [9], and indeed FPT = paraP. One should keep in mind though that this construction does not yield the usual parameterized classes including BPFPT, as these are furthermore characterized by using at most $f(k) \log |x|$ random bits. (cf. [4]). In fact, as observed in [18], BPFPT = paraBPP if and only if P = BPP. As we shall see in the next section, a similar statement remains true when we replace paraBPP by the class PPPT.

2   Error Parameterization

We now provide a formal definition of the class PPPT:

Definition 5. *We say that a parameterized decision problem* A *is in* PPPT *if there exist a computable function* $f : \mathbb{N} \to (0, \frac{1}{2}]$, *a constant* $c \in \mathbb{N}$ *and a probabilistic Turing machine* $\mathcal{M}$ *which on input* (x, k) *halts in time* $(|x| + k)^c$ *with probability at least* $\frac{1}{2} + f(k)$ *of giving the correct answer.*

Based on the convention that the parameter value k is given as a unary string along with the rest of the input x, from this definition it is immediate that PPPT is a subclass of PP. Moreover, it can be shown that this definition is robust in two ways, which makes it easy to see that PPPT is also a subclass of paraBPP.

Proposition 1. *The class* PPPT *in Definition 5 remains the same if*

(i) *the error bound on a correct decision is* $f(k)|x|^{-d}$ *or* $\min(f(k), |x|^{-d})$ *instead;*
(ii) *the probability of a false positive is not bounded away from* $\frac{1}{2}$.

Proof. For (i), note that $\min(f(k)^2, |x|^{-2d}) \leq f(k)|x|^{-d} \leq \min(f(k), |x|^{-d})$, which shows that such error bounds may be used interchangeably. For independent identically distributed Bernoulli variables with $p = \frac{1}{2} + \epsilon$, Hoeffding's inequality states that the probability of the average over n trials being no greater than $\frac{1}{2}$ is at most $e^{-2n\epsilon^2}$. Thus for $\epsilon = \min(f(k)^2, |x|^{-2d})$ we may take the majority vote over $n = |x|^{4d}$ runs to obtain a probability of at least $\frac{1}{2} + f(k)^2$ of giving the right answer. To see this, observe that the claim is trivially true whenever $\min(f(k)^2, |x|^{-2d}) = f(k)^2$, while if $\min(f(k)^2, |x|^{-2d}) = |x|^{-2d}$ then we should have $\frac{1}{2} - f(k)^2 \geq e^{-2}$, which is always the case since $f(k)^2 \leq \frac{1}{4}$ by definition. As this is only a polynomial number of repetitions, any error bound of either of these two alternative forms can be amplified to conform to the one required by Definition 5 without exceeding the time restrictions.

For (ii), we provide essentially the same argument as can be used to show that the class PP remains the same under this less strict requirement. Suppose that $\mathcal{M}$ is a probabilistic Turing machine which decides the problem A in time $(|x| + k)^c$ for some c, with suitable f and d given such that Yes-instances are accepted with probability at least $\frac{1}{2} + f(k)|x|^{-d}$ and No-instances with probability at most $\frac{1}{2}$. Then we may consider a probabilistic Turing machine $\mathcal{M}'$ which on input (x, k) operates the same up to where $\mathcal{M}$ would halt, at which point the outcome of $\mathcal{M}$ is processed as follows. In case of a rejection, the outcome is simply preserved, while an acceptance is rejected with probability $\frac{1}{2}f(k)|x|^{-d}$ (which is possible within the original polynomial time bound). Now a Yes-instance (x, k) of $\mathcal{M}$ is accepted by $\mathcal{M}'$ with probability at least $(\frac{1}{2} + f(k)|x|^{-d})(1 - \frac{1}{2}f(k)|x|^{-d}) \geq \frac{1}{2} + \frac{1}{4}f(k)|x|^{-d}$, while No-instances are accepted with probability at most $\frac{1}{2} - \frac{1}{4}f(k)|x|^{-d}$. Thus $\mathcal{M}'$ halts in time polynomial in $|x| + k$ and bounds the probability of a false positive away from $\frac{1}{2}$ by some $f(k)|x|^{-d}$ as required, which concludes the proof.

One can extend the probabilistic amplification used in the proof of Proposition 1(i) to (instead) remove the term $f(k)$ from the error bound, at the cost of introducing this factor into the runtime. This is generally not allowed for PPPT, but permissible for paraBPP, hence PPPT is a subclass of paraBPP. In fact, this inclusion is easily shown to be strict.

Proposition 2. PPPT $\subsetneq$ paraBPP, *i.e.* PPPT *is a strict subclass of* paraBPP.

Proof. Let A be any decidable problem not in PP, and parameterize this problem by the input size, so that we obtain the problem $|x|$-A. It is immediate that $|x|$-A cannot be in PPPT since A is not in PP. On the other hand, $|x|$-A is in paraBPP as the parameterization permits an arbitrary runtime for the decision procedure for A, hence the inclusion PPPT $\subseteq$ paraBPP is strict.

Note that what the proof above actually shows is that there are problems in FPT which are not in PPPT. This tells us that the inclusion in the statement remains strict even if BPP = P, in which case we have that paraBPP = FPT: see Proposition 5.1 of [18]. Similar arguments can be used to show that PPPT $\subseteq$ FPT implies BPP = P. Most importantly however, we can extend these results by providing an exact characterization of PPPT, the proof of which relies on essentially the same strategy used to show that every problem in FPT is kernelizable.

Theorem 1. PPPT $=$ paraBPP $\cap$ PP. *In particular, if a problem k-A is in* paraBPP *and also in* PP *as an unparameterized problem, then k-A is in* PPPT.

Proof. Suppose k-A is as in the statement of the theorem. Then there is a probabilistic Turing machine $\mathcal{M}$ which on input (x, k) halts in time $f(k)|x|^c$ and makes the correct decision with probability at least $\frac{1}{2} + |x|^{-d}$. Furthermore, there is a probabilistic Turing machine $\mathcal{M}'$ which on input x halts in time $|x|^{c'}$ and makes the correct decision with probability at least $\frac{1}{2} + 2^{-r|x|^{c'}}$. Then on any given (x, k), we can first run $\mathcal{M}$ for $|x|^{c+1}$ steps, adopting its decision if it halts within that time. If it does, then we have given the correct answer with probability at least $\frac{1}{2} + |x|^{-d}$. If it does not, then we may conclude that $f(k) > |x|$, in which case we proceed by running $\mathcal{M}'$ which will halt in time $|x|^{c'}$ with a probability of at least $\frac{1}{2} + 2^{-r \cdot f(k)^{c'}} = \frac{1}{2} + g(k)$ of giving the correct answer. Thus in time $\mathcal{O}(|x|^{c+c'+1})$ we can decide with probability at least $\frac{1}{2} + \min(g(k), |x|^{-d})$ whether (x, k) is in k-A, which means k-A is in PPPT by Proposition 1. As we already observed that PPPT is a subclass of both paraBPP and PP, this yields the conclusion that PPPT = paraBPP ∩ PP as stated.

For context it may be valuable to note that the idea of a complexity class being the intersection of a parameterized and a classical one has been considered before: while [2] looked into W[P] ∩ NP, [18] briefly discussed BPFPT ∩ BPP. However, we believe it is important to note first of all that the class PPPT arose from a reasonably natural definition intended to capture a slightly weaker kind of parameterization, instead of its definition being explicitly constructed to ensure a correspondence to the intersection of paraBPP and PP. Furthermore, and perhaps most importantly, we can actually exhibit natural problems which are complete for (the promise version of) the class PPPT, something which has not yet been done for BPP. In the remainder of this section we shall describe these problems, which originate from the domain of approximate Bayesian inference. We subsequently prove completeness for these problems in Sect. 3, which sets us up to explore the main implications of our results in Sect. 4.

Below we provide a definition of a Bayesian network, so that we may introduce the problem of inference within such networks; for the reader interested in a more detailed treatment, we refer to a standard textbook such as [14].

Definition 6. *A Bayesian network is a pair $\mathcal{B} = (G, \mathrm{Pr})$, where $G = (V, A)$ is a directed acyclic graph whose nodes represent statistical variables, and Pr is a set of families of probability distributions containing for each node V, and each possible configuration $c_{\rho(V)}$ of the variables represented by its parents, a distribution $\mathrm{Pr}(c_V \mid c_{\rho(V)})$ over the possible outcomes of its represented variable.*

As reflected by the notation, one typically blurs the distinction between the node V and the statistical variable which it represents, so that one may use $\Omega(V)$ to refer directly to the set of possible outcomes of this variable.

One of the main computational problems associated with Bayesian networks is that of *inference*, which is to determine what the likelihood is of some given combination of outcomes, possibly conditioned on certain specified outcomes for another set of variables. Below we describe the corresponding decision problem.

BAYESIAN INFERENCE
Input: A Bayesian network $\mathcal{B} = (G, \mathrm{Pr})$, two sets of variables $H, E \subseteq V(G)$, joint value assignments $h \in \Omega(H)$ and $e \in \Omega(E)$, a rational threshold value $0 \leq q \leq 1$.
Question: Is $\mathrm{Pr}(h \mid e) > q$?

Based on whether $E = \emptyset$, we shall refer to this problem as INFERENCE or CONDITIONAL INFERENCE respectively, both of which are PP-complete (see also [19]). Despite their equivalence from a classical perspective, in terms of parameterized approximability the latter is more difficult: [16] has an overview of the main results known thus far. We consider the following specific parameterizations:

ε-INFERENCE

Input: A Bayesian network $\mathcal{B} = (G, \mathrm{Pr})$, a set of variables $H \subseteq V(G)$, a joint value assignment $h \in \Omega(H)$, and rational values $0 \leq q \leq 1$ and $0 < \epsilon \leq \frac{1}{2}$.
Parameter: $\lceil -\log \epsilon \rceil$.
Promise: $\mathrm{Pr}(h) \notin (q - \epsilon, q + \epsilon)$.
Question: Is $\mathrm{Pr}(h) > q$?

$\{\epsilon, \mathrm{Pr}(h), \mathrm{Pr}(e)\}$-CONDITIONAL INFERENCE

Input: A Bayesian network $\mathcal{B} = (G, \mathrm{Pr})$, two sets of variables $H, E \subseteq V(G)$, joint value assignments $h \in \Omega(H)$, $e \in \Omega(E)$, rational values $0 \leq q \leq 1$, $\epsilon > 0$.
Parameters: ϵ, $\mathrm{Pr}(h)$, and $\mathrm{Pr}(e)$.
Promise: $\mathrm{Pr}(h) \notin (q(1 + \epsilon)^{-1}, q(1 + \epsilon))$.
Question: Is $\mathrm{Pr}(h \mid e) > q$?

In what follows we use $\mathrm{Pr}(h) \pm \epsilon \geq q$ as a shorthand for $\mathrm{Pr}(h) > q$ with the promise that $\mathrm{Pr}(h) \notin (q - \epsilon, q + \epsilon)$, and similarly for the relative approximation; for more on this approach, see [17]. For completeness' sake, we remind the reader of the definition of a promise problem, based on [11].

Definition 7. *A promise problem A consists of disjoint sets A_{Yes} and A_{No} of Yes and No-instances respectively, where $A_{Yes} \cup A_{No}$ may be a strict subset of the set of all inputs; $A_{Yes} \cup A_{No}$ is called the promise of the problem A.*

We can separate promise problems into complexity classes just as we do for decision problems by converting the familiar definitions in the following way.

Definition 8. *Let C be any complexity class of decision problems. The class pC of promise problems is defined by applying C's criteria of membership to promise problems instead, e.g. pP is the class of promise problems A for which there exists a polynomial-time algorithm which answers correctly on its promise.*

We now show for both of the approximate inference problems given above that they are in the promise version of PPPT, the former by an explicit argument, the latter by an appeal to Theorem 1.

Proposition 3. ε-INFERENCE *is in* pPPPT.

Proof. By forward sampling the network (i.e. generating outcomes according to the distributions of each variable, following some topological ordering of the graph) and accepting with probability $1 - \frac{q}{2}$ if the sample agrees with h, and with probability $\frac{1}{2} - \frac{q}{2}$ if it does not, we arrive at a probability of acceptance of $\frac{1}{2} + \frac{1}{2}(\mathrm{Pr}(h) - q)$. Under the promise that $\mathrm{Pr}(h) \notin (q - \epsilon, q + \epsilon)$, the probability of giving the correct answer is now at least $\frac{1}{2} + \frac{\epsilon}{2}$, hence the problem is in pPPPT.

Proposition 4. $\{\epsilon, \mathrm{Pr}(h), \mathrm{Pr}(e)\}$-CONDITIONAL INFERENCE *is in* pPPPT.

Proof. In [12] it is shown that rejection sampling (which is forward sampling and dismissing the outcome if it does not agree with e) provides an algorithm which places $\{\epsilon, \Pr(h), \Pr(e)\}$-CONDITIONAL INFERENCE in paraBPP.[3] Because CONDITIONAL INFERENCE is itself in PP, by Theorem 1 we conclude that the parameterized problem is in pPPPT.

3   Reductions and Completeness

At this point we wish to show that the two parameterized problems considered in the previous section are actually complete for the class pPPPT. In order to do this, we first determine which notion of reduction is the most suitable with respect to the class PPPT, after which we identify the machine acceptance problem for pPPPT and demonstrate its completeness for the class under these reductions. We then construct an explicit reduction from this problem to ϵ-INFERENCE, and in turn reduce the latter to $\{\epsilon, \Pr(h), \Pr(e)\}$-CONDITIONAL INFERENCE, thereby establishing completeness for both of these problems.

First of all, it is evident that while some form of parameterized reduction is required, the usual fpt-reductions are unsuitable because they allow the runtime to contain a factor superpolynomial in the parameter value and so PPPT is not closed under these. Furthermore, the reductions cannot be probabilistic either: while this is possible for BPP since the error can be reduced to constant using probabilistic amplification, mitigating the parameterized error bound is generally impossible without parameterized runtime (unless P = PP). Thus in this context it makes sense to consider the notion of a *ppt-reduction*,[4] which was formally introduced in [1]:

Definition 9. *A* ppt-reduction *from A to B is a polynomial-time computable function* $h : (x, k) \mapsto (x', k')$ *such that there exists a polynomial g with the property that* $k' \leq g(k)$, *and furthermore* $(x, k) \in A$ *if and only if* $(x', k') \in B$.

However, we can remove the constraint that k' is bounded by a polynomial in k, and instead simply demand that its value depends only on k.[5] The resulting class of reductions is a slightly broader one for which we introduce the name *ppt-reduction*, for which it is easy to see that PPPT is closed under ppt-reductions. Thus we would like to exhibit a parameterized problem which is complete for PPPT under ppt-reductions; yet here we run into the same issue as with the class BPP, namely that it may be impossible to effectively decide whether a given probabilistic Turing machine has a suitably lower-bounded probability to be correct on all possible inputs. Hence we have to add this explicit requirement in the form of a promise, and so we arrive at the machine acceptance problem stated below and instead study completeness for the promise class pPPPT.

---

[3] Note that the parameter $\Pr(h)$ is only necessary here because we ask for a relative approximation: the same holds when considering INFERENCE instead.

[4] As [7] observes, the acronym 'ppt' can be read equally well as either "polynomial parameter transformation" or "parameterized polynomial transformation".

[5] I hereby express my gratitude to the anonymous reviewer who raised this point.

ERROR PTM ACCEPTANCE

Input: Two strings x and 1^n and a probabilistic Turing machine $\mathcal{M}$.

Parameter: A positive integer k.

Promise: For any valid input to $\mathcal{M}$ and after any number of steps, the probability of acceptance does not lie between $\frac{1}{2} - 2^{-k}$ and $\frac{1}{2} + 2^{-k}$.

Question: After n steps, does $\mathcal{M}$ accept x more often than it rejects?

Proposition 5. ERROR PTM ACCEPTANCE *is complete for* pPPPT.

Proof. The problem ERROR PTM ACCEPTANCE is straightforwardly seen to lie in pPPPT, as it can be decided by running $\mathcal{M}$ for n steps on input x and accepting only if it halts in an accepting state. Based on the promise this has a probability of at least $\frac{1}{2} + 2^{-k}$ of making the correct decision, while taking time polynomial in the input size, which satisfies the requirements for being in PPPT.

In turn, presenting a reduction from a problem $A \in$ pPPPT to ERROR PTM ACCEPTANCE may be done in the following way. Suppose that f, c and $\mathcal{M}$ witness that $A \in$ pPPPT, i.e. on input (x, k) the machine $\mathcal{M}$ will run in time at most $(|x| + k)^c$ and accept or reject based on whether $(x, k) \in A$ with probability at least $\frac{1}{2} + f(k)$. Then we can construct $\mathcal{M}'$ in polynomial time which on input (x, k) simulates the machine $\mathcal{M}$ for exactly $(|x| + k)^c$ steps, deferring the decision until then: this will itself take time at most $(|x| + k)^a$ for some constant $a \geq c$.

Using the above, we find that $(x, k) \in A$ precisely when after $|(x, k)|^a$ steps $\mathcal{M}'$ accepts (x, k) more often than it rejects, hence $(x, k) \in A$ if and only if $((x, k), 1^{(|x|+k)^a}, \mathcal{M}', \lceil -\log f(k) \rceil) \in$ ERROR PTM ACCEPTANCE. Since the remaining data (i.e. other than the machine $\mathcal{M}'$) can also be given in polynomial time, this describes a pppt-reduction as required to show that ERROR PTM ACCEPTANCE is complete for pPPPT under pppt-reductions.

We can now show the problem ϵ-INFERENCE to be complete for pPPPT as well, by providing what is essentially a Cook-style construction of a Bayesian network from the probabilistic Turing machine specification and number of steps and the input which together make up an instance of ERROR PTM ACCEPTANCE. In contrast to the previous result, the reduction resulting from this construction is a ppt-reduction rather than a pppt-reduction.

Theorem 2. ϵ-INFERENCE *is complete for* pPPPT.

Proof. Given an instance $(x, 1^n, \mathcal{M}, k)$ of ERROR PTM ACCEPTANCE we describe a ppt-reduction to ϵ-INFERENCE as follows. First, we construct the underlying graph of the Bayesian network $\mathcal{B}$ by stacking $n + 1$ layers of nodes and connecting these using an intermediate gadget. Any such layer i consists of $n + 1$ nodes $X_{i,0}, \ldots, X_{i,n}$ representing the potentially reachable cells of the machine tape, a pair of nodes TH_i and MS_i which track the current tape head position and machine state respectively, and a series of r nodes $B_{i,1}, \ldots, B_{i,r}$ which act as the random bits which the machine uses to determine its next step. This means that $\Omega(X_{i,j})$ consists of the tape alphabet (including blanks), $\Omega(TH_i) = \{0, \ldots, n\}$, $\Omega(MS_i)$ is the set of machine states, and finally $\Omega(B_{i,j}) = \{0, 1\}$.

Such a layer of nodes i is connected to its successor through a gadget consisting again of $n+1$ nodes $Y_{i,0}, \ldots, Y_{i,n}$, with the parents $\rho(Y_{i,0})$ of $Y_{i,0}$ being TH_i and $X_{i,0}$ and $\rho(Y_{i,j+1}) = \{Y_{i,j}, X_{i,j+1}\}$. These nodes $Y_{i,j}$ can be interpreted as storing the position of the tape head at step i and reading off the tape until the correct cell $X_{i,j}$ is encountered, after which its symbol is copied and carried over all the way to $Y_{i,n}$. To achieve this, we require $\Omega(Y_{i,j})$ to be the disjoint union of $\Omega(TH_i)$ and $\Omega(X_{i,j})$. Now the layer i combined with its gadget is connected to the next one by setting $\rho(X_{i+1,j}) = \{X_{i,j}, MS_i, TH_i, Y_{i,n}, B_{i,1}, \ldots, B_{i,n}\}$, $\rho(TH_{i+1}) = \{MS_i, TH_i, Y_{i,n}, B_{i,1}, \ldots, B_{i,n}\}$ and $\rho(MS_{i+1}) = \{MS_i, Y_{i,n}, B_{i,1}, \ldots, B_{i,n}\}$.

We now have to assign probability distributions to each of these nodes such that they fulfill their intended purposes. First of all, the nodes $B_{i,j}$ are all uniformly distributed so that they may be correctly regarded as random bits. As for the first row, the remaining nodes are fixed to the first $n + 1$ cells of the tape input, the tape head starting location and the initial state of the machine. All other nodes in the network have similar distributions which are deterministic given the values of their parents. In particular, $Y_{i,j} = X_{i,j}$ if $Y_{i,j-1} = j$ and $Y_{i,j} = Y_{i,j-1}$ otherwise (here TH_i should be read for $Y_{i,-1}$), $X_{i+1,j} = X_{i,j}$ unless $TH_i = j$ in which case $MS_i, Y_{i,n}$ and $B_{i,1}, \ldots, B_{i,n}$ together determine the symbol overwriting the previous one according to the transition function of $\mathcal{M}$, and in general the values of TH_{i+1} and MS_{i+1} follow from those of its parents based on this transition function as well.

The reduction can now be straightforwardly expressed as follows: an instance $(x, 1^n, \mathcal{M}, k)$ is mapped to an instance $(\mathcal{B}, MS_n, s_{accept}, \frac{1}{2}, k)$ of ϵ-INFERENCE, where $\mathcal{B}$ is constructed as above and s_{accept} is the accepting state of $\mathcal{M}$. Then as required we have that after n steps $\mathcal{M}$ accepts x more often than it rejects if and only if $\Pr(MS_n = s_{accept}) \pm 2^{-k} \geq \frac{1}{2}$. Since $\mathcal{B}$ is of size polynomial in n and $|\mathcal{M}|$ (in particular because the conditional probability distribution at every node is of polynomial size) and the parameter remains unchanged, this indeed describes a ppt-reduction, which completes the proof.

In turn, we can reduce ϵ-INFERENCE to $\{\epsilon, \Pr(h), \Pr(e)\}$-CONDITIONAL INFERENCE, thereby extending the hardness and hence completeness to the latter.

Corollary 1. $\{\epsilon, \Pr(h), \Pr(e)\}$-CONDITIONAL INFERENCE *is* pPPPT-*complete.*

Proof. Given an instance $(\mathcal{B}, H, h, q, \epsilon)$ of ϵ-INFERENCE, we can adjust $\mathcal{B}$ by building an inverse binary tree below the nodes in H, with terminal node T_H being h when $H = h$ and $\neg h$ otherwise. We then furthermore add an initial, uniformly distributed binary node R and another binary node S with parents R and T_H, distributed as follows:

$$\Pr(s \mid R, T_H) = \begin{cases} 1 & \text{for } R = r, T_H = h \\ 0 & \text{for } R = \neg r, T_H = h \\ \frac{1}{2} & \text{otherwise} \end{cases}$$

Now $\Pr(r) = \Pr(s) = \frac{1}{2}$, and moreover $\Pr(r \mid s) = \frac{1}{2} + \frac{1}{2}\Pr(h)$, hence we find that $\Pr(r \mid s) \pm \frac{1}{2}\epsilon \geq \frac{1}{2} + \frac{1}{2}q$ if and only if $\Pr(h) \pm \epsilon \geq q$. This therefore describes a pppt-reduction from ϵ-INFERENCE to $\{\epsilon, \Pr(h), \Pr(e)\}$-CONDITIONAL INFERENCE, albeit not a ppt-reduction as an artefact of the particular choice of parameter value corresponding to ϵ. The result then follows from Theorem 2.

To conclude this section, we discuss a question which may have occurred to the reader, namely whether one could simplify this approach by avoiding the inference problems altogether and working instead with the following variant of MAJSAT, which is the satisfiability problem complete for PP.

GAP-MAJSAT
Input: A propositional formula φ.
Parameter: A positive integer k.
Promise: The ratio of satisfying truth assignments of φ does not lie between $\frac{1}{2} - 2^{-k}$ and $\frac{1}{2} + 2^{-k}$.
Question: Is φ satisfied by more than half of its possible truth assignments?

The issue here is that the canonical reduction from ERROR PTM ACCEPTANCE to GAP-MAJSAT requires a number of variables proportional to both n and the size of the machine $\mathcal{M}$, hence the original margin of $\pm 2^{-k}$ will shrink by a factor in the input size. The resulting parameter for the GAP-MAJSAT instance will thus depend on the input size, which means this reduction is not even an fpt-reduction. This points to a phenomenon also observed in the W-hierarchy, where W[SAT] (which is defined in terms of a parameterized satisfiability problem) is believed to a proper subclass of W[P] (which is defined in terms of a parameterized circuit satisfiability or machine acceptance problem). That the reduction does work for the inference problems suggests that Bayesian networks do have the direct expressive power of Turing machines lacked by propositional formulas.

4   Application of Results

Ultimately, one of the main open questions in the area of probabilistic computation is whether P = BPP. In contrast to the more famous open question whether P = NP, the generally accepted view is that BPP is likely to equal P, based on works such as [13]. However, due to the lack of natural problems which are known to be BPP-complete, it has not been possible to focus efforts on proving a particular problem to lie in P in order to demonstrate the collapse of BPP. We believe that our work makes an important contribution in that it indirectly provides a problem which can play this part, namely ϵ-INFERENCE. This relies in part on the following proposition adapted from [18] which we hinted at earlier.

Proposition 6. PPPT $\subseteq$ FPT *if and only if* P = BPP.

Proof. Suppose PPPT $\subseteq$ FPT, and let A be an arbitrary problem in BPP. Then certainly $A \in$ PP, and also $A \in$ paraBPP for any constant parameterization, hence $A \in$ PPPT by Theorem 1. By assumption it follows that $A \in$ FPT, hence there is a deterministic algorithm for A which runs in time $f(k)|x|^c$. But now the factor $f(k)$ is a constant term, which means A is actually in P by this algorithm.

Conversely, suppose P = BPP, and let A be an arbitrary problem in PPPT with corresponding error bound function $f(k)$. Given an instance (x, k) of A we can determine whether $|x| \leq f(k)^{-1}$: for the instances where this is true, the problem is in FPT, while it is in BPP for those where it is false. By assumption the latter problem is now in P, which means the entire problem A is in FPT.

Combined with Theorem 2, we arrive at the following result:

Theorem 3. P = BPP *if and only if there exists an efficient deterministic absolute approximation algorithm for* INFERENCE, *i.e. a deterministic approximation which runs in time* $f(\epsilon^{-1})|x|^c$ *for some constant c and computable function f.*

Proof. Since ϵ-INFERENCE $\in$ paraBPP and paraBPP = FPT whenever P = BPP, this part of the equivalence is already established. By Theorem 2 we know that ϵ-INFERENCE is pPPPT-complete, hence PPPT $\subseteq$ FPT if the problem has an fpt-algorithm, by which it follows from Proposition 6 that P = BPP. $\qquad \blacksquare$

At the same time, we can provide some indication as to the hardness of ϵ-INFERENCE by means of the framework of kernelization lower bounds, which is where the notion of a ppt-reduction originated. Here we consider the following reformulation, inspired by [5], of a theorem by Drucker found in [8].

Theorem 4. *If A is an NP-hard or coNP-hard problem and B is a parameterized problem such that there exists a polynomial-time algorithm which maps any tuple $(x_1, \ldots, x_t)$ of n-sized instances of A to an instance y of B such that*

1. *if all x_i are No-instances of A, then y is a No-instance of B;*
2. *if exactly one x_i is a Yes-instance of A, then y is a Yes-instance of B;*
3. *the parameter k of y is bounded by $t^{o(1)}n^c$ for some constant c;*

then B has no randomized (two-sided constant error) polynomial-sized kernels unless coNP $\subseteq$ NP/poly, *collapsing the polynomial hierarchy to the third level.*

We can use this Theorem to prove that ϵ-INFERENCE has no randomized polynomial-sized kernels unless the polynomial hierarchy collapses.

Proposition 7. ϵ-INFERENCE *has no randomized polynomial-sized kernel unless* coNP $\subseteq$ NP/poly.

Proof. Consider the NP-hard problem SAT, and let $\varphi_1, \ldots, \varphi_t$ be propositional formulas in n variables. We can rename the variables so that every formula uses the same $x_1, \ldots, x_n$ if necessary, introduce a new variable x_0, and take the disjunction $\psi = \bigvee_{i=1}^{t} \varphi_i \vee x_0$. Then ψ is a formula with $n + 1$ variables with a majority of its truth assignments being satisfying if and only if at least one of the φ_i is satisfiable, hence $(\psi, n+1) \in$ GAP-MAJSAT if and only if $\varphi_i \in$ SAT for some i. Thus by Theorem 4 GAP-MAJSAT does not have randomized polynomial-sized kernels unless coNP $\subseteq$ NP/poly. Furthermore, by [1] this property is closed under ppt-reductions, and the usual reduction from GAP-MAJSAT to ϵ-INFERENCE (which amounts to constructing a Boolean circuit out of the given formula) is in fact a ppt-reduction, hence neither does ϵ-INFERENCE have randomized polynomial-sized kernels under the assumption that coNP $\nsubseteq$ NP/poly.

While perhaps unsurprising, this result serves in particular as a reminder that hard problems in PPPT such as ϵ-INFERENCE are not solvable by polynomial kernelization followed by a probabilistic (PP) algorithm.

5   Closing Remarks

In this paper we have explored the proposal made in [15,16] of an alternative parameterized randomized complexity class, which we have called PPPT and of which we have shown that it is identical to the intersection of PP and paraBPP. In the preceding sections we showed that the problem ϵ-INFERENCE is a natural fit for this class, as it is not only a member of the class in a straightforward way (Proposition 3), it is moreover complete for the corresponding problem class (Theorem 2). Because of the close relation between classical and parameterized probabilistic computation (Proposition 6), the class PPPT turns out to have unexpected broader relevance, as finding an efficient deterministic absolute approximation algorithm for INFERENCE is necessary and sufficient for the derandomization of BPP to P (Theorem 3).

In other words, we are in the fortunate circumstances where efforts to address a long-standing open question originating in theory can actually coincide with the search for a novel algorithm capable of solving a practical problem, and most importantly the existence of such an algorithm actually follows from a conjecture supported by other considerations. With this paper we wish to call attention to this opportunity for researchers with theoretical and practical motivations alike to engage with a challenge which is broadly relevant to multiple research communities at once. It is our hope that such focused efforts on the ϵ-INFERENCE problem may lead to a valuable breakthrough in both the fields of structural complexity theory and of probabilistic graphical models.

Acknowledgement. The author thanks Johan Kwisthout and Hans Bodlaender for sharing insightful remarks in his discussions with them, and also Ralph Bottesch for providing useful comments on an early draft of this paper.

References

1. Bodlaender, H.L., Thomassé, S., Yeo, A.: Kernel bounds for disjoint cycles and disjoint paths. Theor. Comput. Sci. **412**, 4570–4578 (2011)
2. Cai, L., Chen, J., Downey, R.G., Fellows, M.R.: On the structure of parameterized problem in NP. In: Enjalbert, P., Mayr, E.W., Wagner, K.W. (eds.) Proceedings of STACS 1994, pp. 507–520 (1994)
3. Chauhan, A., Rao, B.V.R.: Parameterized analogues of probabilistic computation. In: Ganguly, S., Krishnamurti, R. (eds.) Algorithms and Discrete Applied Mathematics, pp. 181–192 (2015)

4. Chen, Y., Flum, J., Grohe, M.: Machine-based methods in parameterized complexity theory. Theor. Comput. Sci. **339**, 167–199 (2005)
5. Dell, H.: AND-compression of NP-complete problems: streamlined proof and minor observations. Algorithmica **75**, 403–423 (2016)
6. Downey, R.G., Fellows, M.R.: Parameterized Complexity. Springer, Heidelberg (1999). https://doi.org/10.1007/978-1-4612-0515-9
7. Downey, R.G., Fellows, M.R.: Fundamentals of Parameterized Complexity. Springer, Heidelberg (2013). https://doi.org/10.1007/978-1-4471-5559-1
8. Drucker, A.: New limits to classical and quantum instance compression. Technical report TR12-112, Electronic Colloquium on Computational Complexity (ECCC) (2014). http://eccc.hpi-web.de/report/2012/112/
9. Flum, J., Grohe, M.: Describing parameterized complexity classes. Inf. Comput. **187**, 291–319 (2003)
10. Gill, J.: Computational complexity of probabilistic turing machines. SIAM J. Comput. **6**(4), 675–695 (1977)
11. Goldreich, O.: On promise problems: a survey. In: Goldreich, O., Rosenberg, A.L., Selman, A.L. (eds.) Theoretical Computer Science. LNCS, vol. 3895, pp. 254–290. Springer, Heidelberg (2006). https://doi.org/10.1007/11685654_12
12. Henrion, M.: Propagating uncertainty in Bayesian networks by probabilistic logic sampling. In: Lemmer, J.F., Kanal, L.N. (eds.) Uncertainty in Artificial Intelligence, Machine Intelligence and Pattern Recognition, vol. 5, pp. 149–163 (1988)
13. Impagliazzo, R., Wigderson, A.: P = BPP if E requires exponential circuits: derandomizing the XOR lemma. In: Proceedings of STOC 1997, pp. 220–229 (1997)
14. Koller, D., Friedman, N.: Probabilistic Graphical Models: Principles and Techniques. MIT Press, Cambridge (2009)
15. Kwisthout, J.: Tree-width and the computational complexity of MAP approximations in Bayesian networks. J. Artif. Intell. Res. **53**, 699–720 (2015)
16. Kwisthout, J.: Approximate inference in Bayesian networks: parameterized complexity results. Int. J. Approx. Reason. **93**, 119–131 (2018)
17. Marx, D.: Parameterized complexity and approximation algorithms. Comput. J. **51**, 60–78 (2008)
18. Montoya, J.A., Müller, M.: Parameterized random complexity. Theory Comput. Syst. **52**, 221–270 (2013)
19. Park, J.D., Darwiche, A.: Complexity results and approximation strategies for MAP explanations. J. Artif. Intell. Res. **21**, 101–133 (2004)

On Matrix Ins-Del Systems of Small Sum-Norm

Henning Fernau[1]([✉]), Lakshmanan Kuppusamy[2], and Indhumathi Raman[3]

[1] Fachbereich 4 - Abteilung Informatikwissenschaften, CIRT,
Universität Trier, 54286 Trier, Germany
fernau@uni-trier.de
[2] School of Computer Science and Engineering, VIT, Vellore 632 014, India
klakshma@vit.ac.in
[3] Department of Applied Mathematics and Computational Sciences,
PSG College of Technology, Coimbatore 641 004, India
ind.amcs@psgtech.ac.in

Abstract. A matrix ins-del system is described by a set of insertion-deletion rules presented in matrix form, which demands all rules of a matrix to be applied in the given order. These systems were introduced to model very simplistic fragments of sequential programs based on insertion and deletion as elementary operations as can be found in biocomputing. We are investigating such systems with limited resources as formalized in descriptional complexity. A traditional descriptional complexity measure of such a system is its ins-del size. Summing up the according numbers, we arrive at the *sum-norm*. We show that matrix ins-del systems with sum-norm 4 and (i) maximum length 3 with only one of insertion or deletion being performed under a one-sided context, or (ii) maximum length 2 with both insertion and deletion being performed under a one-sided context, can describe all recursively enumerable languages. We also show that if a matrix ins-del system of size s can describe the class of linear languages LIN, then without any additional resources, matrix ins-del systems of size s also describe the regular closure of LIN.

Keywords: Matrix ins-del systems · Computational completeness
Regular closure of linear languages

1 Introduction

Inserting or deleting words in between parts of sentences often take place when processing natural languages. Insertion and deletion together were first studied in [11]. The corresponding grammatical mechanism is called an *insertion-deletion system* (abbreviated as ins-del system). Informally, if a string η is inserted between two parts w_1 and w_2 of a string $w_1 w_2$ to get $w_1 \eta w_2$, we call the operation *insertion*, whereas if a substring δ is deleted from a string $w_1 \delta w_2$ to get $w_1 w_2$, we call the operation *deletion*. Suffixes of w_1 and prefixes of w_2 are called *contexts*.

© Springer Nature Switzerland AG 2019
B. Catania et al. (Eds.): SOFSEM 2019, LNCS 11376, pp. 192–205, 2019.
https://doi.org/10.1007/978-3-030-10801-4_16

Several variants of ins-del systems have been considered in the literature, imposing regulation mechanisms on top, motivated by classical formal language theory [1]. We refer to the survey paper [17]. In this paper, we focus on *matrix ins-del systems* [13,15]. Viewing insertions and deletions as elementary operations for biocomputing [14], matrices can be seen as a very simple control mechanism.

In a matrix ins-del system, the insertion-deletion rules are given in matrix form. If a matrix is chosen for derivation, then all the rules in that matrix are applied in order and no rule of the matrix is exempted. In the size $(k; n, i', i''; m, j', j'')$ of a matrix ins-del system, the parameters (from left to right) denote the maximum number of rules (*length*) in any matrix, the maximal length of the inserted string, the maximal length of the left context for insertion, and the maximal length of the right context for insertion; a similar list of three parameters concerning deletion follows. We denote the language classes generated by matrix ins-del systems of size s by MAT(s). The tuple formed by the last six parameters, namely $(n, i', i''; m, j', j'')$, is called the *ins-del size*. We call the sum of its parameters the *sum-norm* of the (matrix) ins-del system.

It is known that ins-del systems are computationally complete, i.e., they characterize the family RE of recursively enumerable languages, which readily transfers to the mentioned variants. Descriptional complexity then aims at investigating which of the resources are really needed to obtain computational completeness. For instance, is it really necessary to permit insertion operations that check out contexts of arbitrary length? For resource restrictions that do not (or are not known to) suffice to achieve computational completeness, one is interested in seeing which known families of languages can be still generated. As in our case, for several families of matrix ins-del systems, it is even unknown if all of CF (the context-free languages) can be generated, we then look at the rather large sub-family $\mathcal{L}_{reg}(\text{LIN})$, the *regular closure* of LIN. In Table 1, we report on what resources are needed for a matrix ins-del system of sum-norm 3 or 4 to generate the class specified there, also giving a short literature survey. Further *races for smaller sizes* are described when we discuss the particularities of our results below. Let us highlight that Theorem 4 solves an open problem stated in [12]. This underlines the interest in this type of research. Also, the open questions that we list are sometimes connected, sometimes not. For instance, by the closure of RE under reversal, MAT($*; 1, 0, 0; 1, 1, 0$) = RE iff MAT($*; 1, 0, 0; 1, 0, 1$) = RE, but the questions if MAT($*; 1, 0, 0; 2, 0, 0$) = RE or if MAT($*; 2, 0, 0; 1, 0, 0$) = RE seem to be independent of each other.

Matrix ins-del systems of sizes $(3; 1, 1, 0; 1, 0, 0), (3; 1, 0, 1; 1, 0, 0), (2; 2, 1, 0; 1, 0, 0), (2; 2, 0, 1; 1, 0, 0)$ are not known to describe RE, not even CF. However, these systems have been shown to describe the class MLIN of metalinear languages [6]. It is known that MAT($1; 1, 1, 0; 1, 0, 0$)$\cup$MAT($3; 1, 0, 0; 1, 0, 0$) $\neq$ RE [2]. We show that if LIN $\subseteq$ MAT(s), then $\mathcal{L}_{reg}(\text{LIN}) \subseteq$ MAT(s). As a consequence, it follows that matrix ins-del systems of the sizes mentioned above contain $\mathcal{L}_{reg}(\text{LIN})$.

A further technical contribution consists in formulating a new normal form, called *time separating special Geffert normal form* (tsSGNF), that allows to

Table 1. Generative power of $MAT(k; n, i', i''; m, j', j'')$ with sum-norm 3 or 4

Sum-norm	ID size $(n, i', i''; m, j', j'')$, where $n, m \in \{1, 2\}$ and $i', i'', j', j'' \in \{0, 1, 2\}$	length k	Language family relation	Remarks
3	$(1, 0, 0; 2, 0, 0)$, $(2, 0, 0; 1, 0, 0)$	≥ 1	?	OPEN
3	$(1, 0, 0; 1, 1, 0)$, $(1, 0, 0; 1, 0, 1)$	≥ 1	?	OPEN
3	$(1, 1, 0; 1, 0, 0)$, $(1, 0, 1; 1, 0, 0)$	3	$\supset$ MLIN	[6]
		3	$\supset \mathcal{L}_{reg}(\text{LIN})$	Corollary 1
4	$(1, 0, 0; 1, 1, 1)$, $(1, 1, 1; 1, 0, 0)$	3	RE	[6]
4	$(1, 0, 0; 2, 1, 0)$, $(1, 0, 0; 2, 0, 1)$	≥ 1	?	OPEN
4	$(1, 0, 0; 1, 2, 0)$, $(1, 0, 0; 1, 0, 2)$	3	RE	Theorem 2
4	$(1, 2, 0; 1, 0, 0)$, $(1, 0, 2; 1, 0, 0)$	3	RE	Theorem 3
4	$(1, 1, 0; 1, 1, 0)$, $(1, 0, 1; 1, 0, 1)$	3	RE	[6, 15]
		2	RE	Theorem 4
4	$(1, 1, 0; 1, 0, 1)$, $(1, 0, 1; 1, 1, 0)$	3	RE	[6, 15]
4	$(1, 1, 0; 2, 0, 0)$, $(1, 0, 1; 2, 0, 0)$	2	RE	[6, 15]
4	$(2, 0, 0; 2, 0, 0)$	≥ 1	?	OPEN
4	$(2, 0, 0; 1, 1, 0)$, $(2, 0, 0; 1, 0, 1)$	2	RE	[6, 15]
4	$(2, 1, 0; 1, 0, 0)$, $(2, 0, 1; 1, 0, 0)$	2	$\supset$ MLIN	[6]
		2	$\supset \mathcal{L}_{reg}(\text{LIN})$	Corollary 1

simplify some arguments, because in particular there is no way to have mixtures of terminals and nonterminals at the right end of a derivable sentential form. Hence, such mixed cases need not be considered when proving correctness of simulation results based on tsSGNF. This is important, as the non-existence of such mixed forms is often tacitly assumed in several proofs that use SGNF; replacing SGNF by tsSGNF should help to easily fix these results.

2 Preliminaries

We assume that the readers are familiar with the standard notations in formal
language theory. We recall a few notations here to keep the paper self-contained.

Let Σ^* denote the free monoid generated by the *alphabet* (finite set) Σ.
The elements of Σ^* are called *strings* or *words*; λ denotes the empty string, L^R
and $\mathcal{L}^R$ denote the *reversal* of language L and language family $\mathcal{L}$, respectively.
RE and LIN denote the families of recursively enumerable languages and linear
languages, respectively. Occasionally, we use the shuffle operator, written as ⧢.

For the computational completeness results, we use the fact that type-0 grammars in Special Geffert Normal Form (SGNF) [9] characterize RE and is extensively used in [3–5,15]. In fact, we slightly extend this notion in order to simplify certain arguments below. These simplifications were often tacitly assumed
in previous works, but the following new definition gives good ground for it.

Definition 1. *A type-0 grammar* $G = (N, T, S, P)$ *is said to be in* time separating special Geffert normal form, *tsSGNF for short, if N is the nonterminal
alphabet, T is the terminal alphabet, $S \in N$ is the start symbol and P is the set
of production rules satisfying the following conditions.*

- *N decomposes as $N = N^{(0)} \cup N' \cup N''$, where $N'' = \{A, B, C, D\}$ and $S \in N^{(0)}$, $S' \in N'$,*
- *the only non-context-free rules in P are $AB \to \lambda$ and $CD \to \lambda$,*
- *the context-free rules are of one of the following forms:*
  (a) $X \to Yb$ or $X \to b'Y$ where $X \in N^{(0)}$, $Y \in N^{(0)} \cup N'$, $X \neq Y$, $b \in T$,
  (b) $X \to Yb''$ or $X \to b'Y$ where $X, Y \in N'$, $X \neq Y$, or $S' \to \lambda$,
  where $b' \in \{A, C\}$ and $b'' \in \{B, D\}$ in (a) and (b);
  (c) possibly, there is also the rule $S \to \lambda$.

Remark 1. Notice that there can be at most one nonterminal from $N^{(0)} \cup N'$
present in the derivation of G. We exploit this observation in our proofs. According to the construction of this normal form described in [9,10], the derivation of
a string is performed in two phases. In Phase I, the context-free rules are applied
repeatedly. More precisely, this phase splits into two stages: in stage one, rules
from (a) have left-hand sides from $N^{(0)}$; this stage produces a string of terminal
symbols to the right side of the only nonterminal from $N^{(0)}$ in the sentential form
and codings thereof are put on the left side of the only nonterminal occurrence
from $N^{(0)}$; the transition to stage two is performed by using rules with left-hand
sides from $N^{(0)}$ and one symbol from N' occurring on the right-hand sides; in
stage two, rules from (b) with left-hand sides from N' are applied; importantly,
here (and later) no further terminal symbols are produced. The two erasing rules
$AB \to \lambda$ and $CD \to \lambda$ are not applicable during the first phase as long as there
is a S (or S') in the middle. All the symbols A and C are generated on the
left side of these middle symbols and the corresponding symbols B and D are
generated on the right side. Phase I is completed by applying the rule $S' \to \lambda$.
In Phase II, only the non-context-free erasing rules are applied repeatedly and
the derivation ends. By induction, it is clear that sentential forms derivable by
tsSGNF grammars belong to $\{A, C\}^* N^{(0)} T^* \cup \{A, C\}^* (N' \cup \{\lambda\})\{B, D\}^* T^*$.

Our reasoning shows in particular the following first result:

Theorem 1. *For any $L \in \mathrm{RE}$, there is a grammar G in tsSGNF with $L = L(G)$.*

2.1 Matrix Insertion-Deletion Systems

In this subsection, we describe *matrix insertion-deletion systems* as in [13, 15].

Definition 2. *A matrix insertion-deletion system is a construct $\Gamma = (V, T, A, R)$ where V is an alphabet, $T \subseteq V$, A is a finite language over V, R is a finite set of matrices $\{r_1, r_2, \ldots r_l\}$, where each r_i, $1 \leq i \leq l$, is a matrix of the form $r_i = [(u_1, \alpha_1, v_1)_{t_1}, (u_2, \alpha_2, v_2)_{t_2}, \ldots, (u_k, \alpha_k, v_k)_{t_k}]$. For $1 \leq j \leq k$, $u_j, v_j \in V^*$, $\alpha_j \in V^+$ and $t_j \in \{ins, del\}$.*

The triplet $(u_j, \alpha_j, v_j)_{t_j}$ is called an ins-del rule and the pair (u_j, v_j) is termed the *context* with u_i as the left and v_i as the right context for α_j in t_j; α_j is called *insertion string* if $t_j = ins$ and *deletion string* if $t_j = del$. The elements of A are called *axioms*. For all contexts of t where $t \in \{ins, del\}$, if $u = \lambda$ or $v = \lambda$, then we call the context to be one-sided. If $u = v = \lambda$ for a rule, then the corresponding insertion/deletion can be done freely anywhere in the string and is called context-free insertion/deletion. An insertion rule is of the form $(u, \eta, v)_{ins}$, which means that η is inserted between u and v. A deletion rule is of the form $(u, \delta, v)_{del}$, which means that δ is deleted between u and v. $(u, \eta, v)_{ins}$ corresponds to the rewrite rule $uv \rightarrow u\eta v$, and $(u, \delta, v)_{del}$ to $u\delta v \rightarrow uv$.

At this point, we make a note that in a derivation, the rules of a matrix are applied sequentially one after another in the given order. For $x, y \in V^*$ we write $x \Rightarrow_{r_i} y$, if y can be obtained from x by applying all the rules of a matrix r_i, $1 \leq i \leq l$, in order. The language $L(\Gamma)$ generated by Γ is defined as $L(\Gamma) = \{w \in T^* \mid x \Rightarrow_* w, \text{ for some } x \in A\}$, where $\Rightarrow_*$ (as usual with matrix ins-del systems) denotes the reflexive and transitive closure of $\Rightarrow := \bigcup_{r \in R} \Rightarrow_r$.

If a matrix ins-del system has at most k rules in a matrix and the size of the underlying ins-del system is $(n, i', i''; m, j', j'')$, then we denote the corresponding class of language by $\mathrm{MAT}(k; n, i', i''; m, j', j'')$.

Example 1. The language $L_1 = \{a^n b^m c^n d^m \mid m, n \geq 1\}$ of cross-serial dependencies can be generated by a binary matrix insertion-deletion system as follows: $\Gamma_1 = (\{a, b, c, d\}, \{a, b, c, d\}, \{abcd\}, R)$, where $R = \{m1, m2\}$ with: $m1 = [(a, a, \lambda)_{ins}, (c, c, \lambda)_{ins}]$, $m2 = [(b, b, \lambda)_{ins}, (d, d, \lambda)_{ins}]$. We note that the matrices $m1' = [(\lambda, a, a)_{ins}, (\lambda, c, c)_{ins}]$, $m2' = [(\lambda, b, b)_{ins}, (\lambda, d, d)_{ins}]$ also generate L_1. Hence, $L_1 \in \mathrm{MAT}(2; 1, 1, 0; 0, 0, 0) \cap \mathrm{MAT}(2; 1, 0, 1; 0, 0, 0)$. See [16] for further variants and a discussion of the linguistic relevance of this example.

2.2 Regular Closure of Linear Languages

Recall that a *linear grammar* is a context-free grammar $G = (N, T, S, P)$ whose productions are of the form $A \rightarrow x$, where A is a nonterminal symbol, and x is a word over $N \cup T$ with at most one occurrence of a nonterminal symbol.

The language class LIN collects all languages that can be described by linear grammars. LIN can be characterized by linear grammars in *normal form*, which means that any rule $A \to x$ either obeys $x \in T \cup \{\lambda\}$ or $x \in TN$ or $x \in NT$. It is well known that LIN is not closed under concatenation and Kleene star. This motivates to consider the class $\mathcal{L}_{reg}(\text{LIN})$ as the smallest class containing LIN that is closed under union, concatenation or Kleene star. Similarly, we can assume that any right-linear grammar that we consider is in *normal form*, i.e., it has only rules $A \to aB$ or $A \to \lambda$, with $A \in N$, $B \in N \setminus \{A\}$ and $a \in T$. The following grammatical characterization for $\mathcal{L}_{reg}(\text{LIN})$ was shown in [8].

Proposition 1 [8]. *Let $L \subseteq T^*$ be some language. Then, $L \in \mathcal{L}_{reg}(\text{LIN})$ if and only if there is a context-free grammar $G = (N, T, S, P)$ with $L(G) = L$ that satisfies the following properties.*

- *N can be partitioned into N_0 and N'.*
- *There is a right-linear grammar $G_R = (N_0, N', S, P_0)$.*
- *N' can be further partitioned into $N_1, \ldots, N_k$ for some k, such that the restriction P_i of P involving symbols from $N_i \cup T$ are only linear rules, with T serving as the terminal alphabet.*
- *P can be partitioned into $P_0, P_1, \ldots, P_k$.*

Clearly, the *linear rules* mentioned in the previous proposition can be assumed to be in normal form. In order to simplify the proofs of some of our main results, the following observations from [6] and [7] are helpful.

Proposition 2 [6]. *Let $\mathcal{L}$ be a language class that is closed under reversal. Then, for all non-negative integers $k, n, i', i'', m, j', j''$, we have that*

- *$\text{MAT}(k; n, i', i''; m, j', j'') = [\text{MAT}(k; n, i'', i'; m, j'', j')]^R$;*
- *$\mathcal{L} = \text{MAT}(k; n, i', i''; m, j', j'')$ if and only if $\mathcal{L} = \text{MAT}(k; n, i'', i'; m, j'', j')$;*
- *$\mathcal{L} \subseteq \text{MAT}(k; n, i', i''; m, j', j'')$ if and only if $\mathcal{L} \subseteq \text{MAT}(k; n, i'', i'; m, j'', j')$.*

Proposition 3 [7]. *$L_{reg}(\text{LIN})$ is closed under reversal.*

3   Computational Completeness Results

In this section, we show the computational completeness of matrix ins-del systems of sizes $(3; 1, 0, 0; 1, 2, 0)$, $(3; 1, 0, 0; 1, 0, 2)$, $(3; 1, 2, 0; 1, 0, 0)$, $(3; 1, 0, 2; 1, 0, 0)$, $(2; 1, 1, 0; 1, 1, 0)$, $(2; 1, 0, 1; 1, 0, 2)$, by providing matrix ins-del systems of the above said sizes that simulate type-0 grammars in tsSGNF. We often use labels from $[1 \ldots |P|]$ to uniquely address the rules of a grammar in tsSGNF. Then, such labels (and possibly also primed version thereof) will be used as *rule markers* that are therefore part of the nonterminal alphabet of the simulating matrix ins-del system. For the ease of reference, we collect in P_{ll} the labels of the context-free rules of the form $X \to Yb$ (which resemble left-linear rules) and in P_{rl} the labels of the context-free rules of the form $X \to bY$ (which resemble right-linear rules).[1]

Some of the key features in our construction of matrix ins-del systems are:

---

[1] The symbol (*) marks situations where (parts of) a proof were omitted.

- There is at least one deletion rule in every simulating matrix.
- In the majority of cases, at least one of the deletion rules of every matrix has a *rule marker* in the left context or the marker itself is deleted. A matrix of this type is said to be *guarded*. The importance of a matrix being guarded is that it can be applied only in the presence of the corresponding rule marker. This will avoid interference of any other matrix application.
- After successful application of every matrix, either a rule marker remains or the intended simulation is completed.
- As discussed in Remark 1, during Phase I, the symbols A and C are on the left of the middle nonterminal S or S' and the corresponding symbols B and D are on the right of S or S'. When S' is deleted from the center, the symbols from $\{A, C\}$ and $\{B, D\}$ may combine to be erased in Phase II.
- In the transition to Phase II, a special symbol Z is introduced that is assumed to stay to the left of AB or CD, whatever substring is to be deleted. Special matrices allow Z to move in the sentential form or to be (finally) deleted. This is our novel idea not used in earlier papers.
- There is a subtlety if $\lambda \in L$. Then, we can assume that $S \to \lambda$ is in the simulated grammar, which would add an erasing matrix, similar to Fig. 2.

$$p1 = [(\lambda, p, \lambda)_{ins}, \ (\lambda, p', \lambda)_{ins}, (p'p, X, \lambda)_{del}]$$
$$p2 = [(\lambda, p'', \lambda)_{ins}, (\lambda, p''', \lambda)_{ins}, (p'''p'', p', \lambda)_{del}]$$
$$p3 = [(\lambda, b, \lambda)_{ins}, \ (p'''b, p'', \lambda)_{del}]$$
$$p4 = [(\lambda, Y, \lambda)_{ins}, \ (bY, p, \lambda)_{del}, \ (\lambda, p''', \lambda)_{del}]$$

$$q1 = [(\lambda, q, \lambda)_{ins}, \ (\lambda, q', \lambda)_{ins}, (q'q, X, \lambda)_{del}]$$
$$q2 = [(\lambda, q'', \lambda)_{ins}, (\lambda, b, \lambda)_{ins}, (q''b, q, \lambda)_{del}]$$
$$q3 = [(\lambda, Y, \lambda)_{ins}, (q'Y, q'', \lambda)_{del}, (\lambda, q', \lambda)_{del}]$$

Fig. 1. Matrices of size $(3; 1, 0, 0; 1, 2, 0)$ for simulating the context-free rules of tsSGNF.

$$h1 = [(\lambda, S', \lambda)_{del}, (\lambda, Z, \lambda)_{ins}] \qquad \text{move-}Z = [(\lambda, Z, \lambda)_{del}, (\lambda, Z, \lambda)_{ins}]$$
$$f1 = [(Z, A, \lambda)_{del}, (Z, B, \lambda)_{del}] \qquad \text{del-}Z \ = [(\lambda, Z, \lambda)_{del}]$$
$$g1 = [(Z, C, \lambda)_{del}, (Z, D, \lambda)_{del}]$$

Fig. 2. Matrices of size $(2; 1, 0, 0; 1, 1, 0)$ for simulating the erasing rules of tsSGNF.

In [15], matrices of maximum length 8 and size $(1, 0, 0; 1, 1, 1)$ were used to describe the class of recursively enumerable languages. This length was reduced already to 3 in [6]. In the following, we give a trade-off result compared with the result of [6], namely $\mathrm{MAT}(3; 1, 0, 0; 1, 1, 1) = \mathrm{RE}$.

Theorem 2. $\mathrm{MAT}(3; 1, 0, 0; 1, 2, 0) = \mathrm{MAT}(3; 1, 0, 0; 1, 0, 2) = \mathrm{RE}$.

Proof. Formally, consider a type-0 grammar $G = (N, T, P, S)$ in tsSGNF. The rules from P are labelled, with label sets P_{ll} and P_{rl} as defined above. The nonterminal alphabet decomposes as $N = N^{(0)} \cup N' \cup N''$, $N'' = \{A, B, C, D\}$, $S \in N^{(0)}, S' \in N'$, according to tsSGNF. We construct a matrix ins-del system $\Gamma = (V, T, \{S\}, M)$ with

$$V = N \cup T \cup \{p, p', p'', p''' \mid p \in P_{rl}\} \cup \{q, q', q'' \mid q \in P_{ll}\} \cup \{Z\}.$$

The set of matrices M of Γ consists of the matrices described in the following. We simulate a rule $p: X \to bY$, $X, Y \in N^{(0)} \cup N'$, $b \in N''$, i.e., $p \in P_{rl}$, by the four matrices displayed on the left-hand side of Fig. 1. Similarly, we simulate the rule $q: X \to Yb$, $X, Y \in N^{(0)} \cup N'$, $b \in N'' \cup T$, i.e., $q \in P_{ll}$, by the three matrices shown on the right-hand side of Fig. 1. Recall that applying the rule $h: S' \to \lambda$ starts Phase II within the working of G. In the simulation, the presence of a new symbol, Z, indicates that we are in Phase II. This motivates the introduction of the five matrices listed in Fig. 2.

We now proceed to prove that $L(\Gamma) = L(G)$. We initially prove that $L(G) \subseteq L(\Gamma)$ by showing that Γ correctly simulates the application of the rules of the types p, q, f, g, h, as discussed above. We explain the working of the simulation matrices for the cases p and f mainly, as the working of q and g simulation matrices are similar, and as the working of the simulation of the h rule is clear. Notice that the transition from Phase I to Phase II (as accomplished by applying h in G) is now carried out by applying $h1$ and hence introducing Z which will be always present when simulating Phase II with the system Γ.

Simulation of $p: X \to bY$: Consider the string $\alpha X \beta$ derivable from S in G, with $X \in N^{(0)} \cup N'$ and $\alpha \in \{A, C\}^*$, $\beta \in \{B, D\}^* T^*$ according to Remark 1. We now show that on applying the matrices introduced for simulating rules from P_{rl}, we can derive $\alpha bY \beta$ within Γ, starting from $\alpha X \beta$. First, we apply the rules of matrix $p1$. The markers p and p' are randomly inserted by the first two rules, leading to a string from $p \sqcup p' \sqcup \alpha X \beta$. However, the third rule of $p1$ is applicable only when $p'p$ is inserted before the nonterminal X. This shows that $\alpha X \beta \Rightarrow_{p1} \gamma_1$ is possible if and only if $\gamma_1 = \alpha p'p\beta$. Now, on applying matrix $p2$, p'' and p''' are inserted anywhere, so intermediately we arrive at a string from $p'' \sqcup p''' \sqcup \alpha p'p\beta$. Then, p' is deleted in the left context of $p'''p''$. So, we now arrive at the string $\alpha p'''p''p\beta$. This shows that $\gamma_1 = \alpha p'p\beta \Rightarrow_{p2} \gamma_2$ is possible if and only if $\gamma_2 = \alpha p'''p''p\beta$. We now apply matrix $p3$. Hence, b is first inserted randomly, leading to a string from $b \sqcup \alpha p'''p''p\beta$. The left context in the second rule of $p3$, enforces that, inevitably, we arrive at $\gamma_3 = \alpha p'''bp\beta$. Finally, we apply matrix $p4$. Here, Y is inserted anywhere by the first rule, but the second one enforces that we now look at $\alpha p'''bY\beta$, which yields $\gamma_4 = \alpha bY\beta$ by the last rule. This shows that $\gamma_3 = \alpha p'''bp\beta \Rightarrow_{p4} \gamma_4$ is possible if and only if $\gamma_4 = \alpha bY\beta$. This completes the simulation of rule p.

Simulation of $f: AB \to \lambda$ or $g: CD \to \lambda$: Consider the sentential form $\alpha AB\beta$ derivable in G. This means that we are in Phase II. As said above, the symbol Z will be present in the corresponding sentential form derivable in Γ. Any string from $Z \sqcup \alpha AB\beta$ can be transformed into $\alpha ZAB\beta$ by using matrix move-Z. Now, $\alpha ZAB\beta \Rightarrow_{f1} \alpha Z\beta$ correctly simulates one application of f.

Now, we prove $L(\Gamma) \subseteq L(G)$. Formally, this is an inductive argument that proves the following properties of a string $w \in V^*$ such that $S \Rightarrow_* w$ in Γ:

1. At most one symbol from $N^{(0)} \cup N'$ is occurring in w.
2. If one symbol X from $N^{(0)}$ occurs in w, then $w = \alpha X u$, where $\alpha \in \{A, C\}^*$ and $u \in T^*$: w is derivable in G;

3. If one symbol X from N' occurs in w, then $w = \alpha X \beta u$, where $\alpha \in \{A, C\}^*$, $\beta \in \{B, D\}^*$ and $u \in T^*$: w is derivable in G;
4. If no symbol from $N^{(0)} \cup N'$ occurs in w, then Z occurs at most once in w.
5. If no symbol from $N^{(0)} \cup N' \cup \{Z\}$ occurs in w, then
   (a) either $w = \alpha r'r\beta u$, where $\alpha \in \{A, C\}^*$, r is some context-free rule from G with left-hand side X, $\beta \in \{B, D\}^*$ and $u \in T^*$: $\alpha X \beta u$ is derivable in G;
   (b) or $w = \alpha p'''p''p\beta u$, where $\alpha \in \{A, C\}^*$, $p \in P_{rl}$ with $p : X \rightarrow bY$, $\beta \in \{B, D\}^*$ and $u \in T^*$: $\alpha X \beta u$ is derivable in G;
   (c) or $w = \alpha p'''bp\beta u$, where $\alpha \in \{A, C\}^*$, $p \in P_{rl}$ with $p : X \rightarrow bY$, $\beta \in \{B, D\}^*$ and $u \in T^*$: $\alpha X \beta u$ is derivable in G;
   (d) or $w = \alpha q'q''b\beta u$, where $\alpha \in \{A, C\}^*$, $q \in P_{ll}$ with $q : X \rightarrow Yb$, $\beta \in \{B, D\}^*$ and $u \in T^*$: $\alpha X \beta u$ is derivable in G.
   (e) or $w \in (\{A, B, C, D\} \cup T)^*$: w is derivable in G.
6. If Z occurs in w, then $w \in Z \sqcup \alpha \beta u$, where $\alpha \in \{A, C\}^*$, $\beta \in \{B, D\}^*$ and $u \in T^*$: $\alpha \beta u$ is derivable in G.

Details of the inductive argument are left to the reader. (*)

These considerations complete the proof due to Condition 5(e) that applies to $w \in T^*$. The second equality follows by Proposition 2. □

In [15], matrices of maximum length 8 and size $(1, 1, 1; 1, 0, 0)$ were used to describe the class of recursively enumerable languages. This length was reduced to 3 in [6]. In the following, we give a trade-off result compared to the result of [6], by considering here only a one-sided context for insertion but increasing its length to 2, yet maintaining the maximum matrix length as 3.

Theorem 3. $\mathrm{MAT}(3; 1, 2, 0; 1, 0, 0) = \mathrm{MAT}(3; 1, 0, 2; 1, 0, 0) = \mathrm{RE}$.

Before we sketch our proof, we highlight the key feature of the markers first. In order to simulate, say, $AB \rightarrow \lambda$, we have to use deletion rules $(\lambda, A, \lambda)_{del}$ and $(\lambda, B, \lambda)_{del}$, as deletions cannot be performed under contexts. However, there is the danger that we are deleting unintended occurrences. So we have to carefully place markers before, after and between the chosen nonterminals A and B in order to check that they are neighbored. Also, the auxiliary nonterminal $\$$, which is present in the axiom itself, serves as a semaphore flag, preventing simulation cycles from being interrupted.

Proof (*). Consider a type-0 grammar $G = (N, T, P, S)$ in tsSGNF, with the rules uniquely labelled with $P_{ll} \cup P_{rl}$. Recall the decomposition $N = N^{(0)} \cup N' \cup N''$ by tsSGNF. We can construct a matrix ins-del system $\Gamma = (V, T, \{\$S\}, M)$ with alphabet $V = N \cup T \cup P_{rl} \cup P_{ll} \cup \{f, f', f'', f''', f'''', g, g', g'', g''', g'''', \$\}$. The set of matrices M is defined as follows. Rules $p : X \rightarrow bY \in P_{rl}$ and $q : X \rightarrow Yb \in P_{ll}$ are simulated by the matrices $p1, p2, q1, q2$ shown in Fig. 4(a). The simulation matrices of p, q are in fact borrowed from [6]. We simulate rule $f : AB \rightarrow \lambda$ by the following matrices. Rule $g : CD \rightarrow \lambda$ is simulated alike.

$$f1 = [(\lambda, \$, \lambda)_{del}, (\lambda, f, \lambda)_{ins}, (fA, f', \lambda)_{ins}]$$
$$f2 = [(\lambda, A, \lambda)_{del}, (ff', f'', \lambda)_{ins}, (\lambda, f', \lambda)_{del}]$$
$$f3 = [(f''B, f''', \lambda)_{ins}, (\lambda, B, \lambda)_{del}, (f''f''', f'''', \lambda)_{ins}]$$
$$f4 = [(\lambda, f''', \lambda)_{del}, (\lambda, f'', \lambda)_{del}]$$
$$f5 = [(ff'''', \$, \lambda)_{ins}, (\lambda, f, \lambda)_{del}, (\lambda, f'''', \lambda)_{del}]$$

Recall that our axiom is $\$S$. We also have matrices $\tau = [(\lambda, \$, \lambda)_{del}]$ for termination and $h1 = [(\lambda, \$, \lambda)_{del}, (\lambda, S', \lambda)_{del}, (\lambda, \$, \lambda)_{ins}]$ for $h : S' \to \lambda$.

We now proceed to prove that $L(\Gamma) = L(G)$, starting with the inclusion $L(G) \subseteq L(\Gamma)$. This means that a derivation $S \Rightarrow^* w$ of G in Phase I can be simulated by applying matrices from Γ, where each rule r is simulated by $r1$, followed by $r2$. Hence, $\$S \Rightarrow_* \$w$ is true in Γ. The phase transition rule $h : S' \to \lambda$ is simulated by applying $h1$, so that we can now speak about Phase II of G. By similarity, we discuss only rule f below. To actually produce a terminal word, Γ has to apply τ at the end.

We now discuss Phase II in detail, focussing on $f : AB \to \lambda$. Let $w = \alpha AB\beta$ be a sentential form derivable in G, with $A, B \in N''$ and $\alpha \in \{A, C\}^*$, $\beta \in \{B, D\}^* T^*$, ensured by tsSGNF. This means that $w' \in \$ \sqcup w$ is derivable in Γ (by induction). We can now see that $\$ \sqcup \alpha AB\beta \Rightarrow_{f1} \alpha f Af'B\beta \Rightarrow_{f2} \alpha ff''B\beta \Rightarrow_{f3} \alpha ff''f'''f''''\beta \Rightarrow_{f4} \alpha ff''''\beta \Rightarrow_{f5} \alpha\$\beta$. The purpose of introducing a $\$$ in $f5$ is to enable another simulation of $AB \to \lambda$ or of $CD \to \lambda$. When all occurrence of AB and CD are deleted by repeated applications of the f and g rules, there is still a $\$$ at the end of every simulation. This $\$$ is deleted by applying rule τ, thereby terminating Phase II of tsSGNF. Inductively, this shows that $L(G) \subseteq L(\Gamma)$.

The second claimed completeness result follows by Proposition 2.   □

In the previous two theorems, the maximum length of the insertion/deletion context was two and the other operation, namely deletion/insertion is done in a context-free manner. If we restrict the parameters in the size to be binary (0 or 1), then we achieve computational completeness using matrices of maximum length two; however, both operations are performed under a context. In [15], matrices of maximum length 3 and size $(1, 1, 0; 1, 1, 0)$ were used to describe RE. In the following, we improve this result by decreasing the matrix length from 3 to 2, yet maintaining the ins-del size. By [12], ins-del systems of size $(1, 1, 0; 1, 1, 0)$ do not achieve computational completeness. This corresponds to matrix length one. So, our result is optimal with respect to the length of matrices for this ins-del size. The following result also solves a conjecture (stated in [12]) in the affirmative.

Theorem 4. $MAT(2; 1, 1, 0; 1, 1, 0) = MAT(2; 1, 0, 1; 1, 0, 1) = RE$.

The key feature of the simulation is a combination of the non-context-free rule simulation from Fig. 2 with new matrices for the context-free rules, given in Fig. 3. Here, it is most important to observe that the rule markers, in particular r'', guarantee that no more than one occurrence of $N^{(0)} \cup N'$ is ever introduced. Therefore, the fact that $r1$ can be used to introduce more than one occurrence of r and r' cannot lead to unintended terminal derivations.

$$p1= [(X,p,\lambda)_{ins}, (\lambda,p',\lambda)_{ins}]$$
$$p2= [(p',X,\lambda)_{del}, (p',p'',\lambda)_{ins}]$$
$$p3= [(p'',p,\lambda)_{del}, (p'',Y,\lambda)_{ins}]$$
$$p4= [(p',b,\lambda)_{ins}, (b,p'',\lambda)_{del}]$$
$$p5= [(\lambda,p',\lambda)_{del}]$$

$$q1= [(X,q,\lambda)_{ins}, (\lambda,q',\lambda)_{ins}]$$
$$q2= [(q',X,\lambda)_{del}, (q',q'',\lambda)_{ins}]$$
$$q3= [(q'',q,\lambda)_{del}, (q'',b,\lambda)_{ins}]$$
$$q4= [(q',Y,\lambda)_{ins}, (Y,q'',\lambda)_{del}]$$
$$q5= [(\lambda,q',\lambda)_{del}]$$

(a) How to simulate $p : X \to bY$ (b) How to simulate $q : X \to Yb$

Fig. 3. Matrices of size $(2; 1, 1, 0; 1, 1, 0)$ for simulating context-free rules of tsSGNF

Proof (*). Consider a type-0 grammar $G = (N, T, P, S)$ in tsSGNF, with the rules uniquely labelled with $P_{ll} \cup P_{rl}$. We can construct a matrix ins-del system $\Gamma = (V, T, S, M)$ with alphabet $V = N \cup T \cup \{r, r', r'' \mid r \in P_{ll} \cup P_{rl}\} \cup \{Z\}$. The set of matrices M is defined as follows. Context-free rules of types $p : X \to bY$ and $q : X \to Yb$ are simulated by the matrices as shown in Figs. 3(a) and (b). We simulate the erasing rules $f : AB \to \lambda$, $g : CD \to \lambda$ and $h : S' \to \lambda$ by matrices shown in Fig. 2; this also explains the role of the special symbol Z. We refer to our previous explanations of how these erasing rules are simulated in order to provide a simulation of Phase II of grammar G.

Since the working of the simulation of q-rules is similar to the simulation of p-rules, we discuss the working of the p-rule simulation only. Hence, consider a sentential form $w = \alpha X \beta$ derivable in Phase I of the grammar G. Assume we are about to apply a concrete rule $X \to bY \in P_{rl}$, with $X, Y \in N^{(0)} \cup N'$, yielding $w' = \alpha b Y \beta$. Hence, the matrices listed in Fig. 3(a) should apply, one after the other, giving: $w \Rightarrow_{p1} \alpha p' X p \beta \Rightarrow_{p2} \alpha p' p'' p \beta \Rightarrow_{p3} \alpha p' p'' Y \beta \Rightarrow_{p4} \alpha p' b Y \beta \Rightarrow_{p5} w'$.

The reasoning we provided so far shows that $L(G) \subseteq L(\Gamma)$.

The second completeness result follows by Proposition 2.   □

4   Describing the Regular Closure of Linear Languages

It is shown in [6] that matrix ins-del systems of size $(3; 1, 1, 1; 1, 0, 0)$ can describe RE and if we have a one-sided context for insertion, then matrix ins-del systems of size $(3; 1, 1, 0; 1, 0, 0)$ or $(3; 1, 0, 1; 1, 0, 0)$ and also $(2; 2, 1, 0; 1, 0, 0)$ or $(2; 2, 0, 1; 1, 0, 0)$ can simulate (meta-)linear grammars. However, whether or not one can simulate general context-free grammars with matrix ins-del systems of the above-mentioned sizes is still open. Example 1 shows that there are non-meta-linear languages that can be described by these matrix ins-del systems. For quick reference, we present the matrix ins-del rules of MAT$(3; 1, 1, 0; 1, 0, 0)$ that simulates the linear rules $p : X \to aY$, $q : X \to Ya$, $f : X \to \lambda$ in Fig. 4(a) and the matrix ins-del rules of MAT$(2; 2, 1, 0; 1, 0, 0)$ in Fig. 4(b).

Theorem 5. *For all integers* $n, m \geq 1$, $t \geq 2$ *and* $i', i'', j', j'' \geq 0$ *with* $t + n \geq 4$ *and* $i' + i'' \geq 1$, *if every* $L \in$ LIN *can be generated by a* MAT$(t; n, i', i''; m, j', j'')$

$$p1 = [(X, p, \lambda)_{ins}, (\lambda, X, \lambda)_{del}]$$
$$p2 = [(p, Y, \lambda)_{ins}, (p, a, \lambda)_{ins}, (\lambda, p, \lambda)_{del}]$$
$$q1 = [(X, q, \lambda)_{ins}, (\lambda, X, \lambda)_{del}]$$
$$q2 = [(q, a, \lambda)_{ins}, (q, Y, \lambda)_{ins}, (\lambda, q, \lambda)_{del}]$$
$$f = [(\lambda, X, \lambda)_{del}]$$
(a) MAT$(3; 1, 1, 0; 1, 0, 0)$

$$p1 = [(X, p, \lambda)_{ins}, (\lambda, X, \lambda)_{del}]$$
$$p2 = [(p, aY, \lambda)_{ins}, (\lambda, p, \lambda)_{del}]$$
$$q1 = [(X, q, \lambda)_{ins}, (\lambda, X, \lambda)_{del}]$$
$$q2 = [(q, Ya, \lambda)_{ins}, (\lambda, q, \lambda)_{del}]$$
$$f = [(\lambda, X, \lambda)_{del}]$$
(b) MAT$(2; 2, 1, 0; 1, 0, 0)$

Fig. 4. Matrix ins-del system describing LIN [6]

system with a single axiom that is identical to the start symbol S of a linear grammar describing L, then $\mathcal{L}_{reg}(\text{LIN}) \subseteq \text{MAT}(t; n, i', i''; m, j', j'')$, as well.[2]

Proof (*). Let $L \in \mathcal{L}_{reg}(\text{LIN})$ for some $L \subseteq T^*$. By Proposition 1, we can assume that L is described by a context-free grammar $G = (N, T, S, P)$ that basically consists of a right-linear grammar $G_R = (N_0, N'', S, P_0)$ and linear grammars $G_i = (N_i, T, S_i, P_i)$ for $1 \leq i \leq k$. For technical reasons that should become clear soon, we rather consider $G_i' = (N_i', T, S_i, P_i')$, where $N_i' = N_i \cup \{\langle S_i, A\rangle \mid A \in N_0\}$ and P_i' contains, besides all rules from P_i, rules of the form $\langle S_i, A\rangle \to w$ whenever $S_i \to w \in P_i$ for some $w \in (N_i \cup T)^*$. This means, apart from $L(G_i') = L(G_i)$ (as the new nonterminals will never be used in terminating derivations), that also $L((N_i', T, \langle S_i, A\rangle, P_i')) = L(G_i)$ for any $A \in N_0$.

Since LIN $\subseteq$ MAT$(t; n, i', i''; m, j', j'')$, each G_i' can be simulated by a matrix ins-del system $\Gamma_i = (V_i, T, \{S_i\}, R_i)$ for $1 \leq i \leq k$, each of size $(t; n, i', i''; m, j', j'')$. We assume, without loss of generality, that $V_i \cap V_j = T$ if $1 \leq i < j \leq k$. Let us first consider the case $i' \geq 1$ and $i'' = 0$. We construct a matrix ins-del system Γ for G as follows:[3] $\Gamma = (V, T, \{\langle S_i, A\rangle A' \mid S \to S_i A \in P\}, R \cup R')$, where $V = \left(\bigcup_{i=1}^{k}(V_i \cup \{\langle S_i, A\rangle \mid A \in N_0\})\right) \cup N_0 \cup \{A' \mid A \in N_0\}$; $R = \bigcup_{i=1}^{k} R_i$; and for $t \geq 3$, R' is the set $\{m_p \mid p \in P_0\}$, where:

- $m_p = [(A', \langle S_i, B\rangle, \lambda)_{ins}, (\langle S_i, B\rangle, B', \lambda)_{ins}, (\lambda, A', \lambda)_{del})]$ if $p = A \to S_i B \in P_0$,
- $m_p = [(\lambda, A', \lambda)_{del})]$ if $p = A \to \lambda \in P_0$ (terminating matrix).

For $t = 2$ and $n \geq 2$, we add the following matrix m_p instead of the above-defined matrix m_p into R': $m_p = [(A', \langle S_i, B\rangle B', \lambda)_{ins}, (\lambda, A', \lambda)_{del})]$ for $p = A \to S_i B \in P_0$. Notice that $A \neq B$, which is important both for variants of m_p.

The case when $i' = 0$ and $i'' \geq 1$ follows from Propositions 2 and 3. $\qquad\square$

Combining Theorem 5 with results from [6], we have the following corollary.

---

[2] The technical condition on MAT ins-del systems is not that severe, as we can always take a new start symbol and first generate any finite set with the resources at hand.

[3] There is one subtlety with the case when $\lambda \in L(G)$: in that case, λ should be added as an axiom of Γ.

Corollary 1. *The following assertions are true.*

- $\mathcal{L}_{reg}(\text{LIN}) \subsetneq \text{MAT}(3; 1, 1, 0; 1, 0, 0) \cap \text{MAT}(3; 1, 0, 1; 1, 0, 0)$
- $\mathcal{L}_{reg}(\text{LIN}) \subsetneq \text{MAT}(2; 2, 1, 0; 1, 0, 0) \cap \text{MAT}(2; 2, 0, 1; 1, 0, 0)$.   □

5   Conclusion

In this paper, using matrix ins-del systems, we have obtained some (improved) computational completeness results and described the regular closure of linear languages with small resource needs. It is interesting to note that if one could describe linear languages by a matrix insertion-deletion system of size s, then with the same size s, we could describe the regular closure of linear languages, as well. We have also given a complete picture of the state of the art of the generative power of the matrix ins-del systems with sum-norm 3 or 4 in Table 1. Finally, we believe that tsSGNF offers some features that could be used in other computational completeness proofs. In particular, no substrings with nonterminals to the right of terminals are derivable in this normal form. We now present some further concrete research questions. It would be interesting to explore closure properties for matrix ins-del systems of small sizes. For instance, is the family $\text{MAT}(2; 2, 1, 0; 1, 0, 0)$ closed under reversal? If this would be true, then $\text{MAT}(2; 2, 1, 0; 1, 0, 0) = \text{MAT}(2; 2, 0, 1; 1, 0, 0)$, which would also mean that the statement of Corollary 1 could be simplified. We are working on the question if matrix ins-del systems of length 2 are computationally complete with ins-del sizes $(1, 1, 0; 1, 0, 1)$ or $(1, 0, 1; 1, 1, 0)$. We conjecture that this is indeed the case.

References

1. Dassow, J., Păun, G.: Regulated Rewriting in Formal Language Theory. EATCS Monographs in Theoretical Computer Science, vol. 18. Springer, Heidelberg (1989)
2. Fernau, H., Kuppusamy, L.: Parikh images of matrix ins-del systems. In: Gopal, T.V., Jäger, G., Steila, S. (eds.) TAMC 2017. LNCS, vol. 10185, pp. 201–215. Springer, Cham (2017). https://doi.org/10.1007/978-3-319-55911-7_15
3. Fernau, H., Kuppusamy, L., Raman, I.: Computational completeness of path-structured graph-controlled insertion-deletion systems. In: Carayol, A., Nicaud, C. (eds.) CIAA 2017. LNCS, vol. 10329, pp. 89–100. Springer, Cham (2017). https://doi.org/10.1007/978-3-319-60134-2_8
4. Fernau, H., Kuppusamy, L., Raman, I.: On the computational completeness of graph-controlled insertion-deletion systems with binary sizes. Theor. Comput. Sci. **682**, 100–121 (2017). Special Issue on Languages and Combinatorics in Theory and Nature
5. Fernau, H., Kuppusamy, L., Raman, I.: Computational completeness of simple semi-conditional insertion-deletion systems. In: Stepney, S., Verlan, S. (eds.) UCNC 2018. LNCS, vol. 10867, pp. 86–100. Springer, Cham (2018). https://doi.org/10.1007/978-3-319-92435-9_7
6. Fernau, H., Kuppusamy, L., Raman, I.: Investigations on the power of matrix insertion-deletion systems with small sizes. Nat. Comput. **17**(2), 249–269 (2018)

7. Fernau, H., Kuppusamy, L., Raman, I.: On describing the regular closure of the linear languages with graph-controlled insertion-deletion systems. RAIRO Inf. théor. et Appl./Theor. Inf. Appl. **52**(1), 1–21 (2018)

8. Fernau, H., Kuppusamy, L., Raman, I.: Properties of language classes between linear and context-free. J. Autom. Lang. Combin. **23**(4), 329–360 (2018)

9. Freund, R., Kogler, M., Rogozhin, Y., Verlan, S.: Graph-controlled insertion-deletion systems. In: McQuillan, I., Pighizzini, G. (eds.) Proceedings Twelfth Annual Workshop on Descriptional Complexity of Formal Systems, DCFS, vol. 31 of EPTCS, pp. 88–98 (2010)

10. Geffert, V.: How to generate languages using only two pairs of parentheses. J. Inf. Process. Cybern. EIK **27**(5/6), 303–315 (1991)

11. Kari, L., Thierrin, G.: Contextual insertions/deletions and computability. Inf. Comput. **131**(1), 47–61 (1996)

12. Krassovitskiy, A., Rogozhin, Y., Verlan, S.: Computational power of insertion-deletion (P) systems with rules of size two. Nat. Comput. **10**, 835–852 (2011)

13. Kuppusamy, L., Mahendran, A.: Modelling DNA and RNA secondary structures using matrix insertion-deletion systems. Int. J. Appl. Math. Comput. Sci. **26**(1), 245–258 (2016)

14. Păun, G., Rozenberg, G., Salomaa, A.: DNA Computing: New Computing Paradigms. Springer, Heidelberg (1998). https://doi.org/10.1007/978-3-662-03563-4

15. Petre, I., Verlan, S.: Matrix insertion-deletion systems. Theor. Comput. Sci. **456**, 80–88 (2012)

16. Stabler, E.: Varieties of crossing dependencies: structure dependence and mild context sensitivity. Cogn. Sci. **28**, 699–720 (2004)

17. Verlan, S.: Recent developments on insertion-deletion systems. Comput. Sci. J. Moldova **18**(2), 210–245 (2010)

Separation Logic with Linearly Compositional Inductive Predicates and Set Data Constraints

Chong Gao[1,2], Taolue Chen[3], and Zhilin Wu[1(✉)]

[1] State Key Laboratory of Computer Science, Institute of Software,
Chinese Academy of Sciences, Beijing, China
wuzl@ios.ac.cn
[2] University of Chinese Academy of Sciences, Beijing, China
[3] Department of Computer Science and Information Systems,
Birkbeck, University of London, London, UK

Abstract. We identify difference-bound set constraints (DBS), an analogy of difference-bound arithmetic constraints for sets. DBS can express not only set constraints but also arithmetic constraints over set elements. We integrate DBS into separation logic with linearly compositional inductive predicates, obtaining a logic thereof where set data constraints of linear data structures can be specified. We show that the satisfiability of this logic is decidable. A crucial step of the decision procedure is to compute the transitive closure of DBS-definable set relations, to capture which we propose an extension of quantified set constraints with Presburger Arithmetic (RQSPA). The satisfiability of RQSPA is then shown to be decidable by harnessing advanced automata-theoretic techniques.

1 Introduction

Separation Logic (SL) is a well-established approach for deductive verification of programs that manipulate dynamic data structures [25, 28]. Typically, SL is used in combination with inductive definitions (SLID), which provides a natural and convenient means to specify dynamic data structures. To reason about the property (e.g. sortedness) of data values stored in data structures, it is also necessary to incorporate data constraints into the inductive definitions.

One of the most fundamental questions for a logical theory is whether its satisfiability is decidable. SLID with data constraints is no exception. This problem becomes more challenging than one would probably expect, partially due to the inherent intricacy brought up by inductive definitions and data constraints. It is somewhat surprising that only disproportional research has addressed this question (cf. *Related work*). In practice, most available tools based on SLID only support heuristics without giving completeness guarantees, especially when data

Partially supported by the NSFC grants (No. 61472474, 61572478, 61872340), UK EPSRC grant (EP/P00430X/1), and the INRIA-CAS joint research project VIP.

B. Catania et al. (Eds.): SOFSEM 2019, LNCS 11376, pp. 206–220, 2019.
https://doi.org/10.1007/978-3-030-10801-4_17

constraints are involved. *Complete* decision procedures for satisfiability, however, have been found important in software engineering tasks such as symbolic execution, specification debugging, counterexample generation, etc., let along the theoretical insights they usually shed on the logic system.

The dearth of complete decision procedures for SLID with data constraints has prompted us to launch a research program as of 2015, aiming to identify decidable *and* sufficiently expressive instances. We have made encouraging progress insofar. In [15], we set up a general framework, but could only tackle linear data structures with data constraints in difference-bound arithmetic. In [34], we were able to tackle tree data structures by exploiting machineries such as order graphs and counter machines, though the data constraints therein remained to be in difference-bound arithmetic.

An important class of data constraints that is currently elusive in our investigations is set constraints. They are mandatory for reasoning about, e.g., invariants of data collections stored in data structures. For instance, when specifying the correctness of a sorting algorithm on input lists, whilst the sortedness of the list can be described by difference-bound arithmetic constraints, the property that the sorting algorithm does not change the set of data values on the list requires inductive definitions with *set* data constraints. Indeed, reviewers of the papers [15,34] constantly raised the challenge of set constraints, which compelled us to write the current paper.

Main Contributions. Our *first* contribution is to carefully design the difference-bound set constraints ($\mathcal{DBS}$), and to integrate them into the linearly compositional inductive predicates introduced in [15], yielding $\mathsf{SLID}_{\mathsf{LC}}^{\mathsf{S}}$: *SL with linearly compositional inductive predicates and set data constraints*. The rationale of $\mathcal{DBS}$ is two-fold: (1) it must be sufficiently expressive to represent common set data constraints as well as arithmetic constraints over set elements one usually needs when specifying linear data structures, (2) because of the inductive predicates, it must be sufficiently "simple" to be able to capture the *transitive closure* of $\mathcal{DBS}$-definable set relations[1] in an effective means, in order to render the satisfiability of $\mathsf{SLID}_{\mathsf{LC}}^{\mathsf{S}}$ decidable. As the *second* contribution, we show that the transitive closure of $\mathcal{DBS}$ can indeed be captured in the *restricted extension of quantified set constraints with Presburger arithmetic* ($\mathcal{RQSPA}$) introduced in this paper. Finally, our *third* contribution is to show that the satisfiability of $\mathcal{RQSPA}$ is decidable by establishing a connection of $\mathcal{RQSPA}$ with Presburger automata [29]. This extends the well-known connection of Monadic Second-Order logic on words (MSOW) and finite-state automata *a la* Büchi and Elgot [5,11]. These contributions, together with a procedure which constructs an abstraction (as an $\mathcal{RQSPA}$ formula) from a given $\mathsf{SLID}_{\mathsf{LC}}^{\mathsf{S}}$ formula and which we adapt from our previous work [15], show the satisfiability of $\mathsf{SLID}_{\mathsf{LC}}^{\mathsf{S}}$ is decidable.

We remark that sets are conceptually related to second—rather than first—order logics. While the transitive closure of logic formulae with *first-order variables* is somehow well-studied (especially for simple arithmetic; cf. *Related Work*), the transitive closure of logic formulae with *second-order variables* is

---

[1] This shall be usually referred to as "transitive closure of $\mathcal{DBS}$" to avoid clumsiness.

rarely addressed in literature. (They easily lead to undecidability.) To our best knowledge, the computation of transitive closures of $\mathcal{DBS}$ here represents one of the first practically relevant examples of the computation of this type for a class of logic formulae with second-order variables, which may be of independent interests.

Related Work. We first review the work on SLID with data constraints. (Due to space limit, the work on SLID *without* data constraints will be skipped.) In [7,8,23], SLID with set/multiset/size data constraints were considered, but only (incomplete) heuristics were provided. To reason about invariants of data values stored in lists, SL with list segment predicates and data constraints in universally quantified Presburger arithmetic was considered [1]. The work [26,27] provided decision procedures for SLID with data constraints by translating into many-sorted first-order logic with reachability predicates. In particular, in [27, Section 6], extensions of basic logic GRIT are given to cover set data constraints as well as order constraints over set elements. However, it seems that this app-roach does not address arithmetic constraints over set elements (cf. the "Limi-tations" paragraph in the end of Sect. 6 in [27]). For instance, a list where the data values in adjacent positions are consecutive can be captured in $\mathsf{SLID}^{\mathsf{S}}_{\mathsf{LC}}$ (see the predicate *plseg* in Sect. 3), but appears to go beyond the work [26,27]. Moreover, there is no precise characterisation of the limit of extensions under which the decidability retains. The work [13] introduced the concept of composi-tional inductive predicates, which may alleviate the difficulties of the entailment problem for SLID. Nevertheless, [13] only provided sound heuristics rather than decision procedures. More recently, the work [21,31] investigated SLID with Presburger arithmetic data constraints.

Furthermore, several logics other than separation logic have been consid-ered to reason about both shape properties and data constraints of data struc-tures. The work [30] proposed a generic decision procedure for recursive algebraic data types with abstraction functions encompassing lengths (sizes) of data struc-tures, sets or multisets of data values as special cases. Nevertheless, the work [30] focused on functional programs while this work aims to verify imperative programs, which requires to reason about *partial* data structures such as list seg-ments (rather than complete data structures such as lists). It is unclear how the decision procedure in [30] can be generalised to partial data structures. The work [22] introduced STRAND, a fragment of monadic second-order logic, to reason about tree structures. Being undecidable in general, several decidable fragments were identified. STRAND does not provide an explicit means to describe sets of data values, although it allows using set variables to represent sets of locations.

Our work is also related to classical logics with set constraints, for which we can only give a brief (but by no means comprehensive) summary. Presburger arithmetic extended with sets was studied dating back to 80's, with highly unde-cidability results [6,16]. However, decidable fragments do exist: [33] studied the non-disjoint combination of theories that share set variables and set operations. [20] considered $\mathsf{QFBAPA}^{\leq}_{\infty}$, a quantifier-free logic of sets of real numbers sup-porting integer sets and variables, linear arithmetic, the cardinality operator, infimum and supremum. [17,32] investigated two extensions of the Bernays-Schönfinkel-Ramsey fragment of first-order predicate logic (BSR) with simple linear arithmetic over integers and difference-bound constraints over reals (but crucially, the ranges of the universally quantified variables must be bounded). Since the unary predicate symbols in BSR are uninterpreted and represent sets over integers or reals, the two extensions of BSR can also be used to specify the set constraints on integers or reals. [10] presented a decision procedure for quantifier-free constraints on restricted intensional sets (i.e., sets given by a property rather than by enumerating their elements). None of these logics are able to capture the transitive closure of $\mathcal{DBS}$ as $\mathcal{RQSPA}$ does. MSOW extended with linear cardinality constraints was investigated in [18]. Roughly speaking, $\mathcal{RQSPA}$ can be considered as an extension of MSOW with linear arithmetic expressions on the maximum or minimum value of free set variables. Therefore, the two extensions in [18] and this paper are largely incomparable.

In contrast to set constraints, the computation of transitive closures of relations definable in first-order logic (in particular, difference-bound and octagonal arithmetic constraints) has been considered in for instance, [2–4,9,19].

2 Logics for Sets

We write $\mathbb{Z}$, $\mathbb{N}$ for the set of integers and natural numbers; $\mathbb{S}_{\mathbb{Z}}$ and $\mathbb{S}_{\mathbb{N}}$ for *finite* subsets of $\mathbb{Z}$ and $\mathbb{N}$. For $n \in \mathbb{N}$, $[n]$ stands for $\{1, \cdots, n\}$. We shall work exclusively on finite subsets of $\mathbb{Z}$ or $\mathbb{N}$ unless otherwise stated. For any finite $A \neq \varnothing$, we write $\min(A)$ and $\max(A)$ for the minimum and maximum element of A. These functions, however, are *not* defined over empty sets.

In the sequel, we introduce a handful of logics for sets which will be used later in this paper. We mainly consider two data types, i.e., integer type $\mathbb{Z}$ and (finite) set type $\mathbb{S}_{\mathbb{Z}}$. Typically, $c, c', \cdots \in \mathbb{Z}$ and $A, A', \cdots \in \mathbb{S}_{\mathbb{Z}}$. Accordingly, two types of variables occur: integer variables (ranged over by $x, y, \cdots$) and set variables (ranged over by $S, S', \cdots$). Furthermore, we reserve $\bowtie \in \{=, \leq, \geq\}$ for comparison operators between integers,[2] and $\asymp \in \{=, \subseteq, \supseteq, \subset, \supset\}$ for comparison operators between sets. We start with *difference-bound* set constraints ($\mathcal{DBS}$).

Definition 1 (Difference-bound set constraints). *Formulae of $\mathcal{DBS}$ are defined by the rules:*

$$\begin{aligned}
\varphi &::= S = S' \cup T_s \mid T_i \bowtie T_i + c \mid \varphi \wedge \varphi \\
T_s &::= \varnothing \mid \{\min(S)\} \mid \{\max(S)\} \mid T_s \cup T_s \quad \text{(set terms)} \\
T_i &::= \min(S) \mid \max(S) \quad \text{(integer terms)}
\end{aligned}$$

Remark. $\mathcal{DBS}$ is a rather limited logic, but it has been carefully devised to serve the data formulae in inductive predicates of $\mathsf{SLID}^{\mathsf{S}}_{\mathsf{LC}}[P]$ (cf. Sect. 3). In particular, we remark that only conjunction, but not disjunction, of atomic constraints is

---

[2] The operators $<$ and $>$ can be seen as abbreviations, for instance, $x < y$ is equivalent to $x \leq y - 1$, which will be used later on as well.

allowed. The main reason is, once the disjunction is introduced, the computation of transitive closures becomes infeasible simply because one would be able to encode the computation of Minsky's two-counter machines.   □

To capture the transitive closure of $\mathcal{DBS}$, we introduce *Restricted extension of Quantified Set constraints with Presburger Arithmetic*[3] ($\mathcal{RQSPA}$). Intuitively, an $\mathcal{RQSPA}$ formula is a *quantified* set constraint extended with Presburger Arithmetic satisfying the following restriction: each atomic formula containing *quantified* variables must be a difference-bound arithmetic constraint.

Definition 2 (Restricted extension of Quantified Set constraints with Presburger Arithmetic). *Formulae of $\mathcal{RQSPA}$ are defined by the rules:*

$$\Phi ::= T_s \asymp T_s \mid T_i \bowtie T_i + c \mid T_m \bowtie 0 \mid \Phi \wedge \Phi \mid \neg\Phi \mid \forall x.\ \Phi \mid \forall S.\ \Phi,$$
$$T_s ::= \varnothing \mid S \mid \{T_i\} \mid T_s \cup T_s \mid T_s \cap T_s \mid T_s \setminus T_s,$$
$$T_i ::= c \mid x \mid \min(T_s) \mid \max(T_s),$$
$$T_m ::= c \mid x \mid \max(T_s) \mid \min(T_s) \mid T_m + T_m \mid T_m - T_m.$$

Here, T_s (resp. T_i) represents set (resp. integer) terms which are more general than those in $\mathcal{DBS}$, and T_m terms are *Presburger arithmetic expressions*. Let $\mathsf{Vars}(\Phi)$ (resp. $\mathsf{free}(\Phi)$) denote the set of variables (resp. free variables) occurring in Φ. We require that **all set variables in atomic formulae $T_m \bowtie 0$ are free.** To make the free variables explicit, we usually write $\Phi(\boldsymbol{x}, \boldsymbol{S})$ for a $\mathcal{RQSPA}$ formula Φ. Free variable names are assumed *not* to clash with the quantified ones.

Example 1. $\max(S_1 \cup S_2) - \min(S_1) - \max(S_2) < 0$ and $\forall S_1 \forall S_2.(S_2 \neq \varnothing \rightarrow \max(S_2) \leq \max(S_1 \cup S_2))$ are $\mathcal{RQSPA}$ formulae, while $\forall S_2.\ \max(S_1 \cup S_2) - \min(S_1) - \max(S_2) < 0$ is *not*.   □

The work [6], among others, studied *Presburger arithmetic extended with Sets* ($\mathcal{PS}$), which is *quantifier-free* $\mathcal{RQSPA}$ formulae. In this paper, $\mathcal{PS}$ will serve the data formula part of $\mathsf{SLID}^{\mathsf{S}}_{\mathsf{LC}}[P]$, and we reserve $\Delta, \Delta', \ldots$ to denote formulae from $\mathcal{PS}$ (see Sect. 3).

Semantics. All of these logics ($\mathcal{DBS}$, $\mathcal{RQSPA}$, $\mathcal{PS}$) can be considered as instances of weak monadic second-order logic, and thus their semantics are largely self-explanatory. In particular, set variables are interpreted as *finite* subsets of $\mathbb{Z}$ and integer variables are interpreted as integers. We emphasize that, if a set term T_s is interpreted as $\varnothing$, $\min(T_s)$ and $\max(T_s)$ are undefined. As a result, we stipulate that **any atomic formula containing an undefined term is interpreted as false.**

For an $\mathcal{RQSPA}$ formula $\Phi(\boldsymbol{x}, \boldsymbol{S})$ with $\boldsymbol{x} = (x_1, \cdots, x_k)$ and $\boldsymbol{S} = (S_1, \cdots, S_l)$, $\mathcal{L}(\Phi(\boldsymbol{x}, \boldsymbol{S}))$ denotes

$$\{(n_1, \cdots, n_k, A_1, \cdots, A_l) \in \mathbb{Z}^k \times \mathbb{S}_{\mathbb{Z}}^l \mid \Phi(n_1, \cdots, n_k, A_1, \cdots, A_l)\}.$$

---

[3] An unrestricted extension of quantified set constraints with Presburger Arithmetic is undecidable, as shown in [6].

As expected, typically we use $\mathcal{DBS}$ formulae to define relations between (tuples of) sets from $\mathbb{S}_{\mathbb{Z}}^k$. We say a relation $R \subseteq \mathbb{S}_{\mathbb{Z}}^k \times \mathbb{S}_{\mathbb{Z}}^k$ a *difference-bound set relation* if there is a $\mathcal{DBS}$ formula $\varphi(\boldsymbol{S}, \boldsymbol{S}')$ over set variables $\boldsymbol{S}$ and $\boldsymbol{S}'$ such that $R = \{(\boldsymbol{A}, \boldsymbol{A}') \in \mathbb{S}_{\mathbb{Z}}^k \times \mathbb{S}_{\mathbb{Z}}^k \mid \varphi(\boldsymbol{A}, \boldsymbol{A}')\}$. The *transitive closure* of R is defined in a standard way, viz., $\bigcup_{i \geq 0} R^i$, where $R^0 = \{(\boldsymbol{A}, \boldsymbol{A}) \mid \boldsymbol{A} \in \mathbb{S}_{\mathbb{Z}}^k\}$ and $R^{i+1} = R^i \cdot R$.

3   Linearly Compositional SLID with Set Data Constraints

In this section, we introduce separation logic with *linearly compositional* inductive predicates and *set data* constraints, denoted by $\mathsf{SLID}_{\mathsf{LC}}^{\mathsf{S}}[P]$, where P is an *inductive predicate*. In addition to the integer and set data types introduced in Sect. 2, we also consider the *location* data type $\mathbb{L}$. As a convention, $l, l', \cdots \in \mathbb{L}$ denote locations and $E, F, X, Y, \cdots$ range over location variables. We consider location fields associated with $\mathbb{L}$ and data fields associated with $\mathbb{Z}$.

$\mathsf{SLID}_{\mathsf{LC}}^{\mathsf{S}}[P]$ formulae may contain inductive predicates, each of which is of the form $P(\boldsymbol{\alpha}; \boldsymbol{\beta}; \boldsymbol{\xi})$ and has an associated inductive definition. The parameters are classified into three groups: *source parameters* $\boldsymbol{\alpha}$, *destination parameters* $\boldsymbol{\beta}$, and *static parameters* $\boldsymbol{\xi}$. We require that the source parameters $\boldsymbol{\alpha}$ and the destination parameters $\boldsymbol{\beta}$ are *matched* in type, namely, the two tuples have the same length $\ell > 0$ and for each $i \in [\ell]$, α_i and β_i have the same data type. Static parameters are typically used to store some static (global) information of dynamic data structures, e.g., the target location of tail pointers. Moreover, we assume that for each $i \in [\ell]$, α_i is of either the location type, or the *set type*. (There are no parameters of the integer type.) Without loss of generality, it is assumed that the first components of $\boldsymbol{\alpha}$ and $\boldsymbol{\beta}$ are location variables; we usually explicitly write $E, \boldsymbol{\alpha}$ and $F, \boldsymbol{\beta}$.

$\mathsf{SLID}_{\mathsf{LC}}^{\mathsf{S}}[P]$ formulae comprise three types of formulae: *pure formulae* Π, *data formulae* Δ, and *spatial formulae* Σ. The data formulae are simply $\mathcal{PS}$ introduced in Sect. 2, while Π and Σ are defined by the following rules,

$$\Pi ::= E = F \mid E \neq F \mid \Pi \wedge \Pi \qquad \text{(pure formulae)}$$
$$\Sigma ::= \mathsf{emp} \mid E \mapsto (\rho) \mid P(E, \boldsymbol{\alpha}; F, \boldsymbol{\beta}; \boldsymbol{\xi}) \mid \Sigma * \Sigma \quad \text{(spatial formulae)}$$
$$\rho ::= (f, X) \mid (d, T_i) \mid \rho, \rho \qquad \text{(fields)}$$

where T_i is an integer term as in Definition 2, and f (resp. d) is a location (resp. data) field. For spatial formulae Σ, formulae of the form emp, $E \mapsto (\rho)$, or $P(E, \boldsymbol{\alpha}; F, \boldsymbol{\beta}; \boldsymbol{\xi})$ are called *spatial atoms*. In particular, formulae of the form $E \mapsto (\rho)$ and $P(E, \boldsymbol{\alpha}; F, \boldsymbol{\beta}; \boldsymbol{\xi})$ are called *points-to* and *predicate* atoms respectively. Moreover, E is *the root* of these points-to or predicate atoms.

Linearly Compositional Inductive Predicates. An inductive predicate P is *linearly compositional* if the inductive definition of P is given by the following two rules,

- base rule $R_0 : P(E, \alpha; F, \beta; \xi) ::= E = F \wedge \alpha = \beta \wedge \mathsf{emp}$,
- inductive rule $R_1 : P(E, \alpha; F, \beta; \xi) ::= \exists \boldsymbol{X} \exists \boldsymbol{S}. \; \varphi \wedge E \mapsto (\rho) * P(Y, \gamma; F, \beta; \xi)$.

The left-hand (resp. right-hand) side of a rule is called the *head* (resp. *body*) of the rule.

In the sequel, we specify some constraints on the inductive rule R_1 which are vital to obtain *complete* decision procedures for the satisfiability problem.

C1 None of the variables from F, β occur elsewhere in the right-hand side of R_1, that is, in φ, $E \mapsto (\rho)$.

C2 The data constraint φ in the body of R_1 is a $\mathcal{DBS}$ formula.

C3 For each atomic formula in φ, there is i such that all the variables in the atomic formula are from $\{\alpha_i, \gamma_i\}$.

C4 Each variable occurs in each of $P(Y, \gamma; F, \beta; \xi)$ and ρ at most once.

C5 ξ contains *only* location variables and all location variables from $\alpha \cup \xi \cup \boldsymbol{X}$ occur in ρ.

C6 $Y \in \boldsymbol{X}$ and $\gamma \subseteq \{E\} \cup \boldsymbol{X} \cup \boldsymbol{S}$.

Note that, by **C6**, none of the variables from $\alpha \cup \xi$ occur in γ. Moreover, from **C5** and **C6**, Y occurs in ρ, which guarantees that in each model of $P(E, \alpha; F, \beta; \xi)$, the sub-heap represented by $P(E, \alpha; F, \beta; \xi)$, seen as a directed graph, is connected. We also note that the body of R_1 does *not* contain pure formulae. We remark that these constraints are undeniably technical. However, in practice the inductive predicates satisfying these constraints are usually sufficient to define linear data structures with set data constraints, cf. Example 2.

For an inductive predicate P, let $\mathrm{Flds}(P)$ denote the set of all fields occurring in the inductive rules of P. For a spatial atom a, let $\mathrm{Flds}(a)$ denote the set of fields that a refers to: if $a = E \mapsto (\rho)$, then $\mathrm{Flds}(a)$ is the set of fields occurring in ρ; if $a = P(-)$, then $\mathrm{Flds}(a) = \mathrm{Flds}(P)$.

We write $\mathsf{SLID}^S_{\mathsf{LC}}[P]$ for the collection of separation logic formulae $\phi = \Pi \wedge \Delta \wedge \Sigma$ satisfying the following constraints: (1) P is a linearly compositional inductive predicate, and (2) each predicate atom of Σ is of the form $P(-)$, and for each points-to atom occurring in Σ, the set of fields of this atom is $\mathrm{Flds}(P)$.

For an $\mathsf{SLID}^S_{\mathsf{LC}}[P]$ formula ϕ, let $\mathsf{Vars}(\phi)$ (resp. $\mathsf{LVars}(\phi)$, resp. $\mathsf{DVars}(\phi)$, resp. $\mathsf{SVars}(\phi)$) denote the set of (resp. location, resp. integer, resp. set) variables occurring in ϕ. Moreover, we use $\phi[\boldsymbol{\mu}/\boldsymbol{\alpha}]$ to denote the simultaneous replacement of the variables α_j by μ_j in ϕ. We adopt the standard *classic, precise semantics* of $\mathsf{SLID}^S_{\mathsf{LC}}[P]$ in terms of *states*. In particular, a *state* is a pair (s, h), where s is an assignment and h is a heap. The details can be found in [14].

Example 2. We collect a few examples of linear data structures with set data constraints definable in $\mathsf{SLID}^S_{\mathsf{LC}}[P]$:

sdllseg for sorted doubly linked list segments, $sdllseg(E, P, S; F, L, S') ::= E = F \land P = L \land S = S' \land \text{emp},$ $sdllseg(E, P, S; F, L, S') ::= \exists X, S''. \ S = S'' \cup \{\min(S)\} \land$ $\qquad E \mapsto ((\text{next}, X), (\text{prev}, P), (\text{data}, \min(S))) * sdllseg(X, E, S''; F, L, S').$
plseg for list segments where the data values are consecutive, $plseg(E, S; F, S') ::= E = F \land S = S' \land \text{emp},$ $plseg(E, S; F, S') ::= \exists X, S''. \ S = S'' \cup \{\min(S)\} \land \min(S'') = \min(S) + 1 \land$ $\qquad E \mapsto ((\text{next}, X), (\text{data}, \min(S))) * plseg(X, S''; F, S').$
ldllseg for doubly list segments, to mimic lengths with sets, $ldllseg(E, P, S; F, L, S') ::= E = F \land P = L \land S = S' \land \text{emp},$ $ldllseg(E, P, S; F, L, S') ::= \exists X, S''. \ S = S'' \cup \{\max(S)\} \land \max(S'') = \max(S) - 1 \land$ $\qquad E \mapsto ((\text{next}, X), (\text{prev}, P)) * ldllseg(X, E, S''; F, L, S').$

4   Satisfiability of $\mathsf{SLID}^{\mathsf{S}}_{\mathsf{LC}}[P]$

The satisfiability problem is to decide whether there is a state (an assignment-heap pair) satisfying ϕ for a given $\mathsf{SLID}^{\mathsf{S}}_{\mathsf{LC}}[P]$ formula ϕ. We shall follow the approach adopted in [12,15], i.e., to construct $\mathsf{Abs}(\phi)$, an abstraction of ϕ that is equisatisfiable to ϕ. The key ingredient of the construction is to compute the transitive closure of the data constraints extracted from the inductive rule of P.

Let $\phi = \Pi \land \Delta \land \Sigma$ be an $\mathsf{SLID}^{\mathsf{S}}_{\mathsf{LC}}[P]$ formula. Suppose $\Sigma = a_1 * \cdots * a_n$, where each a_i is either a points-to atom or a predicate atom. For predicate atom $a_i = P(Z_1, \boldsymbol{\mu}; Z_2, \boldsymbol{\nu}; \boldsymbol{\chi})$ we assume that the inductive rule for P is

$$R_1 : P(E, \boldsymbol{\alpha}; F, \boldsymbol{\beta}; \boldsymbol{\xi}) ::= \exists \boldsymbol{X} \exists \boldsymbol{S}. \ \varphi \land E \mapsto (\rho) * P(Y, \boldsymbol{\gamma}; F, \boldsymbol{\beta}; \boldsymbol{\xi}). \qquad (1)$$

We extract the data constraint $\varphi_P(\mathsf{dt}(\boldsymbol{\alpha}), \mathsf{dt}(\boldsymbol{\beta}))$ out of R_1. Formally, we define $\varphi_P(\mathsf{dt}(\boldsymbol{\alpha}), \mathsf{dt}(\boldsymbol{\beta}))$ as $\varphi[\mathsf{dt}(\boldsymbol{\beta})/\mathsf{dt}(\boldsymbol{\gamma})]$, where $\mathsf{dt}(\boldsymbol{\alpha})$ (resp. $\mathsf{dt}(\boldsymbol{\gamma})$, $\mathsf{dt}(\boldsymbol{\beta})$) is the projection of $\boldsymbol{\alpha}$ (resp. $\boldsymbol{\gamma}$, $\boldsymbol{\beta}$) to data variables. For instance, $\varphi_{ldllseg}(S, S') :=$ $(S = S'' \cup \{\max(S)\} \land \max(S'') = \max(S) - 1) [S'/S''] = S = S' \cup \{\max(S)\} \land$ $\max(S') = \max(S) - 1$.

We can construct $\mathsf{Abs}(\phi)$ with necessary adaptations from [15]. For each spatial atom a_i, $\mathsf{Abs}(\phi)$ introduces a Boolean variable to denote whether a_i corresponds to a nonempty heap or not. With these Boolean variables, the semantics of separating conjunction are encoded in $\mathsf{Abs}(\phi)$. Moreover, for each predicate atom a_i, $\mathsf{Abs}(\phi)$ contains an abstraction of a_i, where the formulae $\mathsf{Ufld}_1(a_i)$ and $\mathsf{Ufld}_{\geq 2}(a_i)$ are used. Intuitively, $\mathsf{Ufld}_1(a_i)$ and $\mathsf{Ufld}_{\geq 2}(a_i)$ correspond to the separation logic formulae obtained by unfolding the rule R_1 *once* and *at least twice* respectively. We include the construction here so one can see the role of the transitive closure in $\mathsf{Abs}(\phi)$. The details of $\mathsf{Abs}(\phi)$ can be found in [14].

Let $a_i = P(Z_1, \boldsymbol{\mu}; Z_2, \boldsymbol{\nu}; \boldsymbol{\chi})$ and R_1 be the inductive rule in Eq. (1). If E occurs in $\boldsymbol{\gamma}$ in the body of R_1, we use $\mathsf{idx}_{(P, \boldsymbol{\gamma}, E)}$ to denote the unique index j such that $\gamma_j = E$. (The uniqueness follows from **C4**.)

Definition 3 ($\mathsf{Ufld}_1(a_i)$ and $\mathsf{Ufld}_{\geq 2}(a_i)$). $\mathsf{Ufld}_1(a_i)$ *and* $\mathsf{Ufld}_{\geq 2}(a_i)$ *are defined by distinguishing the following two cases:*

- If E occurs in γ in the body of R_1, then $\mathsf{Ufld}_1(a_i) := (E = \beta_{\mathsf{idx}(P,\gamma,E)} \wedge \varphi_P(\mathsf{dt}(\alpha), \mathsf{dt}(\beta)))[Z_1/E, \mu/\alpha, Z_2/F, \nu/\beta, \chi/\xi]$ and $\mathsf{Ufld}_{\geq 2}(a_i) :=$

$$\begin{pmatrix} E \neq \beta_{\mathsf{idx}(P,\gamma,E)} \wedge E \neq \gamma_{2,\mathsf{idx}(P,\gamma,E)} \wedge \\ \varphi_P[\mathsf{dt}(\gamma_1)/\mathsf{dt}(\beta)] \wedge \varphi_P[\mathsf{dt}(\gamma_1)/\mathsf{dt}(\alpha), \mathsf{dt}(\gamma_2)/\mathsf{dt}(\beta)] \wedge \\ (TC[\varphi_P])[\mathsf{dt}(\gamma_2)/\mathsf{dt}(\alpha)] \end{pmatrix} [Z_1/E, \mu/\alpha, Z_2/F, \nu/\beta, \chi/\xi],$$

where γ_1 and γ_2 are fresh variables.
- Otherwise, let $\mathsf{Ufld}_1(a_i) := \varphi_P[Z_1/E, \mu/\alpha, Z_2/F, \nu/\beta, \chi/\xi]$ and

$$\mathsf{Ufld}_{\geq 2}(a_i) := \begin{pmatrix} \varphi_P[\mathsf{dt}(\gamma_1)/\mathsf{dt}(\beta)] \wedge \\ \varphi_P[\mathsf{dt}(\gamma_1)/\mathsf{dt}(\alpha), \mathsf{dt}(\gamma_2)/\mathsf{dt}(\beta)] \wedge \\ (TC[\varphi_P])[\mathsf{dt}(\gamma_2)/\mathsf{dt}(\alpha)] \end{pmatrix} [Z_1/E, \mu/\alpha, Z_2/F, \nu/\beta, \chi/\xi],$$

where γ_1 and γ_2 are fresh variables.

Here, $TC[\varphi_P](\mathsf{dt}(\alpha), \mathsf{dt}(\beta))$ denotes the transitive closure of φ_P. In Sect. 5, it will be shown that $TC[\varphi_P](\mathsf{dt}(\alpha), \mathsf{dt}(\beta))$ can be written as an $\mathcal{RQSPA}$ formula. As a result, since we are only concerned with satisfiability and can treat the location data type $\mathbb{L}$ simply as integers $\mathbb{Z}$, $\mathsf{Abs}(\phi)$ can also be read as an $\mathcal{RQSPA}$ formula. In Sect. 6, we shall show that the satisfiability of $\mathcal{RQSPA}$ is decidable. Following this chain of reasoning, we conclude that *the satisfiability of* $\mathsf{SLID}^S_{\mathsf{LC}}[P]$ *formulae is decidable.*

5   Transitive Closure of Difference-Bound Set Relations

In this section, we show how to compute the transitive closure of the difference-bound set relation R given by a $\mathcal{DBS}$ formula $\varphi_R(S, S')$. Our approach is, in a nutshell, to encode $TC[\varphi_R](S, S')$ into $\mathcal{RQSPA}$. We shall only sketch part of a simple case, i.e., in $\varphi_R(S, S')$ only one source and destination set parameter are present. The details are however given in [14].

Recall that, owing to the simplicity of $\mathcal{DBS}$, the integer terms T_i in $\varphi_R(S, S')$ can only be $\min(S)$, $\max(S)$, $\min(S')$ or $\max(S')$, whereas the set terms T_s are $\varnothing$, $\{\min(S)\}$, $\{\min(S')\}$, $\{\max(S)\}$, $\{\max(S')\}$, or their union. For reference, we write $\varphi_R(S, S') = \varphi_{R,1} \wedge \varphi_{R,2}$, where $\varphi_{R,1}$ is an equality of set terms (i.e., they are of the form $S = S' \cup T_s$ or $S' = S \cup T_s$), and $\varphi_{R,2}$ is a conjunction of constraints over integer terms (i.e., a conjunction of formulae $T_i \leq T_i + c$). $\varphi_{R,1}$ and $\varphi_{R,2}$ will be referred to as the *set* and *integer subformula* of $\varphi_R(S, S')$ respectively. We shall focus on the case $\varphi_{R,1} := S = S' \cup T_s$. The symmetrical case $\varphi_{R,1} := S' = S \cup T_s$ can be adapted easily.

The integer subformula $\varphi_{R,2}$ can be represented by an *edge-weighted directed graph* $\mathcal{G}(\varphi_{R,2})$, where the vertices are all integer terms appearing in $\varphi_{R,2}$, and there is an edge from T_1 to T_2 with weight c iff $T_1 = T_2 + c$ (equivalent to $T_2 = T_1 - c$), or $T_1 \leq T_2 + c$, or $T_2 + c \geq T_1$ appears in $\varphi_{R,2}$. The weight of a path in $\mathcal{G}(\varphi_{R,2})$ is the sum of the weights of the edges along the path. A *negative cycle* in $\mathcal{G}(\varphi_{R,2})$ is a cycle with negative weight. It is known that $\varphi_{R,2}$ is satisfiable iff $\mathcal{G}(\varphi_{R,2})$ contains no negative cycles [24]. Suppose $\varphi_{R,2}$ is satisfiable. We define the *normal form* of $\varphi_{R,2}$, denoted by $\mathsf{Norm}(\varphi_{R,2})$, as the conjunction of the formulae $T_1 \leq T_2 + c$ such that $T_1 \neq T_2$, T_2 is reachable from T_1 in $\mathcal{G}(\varphi_{R,2})$, and c is path from T_1 to T_2 with the minimal weight in $\mathcal{G}(\varphi_{R,2})$.

S (resp. S') is said to be *surely nonempty* in φ_R if $\min(S)$ or $\max(S)$ (resp. $\min(S')$ or $\max(S')$) occurs in φ_R; otherwise, S (resp. S') is *possibly empty* in φ_R. Recall that, according to the semantics, an occurrence of $\min(S)$ or $\max(S)$ (resp. $\min(S')$ or $\max(S')$) in φ_R implies that S (resp. S') is interpreted as a nonempty set in every satisfiable assignment. Provided that S' is nonempty, we know that $\min(S')$ and $\max(S')$ belong to S'. Therefore, for simplicity, here we assume that in $S = S' \cup T_s$, T_s contains neither $\min(S')$ nor $\max(S')$. The situation that T_s contains $\min(S')$ and $\max(S')$ can be dealt with in a similar way.

Saturation. For technical convenience, we introduce a concept of saturation. The main purpose of saturation is to regularise T_s and $\varphi_{R,2}$, which would make the transitive closure construction more "syntactic".

Definition 4. Let $\varphi_R(S, S') := S = S' \cup T_s \wedge \varphi_{R,2}$ be a DBS formula. Then $\varphi_R(S, S')$ is saturated if $\varphi_R(S, S')$ satisfies the following conditions

- $\varphi_{R,2}$ is satisfiable and in normal forms,
- $T_s \subseteq \{\max(S), \min(S)\}$,
- if S (resp. S') is surely nonempty in φ_R, then $\varphi_{R,2}$ contains a conjunct $\min(S) \leq \max(S) - c$ for some $c \geq 0$ (resp. $\min(S') \leq \max(S') - c'$ for some $c' \geq 0$),
- if both S and S' are surely nonempty in φ_R, then
  - $\varphi_{R,2}$ contains two conjuncts $\min(S) \leq \min(S') - c$ and $\max(S') \leq \max(S) - c'$ for some $c, c' \geq 0$,
  - $\min(S) \notin T_s$ iff $\varphi_{R,2}$ contains the conjuncts $\min(S) \leq \min(S')$ and $\min(S') \leq \min(S)$,
  - $\max(S) \notin T_s$ iff $\varphi_{R,2}$ contains the conjuncts $\max(S') \leq \max(S)$ and $\max(S) \leq \max(S')$,
- if $\varphi_{R,2}$ contains the conjuncts $\min(S) \leq \max(S)$ and $\max(S) \leq \min(S)$, then $\max(S) \notin T_s$ (possibly $\min(S) \in T_s$).

For a formula $\varphi_R(S, S') := S = S' \cup T_s \wedge \varphi_{R,2}$, one can easily saturate φ_R, yielding a saturated formula $\mathsf{Strt}(\varphi_R(S, S'))$. (It is possible, however, to arrive at an unsatisfiable formula, then we are done.)

Proposition 1. Let $\varphi_R(S, S') := \varphi_{R,1} \wedge \varphi_{R,2}$ be a DBS formula such that $\varphi_{R,1} := S = S' \cup T_s$ and $\varphi_{R,2}$ is satisfiable. Then φ_R can be transformed, in polynomial time, to an equisatisfiable formula $\mathsf{Strt}(\varphi_R(S, S'))$, and if the integer subformula of $\mathsf{Strt}(\varphi_R(S, S'))$ is satisfiable, then $\mathsf{Strt}(\varphi_R(S, S'))$ is saturated.

In the sequel, we assume that $\varphi_R(S, S') := \varphi_{R,1} \wedge \varphi_{R,2}$ is satisfiable and *saturated*. For notational convenience, for $A \subseteq \{\min(S), \max(S), \min(S'), \max(S')\}$ with $|A| = 2$, let $\lfloor \varphi_{R,2} \rfloor_A$ denote the conjunction of atomic formulae in $\varphi_{R,2}$ where *all* the elements of A occur.

Evidently, $\lfloor\varphi_{R,2}\rfloor_A$ gives a partition of atomic formulae of $\varphi_{R,2}$. Namely,

$$\varphi_{R,2} = \bigwedge\nolimits_{A\subseteq\{\min(S),\max(S),\min(S'),\max(S')\},|A|=2} \lfloor\varphi_{R,2}\rfloor_A.$$

We proceed by a case-by-case analysis of $\varphi_{R,1}$. There are four cases: (I) $\varphi_{R,1} := S = S'$, (II) $\varphi_{R,1} := S = S' \cup \{\min(S)\}$, (III) $\varphi_{R,1} = S = S' \cup \{\max(S)\}$ and (IV) $\varphi_{R,1} = S = S' \cup \{\min(S),\max(S)\}$. Case (I) is trivial, and Case (III) is symmetrical to (II). However, both (II) and (IV) are technically involved. We shall only give a "sample" treatment of these cases, i.e., part of arguments for Case (II); the full account of Case (II) and (IV) are given in [14].

To start with, Case (II) can be illustrated schematically

as $\underbrace{|-\overbrace{|------|}^{S'}}_{S}$. We observe that S is surely nonempty in φ_R. We then

distinguish two subcases depending on whether S' is possibly empty or surely nonempty in φ_R. Here we give the details of the latter subcase because it is more interesting. In this case, both S and S' are surely nonempty in φ_R. By Definition 4 (4–5), $\varphi_{R,2}$ contains a conjunct $\min(S) \leq \min(S') - c$ for some $c \geq 0$, as well as $\max(S') \leq \max(S)$ and $\max(S) \leq \max(S')$ (i.e., $\max(S') = \max(S)$). Therefore, we can assume

$$\varphi_{R,2} = \max(S') \leq \max(S) \wedge \max(S) \leq \max(S') \wedge \lfloor\varphi_{R,2}\rfloor_{\min(S),\min(S')} \wedge$$
$$\lfloor\varphi_{R,2}\rfloor_{\min(S),\max(S)} \wedge \lfloor\varphi_{R,2}\rfloor_{\min(S'),\max(S')}.$$

Note that in $\varphi_{R,2}$ above, the redundant subformulae $\lfloor\varphi_{R,2}\rfloor_{\min(S),\max(S')}$ and $\lfloor\varphi_{R,2}\rfloor_{\min(S'),\max(S)}$ have been omitted.

The formula $\lfloor\varphi_{R,2}\rfloor_{\min(S),\min(S')}$ is said to be *strict* if it contains a conjunct $\min(S) \leq \min(S') - c$ for some $c > 0$. Otherwise, it is said to be *non-strict*. Intuitively, if $\lfloor\varphi_{R,2}\rfloor_{\min(S),\min(S')}$ is strict, then for $n, n' \in \mathbb{Z}$, the validity of $(\lfloor\varphi_{R,2}\rfloor_{\min(S),\min(S')})[n/\min(S), n'/\min(S')]$ implies that $n < n'$. For the sketch we only present *the case that $\lfloor\varphi_{R,2}\rfloor_{\min(S),\min(S')}$ is strict*; the other cases are similar and can be found in [14].

Evidently, $\mathsf{TC}[\varphi_R](S,S')$ can be written as $(S = S') \vee \bigvee\limits_{n\geq 1} \varphi_R^{(n)}$, where $\varphi_R^{(n)}$

is obtained by unfolding φ_R for n times, that is,

$$\varphi_R^{(n)} = \exists S_1, \cdots, S_{n+1}. \left(\begin{array}{l} S_1 = S \wedge S_{n+1} = S' \wedge \\ \bigwedge\limits_{i\in[n]} (S_i = S_{i+1} \cup \{\min(S_i)\} \wedge \varphi_{R,2}[S_i/S, S_{i+1}/S']) \end{array} \right),$$

where $\varphi_{R,2}[S_i/S, S_{i+1}/S']$ is obtained from $\varphi_{R,2}$ by replacing S (resp. S') with S_i (resp. S_{i+1}).

Clearly, $\varphi_R^{(1)} = \varphi_R$, and

$$\varphi_R^{(2)} = \exists S_2. \, (S = S_2 \cup \{\min(S)\} \wedge S_2 = S' \cup \{\min(S_2)\} \wedge \varphi_{R,2}[S_2/S'] \wedge \varphi_{R,2}[S_2/S]).$$

For $\varphi_R^{(n)}$ where $n \geq 3$, we first simplify $\varphi_R^{(n)}$ to construct a finite formula for $\mathsf{TC}[\varphi_R](S, S')$. The subformula $\bigwedge_{i \in [n]} (S_i = S_{i+1} \cup \{\min(S_i)\} \wedge \varphi_{R,2}[S_i/S, S_{i+1}/S'])$ can be rewritten as

$$\bigwedge_{i \in [n]} \begin{pmatrix} S_i = S_{i+1} \cup \{\min(S_i)\} \wedge \max(S_i) = \max(S_{i+1}) \wedge \\ (\lfloor \varphi_{R,2} \rfloor_{\min(S),\min(S')}[S_i/S, S_{i+1}/S']) \wedge (\lfloor \varphi_{R,2} \rfloor_{\min(S),\max(S)}[S_i/S]) \wedge \\ (\lfloor \varphi_{R,2} \rfloor_{\min(S'),\max(S')}[S_{i+1}/S']) \end{pmatrix}.$$

Because $S_i = S_{i+1} \cup \{\min(S_i)\}$ for each $i \in [n]$, we have $\max(S_1) = \cdots = \max(S_n)$ and $\min(S_1) \leq \cdots \leq \min(S_n)$. Since $\lfloor \varphi_{R,2} \rfloor_{\min(S),\max(S)}$ is a conjunction of difference-bound constraints involving $\min(S)$ and $\max(S)$ only, we have $\bigwedge_{i \in [n]} \lfloor \varphi_{R,2} \rfloor_{\min(S),\max(S)}[S_i/S]$ is equivalent to $\lfloor \varphi_{R,2} \rfloor_{\min(S),\max(S)}[S_1/S] \wedge \lfloor \varphi_{R,2} \rfloor_{\min(S),\max(S)}[S_n/S]$. To see this, assume, for instance,

$$\lfloor \varphi_{R,2} \rfloor_{\min(S),\max(S)} \equiv c \leq \max(S) - \min(S) \leq c'$$

for some constants $c, c' \geq 0$ with $c \leq c'$. Then $\max(S_1) - \min(S_1) \leq c'$ implies $\max(S_i) - \min(S_i) \leq c'$ for each $i \in [n]$, and $c \leq \max(S_n) - \min(S_n)$ implies $c \leq \max(S_i) - \min(S_i)$ for each $i \in [n]$. Therefore, $\lfloor \varphi_{R,2} \rfloor_{\min(S),\max(S)} [S_1/S] \wedge \lfloor \varphi_{R,2} \rfloor_{\min(S),\max(S)}[S_n/S] \equiv c \leq \max(S_1) - \min(S_1) \leq c' \wedge c \leq \max(S_n) - \min(S_n) \leq c'$ implies that $\bigwedge_{i \in [n]} \lfloor \varphi_{R,2} \rfloor_{\min(S),\max(S)}[S_i/S]$, thus they are equivalent. (The other direction is trivial.) Likewise, one has $\lfloor \varphi_{R,2} \rfloor_{\min(S'),\max(S')}[S_2/S'] \wedge \lfloor \varphi_{R,2} \rfloor_{\min(S'),\max(S')}[S_{n+1}/S']$ implies $\bigwedge_{i \in [n]} \lfloor \varphi_{R,2} \rfloor_{\min(S'),\max(S')}[S_{i+1}/S']$, thus they are equivalent. Therefore, $\varphi_R^{(n)}$ can be transformed into

$$\exists S_2, S_n. \begin{pmatrix} \lfloor \varphi_{R,2} \rfloor_{\min(S),\max(S)} \wedge (\lfloor \varphi_{R,2} \rfloor_{\min(S),\max(S)}[S_n/S]) \wedge \\ (\lfloor \varphi_{R,2} \rfloor_{\min(S'),\max(S')}[S_2/S']) \wedge \lfloor \varphi_{R,2} \rfloor_{\min(S'),\max(S')} \wedge S = S_2 \cup \{\min(S)\} \wedge \\ S_n = S' \cup \{\min(S_n)\} \wedge \max(S) = \max(S_2) \wedge \max(S_n) = \max(S') \wedge \\ (\lfloor \varphi_{R,2} \rfloor_{\min(S),\min(S')}[S_2/S']) \wedge (\lfloor \varphi_{R,2} \rfloor_{\min(S),\min(S')}[S_n/S]) \wedge \\ \exists S_3, \cdots, S_{n-1}. \bigwedge_{2 \leq i \leq n-1} \begin{pmatrix} S_i = S_{i+1} \cup \{\min(S_i)\} \wedge \max(S_i) = \max(S_{i+1}) \wedge \\ (\lfloor \varphi_{R,2} \rfloor_{\min(S),\min(S')}[S_i/S, S_{i+1}/S']) \end{pmatrix} \end{pmatrix}.$$

Claim. *Suppose $n \geq 3$ and $\lfloor \varphi_{R,2} \rfloor_{\min(S),\min(S')}$ is strict. Then*

$$\exists S_3, \cdots, S_{n-1}. \bigwedge_{2 \leq i \leq n-1} \begin{pmatrix} S_i = S_{i+1} \cup \{\min(S_i)\} \wedge \max(S_i) = \max(S_{i+1}) \wedge \\ (\lfloor \varphi_{R,2} \rfloor_{\min(S),\min(S')}[S_i/S, S_{i+1}/S']) \end{pmatrix}$$

is equivalent to
$S_n \neq \emptyset \wedge S_2 \setminus S_n \neq \emptyset \wedge S_n \subseteq S_2 \wedge |S_2 \setminus S_n| = n - 2 \wedge \max(S_2 \setminus S_n) < \min(S_n) \wedge \forall y, z. \, \mathsf{succ}((S_2 \setminus S_n) \cup \{\min(S_n)\}, y, z) \to (\lfloor \varphi_{R,2} \rfloor_{\min(S),\min(S')}[y/\min(S), z/\min(S')]),$

where $\mathsf{succ}(S, x, y)$ specifies intuitively that y is the successor of x in S, that is,

$$\mathsf{succ}(S, x, y) = x \in S \wedge y \in S \wedge x < y \wedge \forall z \in S. \, (z \leq x \vee y \leq z).$$

Note that $|\cdot|$ denotes the set cardinality which can be easily encoded into $\mathcal{RQSPA}$. ([14] gives the proof of the claim.) It follows that $\mathsf{TC}[\varphi_R](S, S') =$

$$(S = S') \lor \varphi_R(S, S') \lor \varphi_R^{(2)}(S, S') \lor$$
$$\exists S_1, S_2. \left(\begin{array}{l} S = S_1 \cup \{\min(S)\} \land S_2 = S' \cup \{\min(S_2)\} \land \\ \max(S) = \max(S_1) \land \max(S_2) = \max(S') \land \\ S_2 \neq \varnothing \land S_1 \setminus S_2 \neq \varnothing \land S_2 \subseteq S_1 \land \max(S_1 \setminus S_2) < \min(S_2) \land \\ \lfloor \varphi_{R,2} \rfloor_{\min(S),\max(S)} \land (\lfloor \varphi_{R,2} \rfloor_{\min(S),\max(S)}[S_2/S]) \land \\ (\lfloor \varphi_{R,2} \rfloor_{\min(S'),\max(S')}[S_1/S']) \land \lfloor \varphi_{R,2} \rfloor_{\min(S'),\max(S')} \land \\ (\lfloor \varphi_{R,2} \rfloor_{\min(S),\min(S')}[S_1/S']) \land (\lfloor \varphi_{R,2} \rfloor_{\min(S),\min(S')}[S_2/S]) \land \\ \forall y, z. \left(\begin{array}{l} \mathsf{succ}((S_1 \setminus S_2) \cup \{\min(S_2)\}, y, z) \to \\ (\lfloor \varphi_{R,2} \rfloor_{\min(S),\min(S')}[y/\min(S), z/\min(S')]) \end{array} \right) \end{array} \right).$$

6 Satisfiability of $\mathcal{RQSPA}$

In this section, we focus on the second ingredient of the procedure for deciding satisfiability of $\mathsf{SLID}^S_{\mathsf{LC}}[P]$, i.e., the satisfiability of $\mathcal{RQSPA}$. We first note that $\mathcal{RQSPA}$ is defined over $\mathbb{Z}$. To show the decidability, it turns to be much easier to work on $\mathbb{N}$. We shall write $\mathcal{RQSPA}_{\mathbb{Z}}$ and $\mathcal{RQSPA}_{\mathbb{N}}$ to differentiate them when necessary. Moreover, for technical reasons, we also introduce $\mathcal{RQSPA}^-$, the fragment of $\mathcal{RQSPA}$ excluding formulae of the form $T_m \bowtie 0$.

The decision procedure for the satisfiability of $\mathcal{RQSPA}$ proceeds with the following three steps:

Step I. Translate $\mathcal{RQSPA}_{\mathbb{Z}}$ to $\mathcal{RQSPA}_{\mathbb{N}}$,

Step II. Normalize an $\mathcal{RQSPA}_{\mathbb{N}}$ formula $\Phi(\boldsymbol{x}, \boldsymbol{S})$ into $\bigvee_i (\Phi^{(i)}_{\mathrm{core}} \land \Phi^{(i)}_{\mathrm{count}})$, where $\Phi^{(i)}_{\mathrm{core}}$ is an $\mathcal{RQSPA}^-_{\mathbb{N}}$ formula, and $\Phi^{(i)}_{\mathrm{count}}$ is a conjunction of formulae of the form $T_m \bowtie 0$ which contain only variables from $\boldsymbol{x} \cup \boldsymbol{S}$,

Step III. For each disjunct $\Phi^{(i)}_{\mathrm{core}} \land \Phi^{(i)}_{\mathrm{count}}$, construct a Presburger automaton (PA) $\mathcal{A}^{(i)}_{\Phi}$ which captures the models of $\Phi^{(i)}_{\mathrm{core}} \land \Phi^{(i)}_{\mathrm{count}}$. Satisfiability is thus reducible to the nonemptiness of PA, which is decidable [29].

These steps are technically involved. In particular, the third step requires exploiting Presburger automata [29]. The details can be found in [14].

7 Conclusion

In this paper, we have defined $\mathsf{SLID}^S_{\mathsf{LC}}$, SL with linearly compositional inductive predicates and set data constraints. The main feature is to identify $\mathcal{DBS}$ as a special class of set data constraints in the inductive definitions. We encoded the transitive closure of $\mathcal{DBS}$ into $\mathcal{RQSPA}$, which was shown to be decidable. These together yield a complete decision procedure for the satisfiability of $\mathsf{SLID}^S_{\mathsf{LC}}$.

The precise complexity of the decision procedure—NONELEMENTARY is the best upper-bound we have now—is left open for further studies. Furthermore, the entailment problem of $\mathsf{SLID}^S_{\mathsf{LC}}$ is an immediate future work.

References

1. Bouajjani, A., Drăgoi, C., Enea, C., Sighireanu, M.: Accurate invariant checking for programs manipulating lists and arrays with infinite data. In: Chakraborty, S., Mukund, M. (eds.) ATVA 2012. LNCS, vol. 7561, pp. 167–182. Springer, Heidelberg (2012). https://doi.org/10.1007/978-3-642-33386-6_14
2. Bozga, M., Gîrlea, C., Iosif, R.: Iterating octagons. In: TACAS, pp. 337–351 (2009)
3. Bozga, M., Iosif, R., Konecný, F.: Fast acceleration of ultimately periodic relations. In: CAV, pp. 227–242 (2010)
4. Bozga, M., Iosif, R., Lakhnech, Y.: Flat parametric counter automata. Fundam. Inf. **91**(2), 275–303 (2009)
5. Büchi, R.J.: Weak Second-Order arithmetic and finite automata. Zeitschrift für Mathematische Logik und Grundlagen der Mathematik **6**(1–6), 66–92 (1960)
6. Cantone, D., Cutello, V., Schwartz, J.T.: Decision problems for tarski and presburger arithmetics extended with sets. In: Börger, E., Kleine Büning, H., Richter, M.M., Schönfeld, W. (eds.) CSL 1990. LNCS, vol. 533, pp. 95–109. Springer, Heidelberg (1991). https://doi.org/10.1007/3-540-54487-9_54
7. Chin, W.-N., David, C., Nguyen, H.H., Qin, S.: Automated verification of shape, size and bag properties via user-defined predicates in separation logic. Sci. Comput. Program. **77**(9), 1006–1036 (2012)
8. Chu, D.-H., Jaffar, J., Trinh, M.-T.: Automatic induction proofs of data-structures in imperative programs. In: PLDI, pp. 457–466 (2015)
9. Comon, H., Jurski, Y.: Multiple counters automata, safety analysis and presburger arithmetic. In: Hu, A.J., Vardi, M.Y. (eds.) CAV 1998. LNCS, vol. 1427, pp. 268–279. Springer, Heidelberg (1998). https://doi.org/10.1007/BFb0028751
10. Cristiá, M., Rossi, G.: A decision procedure for restricted intensional sets. In: de Moura, L. (ed.) CADE 2017. LNCS (LNAI), vol. 10395, pp. 185–201. Springer, Cham (2017). https://doi.org/10.1007/978-3-319-63046-5_12
11. Elgot, C.C.: Decision problems of finite automata design and related arithmetics. Trans. Am. Math. Soc. **98**(1), 21–51 (1961)
12. Enea, C., Lengál, O., Sighireanu, M., Vojnar, T.: Compositional entailment checking for a fragment of separation logic. In: APLAS, pp. 314–333 (2014)
13. Enea, C., Sighireanu, M., Wu, Z.: On automated lemma generation for separation logic with inductive definitions. In: Finkbeiner, B., Pu, G., Zhang, L. (eds.) ATVA 2015. LNCS, vol. 9364, pp. 80–96. Springer, Cham (2015). https://doi.org/10.1007/978-3-319-24953-7_7
14. Gao, C., Chen, T., Wu, Z.: Separation logic with linearly compositional inductive predicates and set data constraints (full version). http://arxiv.org/abs/1811.00699
15. Gu, X., Chen, T., Wu, Z.: A complete decision procedure for linearly compositional separation logic with data constraints. In: Olivetti, N., Tiwari, A. (eds.) IJCAR 2016. LNCS (LNAI), vol. 9706, pp. 532–549. Springer, Cham (2016). https://doi.org/10.1007/978-3-319-40229-1_36
16. Halpern, J.Y.: Presburger arithmetic with unary predicates is Π_1^1-complete. J. Symb. Logic **56**(2), 637–642 (1991)
17. Horbach, M., Voigt, M., Weidenbach, C.: On the combination of the Bernays–Schönfinkel–Ramsey fragment with simple linear integer arithmetic. In: de Moura, L. (ed.) CADE 2017. LNCS (LNAI), vol. 10395, pp. 77–94. Springer, Cham (2017). https://doi.org/10.1007/978-3-319-63046-5_6
18. Klaedtke, F., Rueß, H.: Monadic second-order logics with cardinalities. In: Baeten, J.C.M., Lenstra, J.K., Parrow, J., Woeginger, G.J. (eds.) ICALP 2003. LNCS, vol. 2719, pp. 681–696. Springer, Heidelberg (2003). https://doi.org/10.1007/3-540-45061-0_54

19. Konečný, F.: PTIME computation of transitive closures of octagonal relations. In: Chechik, M., Raskin, J.-F. (eds.) TACAS 2016. LNCS, vol. 9636, pp. 645–661. Springer, Heidelberg (2016). https://doi.org/10.1007/978-3-662-49674-9_42

20. Kuncak, V., Piskac, R., Suter, P.: Ordered sets in the calculus of data structures. In: Dawar, A., Veith, H. (eds.) CSL 2010. LNCS, vol. 6247, pp. 34–48. Springer, Heidelberg (2010). https://doi.org/10.1007/978-3-642-15205-4_5

21. Le, Q.L., Sun, J., Chin, W.-N.: Satisfiability modulo heap-based programs. In: Chaudhuri, S., Farzan, A. (eds.) CAV 2016. LNCS, vol. 9779, pp. 382–404. Springer, Cham (2016). https://doi.org/10.1007/978-3-319-41528-4_21

22. Madhusudan, P., Parlato, G., Qiu, X.: Decidable logics combining heap structures and data. In: POPL 2011, pp. 611–622. ACM (2011)

23. Madhusudan, P., Qiu, X., Stefanescu, A.: Recursive proofs for inductive tree data-structures. In: POPL, pp. 123–136 (2012)

24. Miné, A.: A new numerical abstract domain based on difference-bound matrices. In: Danvy, O., Filinski, A. (eds.) PADO 2001. LNCS, vol. 2053, pp. 155–172. Springer, Heidelberg (2001). https://doi.org/10.1007/3-540-44978-7_10

25. O'Hearn, P., Reynolds, J., Yang, H.: Local reasoning about programs that alter data structures. In: Fribourg, L. (ed.) CSL 2001. LNCS, vol. 2142, pp. 1–19. Springer, Heidelberg (2001). https://doi.org/10.1007/3-540-44802-0_1

26. Piskac, R., Wies, T., Zufferey, D.: Automating separation logic using SMT. In: Sharygina, N., Veith, H. (eds.) CAV 2013. LNCS, vol. 8044, pp. 773–789. Springer, Heidelberg (2013). https://doi.org/10.1007/978-3-642-39799-8_54

27. Piskac, R., Wies, T., Zufferey, D.: Automating separation logic with trees and data. In: Biere, A., Bloem, R. (eds.) CAV 2014. LNCS, vol. 8559, pp. 711–728. Springer, Cham (2014). https://doi.org/10.1007/978-3-319-08867-9_47

28. Reynolds, J.C.: Separation logic: a logic for shared mutable data structures. In: LICS, pp. 55–74 (2002)

29. Seidl, H., Schwentick, T., Muscholl, A., Habermehl, P.: Counting in trees for free. In: Díaz, J., Karhumäki, J., Lepistö, A., Sannella, D. (eds.) ICALP 2004. LNCS, vol. 3142, pp. 1136–1149. Springer, Heidelberg (2004). https://doi.org/10.1007/978-3-540-27836-8_94

30. Suter, P., Dotta, M., Kuncak, V.: Decision procedures for algebraic data types with abstractions. In: POPL 2010, pp. 199–210. ACM (2010)

31. Tatsuta, M., Le, Q.L., Chin, W.-N.: Decision procedure for separation logic with inductive definitions and presburger arithmetic. In: Igarashi, A. (ed.) APLAS 2016. LNCS, vol. 10017, pp. 423–443. Springer, Cham (2016). https://doi.org/10.1007/978-3-319-47958-3_22

32. Voigt, M.: The Bernays–Schönfinkel–Ramsey fragment with bounded difference constraints over the reals is decidable. In: Dixon, C., Finger, M. (eds.) FroCoS 2017. LNCS (LNAI), vol. 10483, pp. 244–261. Springer, Cham (2017). https://doi.org/10.1007/978-3-319-66167-4_14

33. Wies, T., Piskac, R., Kuncak, V.: Combining theories with shared set operations. In: Ghilardi, S., Sebastiani, R. (eds.) FroCoS 2009. LNCS (LNAI), vol. 5749, pp. 366–382. Springer, Heidelberg (2009). https://doi.org/10.1007/978-3-642-04222-5_23

34. Xu, Z., Chen, T., Wu, Z.: Satisfiability of compositional separation logic with tree predicates and data constraints. In: de Moura, L. (ed.) CADE 2017. LNCS (LNAI), vol. 10395, pp. 509–527. Springer, Cham (2017). https://doi.org/10.1007/978-3-319-63046-5_31

On the Complexity of Optimal Matching Reconfiguration

Manoj Gupta[1], Hitesh Kumar[2], and Neeldhara Misra[1(✉)]

[1] Indian Institute of Technology, Gandhinagar, Gandhinagar, India
{gmanoj,neeldhara.m}@iitgn.ac.in
[2] NISER Bhubaneswar, Bhubaneswar, India
hitesh.kumar@niser.ac.in

Abstract. We consider the problem of matching reconfiguration, where we are given two matchings M_s and M_t in a graph G and the goal is to determine if there exists a sequence of matchings $M_0, M_1, \ldots, M_\ell$, such that $M_0 = M_s$, all consecutive matchings differ by exactly two edges (specifically, any matching is obtained from the previous one by the addition and deletion of one edge), and $M_\ell = M_t$. It is known that the existence of such a sequence can be determined in polynomial time [5].

We extend the study of reconfiguring matchings to account for the length of the reconfiguration sequence. We show that checking if we can reconfigure M_s to M_t in at most ℓ steps is NP-hard, even when the graph is unweighted, bipartite, and the maximum degree is four, and the matchings M_s and M_t are maximum matchings. We propose two simple algorithmic approaches, one of which improves on the brute-force running time while the other is a SAT formulation that we expect will be useful in practice.

Keywords: Graph theory · Reconfiguration · Matchings
NP-hardness

1 Introduction

A *reconfiguration problem* typically is a reachability question setup in the solution space of some problem, with some appropriate notion of adjacency between solutions. More precisely, given a set of solutions to some fixed problem, and a notion of when two solutions are one step apart, a reconfiguration problem usually asks if there is a path from one solution to another, and often one might be interested in the shortest possible path. As a concrete example, consider the satisfiability problem, and let us say that we are dealing with an arbitrary but fixed instance of SAT over n variables. Every solution can be thought of as a binary vector with n coordinates. One might say that two assignments are adjacent if the Hamming distance between their corresponding vectors is one. Now, given two satisfying assignments τ_1 and τ_2, the reconfiguration problem for SAT would ask if there is a path from τ_1 to τ_2 in the solution graph. In other words, is

© Springer Nature Switzerland AG 2019
B. Catania et al. (Eds.): SOFSEM 2019, LNCS 11376, pp. 221–233, 2019.
https://doi.org/10.1007/978-3-030-10801-4_18

it possible morph τ_1 into τ_2 by flipping the state of one variable at a time while ensuring that each interim assignment is also satisfying? We note that reconfiguration is a broad theme and there are other aspects of the solution space that one may also be interested in, such as structural issues about the graph of solutions. In this work, however, we are focused on the complexity of the reachability question. Notably, the tractability of the "base problem" does not seem to have any predictable influence on the complexity of the corresponding reconfiguration problem: in particular, there are examples of hard problems for which the reconfiguration question is tractable, and vice-versa. For a comprehensive and recent introduction to the general themes in the study of reconfiguration problems, we refer the reader to the survey [8].

In an early study of reconfiguration problems [4,5], a number of tractable cases were identified, most notably, these included the MATCHING RECONFIGU-RATION problem, and the reconfiguration problem for tractable classes of SAT. In an instance of MATCHING RECONFIGURATION, we are given two matchings M_s and M_t in a graph G and the goal is to determine if there exists a sequence of matchings $M_0, M_1, \ldots, M_\ell$, such that $M_0 = M_s$, all consecutive matchings differ by exactly two edges, and $M_\ell = M_t$. It turns out that the exact circumstances in which such a sequence exists admits a neat characterization [5] even for more general notions of reconfiguration. Our work focuses on the natural optimization version of this question, where we are interested not only in the existence of such a sequence, but also one whose length is at most a specified budget.

Our Contributions. We discover that the complexity of the problem is quite different—compared to the issue of feasibility—once we impose a goal on the length of the reconfiguration sequence. In particular, we show that checking if we can reconfigure M_s to M_t in at most ℓ steps is NP-hard, even when the graph is unweighted, bipartite, and the maximum degree is four, and the matchings M_s and M_t are maximum matchings.

We also propose two simple algorithmic approaches. The first is based on a simple dynamic programming formulation that improves on the brute-force running time while the other is by a reduction to a SAT instance where all the clauses have at most four literals. While the running time of the dynamic programming approach is only a mild improvement on the natural brute-force algorithm, the approach is quite generic and works for many other variants of the problem. Similarly, we hope that the SAT formulation will be useful in practice.

Related Work. The MATCHING RECONFIGURATION problem is studied in [5] with the setting of TAR rules of reconfiguration: every step involves the addition or deletion of an edge, and every intermediate matching has size at least k, where k is a fixed threshold. More recently, [6] initiates a study of reconfiguring b-matchings under a more restricted specification of permitted reconfiguration steps, namely that every "step" involves the deletion of an edge *and* the addition of an edge. We recall that b-matchings are a more general notion of matchings, where every vertex has a capacity, and a b-matching M is a subset of edges which are such that the number of edges in M that are incident on a vertex v is at most the capacity of v. The results in [6] establish that the feasibility of reconfiguring b-matchings in this setting can be determined in polynomial time, even when we have an edge-weighted graph, for certain classes of graphs (that include bipartite graphs). However, these results also do not attempt to optimize the total length of the reconfiguration sequence. On the other hand, studies of reconfiguration in the context of other problems have often focused on the shortest path in the reachability graph. As an illustration, we refer the reader to this comprehensive account of vertex cover reconfiguration [7] and also the survey [8].

2   Preliminaries

We employ standard graph-theoretic notation, see, for example, Diestel [3]. We recall some key definitions that will be useful in the subsequent sections. A *matching* is a subset M of edges of G such that no two edges in M share an endpoint. A vertex v in G is said to be *M-covered* or *saturated* by M if v is incident to an edge in M and *M-exposed* or *unsaturated* otherwise. A vertex cover of a graph $G = (V, E)$ is a subset $S \subseteq V$ such that there are no edges in the graph $G \setminus S$. A graph is *bipartite* if its vertex set can be partitioned into two parts (A, B) such that all edges have one of their endpoints in A and the other in B. We recall the following fundamental connection between the size of the smallest vertex cover and the size of a maximum matching in bipartite graphs.

Lemma 1 (König's Lemma [2,3]). *The size of a maximum matching in a bipartite graph G is equal to the size of a minimum vertex cover in G.*

We consider the OPTIMAL MATCHING RECONFIGURATION problem: here, we are given two matchings M_s and M_t of the same size in an unweighted graph G and a budget ℓ. The goal is to determine if there exists a sequence of matchings $M_0, M_1, \ldots, M_\ell$, such that $M_0 = M_s$, all consecutive matchings differ by exactly two edges, that is:

$$|M_{i-1} \setminus M_i| = |M_i \setminus M_{i-1}| = 1 \text{ for all } i \in [\ell],$$

and $M_\ell = M_t$. In other words, every step involves the deletion and addition of one edge. We refer to such a sequence as a *valid $X \leftrightsquigarrow Y$ reconfiguration sequence*. We summarize this problem below.

OPTIMAL MATCHING RECONFIGURATION
Input: A graph G, two matchings M_s and M_t where $|M_s| = |M_t| = k$, and a positive integer ℓ.
Question: Is there a valid reconfiguration sequence of length at most ℓ that starts at M_s and ends at M_t?

We recall that the MATCHING RECONFIGURATION problem as defined in [5] is the following related problem: here we are given an unweighted graph G, a threshold k, two matchings M_s and M_t of size at least k, and the question is if there is a sequence of matchings in G that starts at M_s and ends at M_t, each matching in the sequence is of size at least k and is obtained from the previous one by either the addition or deletion of an edge (we refer to this as the TAR reconfiguration model). Note that in our setting, the emphasis is on the total length of the reconfiguration sequence. Also, for simplicity, our reconfiguration rules are simpler than the TAR model, because we insist on adding one edge and deleting one edge in every step. However, we note that this is without loss of generality, since the two models are essentially the same if the given matchings are of maximum size (as will be the case for our hardness result), while our algorithmic result can be easily amended to account for the TAR rules as well.

We note that any YES-instance of OPTIMAL MATCHING RECONFIGURATION in fact admits a reconfiguration sequence of length $O(n^2)$, and this can be derived by a careful analysis of the characterization given by [5]. Roughly speaking, the greedy reconfiguration algorithm proposed by [5] considers the components of the graph $M_s \Delta M_t$—the main challenge is the components C that form cycles in this graph. It turns out that the M_s edges on C can be reconfigured into the M_t edges precisely when there is a M_s-alternating path starting at an unsaturated vertex of M_s and ending at a vertex of C. So if we were to simply reconfigure every cycle in the natural way, we will have a reconfiguration sequence whose length is at most $O(n^2)$. This also establishes that the problem is in NP: indeed, if $\ell > n^2$, then this problem is equivalent to determining the feasibility of the reconfiguration, and the problem can be resolved in polynomial time. Otherwise, the reconfiguration sequence has bounded length and is the desired certificate.

Proposition 1. *If (G, M_s, M_t, ℓ) is a* YES-*instance of* OPTIMAL MATCHING RECONFIGURATION*, then there exists a valid reconfiguration sequence of length at most $O(n^2)$. In particular, therefore,* OPTIMAL MATCHING RECONFIGURATION *is in* NP.

We also frequently invoke the following fact about reconfigurable pairs of matchings, which is a part of the characterization by [5]. For two matchings M_s and M_t, cycles C in the graph induced by the edges of $M_s \Delta M_t$ are called (M_s, M_t)-alternating cycles.

Proposition 2 (Lemma 1, [5]). *Suppose that both M_s and M_t are maximum matchings of G, and let $k = |M_s| = |M_t|$. Then, there exists a sequence of matchings which transforms M_s into M_t so that all intermediate matchings have size at least $(k-1)$ if and only if, for every (M_s, M_t)-alternating cycle C, there exists an M_s-alternating path in G starting with an M_s-exposed vertex and ending at a vertex in C.*

Note that although the characterization above is stated in the context of the TAR reconfiguration rules, it applies in its stated form to our reconfiguration setup, since—as we have mentioned previously—the two models are equivalent in the context of matchings of maximum size.

3   NP-Hardness of Optimal Reconfiguration

We show that OPTIMAL MATCHING RECONFIGURATION is NP-hard by a reduction from 3-SAT. Recall that an instance of 3-SAT consists of variables and clauses in conjunctive normal form. We note that 3-SAT is hard even when every literal occurs in exactly two clauses, and every clause has exactly three literals [1]. This is the version of 3-SAT that we will reduce from. We refer to this version of the problem as (3, B2)-SAT. We need the restricted version to deduce that the maximum degree of the reduced instance is at most four. In particular, the result that we establish in this section is the following.

Theorem 1. *The* OPTIMAL MATCHING RECONFIGURATION *problem is NP-hard, even on bipartite graphs of maximum degree four.*

Before giving the formal proof of Theorem 1, we give some intuition for why one might expect this problem to be NP-hard. In the setting of determining feasibility, the cycles in $M_s \Delta M_t$ needed "outside help" to be reconfigured: which is to say that we needed to free up a vertex on the cycle by pushing one of them out along an alternating M_s-path to an M_s-exposed vertex. Typically such a reconfiguration sequence would be like the one shown in Fig. 1.

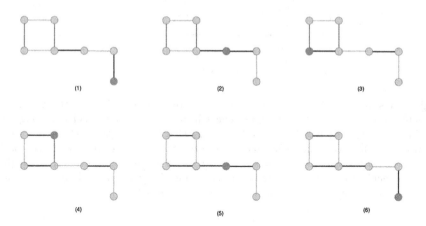

Fig. 1. A typical reconfiguration sequence that handles a cycle in $M_s \Delta M_t$. The edges in red depict the current matching at every stage of the reconfiguration. The matching M_s, therefore, is the collection of red edges in the first step. The target matching M_t comprises of the blue edges in the first step along with the red edge that is not on the cycle in the first step. (Color figure online)

When we are constrained on the length of the reconfiguration sequence, then the choice of *which M_s-alternating path* we use is crucial. As an illustration, consider the example in Fig. 2. Here, we have at least two distinct ways of reconfiguring the two cycles shown and one of them is shorter than the other. It turns out that the difficulty of making these choices can in fact be used to encode an instance of SAT. In particular, we will setup our initial and target matchings M_s and M_0 in such a way that the graph $M_s \Delta M_0$ will have a cycle corresponding to every clause of the SAT formula and the only way to reconfigure this cycle will be via an alternating path ending at M_s-exposed vertices corresponding to variables that appear in the clause. Looking at this from the perspective of the variables, on the other hand: the setup ensures that every variable can "provide" for an alternating path that "fixes" cycles corresponding to all the clauses that the variable can satisfy: however, the budget on the total length of the reconfiguration sequence demanded is designed to ensure that only one set of alternating paths can be "triggered" for any variable—either the ones leading up to the clauses where the variable appears positively, *or* to the ones where it appears negatively. We now turn to a detailed description of our reduction.

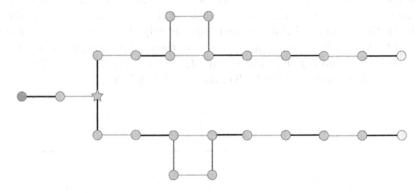

Fig. 2. A situation where we have two possibilities for reconfiguring the two cycles depicted in the picture. As before, the edges in red depict the current matching, and the blue edges are from M_t. Also, any red edge that is not on the cycle is also a part of M_t. Note that we could either reconfigure the two cycles separately via the paths leading up to the exposed yellow vertices, or we could use the green vertex to shift the edge incident on the star vertex on the left, and then reconfigure both cycles using paths starting from the star vertex. (Color figure online)

Let ϕ be an instance of (3, B2)-SAT over variables $V = \{x_1, \ldots, x_n\}$ and clauses $C = \{C_1, \ldots, C_m\}$. We use (G_ϕ, M_s, M_t, ℓ) to denote the OPTIMAL MATCHING RECONFIGURATION instance that we will construct based on ϕ. We begin with a description of the graph G_ϕ, after which we will define the matchings M_s and M_t, and finally, we will specify ℓ. Note that throughout this discussion, we will continue to use n and m to denote, respectively, the number of variables and clauses in ϕ, rather than the number of vertices and edges in the reduced graph.

We begin by introducing variable and clause gadgets as depicted in Fig. 3. Note that the graph constructed so far (by introducing the variable and clause gadgets for each variable and clause in V and C respectively) has $9n+6m$ vertices and $10n + 6m$ edges. The graph is also bipartite with the following[1] bipartition:

$$A = \{t_i, x_i, f_i, x_i^1, x_i^3 \mid i \in [n]\} \cup \{D_j^p, D_j^q, D_j^r \mid j \in [m]\},$$

and:

$$B = \{a_i, b_i, x_i^2, x_i^4 \mid i \in [n]\} \cup \{C_j^p, C_j^q, C_j^r \mid j \in [m]\}.$$

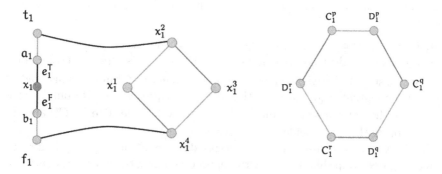

Fig. 3. Left: a variable gadget corresponding to the variable x_1. Right: a clause gadget corresponding to the clause $C_1 := (x_p, x_q, x_r)$. For simplicity of notation, we do not distinguish the polarity (positive or negative occurrence) of the literals in the labels of the vertices of the clause gadget, although we will make this explicit when we describe the connection between the clause and the literal gadgets. (Color figure online)

We will now add edges between the clause and the variable gadgets while maintaining this bipartition. In particular, let p_i and q_i be the indices of the clauses where x_i appears positively. Then we add an edge from t_i to the vertices C_j^i for $j \in \{p_i, q_i\}$. Analogously, let r_i and s_i be the indices of the clauses where x_i appears negatively. Then we add an edge from f_i to the vertices C_j^i for $j \in \{r_i, s_i\}$. It is easy to see that the resulting graph is bipartite.

We now describe the initial and final matchings of our reconfiguration instance. The matching M_s is given by all the red edges shown in the variable and clauses gadgets, and the matching M_t is given by all the blue edges from the four-cycles $(x_i^1, x_i^2, x_i^3, x_i^4)$ in the variable gadget, the blue edges in the cycles of the clause gadgets, and all the remaining red edges that do not involve the vertices $(x_i^1, x_i^2, x_i^3, x_i^4)$. We fix $\ell := 5n + 4m$, and this completes the description of the reduced instance. Note that the sizes of the matchings M_s and M_t are $4n + 3m$. Also note that the vertices:

$$S := \{a_i, b_i, x_i^2, x_i^4 \mid i \in [n]\} \cup \{C_j^p, C_j^q, C_j^r \mid j \in [m]\}$$

---

[1] There is a slight abuse of notation here: for every $j \in [m]$, we are using the superscripts p, q, and r to denote the indices of the variables whose literals appear in the clause C_j. These superscripts also be indexed by j, but these are omitted for clarity.

form a vertex cover of size $(4n + 3m)$. This establishes, by König's Lemma (c.f. Lemma 1), that both the matchings are, in fact, maximum-sized matchings of the graph G_ϕ. It is also easy to see that the maximum degree of any vertex in this graph is at most four, given that every literal appears in exactly two clauses. We now argue the equivalences of these two instances.

The Forward Direction. Let τ be a satisfying assignment for ϕ. We propose a reconfiguration sequence S_τ based on τ. Recall that p_i and q_i are the indices of the clauses where x_i appears positively, while r_i and s_i are the indices of the clauses where x_i appears negatively. For each variable x_i, perform the following steps.

1. If $\tau(x_i) = 1$, then remove the edge (a_i, t_i) and add the edge e_i^T.
2. If the clause gadget corresponding to C_j^i for $j = p_i$ is already reconfigured, skip this step. Otherwise, remove the edge incident to C_i^i in the clause gadget from the matching M_s and add the edge (t_i, C_i^i), for $j = p_i$. Reconfigure the cycle in the natural way, ending with the removal of the edge (t_i, C_j^i) and the addition of the edge incident to C_j^i from M_t.
3. Similarly, if the clause gadget corresponding to C_j^i for $j = q_i$ is already reconfigured, skip this step. Otherwise, proceed along the lines of the previous step, this time for the clause gadget C_j^i for $j = q_i$.
4. Reconfigure the cycle $(x_i^1, x_i^2, x_i^3, x_i^4)$ by removing the edge (x_i^1, x_i^2) and adding the edge (t_i, x_i^2); removing the edge (x_i^4, x_i^1) and adding the edge (x_i^4, x_i^1); and finally, removing the edge (t_i, x_i^2) and adding the edge (x_i^2, x_i^3).
5. Remove e_i^T and add (a_i, t_i), restoring the state of the edge e_i^T.

The steps above are specified similarly for the case when, in the first step, $\tau(x_i) = 0$ rather than $\tau(x_i) = 1$. The only changes are that $j = r_i$ and $j = s_i$ in steps 2 and 3, and the order of addition and removal of edges will be different in the last step. Observe that these steps, once executed, do form a valid $M_s \leadsto M_t$ reconfiguration sequence: the states of all edges common to M_s and M_t are seen to be restored, the edges on the cycles $(x_i^1, x_i^2, x_i^3, x_i^4)$ are explicitly reconfigured for all $i \in [n]$. The only edges left to consider are those from the clause gadgets: but since we started from a satisfying assignment, every clause gadget must have been explicitly handled by the process above. It is straightforward to verify that the sequence described above has at most $5n + 4m$ steps. This completes the argument in the forward direction.

The Reverse Direction. In the reverse direction, let S be a valid $M_s \leadsto M_t$ reconfiguration sequence of length at most ℓ. Our argument is based on how the budget ℓ restricts the reconfiguration sequence in manner that leads us to a well-defined assignment of the variables, which will also turn out to be satisfying if the reconfiguration sequence was valid. We begin by making the following claim:

Claim. In any valid $M_s \leadsto M_t$ reconfiguration sequence of length at most ℓ, for any $i \in [n]$, either e_i^T or e_i^F are present in some matching of the sequence. Consequently, in such matchings, the edge (a_i, t_i) (respectively, (b_i, f_i)) are not present.

This claim follows from the fact that the only M_s-alternating paths starting at an M_s-exposed vertex and ending at the cycle $(x_i^1, x_i^2, x_i^3, x_i^4)$ are:

$$(x_i, a_i) - (a_i, t_i) - (t_i, x_i^2) \text{ and } (x_i, b_i) - (b_i, f_i) - (f_i, x_i^4).$$

Note that $(x_i^1, x_i^2, x_i^3, x_i^4)$ is, in particular, a cycle in the graph induced by the edges of $M_s \Delta M_t$, and therefore the only way to reconfigure it is along one of the two paths above. Therefore, if neither e_i^T or e_i^F are present in any matching of a valid $M_s \rightsquigarrow M_t$ reconfiguration sequence, then we have a contradiction because adding these edges are the only way to remove the edges (x_i^1, x_i^2) and (x_i^3, x_i^4), which are not in M_t. Our second claim is the following:

Claim. In any valid $M_s \rightsquigarrow M_t$ reconfiguration sequence of length at most ℓ, for any $i \in [n]$, if e_i^T appears in some matching of the sequence, then e_i^F does not appear in any matching of the sequence, and vice-versa.

This follows from the following counting argument. Note that we spend at least two steps per vertex gadget (one in adding of the edges e_i^T or e_i^F and another to delete whichever edge was added—recall that these edges do not belong to M_t and have to be removed). Further, we spend three additional steps per vertex gadget in reconfiguring the cycle $(x_i^1, x_i^2, x_i^3, x_i^4)$, and similarly, we need at least four steps to reconfigure the cycles corresponding to the clause gadgets. It is easy to check that all of these requirements are disjoint. Therefore, we have already used up $5n + 4m$ steps assuming that exactly one of the edges e_i^T or e_i^F were added throughout the course of the reconfiguration sequence. In other words, there is "no room", because of the budget ℓ, to add both e_i^T or e_i^F during the course of our reconfiguration.

We are now ready to propose an assignment τ_S to the variables. We set x_i to 1 if e_i^T was added in some matching of S, and 0 otherwise. We claim that this is a satisfying assignment. Suppose not, and in particular, let $C = (\ell_p, \ell_q, \ell_r)$ be a clause that is not satisfied by τ, where ℓ_j is a positive or negative occurrence of the variable x_j, for $j \in \{p, q, r\}$. To make this discussion concrete, let $C = (x_p, \overline{x_q}, x_r)$ (all the other cases have analogous arguments). Since this clause is not satisfied, we have that the edges e_p^T, e_q^F and e_r^T were not added to any matching in the sequence S. However, note that the only M_s-alternating paths starting at M_s-exposed vertices that end at the cycle of the clause gadget corresponding to C are those that start at these edges. Therefore, without adding these edges, it is not possible to reconfigure C, which contradicts the fact that we started with a valid $M_s \rightsquigarrow M_t$ reconfiguration sequence. Therefore, τ_S is in fact a satisfying assignment and this concludes the argument in the reverse direction and of the NP-hardness of OPTIMAL MATCHING RECONFIGURATION.

4 An Exact Algorithm for Optimal Reconfiguration

In this section, we describe an exact algorithm for the OPTIMAL MATCHING RECONFIGURATION problem. Throughout this section, for an instance

(G, M_s, M_t, ℓ) of OPTIMAL MATCHING RECONFIGURATION, we use n and m to denote, respectively, the number of vertices and edges in G. Observe that the natural brute-force approach would be to guess all possible reconfiguration sequences and check if any of them is a valid reconfiguration sequence. This amounts to guessing ℓ pairs of edges $\{(e_i, f_i)\}_{i=1}^{\ell}$ such that e_i is an edge in the matching M_i, while f_i is the edge that is added. Note that if the reconfiguration is between matchings of maximum size, then f_i must be incident on one of the endpoints of e_i. Recalling that k denotes the size of the matchings M_s and M_t, note that the running time of the brute-force algorithm is $(km)^{\ell}$, which is $n^{O(\ell)}$. Using the bound on ℓ that we have by Proposition 2, we note that this running time is bounded by $n^{O(n^2)}$.

We propose two approaches that improve on the brute-force search. One is by a straightforward dynamic programming formulation, and the other is by a reduction to a SAT instance with $O(m\ell)$ variables. While the dynamic programming approach has the better running time, we describe the SAT formulation of the problem as an alternative that may be useful in practice given the availability of SAT solvers. We also note that both approaches are fairly generic and may prove to be relevant for other reconfiguration problems as well.

Dynamic Programming

Let (G, M_s, M_t, ℓ) be an instance of OPTIMAL MATCHING RECONFIGURATION, where $|M_s| = |M_t| = k$. Consider a dynamic programming table with binary entries defined as follows:

$\mathbb{T}(X, Y, d) = 1$ iff there exists a valid $X \leftrightsquigarrow Y$ reconfiguration sequence of length at most d,

where X and Y are matchings of size k and $1 \leq d \leq \ell$ is a positive integer. Note that the size of the DP table is $O(m^{2k} \cdot \ell)$, where m denotes the number of edges in G. By the semantics of the table, we have a YES-instance of OPTIMAL MATCHING RECONFIGURATION if and only if $\mathbb{T}(M_s, M_t, \ell) = 1$. Observe that the entries of the table are easy to populate when $d = 1$: indeed, if $|X \Delta Y|$ is more than one, the corresponding entry is zero, and it is one otherwise (it is clearly one when $|X \Delta Y| = 0$ and if $|X \Delta Y| = 1$, then note that removing the edge in $X \setminus Y$ and adding the edge in $Y \setminus X$ is always a valid reconfiguration step).

Now we turn to the recurrence for the computation of $\mathbb{T}(X, Y, d)$. Let $\mathcal{M}(Y, 1)$ denote the set of all matchings Z of size k such that Y can be obtained from Z in one valid reconfiguration step. Then, we have:

$$\mathbb{T}(X, Y, d) = \vee_{Z \in \mathcal{M}(Y,1)} \mathbb{T}(X, Z, d - 1) \tag{1}$$

Observe that we are, in effect, "guessing" the edge that was added and removed in the last step of the reconfiguration sequence. The dependence graph of this recurrence is clearly a DAG, since we compute the entries in order of increasing values of d. We briefly argue the correctness of the recurrence by induction on d. To this end, we assume that the entries of $\mathbb{T}(X, Y, d')$ for all $d' \leq d$ are computed accurately by the recurrence (1). On the one hand, suppose there exists a valid $X \leftrightsquigarrow Y$ reconfiguration sequence of length at most d:

$$\mathcal{S}_{X,Y} : M_0 := X, M_1, \ldots, M_{d-1}, M_d := Y,$$

Now, note that $\mathcal{S}_{X,Y}$ truncated at the first $d-1$ steps is a valid $X \leftrightsquigarrow M_{d-1}$ reconfiguration sequence of length at most $d-1$ and that $M_{d-1} \in \mathcal{M}(Y, 1)$. On the other hand, if the R.H.S. of (1) is 1, then, by the induction hypothesis, there is a valid $X \leftrightsquigarrow Z$ reconfiguration sequence of length at most $d-1$ for some $Z \in \mathcal{M}(Y, 1)$. By the definition of $\mathcal{M}(Y, 1)$, this sequence can be extended to a valid $X \leftrightsquigarrow Y$ reconfiguration sequence in one step, so that the length of the overall sequence is at most d.

We claim that overall running time of this algorithm is $n^{O(n)}$. The bound on the running time follows from the fact that the number of edges is at most n^2 and $k \leq n$. To compute any entry of the DP table (including the ones corresponding to the base cases), we need polynomial time overall: first we need to compute the matchings in $\mathcal{M}(Y, 1)$, and it is easy to see that there are at most n^2 such matchings, and then we need polynomial time to perform the corresponding lookups.[2] This concludes the argument and gives us the following.

Lemma 2. *The* OPTIMAL MATCHING RECONFIGURATION *problem can be decided by an algorithm whose running time is bounded by* $n^{O(n)}$.

We remark that this approach can be adapted to work with the TAR model of reconfiguration, and also to weighted graphs. We now turn to our second approach, which encodes this problem as a SAT instance.

Reduction to SAT

To create a SAT instance that captures an instance (G, M_s, M_t, ℓ) of OPTIMAL MATCHING RECONFIGURATION, we introduce the following $m\ell$ variables:

$$V_G := \{x_{e,i} \mid e \in E(G), i \in [\ell] \cup \{0\}\}$$

Intuitively, setting the variable $x_{e,i}$ to one represents the fact that the edge e is present in M_i, the ith matching in a hypothetical valid $M_s \leftrightsquigarrow M_t$ reconfiguration sequence. To ensure that an assignment to the variables V_G corresponds to a valid reconfiguration sequence (and vice-versa), we introduce the following clauses.

1. **The chosen edges at every step form a matching.** For all $i \in [\ell] \cup \{0\}$, and for each vertex $v \in G$, introduce the following clause:

$$\bigwedge_{e,f \text{ inc to } v, e \neq f} (\overline{x_{e,i}} \vee \overline{x_{f,i}})$$

---

[2] We are not explicitly emphasizing the polynomial factors here because the exponential term, as stated, is already dominant over them.

2. **Encoding the reconfiguration rule.** For every pair of edges e and f and for every $i \in [\ell - 1] \cup \{0\}$, introduce the clauses:

$$(\overline{x_{e,i}} \vee \overline{x_{f,i}} \vee x_{e,i+1} \vee x_{f,i+1})$$

and also the clauses:

$$(x_{e,i} \vee x_{f,i} \vee \overline{x_{e,i+1}} \vee \overline{x_{f,i+1}})$$

If the clauses above are satisfied, then in the natural interpretation of the assignment of the variables as a sequence of matchings, note that we have neither removed more than one edge in moving from one step to the next nor do we add more than one edge to the matching.

3. **The start and end conditions.** We encode the fact that the initial matching is M_s and the final matching is M_t by introducing singleton clauses corresponding to the edges in these matchings:

$$\left(\bigwedge_{e \in M_s} (x_{e,0}) \right) \wedge \left(\bigwedge_{e \in M_t} (x_{e,\ell}) \right).$$

This concludes the description of the clauses. It is easily checked that satisfying assignments of this formula are in one-to-one correspondence with valid reconfiguration $M_s \rightsquigarrow M_t$ sequences of length ℓ.

5 Concluding Remarks

There are several questions of interest that remain open. Most notably, it would be interesting to improve the exact exponential complexity of OPTIMAL MATCHING RECONFIGURATION. Also, it would be interesting to investigate the parameterized complexity of the problem with respect to two natural parameters: the common size of the matchings (k), and the length of the reconfiguration sequence (ℓ). In the setting of weighted graphs, even the question of the feasibility of the reconfiguration remains open. Some recent work [6] demonstrates polynomial time algorithms for special classes of graphs (which include the class of bipartite graphs) in this setting. Generalizing the characterization obtained by [5] to the case of weighted graphs is an interesting issue to pursue.

References

1. Berman, P., Karpinski, M., Scott, A.D.: Approximation hardness of short symmetric instances of MAX-3SAT. In: Electronic Colloquium on Computational Complexity (ECCC), no. 049 (2003)
2. Dénes, K.: Graphok és mátrixok. Matematikai és Fizikai Lapok, **38** (1931)
3. Diestel, R.: Graph Theory. GTM, vol. 173, 5th edn. Springer, Heidelberg (2017). https://doi.org/10.1007/978-3-662-53622-3

4. Gopalan, P., Kolaitis, P.G., Maneva, E.N., Papadimitriou, C.H.: The connectivity of boolean satisfiability: computational and structural dichotomies. SIAM J. Comput. **38**(6), 2330–2355 (2009)
5. Ito, T., et al.: On the complexity of reconfiguration problems. Theoret. Comput. Sci. **412**, 1054–1065 (2011)
6. Ito, T., Kakimura, N., Kamiyama, N., Kobayashi, Y., Okamoto, Y.: Reconfiguration of maximum-weight b-matchings in a graph. J. Comb. Optim. (2018)
7. Mouawad, A.E., Nishimura, N., Raman, V., Siebertz, S.: Vertex cover reconfiguration and beyond. Algorithms **11**(2), 20 (2018)
8. Nishimura, N.: Introduction to reconfiguration. Algorithms **11**(4), 52 (2018)

Forbidden Directed Minors, Directed Path-Width and Directed Tree-Width of Tree-Like Digraphs

Frank Gurski and Carolin Rehs[(⊠)]

Institute of Computer Science, Algorithmics for Hard Problems Group,
Heinrich-Heine-University Düsseldorf, 40225 Düsseldorf, Germany
carolin.rehs@hhu.de

Abstract. There have been many attempts to find directed graph classes with bounded directed path-width and bounded directed tree-width. Right now, the only known directed tree-width-/path-width-bounded graphs are cycle-free graphs with directed path-width and directed tree-width 0. In this paper, we introduce directed versions of cactus trees and pseudotrees and -forests and characterize them by at most three forbidden directed graph minors. Furthermore, we show that directed cactus trees and forests have a directed tree-width of at most 1 and directed pseudotrees and -forests even have a directed path-width of at most 1.

Keywords: Directed cactus trees · Directed pseudoforests
Directed graph minors · Directed path-width · Directed tree-width

1 Introduction

Cactus trees and pseudotrees are well-known undirected graph classes, which are an attempt to define tree-like graphs that are not exactly trees. There are many problems which are NP-hard on graphs but can be solved in polynomial time for cactus trees. The best known use is for genome comparisons, as for example in [11] and [12]. Cactus trees have bounded undirected tree-width 2.

Pseudotrees are a superclass of the often considered sunlet graphs. They are graphs which contain at most one cycle, so a small extension of trees. Pseudoforests are graphs with only few cycles, technically with at most one cycle per connected component. Many problems, which are NP-hard on graphs, but solvable in polynomial time on trees, are still polynomial on pseudotrees and -forests. Pseudoforests have bounded undirected path-width 2.

As both undirected graph classes have many applications and bounded tree-width or even bouned path-width, it is interesting to define directed versions of these classes and consider their directed tree-width and directed path-width.

The work of the second author was supported by the German Research Association (DFG) grant GU 970/7-1.

B. Catania et al. (Eds.): SOFSEM 2019, LNCS 11376, pp. 234–246, 2019.
https://doi.org/10.1007/978-3-030-10801-4_19

During the last years, width parameters for directed graphs have received a lot of attention [4]. Although among these directed tree-width and directed path-width were much considered, no graph class with bounded directed tree-width or path-width was found except directed acyclic graphs (DAGs) and complete biorientations of graphs of bounded undirected tree-width or path-width. In this paper, we define for the first time directed graph classes, which are bounded in those parameters and further related to some well-known undirected graph classes.

As finding forbidden graph minors for some digraph class of bounded directed path-width seems to be really difficult [9], it is already a step in the right direction to find special class of bounded directed path-width or bounded directed tree-width, which is characterizable by forbidden minors.

2 Preliminaries

Cactus trees are well-known in graph theory. The name "cactus" has been introduced by Harary and Uhlenbeck in 1953 [7]. The definition has slightly changed since then, whereas in the original definition cacti where requested to consist only of triangles, today's more common definition is as follows:

Definition 1 (Cactus tree). *A* cactus tree *is a connected graph* $G = (V, E)$, *where for any two cycles* C_1 *and* C_2 *it holds that they have at most one joint vertex.*

The set of cactus graphs is a superset of the pseudotrees, which are again a superset of the well-known sunlet graphs.

Definition 2 (Pseudotree). *A* pseudotree *is a connected graph which contains at most one cycle.*

It is possible to extend these definitions to forests, which means that they are not necessarily connected. A cactus forest is a graph where any two cycles have at most one joint vertex, and a pseudoforest is a graph where every connected component contains at most one cycle.

Definition 3 (Edge contraction). *Let* $G = (V, E)$ *be a graph with* $e = \{u, v\} \in E$, $u \neq v$. *The* contraction *of* e *leads to a new graph* $G' = (V', E')$ *with* $V' = V \setminus \{u, v\} \cup \{w\}$ *with* $w \notin V$ *and* $E' = \{\{a, b\} \mid a, b \in V \cap V', \{a, b\} \in E$ *or* $a = w, \{u, b\}$ *or* $\{v, b\} \in E$ *or* $b = w, \{a, u\}$ *or* $\{a, v\} \in E\}$.[1]

A graph minor of a graph $G = (V, E)$ is a graph $G' = (V', E')$, if G' can be obtained by forming subgraphs and edge contraction of G. Formally, we write $G' \preceq G$.

---

[1] This means, in graph G' the edge e and its two incident vertices u and v are replaced by the vertex w and all other edges in G incident with u or v are adjacent with w in G'.

Whereas the set of cactus trees and the set of pseudotrees are not closed under the graph minor operation, as subgraphs could create unconnected graphs, cactus forests can be characterized by one forbidden graph minor, the diamond graph D_4 with four vertices, which is the K_4 with one edge less [3]. This means, that every graph, which does not have D_4 as a graph minor, is a cactus forest. As pseudoforests are a subset of cactus forests, D_4 is also a forbidden minor for them, as well as the butterfly graph B_5. Every graph, which has neither D_4 nor B_5 as a graph minor, is a pseudoforest.

Cactus forests are of bounded tree-width and pseudoforests are even of bounded path-width.

3   Directed Cactus Forests and Pseudoforests

Now we want to apply the definitions of cactus trees and forests as well as pseudotrees and -forests to directed graphs. For directed cactus trees, it is possible to use nearly the same definition as for undirected cactus trees:

Definition 4 (Directed cactus tree). *A directed cactus tree is a strongly connected digraph $G = (V, E)$, where for any two directed cycles C_1 and C_2 it holds that they have at most one joint vertex.*

This definition remains equal if C_1 and C_2 must have exactly one joint vertex, and it is equal to the definition given in [1]:

Definition 5 (Directed cactus tree). *A directed cactus tree is a strongly connected digraph in which each arc is contained in exactly one directed cycle. The class of all directed cactus trees is named DCT.*

It would also be possible to define cactus tree as weakly connected subgraphs, where two directed cycles have at most one joint vertex. This would lead to a superset of Definition 4 and a subset of directed cactus forests, which can be defined as follows:

Definition 6 (Directed cactus forest). *A directed cactus forest is a digraph $G = (V, E)$, where for any two directed cycles C_1 and C_2 it holds that they have at most one joint vertex. The class of all directed cactus forests is named DCF.*

Note that if G does not need to be strongly connected, it is not equal if C_1 and C_2 have exactly one directed cycle. It though holds that a graph is a directed pseudoforest, if and only if each arc is contained in at most one cycle. It further holds that if G is a directed cactus tree, then its underlying (undirected) graph $u(G)$ is a cactus tree. But if G is a directed cactus forest, the underlying graph does not need to be neither a cactus tree nor a graph of which every connected component is a cactus tree. The other way around is only true if we use an orientation where no bioriented arcs are allowed. Then if G is an undirected cactus tree or a graph of which every connected component is a cactus tree, then every orientation of G is a directed cactus forest.

For pseudotrees, there are also different ideas of defining a directed version, depending on whether strong or weak connectivity is used. Here it is more sensible to use weak connectivity, because a strongly connected graph containing at most one cycle is exactly a cycle.

Definition 7 (Directed pseudotree). *A directed pseudotree is a weakly connected digraph which contains at most one directed cycle. The class of all directed pseudotrees is named DPT.*

In contrast to directed cactus forests, it does matter for directed pseudoforests if we consider strong or weak connectivity:

Definition 8 (Directed weak pseudoforest). *A directed weak pseudoforest is a digraph, in which every weakly connected component contains at most one directed cycle. The class of all directed weak pseudoforests is named DWPF.*

Definition 9 (Directed strong pseudoforest). *A directed strong pseudoforest is a digraph, in which every strongly connected component contains at most one directed cycle, i.e. every nontrivial component contains exactly one directed cycle. The class of all directed strong pseudoforests is named DSPF.*

Then directed strong pseudoforests are a superclass of directed weak pseudoforest, as every strongly connected component is also a weakly connected component. It further holds, that directed strong pseudoforests are exactly those graphs, where any two directed cycles have no joint vertex, or where every vertex is in at most one cycle.

Note that here as well it holds that if G is a directed pseudotree, the underlying graph $u(G)$ is an undirected pseudotree. For directed weak pseudoforests the underlying undirected graphs are undirected pseudoforests, but for directed strong pseudoforests this is not generally true. But it holds that every orientation of an undirected pseudoforest, is a directed strong pseudoforest.

Proposition 1. *We have the following inclusions for tree-like digraphs.*

$$DPT \subset DCT \subset DCF \tag{1}$$
$$DPT \subset DWPF \subset DSPF \subset DCF \tag{2}$$

4 Directed Graph Minors of Tree-Like Digraphs

As cactus forests and pseudoforests are characterizable by forbidden graph minors, we want to characterize their directed versions by forbidden directed graph minors. Therefore, we first need to define directed graph minors.

Definition 10 (Directed edge contraction). *Let $G = (V, E)$ be a digraph with $e = (u, v) \in E$. The contraction of e leads to a new digraph $G' = (V', E')$ with $V' = V \setminus \{u, v\} \cup \{w\}$ with $w \notin V$ and $E' = \{(a, b) \mid a, b \in V \cap V', (a, b) \in E$ or $a = w, (u, b)$ or $(v, b) \in E$ or $b = w, (a, u)$ or $(a, v) \in E\}$.*[2]

---

[2] This means, in digraph G' the edge e and its two incident vertices u and v are replaced by the vertex w and all other edges in G incident with u or v are incident with w in G'.

There are different ways of defining graph minors using directed edge contraction. As directed path-width and directed tree-width are not monotone under the directed edge contraction on every edge, it is sensible to restrict the edges, on which directed edge contraction can be used. We introduce an equivalent definition to the one introduced by Kintali and Zhang in [10]. Therefore we need to define cycle contraction:

Definition 11 (Directed cycle contraction). *Let* $G = (V, E)$ *be a digraph with* $C = \{v_1, \ldots, v_\ell\}$ *a cycle. The* contraction *of* C *leads to a new digraph* $G' = (V', E')$ *with* $V' = V \setminus C \cup \{w\}$ *with* $w \notin V$ *and* $E' = \{(a, b) \mid a, b \in V \cap V', (a, b) \in E$ *or* $a = w, (v_i, b) \in E$ *for* $1 \le i \le \ell$ *or* $b = w, (a, v_i) \in E$ *for* $1 \le i \le \ell\}$.[3]

Butterfly contractions are defined by Johnson et al. in [8] as directed edge contractions of an edge $e = (u, v)$, where either e is the only outgoing edge of u or e is the only incoming edge of v. The definition of out-contraction of [10] is equal to deleting all outgoing edges of u but e and then doing a butterfly contraction, the definition of in-contraction is equal to deleting all incoming edges of v but e and doing a butterfly contraction of e. Therefore, the following definition of directed graph minors is equal to the one given in [10]:

Definition 12 (Directed graph minor). *Let* $G = (V, E)$ *be a digraph. A digraph* $G' = (V', E')$ *is a* directed minor *of* G, *i.e.* $G' \preceq G$, *if* G' *can be obtained by creating subgraphs, performing cycle contractions and performing butterfly contractions on* G.

Furthermore, the directed graph minor relation is transitive, reflexive and antisymmetric, but not symmetric.

Directed pseudotrees and directed cactus trees can not be closed under directed minor operations, as they are not even closed under the subgraph operation. Directed cactus forests and directed strong/weak pseudoforests are closed under directed graph minor operations by the following results.

Lemma 1. *Directed cactus forests are closed under directed graph minor operations.*

Proof. Let $G = (V, E)$ be a directed cactus forest. Then it holds for every two cycles C_1, C_2 that they have at most one joint vertex. That is, for all $e \in E$ holds that e is part of at most one cycle.

Subgraphs. By deleting vertices or arcs, no edge can become part of another cycle.

Butterfly contraction. Let $e = (u, v)$ be an arc in G such that e is the only outgoing edge of u or e is the only incoming edge of v. Then there is no path from u to v in $G - (u, v)$. Then no additional cycle can be created

---

[3] This means, in digraph G' the cycle C is replaced by the vertex w and all other edges in G incident with a vertex in C are incident with w in G'.

by contraction of e, as no additional arc is created and therefore the only additional possibility to create a new cycle would be containing the new vertex w and a path from w to w, in G, which has not been a path from v to u in G. This is a contradiction to that there is no path from u to v in G. It follows that every arc is still only in at most one cycle in G'.

Cycle contraction. Let C be a cycle in G. By contracting C, no new cycle can be created, as no additional arc is created and there has already been a path from u to v and from v to u for all $u, v \in C$. Therefore, assigning C to only one vertex w does not create new path from w to w. Thus, every arc is still only in at most one cycle in G'.   □

Lemma 2. *Directed strong/weak pseudoforests are closed under directed graph minor operations.*

Proof. We use the same argument as in Lemma 1. For subgraphs, by deleting vertices or arcs, no edge can become part of another cycle. By butterfly and cycle contraction, no additional cycles can be created. From this also follows that these contractions can not create additional strongly connected components. It is easy to see that both contractions can not create weakly connected components, as only arcs are considered, and there are no arcs between two weakly connected components. As directed strong/weak pseudoforests are defined as graphs, where each strong/weakly connected component contains at most one cycle, this means that directed strong/weak pseudoforests are closed under directed graph minor operations.   □

So directed cactus forests and directed strong and weak pseudoforests are closed under graph minor operation. But even more, it is possible to characterize those classes by a finite number of forbidden directed graph minors (Fig. 1):

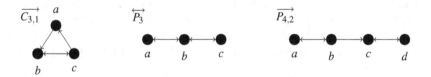

Fig. 1. The forbidden directed minors $\overrightarrow{C_{3,1}}$, $\overleftrightarrow{P_3}$ and $\overrightarrow{P_{4,2}}$.

Theorem 1. *Digraph G is a directed cactus forest if and only if it does not contain the digraph $\overrightarrow{C_{3,1}}$, the directed cycle $\overrightarrow{C_3}$ with one additional arc, as a directed graph minor.*

Proof. ⊆ Let G be a directed cactus forest. Assume that $\overrightarrow{C_{3,1}}$ is a minor of G. As there are two cycles in $\overrightarrow{C_{3,1}}$, $C_1 = \{a, b, c\}$ and $C_2 = \{b, c\}$ containing the vertex b as well as the vertex c, $\overrightarrow{C_{3,1}}$ is not a directed cactus forest. Then Lemma 1, leads to a contradiction.

$\supseteq$ Let G be a digraph with no $\overrightarrow{C_{3,1}}$ as a directed minor. Assume, that G is not a cactus graph. Then there is an arc $e = (u,v)$ in G, such that there are two cycles C_1, C_2 with $u, v \in C_1$ and $u, v \in C_2$. By subgraph operations we obtain G' which contains only of C_1 and C_2 as a graph minor of G. Using then butterfly minor operations on all arcs of C_2 but e and on all arcs of C_1 but e and two other arcs, we obtain $\overrightarrow{C_{3,1}}$ as a directed minor of G'. Then $\overrightarrow{C_{3,1}}$ is a directed minor of G, which leads to a contradiction. $\square$

Further, it holds that $\overrightarrow{C_{3,1}}$ is the minimal forbidden minor for directed cactus forests, as every further minor operation would lead to a graph with only one cycle, so every graph minor of $\overrightarrow{C_{3,1}}$ is a directed cactus forest.

Theorem 2. *Digraph G is a directed strong pseudoforest if and only if it does not contain the digraph $\overrightarrow{C_{3,1}}$ or the digraph $\overleftrightarrow{P_3}$ as a directed graph minor.*

Proof. $\subseteq$ Let G be a directed strong pseudoforest. Assume that $\overrightarrow{C_{3,1}}$ or $\overleftrightarrow{P_3}$ is a minor of G. Both $\overrightarrow{C_{3,1}}$ and $\overleftrightarrow{P_3}$ consist of only one strongly connected component, but include two cycles $\{a, b, c\}$ and $\{b, c\}$ for $\overrightarrow{C_{3,1}}$ and $\{a, b\}$ and $\{b, c\}$ for $\overleftrightarrow{P_3}$, both graphs are no directed strong pseudoforest. Thus Lemma 2, leads to a contradiction.

$\supseteq$ Let G be a digraph with no $\overrightarrow{C_{3,1}}$ or $\overleftrightarrow{P_3}$ as directed minor. Assume that G is not a directed strong pseudoforest. Then G includes a strongly connected component, which has at least two cycles. Let G' be the subgraph of G which only consists of this strongly connected component.

Case 1. Assume that the two cycles in G' have a joint arc. Then, as in the proof of Theorem 1, G' and therefore G has $\overrightarrow{C_{3,1}}$ as a directed minor. This is a contradiction.

Case 2. Assume that the two cycles in G' do not join an arc. As G' is strongly connected, there are two cycles C_1 and C_2 in G' which have a joint vertex. By subgraph operations, delete all arcs and vertices except these two cycles. Then use butterfly contractions to transform these cycles to cycles of size 2. By this, $\overleftrightarrow{P_3}$ results as a directed minor of G. This leads to a contradiction. $\square$

Theorem 3. *Digraph G is a directed weak pseudoforest if and only if it does not contain the digraph $\overrightarrow{C_{3,1}}$, the digraph $\overleftrightarrow{P_3}$ or the digraph $\overrightarrow{P_{4,2}}$ as a directed graph minor.*

Proof. $\subseteq$ Let G be a directed weak pseudoforest. Assume that $\overrightarrow{C_{3,1}}$, the graph $\overleftrightarrow{P_3}$ or the digraph $\overrightarrow{P_{4,2}}$ is a minor of G. As all three graphs contain only one weakly connected component, but two cycles, they are no directed strong pseudoforests. Then Lemma 2 leads to a contradiction.

$\supseteq$ Let G be a digraph with no $\overrightarrow{C_{3,1}}$, $\overleftrightarrow{P_3}$ or $\overrightarrow{P_{4,2}}$ as directed minor. Assume that G is not a directed weak pseudoforest. Then G includes a weakly connected component, which has at least two cycles. Let G' be the subgraph of G which only consists of this weakly connected component.

Case 1. Assume that any two cycles in G' have a joint arc. Then, as in the proof of Theorem 1, G' and therefore G has $\overrightarrow{C_{3,1}}$ as a directed minor. This is a contradiction.

Case 2. Assume that all two cycles in G' do not join an arc, but there are two cycles which have a joint vertex. Then, as in the proof of Theorem 2, G' and therefore G has $\overleftrightarrow{P_3}$ as a directed minor. This is a contradiction.

Case 3. Assume that any two cycles in G' do not have a joint vertex. Let C_1, C_2 be two cycles in G'. By subgraph operations, delete all arcs and vertices except C_1 and C_2 and a directed path connecting C_1 and C_2. Then use butterfly contractions to transform C_1 and C_2 to cycles of size 2 and the path connecting them to a path of length 1. By this, $\overrightarrow{P_{4,2}}$ results as a directed minor of G. This is a contradiction. $\qquad\square$

5   Directed Path-Width of Tree-Like Digraphs

According to Barát [2], the notation of directed path-width was introduced by Reed, Seymour, and Thomas around 1995 and relates to directed tree-width introduced by Johnson, Robertson, Seymour, and Thomas in [8].

Definition 13 (Directed path-width). *A directed path-decomposition of a digraph $G = (V, E)$ is a sequence $(X_1, \ldots, X_r)$ of subsets of V, called bags, such that the following three conditions hold true.*

(dpw-1) $X_1 \cup \ldots \cup X_r = V$.
(dpw-2) For each $(u, v) \in E$ there is a pair $i \leq j$ such that $u \in X_i$ and $v \in X_j$.
(dpw-3) If $u \in X_i$ and $u \in X_j$ for some $u \in V$ and two indices i, j with $i \leq j$,
then $u \in X_\ell$ for all indices ℓ with $i \leq \ell \leq j$.

The width *of a directed path-decomposition $\mathcal{X} = (X_1, \ldots, X_r)$ is*

$$\max_{1 \leq i \leq r} |X_i| - 1.$$

The directed path-width *of G, $d\text{-}pw(G)$ for short, is the smallest integer w such that there is a directed path-decomposition for G of width w.*

Lemma 3 ([2]). *Let G be some complete bioriented digraph, then it holds $d\text{-}pw(G) = pw(u(G))$.*

In order to process the strong components of a digraph we recall the following definition. The *acyclic condensation* of a digraph G, $AC(G)$ for short, is the digraph whose vertices are the strongly connected components $V_1, \ldots, V_c$ of G and there is an edge from V_i to V_j if there is an edge (v_i, v_j) in G such that $v_i \in V_i$ and $v_j \in V_j$. Obviously for every digraph G the digraph $AC(G)$ is always acyclic.

Let $G_1 = (V_1, E_1), \ldots, G_k = (V_k, E_k)$ be k vertex-disjoint digraphs. The *directed union* of $G_1, \ldots, G_k$, denoted by $G_1 \ominus \ldots \ominus G_k$, is defined by their disjoint union plus possible arcs from vertices of G_i to vertices of G_j for all $1 \leq i < j \leq k$.

Lemma 4. *Let G be a digraph, then the directed path-width of G is the maximum directed path-width of its strong components.*

Proof. Let G be a digraph, $AC(G)$ be the acyclic condensation of G, and $v_1, \ldots, v_c$ be a topological ordering of $AC(G)$, i.e. for every edge (v_i, v_j) in $AC(G)$ it holds $i < j$. Further let $V_1, \ldots, V_c$ be the vertex sets of its strong components ordered by the topological ordering. Then G can be obtained by $G = G[V_1] \ominus \ldots \ominus G[V_c]$. Since we have shown in [5] that $d\text{-pw}(G_1 \ominus G_2) = \max\{d\text{-pw}(G_1), d\text{-pw}(G_2)\}$, the statement of the lemma follows.   □

Theorem 4. *Directed cactus trees have unbounded directed path-width.*

Proof. Let G be the complete biorientation of the undirected, binary tree of height h. We know that the path-width of perfect binary trees of hight h is $\lceil h/2 \rceil$ (cf. [14]). Then, by Lemma 3 it follows that $d\text{-pw}(G) = \text{pw}(u(G)) = \lceil h/2 \rceil$. As all complete biorientations of binary trees are directed cactus trees, it follows that directed path-width is not bounded for directed cactus trees.   □

As all directed cactus trees are directed cactus forests, it follows directly:

Corollary 1. *Directed cactus forests have unbounded directed path-width.*

This is not true for directed strong or weak pseudoforests. As complete biorientations of binary trees are no directed pseudoforests, neither strong or weak, as they consist of only one strongly connected component, but contain lots of cycles, the counterexample from the proof of Theorem 4 does not work here. Further, it holds that this graph class has bounded directed path-width:

Theorem 5. *Directed strong pseudoforests have directed path-width at most 1.*

Proof. Let $G = (V, E)$ be a directed strong pseudoforest. Every strong component has at least size one, so the smallest strong components could be single vertices. Let C be a strongly connected component of G. As G is a pseudoforest, C is exactly a directed cycle. For every directed cycle $C = \{c_1, \ldots, c_r\}$ with arcs (c_i, c_{i+1}) for $1 \leq i \leq r - 1$ and (c_r, c_1) we give a directed path-decomposition as follows: For the cycle with $r = 1$ vertex a path-decomposition consists of only one bag, which only contains this single vertex. This is obviously a directed path-decomposition of width 0. For cycles with $r > 1$ vertices, we construct $X_1, \ldots, X_{r-1}$ with $X_1 = \{c_1, c_2\}$, $X_2 = \{c_1, c_3\}$, $\ldots$, $X_{r-1} = \{c_1, c_r\}$. Then $\mathcal{X} = (X_1, \ldots, X_{r-1})$ is a directed path-decomposition of C of width 1. As each strong component of G has directed path-width at most 1, by Lemma 4 the digraph G also has directed path-width at most 1.   □

Since the proof of Lemma 4 using the results of [5] is constructive, we even can give a directed path-decomposition of width 1 for every (not strongly connected) directed pseudoforest. As directed strong pseudoforests are a superclass of weak pseudoforests and directed weak pseudoforests are a superclass of directed pseudotrees, it follows directly:

Corollary 2. *Directed weak pseudoforests and directed pseudotrees have directed path-width at most 1.*

6   Directed Tree-Width of Tree-Like Digraphs

An *acyclic* digraph (*DAG* for short) is a digraph without any cycles as subdigraph. An *out-tree* is a digraph with a distinguished root such that all arcs are directed away from the root. For two vertices u, v of an out-tree T the notation $u \leq v$ means that there is a directed path on ≥ 0 arcs from u to v and $u < v$ means that there is a directed path on ≥ 1 arcs from u to v.

Let $G = (V, E)$ be some digraph and $Z \subseteq V$. A vertex set $S \subseteq V$ is *Z-normal*, if there is no directed walk in $G - Z$ with first and last vertices in S that uses a vertex of $G - (Z \cup S)$. That is, a set $S \subseteq V$ is Z-normal, if every directed walk which leaves and again enters S in $G - Z$ must contain only vertices from $Z \cup S$. Or, a set $S \subseteq V$ is Z-normal, if every directed walk which leaves and again enters S must contain a vertex from Z.

Definition 14 (Directed tree-width, [8]). *A (arboreal) tree-decomposition of a digraph $G = (V_G, E_G)$ is a triple $(T, \mathcal{X}, \mathcal{W})$. Here, $T = (V_T, E_T)$ is an out-tree, $\mathcal{X} = \{X_e \mid e \in E_T\}$ and $\mathcal{W} = \{W_r \mid r \in V_T\}$ are sets of subsets of V_G, such that the following two conditions hold true.*

(dtw-1) $\mathcal{W} = \{W_r \mid r \in V_T\}$ is a partition of V_G into nonempty subsets.[4]
(dtw-2) For every $(u, v) \in E_T$ the set $\bigcup\{W_r \mid r \in V_T, v \leq r\}$ is $X_{(u,v)}$-normal.

The width of a (arboreal) tree-decomposition $(T, \mathcal{X}, \mathcal{W})$ is

$$\max_{r \in V_T} |W_r \cup \bigcup_{e \sim r} X_e| - 1.$$

Here $e \sim r$ means that r is one of the two vertices of arc e. The directed tree-width of G, d-tw(G) for short, is the smallest integer k such that there is a (arboreal) tree-decomposition $(T, \mathcal{X}, \mathcal{W})$ of G of width k.

Lemma 5. *Let G be a digraph, then the directed tree-width of G is the maximum directed tree-width of its strong components.*

Proof. The proof can be done similar to the proof of Lemma 4 using the result in [5] for directed tree-width d-tw($G_1 \ominus G_2$) = max{d-tw(G_1), d-tw(G_2)}, the statement of the lemma follows.   □

Remark 1. Every strong component of a directed cactus forest G consists of r cycles $C_1, \ldots, C_r$ such that for every C_i, $1 \leq i \leq r$, there is a C_j with $i \neq j$, $1 \leq j \leq r$ such that C_i and C_j have exactly one joint vertex. Further, there is a C_i, $1 \leq i \leq r$ such that there is exactly one other cycle C_j with $i \neq j$, $1 \leq j \leq r$ such that C_i and C_j have exactly one joint vertex.

Theorem 6. *Directed cactus forest have directed tree-width at most 1.*

---

[4] A remarkable difference to the undirected tree-width [13] is that the sets W_r have to be disjoint and non-empty.

Proof. Let G be a directed cactus forest. By Lemma 5, the directed tree-width of G is the maximum directed tree-width of the strong components of G. So we only need to consider the strong components of G. Let G' be a strong component of G. By Remark 1, G' consists of r cycles $C_1, \ldots, C_r$ and there is a C_i, $1 \leq i \leq r$ such that there is exactly one other cycle C_j with $i \neq j$, $1 \leq j \leq r$ such that C_i and C_j have exactly one joint vertex. To give a directed tree-decomposition $(T, \mathcal{X}, \mathcal{W})$ for the strong component of G, we start with a vertex of this C_i. A directed tree decomposition of a cycle $C_i = \{c_{i,1}, \ldots, c_{i,\ell}\}$ is always given by a path T and bags $W_{i,t} = \{c_{i,t}\}$ for all $1 \leq t \leq \ell$ and edge sets $X_{(c_{i,t}, c_{i,t+1})} = \{c_{i,t}\}$. Since the order of the vertices in C_i is not unique, our construction leads to a directed tree-decomposition for any order of the vertices in C_i. So we can start with any vertex in C_i and create a directed tree-decomposition for this cycle.

By Remark 1, there is at least one cycle C_j, $i \neq j$, $1 \leq j \leq r$ which has a joint vertex with C_i. So for $C_j = \{c_{j,1}, \ldots, c_{j,k}\}$ there is some $c_{i,q}$, $1 \leq q \leq \ell$ such that $c_{j,1} = c_{i,q}$. (Without loss of generality order C_j in a way such that $c_{j,1}$ is the joint vertex with C_i.) Then append the vertices of C_j to the directed tree-decomposition by creating new bags $W_{j,t} = \{c_{j,t}\}$ for all $1 < t \leq k$ and edges $X_{(c_{j,t}, c_{j,r})} = \{c_{i,t}\}$ for $2 \leq i < k$, $2 < r \leq k$ and $X_{(c_{j,1}, c_{j,2})} = X_{(c_{i,q}, c_{j,2})} = \{c_{i,q}\} = \{c_{j,1}\}$.

By Remark 1 and as of course the strong components of G are strongly connected, there is always a next cycle to insert in the same way somewhere in the tree structure T of our tree-decomposition, till all vertices of the strong component are in a bag of the directed tree-decomposition $(T, \mathcal{X}, \mathcal{W})$. It remains to show that $(T, \mathcal{X}, \mathcal{W})$ really is a directed tree decomposition of width 1 for a strong component of G:

(dtw-1) $\mathcal{W} = \{W_r \mid r \in V_T\}$ is a partition of V_G into nonempty subsets. As already said, all vertices of G are inserted one by one in bags $\mathcal{W}$ by the fact that they are all strongly connected and share a vertex with another cycle. Further, no vertex occurs twice, as in a cactus forest all cycles share at most one joint vertex, and this joint vertex is not added a second time in a W-set.

(dtw-2) For every $(u, v) \in E_T$ the set $\bigcup\{W_r \mid r \in V_T, v \leq r\}$ is $X_{(u,v)}$-normal. Let $(u, v) \in E_T$. Then it holds, by the definition of T that there is a cycle C_j in G such that $(u, v) = (c_{j,t}, c_{j,t+1})$ for $c_{j,t}, c_{j,t+1}$ are elements of the cycle $C_j = (C_{j,1}, \ldots, C_{j,k})$. Further, it holds that $X_{(u,v)} = \{c_{j,t}\}$. By the definition of $(T, \mathcal{X}, \mathcal{W})$ the set $\bigcup\{W_r \mid r \in V_T, v \leq r\}$ consists of a number of cycles, lets say $C_{j+1}, \ldots, C_r$ and the vertices $\{c_{j,t+1}, \ldots, c_{j,k}\}$. As any to cycles in G have at most one vertex in common, it is not possible that there is an arc from one of those cycles to one of the cycles in $C_1, \ldots, C_{j-1}$, as this would create a big cycle including lots of vertices and edges from the cycles this arc would connect. So the only way to get a path from $\bigcup\{W_r \mid r \in V_T, v \leq r\}$ out and back in this set is by using the cycle C_j. It follows that in $G' - X_{(u,v)} = G' - \{c_{j,t}\}$ there is no path out and back in the set $\bigcup\{W_r \mid r \in V_T, v \leq r\}$, which means that this set is $X_{(u,v)}$-normal.

It further holds that $W_r \cup \bigcup_{e \sim r} X_e = \{c_{j,t}\} \cup \{c_{j,t}\} \cup \{c_{j,t-1}\} = \{c_{j,t}, c_{j,t+1}\}$ for all W_r for some C_j cycle of G' and $t > 1$. It then follows that $\max_{r \in V_T} |W_r \cup \bigcup_{e \sim r} X_e| - 1 = 2 - 1 = 1$, so the directed tree-decomposition $(T, \mathcal{X}, \mathcal{W})$ of G' has width 1. It therefore follows that each strong component of G has directed tree-width at most 1, so G has directed tree-width at most 1.   □

Since the proof of Lemma 5 using the results of [5] is constructive, we even can give a directed tree-decomposition of width 1 for every (not strongly connected) directed cactus forest.

As directed pseudoforests and directed cactus trees are both subclasses of directed cactus forests, we can conclude the following corollaries. The first statement also follows by Theorem 5, as the directed tree-width of a graph is always smaller or equal to the directed path-width of this graph [6].

Corollary 3. *Directed strong/weak pseudoforests have directed tree-width at most 1.*

Corollary 4. *Directed cactus trees have directed path-width at most 1.*

The other direction of Theorem 6 does not hold true. There are graphs of directed tree-width 1 which are not directed cactus forests, as for example their forbidden directed graph minor $\overrightarrow{C_{3,1}}$. This graph has directed path-width 1 by the directed path-decomposition $\mathcal{X} = (X_1, X_2)$ with $X_1 = \{a, c\}$ and $X_2 = \{b, c\}$. It then follows that it also has directed tree-width at most 1 and as it includes a cycle, it has directed tree-width exactly 1.

7   Conclusion and Outlook

In this paper we introduced directed cactus trees (DCT) and forests (DCF), directed pseudotrees (DPT) and directed strong (DSPF) and weak pseudoforests (DWPF). We could prove that DCF, DSPF, and DWPF can be characterized by at most three forbidden digraph minors, using a graph minor operation for which directed path-width is monotone. Furthermore, we showed that DCF and its subclasses have directed tree-width at most 1 and DSPF, DCT and their subclasses even have directed path-width at most 1.

We also considered an oriented version of Halin graphs by connecting the leaves within a planar embedding of an out-tree in their clockwise ordering. This leads to a subclass of DWPF as well as DAGs. But these graphs can not be closed under directed minor operations, as they are not even closed under the subgraph operation.

Our results should be a first step on the way to find forbidden directed graph minors for the classes of directed tree-width at most 1 and classes of directed path-width at most 1. The latter have already been proven to have a countable number of forbidden directed graph minors [9], but these minors could not be found yet. Finding them could be an issue of future work, as well as checking if there is a countable number of forbidden digraph minors for the set of digraphs of directed tree-width at most 1 and to find them.

References

1. Bang-Jensen, J., Gutin, G. (eds.): Classes of Directed Graphs. SMM. Springer, Cham (2018). https://doi.org/10.1007/978-3-319-71840-8
2. Barát, J.: Directed pathwidth and monotonicity in digraph searching. Graphs Comb. **22**, 161–172 (2006)
3. El-Mallah, E., Colbourn, C.J.: The complexity of some edge deletion problems. IEEE Trans. Circuits Syst. **35**(3), 354–362 (1988)
4. Ganian, R., et al.: Are there any good digraph width measures? J. Comb. Theory Ser. B **116**, 250–286 (2016)
5. Gurski, F., Rehs, C.: Computing directed path-width and directed tree-width of recursively defined digraphs. ACM Computing Research Repository, abs/1806.04457, p. 16 (2018)
6. Gurski, F., Rehs, C.: Directed path-width and directed tree-width of directed co-graphs. In: Wang, L., Zhu, D. (eds.) COCOON 2018. LNCS, vol. 10976, pp. 255–267. Springer, Cham (2018). https://doi.org/10.1007/978-3-319-94776-1_22
7. Harary, F., Uhlenbeck, G.E.: On the number of husimi trees: I. Proc. Nat. Acad. Sci. **39**(4), 315–322 (1953)
8. Johnson, T., Robertson, N., Seymour, P.D., Thomas, R.: Directed tree-width. J. Comb. Theory Ser. B **82**, 138–155 (2001)
9. Kintali, S., Zhang, Q.: Forbidden directed minors and directed pathwidth. Reseach report (2015)
10. Kintali, S., Zhang, Q.: Forbidden directed minors and Kelly-width. Theor. Comput. Sci. **662**, 40–47 (2017)
11. Paten, B., et al.: Cactus graphs for genome comparisons. J. Comput. Biol. **18**(3), 469–481 (2011)
12. Paten, B., Earl, D., Nguyen, N., Diekhans, M., Zerbino, D., Haussler, D.: Cactus: algorithms for genome multiple sequence alignment. Genome Res. **21**(9), 1512–11528 (2011)
13. Robertson, N., Seymour, P.D.: Graph minors II. Algorithmic aspects of tree width. J. Algorithms **7**, 309–322 (1986)
14. Scheffler, P.: Die baumweite von graphen als mass für die kompliziertheit algorithmischer probleme. Ph.D. thesis, Akademie der Wissenschaften in der DDR, Berlin (1989)

Existence Versus Exploitation: The Opacity of Backdoors and Backbones Under a Weak Assumption

Lane A. Hemaspaandra[1] and David E. Narváez[2(✉)]

[1] Department of Computer Science, University of Rochester,
Rochester, NY 14627, USA
lane.hemaspaandra@icloud.com
[2] College of Computing and Information Sciences, RIT,
Rochester, NY 14623, USA
den9562@rit.edu

Abstract. Backdoors and backbones of Boolean formulas are hidden structural properties. A natural goal, already in part realized, is that solver algorithms seek to obtain substantially better performance by exploiting these structures.

However, the present paper is not intended to improve the performance of SAT solvers, but rather is a cautionary paper. In particular, the theme of this paper is that there is a potential chasm between the existence of such structures in the Boolean formula and being able to effectively exploit them. This does not mean that these structures are not useful to solvers. It does mean that one must be very careful not to assume that it is computationally easy to go from the existence of a structure to being able to get one's hands on it and/or being able to exploit the structure.

For example, in this paper we show that, under the assumption that P ≠ NP, there are easily recognizable families of Boolean formulas with strong backdoors that are easy to find, yet for which it is hard (in fact, NP-complete) to determine whether the formulas are satisfiable. We also show that, also under the assumption P ≠ NP, there are easily recognizable sets of Boolean formulas for which it is hard (in fact, NP-complete) to determine whether they have a large backbone.

1 Introduction

Many algorithms for the Boolean satisfiability problem exploit hidden structural properties of formulas in order to find a satisfying assignment or prove that no such assignment exists. These structural properties are called hidden because they are not explicit in the input formula. A natural question that arises then is what is the computational complexity associated with these hidden structures. In this paper we focus on two hidden structures: backbones and strong backdoors [11].

© Springer Nature Switzerland AG 2019
B. Catania et al. (Eds.): SOFSEM 2019, LNCS 11376, pp. 247–259, 2019.
https://doi.org/10.1007/978-3-030-10801-4_20

The complexity of decision problems associated with backdoors and back-bones has been studied by Nishimura, Ragde, and Szeider [9], Kilby, Slaney, Thiébaux, and Walsh [8], and Dilkina, Gomes, and Sabharwal [3], among others.

In the present paper, we show that, under the assumption that P ≠ NP, there are easily recognizable families of formulas with strong backdoors that are easy to find, yet the problem of determining whether these formulas are satisfiable remains hard (in fact, NP-complete).

Hemaspaandra and Narváez [6] showed, under the (rather strong) assumption that P ≠ NP ∩ coNP, a separation between the complexity of finding backbones and that of finding the values to which the backbone variables must be set. In the present paper, we also add to that line of research by showing that, under the (less demanding) assumption that P ≠ NP, there are families of formulas that are easy to recognize (i.e., they can be recognized by polynomial-time algorithms) yet no polynomial-time algorithm can, given a formula from the family, decide whether the formula has a large backbone (doing so is NP-complete).

Far from being a paper that is intended to speed up SAT solvers, this is a paper trying to get a better sense of the (potential lack of) connection between properties existing and being able to get one's hands on the variables or variable settings that are the ones expressing the property's existence. That is, the paper's point is that there is a potential gap between on one hand the existence of small backdoors and large backbones, and on the other hand using those to find satisfying assignments. Indeed, the paper establishes not just that (if P ≠ NP) such gaps exist, but even rigorously proves that if any NP set exists that is frequently hard (with respect to polynomial-time heuristics), then sets of our sort exist that are essentially just as frequently hard; we in effect prove an inheritance of frequency-of-hardness result, under which our sets are guaranteed to be essentially as frequently hard as any set in NP is.

Our results admittedly are theoretical results, but they speak both to the importance of not viewing backdoors or backbones as magically transparent—we prove that they are in some cases rather opaque—and to the fact that the behavior we mention likely happens on quite dense sets; and, further, since we tie this to whether any set is densely hard, these SAT-solver issues due to this paper have now become inextricably linked to the extremely important, long-open question of how resistant to polynomial-time heuristics the hardest sets in NP can be.[1] We are claiming that these important hidden properties—backdoors and backbones—have some rather challenging behaviors that one must at least be aware of. Indeed, what is most interesting about this paper is likely not the theoretical constructions themselves, but rather the behaviors that those

---

[1] We mention in passing that there are relativized worlds (aka black-box models) in which NP sets exist for which all polynomial-time heuristics are asymptotically wrong half the time [7]; heuristics basically do no better than one would do by flipping a coin to give one's answer. Indeed, that is known to hold with probability one relative to a random oracle, i.e., it holds in all but a measure zero set of possible worlds [7]. Although many suspect that the same holds in the real world, proving that would separate NP from P in an extraordinarily strong way, and currently even proving that P and NP differ is viewed as likely being decades (or worse) away [4].

constructions prove must exist unless P = NP. We feel that knowing that those behaviors cannot be avoided unless P = NP is of potential interest to both AI and theory. Additionally, the behavior in one of our results is closely connected to the deterministic time complexity of SAT; in our result (Theorem 1) about easy-to-find hard-to-assign-values-to backdoors, we show that the backdoor size bound in our theorem cannot be improved even slightly unless NP is contained in subexponential time.

The rest of this paper is organized as follows. Section 2 defines the notation we will use throughout this paper. Sections 3 and 4 contain our results related to backdoors and backbones, respectively. Finally, Sect. 5 adds some concluding remarks.

2 Definitions and Notations

For Boolean formula F, we denote by $V(F)$ the set of variables appearing in F.

Adopting the notations of Williams, Gomes, and Selman [11], we use the following. A partial assignment of F is a function $a_S : S \to \{\text{TRUE}, \text{FALSE}\}$ that assigns Boolean values to the variables in a set $S \subseteq V(F)$. For a Boolean value $v \in \{\text{TRUE}, \text{FALSE}\}$ and a variable $x \in V(F)$, the notation $F[x/v]$ denotes the formula F after replacing every occurrence of x by v and simplifying. This extends to partial assignments, e.g., to $F[a_S]$, in the natural way.

For a finite set A, $\|A\|$ denotes A's cardinality. For any string x, $|x|$ denotes the length of (number of characters of) x.

For each set T and each natural number n, $T^{\leq n}$ denotes the set of all strings in T whose length is less than or equal to n. In particular, $(\Sigma^*)^{\leq n}$ denotes the strings of length at most n, over the alphabet Σ.

3 Results on Backdoors to CNF Formulas

In this section we focus on Boolean formulas in conjunctive normal form, or CNF. A CNF formula is a conjunction of disjunctions, and the disjunctions are called the *clauses* of the formula. Following Dilkina, Gomes, and Sabharwal [3], we define satisfiability of CNF formulas using the language of set theory. This is done by formalizing the intuition that, in order for an assignment to satisfy a CNF formula, it must set at least one literal in every clause to TRUE. One can then define a CNF formula F to be a collection of clauses, each clause being a set of literals. $F \in$ SAT if and only if there exists an assignment $a_{V(F)}$ such that for all clauses $C \in F$ there exists a literal $l \in C$ such that $a_{V(F)}$ assigns l to TRUE. Under this formalization, to be in harmony with the standard conventions that the truth value of the empty conjunctive (resp., disjunctive) formula is TRUE (resp., FALSE), F must be taken to be in SAT if F is empty (since the empty CNF formula must be taken to be TRUE as a consequence of the fact that the empty conjunctive formula is taken to be TRUE) and F must be taken to be in $\overline{\text{SAT}}$ if $\emptyset \in F$ (since that empty clause is an empty disjunctive formula and so by convention is FALSE, and thus F evaluates to FALSE); these two cases are called, respectively, F being trivially TRUE and F being trivially FALSE (as the conventions as just mentioned put these cases not just in SAT and $\overline{\text{SAT}}$ but fix the truth values of the represented formulas to be TRUE and FALSE). We can also formalize simplification using this notation: after assigning a variable x to TRUE (resp., FALSE), the formula is simplified by removing all clauses that contain the literal x (resp., $\overline{x}$) and removing the literal $\overline{x}$ (resp., x) from the remaining clauses. This formalization extends to simplification of a formula over a partial assignment in the natural way.

Example 1. Consider the CNF formula $F = (x_1 \vee \overline{x_2} \vee \overline{x_3} \vee x_5) \wedge (x_1 \vee x_2 \vee x_4 \vee x_5) \wedge (x_3 \vee \overline{x_4}) \wedge (\overline{x_1} \vee x_2 \vee x_3 \vee x_5)$. We can express this formula in our set theory notation as $F = \{\{x_1, \overline{x_2}, \overline{x_3}, x_5\}, \{x_1, x_2, x_4, x_5\}, \{x_3, \overline{x_4}\}, \{\overline{x_1}, x_2, x_3, x_5\}\}$. Suppose we assign x_3 to FALSE and x_4 to TRUE, we have $F[x_3/\text{FALSE}, x_4/\text{TRUE}] = \{\emptyset, \{\overline{x_1}, x_2, x_5\}\}$, which is unsatisfiable because it contains the empty set.

Since CNF-SAT (the satisfiability problem restricted to CNF formulas) is well-known to be NP-complete, a polynomial-time algorithm to determine the satisfiability of CNF formulas is unlikely to exist. Nevertheless, there are several restrictions of CNF formulas for which satisfiability can be decided in polynomial time. When a formula does not belong to any of these restrictions, it may have a set of variables that, once the formula is simplified over a partial assignment of these variables, the resulting formula belongs to one of these tractable restrictions. A formalization of this idea is the concept of backdoors.

Definition 1 (Subsolver [11]). *A polynomial-time algorithm A is a* subsolver *if, for each input formula F, A satisfies the following conditions.*

1. *A either rejects the input F (this indicates that it declines to make a statement as to whether F is satisfiable) or determines F (i.e., A returns a satisfying assignment if F is satisfiable and A proclaims F's unsatisfiability if F is unsatisfiable).*
2. *If F is trivially TRUE A determines F, and if F is trivially FALSE A determines F.*
3. *If A determines F, then for each variable x and each value v, A determines $F[x/v]$.*

Definition 2 (Strong Backdoor [11]). *For a Boolean formula F, a nonempty subset S of its variables is a* strong backdoor *for a subsolver A if, for all partial assignments a_S, A determines $F[a_S]$ (i.e., if $F[a_S]$ is satisfiable A returns a satisfying assignment and if $F[a_S]$ is unsatisfiable A proclaims its unsatisfiability).*

Many examples of subsolvers can be found in the literature (for instance, in Table 1 of [3]). The subsolver that is of particular relevance to this paper is the *unit propagation subsolver*, which focuses on *unit clauses*. Unit clauses are clauses with just one literal. They play an important role in the process of finding models (i.e., satisfying assignments) because the literal in that clause must be set to TRUE in order to find a satisfying assignment. The process of finding a model by searching for a unit clause (for specificity and to ensure that it runs in polynomial time, let us say that our unit propagation subsolver always focuses on the unit clause in the current formula whose encoding is the lexicographically least among the encodings of all unit clauses in the current formula), fixing the value of the variable in the unit clause, and simplifying the formula resulting from that assignment is known in the satisfiability literature as unit propagation. Unit propagation is an important building block in the seminal DPLL algorithm for SAT [1,2]. Notice that the CNF formulas whose satisfiability can be decided by just applying unit propagation iteratively constitute a tractable restriction of SAT. The unit propagation subsolver attempts to decide the satisfiability of an input formula by using only unit propagation and empty clause detection. If satisfiability cannot be decided this way, the subsolver rejects the input formula. Szeider [10] has classified the parameterized complexity of finding backdoors with respect to the unit propagation subsolver.

Example 2. Consider the formula F from Example 1. We will show that $\{x_1, x_3, x_5\}$ is a strong backdoor of F with respect to the unit propagation subsolver by analyzing the possible assignments of these variables. Suppose x_1 is assigned to TRUE and notice $F[x_1/\text{TRUE}] = \{\{x_3, \overline{x_4}\}, \{x_2, x_3, x_5\}\}$. From there it is easy to see that if x_3 is set to TRUE, the resulting formula after simplification is trivially satisfiable. If x_3 is set to FALSE, assigning x_5 to TRUE yields the formula $\{\{\overline{x_4}\}\}$ after simplification and the satisfiability of this formula can be determined by the unit propagation subsolver. Assigning x_5 to FALSE yields a formula with two unit clauses, $\{\{\overline{x_4}\}, \{x_2\}\}$. The unit propagation subsolver will (here we assume that a clause $\{x\}$ precedes a clause $\{y\}$ in lexicographical order if x precedes y in lexicographical order.) pick the unit clause $\{x_2\}$, assign the truth value of x_2 and simplify, and will then pick the (sole) remaining unit clause, $\{\overline{x_4}\}$, and assign the truth value of x_4 and simplify to obtain a trivially satisfiable formula. Now suppose x_1 is assigned to FALSE and notice $F[x_1/\text{FALSE}] = \{\{\overline{x_2}, \overline{x_3}, x_5\}, \{x_2, x_4, x_5\}, \{x_3, \overline{x_4}\}\}$. If we now assign x_3 to TRUE, notice $F[x_1/\text{FALSE}, x_3/\text{TRUE}] = \{\{\overline{x_2}, x_5\}, \{x_2, x_4, x_5\}\}$. If we assign x_5 to TRUE F simplifies to a trivially satisfiable formula. If we assign x_5 to FALSE, the formula simplifies to $\{\{\overline{x_2}\}, \{x_2, x_4\}\}$. The unit propagation subsolver will pick the unit clause $\{\overline{x_2}\}$, assign the truth value of x_2, and the resulting formula after simplification will be $\{\{x_4\}\}$ whose satisfiability can be determined by the unit propagation subsolver. If we assign x_3 to FALSE, notice $F[x_1/\text{FALSE}, x_3/\text{FALSE}] = \{\{x_2, x_4, x_5\}, \{\overline{x_4}\}\}$. If we now assign x_5 to TRUE and simplify, the resulting formula would be $\{\{\overline{x_4}\}\}$ whose satisfiability can be determined by the unit propagation subsolver. If we assign x_5 to FALSE and simplify, the resulting formula would contain the unit clause $\{\overline{x_4}\}$. The unit propagation subsolver would then set the value of x_4 to FALSE and simplify, yielding the formula $\{\{x_2\}\}$, whose satisfiability can also be determined by the unit propagation subsolver. It should be clear from the case analysis above that just setting the values of x_1 and x_3 is not enough for the unit propagation subsolver to always be able to determine the satisfiability of the resulting formula. In fact, a similar analysis done on every 2-element subset and every 3-element subset of $V(F)$—which we do not write out here—shows that $\{x_1, x_3, x_5\}$ is actually the smallest strong backdoor of F with respect to the unit propagation subsolver.

We're ready to prove our main result about backdoors: Under the assumption that $P \neq NP$, there are families of Boolean formulas that are easy to recognize and have strong unit propagation backdoors that are easy to find, yet deciding whether the formulas in these families are satisfiable remains NP-complete.

Theorem 1. *If* $P \neq NP$, *for each* $k \in \{1, 2, 3, \ldots\}$ *there is a set* A *of Boolean formulas such that all the following hold.*

1. $A \in P$ *and* $A \cap SAT$ *is* NP-*complete.*
2. *Each formula* G *in* A *has a strong backdoor* S *with respect to the unit propagation subsolver, with* $\|S\| \leq \|V(G)\|^{\frac{1}{k}}$.
3. *There is a polynomial-time algorithm that, given* $G \in A$, *finds a strong backdoor having the property stated in item 2 of this theorem.*

Proof. For $k = 1$ the theorem is trivial, so we henceforward consider just the case where $k \in \{2, 3, \ldots\}$. Consider (since in the following set definition F is specified as being in CNF, we can safely start the following with "$F\wedge$" rather than for example "$(F)\wedge$") $A \in P$ defined by $A = \{F \wedge (\mathbf{new}_1 \wedge \cdots \wedge \mathbf{new}_{\|V(F)\|^k - \|V(F)\|}) \mid F$ is a CNF formula$\}$, where $\mathbf{new}_i$ is the ith (in lexicographical order) legal variable name that does not appear in F. For instance, if F contains literals $\overline{x_1}$, x_2, x_3, and $\overline{x_3}$, and if our legal variable universe is $x_1, x_2, x_3, x_4, \ldots$, then $\mathbf{new}_1$ would be x_4. The backdoor is the set of variables of F, which can be found in polynomial time by parsing. It is clear that the formula resulting from simplification after assigning values to all the variables of F only has unit clauses and potentially an empty clause, so satisfiability for this formula can be decided by the unit propagation subsolver. Finally, it is easy to see that $F \wedge (\mathbf{new}_1 \wedge \cdots \wedge \mathbf{new}_{\|V(F)\|^k - \|V(F)\|}) \in SAT \Leftrightarrow F \in SAT$ so, since the formula-part that is being postponed to F can easily be polynomial-time constructed given F, under the assumption that $P \neq NP$ deciding satisfiability for the formulas in A is hard. (Note: One can add "CNF" after "Boolean" in the theorem statement with just a minor proof adjustment.) $\square$

Let us address two natural worries the reader might have regarding Theorem 1. First, the reader might worry that the hardness spoken of in the theorem occurs very infrequently (e.g., perhaps except for just one string at every double-exponentially spaced length everything is easy). That is, are we giving a worst-case result that deceptively hides a low typical-case complexity? We are not (unless all of NP has easy typical-case complexity): we show that if any set in NP is frequently hard with respect to polynomial-time heuristics, then a set of our sort is almost as frequently hard with respect to polynomial-time heuristics. We will show this as Theorem 3.

But first let us address a different worry. Perhaps some readers will feel that the fact that Theorem 1 speaks of backdoors of size bounded by a fixed kth root in size is a weakness, and that it is disappointing that the theorem does not establish its same result for a stronger bound such as "constant-sized backdoors", or if not that then polylogarithmic-sized, or if not that then at least ensuring that not just each fixed root is handled in a separate construction/set but that a single construction/set should assert/achieve the case of a growth rate that is asymptotically less than every root. Those are all fair and natural to wonder about. However, we claim that not one of those improvements of Theorem 1 can be proven without revolutionizing the deterministic speed of SAT. In particular, the following result holds, showing that those three cases would respectively put NP into P, quasipolynomial time, and subexponential time.

Theorem 2. 1. *[Constant case] Suppose there is a $k \in \{1, 2, 3, \ldots\}$ and a set A of Boolean formulas such that all the following hold: (a) $A \in$ P and $A \cap$ SAT is NP-complete; (b) each formula G in A has a strong backdoor S with respect to the unit propagation subsolver, with $\|S\| \leq k$; and (c) there is a polynomial-time algorithm that, given $G \in A$, finds a strong backdoor having the property stated in item (b). Then $P = NP$.*

2. *[Polylogarithmic case] Suppose there is a function $s(n)$, with $s(n) = (\log n)^{\mathcal{O}(1)}$, and a set A of Boolean formulas such that all the following hold: (a) $A \in$ P and $A \cap$ SAT is NP-complete; (b) each formula G in A has a strong backdoor S with respect to the unit propagation subsolver, with $\|S\| \leq s(\|V(G)\|)$; and (c) there is a polynomial-time algorithm that, given $G \in A$, finds a strong backdoor having the property stated in item (b). Then NP is in quasipolynomial time, i.e., $NP \subseteq \bigcup_{c>0} DTIME[2^{(\log n)^c}]$.*

3. *[Subpolynomial case] Suppose there is a polynomial-time computable function r and a set A of Boolean formulas such that all the following hold: (a) for each $k \in \{1, 2, 3, \ldots\}$, $r(0^n) = \mathcal{O}(n^{\frac{1}{k}})$; (b) $A \in$ P and $A \cap$ SAT is NP-complete; (c) each formula G in A has a strong backdoor S with respect to the unit propagation subsolver, with $\|S\| \leq r(0^{\|V(G)\|})$; and (d) there is a polynomial-time algorithm that, given $G \in A$, finds a strong backdoor having the property stated in item (c). Then NP is in subexponential time, i.e., $NP \subseteq \bigcap_{\epsilon>0} DTIME[2^{n^\epsilon}]$.*

We can see this as follows. Consider the "Constant case"—the first part—of the above theorem. Let k be the constant of that part. Then there are at most $\binom{N}{k}$ ways of choosing k of the variables of a given Boolean formula of N bits (and thus of at most N variables). And for each of those ways, we can try all 2^k possible ways of setting those variables. This is $\mathcal{O}(N^k)$ items to test—a polynomial number of items. If the formula is satisfiable, then via unit propagation one of these must yield a satisfying assignment (in polynomial time). Yet the set $A \cap$ SAT was NP-complete by the first condition of the theorem. So we have that $P = NP$, since we just gave a polynomial-time algorithm for $A \cap$ SAT. The other three cases are analogous (except in the final case, we in the theorem needed to put in the indicated polynomial-time constraint on the bounding function r since otherwise it could be badly behaved; that issue doesn't affect the second part of the theorem since even a badly behaved function s of the second part is bounded above by a simple-to-compute function s' satisfying $s'(n) = (\log n)^{\mathcal{O}(1)}$ and we can use s' in place of s in the proof).

Even the final part of the above theorem, which is the part that has the weakest hypothesis, implies that NP is in subexponential time. However, it is widely suspected that the NP-complete sets lack subexponential-time algorithms. And so we have established that the $n^{1/k}$ growth, which we *do* prove in Theorem 1, is the smallest bound in part 2 of that result that one can hope to prove Theorem 1 for without having to as a side effect put NP into a deterministic time class so small that we would have a revolutionarily fast deterministic algorithm for SAT.

Moving on, we now, as promised above, address the frequency of hardness of the sets we define in Theorem 1, and show that if any set in NP is frequently hard then a set of our type is almost-as-frequently hard. (Recall that, when n's universe is the naturals as it is in the following theorem, "for almost every n" means "for all but at most a finite number of natural numbers n".) We will say that a (decision) algorithm errs with respect to B on an input x if the algorithm disagrees with B on x, i.e., if the algorithm accepts x yet $x \notin B$ or the algorithm rejects x yet $x \in B$.

Theorem 3. *If h is any nondecreasing function and for some set $B \in$ NP it holds that each polynomial-time algorithm errs with respect to B, at infinitely many lengths n (resp., for almost every length n), on at least $h(n)$ of the inputs up to that length, then there will exist an $\epsilon > 0$ and a set $A \in$ P of Boolean formulas satisfying the conditions of Theorem 1, yet being such that each polynomial-time algorithm g, at infinitely many lengths n (resp., for almost every length n), will fail to determine membership in $A \cap$ SAT for at least $h(n^\epsilon)$ inputs of length at most n.*

Before getting to the proof of this theorem, let us give concrete examples that give a sense about what the theorem is saying about density transference. It follows from Theorem 3 that if there exists even one NP set such that each polynomial-time heuristic algorithm asymptotically errs exponentially often up to each length (i.e., has $2^{n^{\Omega(1)}}$ errors), then there are sets of our form that in the same sense fool each polynomial-time heuristic algorithm exponentially often. As a second example, it follows from Theorem 3 that if there exists even one NP set such that each polynomial-time heuristic algorithm asymptotically errs quasipolynomially often up to each length (i.e., has $n^{(\log n)^{\Omega(1)}}$ errors), then there are sets of our form that in the same sense fool each polynomial-time heuristic algorithm quasipolynomially often. Since almost everyone suspects that some NP sets are quasipolynomially and indeed even exponentially densely hard, one must with equal strength of belief suspect that there are sets of our form that are exponentially densely hard.

Proof of Theorem 3. For conciseness and to avoid repetition, we build this proof on top of a proof (namely, of Theorem 6) that we will give later in the paper. That later proof does not rely directly or indirectly on the present theorem/proof, so there is no circularity at issue here. However, readers wishing to read the present proof should probably delay doing that until after they have first read that later proof.

We define r_B as in the proof of Theorem 6 (the r_B given there draws on a construction from Appendix A of [5], and due to that construction's properties outputs only conjunctive normal form formulas). For a given k, we define

$A = \{r_B(x) \wedge (\text{new}_1 \wedge \cdots \wedge \text{new}_{\|V(r_B(x))\|^k - \|V(r_B(x))\|}) \mid x \in \Sigma^*\}$, and since $r_B(x) \wedge (\text{new}_1 \wedge \cdots \wedge \text{new}_{\|V(r_B(x))\|^k - \|V(r_B(x))\|}) \in \text{SAT} \Leftrightarrow r_B(x) \in \text{SAT}$ and $r_B(x) \in \text{SAT} \Leftrightarrow x \in B$, we can now proceed as in the proof of Theorem 6, since here too the tail's length is polynomially bounded. $\qquad\square$

4   Results on Backbones

For completeness, we start this section by restating the definition of backbones as presented by Williams, Gomes, and Selman [11]. We restrict ourselves to the Boolean domain, since we only deal with Boolean formulas in this paper.

Definition 3 (Backbone [11]). *For a Boolean formula F, a subset S of its variables is a* backbone *if there is a unique partial assignment a_S such that $F[a_S]$ is satisfiable.*

The *size* of a backbone S is the number of variables in S. One can readily see from Definition 3 that all satisfiable formulas have at least one backbone, namely, the empty set. This backbone is called the *trivial* backbone, while backbones of size at least one are called *nontrivial* backbones. It follows from Definition 3 that unsatisfiable formulas do not have backbones. Note also that some satisfiable formulas have no nontrivial backbones, e.g., $x_1 \vee x_2 \vee x_3$ is satisfiable but has no nontrivial backbone.

Example 3. Consider $F = x_1 \wedge (x_1 \leftrightarrow \overline{x_2}) \wedge (x_2 \leftrightarrow x_3) \wedge (x_2 \vee x_4 \vee x_5)$. Any satisfying assignment of F must have x_1 set to TRUE, which in turn constrains x_2 and x_3. Then $\{x_1, x_2, x_3\}$ is a backbone of F, as is any subset of this backbone. It is also easy to see that $\{x_1, x_2, x_3\}$ is the largest backbone of this formula since the truth values of x_4 and x_5 are not entirely constrained in F (since F in effect is—once one applies the just-mentioned forced assignments—$x_4 \vee x_5$).

Our first result states that if P $\neq$ NP then there are families of Boolean formulas that are easy to recognize, with the property that deciding whether a formula in these families has a large backbone is NP-complete (and so is hard). As a corollary to its proof, we have that if P $\neq$ NP then there are families of Boolean formulas that are easy to recognize, with the property that deciding whether a formula in these families has a nontrivial backbone is NP-complete (and so is hard).[2]

---

[2] We have not been able to find Corollary (to the Proof) 5 in the literature. Certainly, two things that on their surface might seem to be the claim we are making in Corollary (to the Proof) 5 are either trivially true or are in the literature. However, upon closer inspection they turn out to be quite different from our claim.

In particular, if one removes the word "nontrivial" from Corollary (to the Proof) 5's statement, and one is in the model in which every satisfiable formula is considered to have the empty collection of variables as a backbone and every unsatisfiable formula is considered to have no backbones, then the thus-altered version of Corollary (to the Proof) 5 is clearly true, since if one with those changes takes A to

Theorem 4. *For any real number* $0 < \beta < 1$, *there is a set* $A \in P$ *of Boolean formulas such that the language* $L_A = \{F \mid F \in A \text{ and } F \text{ has a backbone } S \text{ with } \|S\| \geq \beta\|V(F)\|\}$ *is NP-complete (and so if* $P \neq NP$ *then* L_A *is not in* P).

Corollary (to the Proof) 5. *There is a set* $A \in P$ *of Boolean formulas such that* $L_A = \{F \mid F \in A \text{ and } F \text{ has a nontrivial backbone } S\}$ *is NP-complete (and so if* $P \neq NP$ *then* L_A *is not in* P).

Proof of Theorem 4 and Corollary 5. We will first prove Theorem 4, and then will note that Corollary 5 follows easily as a corollary to the proof/construction.

So fix a β from Theorem 4's statement. For each Boolean formula G, let $q(G) = \left\lceil \frac{\beta\|V(G)\|}{1-\beta} \right\rceil$. Define $A = \{(G) \wedge (\text{new}_1 \wedge \text{new}_2 \wedge \cdots \wedge \text{new}_{q(G)}) \mid G$ is a Boolean formula having at least one variable$\}$, where, as in the proof of Theorem 1, we define new_i as the ith variable that does not appear in G. Note that $\text{new}_1 \wedge \text{new}_2 \wedge \cdots \wedge \text{new}_{q(G)}$ is a backbone if and only if $G \in \text{SAT}$, thus under the assumption that $P \neq NP$ and keeping in mind that for zero-variable formulas satisfiability is easy to decide, it follows that no polynomial-time algorithm can decide L_A, since the size of this backbone is $q(G) > 0$, which by our definition of q will satisfy the condition $\|S\| \geq \beta\|V(F)\|$. Why does it satisfy that condition? $\|S\|$ here is $q(G)$. And $\|V(F)\|$ here, since F is the formula $(G) \wedge (\text{new}_1 \wedge \text{new}_2 \wedge \cdots \wedge \text{new}_{q(G)})$, equals $\|V(G)\| + q(G)$. So the condition is claiming that $q(G) \geq \beta(\|V(G)\| + q(G))$, i.e., that $q(G) \geq \frac{\beta}{(1-\beta)}\|V(G)\|$, which indeed holds in light of the definition of q. And why do we claim that no polynomial-time algorithm can decide L_A? Well, note that SAT many-one polynomial-time reduces to L_A via the reduction $g(H)$ that equals some fixed string in L_A if H is in SAT and H has zero variables and that equals some fixed string in $\overline{L_A}$ if H is not in SAT and H has zero variables (these two cases are included merely to handle degenerate things such as TRUE $\vee$ FALSE that can occur if we allow TRUE and FALSE as atoms in our propositional formulas), and that equals $(H) \wedge (\text{new}_1 \wedge \text{new}_2 \wedge \cdots \wedge \text{new}_{q(H)})$ otherwise (the above formula is H conjoined with a large number of new variables). Since L_A is in NP, we have that it is NP-complete, and since $P \neq NP$ was part of the theorem's hypothesis, L_A cannot be in P.

---

be the set of all Boolean formulas, then the theorem degenerates to the statement that if $P \neq NP$, then SAT is (NP-complete, and) not in P.

Also, it is stated in Kilby et al. [8] that finding a backbone of CNF formulas is NP-hard. However, though this might seem to be our result, their claim and model differ from ours in many ways, making this a quite different issue. First, their hardness refers to Turing reductions (and in contrast our paper is about many-one reductions and many-one completeness). Second, they are not even speaking of NP-Turing-hardness—much less NP-Turing-completeness—in the standard sense since their model is assuming a function reply from the oracle rather than having a set as the oracle. Third, even their notion of backbones is quite different as it (unlike the influential Williams, Gomes, and Selman 2003 paper [11] and our paper) in effect requires that the function-oracle gives back both a variable *and its setting*. Fourth, our claim is about *nontrivial* backbones.

The above proof establishes Theorem 4. Corollary 5 follows immediately from the proof/construction of Theorem 4. Why? The set A from the proof of Theorem 4 is constructed in such a way that each of its potential members $(G) \wedge (\mathbf{new}_1 \wedge \mathbf{new}_2 \wedge \cdots \wedge \mathbf{new}_{q(G)})$ (where G is a Boolean formula having at least one variable) either has no nontrivial backbone (indeed, no backbone) or has a backbone of size at least $\beta(\|V(G)\|)$. Thus the issue of backbones that are nontrivial but smaller than $\beta(\|V(F)\|)$, where F is $(G) \wedge (\mathbf{new}_1 \wedge \mathbf{new}_2 \wedge \cdots \wedge \mathbf{new}_{q(G)})$, does not cause a problem under the construction. That is, our A (which itself is dependent on the value of β one is interested in) is such that we have ensured that $\{F \mid F \in A$ and F has a nontrivial backbone $S\} = \{F \mid F \in A$ and F has a backbone S with $\|S\| \geq \beta\|V(F)\|\}$.   □

We now address the potential concern that the hard instances for the decision problems we just introduced may be so infrequent that the relevance of Theorem 4 and Corollary 5 is undercut. The following theorem argues against that possibility by proving that, unless not a single NP set is frequently hard (in the sense made rigorous in the theorem's statement), there exist sets of our form that are frequently hard. (This result is making for backbones a point analogous to the one our Theorem 3 makes for backdoors. Hemaspaandra and Narváez [6] looks at frequency of hardness result for backbones, but with results focused on $\text{NP} \cap \text{coNP}$ rather than NP.)

Theorem 6. *If h is any nondecreasing function and for some set $B \in \text{NP}$ it holds that each polynomial-time algorithm errs with respect to B, at infinitely many lengths n (resp., for almost every length n), on at least $h(n)$ of the inputs up to that length, then there will exist an $\epsilon > 0$ and a set $A \in \text{P}$ of Boolean formulas satisfying the conditions of Theorem 4, yet being such that each polynomial-time algorithm g, at infinitely many lengths n (resp., for almost every length n), will fail to correctly determine membership in L_A for at least $h(n^\epsilon)$ inputs of length at most n. The same claim also holds for Corollary 5.*

Proof. We will prove the theorem's statement regarding Theorem 4. It is not hard to also then see that the analogous claim holds regarding Corollary 5.

$B \in \text{NP}$ and SAT is NP-complete. So let r_B be a polynomial-time function, transforming strings into Boolean formulas, such that (a) $r_B(x) \in \text{SAT} \Leftrightarrow x \in B$, and (b) r_B is one-to-one. (A construction of such a function is given in Appendix A of [5], and let us assume that construction is used.) As in the proof of Theorem 4, if F is a Boolean formula we define $q(F) = \left\lceil \frac{\beta\|V(F)\|}{1-\beta} \right\rceil$.

Without loss of generality, we assume that r_B outputs only formulas having at least one variable. Note that throughout this proof, q is applied only to outputs of r_B. Thus we have ensured that none of the logarithms in this proof have a zero as their argument.

Set $A = \{(r_B(x)) \wedge (\mathbf{new}_1 \wedge \mathbf{new}_2 \wedge \cdots \wedge \mathbf{new}_{q(r_B(x))}) \mid x \in \Sigma^*\}$. Because r_B is computable in polynomial time, there is a polynomial b such that for every input x of length at most n, the length of $r_B(x)$ is at most $b(n)$. Fix some such polynomial b, and let k denote its degree. In order to find a bound for the length of the added "tail" $\mathbf{new}_1 \wedge \mathbf{new}_2 \wedge \cdots \wedge \mathbf{new}_{q(r_B(x))}$ in terms of $b(n)$, notice that the length of the tail is less than some constant (that holds over all x and n, $|x| \leq n$) times $q(r_B(x)) \log q(r_B(x))$. Since $q(r_B(x)) = \left\lceil \frac{\beta \|V(F)\|}{1-\beta} \right\rceil$ and the length of $r_B(x)$ is at least a constant times the number of its variables, our assumption that $|r_B(x)| \leq b(n)$ implies the existence of a constant c such that, for all x and n, $|x| \leq n$, we have $q(r_B(x)) \leq c \cdot b(n)$. Taken together, the two previous sentences imply the existence of a constant d such that, for all x and n, $|x| \leq n$, we have that the length of $\mathrm{new}_1 \wedge \mathrm{new}_2 \wedge \cdots \wedge \mathrm{new}_{q(r_B(x))}$ is at most $d \cdot b(n) \log(b(n))$, and so certainly is less than $d \cdot b^2(n)$. Let N be a natural number such that, for all $n \geq N$ and all x, $|x| \leq n$ implies that $|(r_B(x)) \wedge (\mathrm{new}_1 \wedge \mathrm{new}_2 \wedge \cdots \wedge \mathrm{new}_{q(r_B(x))})| \leq n^{2k+1}$; by the previous sentence and the fact that b is of degree k, such an N will exist. Let g be a polynomial-time heuristic for L_A. Notice that $g \circ r_B$—i.e., $g(r_B(\cdot))$—is a polynomial-time heuristic for B, since $(r_B(x)) \wedge (\mathrm{new}_1 \wedge \mathrm{new}_2 \wedge \cdots \wedge \mathrm{new}_{q(r_B(x))}) \in L_A \Leftrightarrow r_B(x) \in$ SAT and $r_B(x) \in \mathrm{SAT} \Leftrightarrow x \in B$. Let $n_B \geq N$ be such that there is a set of strings $S_{n_B} \subseteq (\Sigma^*)^{\leq n_B}$, $\|S_{n_B}\| \geq h(n_B)$, having the property that for all $x \in S_{n_B}$, $g \circ r_B$ fails to correctly determine the membership of x in B. Consequently, there is a set of strings $T_{n_B} \subseteq (\Sigma^*)^{\leq (n_B)^{2k+1}}$, $\|T_{n_B}\| \geq h(n_B)$, such that for all $x \in T_{n_B}$, g fails to correctly determine the membership of x in L_A; in particular the set $T_{n_B} = \{(r_B(x)) \wedge (\mathrm{new}_1 \wedge \mathrm{new}_2 \wedge \cdots \wedge \mathrm{new}_{q(r_B(x))}) \mid x \in S_{n_B}\}$ has this property.

Using the variable renaming $n_A = (n_B)^{2k+1}$, it is now easy to see that we have proven that every length $n_B \geq N$ at which $g \circ r_B$ (viewed as a heuristic for B) errs on at least $h(n_B)$ inputs of length up to n_B has a corresponding length n_A at which g (viewed as a heuristic for L_A) errs on at least $h((n_A)^{\frac{1}{2k+1}})$ inputs of length up to n_A. Our hypothesis guarantees the existence of infinitely many such $n_B \geq N$ (resp., almost all $n \geq N$ can take the role of n_B), each with a corresponding n_A. Setting $\epsilon = \frac{1}{2k+1}$, our theorem is now proven.   □

5   Conclusions

We constructed easily recognizable families of Boolean formulas that provide hard instances for decision problems related to backdoors and backbones under the assumption that P $\neq$ NP. In particular, we have shown that, under the assumption P $\neq$ NP, there exist easily recognizable families of Boolean formulas with easy-to-find strong backdoors yet for which it is hard to determine whether the formulas are satisfiable. Under the same P $\neq$ NP assumption, we have shown that there exist easily recognizable collections of Boolean formulas for which it is hard (in fact, NP-complete) to determine whether they have a backbone, and that there exist easily recognizable collections of Boolean formulas for which it is hard (in fact, NP-complete) to determine whether they have a large backbone. (These results can be taken as indicating that, under the very plausible assumption that P $\neq$ NP, search and decision shear apart in complexity for backdoors and backbones. That makes it particularly unfortunate that their definitions in the literature are framed in terms of decision rather than search, especially since when one tries to put these to work in SAT solvers, it is the search case that one typically tries to use and leverage.)

For both our backdoor and backbone results, we have shown that if *any* problem B in NP is frequently hard, then there exist families of Boolean formulas of the sort we describe that are hard almost as frequently as B.

Acknowledgments. We thank the SOFSEM referees for helpful comments. Work done in part while L. Hemaspaandra was visiting ETH-Zürich and U-Düsseldorf.

References

1. Davis, M., Logemann, G., Loveland, D.: A machine program for theorem-proving. Commun. ACM **5**, 394–397 (1962)
2. Davis, M., Putnam, H.: A computing procedure for quantification theory. J. ACM **7**(3), 201–215 (1960)
3. Dilkina, B., Gomes, C., Sabharwal, A.: Tradeoffs in the complexity of backdoors to satisfiability: dynamic sub-solvers and learning during search. Ann. Math. Artif. Intell. **70**(4), 399–431 (2014)
4. Gasarch, W.: The second P =? NP poll. SIGACT News **43**(2), 53–77 (2012)
5. Hemaspaandra, L., Narváez, D.: The opacity of backbones. Technical report, June 2016. arXiv:1606.03634 [cs.AI], Computing Research Repository, arXiv.org/corr/. Accessed June 2017 to December 2018. Revised January 2017
6. Hemaspaandra, L., Narváez, D.: The opacity of backbones. In: Proceedings of the 31st AAAI Conference on Artificial Intelligence, pp. 3900–3906. AAAI Press, February 2017
7. Hemaspaandra, L., Zimand, M.: Strong self-reducibility precludes strong immunity. Math. Syst. Theory **29**(5), 535–548 (1996)
8. Kilby, P., Slaney, J., Thiébaux, S., Walsh, T.: Backbones and backdoors in satisfiability. In: Proceedings of the 20th National Conference on Artificial Intelligence, pp. 1368–1373. AAAI Press (2005)
9. Nishimura, N., Ragde, P., Szeider, S.: Detecting backdoor sets with respect to Horn and binary clauses. In: Informal Proceedings of the 7th International Conference on Theory and Applications of Satisfiability Testing, pp. 96–103, May 2004
10. Szeider, S.: Backdoor sets for DLL subsolvers. J. Autom. Reasoning **35**(1–3), 73–88 (2005)
11. Willams, R., Gomes, C., Selman, B.: Backdoors to typical case complexity. In: Proceedings of the 18th International Joint Conference on Artificial Intelligence, pp. 1173–1178. Morgan Kaufmann, August 2003

On Point Set Embeddings for k-Planar Graphs with Few Bends per Edge

Michael Kaufmann[✉]

Wilhelm-Schickhard-Institut für Informatik, Universität Tübingen,
Tübingen, Germany
mk@informatik.uni-tuebingen.de

Abstract. We consider the point set embedding problem (PSE) for 1-, 2- and k-planar graphs where at most 1, 2, or k crossings resp. are allowed for each edge which greatly extends the well-researched class of planar graphs. For any set of n points and any given embedded graph that belongs to one of the above graph classes, we compute a 1-to-1 mapping of the vertices to the points such that the edges can be routed using only a limited number of bends according to the given embedding and the sequences of crossings. Surprisingly, for the class of 1-planar graphs the same results can be achieved as the best known results for planar graphs. Additionally for k-planar graphs, the bounds are also much better than expected from the first sight.

1 Introduction

Graph embeddings have been popular since quite some time in the area of combinatorics, graph algorithms and graph drawing. Variants include the bandwidth minimization problem [15] where the vertices should be ordered in unit distance on a line such that the total sum of the distances is minimized, but also graph embeddings, where a guest graph has to be embedded to a host graph under certain parameters like dilation, congestion, expansion, etc. [24].

In the geometric variant of the point set embedding (PSE) problem, a set S of points is assumed to be given, together with an input graph with certain properties. In the most simple version, S has exactly n points and the problem is to find a 1-to-1 mapping of the vertices to the points such that the corresponding straight-line edges are crossing-free. Techniques have been developed for the cases that the input graph is a tree [8,20], an outerplanar graph [7,29]. For the case of general planar graphs, the problem has been shown to be NP-hard. [9]. Another research direction concentrates on restricting the number of additional points that is needed to guarantee a planar straight-line embedding for any given planar graph. This has been called a 'universal point set'. Unfortunately, the best upper bound for the general case is $O(n^2)$ points as the universal point set [17], very far away from the lower bound of $1.098n$ [12]. Recently, smaller universal point sets of subquadratic size have been found for planar 3-trees [18], graphs of bounded pathwidth [4] and for the quite general class of so-called k-outerplanar graphs [3].

© Springer Nature Switzerland AG 2019
B. Catania et al. (Eds.): SOFSEM 2019, LNCS 11376, pp. 260–271, 2019.
https://doi.org/10.1007/978-3-030-10801-4_21

Another variant of the point set embedding problem with several applications assumes a given mapping of the vertices to the points. A linear number of bends per edge is sufficient and sometimes almost necessary [28] .

A well-recognized approach without a prescribed mapping but allowing bends on the edges has been developed by Kaufmann and Wiese [23], who developed an efficient PSE algorithm for any pointset of size n with only one bend per edge if the given plane graph is hamiltonian, and with only two bends per edge for general plane graphs. We will recall their algorithm in Sect. 2. We extend their technique and apply it to graph classes from a research direction which is called 'beyond planarity'.

'Beyond planarity' is an informal term for a recent research direction in Graph Drawing, which is currently receiving a great deal of attention [19,21,25]. It mainly focuses on combinatorial and algorithmic aspects for classes of graphs that can be drawn on the plane while avoiding specific kinds of edge crossings; see, e.g., [14] for a survey.

Even quite long time ago, 1-planar graphs [30] have been considered as the family of graphs that admit drawings with at most one crossing per edge. Later generalizations include 2-planar graphs [27], 3-planar graphs [5,26], k-planar graphs, where k crossings per edge are allowed. Others have more complicated restrictions like fan-planar graphs [22], where an edge is only allowed to be crossed by a set of adjacent edges from the same side (a fan), or fan-crossing-free graphs [10], where exactly this configuration was forbidden and only crossings by pairwise independent edges are allowed. Quasiplanar and k-quasiplanar graphs where no three resp. no k mutually crossing edges are allowed, received special considerations in the past due to their combinatorial structure [1,2,16] We finally mention the RAC graphs [13], i.e. the family of graphs that admit polyline drawings, with few bends per edge, in which the angles formed at the edge crossings are 90°.

Our Contribution: In this paper, we consider the PSE problem for 1-, 2- and k-plane graphs in the model, where the vertices-to-points mapping is not given and bends on the edges are allowed and should be minimized. In particular, after reviewing some theorems from the literature and the basic algorithm by Kaufmann and Wiese in Sect. 2, we show that an n-vertex 1-plane graph can be embedded on any given point set of size n with only 2 bends per edge in Sect. 3, matching the previously known best bound for plane graphs. In Sect. 4 we extend these techniques to 2-plane graphs where 2 crossings are allowed on each edge. Finally we demonstrate in Sect. 5 that even for general k-plane graphs very good results on the number of bends per edge can be achieved when the input graph is mapped on the given set of n points. We discuss open problems in Sect. 6.

2   Preliminaries

In this section, we introduce again the problem for our model. Then we recall some theorems and techniques from the literature, that we will use to obtain our results.

Let S be a set of n given disjoint points $p_1 = (x_1, y_1), p_2 = (x_2, y_2), ..., p_n = (x_n, y_n)$ in the plane, and let $G = (V, E)$ be an n vertex graph. The graph G is assumed to be simple and given as a topological drawing, i.e. a drawing of disjoint points as vertices and simple curves, representing the edges, such that there are no self-intersections, self-loops nor intersections of adjacent edges. We have to find a mapping of the vertices to the points and realizations of the edges as polygonal lines such that the topological drawing of the graph is preserved and the maximal number of bends on each edge is as small as possible.

For our approach, we will use the following three theorems, in particular Theorem 3.

Theorem 1 *(Biedl, Kant, Kaufmann [6]). Any 2-connected planar graph without separating triangles can be triangulated efficiently without introducing new separating triangles. The new graph will be 4-connected.*

Theorem 2 *(Chiba, Nishizeki [11]). Any 4-connected planar graph has a hamiltonian cycle, which can be constructed in linear time.*

Theorem 3 *(Kaufmann, Wiese [23]). Any 4-connected plane graph can be embedded on any point set of the same size with only one bend per edge. For 3-connected plane graphs a corresponding result holds but the edges might have at most two bends.*

We recall the algorithm of Kaufmann and Wiese [23]:

Let S be the given set of n points in the plane rotated such that the x-coordinates are disjoint and at least a certain constant space in x-direction exists between subsequent points. Let G a plane graph which can be assumed to be triangulated. If G is four-connected, we know that there exists a hamiltonian cycle $C = v_1, ..., v_n, v_1$, which can be efficiently computed [11]. We subsequently map the vertices v_i along the cycle to the points of S ordered from left to right. Consequently, the edges of the cycle C from v_1 to v_n form an x-monotone polygonal line P. The closing edge (v_n, v_1) is drawn with one bend above the polygonal line, such that the slopes of the ascending and the decreasing segment of the edge are drawn with the same slopes but with opposite sign. All the other edges can now be drawn easily either above the polygonal line inside of the cycle in a similar way as (v_n, v_1) with one bend each, or below the line such that the bend is the lowest point of the edge which is drawn symmetrically as well.

More concretely, the slope is computed as follows: Define the parameter $\rho = max_i \{ \frac{|y_{i+1} - y_i|}{x_{i+1} - x_i} \}$ which denotes the steepest slope of a polygonal segment along the polygonal line P. To draw the edges not on the path with one bend symmetrically, we use slopes that are just slightly larger than ρ for ascending segments resp. slightly smaller than $-\rho$ for descending segments. Simple adjustment of the slopes ensures the absence of crossings. Details can be found in [23]. This algorithm completes the first part.

The second part is more tricky. We assume that the graph is triangulated but not 4-connected, which implies that it has some separating triangles. Let $T = \{u, v, w\}$ be one of the separating triangles with corresponding edges $(u, v), (v, w), (w, u)$. We break T by taking one of the edges, say (u, v) and subdivide it by a dummy vertex d, a so-called b-dummy. Additionally, we triangulate the two quadrangular faces adjacent to d by two new so-called breaking edges incident to b. We apply this step to all the separating triangles and get a triangulated graph with some dummy vertices which has no separating triangles, i.e. which is 4-connected and hence has a hamiltonian cycle C. Clearly C also passes through the dummy vertex d. If it does not use both corresponding breaking edges, the dummy vertex is useless and can be removed without changing the remaining hamiltonian cycle C. Only if C passes through both breaking edges the dummy vertex is useful as the edge (u, v) now consists of two subedges (u, d) and (d, v) which lie on different sides of the polygonal line P. If we apply now the basic algorithm of the four-connected case to the plane graph extended by the dummy vertices, we get at most 3 bends per edge, one for each part, and one at the dummy vertex. By changing the shape of the edges such that the edge segment incident to a dummy vertex consist of two vertical segments, the bend at the dummy vertex can be saved, such that each edge finally has at most 2 bends. Figure 1 shows the effect of the bend-saving technique. As it is explained in [23], the area consumption might be exponential in the size of the bounding rectangle of the point set S, see also Sect. 6.

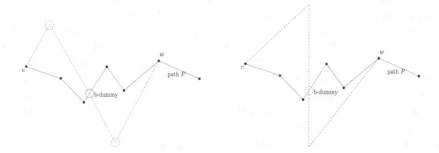

Fig. 1. The left figure shows the 3-bend version with symmetrical slopes. The bend at the dummy vertex can be saved by aligning the two middle segments vertically (right figure). Note that w might be to the left of the b-dummy, such that three bends seems unavoidable for the case of symmetrical slopes without the bend-saving technique.

In summary, edges that are incident to two original vertices have only one bend, and the edges that consist of two parts both incident to a dummy vertex have 3 bends, which can be reduced to two bends. In the following, we will heavily use this basic scheme, which will be extended appropriately.

We will planarize our non-planar input graph which will be given as a topological graph drawn in the plane by replacing the crossings by so-called *c-dummy* vertices. The corresponding edges will now consist of subedges between 'real' vertices and c-dummies. Those subedges will be called *c-c edges* if they are incident to two c-dummies. In case, the subedge is between a real vertex and a c-dummy, it is an *r-c edge*. Edges between a real and a b-dummy, are called *r-b edges*. If an edge is incident to two 'real' vertices, we call it an *r-r edge*.

3   1-Plane Graphs

In this section, we consider 1-plane graphs, i.e. graphs with at most one crossing per edge embedded in the plane. Let $G = (V, E)$ be a 1-plane graph and S be a set of n points in the plane which is rotated such that the x-coordinates of the points are disjoint. As a first step, we replace the crossings of G by c-dummies and achieve a plane graph G'. Now, we want to apply the algorithm of Kaufmann and Wiese to G'. Because of 1-planarity, the edges in G might correspond to two r-c edges each in G'. So, a first estimate indicates that the edges in G might have at most 5 bends, namely 2 bends for each subedge plus a bend at the corresponding c-dummy vertex. Unfortunately, the trick to reduce the bend at the dummy vertex cannot be applied directly.

Nevertheless we will prove the following theorem achieving a much better bound:

Theorem 4. *Any 2-connected 1-plane graph can be embedded on any point set of the same size with at most two bends per edge.*

Proof. After the replacement step of the crossings, we obtain the plane graph G'. At each c-dummy, we have 4 incident r-c subedges. We triangulate G', e.g. by the algorithm of Biedl et al. [6] which has the property that new separating triangles are produced only if it is necessary. Next, we want to break the separating triangles. For each such triangle T, we select one of the edges to add a b-dummy vertex, which breaks the triangle T. Note that by 1-planarity, there is no edge between two c-dummies. So, if one of the edges is a r-c edge, then there are exactly two r-c edges in T. Clearly, the third edge is an r-r edge and we choose that edge to place the b-dummy vertex and which then is split into two r-b edges. Hence we never place an b-dummy on an r-c edges.

After the triangulation and breaking all the separation triangles by the b-dummies, we obtain G''. G'' finally is 4-connected such that a hamiltonian cycle C exists and we can apply approach of Kaufmann and Wiese. The hamiltonian cycle forms a monotone polygonal line P of straight-line segments between the points. Auxiliary points at appropriate positions, e.g. equidistantly spaced on the corresponding straight-line segment of the polygonal line P host the intermediate dummy vertices. Since only the edges of G' between two real vertices are split into at most two parts, we have on each original edge at most one dummy vertex, either a c-dummy or a b-dummy. Hence we have achieved already a bound of 3 for the number of bends per edge.

Next, we show how to save one of the three bends. If the dummy is a b-dummy, then the original technique of Kaufmann and Wiese can be applied and the middle bend can be saved. Note that this is quite straightforward also because b-dummies have degree 2. For the c-dummies, the four incident r-c edges make it more involved. In principle, we route the r-c edges such that the second segments (those that incident to the c-dummy) are nearly vertical, deviating from the vertical by a very small ($\epsilon < 0$) angle to avoid unnecessary overlaps. Note that by chosing the bend points appropriately, this can be done without violating planarity. The preliminary crossing point is placed at the position of the c-dummy.

We distinguish 3 configurations, depending on how many of those four r-c edges are on the same sides of the polygonal line P defined by the hamiltonian path $(v_1, ..., v_n)$. We have either 2, 3 or 4 of the r-c edges on different sides. The next figure shows some different configurations and our corresponding solutions.

(a) Two of the r-c edges are on each side of P.

For this case, we see that the bend-saving technique of Kaufmann and Wiese can be applied directly. We leave the crossing at the c-dummy, and ensure that there are no bends there. Hence the claim to achieve 2 bends holds for this case.

(b) Three of the r-c edges are on the same side of P.

This means that one of the two edges, call it e, consisting of two of the r-c edges does not cross P. Let b_1, b_2 and b_3 be the bends along e, where b_2 is the c-dummy. Clearly, the triangle b_1, b_2, b_3 is empty and is only crossed by one of the other two r-c edges incident to b_2. Now, we omit b_2 from e and add a straight-line segment from b_1 to b_3. Hence we save the middle bend. The crossing point is now at another position, but since its position is not prescribed, this is not problematic. The other edge which crosses P at the c-dummy, can be routed with only 2 bends as in Case (a). Note that the topology of the crossing is preserved.

(c) All four r-c edges are on the same side of the polygonal line.

We apply the same bend-saving trick as in the case (b) but now for both edges. The properties that the corresponding triangles described by the 3 bends are empty hold as well. Hence we can save the two middle bends achieving the two-bends bound. See also the right-hand side of Fig. 2. Clearly, the crossing still exists since we do not change the beginning and end of the two edges.   □

In the next section, we examine how much of the techniques can be transfered to the 2-planarity case, where there might be two crossings on each edge.

4   2-Plane Graphs

Let $G = (V, E)$ be a 2-connected 2-plane graph and S be a set of n points in the plane which is rotated to provide disjoint x-coordinates of the points. As before,

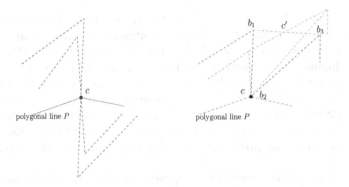

Fig. 2. Two different configurations (a) and (c) for a crossing c with the corresponding incident edge segments. On the right-hand side, bend-saving is applied, and the dashed segments denote the new middle segments

we replace the crossings of G by so-called c-dummies and achieve a plane graph G'. Before executing the triangulation step, we analyse the existing separating triangles and restrict the set of edges that will be split by b-dummy vertices.

Lemma 1. *In a separating triangle in G', there is an r-r edge.*

Proof. We first assume that there is a separating triangle $T = \{u, v, w\}$ in G' such that it contains only c-c edges, hence u, v and w are c-dummies. Hence the component inside of the separating triangle has either no or only one edge connecting it to the three crossings, which is a contradiction to the 2-planarity assumption, as the edge might have 3 crossings, or there are two edges from the inside of T to the crossings, then T forms a selfloop of an edge consisting of at least three parts.

Second, we assume that there is a separating triangle T, that has an r-c edge, but no r-r edge. Note that then it has exactly two incident r-c edges, and one c-c edge. Let v be the real vertex in T. By 2-connectivity, the component has at least two different connections to T. If both c-dummies are adjacent to the inside component, then T is a selfloop edge consisting of three subedges and incident to vertex v. If only one c-dummy, say w, is adjacent to the inside component, then the subedges (v, w) and (w, u) belong to the same edge, and this edge crosses the other edge that starts at the real vertex with the subedge (v, u), which is then a contradiction to the simplicity of the drawing, as two incident edges intersect. Figure 3 gives an overview of the different cases.   □

Now we extend G' by new triangulation edges as planned before [6]. There might occur some separating triangles that existed already in G', or some new separating triangles that include some new (fictitious) triangulation edges. To break the later separating triangles, we use the fictitious edges to place the b-dummies onto, since those edges will not be drawn anyway and hence we do not care about the number of bends they will have.

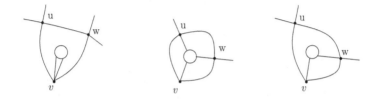

Fig. 3. Different configurations for a separating triangle with real vertex v and c-dummies u and w. The circle inside the triangle denotes an inside component.

We only focus on the 'original' separating triangles in G' and we apply Lemma 1. Hence we split these separating triangles only by b-dummy vertices on r-r edges. We again have a 4-connected graph G'' with a hamiltonian cycle C which then implies the order of the real vertices in which those vertices will be mapped along the polygonal monotone path P. The dummy vertices are placed in between respectively. Since b-dummies are only placed at r-r edges and fictitious edges, the bend-saving technique of Kaufmann and Wiese can be applied to make sure that the number of bends on the real edges is at most two. The most critical edges are those edges with two crossings, as they are subdivided by two c-dummies. They consist of three subedges with two segments respectively with slopes slightly larger than parameter ρ for ascending segments, and slightly smaller than $-\rho$ for descending segments. Without the bend-saving technique, the intermediate c-dummies would imply two more bends, and hence 5 bends in total.

Applying the bend-saving technique realizes each subedge by a nearly-vertical segment and the second segment with a large (positive or negative) slope. Unfortunately, not both of the bends at the c-dummies can be saved by the technique if the middle subedge lies on the other side of the polygonal path P, i.e. if we have to cross P twice at the c-dummies. Corresponding bends can only be saved if we have (almost)-vertical segments at the c-dummies. This can only be realized at one of them. Hence the number of bends on each edge can be upperbounded by 4.

Theorem 5. *Any 2-connected 2-plane graph can be embedded on any point set of the same size with at most four bends per edge.*

In Fig. 4 we give an example of an edge with two crossings $c1$ and $c2$, where the hamiltonian cycle is crossed twice and the applicability of the bend-saving technique is limited. This effect generalizes also to the case of higher number of crossings per edge (see Sect. 5) if the edge crosses the hamiltonian cycle upto k times. In the next section, we will transfer the insights about the problems in 2-planar graphs to general k-planar graphs for $k \geq 2$.

5   k-Plane Graphs

Let G be a k-plane graph for $k \geq 2$ given as a topological drawing with at most k crossings on each edge and let S be a set of n points in the plane which is

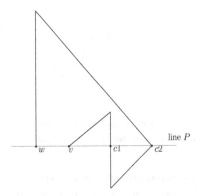

Fig. 4. A sketch of the routing of an edge (v, w) which is crossed twice, and the vertices and c-dummies come in a certain order along the hamiltonian cycle. The bend at $c1$ can be saved, but the one at $c2$ cannot. This indicates that saving the bends can eventually be applied only at every other c-dummy.

rotated to provide disjoint x-coordinates of the points. Again, the crossings of G are replaced by the c-dummies such that a plane graph G' has been constructed. Note that each edge which has l crossings is subdivided into two r-c edges and $l - 1$ c-c edges. Before executing the triangulation step, we will try to make similar observations as in the 2-planarity case such that we restrict the set of edges that will be split by b-dummy vertices to remove the separating triangles.

Lemma 2. *In a separating triangle in G', there is an r-r edge.*

Proof. We assume that there is a separating triangle $T = \{u, v, w\}$ in G' without any r-r edge.

By 2-connectivity, at least two of the (dummy) vertices have one or more connections to the inside of T. As there are no r-r edges in T, there are at least two c-dummies in T. Only one of three cases might occur: Either the three subedges of T form a loop in G, or there are two edges in G that cross each other twice in G, or one of the vertices of T is a real vertex v, and there are two edges incident to v that cross each other. All three cases lead to a contradiction to simplicity assumptions for the topological embedding of G. Some of the cases can be checked in Fig. 3.   □

Next we apply the triangulation procedure adding some (fictitious) edges. Then we extend the graph by breaking separating triangles either placing b-dummy vertices on fictitious edges (if they exist), or we can use the r-r edges within the 'original' separating triangles to destroy those separating triangles by placing b-dummies. Then we proceed with the algorithm of Kaufmann and Wiese for 4-connected plane graphs, guided by the hamiltonian cycle C which we determine by the algorithm of Chiba/Nishizeki [11], and parameter ρ. Without the bend-saving tricks, we achieve for an edge with k crossings, hence with k c-dummies and $k + 1$ subedges at most $2k + 1$ bends.

Theorem 6. *Any 2-connected k-plane graph can be embedded on any point set of the same size with at most $2k + 1$ bends per edge.*

Further improvements can be achieved applying the bend-saving techniques that we developed before. Unfortunately, we were not able to save all the bends at the c-vertices. It can successfully be done when two of the subedges consequently are on the same side of the polygonal line formed by the monotone path P along the hamiltonian cycle. The worst case appears if subsequent subedges alternate between the sides of P. In that case, we can save only every second bend at a c-vertex, as we have to use two nearly-vertical segments at this vertex. When we have fulfilled this condition for a c-vertex, we cannot fulfill it for the next c-vertex, or we have to use an additional bend, cf. Fig. 4 where the case for 2 crossings is sketched. Hence we can conclude

Theorem 7. *Any 2-connected k-plane graph can be embedded on any point set of the same size with at most $k + 1 + \lfloor k/2 \rfloor \leq \frac{3}{2}k + 1$ bends per edge.*

6 Discussions and Conclusions

We have shown how to effectively apply the algorithm of Kaufmann and Wiese [23] to 1-planar, 2-planar and finally k-planar embedded graphs for $k \geq 2$. Note that the area consumption is exponential even in the basic algorithm for plane graphs if the bend-saving techniques are applied and the bound of only two bends should be maintained. For the case of three bends where the parameter ρ is used to control the size of the slopes, and using the assumption of an integer grid, the authors mention an upper bound of $O(W^3)$ for the area, where W denotes the width of the bounding box of the points. We give the corresponding bounds for the graph classes here:

Theorem 8. *Any 2-connected k-plane graph can be embedded on any point set of the same size with at most $2k + 1$ bends per edge using only a drawing area of polynomial size. In particular, for $k = 1$, we have 3 bends per edge, and for $k = 2$, we have 5 bends per edge.*

We conclude with some open problems:

1. Improve the bounds for the number of bends per edge for 2-plane graphs from four to three. This then might imply an improvement of the bounds for general k-plane graphs as well.
2. The area issue should be discussed in more detail, even for the crossing-free case. Is it possible to keep the area of polynomial size and still have only two bends per edge. Develop lower bounds.
3. Consider the special case of outer-1-plane graphs. Note that there are efficient algorithms for straight-line embeddings of outerplanar graphs on pointsets in general positions. Can this result be transfered to outer-1-plane graphs? What are the corresponding bounds for outer-2-plane and outer-k-planar graphs?

Acknowledgement. The author wishes to thanks the participants of the GNV workshop in Heiligkreuztal 2018 for inspiring discussions.

References

1. Ackerman, E., Tardos, G.: On the maximum number of edges in quasi-planar graphs. J. Comb. Theor. Ser. A **114**(3), 563–571 (2007)
2. Agarwal, P.K., Aronov, B., Pach, J., Pollack, R., Sharir, M.: Quasi-planar graphs have a linear number of edges. Combinatorica **17**(1), 1–9 (1997)
3. Angelini, P., et al.: Small universal point sets for k-outerplanar graphs. Discret. Comput. Geom. **60**(2), 430–470 (2018). https://doi.org/10.1007/s00454-018-0009-x
4. Bannister, M.J., Cheng, Z., Devanny, W.E., Eppstein, D.: Superpatterns and universal point sets. J. Graph Algorithms Appl. **18**(2), 177–209 (2014). https://doi.org/10.7155/jgaa.00318
5. Bekos, M.A., Kaufmann, M., Raftopoulou, C.N.: On optimal 2- and 3-planar graphs. In: Aronov, B., Katz, M.J. (eds.) Symposium on Computational Geometry. LIPIcs, vol. 77, pp. 16:1–16:16. Schloss Dagstuhl - Leibniz-Zentrum fuer Informatik (2017)
6. Biedl, T.C., Kant, G., Kaufmann, M.: On triangulating planar graphs under the four-connectivity constraint. Algorithmica **19**(4), 427–446 (1997). https://doi.org/10.1007/PL00009182
7. Bose, P.: On embedding an outer-planar graph in a point set. In: Proceedings of Graph Drawing, 5th International Symposium, GD 1997, 18–20 September 1997, Rome, Italy, pp. 25–36 (1997). https://doi.org/10.1007/3-540-63938-1_47
8. Bose, P., McAllister, M., Snoeyink, J.: Optimal algorithms to embed trees in a point set. In: Proceedings of Graph Drawing, Symposium on Graph Drawing, GD 1995, 20–22 September 1995, Passau, Germany, pp. 64–75 (1995). https://doi.org/10.1007/BFb0021791
9. Cabello, S.: Planar embeddability of the vertices of a graph using a fixed point set is NP-hard. J. Graph Algorithms Appl. 10(2), 353–363 (2006). http://jgaa.info/accepted/2006/Cabello2006.10.2.pdf
10. Cheong, O., Har-Peled, S., Kim, H., Kim, H.: On the number of edges of fan-crossing free graphs. Algorithmica **73**(4), 673–695 (2015)
11. Chiba, N., Nishizeki, T.: The hamiltonian cycle problem is linear-time solvable for 4-connected planar graphs. J. Algorithms **10**(2), 187–211 (1989). https://doi.org/10.1016/0196-6774(89)90012-6
12. Chrobak, M., Karloff, H.J.: A lower bound on the size of universal sets for planar graphs. SIGACT News **20**(4), 83–86 (1989). https://doi.org/10.1145/74074.74088
13. Didimo, W., Eades, P., Liotta, G.: Drawing graphs with right angle crossings. Theor. Comput. Sci. **412**(39), 5156–5166 (2011)
14. Didimo, W., Liotta, G., Montecchiani, F.: A survey on graph drawing beyond planarity. CoRR abs/1804.07257 (2018)
15. Feige, U.: Approximating the bandwidth via volume respecting embeddings (extended abstract). In: Proceedings of the Thirtieth Annual ACM Symposium on the Theory of Computing, 23–26 May 1998, Dallas, Texas, USA, pp. 90–99 (1998). https://doi.org/10.1145/276698.276716
16. Fox, J., Pach, J., Suk, A.: The number of edges in k-quasi-planar graphs. SIAM J. Discret. Math. **27**(1), 550–561 (2013)
17. de Fraysseix, H., Pach, J., Pollack, R.: How to draw a planar graph on a grid. Combinatorica **10**(1), 41–51 (1990). https://doi.org/10.1007/BF02122694
18. Fulek, R., Tóth, C.D.: Universal point sets for planar three-trees. J. Discret. Algorithms **30**, 101–112 (2015). https://doi.org/10.1016/j.jda.2014.12.005

19. Hong, S., Tokuyama, T.: Algorithmics for beyond planar graphs. NII Shonan Meeting Seminar 089, 27 November–1 December 2016

20. Ikebe, Y., Perles, M.A., Tamura, A., Tokunaga, S.: The rooted tree embedding problem into points in the plane. Discret. Comput. Geom. **11**, 51–63 (1994). https://doi.org/10.1007/BF02573994

21. Kaufmann, M., Kobourov, S., Pach, J., Hong, S.: Beyond planar graphs: algorithmics and combinatorics. Dagstuhl Seminar 16452, 6–11 November 2016

22. Kaufmann, M., Ueckerdt, T.: The density of fan-planar graphs. CoRR abs/1403.6184 (2014)

23. Kaufmann, M., Wiese, R.: Embedding vertices at points: few bends suffice for planar graphs. J. Graph Algorithms Appl. **6**(1), 115–129 (2002). http://www.cs.brown.edu/publications/jgaa/accepted/2002/KaufmannWiese2002.6.1.pdf

24. Leighton, F.T.: Introduction to Parallel Algorithms and Architectures: Array, Trees, Hypercubes. Morgan Kaufmann Publishers Inc., San Francisco (1992)

25. Liotta, G.: Graph drawing beyond planarity: some results and open problems. SoCG Week, Invited talk, 4 July 2017

26. Pach, J., Radoičić, R., Tardos, G., Tóth, G.: Improving the crossing lemma by finding more crossings in sparse graphs. Discret. Comput. Geom. **36**(4), 527–552 (2006)

27. Pach, J., Tóth, G.: Graphs drawn with few crossings per edge. Combinatorica **17**(3), 427–439 (1997)

28. Pach, J., Wenger, R.: Embedding planar graphs at fixed vertex locations. In: Proceedings of Graph Drawing, 6th International Symposium, GD 1998, August 1998, Montréal, Canada, pp. 263–274 (1998). https://doi.org/10.1007/3-540-37623-2_20

29. Pach, J., Gritzmann, P., Mohar, B., Pollack, R.: Embedding a planar triangulation with vertices at specified points. Am. Math. Mon. **98**, 165–166 (1991). Professor Pach's number: [065]

30. Ringel, G.: Ein Sechsfarbenproblem auf der Kugel. Abh. Math. Sem. Univ. Hamb. **29**, 107–117 (1965)

Enumerating Connected Induced Subgraphs: Improved Delay and Experimental Comparison

Christian Komusiewicz and Frank Sommer$^{(\boxtimes)}$

Fachbereich Mathematik und Informatik, Philipps-Universität Marburg,
Marburg, Germany
{komusiewicz,fsommer}@informatik.uni-marburg.de

Abstract. We consider the problem of enumerating all connected induced subgraphs of order k in an undirected graph $G = (V, E)$. Our main results are two enumeration algorithms with a delay of $\mathcal{O}(k^2\Delta)$ where Δ is the maximum degree in the input graph. This improves upon a previous delay bound [Elbassioni, JGAA 2015] for this problem. In addition, we give improved worst-case running time bounds and delay bounds for several known algorithms and perform an experimental comparison of these algorithms for $k \leq 10$ and $k \geq |V| - 3$.

1 Introduction

We study algorithms for the following fundamental graph problem.

CONNECTED INDUCED SUBGRAPH ENUMERATION (CISE)
Input: An undirected graph $G = (V, E)$ and an integer k.
Task: Enumerate all connected induced subgraphs of order k.

We call a connected subgraph of order k a *solution* in the following. The enumeration of connected subgraphs is important in many applications, such as the identification of network motifs (statistically overrepresented induced subgraphs of small size). A straightforward algorithm to find such motifs is to enumerate all connected induced subgraphs and to count how often each subgraph of order k occurs [6,14]. A further application arises when semantic web data is searched using only keywords instead of structured queries [5]. Finally, many fixed-cardinality optimization problems can be solved by an algorithm whose first step is to enumerate connected induced subgraphs of order k [8]. This algorithm can solve for example CONNECTED DENSEST-k-SUBGRAPH, the problem of finding a connected subgraph of order k with a maximum number of edges. Experiments showed that enumeration-based algorithms can be competitive with other algorithmic approaches [9].

At first sight, providing any nontrivial upper bounds on the running time of CISE seems hopeless: As evidenced by a clique on n vertices, graphs may have up to $\binom{n}{k}$ CISE solutions. Even very sparse graphs may have $\binom{n-1}{k-1}$ CISE

© Springer Nature Switzerland AG 2019
B. Catania et al. (Eds.): SOFSEM 2019, LNCS 11376, pp. 272–284, 2019.
https://doi.org/10.1007/978-3-030-10801-4_22

solutions as evidenced by a star graph with $n-1$ leaves. It is maybe due to these lower bounds that, despite its importance, CISE has not received too much attention from the viewpoint of worst-case running time analysis.

One way to achieve relevant running time bounds is to consider degree-bounded graphs where the number of solutions is much smaller than in general.

Lemma 1 ([3, Eq. 7]). *Let G be a graph with maximum degree Δ. Then the number of connected induced subgraphs of order k that contain some vertex v is at most $(e(\Delta-1))^{(k-1)}$. Hence, the overall number of connected induced subgraphs of order k in G is $\mathcal{O}((e(\Delta-1))^{(k-1)} \cdot (n/k))$.*

This observation can be exploited to obtain an algorithm for CISE that runs in $\mathcal{O}((e(\Delta-1))^{(k-1)} \cdot (\Delta+k) \cdot (n/k))$ time [8].

A second approach to provide nontrivial running time bounds is to prove upper bounds on the *delay* of the enumeration. The delay is the maximal time that the algorithm spends between the output of consecutive solutions. The *reverse search* framework is a general paradigm for enumeration algorithms with bounded delay. The basic idea is to construct a tree where each node represents a unique element of the enumeration process. By traversing this tree from the root, each element is enumerated exactly once. By using reverse search, one can enumerate all induced subgraphs of order *at most* k with polynomial delay [1]. When we are interested only in solutions of order *exactly* k [4], this algorithm is *not* output polynomial, that is, the running time is not bounded by a polynomial in the input and output size. Hence, it does not achieve polynomial delay either. A different reverse search algorithm, however, achieves delay $\mathcal{O}(k \min(n-k, k\Delta)(k(\Delta+\log k)+\log n))$ [4].

Thus, k and Δ appear to be central parameters governing the complexity of CISE. Motivated by this observation, we aim to make further progress at exploiting small values of Δ and k.

Related Work. Most known CISE algorithms follow the same strategy: starting from an initial vertex set $S := \{v\}$ for some vertex v, build successively larger connected induced subgraphs $G[S]$ until an order-k subgraph is found. Wernicke [13] describes a procedure following this paradigm, which we refer to as *Simple*. The idea is to branch into the different possibilities to add one vertex u from $N(S)$. Another popular enumeration algorithm is *Kavosh* [6] which also considers adding vertices of $N(S)$ but creates one branch for each subset of $N(S)$ that has size at most $k - |S|$.

A slightly different strategy is to first pick a vertex p of the current set S whose neighbors are added in the next step and then branch on the up to $(\Delta-1)$ possibilities for adding a neighbor of this vertex. The vertex p is called the *active* vertex of the enumeration. The corresponding algorithm, which we call *Pivot*, has a worst-case running time of $\mathcal{O}((4(\Delta-1))^k \cdot (\Delta+k) \cdot n)$ [7]. A further variant of *Pivot* achieves the running time of $\mathcal{O}(e((\Delta-1))^{(k-1)} \cdot (\Delta+k) \cdot n/k)$ mentioned above [8]. This variant, which we call *Exgen*, generates *exhaustively* all subsets S' of $N(p) \setminus S$ of size at most $k - |S|$ and creates for each such set S' one branch in which S' is added to S. The final variant that we consider is $BDDE$ [11]. For a fixed vertex v, $BDDE$ enumerates the connected subgraphs containing v for increasing subgraph orders. The main idea is to use two functions, one to discover new graph edges and one to copy siblings in the enumeration tree.

The above-mentioned algorithms with polynomial delay [4] work differently. They use reverse search and, more generally, the supergraph method [1]. There, for a given graph G and parameter k, the supergraph $\mathcal{G}$ contains a node for each CISE solution in G. Furthermore, two nodes in $\mathcal{G}$ are connected if and only if the corresponding connected subgraphs differ in exactly one vertex. Let $|\mathcal{G}|$ denote the number of vertices in $\mathcal{G}$, that is, the number of CISE solutions. The basic idea is to explore the supergraph $\mathcal{G}$ efficiently. The first variant, which we refer to as RwD (*Reverse Search with Dictionary*) has a delay of $\mathcal{O}(k \min{(n-k, k\Delta)}(k(\Delta+\log k)+\log n))$ and requires $\mathcal{O}(n+m+k|\mathcal{G}|)$ space where m is the number of edges in the input graph G. The second variant, which we refer to as RwP (*Reverse Search with Predecessor*), has a delay of $\mathcal{O}((k \min{(n-k, k\Delta)})^2(\Delta+\log k))$ and requires $\mathcal{O}(n+m)$ space [4]. Hence, algorithm RwD admits a better delay but requires exponential space, since $\mathcal{G}$ may grow exponentially with the size of G.

Our Results. We show how to adapt *Simple* and *Pivot* in such a way that the worst-case delay between the output of two solutions is $\mathcal{O}(k^2\Delta)$ and the algorithms requires $\mathcal{O}(n+m)$ space. This improves over the previous best delay bound of RwD [4] while requiring only linear space. As a side result, we show that these variants of *Simple* and *Pivot* achieve an overall running time of $\mathcal{O}(e((\Delta-1))^{(k-1)} \cdot (\Delta+k) \cdot n/k)$ and $\mathcal{O}((e(\Delta-1))^{k-1} \cdot \Delta \cdot n)$, respectively. For *Simple* this is the first running time bound, for *Pivot*, this is a substantial improvement over the previous running time bound.

Finally, we compare these algorithms experimentally with implementations of *Kavosh* [6], *Exgen* [8], and *BDDE* [11]. For $k \leq 10$, we observe that RwD and RwP are significantly slower than the other algorithms. The *Simple* algorithm is faster than RwD and RwP but substantially slower than the other algorithms. *Kavosh* [6] is the fastest with *Pivot* being surprisingly competitive. For k close to the order of the largest connected component, we observe that our adaptions are necessary to solve these instances. Again, RwD and RwP are slower than the other algorithms and again, *Kavosh* is the fastest algorithm with *Simple* being second-best but not competitive with *Kavosh*.

Due to lack of space, several proofs are deferred to a long version of the article.

2   Preliminaries and Main Algorithm

Graph Notation. We consider undirected simple graphs $G = (V, E)$. The order $n := |V|$ denotes the number of vertices in G and $m := |E|$ denotes the number of edges in G. For a vertex v, $N(v) := \{u \mid \{u, v\} \in E\}$ denotes the *open neighborhood* of v, and $N[v] := N(v) \cup \{v\}$ denotes the *closed neighborhood* of v. For a vertex set $W \subseteq V$, $N(W) := \bigcup_{v \in W} N(v) \setminus W$ denotes the *open neighborhood of W* and $N[W] := N(W) \cup W$ denotes the *closed neighborhood of W*.

Algorithm 1. The main loop for calling the enumeration algorithms; *Enum-Algo* can be any of *Simple, Pivot, Exgen, Kavosh,* and *BDDE.*

1: **procedure** *Enumerate*$(G = (V, E))$
2:    **while** $|V(G)| \geq k$ **do**
3:        choose vertex v from $V(G)$
4:        enumerate all *CISE* solutions containing v with *Enum-Algo*
5:        remove v from G

The graph $G[W] := (W, \{\{u, v\} \in E \mid u, v \in W\})$ is the *subgraph induced by* W. For a set W the graph $G - W := G[V \setminus W]$ is the subgraph of G obtained by deleting the vertices of W. A connected component of G is a maximal subgraph where any two vertices are connected to each other by paths.

Enumeration Trees and the Main Algorithm Loop. With the exception of *RwD* and *RwP*, the enumeration algorithms use a search tree method which is called from a main loop whose pseudo code is given in Algorithm 1. Different algorithms, for example *Simple* or *Pivot*, can be used as *Enum-Algo* in Line 4 in Algorithm 1. For each vertex in the graph, Algorithm 1 creates a unique *enumeration tree*. In other words, Algorithm 1 produces a forest consisting of $|V|$ enumeration trees. To avoid confusion, we refer to the vertices of the enumeration trees as *nodes*. Each node represents a connected subgraph $G[S]$ of order at most k. Roughly speaking, a node N is a child of another node M if the subgraph corresponding to M is a subgraph of the subgraph corresponding to N. The exact definition of child depends on the choice of *Enum-Algo*. A *leaf* is a node without any children. Further, a leaf is *interesting* if S has size k; otherwise it is *boring*. A node leads to an interesting leaf, if at least one of its descendants is an interesting leaf.

In the main algorithm loop, we enumerate for each vertex of the input graph all *CISE* solutions containing the vertex v by calling the respective enumeration procedures; the first call of the enumeration procedure is the *root* of the enumeration tree and it represents the connected subgraph $G[\{v\}]$. After enumerating all solutions containing v, the vertex v is removed from the graph.

Cleaning the Graph. The removal of v may create connected components of order less than k. If *Enumerate* chooses all vertices from such connected components, then we will not achieve the claimed delays. Hence, we show how to remove these connected components quickly.

Lemma 2. *Let G be a graph such that each connected component has order at least k and let v be an arbitrary vertex of G. In $\mathcal{O}(k^2 \Delta)$ time we can delete every vertex of $G - \{v\}$ that is in a connected component of order less than k.*

3   Polynomial Delay with Simple

We now adapt *Simple* to obtain a polynomial delay algorithm; the pseudo code is shown in Algorithm 2. In *Simple*, we start with a single vertex v and find successively larger connected subgraphs containing v. The vertex set of a subgraph set is denoted by P. Further, the set X, called *extension set*, contains those neighbors of P which can be added to P to enlarge this subgraph. When putting u in the set P, we remove u from X and add to X each neighbor of u which is not in $N[P]$. Lines 10 and 11 of Algorithm 2 are not part of the plain version of *Simple* [13]. Without these two lines *Simple* is not a polynomial delay algorithm for CISE.

We now present a pruning rule (Lines 10 and 11 of Algorithm 2) that will establish polynomial delay. Consider a path $T_1, \ldots, T_i$ from the root T_1 to a node T_i of an enumeration tree. We denote the subgraph set of a node T_i by P_i and its extension set by X_i. To avoid some unnecessary recursions, we check after each recursive call of *Simple* in node T_i whether this call reported a new solution. If not, we return in T_i to its parent T_{i-1}. First, we prove that this pruning rule is correct. Recall that a leaf T_j is called interesting if the corresponding subgraph set P_j is a solution for CISE and that T_j is called boring otherwise.

Algorithm 2. The *Simple* algorithm; the initial call is $Simple(\{v\}, N(v))$.

```
1:  procedure SIMPLE(P, X)
2:      if |P| = k then
3:          output P
4:          return
5:      while X ≠ ∅ do
6:          u := choose arbitrary vertex from X
7:          delete u from X                    ▷ The current set P will be extended
8:          X' := X ∪ (N(u) \ N[P])
9:          Simple(P ∪ {u}, X')
10:         if output of Simple(P ∪ {u}, X') was empty then
11:             return                         ▷ Stop recursion if no new solution found
12:     return
```

Lemma 3. *If the output of a recursive call of Simple in node T_i is empty, then no subsequent recursive call of Simple in node T_i leads to an interesting leaf.*

Now we prove that *Simple* achieves a polynomial delay. To this end, we present a new data structure to store the extension set during the algorithm. In the following, we denote by p_i the vertex which was added to the subgraph set P_i when T_i is created. In other words, if T_{i-1} is the parent of T_i, then $p_i \in P_i \setminus P_{i-1}$. First, we prove that for a node T_i in the enumeration tree we need $\mathcal{O}(\Delta)$ time to either compute its next child T_{i+1} or to restore its parent T_{i-1}.

Lemma 4. *Simple can be implemented in such a way that for every node T_i of the enumeration tree, we need $\mathcal{O}(\Delta)$ time to either compute the next child T_{i+1} or to restore the parent T_{i-1} and that the overall space needed is $\mathcal{O}(n+m)$.*

Proof. We describe the data structures that we use to fulfill the running time and space bounds of the lemma. To check whether a vertex is in some extension

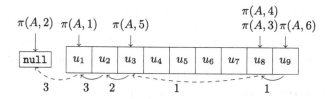

Fig. 1. An example for the pointer movement: Pointer $\pi(A, 6)$ points to u_9, an exclusive neighbor of p_6. Before adding u_9 to the subgraph set P_6, we move pointer $\pi(A, 6)$ to the left to p_8, an exclusive neighbor of vertex $\pi(A, 3)$. Hence, we move $\pi(A, 6)$ to the position of pointer $\pi(A, 5)$, since T_5 is the parent of T_6. Next, we create a child of T_6 by adding u_9 to the subgraph set P_6. The next time we are in node T_6, we move $\pi(A, 6)$ one to the left to vertex u_2 create a child of T_6 by adding u_3 to P_6. After returning from this child, we move $\pi(A, 6)$ to vertex u_1 which is an exclusive neighbor of vertex p_1. Hence, we move $\pi(A, 6)$ to the position of $\pi(A, 2)$, since T_2 is the parent of T_3. Afterwards, we create a child by adding u_2 to P_6. The next time we come back to node T_6, we delete pointer $\pi(A, 6)$, since $\pi(A, 6)$ points to `null`, and return to the parent T_5 of node T_6.

set, we color some vertices of with $k+1$ colors $c_0, \ldots, c_k$ as follows. For a node T_i, we call the *exclusive neighbors of* p_i the vertices which are in $N[P_i] \setminus N[P_{i-1}]$ where T_{i-1} is the parent of T_i. These are exactly the vertices that are added to X_{i-1} in Line 7 of Algorithm 2 to construct the set X_i for the node T_i. Throughout the algorithm we maintain the following invariant: The vertex p_1 has color c_0. A vertex has color c_i, $i \geq 1$, if and only if it is an exclusive neighbor of p_i. In a nutshell, the colors $c_0, \ldots, c_j$ represent the vertices in $N[P_j]$. It is necessary to use $k + 1$ different colors to determine in which node a vertex was added to the extension set. Note that every vertex may have at most one color.

The extension sets of all nodes on the path from the root T_1 to an enumeration tree node T_i are represented by an array A of length $k\Delta$ with up to k pointers pointing to positions of A. There is one pointer $\pi(A, i)$ corresponding to T_i and one pointer $\pi(A, j)$ for each ancestor T_j of T_i. An entry of A is either empty or contains a pointer to a vertex of the extension set X_i. New vertices for the extension set replace empty entries in the back. Pointer $\pi(A, i)$ points to the vertex x in the extension set X_i which will be added to P_T in the *next* recursive call of *Simple* in node T_i. If at node T_i already all children of T_i have been created, then $\pi(A, i)$ points to `null`. Hence, we may check in constant time whether T_i has further children and return to the parent of T_i if this is not the case.

In addition to A, we use two further simple data structures: The subgraph set P_i at a node T_i is implemented as stack Q that is modified in the course of the algorithm with the top element of the stack being p_i. Also, for each node T_i, we create a list L_i of its exclusive neighbors. This list is necessary to undo some later operations. We now describe how these data structures are maintained throughout the traversal of the enumeration tree.

Initialization. At the root T_1 of the enumeration tree, we initialize A as follows: add all neighbors of the start vertex $p_1 := v$ to A, set pointer $\pi(A, 1)$ to the last non-empty position in A. Hence, the initial extension set is represented by all vertices from the first vertex in A to the initial position of pointer $\pi(A, 1)$. These are precisely the vertices of the exclusive neighborhood of v. The stack Q consists of the vertex v and L_1 contains all neighbors of v.

Creation of New Children. As discussed above, a node T_i has a further child T_{i+1} if it points to an index containing some vertex x. We create child T_{i+1} as follows:

1. move the pointer $\pi(A, i)$ to the left,
2. check whether x is an exclusive neighbor of p_i, and remove x from A if this is the case, and
3. create the child T_{i+1} with $p_{i+1} = x$ and enter the recursive call for T_{i+1}.

We now specify how to *move the pointer $\pi(A, i)$ to the left* when it currently points to vertex x of color c_ℓ. For an example of the pointer movement, see Fig. 1. Note that if x is an exclusive neighbor of p_i, we have $i = \ell$. If x is contained in the first entry of A, then redirect $\pi(A, i)$ to null. Otherwise, decrease the position of $\pi(A, i)$ by one. If $\pi(A, i)$ now points to a position containing a vertex y of color c_j such that $\pi(A, j)$ also points to y, then move $\pi(A, i)$ to the position that $\pi(A, \ell - 1)$ points to. Observe that if $j = \ell - 1$ this means that the pointer does not move in the second step.

We now describe how the algorithm creates a child T_{i+1} of T_i after fixing $p_{i+1} := x$ as described above. If node T_{i+1} is an interesting leaf, that is, if $i = k - 1$, we output $P_{i+1} \cup \{x\}$ and return to node T_i. Otherwise, we add vertex x to the stack Q representing the subgraph set and create an initially empty list L_{i+1}. Then we update A so that it represents X_{i+1}. For each neighbor u of x, check if u has some color c_j. If this is not the case, then color u with color c_{i+1} and add u to L_{i+1}. Now store the vertices of L_{i+1} in the left-most non-empty entries of A. Finally, create the pointer $\pi(A, i + 1)$ and let it point to the last non-empty position in A. Observe that this procedure runs in $\mathcal{O}(\Delta)$ time.

Restoring the Parent. Finally, we describe how the algorithm returns to the parent T_{i-1} of a node T_i. Note that the case that T_i is an interesting leaf was already handled above, hence, assume that T_i is not an interesting leaf. When returning to T_{i-1}, first delete the last element of stack Q. Then, for each vertex in L_i, we remove its color c_i. Finally, remove pointer $\pi(A, i)$ from array A. Observe that this can be done in $\mathcal{O}(\Delta)$ time as well. Hence, the overall running time is $\mathcal{O}(\Delta)$ as claimed. Moreover, the size of stack Q is bounded by k, array A has a length of $\min(k\Delta, n)$, and the sum of the sizes of all lists L_i is at most $\min(k\Delta, n)$. Hence, *Simple* needs $\mathcal{O}(n + m)$ space. The proof of the correctness of the algorithm is deferred to a long version of the article.   □

With this running time bound we may now prove the claimed delay.

Theorem 1. *Enumerate with Simple solves CISE for any graph G where each connected component has order at least k and the maximum degree is Δ with delay $\mathcal{O}(k^2 \Delta)$ and space $\mathcal{O}(n + m)$.*

Proof. Enumerate chooses an arbitrary start vertex v. According to Lemma 2, after the deletion of vertex v, we can delete every vertex of each connected component with less than k vertices in $\mathcal{O}(k^2 \Delta)$ time. Thus it is sufficient to bound the time which is needed to output the next solution within *Simple*.

Consider a node T_i in the enumeration tree of one call of *Enumerate* with *Simple* and its associated sets P_i (the subgraph set of node T_i) and X_i (the extension set of node T_i). Every time we call *Simple* recursively, we add exactly one vertex to the subgraph set. Hence, we need at most k iterations to reach a leaf T_j. If T_j is interesting, that is, if we find a solution for CISE, then we have a delay of $\mathcal{O}(k\Delta)$. If T_j is boring, then according to Lemma 3 the pruning rule applies to each node T_ℓ on the path from T_j to T_i since no other subsequent child of node T_ℓ yields a path to an interesting leaf. Hence, we will return in altogether $\mathcal{O}(k\Delta)$ time to the parent T_{i-1} of node T_i. Now, we are in the same situation as above. Either the first path from node T_{i-1} to a leaf leads to a solution for CISE or the pruning rule applies and we return to the parent of T_{i-1}. The crucial difference is that the depth of node T_{i-1} in the enumeration tree is one less than the depth of node T_i. Since the depth of the enumeration tree is bounded by k, we can go up at most k times until we return from the root (which finishes this call to *Simple*). Each time, we either report a new solution in $\mathcal{O}(k\Delta)$ time or go up once more. Hence, the overall delay is $\mathcal{O}(k^2 \Delta)$. The space complexity follows from Lemma 3. □

We can use Lemma 4 also to bound the overall running time of the algorithm.

Proposition 1. *Enumerate with Simple has running time* $\mathcal{O}((e(\Delta - 1))^{k-1} \cdot (\Delta + k) \cdot n/k)$.

4   Polynomial Delay with Pivot

We now adapt *Pivot* of Komusiewicz and Sorge [7] to obtain polynomial delay and a better running time bound. In *Pivot*, in each enumeration tree node, the vertex set of the subgraph set is partitioned into two sets P and S. The set P contains those vertices whose neighbors may still be added to extend the subgraph set and set S contains the other vertices of this subgraph, that is, no neighbor of S may be added to the subgraph. Moreover, we have a set F containing further vertices that may not be added to the connected subgraph. In the original algorithm [7] each node in the enumeration tree has an *active* vertex of the set P whose neighbors will be added to the subgraph. After adding each possible neighbor, the vertex becomes *inactive* and is added to set S. This version of the algorithm has a running time of $\mathcal{O}(4^k(\Delta - 1)^k n(n + m))$ [7] and no polynomial delay.

We improve this algorithm such that the number of enumeration tree nodes will be worst-case optimal and the algorithm has polynomial delay. The pseudo code of *Pivot* with improved running time and with pruning rule can be found in Algorithm 3. Consider a path $T_1, \ldots, T_i$ from the root T_1 to a node T_i of the enumeration tree. We will not associate enumeration tree nodes with active vertices. Instead, with each node T_i we associate P_i which is the subset of the subgraph set which can have further neighbors, S_i which is the remaining subgraph set, and F_i which is the set of forbidden vertices. Hence, we are using a Line 5 instead of creating a new child for each new active vertex. Now we do the following until P_i is empty: Pick an arbitrary $p \in P_i$. Next, for each neighbor v of p that is not in $P_i \cup S_i \cup F_i$, create a child node T_{i+1} in which v is added to P_i. After recursively solving the subproblem of T_{i+1}, move v to F_i. Consequently, v is contained in F_i in all subsequent children of T_i. Finally, after creating a child for each neighbor of p, remove p from P_i and put it into S_i. With this simple improvement, the number of enumeration tree nodes is now exactly the number of connected subgraphs of order at most k.

Algorithm 3. The *Pivot* algorithm; the initial call is $Pivot(\{v\}, \emptyset, \emptyset)$.

1: **procedure** $Pivot(P, S, F)$
2:     **if** $|P \cup S| = k$ **then**
3:         **output** $P \cup S$
4:         **return**
5:     **while** $P \neq \emptyset$ **do**
6:         $p :=$ choose element of P
7:         **for each** $z \in N(p) \setminus (P \cup S \cup F)$ **do**
8:             $Pivot(P \cup \{z\}, S, F)$
9:             $F := F \cup \{z\}$
10:            **if** output of $Pivot(P \cup \{z\}, S, F)$ was empty **then**
11:                **return**   ▷ Stop recursion if no solution was found
12:         $P := P \setminus \{p\}$
13:         $S := S \cup \{p\}$
14:     **return**

Lemma 5. *For each connected induced subgraph $G[U]$ of order at most k containing v, there is exactly one node T of the enumeration tree created by $Pivot(\{v\}, \emptyset, \emptyset)$ such that $P_T \cup S_T = U$.*

To obtain polynomial delay we add in Lines 10 and 11 a similar pruning rule to *Pivot* as for *Simple*: After each recursive call of *Pivot* in node T_i we check whether the call of node T_{i+1} outputs at least one solution for CISE. If not, we return in node T_i to its parent T_{i-1} of the enumeration tree. These two lines were not part of the original algorithm.

Lemma 6. *Let T_i be a node in the enumeration tree in a call of Pivot. If the output of a recursive call of Pivot in node T_i is empty, then no subsequent recursive call of Pivot in node T_i yields a path to an interesting leaf.*

Next, we prove that with suitable data structures for maintaining the sets P, S, and F during the enumeration, we can quickly traverse the enumeration tree.

Lemma 7. *Pivot can be implemented in such a way that for every node T_i of the enumeration tree, we need $\mathcal{O}(k\Delta)$ time to either compute the next child T_{i+1} or to restore the parent T_{i-1} and that the overall space needed is $\mathcal{O}(n+m)$.*

Proof. To check in constant time whether a vertex belongs to P_i, S_i, or F_i at an enumeration tree node T_i, we color some vertices of the graph with colors c_F, c_P, and c_S. For a node T_i the set of c_F-colored vertices represents the forbidden vertices F_i, the set of c_P-colored vertices represents the set of vertices P_i which can have new neighbors, and the set of c_S-colored vertices represents the set of vertices S_i which have no new neighbors. At the root of the enumeration tree, no vertex has color c_F or c_S. Only the single vertex v in P has color c_P. Testing if a vertex has color c_P, c_S, or c_F can be done in constant time.

To represent the partition of the subgraph set of node T_i into P_i and S_i we use an array A of length k. The array A contains $i = |P_i \cup S_i|$ nonempty elements. In A, we first save all vertices of S_i. Then the vertices of P_i follow. Further, a pointer $\pi(A,i)$ points to the vertex p of P_i with minimal index in A. Vertex p is the vertex which was chosen in Line 6 of *Pivot* and the vertex one position to the right of p will be chosen next. Hence, in node T_i altogether $|P_i \cup S_i|$ many pointers (one for node T_i and one for each of its ancestors) point to positions of A. To represent the set of forbidden vertices, we use a list L_i for each node T_i. The union of all vertices in lists $L_1, \ldots, L_i$ represents the set F_i of forbidden vertices in node T_i. List L_i contains all vertices in $F_i \setminus F_{i-1}$. List L_i is used to restore F_{i-1} when we return from node T_i to its parent T_{i-1}.

Initialization. If we call *Pivot* with the chosen start vertex v we create the root T_1 of the enumeration tree. The first and only non-empty entry of A contains v, pointer $\pi(A,1)$ points to v, and list L_1 is empty. Now, we describe how to update these data structures in order to the next child T_{i+1}, or restore the parent T_{i-1} of any node T_i in $\mathcal{O}(k\Delta)$ time.

Determining the Next Child of T_i. Do the following while $\pi(A,i)$ points to a vertex p. Check in $\mathcal{O}(\Delta)$ time whether p has a neighbor u which has none of the colors c_P, c_S, or c_F. If yes, we have determined that by adding u to P_i we can create a new child T_{i+1} of node T_i. Otherwise, all neighbors of p have some color, and we remove color c_P from p, recolor p with c_S, and move pointer $\pi(A,i)$ one position to the right. If pointer $\pi(A,i)$ points to an empty entry of A, then P_i is empty and T_i contains no more children. Overall, we need $\mathcal{O}(k\Delta)$ time to determine the vertex to add for the next child T_{i+1} of T_i.

Creating a New Child. To create T_{i+1}, we update the data structure to represent the sets P_{i+1}, S_{i+1}, and F_{i+1}: We replace the empty entry with minimal index in A by vertex u, we color u with c_P, and create pointer $\pi(A, i+1)$ which points to the same vertex as $\pi(A, i)$. Further, we create the list L_{i+1}. This list is empty since $F_{i+1} = F_i$. Thus, the child can be created in constant time.

Restoring a Parent. Now, we prove that we can restore the parent T_{i-1} in $\mathcal{O}(k\Delta)$ time when we have determined that T_i has no further children: We need to restore the sets P_{i-1}, S_{i-1}, and F_{i-1} of the parent T_{i-1} of node T_i. All vertices in list L_i are forbidden vertices which were added in node T_i. In other words: $L_i = F_i \setminus F_{i-1}$. Removing color c_F from these vertices and deleting list L_i afterwards needs $\mathcal{O}(k\Delta)$ time, since the set P_i can have at most $k\Delta$ neighbors and hence, node T_i can have at most $k\Delta$ children. Next, we remove pointer $\pi(A, i)$ from A in constant time. Afterwards, we remove the last non-empty vertex x from A, add x to list L_{i-1}, and change the color of x to color c_F. To restore the coloring of P_{i-1} and S_{i-1} we use the position of pointer $\pi(A, i-1)$. More precisely, all vertices from $\pi(A, i-1)$ to the last non-empty entry of A get color c_P, and all other vertices of A get color c_S. Overall, we need $\mathcal{O}(k\Delta)$ time for this step.

As shown above, the algorithm has the claimed running time. Moreover, array A has length k and the sum of the list sizes is $\min(k\Delta, n)$. Hence, the algorithm needs $\mathcal{O}(n + m)$ space.   □

Together with the pruning rule, the above gives a delay of $\mathcal{O}(k^3\Delta)$.

Proposition 2. *Pivot can be implemented in such a way that Enumerate with Pivot solves* CISE *for any graph G where each connected component has order at least k and the maximum degree is Δ with delay $\mathcal{O}(k^3\Delta)$ and space $\mathcal{O}(n+m)$.*

Next, we will improve the delay to $\mathcal{O}(k^2\Delta)$. The bottleneck in the delay provided by Proposition 2 is that when we have a node T_i that does not lead to an interesting leaf, we may have to go up $\Theta(k)$ levels before reaching a node that leads to an interesting leaf, each time needing $\Theta(k^2\Delta)$ time to check if the current node T_i leads to an interesting leaf. We will do the following: Before generating child T_{i+1} of node T_i, we invest $\mathcal{O}(k\Delta)$ time to check if the next child T'_{i+1} of T_i yields a path to an interesting leaf. This will be done by coloring at most $k - i$ vertices with a new color c_t. If and only if $k - i$ vertices received color c_t the next child T'_{i+1} yields a path to an interesting leaf. Afterwards, color c_t will be removed from each vertex to use color c_t for the next node in the enumeration tree. With this we can prove the following delay bound.

Theorem 2. *Pivot can be implemented in such a way that Enumerate with Pivot solves* CISE *for any graph G where each connected component has order at least k and the maximum degree is Δ with delay $\mathcal{O}(k^2\Delta)$ and space $\mathcal{O}(n + m)$.*

Finally, we can prove a better running time bound for *Pivot*.

Proposition 3. *Enumerate with Pivot has running time $\mathcal{O}((e(\Delta-1))^{k-1} \cdot \Delta \cdot n)$.*

5   An Experimental Comparison

We implemented *Simple*, *Pivot*, *Exgen*, and *Kavosh* with and without the pruning rules. Note that adding the pruning rule to *Exgen* and *Kavosh* does not make them polynomial delay algorithms. We also implemented *BDDE* [11], the Reverse Search with dictionary (*RwD Old*), and the Reverse Search with predecessor (*RwP Old*) algorithm [4]. For reverse search-based algorithms we also implemented another method to determine neighbors in the supergraph (*RwD New* and *RwP New*).

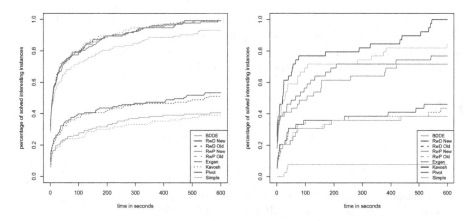

Fig. 2. Comparison for $k \in \{3, \ldots, 10\}$ (left) and $k \in \{n_c - 1, n_c - 2, n_c - 3\}$ (right) on interesting instances.

Each experiment was performed on a single thread of an Intel(R) Xeon(R) Silver 4116 CPU with 2.1 GHz, 24 CPUs and 128 GB RAM running Python 2.7.14 with *igraph* (http://igraph.org/python/) as the general graph data structure and *NetworkX* (https://networkx.github.io/) as the data structure for maintaining the enumeration tree in *BDDE*. [1] As benchmark data set we used 30 sparse social, biological, and technical networks obtained from the Network Repository [12], KONECT [10], and the 10th DIMACS challenge [2] and 20 random graphs generated in the $G_{n,p}$ model with $n \in \{100, 200, \ldots, 1000\}$ and $p \in \{0.1, 0.2\}$. The real-world networks range from very small (up to 500 vertices) to very large networks (up to 500 000 vertices).

Each algorithm was run on each instance with a time limit of 600 s. An instance is *interesting* if at least one of the 14 algorithms solved it within the time limit. For *Simple*, *Pivot*, *Exgen*, and *Kavosh* only the variant with the pruning rule is plotted in Fig. 2 since these variants were the fastest. Figure 2 shows the result for $k \in \{3, \ldots, 10\}$. Both versions of *RwD* and *RwP* only solve half as many instances as the other algorithms. All instances solved by *RwD* were solved by the remaining algorithms in 20 s. *Simple* is a factor 2 slower than *Pivot*, *BDDE*, *Exgen*, and *Kavosh*; *Kavosh* is slightly faster than *Pivot*, *BDDE*, and *Exgen*. Hence, for small k, one should use *Kavosh*.

Figure 2 shows the result for $k \in \{n_c - 1, n_c - 2, n_c - 3\}$ where n_c is the order of the largest connected component in the graph. Since *BDDE* stores the enumeration tree, it produced many memory errors and solved only the smallest instances. All instances solved by *RwD* or *RwP* were solved by *Pivot*, *Simple*, *Exgen*, and *Kavosh* with pruning rules in less than 100 s. The versions of the algorithms without the pruning rule only solved the same number of instances as *BDDE*. Hence, adding these pruning rules was necessary to solve CISE for

---

[1] The source code of our program *Enucon* is available at www.uni-marburg.de/fb12/ arbeitsgruppen/algorithmik/software/.

large k. Again, *Kavosh* is the fastest algorithm, despite the fact that adding the pruning rule to *Kavosh* does not yield polynomial delay. Hence, for large k, also *Kavosh* should be used. It seems that *Pivot* is slower for large k because it may spend $\Theta(k\Delta)$ time before creating the next child.

References

1. Avis, D., Fukuda, K.: Reverse search for enumeration. Discret. Appl. Math. **65**(1–3), 21–46 (1996)
2. Bader, D.A., Meyerhenke, H., Sanders, P., Schulz, C., Kappes, A., Wagner, D.: Benchmarking for graph clustering and partitioning. In: Alhajj, R., Rokne, J. (eds.) Encyclopedia of Social Network Analysis and Mining, pp. 73–82. Springer, New York (2014). https://doi.org/10.1007/978-1-4614-6170-8_23
3. Bollobás, B.: The Art of Mathematics - Coffee Time in Memphis. Cambridge University Press, Cambridge (2006)
4. Elbassioni, K.M.: A polynomial delay algorithm for generating connected induced subgraphs of a given cardinality. J. Graph Algorithms Appl. **19**(1), 273–280 (2015)
5. Elbassuoni, S., Blanco, R.: Keyword search over RDF graphs. In: Proceedings of the 20th ACM Conference on Information and Knowledge Management, (CIKM 2011), pp. 237–242. ACM (2011)
6. Kashani, Z.R.M., et al.: Kavosh: a new algorithm for finding network motifs. BMC Bioinform. **10**, 318 (2009)
7. Komusiewicz, C., Sorge, M.: Finding dense subgraphs of sparse graphs. In: Thilikos, D.M., Woeginger, G.J. (eds.) IPEC 2012. LNCS, vol. 7535, pp. 242–251. Springer, Heidelberg (2012). https://doi.org/10.1007/978-3-642-33293-7_23
8. Komusiewicz, C., Sorge, M.: An algorithmic framework for fixed-cardinality optimization in sparse graphs applied to dense subgraph problems. Discret. Appl. Math. **193**, 145–161 (2015)
9. Komusiewicz, C., Sorge, M., Stahl, K.: Finding connected subgraphs of fixed minimum density: implementation and experiments. In: Bampis, E. (ed.) SEA 2015. LNCS, vol. 9125, pp. 82–93. Springer, Cham (2015). https://doi.org/10.1007/978-3-319-20086-6_7
10. Kunegis, J.: KONECT: the Koblenz network collection. In: Proceedings of the 22nd International World Wide Web Conference (WWW 2013), pp. 1343–1350. International World Wide Web Conferences Steering Committee/ACM (2013)
11. Maxwell, S., Chance, M.R., Koyutürk, M.: Efficiently enumerating all connected induced subgraphs of a large molecular network. In: Dediu, A.-H., Martín-Vide, C., Truthe, B. (eds.) AlCoB 2014. LNCS, vol. 8542, pp. 171–182. Springer, Cham (2014). https://doi.org/10.1007/978-3-319-07953-0_14
12. Rossi, R.A., Ahmed, N.K.: The network data repository with interactive graph analytics and visualization. In: Proceedings of the 29th AAAI Conference on Artificial Intelligence (AAAI 2015), pp. 4292–4293. AAAI Press (2015). http://networkrepository.com
13. Wernicke, S.: A faster algorithm for detecting network motifs. In: Casadio, R., Myers, G. (eds.) WABI 2005. LNCS, vol. 3692, pp. 165–177. Springer, Heidelberg (2005). https://doi.org/10.1007/11557067_14
14. Wernicke, S.: Combinatorial algorithms to cope with the complexity of biological networks. Ph.D. thesis, Friedrich Schiller University of Jena (2006). http://d-nb.info/982598882

Multi-stranded String Assembling Systems

Martin Kutrib$^{(\boxtimes)}$ and Matthias Wendlandt

Institut für Informatik, Universität Giessen, Arndtstr. 2, 35392 Giessen, Germany
{kutrib,matthias.wendlandt}@informatik.uni-giessen.de

Abstract. Classical string assembling systems form computational models that generate strings from copies out of a finite set of assembly units. The underlying mechanism is based on piecewise assembly of a double-stranded sequence of symbols, where the upper and lower strand have to match. The generative power of such systems is driven by the power of the matching of the two strands. Here we generalize this approach to multi-stranded systems. The generative capacities and the relative power are our main interest. In particular, we consider briefly one-stranded systems and obtain that they describe a subregular language family. Then we explore the relations with one-way multi-head finite automata and show an infinite, dense, and strict strand hierarchy. Moreover, we consider the relations with the linguistic language families of the Chomsky Hierarchy and consider the unary variants of k-stranded string assembling systems.

1 Introduction

The Post Correspondence Problem is one of the oldest problems for which undecidability has been shown. The comprehensible structure and the multifarious appliance makes it a very useful tool for showing undecidability. The common view on the PCP is that it works somehow like the intersection of two generating mechanisms. A different view would be, to see the pairs as two strands or dominoes

$$\binom{ll}{el}_1, \binom{He}{H}_2, \binom{o}{lo}_3,$$

that can be connected to resulting double-strands

$$\Rightarrow^{2,1,3} \binom{Hello}{Hello}$$

and describe a language in this way.

Nowadays a bunch of publications examine the usability of DNA for computations. DNA consists of double-strands as well. Maybe the basic idea of using DNA sequences for computations goes back to Bennet and Landauer in the 1980's [2] where they describe how DNA could be used to realize a variant of Turing machines, called Brownian Turing machines. But the initial spark for the

© Springer Nature Switzerland AG 2019
B. Catania et al. (Eds.): SOFSEM 2019, LNCS 11376, pp. 285–297, 2019.
https://doi.org/10.1007/978-3-030-10801-4_23

intensive and wide-spreaded investigation of DNA computing was the experiment by Adleman [1] in 1994 where he solves a (small) instance of the Hamilton path problem in the laboratory by using DNA sequences. Although Hartmanis [6] argues that one needs more DNA than the weight of the Earth to solve an instance with a graph of 200 nodes by this technique, the euphoria of investigating DNA based computing is unbroken.

A basic approach to model computations with DNA strands is the generative system of sticker systems. They were introduced in [7] in their basic one-way variant. A sticker system basically consists of dominoes that can be seen as double-stranded molecules, where both strands have to overlap each other. These dominoes are concatenated to form double-strands of arbitrary length. However, due to the overlap the length difference of both strands is bounded by a constant that depends on the set of the dominoes. Sticker systems with such bounded delay are at most as powerful as one-turn pushdown automata which are known to accept linear context-free languages. Different variants of sticker systems have been investigated in [4, 7, 14, 15].

An inspiring generalization is suggested by the Post Correspondence Problem, where the two strings provided need not to overlap. This generalization harbors an enormous increase of generative capacity. So, assembly units are seen as pairs of (not necessarily overlapping) substrings that can be connected to the upper and lower string generated so far synchronously. String assembling systems have two further control mechanisms. First, it is required that the first symbol of a substring that should be connected has to be the same as the last symbol of the strand to which it is connected. One can imagine that both symbols are glued together one at the top of the other and, thus, just one appears in the final string. Second, as for the notion of strictly locally testable languages [12, 18] we distinguish between assembly units that may appear at the beginning, during, and at the end of the assembling process.

String assembling systems have been introduced in [9], where connections to one-way two-head finite automata are shown. In particular they are less powerful than the non-restricted variant of one-way two-head finite automata and they are more powerful than stateless one-way two-head finite automata. On the other hand, it has been shown in [3] that adding nonterminals to the generation increases the capacity to that of one-way two-head finite automata. Moreover, the generative capacity of bidirectional string assembling systems is incomparable with the power of such automata [10]. The impact of the model-inherent control mechanisms is studied in [11].

The computational power of one-way k-head automata increases with each head added to the system [17]. This suggests naturally to examine multi-stranded string assembling systems and to investigate whether a similar hierarchy can be achieved. Here we investigate these generalizations. The generative capacities and the relative power are our main interest. In particular, we consider briefly one-stranded systems and obtain that they describe a subregular language family. Then we explore the relations with one-way multi-head finite automata and show an infinite, dense, and strict strand hierarchy. Moreover, we consider the relations with the linguistic language families of the Chomsky Hierarchy and consider the unary variants of k-stranded string assembling systems.

2   Preliminaries and Definitions

We write Σ^* for the set of all words (strings) over the finite alphabet Σ. The empty word is denoted by λ, and $\Sigma^+ = \Sigma^* \setminus \{\lambda\}$. The reversal of a word w is denoted by w^R and for the length of w we write $|w|$. For the number of occurrences of a symbol a in w we use the notation $|w|_a$. Generally, for a singleton set $\{a\}$ we also simply write a. We use $\subseteq$ for inclusions and $\subset$ for strict inclusions. In order to avoid technical overloading in writing, two languages L and L' are considered to be equal, if they differ at most by the empty word, that is, $L \setminus \{\lambda\} = L' \setminus \{\lambda\}$.

We are especially interested in how string assembling systems can be used to describe languages. To this end, we consider arbitrary alphabets and do not restrict on the natural alphabet $\{A, G, C, T\}$. Clearly, there are ways to encode an arbitrary alphabet in the natural alphabet.

Before we continue with the definition of the systems to be studied, we clarify our notation of k-stranded words. Given some alphabet Σ, a k-stranded word over Σ is a k-tuple $(w_1, w_2, \ldots, w_k)$, where $w_i \in \Sigma^+$, $1 \leq i \leq k$, and w_i is a prefix of w_j or vice versa for all $1 \leq i < j \leq k$. A k-stranded word is said to be complete if $w_1 = w_2 = \cdots = w_k$. The set of all k-stranded words over Σ is denoted by D_Σ^k. Note that all strands of multi-stranded words are nonempty.

A k-stranded string assembling system generates k-stranded words by assembling units. Each unit consists of k substrings that are connected to the k strings generated so far. The corresponding symbols of all strands have to be equal. Moreover, a unit can only be assembled when the first symbols of its substrings match the last symbols of the current strands. In this case the matching symbols are glued together one at the top of the other. The generation has to begin with a unit from the set of initial units. Then it may continue with units from a different set. When a unit from a third set of ending units is applied the process necessarily stops. The generation is said to be valid if and only if all strands are identical when the process stops. More precisely:

Let $k \geq 1$. A k-stranded string assembling system (SAS(k)) is a quadruple $S = \langle \Sigma, A, T, E \rangle$, where Σ is the finite, nonempty set of symbols or letters, $A \subset D_\Sigma^k$ is the finite set of axioms, $T \subset (\Sigma^+)^k$ is the finite set of assembly units, and $E \subset (\Sigma^+)^k$ is the finite set of ending units.

The derivation relation $\Rightarrow$ is defined on $D_\Sigma^k \times D_\Sigma^k$ by

$$(u_1 x_1, u_2 x_2, \ldots, u_k x_k) \Rightarrow (u_1 x_1 v_1, u_2 x_2 v_2, \ldots, u_k x_k v_k)$$

if $(x_1 v_1, x_2 v_2, \ldots, x_k v_k) \in T \cup E$, for $u_i, v_i \in \Sigma^*$, $x_i \in \Sigma$, $1 \leq i \leq k$.

A derivation is said to be successful if it initially starts with an axiom from A, continues with assembling units from T, and ends with assembling an ending unit from E. The process necessarily stops when an ending assembly unit is added. The sets A, T, and E are not necessarily disjoint.

The *language $L(S)$ generated* by S is defined to be the set

$$L(S) = \{\, w \in \Sigma^+ \mid (u_1, u_2, \ldots, u_k) \Rightarrow^* (w, w, \ldots, w) \text{ is a successful derivation} \,\},$$

where $\Rightarrow^*$ refers to the reflexive, transitive closure of the derivation relation $\Rightarrow$.

It is known that the concatenation of the languages $L = \{\, a^n b^n \mid n \geq 1 \,\}$ and $L' = \{\, a^m \mid m \geq 1 \,\}$ cannot be generated by any double-stranded SAS [9]. The following example shows that the concatenation LL' is generated by a 3-stranded SAS.

Example 1. The language $L = \{\, a^n b^n a^m \mid m, n \geq 1 \,\}$ is generated by the SAS(3)$S = \langle \Sigma, A, T, E \rangle$ with the single axiom $A = \{(a, a, a)\}$, the ending units

$$E = \{(aa, aa, a), (aaa, aaa, aa), (a, ba, ba), (a, baa, baa), (ba, aba, ba)\}$$

and the assembling units in T being

1. (aa, a, aa)	5. (b, bb, b)	9. (aaa, aaa, b)
2. (ab, a, ab)	6. (ba, b, b)	10. (a, a, baa)
3. (bb, aa, b)	7. (a, b, bb)	11. (a, a, aaa).
4. (b, ab, b)	8. (a, ba, b)	

The derivation starts with the sole axiom. Then the first a-block is generated in the first and the third strand by applying repeatedly unit 1 and unit 2 once. In a valid derivation it is not possible to apply unit 11 in this phase, since otherwise the number of a's in the first and the third strand will never match again. After applying unit 2, the current 3-stranded word is of the form $(a^n b, a, a^n b)$. Now units 3 and 4 become applicable. The repeated application of unit 3 followed by one application of unit 4 yields a 3-stranded word of the form $(a^n b^n, a^n b, a^n b)$. So, there are as many a's in the first block as b's in the second block of the first strand. After applying unit 4, units 5 and 6 from T become applicable. A repeated application of unit 5 completes the b-block of the second strand. After subsequently applying unit 6 once, a 3-stranded word of the form $(a^n b^n a, a^n b^n, a^n b)$ is derived. Note, if unit 5 is applied too many times, unit 6 and, thus, a successful derivation becomes impossible. Now units 7 and 8 are applicable. The repeated application of unit 7 followed by one application of unit 8 yields a 3-stranded word of the form $(a^n b^n a, a^n b^n a, a^n b^n)$. Next, the third a-block is generated. In order to avoid that it is followed by another b-block, its parity is utilized. So, applying unit 9 repeatedly and unit 10 once gives a 3-stranded word of the form $(a^n b^n a^i, a^n b^n a^i, a^n b^n aa)$, where i is an odd number. Subsequent applications of unit 11 fill the a-block of the third strand but let its length be even. In this way further applications of units 1 or 2 fail. Finally, the derivation is made successful by selecting ending unit (aa, aa, a) or (aaa, aaa, aa). The further ending units are used to derive short blocks. ∎

3   Single-Stranded SAS

As discussed before, the strong connection with the Post Correspondence Problem reflects the power of double-stranded modeling systems, since it goes along with the undecidability of emptiness. Here we first drop the power of the double strands and investigate 1-stranded SAS.

The next theorem shows that the restriction to one strand reduces the power significantly.

Theorem 2. *The family of languages generated by 1-stranded SAS is strictly included in the family of regular languages.*

Proof. Given a single-stranded SAS $S = \langle \Sigma, A, T, E \rangle$ we construct a nondeterministic finite automaton $M = \langle Q, \Sigma, \delta, q_0, F \rangle$ such that $L(S) = L(M)$ (see Example 3).

Since S is single stranded, the units are basically words from Σ^+. Since units in T of length one do not have any effect on SAS(1), they can safely be omitted. So, we assume that each unit in T is at least of length two.

For each axiom $u = x_1 x_2 \cdots x_\ell \in A$ we define the states $\{u_0, u_1, \ldots, u_\ell\}$ and the transitions $\delta(u_{j-1}, x_j) = u_j$, for $1 \leq j \leq \ell$. Similarly, for each unit $u = x_1 x_2 \cdots x_\ell \in T \cup E$ we define the states $\{u_1, u_2, \ldots, u_\ell\}$ and the transitions $\delta(u_{j-1}, x_j) = u_j$, for $2 \leq j \leq \ell$.

So far, we have a different chain of connected states for each unit of S. The chains are interconnected as follows. Let q_0 be a new state that is set to be the initial state. By the λ-transitions $\delta(q_0, \lambda) = u_0$, for all $u \in A$, the NFA guesses an axiom of S. The last state of each chain associated with units $u = x_1 x_2 \cdots x_\ell$ from A or T is connected to a next unit $v = y_1 y_2 \cdots y_m$ from T or E by λ-transitions $\delta(u_\ell, \lambda) = v_1$ if the last symbol of the first chain matches the first symbol of the second chain, that is, if $x_\ell = y_1$. In this way the assembling of units is simulated. Finally, the last state of each chain associated with an ending unit is made accepting, that is, $F = \{ q \mid q = v_m \text{ for some } v = y_1 y_2 \cdots y_m \in E \}$. So, there are no transitions from states at the end of chains associated with ending units.

The NFA constructed simulates derivations of S and vice versa, that is, for each derivation of a word $w \in L(S)$ that applies the units $u_0, u_1, \ldots, u_n$, there is an accepting path in M that runs through q_0 and the chains associated with $u_0, u_1, \ldots, u_n$, and vice versa. This shows that every language generated by some SAS(1) is regular.

The strictness of the inclusion claimed follows since the regular language $L = \{a\} \cup \{ a^{2n} \mid n \geq 2 \}$ is not generated by any even double-stranded SAS [9]. □

Example 3. Given the single-stranded SAS $S = \langle \Sigma, A, T, E \rangle$ with $u^{(1)} = a$, $u^{(2)} = aa$, $u^{(3)} = aab$, $u^{(4)} = bb$, $u^{(5)} = bb$, and $A = \{u^{(1)}\}$, $T = \{u^{(2)}, u^{(3)}, u^{(4)}\}$, and $E = \{u^{(5)}\}$, the construction in the proof of Theorem 2 gives the NFA shown in Fig. 1. ■

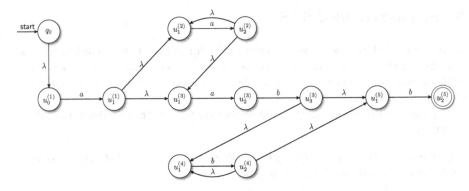

Fig. 1. Example of an NFA constructed from a single-stranded SAS generating the language aaa^*bbb^*.

Next we turn to two observations that are in the spirit of pumping lemmas. They are useful to show that languages are not generated by single-stranded SAS.

Lemma 4. 1. Let $L \subseteq \Sigma^*$ be a language whose words have infinitely many unary factors, that is, for some $a \in \Sigma$, the set

$$\{ a^n \mid n \geq 1 \text{ and there are } u, v \in \Sigma^* \text{ such that } ua^nv \in L \}$$

is infinite. If L is generated by a single-stranded SAS then there is a unit $(a^i) \in T$, for some $i \geq 2$.
2. Let $L \subseteq \Sigma^*$ be a language such that for some $a, b \in \Sigma$ and all $n \geq 1$ the set

$$\{ a^ib^j \mid i,j \geq n \text{ and there are } u, v \in \Sigma^* \text{ such that } ua^ib^jv \in L \}$$

is infinite. If L is generated by a single-stranded SAS then there is a unit $(a^ib^j) \in T$, for some $i, j \geq 1$.

Example 5. The language $L = \{ a^\ell b^m a^n \mid \ell, m, n \geq 1 \}$ is not generated by any single-stranded SAS. Contrarily assume that L is generated by a single-stranded SAS $S = \langle \Sigma, A, T, E \rangle$. Then Lemma 4 says that there are assembling units (a^n), $n \geq 2$, $(b^{n'})$, $n' \geq 2$, (a^ib^j), $i,j \geq 1$, and $(b^{i'}a^{j'})$, $i',j' \geq 1$, in T. Moreover, there must be an axiom (a^r), $r \geq 1$, in A and an ending unit $(a^{r'})$, $r' \geq 1$, in E. Therefore, the word $a^{r+i-1}b^jb^{i'-1}a^{j'}a^{i-1}b^jb^{i'-1}a^{j'+r'-1}$ can be derived as well, a contradiction. ∎

It has been shown that the copy language $\{ \$_1w\$_2w\$_3 \mid w \in \{a,b\}^+ \}$, which is not even context free, is generated by some double-stranded SAS [9]. This result together with Theorem 2 implies that the generative capacity of SAS(k), $k \geq 2$, is strictly larger than that of SAS(1).

4   String Assembling Systems with Multiple Strands

Now we turn to investigate the general case of k-stranded SAS. First we derive an upper bound for the generative capacity. The upper bound is given by the recognition power of nondeterministic one-way k-head finite automata. Their two-way variants characterize the complexity class $\mathsf{NL} = \mathsf{NSPACE}(\log n)$ [5]. So, together with the next theorem, we obtain that the family of languages generated by $SAS(k)$, $k \geq 1$, is properly included in NL.

Theorem 6. *Let $k \geq 1$ and S be a k-stranded SAS. Then there exists a nondeterministic one-way k-head finite automaton M that accepts $L(S)$.*

Proof. The main idea of the construction of M is that each of the k heads is used to check one strand. Initially, M guesses one of the axioms. The guess is verified by reading the k strands of the unit with the k input heads. In this way, it is checked whether the input given fits to the unit guessed. The last symbols are the new overlappings and the heads stay on it for the new guess. Next, M guesses dependent on the currently scanned input symbols (the current overlappings) which assembly unit comes next. Then the guess is verified. After each verification, M determines whether the assembling process is completed and guesses another unit to be assembled otherwise. If an ending unit is guessed then after its verification M moves all heads synchronously one step to the right. If all heads see the right endmarker, M accepts.   □

Theorem 6 allows an immediate generalization of results for double-stranded SAS from [9] to k-stranded SAS. It is known that nondeterministic one-way k-head finite automata cannot accept the deterministic and linear context-free language $\{\, wcw^R \mid w \in \{a,b\}^* \,\}$ [8]. Moreover, as mentioned above, the non-context-free language $\{\, \$_1w\$_2w\$_3 \mid w \in \{a,b\}^+ \,\}$ is generated by some $SAS(2)$. Since NL is strictly included in $\mathsf{NSPACE}(n)$ (see, for example, [13]), which in turn is equal to the family of context-sensitive languages, we have the following relations.

Corollary 7. *For $k \geq 1$, the family of languages generated by $SAS(k)$ is strictly included in NL and, thus, in the family of context-sensitive languages.*

For $k \geq 2$, the family of languages generated by $SAS(k)$ is incomparable with the family of (deterministic) (linear) context-free languages.

Next, we turn to an infinite, dense, and strict strand hierarchy. The question whether there is a proper head hierarchy for one-way multi-head finite automata has been raised in [16]. It has been answered in the affirmative in [17]. The basic witness languages are

$$\{\, w_1 * w_2 * \cdots * w_{2n} \mid w_i \in \{a,b\}^*, w_i = w_{2n+1-i}, 1 \leq i \leq 2n \,\}$$

which can be accepted by a (nondeterministic) one-way k-head finite automaton if and only if $n \leq \binom{k}{2}$. Here we use a slight modification of these languages. Let

$h\colon \{a_i, b_i \mid 1 \leq i \leq 2n\}^* \to \{a, b\}^*$ be a homomorphism with $h(a_i) = a$ and $h(b_i) = b$, for $1 \leq i \leq 2n$. Then we define

$$L_n = \{ \$_0 w_1 \$_1 w_2 \$_2 \cdots \$_{2n-1} w_{2n} \$_{2n} \mid$$
$$w_i \in \{a_i, b_i\}^+, h(w_i) = h(w_{2n+1-i}), 1 \leq i \leq 2n \}.$$

Clearly the head hierarchy for one-way multi-head finite automata is witnessed by the languages L_n as well, since the transduction from the former to L_n can be done by a sequential transducer, that can be simulated by a multi-head finite automaton in parallel. So, a strand hierarchy for SAS would follow if there were a k-stranded SAS generating the language L_n, $n = \binom{k}{2}$. The construction is shown in the proof below. Afterwards the construction is applied to an example to illustrate the details.

Theorem 8. *Let $k \geq 2$ be an integer. The family of languages generated by $SAS(k)$ is strictly included in the family of languages generated by $SAS(k+1)$.*

Proof. It can be concluded from Theorem 2 that the family of languages generated by 1-stranded SAS is a proper subfamily of the family of languages generated by 2-stranded SAS. As mentioned before, for $k \geq 2$ it is sufficient to show that language L_n with $n = \binom{k}{2}$ is generated by some $SAS(k)$. To this end, we now construct the $SAS(k)$ $S = \langle \Sigma, A, T, E \rangle$, where $A = \{(\$_0, \$_0, \ldots, \$_0)\}$ and $E = \{(\$_{2n}, \$_{2n}, \ldots, \$_{2n})\}$. Example 9 illustrates the construction.

Example 1 shows how words between delimiters can be copied to another position of another strand. This is the underlying technique of the construction. However, whenever this technique is used to copy a subword of L_n from one strand to another or to compare two subwords on different strands, clearly, due to the positions of matching subwords these two strands cannot be used for any further pair of matching subwords. The idea of the construction is as follows. Each pair of different strands is used to copy one subword to its matching position. In this way, $\binom{k}{2}$, that is, the required n matching subwords can be generated. The initial phase of the generation is slightly different.

Initially, the first strand is constructed up to block $2n - (k-1)$. In this process the content of the blocks is nondeterministically guessed. For the rest of the proof let $x, y \in \{a, b\}$. For $1 \leq i \leq 2n - (k-1)$ we define the units

1. $(\$_{i-1} x_i, \$_0, \ldots, \$_0)$
2. $(x_i y_i, \$_0, \ldots, \$_0)$
3. $(x_i \$_i, \$_0, \ldots, \$_0)$.

In particular, these units ensure that the format of the word generated is correct up to block $2n - (k-1)$.

Next, the remaining $k-1$ blocks are copied from their matching positions. More precisely, block $i-1$ of strand i is copied to position $2n+2-i$ of strand 1. Here we define the corresponding units. However, in order to apply these units, first the leftmost blocks of strand i have to be generated. Whenever an block of strand i is not copied but has to be completed, units defined below are used.

In order to distinguish between blocks that are copied or just completed, for any strand i, a set ω_i is maintained that, in the end, contains all indices of words that are copied from or to the strand. For $2 \leq i \leq k$ and $z_j \in \Sigma$, $j \in \{1, 2, \ldots, k\} \setminus \{1, i\}$, we define the units

4. $(z_1, z_2, \ldots, z_k)$ with $z_1 = \$_{2n+1-i} x_{2n+2-i}$ and $z_i = \$_{i-2} x_{i-1}$,
5. $(z_1, z_2, \ldots, z_k)$ with $z_1 = x_{2n+2-i} y_{2n+2-i}$ and $z_i = x_{i-1} y_{i-1}$,
6. $(z_1, z_2, \ldots, z_k)$ with $z_1 = x_{2n+2-i} \$_{2n+2-i}$ and $z_i = x_{i-1} \$_{i-1}$,

and define $i - 1 \in \omega_i$ and $2n + 2 - i \in \omega_1$.

Now the first strand is complete, its format is correct, and the first $k - 1$ subwords match their mates at the end of the strand. This completes the initial phase.

In the second phase the remaining pairs of subwords have to be compared. The overall generation may only end successfully if the matching is verified. The comparisons are done by trying to copy one subword to the position of its matching subword. Clearly, the copying can only be successful if the subwords (which are already fixed in the first strand) are equal. To this end, pairs of strands are used. The first strand has already been used together with each other strand to ensure that $k - 1$ subwords match their mates. Next strand 2 is used together with strands 3 to k to compare the subwords k up to $k - 1 + k - 2$ with their matching mates. Subsequently, strand 3 is used together with strands 4 to k to compare the subwords $k - 1 + k - 2 + 1$ up to $k - 1 + k - 2 + k - 3$ with their matching mates, and so on. Formally, for $2 \leq i < j \leq k$, strand i is is used together with strand j to compare the subword at position $\alpha = 2n + (i+1) - j - (k - 1 + k - 2 + \cdots + k - (i - 1))$ with the subword at position $\beta = j - i + (k - 1 + k - 2 + \cdots + k - (i - 1))$. For $2 \leq i < j \leq k$ and $z_\ell \in \Sigma$, $\ell \in \{1, 2, \ldots, k\} \setminus \{i, j\}$, we define the units

7. $(z_1, z_2, \ldots, z_k)$ with $z_i = \$_{\alpha-1} x_\alpha$ and $z_j = \$_{\beta-1} x_\beta$,
8. $(z_1, z_2, \ldots, z_k)$ with $z_i = x_\alpha y_\alpha$ and $z_j = x_\beta y_\beta$,
9. $(z_1, z_2, \ldots, z_k)$ with $z_i = x_\alpha \$_\alpha$ and $z_j = x_\beta \$_\beta$,

and define $\alpha \in \omega_i$ and $\beta \in \omega_j$.

As mentioned before, in order to apply these units, first the leftmost blocks of strands i and j have to be generated. To this end, the blocks are completed with arbitrary symbols unless they are used for comparisons in which case the block number appears in the associated set ω. Actually, completing with arbitrary symbols means to copy the block from the first strand since there is no other possibility to generate a word successfully. Since the first strand is already complete, no further units have to be provided for its blocks. So, for all $2 \leq j \leq k$, $i \in (\{1, 2, \ldots, 2n\} \setminus \omega_j)$, and $z_\ell \in \Sigma$, $\ell \neq j$ we define the units

10. $(z_1, z_2, \ldots, z_k)$ with $z_j = \$_{i-1} x_i$,
11. $(z_1, z_2, \ldots, z_k)$ with $z_j = x_i y_i$,
12. $(z_1, z_2, \ldots, z_k)$ with $z_j = x_i \$_i$.

This completes the construction of the SAS(k). $\square$

In order to clarify the details of the proof we present an example for language L_3 which is generated by a 3-stranded string assembling system.

Example 9. The language

$$L_3 = \{ \$_0 w_1 \$_1 w_2 \$_2 w_3 \$_3 w_4 \$_4 w_5 \$_5 w_6 \$_6 \mid$$
$$w_i \in \{a_i, b_i\}^*, h(w_i) = h(w_{7-i}), 1 \le i \le 6 \}$$

is generated by the 3-stranded string assembling system $S = \langle \Sigma, A, T, E \rangle$ with $A = \{(\$_0, \$_0, \$_0)\}$ and $E = \{(\$_6, \$_6, \$_6)\}$. Let $x, y \in \{a, b\}$. The first assembling units generate the first strand up to $\$_4$:

1. $(\$_0 x_1, \$_0, \$_0)$	5. $(x_2 y_2, \$_0, \$_0)$	9. $(x_3 \$_3, \$_0, \$_0)$
2. $(x_1 y_1, \$_0, \$_0)$	6. $(x_2 \$_2, \$_0, \$_0)$	10. $(\$_3 x_4, \$_0, \$_0)$
3. $(x_1 \$_1, \$_0, \$_0)$	7. $(\$_2 x_3, \$_0, \$_0)$	11. $(x_4 y_4, \$_0, \$_0)$
4. $(\$_1 x_2, \$_0, \$_0)$	8. $(x_3 y_3, \$_0, \$_0)$	12. $(x_4 \$_4, \$_0, \$_0)$.

The units 1–12 are used to derive a multi-strand of the form

$$(\$_0 w_1 \$_1 w_2 \$_2 w_3 \$_3 w_4 \$_4, \$_0, \$_0).$$

The following units are used to generate the subwords w_5 and w_6 of the first strand in parallel to the subwords w_2 and w_1 (after completing the first block of the third strand with the units 25–27).

13. $(\$_4 x_5, \$_0, \$_1 x_2)$	15. $(x_5 \$_5, \$_0, x_2 \$_2)$	17. $(x_6 y_6, x_1 y_1, \$_2)$
14. $(x_5 y_5, \$_0, x_2 y_2)$	16. $(\$_5 x_6, \$_0 x_1, \$_2)$	18. $(x_6 \$_6, x_1 \$_1, \$_2)$.

Units 13–15 are used to complete and copy w_2 on the third strand to w_5 on the first strand. Similarly, units 16–18 are used to complete and copy w_1 on the second strand to w_6 on the first strand. After completing the first block of the third strand with the units 25–27 and applying units 13–18, a multi-strand of the form $(\$_0 w_1 \$_1 w_2 \$_2 w_3 \$_3 w_4 \$_4 w_5 \$_5 w_6 \$_6, \$_0 w_1 \$_1, \$_0 w_1 \$_1 w_2 \$_2)$ is obtained. After this phase, the sets ω are $\omega_1 = \{5, 6\}$, $\omega_2 = \{1\}$ and $\omega_3 = \{2\}$.

With the next units subword w_4 on the second strand is completed and copied on the third strand to w_3 (after completing the missing blocks of the second strand with the units 22–24).

19. $(\$_6, \$_3 x_4, \$_2 x_3)$	20. $(\$_6, x_4 y_4, x_3 y_3)$	21. $(\$_6, x_4 \$_4, x_3 \$_3)$.

Afterwards the sets ω are $\omega_1 = \{5, 6\}$, $\omega_2 = \{1, 4\}$ and $\omega_3 = \{2, 3\}$.

The units that complete the strands remain to be defined. As described in the construction above no blocks of the first strand has to be completed. We define

22. $(z, \$_{i-1}x_i, z')$ 23. (z, x_iy_i, z') 24. $(z, x_i\$_i, z')$

for $i \in \{2, 3, 5, 6\}$ and $z, z' \in \Sigma$, since $\omega_2 = \{1, 4\}$. Furthermore, define

25. $(z, z', \$_{i-1}x_i)$ 26. (z, z', x_iy_i) 27. $(z, z', x_i\$_i)$

for $i \in \{1, 4, 5, 6\}$ and $z, z' \in \Sigma$, since $\omega_3 = \{2, 3\}$.   ∎

The last theorem established an infinite, dense, and tight strand hierarchy of SAS. The hierarchy is based partially on the simulations by one-way k-head finite automata. Further relations with language families have been given in Corollary 7. Still open are the relations of the families of languages generated by SAS(k) with the families of languages accepted by one-way k-head finite automata (is the inclusion proper?) and with the regular languages. The next results show that there is in fact a very simple, that is, unary regular language that cannot be generated by any SAS(k). For the proof we utilize the observation that the ordering of applied units in the generation of unary words does not matter. Any permutations of the units in T yield the same word. This observation is formalized in the next lemma.

Lemma 10. *Let $k \geq 1$ be an integer and $S = \langle \Sigma, A, T, E \rangle$ be an SAS(k). If some unary word x^n, for $x \in \Sigma$, is generated by S by using an axiom $u_a \in A$ and applying successively units $u_1, u_2, \ldots, u_m$ and an ending unit $u_e \in E$, then the same word x^n is generated by S by using u_a and applying successively units $u_{i_1}, u_{i_2}, \ldots, u_{i_m}$ and ending unit u_e, where $i_1, i_2, \ldots, i_m$ is an arbitrary permutation of $1, 2, \ldots, m$.*

Theorem 11. *For any $k \geq 1$, the regular language $L = \{a\} \cup \{a^{2n} \mid n \geq 2\}$ is not generated by any SAS(k).*

Proof. In contrast to the assertion assume that L is generated by some SAS(k) $S = \langle \Sigma, A, T, E \rangle$. Let $T = \{u_1, u_2, \ldots, u_n\}$.

Lemma 10 reveals that any successful generation of S can be represented by its axiom u_a, its ending unit u_e and the multiplicities of applications of units from T, that is, by the n numbers $i_1, i_2, \ldots, i_n$, where $i_j \geq 0$ gives the number of applications of unit u_j, $1 \leq j \leq n$.

Since L is infinite, there is at least one pair of axiom u_a and ending unit u_e such that there are infinitely many different words generated by S by using u_a and u_e. Let $L_{u_a, u_e} \subseteq L(S)$ be the language of these words.

Since L_{u_a, u_e} is infinite, there exist two different multiplicities of applications of units $i_1, i_2, \ldots, i_n$ and $j_1, j_2, \ldots, j_n$ that represent successful generations of two different words $v, w \in L_{u_a, u_e}$, where $4 \leq |v| < |w|$, $i_\ell \leq j_\ell$, for all $1 \leq \ell \leq n$. We conclude that the multiplicities $j_1 - i_1, j_2 - i_2, \ldots, j_n - i_n$ represent the successful generation of $a^{|w|-|v|} \in L_{u_a, u_e}$. Moreover, applying the units from T as often as given by $j_1 - i_1, j_2 - i_2, \ldots, j_n - i_n$ extends any of the k strands by the same number of symbols. Since $|v|$ and $|w|$ have to be even, $|w| - |v|$ is even.

For the extension by an even number of symbols, an odd number of symbols is needed since one symbol is lost by gluing the strands together.

Since $a \in L$ there is an axiom as well as an ending unit of the form $(a, a, \ldots, a)$ in S. Now we consider the word w' whose generation is described by this axiom and ending unit and the multiplicities $j_1 - i_1, j_2 - i_2, \ldots, j_n - i_n$. The generation is successful and $|w'| > 2$ is odd, a contradiction.   □

It has been shown that k-stranded SAS are able to generate non-context-free languages. On the other hand the unary regular language L cannot be generated by any k-stranded SAS.

Thus it can be concluded that the family of languages generated by k-stranded SAS is incomparable with the context-free languages as well as the regular languages.

Corollary 12. *Let $k \geq 1$ be an integer. (i) The family of languages generated by $SAS(k)$ is incomparable with the families of languages accepted by one-way k'-head finite automata if $k' < k$. (ii) The family of languages generated by $SAS(k)$ is strictly included in the families of languages accepted by one-way k'-head finite automata if $k' \geq k$. (iii) The family of languages generated by $SAS(k)$ is incomparable with the (unary) regular languages if $k \geq 2$.*

The results concerning the generative power of $SAS(k)$ are summarized in Fig. 2.

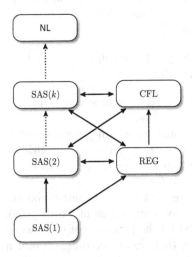

Fig. 2. Hierarchy of language families. A single arrow indicates strict inclusion, a double arrow incomparability, and a dotted line infinitely many hierarchy levels.

References

1. Adleman, L.M.: Molecular computation of solutions to combinatorial problems. Science **266**, 1021–1024 (1994)
2. Bennett, C.H., Landauer, R.: The fundamental physical limits of computation. Sci. Am. **253**, 48–56 (1985)
3. Bordihn, H., Kutrib, M., Wendlandt, M.: Nonterminal controlled string assembling systems. J. Autom. Lang. Comb. **19**, 33–44 (2014)
4. Freund, R., Păun, G., Rozenberg, G., Salomaa, A.: Bidirectional sticker systems. In: Pacific Symposium on Biocomputing (PSB 1998), pp. 535–546. World Scientific, Singapore (1998)
5. Hartmanis, J.: On non-determinancy in simple computing devices. Acta Inform. **1**, 336–344 (1972)
6. Hartmanis, J.: On the weight of computations. Bull. EATCS **55**, 136–138 (1995)
7. Kari, L., Păun, G., Rozenberg, G., Salomaa, A., Yu, S.: DNA computing, sticker systems, and universality. Acta Inform. **35**, 401–420 (1998)
8. Kutrib, M., Malcher, A., Wendlandt, M.: Set automata. Int. J. Found. Comput. Sci. **27**, 187–214 (2016)
9. Kutrib, M., Wendlandt, M.: String assembling systems. RAIRO Inform. Théor. **46**, 593–613 (2012)
10. Kutrib, M., Wendlandt, M.: Bidirectional string assembling systems. RAIRO Inform. Théor. **48**, 39–59 (2014)
11. Kutrib, M., Wendlandt, M.: Parametrizing string assembling systems. In: Câmpeanu, C. (ed.) CIAA 2018. LNCS, vol. 10977, pp. 236–247. Springer, Cham (2018). https://doi.org/10.1007/978-3-319-94812-6_20
12. McNaughton, R.: Algebraic decision procedures for local testability. Math. Syst. Theory **8**, 60–76 (1974)
13. Papadimitriou, C.H.: Computational Complexity. Addison-Wesley, Boston (1994)
14. Păun, G., Rozenberg, G.: Sticker systems. Theor. Comput. Sci. **204**, 183–203 (1998)
15. Păun, G., Rozenberg, G., Salomaa, A.: DNA Computing: New Computing Paradigms. Texts in Theoretical Computer Science. Springer, Heidelberg (1998)
16. Rosenberg, A.L.: On multi-head finite automata. IBM J. Res. Dev. **10**, 388–394 (1966)
17. Yao, A.C., Rivest, R.L.: $k+1$ heads are better than k. J. ACM **25**, 337–340 (1978)
18. Zalcstein, Y.: Locally testable languages. J. Comput. System Sci. **6**, 151–167 (1972)

Towards Automatic Comparison of Cloud Service Security Certifications

Martin Labaj[(✉)] ⓘ, Karol Rástočný ⓘ, and Daniela Chudá ⓘ

Institute of Informatics, Information Systems and Software Engineering,
Faculty of Informatics and Information Technologies,
Slovak University of Technology in Bratislava,
Ilkovičova 2, 842 16 Bratislava 4, Slovak Republic
{martin.labaj,karol.rastocny,daniela.chuda}@stuba.sk

Abstract. Cloud service providers who offer services to their users traditionally signal security of their offerings through certifications based on various certification schemes. Currently, a vast number of schemes and standards exists on one side (cloud service certifications), while another large set of security requirements stemming from internal needs or laws and regulations stand on the other side (users of cloud services). Determining whether a service with an arbitrary certificate in one country fulfills requirements imposed by the user in another country is a difficult task and therefore a project (EU-SEC) was started focusing on allowing cross-border usage of cloud services. In this paper, we propose automated comparison of cloud service security certification schemes and, subsequently, security of cloud services certified using these schemes. In the presented method, we map requirements in schemes, standards, laws, and regulations into a proposed cloud service security ontology. Due to the free-form text nature of these items, we also describe a supporting method for semi-automated conversion of free text into this ontology using natural language processing. The requirements described in ontology format are then easily compared against each other. We also describe an implementation of a prototype system supporting the conversion and comparison with preliminary results on describing and comparing two well-known schemes.

Keywords: Cloud service certification · Natural language processing Certification scheme ontology

1 The Problem Domain

Security in information technology is one the most important topics today and as applications, platforms, and infrastructures were moved to the cloud, the security of cloud services becomes as important. *Cloud Service Providers* (CSP) offer their services to customers and in just the same way as in any other line of business when selling a product, they need to describe their product appropriately. With quantitative technical properties, this is straightforward, one would advertise/describe that a computing service is offered with 10 processor cores equivalent to a given processor model or that a messaging service allows processing 100.000 events a month. Also, the customer can verify such physical attributes easily.

© Springer Nature Switzerland AG 2019
B. Catania et al. (Eds.): SOFSEM 2019, LNCS 11376, pp. 298–309, 2019.
https://doi.org/10.1007/978-3-030-10801-4_24

The situation is far more difficult in the case of security properties of the service. How does the provider describe that the service is secure? One might vaguely state that some steps are taken, e.g., to not use customer production data in development environments for testing. Then, how does the customer verify this? Traditionally, providers signal such attributes of their services by certifying themselves and their services under a certification scheme or standard through auditors and certifications bodies. However, there are various schemes, frameworks and standards: ISO/IEC 27000-series, CSA CCM, BSI C5 and others, used in different industries, different countries or for different customers. This creates a barrier for offering a procuring cloud services across borders (both state borders and market niches).

The cloud service provider would need to undergo multiple certifications with often repeated controls to capture a foreign market. At the same time, due diligence processes of cloud service users when obtaining a service are increasingly more complicated. When the prospective provider does not hold the "right" certificate, it may prevent using the service at all. This is not limited to certification schemes and certificates. Users of cloud services may have other requirements arising from internal needs or being imposed on them, e.g., by national laws and regulations, and those may not be mapped to known certification schemes used abroad and the task of determining a suitable provider matching those requirements becomes a mundane task of comparing certificates against each other and against laws manually. And when a scheme or law gets updated, this needs to be done again.

1.1 Proposed Automatic Comparison

In order to alleviate the aforementioned problems, we propose a method for automatic comparison of cloud service security requirements defined in certification schemes, user needs, and regulations by describing them in a proposed cloud service security ontology. The machine-understandable description using ontology then allows for quick comparison of schemes against each other, schemes against regulations, etc. This benefits all the personas involved in the process:

- Provider – The cloud service provider can see a differential comparison of their currently held certifications against other possible certifications and see what amount of work it would take to certify against another certification when entering a new market. Moreover, the following advantages for cloud service users create more demand and bring more customers for the provider.
- User – The cloud service user can see whether a given service is suitable for their requirements, regardless whether the service is offered across the border and holds only a set of certificates not exactly matching user's expected certificates. This means larger supply for their demand.
- Auditor/certification body – The entity issuing the certificate upon performing an audit can benefit from scheme comparison too, by possibly performing only a differential audit without going over the previously verified and possibly stored audit evidence. Only the different controls of a new/updated scheme need to be audited primarily. This also benefits the providers by offering less exhausting and less expensive certifications.

When converting the often complex free-form text[1] into the ontology, the hard task of having to understand these requirements does not disappear, it still needs to be done for the first time when the requirement is transcribed into ontology notation. The benefit is that this needs to be done only once while transcribing each separate scheme (with N schemes, we need to do N transcriptions – comprehensions) instead of comparing each scheme with each other (N^2 comparisons – comprehensions). In the case of updates to a single scheme, only description of this scheme needs to be updated (*1* transcription) and the resulting comparison using above method is fully automatic, instead of comparing it to all other existing schemes, regulations, user requirements (N comparisons).

Still, this task can be supported using methods of natural language processing (NLP) to preprocess the free-text requirements and even propose its ontological description, which is then to be edited and confirmed by the domain expert. Overview of the proposed approach is show on Fig. 1.

1.2   Organization of This Paper

The rest of this paper is organized as follows. In Sect. 2, we describe the proposed ontology for cloud security requirements and the process of automatic requirement comparison. In Sect. 3, a supporting method for requirement conversion to ontology using NLP is described. We describe a prototype implementation of the proposed methods in a web-based system and the results of describing and comparing two widely used schemes in Sect. 4. Other possible application domains for both requirement and non-requirement comparisons are explored at the end of this paper.

2   Formalizing Certification Schemes Using Ontology

Several cloud security ontologies have been proposed already [1, 2]. Each of these ontologies are focused on different levels of cloud services (e.g. IaaS in the CoCoOn ontology [3]) and on different aspects of cloud security [2], while having various complexity (see Table 1). Therefore, in order to devise a cloud security ontology we considered the following possibilities:

1. Uniforming ontology built by mapping of the selected cloud security ontologies. This approach grants reusability of existing ontological descriptions of cloud services, but it must deal with problems of ontology mapping techniques [4, 5].
2. Extended ontology based on one selected ontology which should be selected with respect to its complexity, recency and usability. The Linked USDL ontology [6] and the taxonomy defined in [7] seem to be promising.
3. Ontology built from scratch, which should be defined in parallel with description of cloud security schemes and standards by cloud security experts.

---

[1] Example: *"Prior to granting customers access to data, assets, and information systems, identified security, contractual, and regulatory requirements for customer access shall be addressed."* (AIS-02, CSA CCM v3.0.1).

Current (naïve) approach

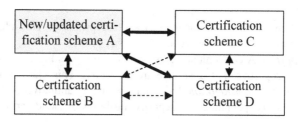

Manual evaluation against
all known schemes must be
performed

Proposed method

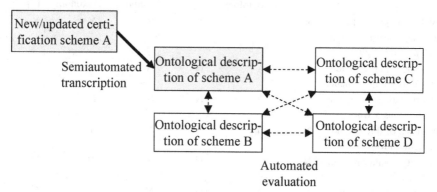

Automated
evaluation

Fig. 1. Overview of the proposed method for security requirements comparison showing fully automated and semi-automated parts, compared to naïve (manual) approach.

Our proposed approach uses ontology of cloud service security requirements based on Linked USDL format which models service level contract [6]. This approach, the second option, does not require intense work of cloud security experts and allows reusing the state-of-the-art in the cloud security formalization domain. The Linked USDL ontology was chosen for availability of formal descriptions. It is also used in current research projects [8, 9] and is based on Unified Service Description Language[2] which was developed as W3C incubator activity[3]. These facts give good presumption of overall acceptance of this ontology.

We also considered Takahashi et al.'s ontology [10]. Properties of this ontology look promising, but the ontology is only described in research papers and its formal description is not publicly available. Using this ontology would require reconstructing it from research papers, heightening the barriers to entry.

---

[2] https://www.w3.org/2005/Incubator/usdl/XGR-usdl-20111027/.

[3] https://www.w3.org/2005/Incubator/usdl/.

Linked USDL already defines necessary building blocks in its namespace *usdl-agreement*:

- *Metric* – A metric used to measure a service property. This class is used for definition of cloud security metrics.
- *ServiceProperty* – A convenience class which represents the class of qualitative or quantitative properties that a service may specify. This class is defined over a specialization of *GoodRelations*[4] qualitative and quantitative service properties. *ServiceProperty* is specialized by security attributes of cloud services.

Table 1. Cloud security attributes addressed by cloud ontologies based on [2] extended with Linked USDL.

Attributes	Cloud ontology	Takahashi et al.'s ontology	CoCoOn	Keerthana et al.'s ontology	Gonzalez et al.'s taxonomy	Linked USDL [6]
Well-defined ontology with appropriate design and methodology	Y	Y	Y	Y	Y	Y
Ontological specification of security requirements	Y	Y	N	N	Y	Y
Incorporation of security attributes as a whole	N	N	N	N	N	N
Analysis of threats and vulnerabilities	N	Y	N	N	Y	N
Security countermeasures and controls	N	Y	N	N	N	N
Validation and evaluation	N	Y	N	N	Y	Y
Availability of formal description	N	N	OWL[a]	N	N	TTL[b]

[a]https://sites.google.com/site/molkarekiklaadhar/home/cso-owl
[b]https://github.com/linked-usdl

- *AgreementCondition* – Service property constraint which can be checked within the terms of a service level agreement. This class is used for defining control objectives addressed by controls of cloud service security standards and schemes and for defining service level objectives and service quality objectives.

To extend the Linked USDL ontology in respect to the domain of cloud service security, we propose seven extending classes (Fig. 2):

- *Audit* – Class for definition of audit schemes. Hierarchy between audits is specified by the relation *fulfils*. A stricter audit refers to a less strict one. If a certification is audited by the stricter audit, it transitively fulfils the requirements of other audits in the hierarchy.

---

[4] http://www.heppnetz.de/ontologies/goodrelations/v1.

- *Certificate* – Class for certificates awarded to cloud services. Each certificate has attributes describing auditor (*issuedBy*), number (*hasNumber*), scope (*hasScope*), effective date (*hasEffectiveDate*), and expire date (*hasExpireDate*). Certificates are also related to certifications, based on which are certificates issued.
- *Certification* – Class for definition of cloud security certifications. Each certification is audited by a related audit (relation *auditedBy*) according to a certification scheme (relation *accordingTo*).
- *CertificationScheme* – Class for definition of cloud security certification schemes and standards. Certification schemes are decomposed into their controls (relation *hasControl*).
- *Control* – Class for certification scheme's controls. Each control has defined control objectives via relations to agreement conditions.
- *SecurityAttribute* – Class for definition of cloud service security attributes. The class *SecurityAttribute* is specialization of the class *ServiceProperty*. Each security attribute is categorized to a control domain (relation *hasDomain*).
- *ControlDomain* – Class for definition of control domains.

To illustrate the usage of the extended ontology on an example, we define:

- One instance of the class *ControlDomain*:
  - Application & Interface Security
- Two instances of the class *Metric*:
  - IsPresent – measures whether a property is present in a cloud service
    hasExpression: Is property present
    hasMeasuringInterval: 1 year
  - UsedTool – measures a number of used tools
    hasExpression: A number of used tools
    hasMeasuringInterval: 1 week
- Three instances of the class *SecurityAttribute*:
  - TechnicalVulnerabilitiesObtained – Technical vulnerabilities of information systems are obtained
    hasMetric: IsPresent
    rdfs:range: gr:QualitativeValue
    hasDomain: Application & Interface Security
  - TechnicalVulnerabilitiesEvaluated – Exposure to technical vulnerabilities is evaluated
    hasMetric: IsPresent
    rdfs:range: gr:QualitativeValue
    hasDomain: Application & Interface Security
  - SourceCodeAnalysisTool – Automated source code analysis tool is used
    hasMetric: UsedTool
    rdfs:range: gr:QuantitativeValue
    hasDomain: Application & Interface Security

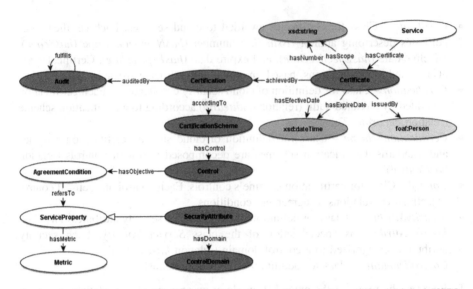

Fig. 2. The extended ontology. White colored classes are requisitioned from Linked USDL ontology. Blue colored classes are extending classes necessary for the framework. (Color figure online)

Now, we can describe control A.12.6.1 from ISO/IEC 27001:2013 by creating:

- Control objective: A.12.6.1 Obtaining
  - rdf:type: GuaranteedValue
  - hasValue: true
  - refersTo: TechnicalVulnerabilitiesObtained
- Control objective: A.12.6.1 Evaluation
  - rdf:type: GuaranteedValue
  - hasValue: true
  - refersTo: TechnicalVulnerabilitiesEvaluated

The control AIS-01 from CSA CCM v3.0.1 could be described as follows:

- Control objective: AIS-01.2
  - rdf:type: MinGuaranteedValue
  - hasValue: 1
  - refersTo: SourceCodeAnalysisTool
- Control objective: AIS-01.5
  - rdf:type: GuaranteedValue
  - hasValue: true
  - refersTo: TechnicalVulnerabilitiesEvaluated

In this manner, we can describe cloud service security schemes and standards independently and detailed comparison can be created automatically using straightforward approach of standard ontological descriptions comparison. In the example above, controls A.12.6.1 and AIS-01 are considered "partially mapped". Both of them require technical vulnerabilities to be evaluated, as both the *A.12.6.1 Evaluation* and *AIS-01.5* refer to the same security attribute, and with the same value. However, a "gap" exists in control objectives *A.12.6.1 Obtaining* (no respective control objective with given security attribute exists in AIS-01) and *AIS-01.2* (no respective control objective in A.12.6.1). Apart from a full mapping (same security attribute, same values) and a full gap (security attribute missing), a partial gap can also exist, where both control objectives refer to the same security attribute, but one requires stricter values, for example, one scheme requires 1 inspection, whereas another scheme requires 3 inspections.

3   Employing Natural Language Processing

Performing the above method by the human on requirements expressed in a free text sentences[5] requires multiple steps to be performed on each requirement:

1. *Normalizing the requirement.* A requirement may be already expressed as a standalone statement, or it could be expressed as a question to check for compliance, or it can refer to previous statements. Each sentence is converted to a standalone statement by rephrasing the question and/or replacing the references.
2. *Splitting the sentence into atomic requirements.* When the sentence expresses multiple requirements (e.g., by listing multiple grammar subjects as in "risks and threats must be assessed" or multiple verbs, or being a compound sentence), it is split into separate requirements (i.e. "risks must be assessed" and "threats must be assessed").
3. *Mapping the atomic requirements to ontology classes.* New instances of *ControlObjective* and other classes are created as described in Sect. 2 with *SecurityAttribute* and *Metric* reused where possible as they are the basis for mapping/gap analysis.

To assist this process and further (semi-)automatize the scheme comparison as a whole, natural language processing methods can be used, if not to perform all these steps automatically, at least to: (a) convert questions to statements in Step 1, (b) propose candidates for split atomic requirements in Step 2, and (c) on each of these candidates, propose candidates for existing and new ontology instances in Step 3.

Normalization. The sentence is parsed into a syntax tree. The tree for requirement expressed as question "Do you provide tenants with guidance on how to create production environments and tests?" is shown on Fig. 3 with respective Part of Speech (POS) tags using Penn Treebank notation [11]. Using heuristics and identifying words in the POS-tagged sentence, the question is converted to statement replacing "Do you provide tenants" with "Tenants must be provided". The rephrased statement is reparsed again.

---

[5] (n 1).

Splitting. Looking for conjunctions and commas, connected clauses are split and each respective part is raised in the syntax tree recursively producing separate sentences. In the example above, "*production environments and tests*" is split into "production environment" and "tests" and both are used to produce atomic requirements "Tenants must be provided with guidance on how to create *production environments*" and "Tenants must be provided with guidance on how to create *tests*". Note that correctly identifying how far to raise/connect the split clause (linguistic phenomenon known as right node raising) is an open research problem and due to ambiguities in complex cases, the compound requirements may be split incorrectly, therefore this method only produces candidates for atomic requirements subject to human approval.

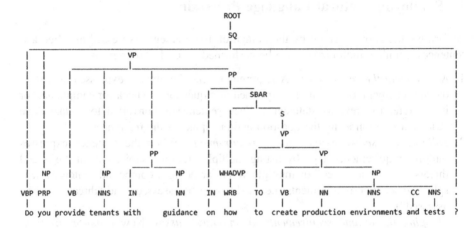

Fig. 3. Example of a parsed control question (IVS-08.2, CSA CCM v3.0.1).

Mapping. By looking for verbs and heuristically identifying objects and subjects, the respective parts of the atomic sentence are compared against known existing security attributes and metrics to aid in mapping the rest of the sentence to respective classes and their properties, producing Control Objective candidates, if no suitable instances are found, candidates of new security attributes and metrics are also produced.

4 Prototype Implementation and Results

To evaluate the feasibility of the proposed method, we implemented a web-based system for basic use-cases:

- Importing certification schemes. A certification scheme can be imported using a common spreadsheet format listing its controls and control questions and additional data. The raw text serves as base for NLP processing or manual description.
- Describing new certification schemes manually. A certification scheme can be also created manually without any import.

- Proposing ontological descriptions using NLP. The scheme text, either imported or created manually, is processed and split into control objectives with associated metrics and security attributes are proposed. This is approved/further edited by the domain expert responsible for the scheme transcription.
- Updating and managing descriptions. When the published scheme becomes updated, its ontological description must be updated in the system.
- Comparing described schemes and providing mapping/gap reports.
- Publishing described schemes in an interoperable RDF format.
- Managing system-wide vocabulary (security attributes and metrics). When describing a new scheme with controls not previously present in the system (a first scheme or a scheme containing additional controls compared to existing schemes), new security attributes and metrics are created by the domain expert at the same time. These are then to be reused in subsequent schemes having similar requirements. Therefore, vocabulary management (approving/editing/rejecting newly created attributes) is an important process in the workflow.

The front-end of the prototype system was implemented in Python using Django framework, the back-end uses combination of Python and Java for background intensive tasks (import, natural language processing). PostgreSQL was used as a datastore, the ontology is stored in relational representation. For natural language processing, we used Stanford CoreNLP [12] for POS-tagging the free text and Nodebox English Linguistics[6] library for improving initially preprocessed sentences and unifying synonymic definitions, verb forms, etc. in various requirements. Because it is crucial to use existing metrics and attributes created from previous requirements, full-text search was provided for users who are transcribing the schemes to easily search through existing elements and this was facilitated through Elasticsearch. Sample screen from the system is shown in Fig. 4, showing sample comparison of two schemes using the same security attribute in their requirements.

Using the above system, 7 persons who have not previously worked with ontologies and untrained in the schemes in question were recruited as scheme editors and tasked with transcribing selected parts of CSA CCM v3.0.1 and ISO/IEC 27001:2013 schemes from scratch using the system both in the manual workflow by breaking down the scheme's requirements manually as described in Sect. 3 and by using the import of spreadsheet files with NLP-assisted pre-processing into control objective and metric candidates, to be further manually edited and confirmed. The work performed is summarized up in Table 2. The preliminary evaluation has shown that both the method for describing individual certification schemes into proposed ontology and the proposed ontology are feasible and allow easier comparison of certification schemes than when comparing schemes directly by experts, even allowing for security non-experts/persons untrained in populating ontologies to be engaged in the process.

---

[6] https://www.nodebox.net/code/index.php/Linguistics.

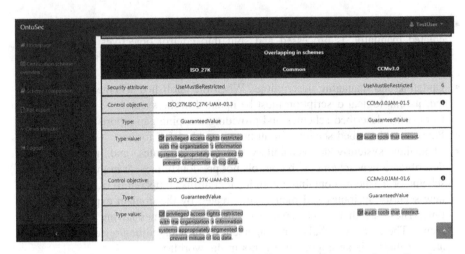

Fig. 4. Screenshot of scheme comparison showing different use of one security attribute in two different schemes, highlighting the difference.

Table 2. (Part of) evaluated schemes as transcribed into the ontology in the prototype system.

Property	Certification scheme	
	CSA CCM v3.0.1	ISO/IES 27001:2013
Controls	133	140
Control objectives	1529	289
Unique security attributes used	278	191
Unique metrics used	111	81

5 Looking Ahead

The proposed ontology and NLP-based assistance was implemented and tested through prototype web-based system showing promise both in the proposed ontology for describing and *fully automatically* comparing certification schemes and in the NLP-assisted *semiautomated* conversion of free text format of schemes into the ontology. The NLP-based method, although only an assistive function to save initial human efforts, can be further improved with advances in text processing, e.g. using deep learning methods.

The core, proposed ontology, approach is not limited only to the domain of certification schemes and even cloud security. A very similar domain, both in the nature of the underlying data (free text requirements/statements) and the need for cross-border comparison, exists in the area of laws. Laws in different countries are harmonized to some extent (e.g. traffic signs unified through Vienna Conventions, or EU member state national laws unified through EU directives, or laws of countries split from previous common state), yet smaller or larger details are incrementally added or nevertheless implemented differently and, in the same way cloud service users and providers benefit from up-to-date and possibly cross-border comparison of security schemes between each other or between security schemes and laws/requirements applicable to them, for example, a motorist travelling abroad would benefit greatly from summarized up-to-date review of differences in traffic rules across the border.

Acknowledgement. This work was partially supported by the project EU 731845 – EU-SEC.

References

1. Androcec, D., Vrcek, N., Seva, J.: Cloud computing ontologies: a systematic review. In: MOPAS 2012: The Third International Conference on Models and Ontology-Based Design of Protocols, Architectures and Services Cloud. IARIA, pp. 9–14 (2012)
2. Singh, V., Pandey, S.K.: A comparative study of cloud security ontologies. In: Proceedings of 3rd International Conference on Reliability, Infocom Technologies and Optimization. IEEE (2014)
3. Zhang, M., Ranjan, R., Haller, A., et al.: An ontology-based system for cloud infrastructure services' discovery. In: 8th International Conference on Collaborative Computing: Networking, Applications and Worksharing (CollaborateCom). IEEE, Pittsburgh, pp. 524–530 (2012)
4. Zhu, J.: Survey on ontology mapping. Phys. Procedia **24**, 1857–1862 (2012). https://doi.org/10.1016/j.phpro.2012.02.273
5. Hooi, Y.K., Hassan, M.F., Shariff, A.M.: A survey on ontology mapping techniques. In: Jeong, H.Y., S. Obaidat, M., Yen, N.Y., Park, J.J.(Jong Hyuk) (eds.) Advances in Computer Science and its Applications. LNEE, vol. 279, pp. 829–836. Springer, Heidelberg (2014). https://doi.org/10.1007/978-3-642-41674-3_118
6. Pedrinaci, C., Cardoso, J., Leidig, T.: Linked USDL: a vocabulary for web-scale service trading. In: Presutti, V., d'Amato, C., Gandon, F., d'Aquin, M., Staab, S., Tordai, A. (eds.) ESWC 2014. LNCS, vol. 8465, pp. 68–82. Springer, Cham (2014). https://doi.org/10.1007/978-3-319-07443-6_6
7. Gonzalez, N., Miers, C., Redígolo, F., et al.: A quantitative analysis of current security concerns and solutions for cloud computing. J. Cloud Comput. Adv. Syst. Appl. **1**, 11 (2012). https://doi.org/10.1186/2192-113x-1-11
8. Veloudis, S., Paraskakis, I.: Ontological templates for modelling security policies in cloud environments. In: Proceedings of the 20th Pan-Hellenic Conference on Informatics - PCI 2016. ACM Press, New York (2016)
9. Garcia, J.M., Fernandez, P., Pedrinaci, C., et al.: Modeling service level agreements with linked USDL agreement. IEEE Trans. Serv. Comput. **10**, 52–65 (2017). https://doi.org/10.1109/TSC.2016.2593925
10. Takahashi, T., Kadobayashi, Y., Fujiwara, H.: Ontological approach toward cybersecurity in cloud computing. In: Proceedings of the 3rd International Conference on Security of Information and Networks - SIN 2010. ACM Press, New York, pp. 100–109 (2010)
11. Marcus, M.P., Marcinkiewicz, M.A., Santorini, B.: Building a large annotated corpus of English: the Penn treebank. Comput. Linguist. **19**, 313–330 (1993)
12. Manning, C., Surdeanu, M., Bauer, J., et al.: The Stanford CoreNLP natural language processing toolkit. In: Proceedings of 52nd Annual Meeting of the Association for Computational Linguistics: System Demonstrations. Association for Computational Linguistics, Stroudsburg, pp. 55–60 (2014)

On the Expressive Power
of GF(2)-Grammars

Vladislav Makarov$^{(\boxtimes)}$ and Alexander Okhotin

St. Petersburg State University, 7/9 Universitetskaya nab.,
Saint Petersburg 199034, Russia
vm450@yandex.ru, alexander.okhotin@spbu.ru

Abstract. GF(2)-grammars, recently introduced by Bakinova et al. ("Formal languages over GF(2)", LATA 2018), are a variant of ordinary context-free grammars, in which the disjunction is replaced by exclusive OR, whereas the classical concatenation is replaced by a new operation called GF(2)-concatenation: $K \odot L$ is the set of all strings with an odd number of partitions into a concatenation of a string in K and a string in L. This paper establishes several results on the family of languages defined by these grammars. Over the unary alphabet, GF(2)-grammars define exactly the 2-automatic sets. No language of the form $\{a^n b^{f(n)} \mid n \geqslant 1\}$, with uniformly superlinear f, can be described by any GF(2)-grammar. The family is not closed under union, intersection, classical concatenation and Kleene star, non-erasing homomorphisms. On the other hand, this family is closed under injective nondeterministic finite transductions, and contains a hardest language under reductions by homomorphisms.

1 Introduction

A new family of formal grammars, the *GF(2)-grammars*, was recently introduced by Bakinova et al. [2]. These grammars differ from the ordinary grammars (Chomsky's "context-free") as follows. In ordinary grammars, the operations are: the *disjunction* of syntactical conditions, expressed by multiple rules for the same nonterminal symbol, and the *concatenation* of languages, which is defined through conjunction and disjunction [17]. In GF(2)-grammars, these operations are modified by replacing the underlying Boolean logic with the GF(2) field. Accordingly, instead of set-theoretic union of languages, GF(2)-grammars feature *symmetric difference*, whereas concatenation of languages is replaced with a new operation called *GF(2)-concatenation*, defined as follows.

$$K \odot L = \{w \mid \text{the number of partitions } w = uv, \text{ with } u \in K \text{ and } v \in L, \text{ is odd}\}$$

GF(2)-grammars were introduced as a part of a general study of GF(2)-concatenation as an operation on formal languages. Their formal definition

---

Supported by Russian Science Foundation, project 18-11-00100.

B. Catania et al. (Eds.): SOFSEM 2019, LNCS 11376, pp. 310–323, 2019.
https://doi.org/10.1007/978-3-030-10801-4_25

is based on parse trees in the corresponding ordinary grammar with classical concatenation and union: assuming that every string has a finite number of parse trees, the GF(2)-grammar defines all strings with an odd number of parse trees. The intuitive correctness of this definition is confirmed by a result that if a grammar is represented by a system of *language equations*, similar to the equations of Ginsburg and Rice [8], but using the operations of GF(2)-concatenation and symmetric difference, then the language defined by this GF(2)-grammar satisfies the system.

A few related, more general grammar models were studied before. Knuth [13] investigated specification of *multisets* by grammars, with every parse tree contributing an element to a multiset. A more general extension are formal languages over multiplicities, that is, mappings from the set of strings to a semiring. Under certain monotonicity assumptions on the semiring, equations in formal power series over a semiring behave similarly to ordinary grammars, and a substantial theory has been developed around them, see the survey by Petre and Salomaa [18]. However, the two-element field does not have the required monotonicity properties, and this general theory does not apply to GF(2)-grammars. Another matter is that languages over multiplicities are, after all, functions, and not languages as such. There are just two cases when actual languages are defined: this is when the semiring is either the Boolean semiring or the GF(2) field. Whereas the former case is classical, the other case deserves investigation.

The study of GF(2)-grammars is a part of the research on formal grammars with different sets of operations [17]. All these grammars are variants of Chomsky's "context-free" model, and particular models include *conjunctive grammars* equipped with a conjunction operator in the rules [15]; *multi-component grammars* [19] that allow substrings with gaps as basic constituents; *grammars with context operators* [3], and a few other models.

Every *unambiguous grammar* is a GF(2)-grammar, and it still defines the same language. In the presence of ambiguity, ordinary grammars assert the *existence* of a parse tree, whereas GF(2)-grammars check the *parity*. For this reason, ordinary grammars and GF(2)-grammars are two different generalizations of unambiguous grammars. These two generalizations share the same complexity upper bound: there is a basic cubic-time parsing algorithm, more efficient parsing by matrix multiplication, and parallel parsing in NC^2 [2]. These practically valuable properties make the class of GF(2)-grammars potentially useful and accordingly deserving further study.

There is some evidence that the formal properties of ordinary and GF(2)-grammars are not symmetric. First, unlike the classical concatenation, the GF(2)-concatenation is *invertible*: to be precise, for every language L containing the empty string, there exists a language L^{-1}, for which $L \odot L^{-1} = L^{-1} \odot L = \{\varepsilon\}$ [2]. How this property affects language specification, remains to be investigated. Second, over a unary alphabet, GF(2)-grammars can describe some non-regular sets, such as $\{a^{2^n} \mid n \geqslant 0\}$. These differences make this family an interesting subject for theoretical research.

The goal of this paper is to investigate the family of GF(2)-grammars and to determine, which languages they can describe and which they cannot. These results shall be used to establish the basic closure properties of GF(2)-grammars. as well as to compare their expressive power with that of the main families of formal grammars.

2   GF(2)-Grammars

Syntactically, a GF(2)-grammar is defined exactly as an ordinary grammar, with a finite sequence of symbols and nonterminal symbols on the right-hand side of each rule. However, every such sequence has semantics of GF(2)-concatenation, whereas multiple rules for the same nonterminal symbol implicitly denote symmetric difference of the given conditions.

Definition 1 ([2]). *A GF(2)-grammar is a quadruple $G = (\Sigma, N, R, S)$, where:*

- *Σ is the alphabet of the language;*
- *N is the set of nonterminal symbols;*
- *every rule in R is of the form $A \to X_1 \odot \ldots \odot X_\ell$, with $\ell \geqslant 0$ and $X_1, \ldots X_\ell \in \Sigma \cup N$, which represents all strings that have an odd number of partitions into $w_1 \ldots w_\ell$, with each w_i representable as X_i;*
- *$S \in N$ is the initial symbol.*

The grammar must satisfy the following condition. Let $\widehat{G} = (\Sigma, N, \widehat{R}, S)$ be the corresponding ordinary grammar, with $\widehat{R} = \{A \to X_1 \ldots X_\ell \mid A \to X_1 \odot \ldots \odot X_\ell \in R\}$. It is assumed that, for every string $w \in \Sigma^*$, the number of parse trees of w in $\widehat{G}$ is finite; if this is not the case, then G is considered ill-formed.

Then, for each $A \in N$, the language $L_G(A)$ is defined as the set of all strings with an odd number of parse trees as A in $\widehat{G}$.

A grammar is GF(2)-linear, if, in each rule, at most one of $X_1, \ldots, X_\ell$ is a nonterminal symbol.

Theorem A ([2]). *Let $G = (\Sigma, N, R, S)$ be a GF(2)-grammar. Then the substitution $A = L_G(A)$ for all $A \in N$ is a solution of the following system of language equations.*

$$A = \bigtriangleup_{A \to X_1 \odot \ldots \odot X_\ell \in R} X_1 \odot \ldots \odot X_\ell \qquad (A \in N)$$

Multiple rules for the same nonterminal symbol can be denoted by separating the alternatives with the "sum modulo two" symbol ($\oplus$), as in the following example.

Example 2 ([2]). The following GF(2)-linear grammar defines the language $\{a^\ell b^m c^n \mid \ell = m$ or $m = n$, but not both$\}$.

$$S \to A \oplus C$$
$$A \to aA \oplus B$$
$$B \to bBc \oplus \varepsilon$$
$$C \to Cc \oplus D$$
$$D \to aDb \oplus \varepsilon$$

Indeed, each string $a^\ell b^m c^n$ with $\ell = m$ or with $m = n$ has a parse tree, and if both equalities hold, then there are accordingly two parse trees, which cancel each other.

Since GF(2)-concatenation with a singleton language is the same as classical concatenation, GF(2)-linear grammars are a special case of *linear Boolean grammars*, in which the allowed operations are all Boolean operations and concatenation with singletons. In the latter grammars, negation can be eliminated, resulting in a *linear conjunctive grammar* [14].

Since linear conjunctive grammar over a unary alphabet define only regular languages, so do the GF(2)-linear grammars. On the other hand, GF(2)-grammars of the general form can define some non-regular unary languages.

Example 3 ([2]). The following grammar describes the language $\{a^{2^n} \mid n \geqslant 0\}$.

$$S \to (S \odot S) \oplus a$$

The main idea behind this grammar is that the GF(2)-square $S \odot S$ over a unary alphabet doubles the length of each string: $L \odot L = \{a^{2\ell} \mid a^\ell \in L\}$. The grammar iterates this doubling to produce all powers of two.

3   GF(2)-Grammars over the Unary Alphabet

Ordinary grammars over the unary alphabet $\Sigma = \{a\}$ define only regular languages [8]. On the other hand, as demonstrated by Example 3, GF(2)-grammars can define some non-regular languages. The question is, which unary languages can be defined? The answer follows from the famous Christol's theorem [5].

A few definitions are necessary.

Definition 4. *A set of natural numbers $S \subseteq \mathbb{N}$ is called k-automatic* [1]*, if there is a finite automaton over the alphabet $\Sigma_k = \{0, 1, \ldots, k-1\}$ recognizing base-k representations of these numbers.*

Let $\mathbb{F}_k[t]$ be the ring of polynomials over the k-element field GF(k), and let $\mathbb{F}_k[[t]]$ denote the ring of formal power series over the same field.

Definition 5. *A formal power series $f \in \mathbb{F}_k[[t]]$ is said to be algebraic, if there exists a non-zero polynomial P with coefficients from $\mathbb{F}_k[t]$, such that $P(f) = 0$.*

Theorem B (Christol's theorem for GF(2)). *A formal power series $\sum_{n=0}^{\infty} f_n t^n \in \mathbb{F}_2[[t]]$ is algebraic if and only if the set $\{n \in \mathbb{N}_0 \mid f_n = 1\}$ is 2-automatic.*

For a unary alphabet, solutions of language equations corresponding to a GF(2)-grammar, as in Theorem A, are algebraic formal power series in $\mathbb{F}_2[[t]]$, which has the following consequence.

Corollary 6. *Every unary language defined by GF(2)-grammar is 2-automatic.*

Inferring the converse characterization from Christol's theorem is not trivial, it is easier to give a direct proof.

Theorem 7. *Every 2-automatic unary language is described by a GF(2)-grammar.*

Proof. Let $\mathcal{A} = (\{0,1\}, Q, q_0, \delta, F)$ be a DFA that recognizes binary representations of natural numbers without leading zeroes. The corresponding GF(2)-grammar is defined as $G = (\{a\}, \{A_q \mid q \in Q\} \cup \{S\}, R, S)$, with the following set of rules.

$$
\begin{array}{ll}
A_q \to A_p \odot A_p & (p \in Q,\ \delta(p,0) = q) \\
A_q \to a \odot A_p \odot A_p & (p \in Q,\ \delta(p,1) = q) \\
A_q \to a & (\delta(q_0, 1) = q) \\
S \to A_q & (q \in F) \\
S \to \varepsilon & (\text{if } 0 \in L(\mathcal{A}))
\end{array}
$$

Here, as in Example 3 the rule for $A_q \to A_p \odot A_p$ produces all strings $a^{2\ell}$, with a^ℓ defined by A_p: this effectively appends zero to the binary representation. The rule $A_q \to A_p \odot A_p \odot a$ doubles the length and adds one, thus appending digit 1.
Then, $L(A_q)$ consists of all strings $a^{(1w)_2}$, with $\delta(q_0, 1w) = q$. ☐

By the above, the unary languages defined by GF(2)-grammars are exactly the 2-automatic languages. This characterization also gives a tool for proving that a given language over a non-unary alphabet cannot be defined by any GF(2)-grammar.

Theorem 8 (Method of unary image). *Let a language L over an alphabet Σ be defined by a GF(2)-grammar, and let $h \colon \Sigma \to \{t\}^*$ be a non-erasing homomorphism that is injective on L, in the sense that $h(u) \neq h(v)$ for any distinct $u, v \in L$. Then, $h(L)$ is a 2-automatic language over the unary alphabet $\{t\}$.*

In the GF(2)-grammar for L, it is sufficient to replace every occurrence of every symbol $a \in \Sigma$ in the rules with $h(a)$. The resulting GF(2)-grammar defines the language $h(L)$, which is then 2-automatic by Corollary 6.

4 Representability of Subsets of a^*b^*

Defining languages of the form $L \subseteq a^*b^*$ in a certain formalism represents its ability to *count*. A particular special case are languages of the form $L_f = \{a^n b^{f(n)} \mid n \geqslant 1\}$, where f is a function $f \colon \mathbb{N} \to \mathbb{N}$.

Finite automata *cannot keep count*, in the sense that L_f is regular only if there is a partition of $\mathbb{N}$ into finitely many pairwise disjoint arithmetic progressions, including singletons, and for each arithmetic progression $\{m_0 + ip \mid i \geqslant 0\}$, the language contains a subset $\{a^{m_0 + ip} b^{n_0} \mid i \geqslant 0\}$, for a fixed number n_0.

For ordinary grammars, the subset may be *linear*, that is, of the form $\{a^{m_0+ip}b^{n_0+iq} \mid i \geqslant 0\}$. Using linear conjunctive grammars, more sophisticated languages with exponential growth can be expressed.

Example 9 (Ibarra and Kim [11]). The language $\{a^n b^{2^n} \mid n \geqslant 1\}$ is recognized by a one-way real-time cellular automaton, and, equivalently, is described by a linear conjunctive grammar.

At the same time, there is the following bound on the growth of f.

Theorem C (Buchholz and Kutrib [4]). *For every function $f \colon \mathbb{N} \to \mathbb{N}$, if the language $\{a^n b^{f(n)} \mid n \geqslant 1\}$ is linear conjunctive, then f is bounded by an exponential function.*

Buchholz and Kutrib [4] further examined the ability to count for several classes of cellular automata. The question investigated in this paper is, what kind of languages of the form L_f can be expressed using GF(2)-grammars? The starting point is the following class of obviously representable languages.

Theorem 10. *Let $\mathbb{N}$ be represented as a disjoint union of finitely many 2-automatic sets: $\mathbb{N} = S_1 \uplus \ldots \uplus S_k$. For each of these sets, S_j, let L_j be a language of the following form: either $L_j = \{a^m b^{n_0} \mid m \in S_j\}$, for some $n_0 \geqslant 0$, or, as long as S_j is an arithmetic progression $\{m_0 + ip \mid i \geqslant 0\}$ with $m_0 \geqslant 0$ and $p \geqslant 1$, a language $L_j = \{a^{m_0+ip} b^{n_0+iq} \mid i \geqslant 0\}$ with $n_0 \geqslant 0$ and $q \geqslant 1$. Then, the languages $L_1, \ldots, L_k$ are pairwise disjoint, their union is a language of the form L_f, and it can be described by a GF(2)-grammar.*

The current conjecture is that no other languages of the form L_f can be represented. The next theorem identifies a class of non-representable languages, which are all those with a superlinearly growing function f, under the following uniformness restriction.

Definition 11. *A function $f \colon \mathbb{N} \to \mathbb{N}$ is called* uniformly superlinear, *if, for every $c > 0$, there exists $N \in \mathbb{N}$, such that $f(n+1) - f(n) > c$ for all $n > N$; in other words, $f(n+1) - f(n)$ is eventually larger than any constant, or $\liminf\limits_{n \to +\infty} f(n+1) - f(n) = +\infty$.*

Theorem 12. *Let $f \colon \mathbb{N} \to \mathbb{N}$ be a monotonically increasing uniformly superlinear function. Then the language $\{a^n b^{f(n)} \mid n \in \mathbb{N}\}$ is not described by any GF(2)-grammar.*

Proof. Proof by contradiction. Suppose that $L := \{a^n b^{f(n)} \mid n \in \mathbb{N}\}$ is described by some GF(2)-grammar. Then $S_1 := \{n + f(n) \mid n \in \mathbb{N}\}$ and $S_2 := \{2n + f(n) \mid n \in \mathbb{N}\}$ are both 2-automatic by virtue of being unary images of L under homomorphisms $a \to t, b \to t$ and $a \to t^2, b \to t$, respectively.

Let ℓ be an integer large enough, so that $2^{\ell/2}$ is greater than the number of states in the minimal NFAs recognizing both S_1 and S_2 in binary notation. By the uniform superlinearity of f, there exists a number M, such that $f(n+1) - f(n) > 2^\ell$ for all $n \geqslant M$. Consider the integers $M + 1, M + 2, \ldots, M + 2^\ell$. Clearly, they all have different remainders modulo 2^ℓ.

Claim. For any function $f \colon \mathbb{N} \to \mathbb{N}$ and for any two numbers $\ell, M \in \mathbb{N}$, there exists a factor $k \in \{1, 2\}$ and a set $X \subseteq \{M + 1, M + 2, \ldots, M + 2^{\ell}\}$, such that $|X| = \lceil 2^{\ell/2} \rceil$ and all residues $kn + f(n)$ modulo 2^{ℓ} for $n \in X$ are distinct.

Proof (of the claim). The first observation is that the mapping $n \mapsto (n + f(n), 2n + f(n))$ (mod 2^{ℓ}) is injective on $\{M + 1, \ldots, M + 2^{\ell}\}$. Indeed, if, for any two arguments $n, n' \in \{M + 1, \ldots, M + 2^{\ell}\}$ the values coincide, that is, $n + f(n) \equiv n' + f(n')$ (mod 2^{ℓ}) and $2n + f(n) \equiv 2n' + f(n')$ (mod 2^{ℓ}), then, subtracting the former equality from the latter yields $n \equiv n'$ (mod 2^{ℓ}), which implies that the arguments must be the same.

Now the statement is proved by contradiction. For $k = 1$, the assumption that no such set X exists means that there are fewer than $2^{\ell/2}$ distinct values $n + f(n)$ modulo 2^{ℓ}, for $n \in \{M + 1, M + 2, \ldots, M + 2^{\ell}\}$. Similarly, for $k = 2$, by assumption, there are fewer than $2^{\ell/2}$ distinct values $2n + f(n)$ modulo 2^{ℓ}, for all $n \in \{M + 1, M + 2, \ldots, M + 2^{\ell}\}$. Therefore, the number of distinct pairs $(n + f(n), 2n + f(n))$ modulo 2^{ℓ}, obtained for different n, is strictly less than $2^{\ell/2} \cdot 2^{\ell/2} = 2^{\ell}$. Since there are 2^{ℓ} different arguments n, the mapping $n \mapsto (n + f(n), 2n + f(n))$ (mod 2^{ℓ}) cannot be injective, which contradicts the above observation. $\square$

Resuming the proof of the theorem, by the lemma, there exist $k \in \{1, 2\}$ and $X \subseteq \{M + 1, M + 2, \ldots, M + 2^{\ell}\}$, such that all numbers $(kn + f(n))$, with $n \in X$, are pairwise distinct modulo 2^{ℓ}. Let $X := \{n_1, n_2, \ldots, n_{|X|}\}$. For each number n_i, the least significant ℓ digits in its binary representation are denoted by $v_i \in \{0, 1\}^{\ell}$; if n_i is less than $2^{\ell-1}$, the string is accordingly padded by zeroes. Similarly, let $u_i \in \{0, 1\}^*$ be the string of all remaining digits, so that $(u_i v_i)_2 = kn_i + f(n_i)$. By the choice of X, all v_i are different.

It is claimed that the set of $|X|$ pairs $(u_1, v_1), \ldots, (u_{|X|}, v_{|X|})$ forms a fooling set for the language of binary representations of S_k. Indeed, $(u_i v_i)_2 = (kn_i + f(n_i)) \in S_k$ for $i = 1, 2, \ldots, |X|$. On the other hand, for any i and j from $\{1, 2, \ldots, |X|\}$, such that $i \neq j$, at least one of the numbers $(u_i v_j)_2$ and $(u_j v_i)_2$ is not in S_k. By choice of X, $v_i \neq v_j$. Without loss of generality, let $(v_i)_2 < (v_j)_2$. For the sake of a contradiction, suppose that $(u_j v_i)_2 \in S_k$, that is, $(u_j v_i)_2 = km + f(m)$ for some $m \in \mathbb{N}$. On the one hand, $(u_j v_i)_2 < (u_j v_j)_2 < (u_j v_i)_2 + 2^{\ell}$, because $(u_j v_j)_2$ and $(u_j v_i)_2$ differ only in ℓ lowest bits, and $(v_i)_2 < (v_j)_2$. On the other hand, since $km + f(m) = (u_j v_i)_2 < (u_j v_j)_2 = kn_j + f(n_j)$ and f is non-decreasing, one can conclude that $m < n_j$ and $(u_j v_j)_2 - (u_j v_i)_2 = (kn_j + f(n_j)) - (km + f(m)) > f(n_j) - f(m) \geqslant f(n_j) - f(n_j - 1) > 2^{\ell}$, because $n_j - 1 \geqslant M$ and $f(n + 1) - f(n) > 2^{\ell}$ for $n \geqslant M$. Together these facts yield $2^{\ell} > (u_j v_j)_2 - (u_j v_i)_2 > 2^{\ell}$, contradiction.

It has thus been proved that the language of binary representations of S_k. has a fooling set of size $2^{\ell/2}$, and therefore every NFA recognizing this language must have at least this many states. This contradicts the assumption that there is a smaller NFA for this language. $\square$

Example 13 (cf. Example 9). The language $\{a^n b^{2^n} \mid n \geqslant 1\}$ is not described by any GF(2)-grammar, because the function $f(n) = 2^n$ is increasing and uniformly superlinear.

5 A Separating Example and the Hierarchy

In order to compare the expressive power of GF(2)-grammars to other grammar families, it is essential to find a simple language which they could not represent, but other kinds of grammars could. Most of the results presented later on are based on the following language over the alphabet $\{a, b\}$.

$$L = \{ba^{2 \cdot 3^n - 1} \ldots ba^{17} \, ba^5 \, ba \, bbb \, a^3 b \, a^{11} b \, a^{35} b \ldots a^{4 \cdot 3^n - 1} b \mid n \geqslant 0\}$$

Lemma 14. *The language L is representable as $L = L_1 \cap L_2$, where both L_1 and L_2 are described by unambiguous linear grammars.*

Furthermore, their complements $\overline{L_1}$ and $\overline{L_2}$ are described by linear grammars, and therefore so is the complement of L.

Proof (a sketch). This is a standard construction, inspired by a proof by Ginsburg and Spanier [9]. Each string in L encodes two sequences of numbers: $1, 5, 17, \ldots, 2 \cdot 3^n - 1$ on the left, and $3, 11, 35, \ldots, 4 \cdot 3^n - 1$ on the right. The language L_1 ensures that for each i-th element m on the left-hand side, the i-th element on the right-hand side must be $2m + 1$; the language L_2 similarly ensures that for each i-th element $2m - 1$ on the right-hand side, the $(i + 1)$-th element on the left-hand side must be $3m - 1$, and also that the first element of the left-hand-side sequence is 1.

Linear grammars for the complements of L_1 and of L_2 simply check that there is at least one error in the above correspondence. □

Lemma 15. *Neither L nor its complement are described by any GF(2)-grammars.*

Proof. Indeed, the unary image of L is $\{a^{3^n} \mid n \geqslant 2\}$, and the latter language is not described by any GF(2)-grammar by Christol's theorem. Since complementation is representable in GF(2)-grammars, there cannot be a grammar for the complement of L either. □

With this last example, the position of GF(2)-grammars in the hierarchy of grammars with different sets of operations can be determined as follows. The hierarchy in Fig. 1 includes the following grammar families: ordinary grammars or Chomsky's context-free (union and concatenation: ORDINARY); unambiguous grammars (disjoint union, unambiguous concatenation: UNAMB); linear grammars (union, concatenation with symbols: LIN); unambiguous linear grammars (disjoint union, concatenation with symbols: UNAMBLIN); linear conjunctive grammars (union, intersection, concatenation with symbols: LINCONJ); conjunctive grammars (union, intersection, concatenation: CONJ).

The families are separated by the following examples.

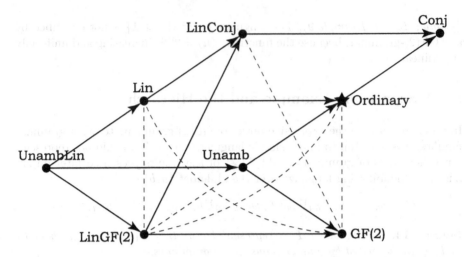

Fig. 1. The hierarchy of grammars: solid lines indicate proper inclusions, dashed lines mark incomparable families (shown only for GF(2)-families).

- GF(2)-linear, but not ordinary (and thus not linear and not unambiguous): $\{a^\ell b^m c^n \mid \ell = m$ or $m = n$, but not both$\}$ (Example 2).
- GF(2), but not ordinary (and thus not unambiguous) and not linear conjunctive (and thus neither linear nor GF(2)-linear): $\{a^{2^n} \mid n \geqslant 0\}$ (Example 3).
- Linear conjunctive but not GF(2) (and thus not GF(2)-linear): $\{a^n b^{2^n} \mid n \geqslant 1\}$ (Example 13).
- Linear (and also ordinary), but not GF(2) (and thus not GF(2)-linear): the complement of L (Lemmata 14–15).
- Conjunctive, but not GF(2): the complement of L applies as well. Furthermore, non-containment is witnessed by the unary language $\{a^{3^n} \mid n \geqslant 0\}$, which has a conjunctive grammar [12], but not a GF(2)-grammar (Corollary 6).
- Unambiguous, but not GF(2)-linear: a language defined by an unambiguous grammar, but not a linear conjunctive grammar, was constructed by Okhotin [16, Lemma 4] using a method of Terrier [20].

The comparison between GF(2)-grammars and conjunctive grammars remains incomplete, because no example of a language defined by a GF(2)-grammar, but not by any conjunctive grammar, is known. The conjectured example is $\{uv \mid u, v \in \{a, b\}^*, |u| = |v|, u$ and v differ in an odd number of positions$\}$ [2]. No way of constructing a conjunctive grammar for this language is known; however, no proof of this could be given due to the general lack of knowledge on conjunctive grammars [15].

6   Closure Properties

Some closure properties of GF(2)-grammars are quite expected, and follow by well-known arguments. Such is the closure under intersection with regular languages: the classical construction by Bar-Hillel et al. applies verbatim, because it preserves multiplicities of parse trees. It makes sense to prove this result in its most general form, for all mappings computed by injective nondeterministic finite tranducers (NFT). First, it is established under the following technical assumption.

Lemma 16. *Let G be a GF(2)-grammar over an alphabet Σ, and let a mapping $T\colon \Sigma^* \to 2^{\Omega^*}$ be computed by an injective NFT, which has the following property: for every pair $(w, x) \in \Sigma^* \times \Omega^*$, there is at most one computation on w that emits x. Then the language $T(L(G))$ is defined by a GF(2)-grammar G'. Furthermore, if G is linear, then so is G'.*

Proof (a sketch). The construction is standard, and the assumptions of injectivity and of the uniqueness of a computation ensure that, whenever the original grammar G defines w and $x \in T(w)$, the number of parse trees of x in the constructed grammar G' is the same as the number of parse trees of w in G. For that reason, $L(G') = T(L(G))$, as desired. □

Using this result, the desired closure property is proved as follows.

Theorem 17. *Let a mapping $T\colon \Sigma^* \to 2^{\Omega^*}$ be computed by an injective NFT. Then the language families defined by GF(2)-grammars and GF(2)-linear grammars are closed under T.*

Proof (a sketch). Since the definition of NFT is symmetric with respect to its input and its output, there is a single-valued NFT implementing a partial mapping $T'\colon \Omega^* \to \Sigma^*$, with $w = T'(x)$ if and only if $x \in T(w)$. As proved by Eilenberg [6, p. 186] a single-valued NFT can be transformed to an *unambiguous NFT*, that is, with at most one accepting computation on every input. Swapping the input and the output again yields an NFT implementing the original mapping T that satisfies the conditions of Lemma 16. Therefore, by Lemma 16, both language families are closed under T. □

Corollary 18. *The families defined by GF(2) and GF(2)-linear grammars are closed under intersection with regular languages, as well as under union with regular languages.*

For all other standard operations on languages, GF(2)-grammars demonstrate non-closure.

Theorem 19. *The family of languages described by GF(2)-grammars is not closed under (a) union, (b) intersection, (c) concatenation, (d) Kleene star, (e) left- and right-quotient with a two-element set, and (f) non-erasing homomorphisms. The same results hold for GF(2)-linear grammars.*

Intuitively, all these operations essentially use conjunction or disjunction in their definitions, and those Boolean operations are not expressible in GF(2).

Proof. The proofs of all cases are based on the languages L, L_1 and L_2 given in Lemmata 14–15. By Lemma 14, L_1 and L_2 are described by unambiguous linear grammars, and hence both these languages and their complements are described by GF(2)-linear grammars.

(a) Both $\overline{L_1}$ and $\overline{L_2}$ are described by GF(2)-linear grammars, but their union $\overline{L_1} \cup \overline{L_2} = \overline{L}$ is not.

(b) Similarly, $L_1 \cap L_2 = L$, where L_1 and L_2 are described by GF(2)-grammars, but L is not.

(c) Let c be a new symbol. By the assumptions, $\{\varepsilon, c\}$ and $c\overline{L_1} \triangle \overline{L_2} = c\overline{L_1} \cup \overline{L_2}$ are described by some GF(2)-grammars. Their concatenation is the following language.

$$(\{\varepsilon, c\} \cdot (c\overline{L_1} \cup \overline{L_2})) = \overline{L_2} \cup c(\overline{L_1} \cup \overline{L_2}) \cup cc\overline{L_1}$$

If it is represented by some GF(2)-grammar, then, by Theorem 17, so is its image under a finite transduction T defined by $T(cw) = w$ for all $w \in \Sigma^*$, and undefined on all other strings. This image is the language $\overline{L}$, which is not described by any GF(2) grammar, contradiction.

(d) By Corollary 18, $(ccc\overline{L_1} \triangle cc\overline{L_2} \triangle c)^* \cap (c^3 \{a, b\}^*) = (ccc\overline{L_1} \cup cc\overline{L_2} \cup c)^* \cap (c^3 \{a, b\}^*) = c^3(\overline{L_1} \cup \overline{L_2} \cup \{\varepsilon\}) = c^3(\overline{L} \cup \{\varepsilon\}) = c^3\overline{L}$. Similarly to (c), this language is not described by any GF(2)-grammar.

(e) For symmetry reasons it suffices to prove only the left-quotient result. Denote $K \backslash M = \{v \mid \exists u \in K : uv \in M\}$. Indeed, $(\{\varepsilon, c\} \backslash (c\overline{L_1} \triangle \overline{L_2})) \triangle c\overline{L_1} = (\{\varepsilon, c\} \backslash (c\overline{L_1} \cup \overline{L_2})) \triangle c\overline{L_1} = (\overline{L_1} \cup \overline{L_2} \cup \overline{L_1} \cup \varnothing) \triangle c\overline{L_1} = (\overline{L_1} \cup \overline{L_2}) = \overline{L}$.

(f) $h(c\overline{L_1} \triangle d\overline{L_2}) = h(c\overline{L_1} \cup d\overline{L_2}) = c(\overline{L_1} \cup \overline{L_2}) = c\overline{L}$, where the images of the letters under homomorphism $h \colon \{a, b, c, d\} \to \{a, b, c\}^*$ are $h(a) = a, h(b) = b, h(c) = h(d) = c$ respectively.   □

In Table 1, the closure properties of GF(2)-grammars and of their linear subclass are summarized and compared with other grammar families. The operations featured in the table are: intersection with regular languages ($\cap$Reg), union ($\cup$), intersection ($\cap$), complementation ($\sim$), concatenation ($\cdot$), Kleene star ($*$), GF(2)-concatenation ($\odot$), GF(2)-inverse ($^{-1}$), quotient with regular languages (/Reg), homomorphisms (h), injective homomorphisms (h_{inj}), inverse homomorphisms (h^{-1}). All closure properties of GF(2) and GF(2)-linear grammars are proved in this paper. Non-closure of the classical families under GF(2)-concatenation and GF(2)-inverse is known [2]; most likely, this non-closure extends to linear conjunctive grammars and could be proved by the method of Terrier [20].

7   Hardest Language

Some formal properties of GF(2)-grammars are the same as for ordinary grammars and are established by the same argument. One such property is Greibach's *hardest language theorem* [10], which has the same statement in the case of GF(2)-grammars.

Table 1. Closure properties of grammars under classical and under GF(2)-operations.

	∩Reg	U	∩	~	·	*	⊙	−1	/Reg	h	h_{inj}	h^{-1}
GF(2)-linear ($\triangle$, LIN·)	+	−	−	+	−	−	−	−	−	−	+	+
Unambiguous ($\uplus$, UNAMB·)	+	−	−	−	−	−	−	−	−	−	+	+
Ordinary (U, ·)	+	+	−	−	+	+	−	−	+	+	+	+
GF(2) ($\triangle$, $\odot$)	+	−	−	+	−	−	+	+	−	−	+	+
Linear conjunctive (U,∩, LIN·)	+	+	+	+	−	−	?	?	−	−	+	+
Conjunctive (U,∩, ·)	+	+	+	?	+	+	?	?	−	−	+	+

Theorem 20. *There exist an alphabet Σ_0 and a GF(2)-grammar $G_0 = (\Sigma_0, N_0, R_0, S_0)$, such that for every GF(2)-grammar over any alphabet Σ, there exists a homomorphism $h\colon \Sigma \to \Sigma_0^*$, such that a non-empty string w over Σ is in $L(G)$ if and only if $h(w)$ is in $L(G_0)$.*

The proof requires a Greibach normal form.

Definition 21. *A GF(2)-grammar is said to be in Greibach normal form (GNF), if all its rules are of the form $A \to a \odot B_1 \odot \ldots \odot B_\ell$, with $a \in \Sigma$, $\ell \geqslant 0$ and $B_1, \ldots, B_\ell \in N$.*

Proposition 22. *For every GF(2)-grammar G with $\varepsilon \notin L(G)$, there exists a GF(2)-grammar in the Greibach normal form that describes the same language.*

It is known that the transformation to the Greibach normal form preserves the number of parse trees [7, Lemma 4]. Taking this modulo two yields Proposition 22.

Proof (of Theorem 20). Greibach's classical construction for ordinary grammars applies here, because it is known to *preserve multiplicities*: for a non-empty string $w \in \Sigma^*$, its image $h(w)$ has the same number of parse trees in G_0 as w has in G [10, p. 307]. Therefore, the number of trees modulo 2 is preserved as well.   □

8   Conclusion

The new negative results for GF(2)-grammars were sufficient to establish their position in the hierarchy and their basic closure properties. However, these methods are still quite limited, and the existence of GF(2)-grammars remains unknown even for some very simple languages. For instance, can the language $\{a^n b^n c^n \mid n \geqslant 0\}$ be defined by these grammars? Exactly which subsets of $a^* b^*$ can be defined? In particular, what is the exact class of functions f, for which the language $L_f = \{a^n b^{f(n)} \mid n \geqslant 0\}$ can be defined—is it any larger than the class in Theorem 10?

References

1. Allouche, J.-P., Shallit, J.: Automatic Sequences: Theory, Applications, Generalizations. Cambridge University Press, Cambridge (2003)
2. Bakinova, E., Basharin, A., Batmanov, I., Lyubort, K., Okhotin, A., Sazhneva, E.: Formal languages over GF(2). In: Klein, S.T., Martín-Vide, C., Shapira, D. (eds.) LATA 2018. LNCS, vol. 10792, pp. 68–79. Springer, Cham (2018). https://doi.org/10.1007/978-3-319-77313-1_5
3. Barash, M., Okhotin, A.: An extension of context-free grammars with one-sided context specifications. Inf. Comput. **237**, 268–293 (2014). https://doi.org/10.1016/j.ic.2014.03.003
4. Buchholz, T., Kutrib, M.: On time computability of functions in one-way cellular automata. Acta Informatica **35**(4), 329–352 (1998). https://doi.org/10.1007/s002360050123
5. Christol, G.: Ensembles presque periodiques k-reconnaissables. Theor. Comput. Sci. **9**, 141–145 (1979). https://doi.org/10.1016/0304-3975(79)90011-2
6. Eilenberg, S.: Automata, Languages and Machines, vol. 1. Academic Press, Cambridge (1974)
7. Forejt, V., Jančar, P., Kiefer, S., Worrell, J.: Language equivalence of probabilistic pushdown automata. Inf. Comput. **237**, 1–11 (2014). https://doi.org/10.1016/j.ic.2014.04.003
8. Ginsburg, S., Rice, H.G.: Two families of languages related to ALGOL. J. ACM **9**, 350–371 (1962). https://doi.org/10.1145/321127.321132
9. Ginsburg, S., Spanier, E.H.: Quotients of context-free languages. J. ACM **10**(4), 487–492 (1963). https://doi.org/10.1145/321186.321191
10. Greibach, S.A.: The hardest context-free language. SIAM J. Comput. **2**(4), 304–310 (1973). https://doi.org/10.1137/0202025
11. Ibarra, O.H., Kim, S.M.: Characterizations and computational complexity of systolic trellis automata. Theor. Comput. Sci. **29**, 123–153 (1984). https://doi.org/10.1016/0304-3975(84)90015-X
12. Jeż, A.: Conjunctive grammars can generate non-regular unary languages. Int. J. Found. Comput. Sci. **19**(3), 597–615 (2008). https://doi.org/10.1142/S012905410800584X
13. Knuth, D.E.: Context-free multilanguages. In: Theoretical Studies in Computer Science, pp. 1–13. Academic Press, Cambridge (1992)
14. Okhotin, A.: On the equivalence of linear conjunctive grammars to trellis automata. RAIRO Informatique Théorique et Applications **38**(1), 69–88 (2004). https://doi.org/10.1051/ita:2004004
15. Okhotin, A.: Conjunctive and Boolean grammars: the true general case of the context-free grammars. Comput. Sci. Rev. **9**, 27–59 (2013). https://doi.org/10.1016/j.cosrev.2013.06.001
16. Okhotin, A.: Input-driven languages are linear conjunctive. Theor. Comput. Sci. **618**, 52–71 (2016). https://doi.org/10.1016/j.tcs.2016.01.007
17. Okhotin, A.: Underlying principles and recurring ideas of formal grammars. In: Klein, S.T., Martín-Vide, C., Shapira, D. (eds.) LATA 2018. LNCS, vol. 10792, pp. 36–59. Springer, Cham (2018). https://doi.org/10.1007/978-3-319-77313-1_3
18. Petre, I., Salomaa, A.: Algebraic systems and pushdown automata. In: Droste, M., Kuich, W., Vogler, H. (eds.) Handbook of Weighted Automata, pp. 257–289. Springer, Heidelberg (2009). https://doi.org/10.1007/978-3-642-01492-5_7

19. Seki, H., Matsumura, T., Fujii, M., Kasami, T.: On multiple context-free grammars. Theor. Comput. Sci. **88**(2), 191–229 (1991). https://doi.org/10.1016/0304-3975(91)90374-B
20. Terrier, V.: On real-time one-way cellular array. Theor. Comput. Sci. **141**(1–2), 331–335 (1995). https://doi.org/10.1016/0304-3975(94)00212-2

An Efficient Algorithm for Combining Verification and Validation Methods

Isela Mendoza[1]($\boxtimes$), Uéverton Souza[1], Marcos Kalinowski[2],
Ruben Interian[1], and Leonado Gresta Paulino Murta[1]

[1] Computer Institute, Fluminense Federal University,
Niterói, Rio de Janeiro, Brazil
{imendoza, ueverton, rinterian, leomurta}@ic.uff.br
[2] Informatics Department, Pontifical Catholic University of Rio de Janeiro,
Rio de Janeiro, Rio de Janeiro, Brazil
kalinowski@inf.puc-rio.br

Abstract. An adequate combination of verification and validation (V&V) methods is important to improve software quality control throughout the development process and to reduce costs. However, to find an appropriate set of V&V methods that properly addresses the desired quality characteristics of a given project is a NP-hard problem. In this paper, we present a novel approach that combines V&V methods efficiently in order to properly cover a set of quality characteristics. We modelled the problem using a bipartite graph to represent the relationships between V&V methods and quality characteristics. Then we interpreted our problem as the Set Cover problem. Although Set Cover is considered hard to be solved, through the theoretical framework of Parameterized Complexity we propose an FPT-Algorithm (fixed-parameter tractable algorithm) that effectively solves the problem, considering the number of quality characteristics to be covered as a fixed parameter. We conclude that the proposed algorithm enables combining V&V methods in a scalable and efficient way, representing a valuable contribution to the community.

Keywords: Combination · Verification · Validation · Software quality
FPT · Set cover · Parameterized Complexity

1 Introduction

Studies suggest high costs related to quality assurance activities in software development projects [1]. The appropriate combination of verification and validation (V&V) methods is seen in the literature as a way to reduce these costs and increase product quality [2]. Over the years, some knowledge has been generated regarding V&V methods when observed in isolation [3]. However, the selection of different V&V methods as well as the interdependencies among them are still not well-understood [4].

A significant part of the software industry is made up of small and medium-sized companies that, given the lack of guidelines for performing the right combination of V&V methods, have difficulties in optimizing this combination for their context, increasing the costs of resources and time and mainly harming the quality of the produced software.

© Springer Nature Switzerland AG 2019
B. Catania et al. (Eds.): SOFSEM 2019, LNCS 11376, pp. 324–340, 2019.
https://doi.org/10.1007/978-3-030-10801-4_26

According to the Guide to the Software Engineering Body of Knowledge (SWE-BOK) [5], verification is used to ensure that the software product is built in the correct way, that is, it complies with the previously defined specifications. On the other hand, validation guarantees that the product is adherent to the user needs. It is known that an adequate combination of V&V methods outperforms any method alone [6]. Most of the studies presented in a systematic mapping [9] do not clearly specify which V&V methods cover which quality characteristics (e.g., considering the quality characteristics described in the ISO 25010 quality model standard).

Finding a set of methods that together properly addresses all quality characteristics of interest can be seen as a Set Cover Problem (SCP) [11]. The SCP is a classic NP-hard problem in the computational complexity area, whose decision version belongs to the list of the 21 Karp's NP-complete problems [11]. This means that when the number of methods or quality characteristics increase, the performance of an algorithm to aiming at combining them in an optimal way would drastically decrease.

The existence of efficient algorithms to solve NP-complete, or otherwise NP-hard, problems is unlikely, if the input parameters are not fixed; all known algorithms that solve these problems require exponential time (or at least super-polynomial time) in terms of the input size. However, some problems can be solved by algorithms for which we can split the running time into two parts: one exponential, but only with respect to the size of a fixed parameter, and another polynomial in the size of the input. Such algorithms are denoted FPT (fixed-parameter tractable) in the *Parameterized Complexity* field, because the problem can be solved efficiently for small values of the fixed parameter [13–15]. This field emerged as a promising alternative for working with NP-hard problems [12].

In this paper, we propose an algorithm to obtain an optimal combination of methods covering software quality characteristics of interest in reasonable computational time. In order to find an optimal solution for the problem (based on the desired software quality characteristics and the relation between those and V&V methods, provided as input), we adopted a parameterized approach, considering the set of quality characteristics as fixed parameter, and obtaining an algorithm classified as FPT. The implemented FPT algorithm is the first of its kind that solves the SCP.

Our proposed algorithm reached its goals: it runs in $O(f(k) \times n)$, where the constant k is the number of quality characteristics, n is the number of methods, and $f(k)$ is some function of k. Considering that the number of quality characteristics of a given quality standard is always constant, the algorithm runs in polynomial time in terms of the number of V&V methods to be combined. As a result, it provides the minimum set of V&V methods addressing all quality characteristics of interest. While this information is surely useful, we are aware that companies may choose to complement these methods with others to further assure the quality of the product (or even chose others) and that other factors, such as cost, should be considered when taking the final decision.

The remainder of this paper is organized in the following sections: Sect. 2 presents the background and related work concerning quality characteristics, V&V methods, and the combination of V&V methods. In Sect. 3 the problem is modeled as a SCP. Section 4 briefly introduces parameterized complexity theory. Section 5 presents the FPT–Algorithm that obtains the optimal combination. Section 6 contains a computational experiment analysis. Section 7 discusses the contributions and limitations of our approach. Section 8 presents the concluding remarks.

2   Background and Related Work

2.1   Quality Characteristics

Concerning software product quality characteristics, the product quality model defined in the ISO 25010 standard includes eight characteristics, for which quality requirements may be defined and measured during software development [10]. The characteristics and short descriptions for them, based on the ISO 25010 standard, can be found in Table 1.

Table 1. ISO 25010 quality characteristics.

Characteristic	Short description
Function suitability	Degree to which a product or system provides functions that meet stated and implied needs when used under specified conditions
Performance efficiency	Represents the performance relative to the amount of resources used under stated conditions
Compatibility	Degree to which a product, system or component can exchange information with other products, systems or components, and/or perform its required functions, while sharing the same hardware or software environment
Usability	Degree to which a product or system can be used by specified users to achieve specified goals with effectiveness, efficiency and satisfaction in a specified context of use
Reliability	Degree to which a system, product or component performs specified functions under specified conditions for a specified period
Security	Degree to which a product or system protects information and data so that persons or other products or systems have the degree of data access appropriate to their types and levels of authorization
Maintainability	Degree of effectiveness and efficiency with which a product or system can be modified to improve it, correct it or adapt it to changes in environment, and in requirements
Portability	Degree of effectiveness and efficiency with which a system, product or component can be transferred from one hardware, software or other operational or usage environment to another

2.2   V&V Methods

Several V&V methods have been proposed over the years. In this paper we concentrate on a subset of V&V methods extracted mainly from the SWEBOK [5] and some other sources [19, 20] to compose our corpus. The list of methods can be found in Table 2. Due to space constraints, a short description of the methods is not provided, but it can be easily obtained in the cited sources. It is noteworthy that there are variations for each of these methods (e.g., different control flow-based criteria, different inspection techniques). Nevertheless, we use this more generic classification as a starting point, given that characterizing all possible variations to obtain a representative input for our algorithm would be hard to accomplish. Thus, a method covering a quality characteristic, in the context of this paper, means that there are ways of appropriately addressing it using the method.

Table 2. V&V methods.

Classification	Method
Based on intuition & experience	*Ad Hoc* Testing
	Exploratory Testing
Input domain-based	Equivalence Partitioning
	Pair wise Testing
	Boundary-Value Analysis
	Random Testing
	Cause-Effect Graphing
Code-based	Control Flow-Based Criteria
	Data Flow-Based Criteria
Fault-based	Error Guessing
	Mutation Testing
Usage-based	Operational Profile
	Usability Inspection Methods
Model-based	Finite-State Machines
	Workflow Models
Reviews	Walkthrough
	Peer Review or desk checking
	Technical Review
	Inspection

2.3   Combination of V&V Methods

It is known that the quality of software products is strongly dependent on the appropriate combination of V&V methods employed during development [2]. Experimental studies have long demonstrated that the use of combinations of different V&V methods to ensure the quality of a software is more effective than using isolated methods [7, 8].

Elbertzhager *et al.* [9] conducted a mapping study concerning the combination of V&V methods. They describe two fundamental approaches: *Compilation* and *Integration*. We focus on the *Compilation* approach since our purpose is purely to combine existing V&V methods (*Compilation* process). We are not focusing on *creating* new techniques by combining different methods into one, nor in using the results of the application of some technique as an instance to apply another one (*Integration* of V&V methods).

In order to establish how other works perform the combination of V&V methods in the *Compilation* approach, Elberzhager *et al.* [9] created a categorization to classify and organize these studies into three subgroups.

In the first subgroup, static and dynamic techniques are combined, focusing on thread escape analysis, atomicity analysis, protocol analysis, vulnerability analysis, concurrent program analysis or on defects in general. All these combinations are supported by open-source or proprietary tools.

The second subgroup compares different testing and inspection techniques discussing advantages and disadvantages among them. In most cases, two or three techniques are compared to each other. Several studies initially perform inspections, followed by some tests, corroborating then the effectiveness of the combination of both techniques.

The last subgroup describes other combinations, such as testing techniques and inspections combined with formal specifications, bug-finding tools, comprehensive quality control processes in industrial environments, comprising several inspections and technical tests, requirements and static analysis, tutorials, simulations, and vision-based approaches.

The most cited papers in the systematic mapping [9] regarding the *Compilation* approach are: Basili [22], Kamsties and Lott [23], and Wagner *et al.* [24]. Basili [22] makes a comparison of three software testing techniques: reading of code by gradual abstraction, functional testing using equivalence partition and border value analysis, and structural testing using total coverage of criticism, according to efficiency, cost, and fault detection classes. Kamsties and Lott [23], evaluate three techniques through a controlled experiment: reading of code by gradual abstraction, functional (black-box) testing and structural (white-box) testing. Wagner [24] describes a case study where several projects are analyzed in an industrial environment. In this project, automatic static analysis, testing, and reviews are used to detect defects. Their results show that these techniques complement each other and that they should be combined.

In the systematic mapping [9], papers were analyzed until 2010. This led us to carry out an update regarding the compilation approach, with the aim of finding more relevant and recent papers from 2010 to present. Due to space constraints the details of our mapping update will not be provided in this paper and we focus directly on the recent related work.

Dwyer and Elbaum [25] suggest an approach based on dividing V&V methods into two main classes: those that make dynamic analyses (or focused on behavior of the system, e.g., testing) and those that use static analysis (typically focused on a single property of the system at a time). Runeson *et al.* [27] compare code inspections and structural unit tests by analyzing three replications of an experiment in order to know which method finds more faults. Olorisade *et al.* [28] investigate the effectiveness of two test techniques (partition of equivalence class and decision coverage) and one review technique (code by abstraction) in terms of their ability to detect faults. Cotroneo *et al.* [29] combine testing techniques adaptively, based on machine learning, during the testing process, by learning from past experience and adapting the technique selection to the current testing session. Bishop *et al.* [30] combine a monotonicity analysis with a defined set of tests, showing that, unlike "independent" dynamic methods, this combination provides a full error coverage. Solari and Matalonga [31] study the behavior of two techniques, equivalence partition and decision coverage, to determine the types of defects that are undetectable for either of them. Finally, Gleirscher *et al.* [32] analyze three different techniques of automated static analysis: code clone detection, bug pattern detection, and architecture conformance analysis. They claim that this combination tends to be affordable in terms of application effort and cost to correct defects.

It is noteworthy that none of the related work has implemented something similar to our proposal, since we focus on covering a set of quality characteristics with few methods, thus obtaining an optimal combination of V&V methods. While applying all available methods represents a solution, this option might not be applicable due to cost constraints.

3 Modeling the Problem

In this section we describe how the problem of finding the smallest combination of methods that cover a specific set of quality characteristics can be modelled as a *Set Cover Problem*.

Consider C as the set of characteristics, and $N(m)$ as the subset of C that is covered by a specific method m. We need to *find the smallest set of subsets that cover C*. The problem is NP-hard in general. The relation between the characteristics and the methods can be modeled as an undirected bipartite graph as shown in Fig. 1.

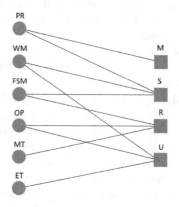

Fig. 1. Undirected bipartite graph. In the left-hand side the methods are positioned, and in the right-hand side the characteristics. Edges reflect the relationship between methods and characteristics. In the depicted instance, the set of methods contains the following elements: Peer Review (*PR*), Workflow Models (*WM*), Finite-State Machines (*FSM*), Operational Profile (*OP*), Mutation Testing (*MT*), and Exploratory Testing (*ET*). The set of characteristics is composed by four characteristics: Usability (**U**); Reliability (**R**); Security (**S**), and Maintainability (**M**). The graph shows a scenario in which method *PR* covers characteristics **M** and **S**, *WM* covers **S** and **U**, *FSM* covers **S** and **R**, *OP* covers **R** and **U**, *MT* covers **R**, and *ET* covers **U**.

The example instance was obtained from the results of a survey [21] that collected the opinion of experts about these V&V methods. The experts answered about their agreement on the suitability of the methods to address the different quality attributes of the ISO 25010 standard. The relationship between some method m and some characteristic c is obtained from the median of the survey answers (1 – disagree, 2 – partially disagree, 3 – partially agree, 4 – agree). In this example we considered that m properly covers c if the median is bigger or equal to 3. I.e., only methods that cover a quality characteristic to a certain degree will have edges in the graph. The outcome of the survey relating the quality characteristics to the V&V method can be seen in more details in [21].

For illustrative purposes we built this graph instance taking a subset of our real data, considering only four quality characteristics and some of the methods that can be used to properly address them according to the answers of the respondents (19 experts from 7 different countries, all with PhDs in software engineering, active in major software engineering and V&V venue committees, and with relevant publications in the area of V&V). The example perfectly serves our illustrative purposes to present the V&V method combination algorithm. Actually, this smaller example allows providing a better understanding of the algorithm's execution and correctness.

4 Parameterized Complexity

The *Parameterized Complexity* field emerged as a promising way to deal with NP-hard problems [12, 26]. It is a branch of the Computational Complexity Theory that focuses on classifying computational problems according to their hardness with respect to different parameters of the input. The complexity of a problem is mainly expressed through a function of these parameters.

The theory of NP-completeness was developed to identify problems that cannot be solved in polynomial time if $P \neq NP$. However, several NP-complete and NP-hard problems still need to be solved in practice.

For many problems, only super-polynomial time algorithms are known when the complexity is measured according to the size of the input, and in general, they are considered "intractable" from the theoretical point of view assuming that P is different from NP. Nevertheless, for several problems we can develop algorithms in which we can split the running time into a part computed in polynomial time with respect to the size of the input and another part computed in at least exponential time, but only with respect to a parameter k. Consequently, if we set the parameter k to a small value and its growth is relatively small we could consider these problems as "manageable" and not "intractable" [13–15].

Thus, an important question arises: *"Do these hard problems admit non-polynomial time algorithms whose exponential complexity part is a function of merely some aspects of the problem?"* [12]. The existence of such algorithms was analyzed by Downey and Fellows in [13], and is briefly discussed in the next section.

4.1 Fixed-Parameter Tractable (FPT) Approach

The fixed-parameter tractable (FPT) approach [13] considers the following format for the problems: *"Given an object x and a non-negative integer k, the goal is to determine whether x has some property that depends on k?"* The parameter k is considered small compared to the size of x. The relevance of these parameters lies precisely in the small range of values they can take, being a very important factor in practice [12].

The FPT-algorithms sacrifice the execution time, which can be exponential, but guarantee that the exponential dependency is restricted to the parameter k, which means that the problem can be solved efficiently for small values of that fixed parameter. The use of these algorithms provides a more rigorous analysis of problem's time complexity since this complexity is generally obtained from the size of the input [12].

Formally, a problem Π belongs to the class *FPT* (it is fixed-parameter tractable) with respect to a parameter k if it admits an algorithm to solve it whose running time is of the form: $f(k) \times n^a$, where a is a constant, and f is an arbitrary computable function. Note that whenever k is bounded by a constant we have $f(k) = O(1)$, hence the running time of the algorithm will be polynomial.

Finally, for the problem of this paper, we present a fixed-parameter tractable algorithm where the size k of the set of characteristics to be covered is the parameter. I.e., we are limiting the complexity by the number of relevant product characteristics to be considered when developing the software.

4.2 Scalability of the FPT-Algorithms

Scalability is the ability of a system or process to handle an increasing amount of data [16]. Computer algorithms can be called scalable if they are efficient when applied to large instances, i.e., instances with a large size of the input [17].

We can say that FPT-algorithms are scalable because they are efficient when executed in large instances. These algorithms take advantage of the specific structure of the instances, which is a differential when compared to exact or exhaustive search algorithms that require high computational time.

It is important to note that the studied problem can handle a large number of V&V methods, given that the number of quality characteristics tends to be relatively small. Therefore, an FPT-algorithm with respect to the number of characteristics to be covered will produce a tool for combination of V&V methods with high scalability.

Indeed, in our problem, the number of quality characteristics is already a known small integer (in the ISO standard this number is 8). Therefore, scalability relies on the ability to find the optimal solution even if the number of considered methods is growing. Our initial set comprises 19 methods, but additional methods have been reported by the survey respondents and our algorithm allows to efficiently work, for example, with 30, 50, or 100 methods.

5 FPT–Algorithm to Combine V&V Methods

The goal of the algorithm, shown in the Fig. 2, is to obtain the optimal combination (smallest number) of V&V methods that properly cover all the relevant quality characteristics for the product to be developed. Certainly, a software organization could complement the resulting set with other V&V methods that cover similar quality characteristics to find more defects and to further enhance quality, but at least they would know about the minimum set of methods to consider in order to address all the quality characteristics that are relevant for the product to be developed. I.e., a combination such that there is a method properly addressing (i.e., with an edge in the graph for) each relevant quality characteristic and none of them remains uncovered.

The objective function is the number of selected methods that properly cover all the characteristics. The parameter to be set is the number of the selected quality characteristics. In this way, we are parameterizing the *Set Cover Problem* by the number of characteristics to be covered by the V&V methods.

Coming up next, we present some definitions that are used in the algorithm presented in Fig. 2:

C – set of characteristics.

M – set of methods.

$N(m)$ – set of characteristics covered by the method m.

$P(c) = \{x \in M : c \in N(x)\}$ – set of methods that cover the characteristic c.

$R(m) = \{x \in M : N(x) \subseteq N(m)\}$ – set of methods that cover a subset of $N(m)$.

The input parameters are the set of characteristics C and the set of methods M. The redundant methods are removed in line 6 by using a simple preprocessing step. It removes methods that cover a subset of characteristics covered by any other method. A characteristic c is selected from the set of characteristics in line 7. The algorithm then focuses on selecting the method that will cover c in the optimal solution. In line 8, the

---

Algorithm 1 Set Cover Algorithm, **Parameters:** sets C, M

1: $M^* \leftarrow M$
2: $f^* = |M|$
3: **if** $C = \emptyset$ **then**
4:     **return** $\emptyset$
5: **else**
6:     $M \leftarrow RemoveSubsets(M)$
7:     $c \leftarrow SelectCharacteristic(C)$
8:     $C_t \leftarrow C \setminus \{c\}$
9:     **for all** $m \in P(c)$ **do**
10:        $M' \leftarrow \{m\}$
11:        $C_t \leftarrow C_t \setminus N(m)$
12:        $M_t \leftarrow M \setminus R(m)$
13:        **while** $\exists c' \in C_t : |P(c') \cap M_t| = 1$ **do**
14:            Let $m' \in P(c')$
15:            $C_t \leftarrow C_t \setminus N(m')$
16:            $M_t \leftarrow M_t \setminus \{m'\}$
17:            $M' \leftarrow M' \cup \{m'\}$
18:        **end while**
19:        $M'^* \leftarrow SetCover(C_t, M_t)$
20:        **if** $|M'^*| + |M'| < f^*$ **then**
21:            $f^* \leftarrow |M'^*| + |M'|$
22:            $M^* \leftarrow M'^* \cup M'$
23:        **end if**
24:        $C_t \leftarrow C \setminus \{c\}$
25:     **end for**
26: **end if**
27: **return** M^*

Fig. 2. Pseudocode of the set cover FPT-algorithm.

variable C_t that contains the characteristics to be covered is initialized. A loop runs through all the methods that cover c in lines 9–25. The set M' that stores the methods that will be part of a feasible solution is initialized with method m in line 10. The set of characteristics to cover C_t is updated in line 11 by removing the characteristics already covered by m. The set M_t, containing the methods available to cover C_t, is initialized in line 12 with all methods of M except those covering a subset of $N(m)$. In lines 13–18 a loop is executed while there are characteristics c' that are covered by a single method m'. The variables C_t, M_t and M' are updated in lines 15–17. The available M_t methods and the characteristics that have not been covered until now are used to obtain an optimal sub-problem solution by recursively calling the *SetCover* algorithm. In line 19, the obtained optimal solution is stored in M'^*. If the methods selected in M' together with the optimal solution M'^* of the sub-problem improve the optimum value found so far (f^*), then f^* and M^* are updated in lines 20–23. The value of C_t is reinitialized in line 24. The best solution found (M^*), is returned as the optimal solution to the problem in line 27.

5.1   Execution of the Set Cover Algorithm

Taking the graph represented in Fig. 1 as the entry of the algorithm, we now illustrate the execution of the pseudocode. After initialization steps 1–5, line 6 removes redundant methods. In this case, methods *MT* and *ET* are removed, because they cover only one characteristic, already covered by other methods. The result is shown in Fig. 3.

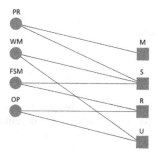

Fig. 3. Algorithm execution. State of the graph after the preprocessing step.

Afterwards, the first characteristic **M** is chosen as c, and all the methods that cover **M** must be considered in the loop that begins on line 9. Therefore, method PR is selected. In line 11, we remove all the characteristics already covered by PR, that is, **M** and **S**. The variable M_t gets the set of methods {*WM, FSM, OP*} in line 12. Since there are no characteristics covered by only one method, the loop on lines 13–18 does not perform any action, and the algorithm is called recursively in line 19 with set of characteristics {**R, U**}, and set of methods {*WM, FSM, OP*} as parameters. Figure 4 illustrates the graph at this stage.

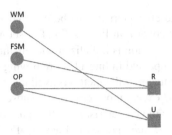

Fig. 4. Algorithm execution. State of the graph after the recursive call.

Finally, the algorithm is executed again from the beginning. Methods *WM* and *FSM* are immediately removed as redundant, and the remaining method *OP* is selected to cover the last two characteristics. The variables C_t and M_t became empty, and in the next recursive call, the stopping criterion is reached. The *OP* method is returned as a solution of the instance represented in Fig. 4, forming the final solution of the whole instance together with already selected method *PR*. The smallest set of methods M^* is set as $\{PR, OP\}$, and the optimal value f^* is set to 2.

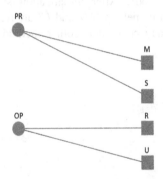

Fig. 5. Methods that form the optimal solution returned by the algorithm when executed in the graph. If **M** is selected as the first characteristic at the beginning of the execution, then the optimal set of methods returned by the algorithm is $\{PR, OP\}$.

As can be observed from the execution, the algorithm considers all the possible ways of covering the quality characteristics, keeping the most efficient ones. In this sense, the obtained solution can be considered optimal for the problem and model we pose. In fact, the algorithm is able to determine the optimal combination (smallest number) of V&V methods that properly cover quality characteristics of interest for a product to be developed based on any initial graph configuration connecting V&V methods to the quality characteristics they properly address (Fig. 5).

5.2   Running Time Analysis

Suppose that there are n methods in the set M, and there are k characteristics in the set C, being k some small integer. We note that a naive (brute force) algorithm would test all solutions (subsets of the set M) and chose which of them cover C having smaller size. Because there are 2^n subsets of the set M, this naive algorithm has a time complexity of $O(2^n)$. This exponential order is intractable even for some relatively small values of n, like 30 or 40 (which could easily be achieved when including specific variations of the V&V methods as input).

Instead, the proposed algorithm tries to determinate which method is the best option to cover each characteristic. After choosing some characteristic c, the algorithm tries to select each method that properly covers c, covering the rest of characteristics recursively. Because the number of methods that cover each characteristic is at most n, the order of this algorithm can be initially bounded by $O(n^k)$. In general, this order is already better than the 'naive' solution.

Nevertheless, we improve the upper bound of our algorithm's running time by refining the actual number of methods it will analyze. In fact, there are only 2^k different ways of covering a set C of k elements. If there is more than 2^k methods, then necessarily there will be two of them that cover exactly the same set of characteristics. That means that these two methods would be indistinguishable to our algorithm; that is, if they cover the same characteristics, any one of them can be used. Using this fact, we can successively preprocess the input, improving the algorithm performance from $O(n^k)$ to $O(f(k) \times n)$, where $f(k) \leq (2^k)^k = 2^{k^2}$. In our case, $k = 8$ and this means that $f(k)$ is bounded above by a constant, i.e., $f(k) = O(1)$. Then, we have a linear algorithm for the problem instead of an exponential or even a $O(n^8)$-time algorithm.

Once again, the upper bound for $f(k)$ is improved (decreased) using the fact that if some method m_1 covers a subset of characteristics covered by some other method m_2, then m_1 can be removed from the set of methods. This is because if m_1 is actually chosen, you can instead choose m_2, since m_2 is 'better' method in the sense that it covers all that m_1 covers, and possibly more. Lubell [18] showed that there are no more than $\binom{k}{\lfloor k/2 \rfloor}$ combinations with the property that no one is a subset of the other. This implies that for $k = 8$, there can be much less than 2^k different methods with the property that there is no method that covers a subset of characteristic of some other method. In particular, for $k = 8$ there can be at most $\binom{8}{4} = 70$ methods satisfying this property, and there can be at most $\binom{7}{3} = 35$ of these methods covering one common characteristic. Therefore, the redundant methods are removed by using a simple preprocessing step that searches for methods that cover a subset of characteristics covered by any other method. At each iteration of the algorithm, the number of characteristics to be covered decreases and the previous steps of the algorithm are repeated considering a decremented k value.

Summarizing, it holds that:

$$f(k) < \binom{k-1}{\lfloor (k-1)/2 \rfloor} \times \binom{k-2}{\lfloor (k-2)/2 \rfloor} \times \cdots \times \binom{2}{1} \tag{1}$$

For $k = 8$, it follows that $f(k) < 35 \times 20 \times 10 \times 6 \times 3 \times 2$, then our $O(f(k) \times n)$-time algorithm is an efficient (linear) algorithm where $f(k)$ is bounded above by a constant. In practice, this constant is even lower, and does not depend on the number of existing methods, which produces scalability with respect to the number of methods to be worked.

6   Computational Experiments

Several experiments were performed to assess the algorithm presented above. The algorithm was implemented in C# programming language and compiled by Roslyn, a reference C# compiler, in an Intel Core i3 machine with a 2.0 GHz processor and 4 GB of random-access memory, running under the Windows 10 operating system.

A number of test problems created by a random generator is considered. Each test problem has two parameters: the number of vertices n and the probability p of a method to cover a characteristic.

The FPT–algorithm is also executed on the instance obtained from the survey described in [21], according to the criteria explained in the *Sect.* 3 when we build the example with a subset of this same data.

Table 3 shows the optimal solution sizes and execution times (in seconds) for the FPT-Algorithm solver with and without instance preprocessing (Fig. 2, line 6) and a naive algorithm (brute force), for each instance. The name of the instance indicates the number of methods, followed by the probability of a characteristic to be covered by a method, in percent. The optimal solution sizes (number of methods returned) are equal for all instances, indicating the correctness of the Algorithms.

The FTP–Algorithm with the preprocessing is more efficient than without preprocessing and the Naive Algorithm, obtaining the result in less than 0:01 s in all cases even for the biggest instances, unlike the other algorithms in which for some instance sizes the solution is not found in a reasonable waiting time (—). The FTP–Algorithm without processing proves to be, in turn, more efficient than the Naive Algorithm, executing more instances with better runtime.

For the instance obtained from the survey [21], the FPT–Algorithm with the processing of the instance returned the methods: 12, 19 and the FPT–Algorithm without the processing and the Naive Algorithm returned the methods: 12, 18. The solution returned by the FPT–Algorithm with the processing contains the method 19 that covers a superset of the characteristics covered by the method 18 in the others algorithm solutions, showing that FPT–Algorithm with preprocessing performs better when concerning coverage, by using this additional comparison criterion.

Table 3. Computational experiments

Instance	Runtime FPT-Alg (with pre-processing)	Optimal solution FPT-Alg (with pre-processing)	Runtime FPT-Alg (no pre-processing)	Optimal solution FPT-Alg (no pre-processing)	Runtime Alg-Naive (brute force)	Optimal Alg-Naive (brute force)
Instance_20_10	0.00031	5	0.00375	5	0.14907	5
Instance_20_20	0.00047	3	0.01094	3	0.17703	3
Instance_20_50	0.00062	2	0.08859	2	0.22297	2
Instance_50_10	0.00031	5	0.14359	5	—	—
Instance_50_20	0.00110	2	4.97453	2	—	—
Instance_50_50	0.00094	2	11.4025	2	—	—
Instance_100_10	0.00094	2	64.7025	2	—	—
Instance_100_20	0.00125	3	—	—	—	—
Instance_100_50	0.00110	2	—	—	—	—
Instance_200_10	0.00344	3	—	—	—	—
Instance_200_20	0.00188	2	—	—	—	—
Instance_200_50	0.00094	1	—	—	—	—
Instance_500_10	0.00359	3	—	—	—	—
Instance_500_20	0.00469	2	—	—	—	—
Instance_500_50	0.00234	1	—	—	—	—
Instance_1000_10	0.00766	3	—	—	—	—
Instance_1000_20	0.00578	2	—	—	—	—
Instance_1000_50	0.00500	1	—	—	—	—

7 Discussion

Our proposed algorithm is effective, being able to provide the optimal combination (smallest number) of V&V methods properly covering a set of chosen quality characteristics to be considered when developing a software product. Additionally, it is more efficient than brute-force or exhaustive search algorithms and its execution time properties match the particularities of the problem well. Indeed, the algorithm can be applied to instances of different sizes, making our approach scalable, i.e., suitable for larger case studies (for instance, considering more V&V methods, including specific variations of the more generic methods used for our sample).

There is, however, a basic assumption for applying the algorithm, which is having a defined input with information on which V&V methods properly address the different quality characteristics. For illustrative purposes, our example was based on initial outcomes of an expert survey. It is also noteworthy that our set of 19 V&V methods represents generic methods for which several variations are available (e.g., applying specific testing criteria or variations of inspection methods). While they perfectly fit our illustrative example and allowed us getting feedback from experts on whether they can be employed to properly address quality characteristics, information on more specific methods could be provided as input to combination algorithm. We highlight that this initial configuration is out of the scope of the intended contribution of this paper and that companies could use an initial configuration based on their own sets of evidence on the V&V methods they typically use or on their own elicited expert beliefs.

Moreover, from a practical point of view, companies might decide to complement the optimal solution provided by the algorithm by applying additional V&V methods that cover similar quality characteristics (e.g., aiming at finding additional defects and further enhancing product quality), in particular for critical projects. However, using our approach at least they would know about a minimum set of methods that would allow them avoiding neglecting quality characteristics that are relevant for the product to be developed.

Also, specialists on software engineering economics might argue that our solution providing the smallest number of V&V methods is not considering the cost of applying each method. However, to address this issue we would need to know the relative cost among the V&V methods and this information is extremely context specific and hard to generalize. We are aware of this limitation and further addressing it is part of our future work. A solution option to handle this issue when using the approach described in this paper would be removing the methods that are cost restrictive from the initial configuration.

8   Concluding Remarks

In this paper, we modeled the problem of finding a combination of V&V methods to cover software quality characteristics as the Set Cover problem, a NP-hard combinatorial optimization problem. We defined a parameterized FPT algorithm that is specially designed for our instances, since typically the number of considered quality characteristics is small. Provided by a valid input, the proposed algorithm is able to efficiently provide an optimal combination (smallest number) of V&V methods properly covering a set of chosen quality characteristics to be considered when developing a software product. Additionally, we showed that it is more efficient than Naive (brute-force) algorithms. Furthermore, the algorithm can be applied to instances of different sizes, making our approach scalable, i.e., suitable for larger studies (for instance, considering more V&V methods).

Our future works consist of development of a support tool that, given a set of selected quality characteristics and an initial configuration (e.g., from the survey results, or any other source such as within-company expert belief elicitation), provide the optimal combination of V&V methods. Finally, for now we focused on product quality, and a next step would be to integrate cost-related issues into the approach. Moreover, we believe that the Fixed-Parameter Tractable algorithm approach can be applied to solve other problems in the software engineering domain and that sharing our V&V method combination experience with the community could foster discussions towards other graph theory-based solutions for relevant software engineering problems.

Acknowledgment. The authors would like to thank CNPq and FAPERJ for the financial support (Project N°. E-26/010.001578/2016, Title: *"Resolution of Critical Problems of the Software Industry through Graph Theory and its algorithms"*). Thanks also to the survey respondents, which provided us the initial configurations to test our approach.

References

1. Meyers, G.J., Badgett, T., Thomas, T., Csandler, C.: The Art of Software Testing, 3rd edn. Wiley, Hoboken (2011). ISBN 978-1118031964
2. Feldt, R., Torkar, R., Ahmad, E., Raza, B.: Challenges with software verification and validation activities in the space industry. In: Third International Conference on Software Testing, Verification and Validation (ICST) (2010)
3. Boehm, B., Basili, V.: Software defect reduction top 10 list. IEEE Softw. **34**(1), 135–137 (2001)
4. Feldt, R., Marculescu, B., Schulte, J., Torkar, R., Preissing, P., Hult, E.: Optimizing verification and validation activities for software in the space industry. In: Data Systems in Aerospace (DASIA), Budapest (2010)
5. Bourque, P., Fairley, R.E.: SWEBOK guide V3.0, guide to the software engineering body of knowledge. IEEE Computer Society (2004)
6. Endres, A., Rombach, D.: A Handbook of Software and Systems Engineering. Addison Wesley, Reading (2003)
7. Myers, G.J.: A controlled experiment in program testing and code walkthroughs/inspections. Commun. ACM **21**(9), 760–768 (1978)
8. Wood, M., Roper, M., Brooks, A., Miller, J.: Comparing and combining software defect detection techniques: a replicated empirical study. In: Jazayeri, M., Schauer, H. (eds.) ESEC/SIGSOFT FSE-1997. LNCS, vol. 1301, pp. 262–277. Springer, Heidelberg (1997). https://doi.org/10.1007/3-540-63531-9_19
9. Elberzhager, F., Münch, J., Nha, V.T.N.: A systematic mapping study on the combination of static and dynamic quality assurance techniques. Inf. Softw. Technol. **54**(1), 1–15 (2012)
10. ISO25000 Software Product Quality, ISO/IEC 25010, Official site (2011). http://iso25000.com/index.php/en/iso-25000-standards/iso-25010
11. Karp, R.M.: Reducibility among combinatorial problems. In: Miller, R.E., Thatcher, J.W. (eds.) Complexity of Computer Computations, pp. 85–103. Plenum, New York (1972)
12. dos Santos, V.F., dos Santos Souza, U.: Uma Introdução à Complexidade Parametrizada. In: Anais da 34° Jornada de Atualização em Informática, CSBC, pp. 232–273 (2015)
13. Downey, R.G., Fellows, M.R.: Parameterized Complexity. Monographs in Computer Science. Springer, New York (1999). https://doi.org/10.1007/978-1-4612-0515-9
14. Flum, J., Grohe, M.: Parameterized Complexity Theory. Springer, Heidelberg (2006). https://doi.org/10.1007/3-540-29953-X
15. Niedermeier, R.: Invitation to Fixed-Parameter Algorithms. Oxford Lecture Series in Mathematics and Its Applications. Oxford University Press, Oxford (2006)
16. Bondi, A.B.: Characteristics of scalability and their impact on performance. In: Proceedings Second International Workshop on Software and Performance WOSP, pp. 195–203 (2000)
17. Laudon, K.C., Traver, C.G.: E-commerce: Business, Technology, Society. Stanford University, Stanford (2008)
18. Lubell, D.: A short proof of Sperner's lemma. J. Comb. Theory **1**(2), 299 (1996)
19. Wagner, S.: Software Product Quality Control. Springer, Heidelberg (2013). https://doi.org/10.1007/978-3-642-38571-1

20. Wiegers, K.E.: Peer Reviews in Software: A Practical Guide, 1st edn. Addison-Wesley Longman Publishing Co., Inc., Boston (2002)

21. Mendoza, I., Kalinowski, M., Souza, U., Felderer, M.: Relating verification and validation methods to software product quality characteristics: results of an expert survey. In: 11th Software Quality Days (SWQD). Lecture Notes on Business Information Processing, Vienna, Austria. Springer (2019, to appear)

22. Basili, V.R.: Comparing the effectiveness of software testing strategies. IEEE Trans. Softw. Eng. 13(12), 1278–1296 (1987)

23. Kamsties, E., Lott, C.M.: An empirical evaluation of three defect-detection techniques. In: Schäfer, W., Botella, P. (eds.) ESEC 1995. LNCS, vol. 989, pp. 362–383. Springer, Heidelberg (1995). https://doi.org/10.1007/3-540-60406-5_25

24. Wagner, S., Jürjens, J., Koller, C., Trischberger, P.: Comparing bug finding tools with reviews and tests. In: Khendek, F., Dssouli, R. (eds.) TestCom 2005. LNCS, vol. 3502, pp. 40–55. Springer, Heidelberg (2005). https://doi.org/10.1007/11430230_4

25. Dwyer, M.B., Elbaum, S.: Unifying verification and validation techniques: relating behavior and properties through partial evidence. In: FSE/SDP Workshop on Future of Software Engineering Research (FOSE), Santa Fe, New Mexico, USA, pp. 93–98 (2010)

26. Cygan, M., et al.: Parameterized Algorithms. Springer, Cham (2015). https://doi.org/10.1007/978-3-319-21275-3

27. Runeson, P., Stefik, A., Andrews, A., Grönblom, S., Porres, I., Siebert, S.: A comparative analysis of three replicated experiments comparing inspection and unit testing. In: Proceedings 2nd International Workshop on Replication in Empirical Software Engineering Research (RESER), Banff, AB, Canada, Article No. 6148335, pp. 35–42 (2012)

28. Olorisade, B.K., Vegas, S., Juristo, N.: Determining the effectiveness of three software evaluation techniques through informal aggregation. Inf. Softw. Technol. 55(9), 1590–1601 (2013)

29. Cotroneo, D., Pietrantuono, R., Russo, S.: A learning-based method for combining testing techniques. In: Proceedings 35th International Conference on Software Engineering (ICSE), San Francisco, CA, USA, Article No. 6606560, pp. 142–151 (2013)

30. Bishop, P., Bloomfield, R., Cyra, L.: Combining testing and proof to gain high assurance in software: a case study. In: IEEE 24th International Symposium on Software Reliability Engineering (ISSRE), Pasadena, CA, USA, Article No. 6698924, pp. 248–257 (2013)

31. Solari, M., Matalonga, S.: A controlled experiment to explore potentially undetectable defects for testing techniques. In: Proceedings of the 26th International Conference on Software Engineering and Knowledge Engineering (SEKE), Canada, pp. 106–109 (2014)

32. Gleirscher, M., Golubitskiy, D., Irlbeck, M., Wagner, S.: Introduction of static quality analysis in small- and medium-sized software enterprises: experiences from technology transfer. Softw. Qual. J. 22(3), 499–542 (2014)

Robustness Radius
for Chamberlin-Courant on Restricted
Domains

Neeldhara Misra[(⊠)] and Chinmay Sonar

Indian Institute of Technology, Gandhinagar, Gandhinagar, India
{neeldhara.m,sonar.chinmay}@iitgn.ac.in

Abstract. The notion of robustness in the context of committee elections was introduced by Bredereck et al. [SAGT 2018] [2] to capture the impact of small changes in the input preference orders, depending on the voting rules used. They show that for certain voting rules, such as Chamberlin-Courant, checking if an election instance is robust, even to the extent of a small constant, is computationally hard. More specifically, it is NP-hard to determine if one swap in any of the votes can change the set of winning committees with respect to the Chamberlin-Courant voting rule. Further, the problem is also W[1]-hard when parameterized by the size of the committee, k. We complement this result by suggesting an algorithm that is in XP with respect to k. We also show that on nearly-structured profiles, the problem of robustness remains NP-hard. We also address the case of approval ballots, where we show a hardness result analogous to the one established in [2] about rankings and again demonstrate an XP algorithm.

Keywords: Robustness radius · Chamberlin-Courant · Single-peaked · Single-crossing · NP-hardness

1 Introduction

A *voting rule* is a function that maps a collection of preferences over a fixed set of alternatives to a set of winning options, where each option could be one or more alternatives—corresponding, respectively, to the scenarios of single-winner and committee elections. A voting rule is *vulnerable to change* if small perturbations in the input profile can cause its outcome to vary wildly. There have been several notions in the contemporary computational social choice literature that captures the degree of vulnerablity of various voting rules.

A recent exercise in this direction was carried out in [2], where the notion of *robustness radius* was introduced as the minimum number of swaps that was required between consecutive alternatives to change the outcome of a multiwinner voting rule. We note here that we are implicitly assuming that preferences are modeled as linear orders over the alternatives, although the notion of swaps can be defined naturally for the situation where the votes are given by approval

© Springer Nature Switzerland AG 2019
B. Catania et al. (Eds.): SOFSEM 2019, LNCS 11376, pp. 341–353, 2019.
https://doi.org/10.1007/978-3-030-10801-4_27

ballots (each vote indicates the set of approved candidates). In the work of [2], several voting rules are considered, and efficient algorithms were proposed for ROBUSTNESS RADIUS for many of these rules. On the other hand, for some voting rules, the problem turned out to be hard: even when the question was to decide if there is *one* swap that influences the outcome. This is the motivation for the present work: we focus on the Chamberlin-Courant voting rule (c.f. Sect. 2 on Preliminaries for the definition), for which ROBUSTNESS RADIUS turns out to be intractable, and look for exact algorithms on general profiles and ask if the problem becomes easier to tackle on structured preferences.

Our Contributions. Our first contribution is an explicit XP algorithm (recall that a problem is XP parameterized by k if there exists an algorithm which solves it in time $\mathcal{O}(n)^{f(k)}$) for the ROBUSTNESS RADIUS problem in the context of the Chamberlin-Courant voting rule. Recall that it is already NP-hard to determine if there exists *one* swap which changes the set of winning committees. Notice that the natural brute-force approach to check if there are at most r swaps which affect the set of winning committees is to simply try all possible ways of executing r swaps and recompute the set of winning committees at every step. This approach, roughly speaking, requires $O((mn)^r \cdot m^k)$ time where m, n are number of candidates and voters (respectively) in the given election instance. We improve this by suggesting an algorithm whose running time can be bounded by $O^\star(m^k)$. We show this result for both the Chamberlin-Courant voting rule with the Borda misrepresentation function as well as for the approval version of the Chamberlin-Courant voting rule. For the latter, we also show that an analogous hardness result holds.

On the other hand, we initiate an exploration of whether the ROBUSTNESS RADIUS problem remains hard on structured preferences. We provide some insights on this issue by demonstrating that the problem remains NP-hard on "nearly-structured" profiles. In particular, we show that:

1. Determining if the robustness radius of a profile is one for the ℓ_1-CC (respectively, ℓ_∞-CC) voting rule, with respect to the Borda misrepresentation score, is NP-hard even when the input profiles are restricted to the six-crossing domain[1] (respectively, the four-crossing domain).
2. Determining if the robustness radius of a profile is one for the ℓ_∞-CC voting rule, with respect to the Borda misrepresentation score, is NP-hard even when the domain is a four-composite single-peaked domain.

Related Work. The notion of robustness is also captured by other closely related notions, such as the margin of victory (MoV) [11] and swap bribery [5]. In the former, the metric of change is the number of voters who need to be influenced, rather than the total number of swaps. On the other hand, in swap bribery, the goal is not to simply influence a change in the set of committees, but to

---

[1] We refer the reader to the section on Preliminaries for the definition of ℓ-single-crossing domains. Some definitions and results are deferred to the full version due to lack of space and are marked with a ($\star$).

ensure that a specific committee does or does not win (corresponding to constructive and destructive versions of the problem, respectively). We note that swap bribery has been mostly studied in the context of single-winner voting rules. Observe that any profile that is a non-trivial YES-instance of swap bribery is also a YES-instance of ROBUSTNESS RADIUS with the same budget, but the converse is not necessarily true. Similarly, any profile that is a YES-instance of ROBUSTNESS RADIUS is also a YES-instance of MoV with the same budget, but again the converse need not be true. However, we remark that in the case of the Approval-CC voting rule, the notions of ROBUSTNESS RADIUS and MoV happen to coincide. Robustness has also been studied for single-winner voting rules in earlier work [10].

2 Preliminaries

In this section, we introduce some key definitions and establish notation. For a comprehensive introduction, we refer the reader to [1,6].

Notation. For a positive integer ℓ, we denote the set $\{1, \ldots, \ell\}$ by $[\ell]$. We first define some general notions related to voting rules. Let $V = \{v_i : i \in [n]\}$ be a set of n *voters* and $C = \{c_j : j \in [m]\}$ be a set of m *candidates*. If not mentioned otherwise, we denote the set of candidates, the set of voters, the number of candidates, and the number of voters by C, V, m, and n respectively.

Every voter v_i has a *preference* $\succ_i$ which is typically a complete order over the set C of candidates (rankings) or a subset of approved candidates (approval ballots). An instance of an election consists of the set of candidates C and the preferences of the voters V, usually denoted as $E = (C, V)$. A *multiwinner committee rule* $\mathcal{R}$ is a function that, given an election E and a committee size k, outputs a family $\mathcal{R}(E, k)$ consisting winning committees of k-sized subsets of C.

We now state some definitions in the context of rankings, although we remark that analogous notions exist also in the setting of approval ballots. We say voter v_i prefers a candidate $x \in C$ over another candidate $y \in C$ if $x \succ_i y$. We denote the set of all preferences over C by $\mathcal{L}(C)$. The n-tuple $(\succ_i)_{i \in [n]} \in \mathcal{L}(C)^n$ of the preferences of all the voters is called a *profile*. Note that a profile, in general, is a multiset of linear orders. For a subset $M \subseteq [n]$, we call $(\succ_i)_{i \in M}$ a sub-profile of $(\succ_i)_{i \in [n]}$. For a subset of candidates $D \subseteq C$, we use $\mathcal{P}|_D$ to denote the projection of the profile on the candidates in D alone. A *domain* is a set of profiles.

Chamberlin-Courant for Rankings. The Chamberlin–Courant voting rule is based on the notion of a *dissatisfaction function* or a *misrepresentation function* (we use these terms interchangeably). This function specifies, for each $i \in [m]$, a voter's dissatisfaction from being represented by candidate she ranks in position i. A popular dissatisfaction function is Borda, given by $\alpha_B^m(i) = \alpha_B(i) = i - 1$, and this will be our measure of dissatisfaction in the setting of rankings.

We now turn to the notion of an assignment function. Let k be a positive integer. A *k-CC-assignment function* for an election $E = (C, V)$ is a mapping

$\Phi: V \to C$ such that $\|\Phi(V)\| = k$, where $\|\Phi(V)\|$ denotes the image of Φ. For a given assignment function Φ, we say that voter $v \in V$ is *represented* by candidate $\Phi(v)$ in the chosen committee. There are several ways to measure the quality of an assignment function Φ with respect to a dissatisfaction function α; we use the following:

1. $\ell_1(\Phi, \alpha) = \sum_{i=1,\ldots,n} \alpha(\mathrm{pos}_{v_i}(\Phi(v_i)))$, and
2. $\ell_\infty(\Phi, \alpha) = \max_{i=1,\ldots,n} \alpha(\mathrm{pos}_{v_i}(\Phi(v_i)))$.

Unless specified otherwise, α will be the Borda dissatisfaction function described above. We are now ready to define the Chamberlin-Courant voting rule.

Definition 1 (Chamberlin-Courant [3]). *For $\ell \in \{\ell_1, \ell_\infty\}$, the ℓ−CC voting rule is a mapping that takes an election $E = (C, V)$ and a positive integer k with $k \leqslant |C|$ as its input, and returns the images of all the k-CC-assignment functions Φ for E that minimizes $\ell(\Phi, \alpha)$.*

Chamberlin Courant for Approval Ballots. Recall that an approval vote v on the set of candidates C is an arbitrary subset S_v of C such that v approves all the candidates in S_v. We define the misrepresentation score for k-sized committee T for an approval voting profile as the number of voters which do not have any of their approved candidates in T (i.e. $T \cap S_v = \phi$). Hence the optimal committees under approval Chamberlin Courant are the committees which maximize the number of voters with at least one approved candidate in the winning committee. This notion of Chamberlin-Courant for the setting of approval ballots was proposed by [8].

Single Crossing Profiles. A preference profile is said to belong to the single crossing domain if it admits a permutation of the voters such that for any pair of candidates a and b, there is an index $j[(a,b)]$ such that either all voters v_j with $j < j[(a,b)]$ prefer a over b and all voters v_j with $j > j[(a,b)]$ prefer b over a, or vice versa. The formal definition is as follows.

Definition 2 (Single Crossing Domain). *A profile $\mathcal{P} = (\succ_i)_{i \in [n]}$ of n preferences over a set C of candidates is called a single crossing profile if there exists a permutation σ of $[n]$ such that, for every pair of distinct candidates $x, y \in C$, whenever we have $x \succ_{\sigma(i)} y$ and $x \succ_{\sigma(j)} y$ for two integers i and j with $1 \leqslant \sigma(i) < \sigma(j) \leqslant n$, we have $x \succ_{\sigma(k)} y$ for every $\sigma(i) \leqslant k \leqslant \sigma(j)$.*

We generalize the notion of single-crossing domains to r-single crossing domains in the following natural way (c.f. [9]): for every pair of candidates (a, b), instead of demanding one index where the preferences "switch" from one way to the other, we allow for r such switches. More formally, a profile is r-single crossing if for every pair of candidates a and b, there exist r indices $j_0[(a,b)], j_1[(a,b)], \ldots j_r[(a,b)], j_{r+1}[(a,b)]$ with $j_0[(a,b)] = 1$ and $j_{r+1}[(a,b)] = n+1$, such that for all $1 \leqslant i \leqslant r+1$, all voters v_j with $j_i[(a,b)] \leqslant j < j_{i+1}[(a,b)]$ are unanimous in their preferences over a and b.

Robustness Radius. Let $\mathcal{R}$ be a multiwinner voting rule. For the given election $E = (C, V)$, a committee size k, and an integer r, in the $\mathcal{R}$-ROBUSTNESS RADIUS problem we ask if it is possible to obtain an election E' by making at most r swaps of adjacent candidates within the rankings in E (or by introducing or removing at most r candidates from the approval sets of voters in case of approval ballots) so that $\mathcal{R}(E', k) \neq \mathcal{R}(E, k)$.

Parameterized Complexity. We occasionally use terminology from parameterized complexity, mainly to describe our results in an appropriate context. A parameterized problem is denoted by a pair $(Q, k) \subseteq \Sigma^* \times \mathbb{N}$. The first component Q is a classical language, and the number k is called the parameter. Such a problem is *fixed–parameter tractable* (FPT) if there exists an algorithm that decides it in time $O(f(k)n^{O(1)})$ on instances of size n. On the other hand, a problem is said to belong to the class XP if there exists an algorithm that decides it in time $n^{O(f(k))}$ on instances of size n. We refer the reader to [4] for a comprehensive introduction to parameterized algorithms.

3   XP Algorithms for Robustness Radius

The ROBUSTNESS RADIUS problem for the ℓ_1-Chamberlin-Courant voting rule with the Borda dissatisfaction function is known to be in FPT when parameterized by either the number of candidates or the number of voters. For the former, the approach involves formulating the problem as an ILP and then using Lenstra's algorithm. In the case of the latter, the algorithm is based on guessing all possible partitions of the voters based on their anticipated representatives and then employing a dynamic programming approach.

In this section, we give a simple but explicit algorithm for the problem which has a XP running time in k, the committee size. This complements the W[1]-hardness of the problem when parameterized by k [2]. We establish this result for both when the votes are rankings as well as when they are approval ballots. First, we address the case when the votes are rankings.

Theorem 1. *On general profiles comprising of rankings over alternatives,* ROBUSTNESS RADIUS *for the ℓ_1-Chamberlin-Courant voting rule with the Borda dissatisfaction function admits a $O^*(m^k)$ algorithm, where m is the number of candidates and k is the committee size.*

Proof. We first determine the set of all optimal committees of size k in time $O(m^k)$. Suppose there are at least two committees, say A and B, that are both optimal. The manner in which this case can be handled is also addressed in [2]. For the sake of completeness, we reproduce the main point here, but in particular we do not address certain edge cases: for example, a slightly different discussion is called for if there are less than k candidates in total occupying the top positions across the votes. We refer the reader to [2] for a more detailed explaination.

Now, note that since A and B are distinct committees, there is at least one voter v whose Chamberlin-Courant representative with respect to A and

B are distinct candidates: say c_a and c_b, respectively. Assume, without loss of generality, that $c_a \succ_v c_b$. Note that swapping the candidate c_b so that its rank in the vote v decreases by one results in a new profile where:

1. the dissatisfication score of the committee B is one less than in the original profile, and,
2. the dissatisfication score of the committee A is at least its score in the original profile (indeed; the dissatisfaction score either stays the same or increases if c_a is adjacent to c_b in the vote v).

Therefore, when there are at least two optimal committees, it is possible to change the set of winning committees with only one swap, making this situation easy to resolve. We now turn to the case when the input profile admits a unique winning committee A. Our overall approach in this case is the following: we "guess" a committee B that belongs to the set of winning committees after r swaps (note that such a committee must exist if we are dealing with a YES-instance). For a fixed choice of B, we determine, greedily, the minimum number swaps required to make B a winning committee. We now turn to a formal description of the algorithm.

Recall that a profile Q is said to be within r swaps of a profile P if Q can be obtained by at most r swaps of consecutive candidates in P. In the following discussion, we say that a committee B is *nearly winning* if there exists a profile Q, within r swaps of P, where B is a winning committee. We refer to Q as the *witness* for B. Note that the existence of a nearly winning committee $B \neq A$ characterizes the YES-instances. Let $\Delta_{B,A}(P)$ denote the difference between the dissatisfaction scores of the committees B and A with respect to the profile P. We begin by making the following observation.

Proposition 1. *Let P and Q be two profiles such that Q can be obtained by making at most r swaps of consecutive candidates in the profile P. Note that:*

$$\Delta_{B,A}(P) - 2r \leqslant \Delta_{B,A}(Q) \leqslant \Delta_{B,A}(P) + 2r.$$

The claim above follows from the fact that if Q is a profile obtained from P by one swap of consecutive candidates in some vote of P, then it is easy to see that $\Delta_{B,A}(P) - 2 \leqslant \Delta_{B,A}(Q) \leqslant \Delta_{B,A}(P) + 2$. Note that if B is nearly winning, then $\Delta_{B,A}(Q) \leqslant 0$, where Q is the witness profile. We now have a case analysis based on $\Delta_{B,A}(P)$.

Case 1. $\Delta_{B,A}(P) > 2r$. In this case, by Proposition 1, we know that in every profile Q within r swaps of P, $\Delta_{B,A}(Q) > 0$, which is to say that B will have a greater Borda dissatisfaction score than A in every profile that is r swaps away from the input profile. Therefore, in this case, we reject the choice of B as a potential nearly winning committee.

Case 2. $\Delta_{B,A}(P) \leqslant r$. An analogous argument can be used to see that B is in fact nearly winning in this case. Indeed, any r swaps that improve the ranks of the candidates in B will result in a profile Q that is within r swaps of P and where $\Delta_{B,A}(\mathcal{Q}) \leqslant 0$. So, B is either nearly winning with witness profile $\mathcal{Q}$, or A is no longer a winning committee in $\mathcal{Q}$. Therefore, in this situation, we output YES.

Case 3. $\Delta_{B,A}(\mathcal{P}) = r + s, 1 \leqslant s \leqslant r$. For a vote v, let $A(v)$ and $B(v)$ denote, respectively, the candidates from A and B with the highest rank in the vote v. Further, let $d_{B,A}(v)$ denote the difference between the ranks of $B(v)$ and $A(v)$. Let $W \subseteq V$ be the subset of votes for which $d_{B,A}(v) > 0$, and let $w_1, w_2, \ldots$ denote an ordering of the votes in W in increasing order of these differences. We now make the following claim.

Proposition 2. *There exists a profile $\mathcal{Q}$ that is r swaps away from $\mathcal{P}$ where $\Delta_{B,A}(\mathcal{Q}) \leqslant 0$ if, and only if:*

$$t := \sum_{i=1}^{s} d_{B,A}(w_i) \leqslant r. \tag{1}$$

Proof. In the forward direction, suppose (1) holds. Then perform swaps in the votes $w_1, \ldots, w_s$ so that for any $i \in [s]$, the candidate $B(w_i)$ is promoted to the position just above $A(w_i)$. In other words, each swap involves $B(w_i)$ and in the profile obtained after the swaps, $B(w_i) \succ A(w_i)$ for all $i \in [s]$, and the difference in the ranks of these pairs is exactly one. Note that a total of t swaps are performed to obtain this profile. Denote this profile by $\mathcal{R}$ and note that $\Delta_{B,A}(\mathcal{R}) = r + s - t - s = r - t$ (since the last swap made on each vote w_i reduces the gap between the dissatisfaction scores of the two committees by two). Also, $(r - t)$ is also exactly the number of remaining swaps we can still make, so a witness profile can be obtained using the argument we made in the previous case. The proof of the other direction is deferred to a full version due to lack of space.   □

To summarize, our algorithm in this case identifies and sorts the votes in W, and returns YES if condition (1) holds, and rejects the choice of B otherwise. Observe that we output No if no choice of B results in a positive outcome in this case analysis. In terms of the running time, we require $O(m^k)$ time in distinguishing whether we have a unique winning committee or not, and if we are in the former situation, we need $O(m^k)$ time to guess a nearly winning committee. For each choice B of a potential winning committee, we spend time $O(mn \log n)$ in the worst case to determine if B is indeed a nearly winning committee. Therefore, hiding polynomial factors, the overall running time of our algorithm is $O^\star(m^k)$ and this concludes the proof.   □

We now turn to the case of approval ballots. First, we show that the robustness radius problem in this setting remains NP-hard even for determining if the robustness radius is one, as was true for the case when the votes were rankings.

Theorem 2. ROBUSTNESS RADIUS *for the Approval Chamberlin-Courant voting rule is* NP*-hard, even when the robustness radius is one and each voter approves at most three candidates. It is also* W[2]*-hard parameterized by the size* of the committee when there are no restrictions on the size of the number of candidates approved by a voter, and the robustness radius is one.

Proof. We reduce from the HITTING SET problem. Note that the NP-hardness in the restricted setting follows from the fact that HITTING SET is already hard for sets of size at most two (recall that this is the VERTEX COVER problem), while the W[2]-hardness follows from the fact that HITTING SET is W[2]-hard when parameterized by the size of the hitting set [4] and our reduction will be parameter-preserving with respect to the parameter of committee size.

Let $(U, \mathcal{F}; k)$ be an instance of HITTING SET. Recall that this is a YES-instance if and only if there exists $S \subseteq U$, with $|S| \leqslant k$ such that $S \cap X \neq \emptyset$ for any $X \in \mathcal{F}$. We construct a profile $\mathcal{P}$ over alternatives $\mathcal{A}$ as follows. Let:

$$\mathcal{A} := \underbrace{\{c_u \mid u \in U\}}_{\mathcal{C}} \cup \underbrace{\{d_1, \ldots, d_k\}}_{\mathcal{D}}$$

Also, for every $1 \leqslant i \leqslant k$, and for every $X \in \mathcal{F}$, introduce a vote $v(X, i)$ that approves the candidates corresponding to the elements in X along with d_i. This completes the construction of the instance. We claim that this instance has a robustness radius of one if and only if $(U, \mathcal{F}; k)$ is a YES-instance of HITTING SET.

Forward Direction. Suppose S is a hitting set for $(U, \mathcal{F})$ of size k. Then the set $C_S := \{c_u \mid u \in S\}$ and $\mathcal{D}$ are two optimal Approval-CC committees with dissatisfaction scores of zero each. Note that removing the candidate d_1 from any vote of the form $v(X, 1)$ will lead to a profile where the set of winning committees contains C_S but does not contain $\mathcal{D}$. Hence, the robustness radius is indeed one.

Reverse Direction. For the reverse direction, suppose the profile $\mathcal{P}$ has robustness radius one. We will now argue the existence of a hitting set of size at most k. Note that $\mathcal{D}$ is already an optimal committee with respect to $\mathcal{P}$ as it has the best possible Approval-CC dissatisfaction score of zero. Now, suppose $\mathcal{P}$ admits another winning committee $\mathcal{W}$ distinct from $\mathcal{D}$. Then notice that the Approval-CC dissatisfaction score of $\mathcal{W}$ must also be zero, and since there is at least one candidate from $\mathcal{D}$ (say d_i) that is not present in $\mathcal{W}$, it is easy to see that the candidates in $\mathcal{C} \cap \mathcal{W}$ form a hitting set for the instance $(U, \mathcal{F}; k)$—indeed, note that every voter in the sub-profile $\{v(X, i) \mid X \in \mathcal{F}\}$ does not approve anyone in $\mathcal{D} \cap \mathcal{W}$, and therefore must approve someone of in $\mathcal{C} \cap \mathcal{W}$, making this a hitting set for $\mathcal{F}$.

Therefore, the interesting case is when $\mathcal{D}$ is the unique winning committee for $\mathcal{P}$. We claim that any other subset of candidates $\mathcal{W}$ of size k has an Approval-CC dissatisfaction score of at least two. This would imply that the robustness radius of $\mathcal{P}$ cannot possibly be one, and therefore there is nothing to prove. To this end, observe that $C_W := \mathcal{W} \cap \mathcal{C}$ is not a hitting set[2] for $\mathcal{F}$: indeed, if C_W

---

[2] Note the slight abuse of terminology here: when referring to C_W as a hitting set, we are referring to the elements of U corresponding to the candidates in C_W. As long as this is clear from the context, we will continue to use this convention.

was a hitting set then it is easy to see that $\mathcal{W}$ is also an optimal committee with respect to $\mathcal{P}$, contradicting the case that we are in. Let X denote a set that is not hit by C_W. Now, we consider two cases:

$\mathcal{W}$ Omits Two Candidates from $\mathcal{D}$. In this case, there are at least two candidates in $\mathcal{D}$—say d_i and d_j—who do not belong to $\mathcal{W}$. Then $\mathcal{W}$ earns a dissatisfaction score of one from each of $v(X, i)$ and $v(X, j)$, which makes its dissatisfaction score at least two, as desired.

$\mathcal{W}$ Omits Exactly One Candidate from $\mathcal{D}$. In this case, notice that $|C_W| = 1$ and that C_W does not hit at least two sets, say X and Y: else C_W along with an arbitrarily chosen element from X and another chosen from Y, along with an arbitrary choice of $k - 3$ additional candidates would constitute a winning committee in $\mathcal{P}$ different from $\mathcal{D}$, again contradicting the case that we are in. Therefore, observe that d_i is the candidate from $\mathcal{D}$ that is not present in $\mathcal{W}$, the votes $v(X, i)$ and $v(Y, i)$ contribute one each to the dissatisfaction score of the committee W.

Overall, therefore, if $\mathcal{D}$ is the unique winning committee in $\mathcal{P}$, then the robustness radius is greater than one, and there is nothing to prove. This concludes our argument in the reverse direction. □

We now turn to $O^\star(m^k)$ algorithm for ROBUSTNESS RADIUS with respect to approval ballots. The general approach is quite analogous to the setting of rankings. However, the notion of swaps is slightly different, and the overall case analysis is, in fact, simpler. Since the main ideas are identical, in the interest of space, we defer a proof of the following claim to a full version of the paper.

Lemma 1 ($\star$). *On general profiles comprising of approval ballots over alternatives, ROBUSTNESS RADIUS for the ℓ_1-Chamberlin-Courant voting rule with the Borda dissatisfaction function admits a $O^\star(m^k)$ algorithm, where m is the number of candidates and k is the committee size.*

4 Hardness for ℓ-Crossing Profiles

In this section, we explore the complexity of ROBUSTNESS RADIUS on nearly-structured preferences. We discover that the problem remains NP-hard parameterized by the size of the committee sought, even on profiles which are 6-crossing even when the robustness radius is one. We note that our overall approach is very similar to the one employed in [2].

Theorem 3. *Determining if the robustness radius of a profile is one for the ℓ_1-CC voting rule, with respect to the Borda misrepresentation score, is NP-hard even when the input profiles are restricted to the six-crossing domain.*

Proof. We reduce from INDEPENDENT SET ON 3-REGULAR GRAPHS. Let (G, t) be an INDEPENDENT SET ON 3-REGULAR GRAPHS [7]. We construct a profile based on G as follows. Our set of candidates $\mathcal{C}$ is given by:

$$\mathcal{C} := \underbrace{\{c_u \mid u \in V(G)\}}_{V} \cup \underbrace{\{d_1, \ldots, d_h\}}_{\mathcal{D}} \cup \underbrace{\{Z_0, Z_1\}}_{\mathcal{Z}} \cup \underbrace{\{x_1, \ldots, x_{t+1}\}}_{X},$$

where h is a parameter that we will specify in due course. We refer to the candidates in X as the *safe candidates* and Z_0 & Z_1 are two special candidates. We will use τ denote a subset of Δ many unique dummy candidates, where $\Delta :=$ $12nt$. Now we describe the votes. Our voters are divided into three categories as follows:

Special Candidate Votes: This group consists of $t + 3$ copies of the vote,

$$Z_0 \succ \tau \succ \cdots$$

These votes ensure that every winning committee must include Z_0.

"Safe Committee" Votes: For each candidate x_i we have $\frac{18t^2}{t+1}$ copies of the vote:

$$v_{x_i} := x_i \succ Z_1 \succ \tau \succ \cdots$$

Independent Set Votes: For every edge $\{u, v\}$ in the graph, we introduce $2t$ copies of following two votes:

$$u \succ v \succ Z_0 \succ \tau \succ \cdots$$
$$v \succ u \succ Z_0 \succ \tau \succ \cdots$$

We denote the block of these $4t$ votes by $V_{u,v}$. The intuition for this is to ensure that if some committee has both the endpoints of some edge then the overall misrepresentation will be more than Δ.

The votes described above together constitute our profile $\mathcal{P}$. By fixing an ordering on $\mathcal{C}$ and respecting it on the unspecified votes, it is straightforward to verify that all pairs of candidates cross at most six times in this profile. We note that the candidates corresponding to the vertices cross at most six times because the construction is based on a three regular graph. Define $k = t + 2$ and $r = 1$. The ℓ_1-CC -ROBUSTNESS RADIUS instance thus constructed is given by $(\mathcal{C}, \mathcal{P}, k = t+2, r = 1)$. This completes the construction of the instance. We now make some observations about the nature of the optimal committees which will help us argue the equivalence subsequently.

Possible Winning Committees. Let T denote the set of candidates corresponding to t-sized independent set in G (whenever it exists). We refer to the subset of candidates given by $\{Z_0, x_1, x_2, \ldots, x_{t+1}\}$ as the *safe committee* and denote it by $\mathcal{S}$.

Lemma 2. *The constructed profile has a unique winning committee if and only if the graph G has no independent set of size t. The safe committee S has a dissatisfaction score of Δ and is always a winning committee. If (G, t) is a* YES *instance, then $\{Z_0, Z_1\} \cup T$ is also an optimal committee, where T denotes an independent set of size t in G. Further, any k-sized committee not of this form will have dissatisfaction strictly greater than $\Delta + 1$.*

Proof. It is easy to see that the dummy candidate will not appear in any optimal committee, since it appears in the top Δ positions for exactly one vote.

Let us compute the dissatisfaction score for the two proposed committees. For the *safe committee*, we get zero dissatisfaction from the special candidate votes and safe committee votes and we get $8t$ dissatisfaction for each edge which gives us a total dissatisfaction score of $8t \cdot \frac{3n}{2} = 12nt = \Delta$. For the committee based on the independent set, we get zero dissatisfaction from the special candidate votes, $18t^2$ from the *safe committee* votes (one per vote) and $(\frac{3n}{2} - 3t) \cdot 8t + 3t \cdot 2t = 12nt - 18t^2$ from the independent set detector votes. Hence, for both the committees the total dissatisfaction is Δ. It is easy to see that this is the best possible dissatisfaction score that can be achieved by any committee of size k.

Note that any optimal winning committee will have candidate Z_0 otherwise, one has to pick $k+1$ dummy candidates (to remain optimal), which would exceed the committee size. With Z_0 in optimal committee if we intend to choose only few of $x_i's$ then candidate Z_1 is forced in the committee. With these constraints, now, we only have two possible structures for any optimal committee. We will analyze both in next part of the proof.

Consider the possible optimal committees which picks Z_0, Z_1, few endpoints of edges which are covered twice and the partial independent set (set of vertices which only has one endpoint with given edge). The edges for which both the endpoints are in committee gives zero dissatisfaction, edges for which one endpoint lies in committee gives $2t$ dissatisfaction and edges for which both the endpoints are not in committees gives $8t$ dissatisfaction. Hence, the non-uniformity in dissatisfaction clearly indicates that it is better to cover maximum number of edges by picking one end-point rather than completely losing an edge which causes very high dissatisfaction. So, with the remaining budget for t-candidates, the committee with all candidates from independent set will cover maximum edges (to represent by one endpoint) and will cause strictly less dissatisfaction from any other committee by at least $2t$ points.

We now consider a possible winning committee which contains Z_0, Z_1, partial independent set and $x_i's$ for the remaining budget. Let's compute the dissatisfaction for this committee. Say we pick p candidates among the $x_i's$ and $(k - 2 - p) = (t - p)$ candidates from the independent set. The dissatisfaction is:

$$(t + 1 - p) \cdot \frac{18t^2}{t + 1} + \left(\frac{3n}{2} - 3(t - p) \right) \cdot 8t + (3 \cdot (t - p) \cdot 2t)$$

which simplifies to: $\Delta + (t - p) \left(\frac{18t^2}{t+1} - 18t \right) + \frac{18t^2}{t+1}$.

For any value of t, it is straightforward to verify the above expression has value strictly greater than $\Delta + 1$. Hence, committees with this structure will also not be optimal, and this proves the claim.   □

Now, we turn to the equivalence of the two instances.

Forward Direction. We need to show that the existence of t-sized independent set in the graph implies the existence of one swap of adjacent candidates which changes the set of winning committees for the new election instance. From the above claim we know that when there exist a t-sized independent set T, we have two winning committees. In this election instance consider the swap of Z_1 with a dummy candidate on right in any of the *safe committee votes*. Now the score for $\{Z_0, Z_1\} \cup T$ is $\Delta + 1$ and it's not optimal anymore. Hence, we have changed the set of winning committees. This completes the argument for forward direction.

Reverse Direction. From Lemma 2, we know that unless independent set exists any k-candidate committee other than the *safe committee* has dissatisfaction score strictly greater than $\Delta + 1$. This implies there does not exist any swap which can introduce a new committee in winning committee set (since a single swap can change the score of any committee by at most one) or can knock off *safe committee* from the set. Hence, in this case robustness radius equal to one forces the existence of required independent set (since this is the only committee that can change the set of winning committees). This concludes the proof.   □

We remark that an analogous result can be established for the ℓ_∞-CC voting rule as well, but exclude the proof due to lack of space.

5   Concluding Remarks and Open Problems

We demonstrated XP algorithms for the ROBUSTNESS RADIUS problem, when parameterized by the size of the committee, for both the ℓ_1-CC and the Approval-CC voting rules, using a greedy approach. This complements the known W[1]-hardness of the problem with respect to this parameter. We also explicitly establish the W[2]-hardness of ROBUSTNESS RADIUS for the Approval-CC voting rule when parameterized by the size of the committee, even when every voter approves at most three candidates, and when the robustness radius is one. We also established that ROBUSTNESS RADIUS for the ℓ_1-CC and ℓ_∞-CC voting rules remains intractable on fairly structured preferences, such as six-crossing profiles.

A natural direction for further thought is if our XP algorithm can be improved to a better running time, especially on structured profiles such as single-peaked or single-crossing domains. A tempting approach is to see if we can exploit the fact that optimal Chamberlin-Courant committees can be computed in polynomial time on these domains. One immediate challenge is the following: if we require our swaps to be such that the resulting profile also remains in the domain that we are working on, then the case when the input profile has multiple winning committees is harder to decide: we can no longer push a committee out of the winning set with one swap, because the said swap may disturb the structure of the profile. We also believe that instead of guessing all possible choices for a nearly winning committee B, on structured profiles one might be able to cleverly anticipate the right choice of B without trying all of them.

References

1. Brandt, F., Conitzer, V., Endriss, U., Lang, J., Procaccia, A.: Handbook of Computational Social Choice. Cambridge University Press, Cambridge (2016)
2. Bredereck, R., Faliszewski, P., Kaczmarczyk, A., Niedermeier, R., Skowron, P., Talmon, N.: Robustness among multiwinner voting rules. In: Bilò, V., Flammini, M. (eds.) SAGT 2017. LNCS, vol. 10504, pp. 80–92. Springer, Cham (2017). https://doi.org/10.1007/978-3-319-66700-3_7
3. Chamberlin, J.R., Courant, P.N.: Representative deliberations and representative decisions: proportional representation and the Borda rule. Am. Polit. Sci. Rev. **77**(03), 718–733 (1983)
4. Cygan, M., et al.: Parameterized Algorithms. Springer, Cham (2015). https://doi.org/10.1007/978-3-319-21275-3
5. Elkind, E., Faliszewski, P., Slinko, A.: Swap bribery. In: Mavronicolas, M., Papadopoulou, V.G. (eds.) SAGT 2009. LNCS, vol. 5814, pp. 299–310. Springer, Heidelberg (2009). https://doi.org/10.1007/978-3-642-04645-2_27
6. Endriss, U.: Trends in Computational Social Choice. lulu.com (2017)
7. Fleischner, H., Sabidussi, G., Sarvanov, V.I.: Maximum independent sets in 3- and 4-regular Hamiltonian graphs. Discrete Math. **310**(20), 2742–2749 (2010). Graph Theory Dedicated to Carsten Thomassen on his 60th Birthday
8. Lackner, M., Skowron, P.: Consistent approval-based multi-winner rules. In: Proceedings of the 2018 ACM Conference on Economics and Computation, pp. 47–48. ACM (2018)
9. Misra, N., Sonar, C., Vaidyanathan, P.R.: On the complexity of Chamberlin-Courant on almost structured profiles. In: Rothe, J. (ed.) ADT 2017. LNCS (LNAI), vol. 10576, pp. 124–138. Springer, Cham (2017). https://doi.org/10.1007/978-3-319-67504-6_9
10. Shiryaev, D., Yu, L., Elkind, E.: On elections with robust winners. In: Proceedings of the International Conference on Autonomous Agents and Multi-Agent Systems, (AAMAS), pp. 415–422. IFAAMAS (2013)
11. Xia, L.: Computing the margin of victory for various voting rules. In: Proceedings of the ACM Conference on Electronic Commerce, (EC), pp. 982–999. ACM (2012)

On the Complexity of Color-Avoiding Site and Bond Percolation

Roland Molontay[1,2(✉)] and Kitti Varga[3]

[1] Department of Stochastics, Budapest University of Technology and Economics,
Budapest, Hungary
[2] MTA-BME Stochastics Research Group, Budapest, Hungary
molontay@math.bme.hu
[3] Department of Computer Science and Information Theory,
Budapest University of Technology and Economics, Budapest, Hungary
vkitti@cs.bme.hu

Abstract. The mathematical analysis of robustness and error-tolerance of complex networks has been in the center of research interest. On the other hand, little work has been done when the attack-tolerance of the vertices or edges are not independent but certain classes of vertices or edges share a mutual vulnerability. In this study, we consider a graph and we assign colors to the vertices or edges, where the color-classes correspond to the shared vulnerabilities. An important problem is to find robustly connected vertex sets: nodes that remain connected to each other by paths providing any type of error (i.e. erasing any vertices or edges of the given color). This is also known as color-avoiding percolation.

In this paper, we study various possible modeling approaches of shared vulnerabilities, we analyze the computational complexity of finding the robustly (color-avoiding) connected components. We find that the presented approaches differ significantly regarding their complexity.

Keywords: Computational complexity · Color-avoiding percolation
Robustly connected components · Attack tolerance
Shared vulnerability

1 Introduction and Related Works

Understanding the attack and error tolerance of complex networks – i.e. the ability to maintain the overall connectivity of the network as the vertices (or edges) are removed – has attracted a great deal of research interest in the last two decades [1–3,18,26]. Most of the works have focused on random error, meaning that the nodes are considered to be homogeneous with respect to their vulnerabilities, or hub-targeted attack, i.e. the nodes fail preferentially according to a structural property such as degree or betweenness centrality. However, real-world networks are typically heterogeneous (not only with respect to their degree distribution): nodes (or edges) can be separated into different classes regarding

© Springer Nature Switzerland AG 2019
B. Catania et al. (Eds.): SOFSEM 2019, LNCS 11376, pp. 354–367, 2019.
https://doi.org/10.1007/978-3-030-10801-4_28

a mutually shared vulnerability within the class. The shared vulnerabilities can be modeled by assigning a color to each class that may represent a shared eavesdropper, a controlling entity or correlated failures. A "color-avoiding" percolation framework was developed by Krause et al. [10,12,13].

Traditional percolation theory can be used to study the behaviour of connected components in a graph if a vertex (site) or edge (bond) failure occurs with a given probability [15,23]. In the traditional approach a single path provides connectivity, however, here connectivity corresponds to the ability to avoid all vulnerable sets of vertices or edges via multiple paths such that no color is required for all paths (color-avoiding connectivity).

This question is different from k-core percolation where any k paths are sufficient between two nodes [5,25]; and also different from k-connectivity, where k mutually independent paths are required [16]. Another related concept is percolation on multiplex networks [9,22] where the layers can be thought of as the colors, but this approach also differs in the definition of connectivity [10].

To study color-avoiding percolation a new framework was needed. The problem was introduced by Krause et al. [12] who analyzed the color-avoiding connectivity of networks with shared vulnerabilities on the vertices. The authors also examined the latent color-avoiding connectivity structure of the AS-level Internet [20] where the color of the node represents the country to which the router is registered. They have found that 26,228 out of 49,743 of the routers are in the largest color-avoiding component, i.e. secure communication can be obtained among them by splitting the message into more pieces and transmitting them on different paths - even if every country is eavesdropping on its traffic [12]. In [13] the theory of color-avoiding percolation has been extended. The authors study analytically and numerically the maximal set of nodes that are color-avoiding connected in random networks with randomly distributed colors. Shekhtman et al. [19] generalize this framework to study secure message-passing in networks with a given community structure and different classes of vulnerabilities. Kadović et al. [10] formulated color-avoiding percolation for colored edges as well and studied color-avoiding bond and site percolation for networks with a power-law degree distribution.

Beside color-avoiding percolation, finding paths in colored graphs has gained interest in other domains as well. Wu [24] introduced the Maximum Colored Disjoint Paths (MAXCDP) problem that is to find the maximum number of vertex-disjoint paths with edges of the same color between two vertices. The complexity of MAXCDP has been investigated: it can be solved in polynomial time if the input graph contains one color but if there are at least two colors the problem is NP-hard [7,24]. An even harder variant of the problem, to find the maximum number of vertex-disjoint and color-disjoint uni-color paths, was introduced by Dondi et al. [4].

Another related problem is finding k-multicolor paths in a graph. Santos et al. [17] show that the problem is NP-hard and it is also hard to approximate. Another interesting problem is finding a path from a vertex that meets all the colors in a graph. This problem is NP-hard considering a properly colored directed graph, as well as finding the shortest or longest such paths [8]. A survey on algorithmic and computational results for other coloring problems can be found in [14].

On the other hand, none of the previously mentioned works have addressed the computational complexity of color-avoiding percolation. In this article we present various modeling approaches to handle shared vulnerabilities along with the analysis regarding computational complexity. Section 2 is devoted to the problem definition considering both vertices and edges as the targets of the attack. In Sect. 3 we conduct complexity analysis and find that, although the presented approaches are seemingly very similar, they differ significantly regarding computational complexity. Section 4 concludes the work.

2   Modeling Shared Vulnerabilities in Networks

There are two main approaches of modeling shared vulnerabilities in networks depending on the target of the attack: either the links between the nodes can be destroyed (or eavesdropped) or the nodes themselves are the subject of a possible failure. The former is modeled by coloring the edges according to the shared vulnerabilities leading to color-avoiding bond percolation, while the latter is represented by assigning a color to the vertices having the same vulnerability resulting in color-avoiding site percolation. The color-avoiding edge- and vertex-connectivity is illustrated in Fig. 1 on an Erdős-Rényi random graph, the exact concepts will be introduced later in this section.

2.1   Coloring the Edges

It is natural to consider the case when the edges with shared vulnerabilities are the subject of the attack and the vertices remain indistinguishable and unharmed. Some possible real-world examples with edges having shared vulnerabilities: different means of transportation (bus, underground, railway etc.) for a traffic network, various metabolic pathways depending on particular biochemical profiles [11].

Every Edge Has One Color

Definition 1. *Let G be a graph, $C = \{c_1, \ldots, c_k\}$ a color set and $c : E(G) \to C$ a function that assign colors to the edges.*

*For any $i \in \{1, \ldots, k\}$ the vertices $u, v \in V(G)$ are called c_i-**avoiding edge-connected** if after the removal of the edges of color c_i, u and v are in the same component in the remaining graph, i.e there exists a path between u and v which does not contain any edges of color c_i – such a path is called a c_i-**avoiding path**.*

*We say that the vertices $u, v \in V(G)$ are **color-avoiding edge-connected** if they are c_i-avoiding edge-connected for all $i \in \{1, \ldots, k\}$.*

The relation of color-avoiding edge-connectivity (see Fig. 2) is an equivalence relation and thus it defines a partition of the vertex set. The equivalence classes are called *color-avoiding edge-connected components*.

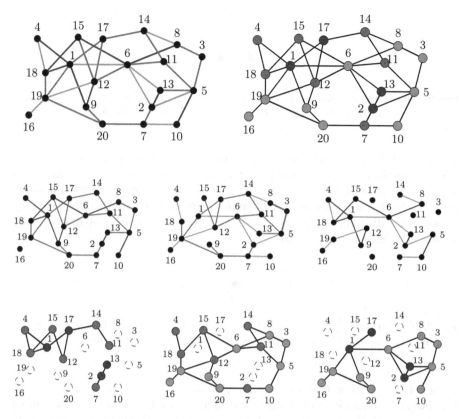

Fig. 1. Illustrating the color-avoiding edge- and vertex-connectivity on an Erdős-Rényi random graph $G(20, 0.15)$. The first row depicts an edge-colored and a vertex-colored graph with three equiprobable colors. The second and third rows show the graphs after removing the green, blue and red edges/vertices, respectively. Note that vertex 5 and vertex 14 are color-avoiding edge-connected since the path 5-11-14 serves as both a green-avoiding and a blue-avoiding path, while the path 5-2-6-8-14 serves as a red-avoiding path. On the other hand, vertex 16 is not color-avoiding edge-connected to any other vertices since without green edges it gets isolated. Considering color-avoiding vertex-connectivity, vertex 6 and vertex 10 are strongly (and therefore weakly) color-avoiding connected since the path 6-11-5-10 serves as a blue-avoiding path, the path 6-13-5-10 serves as a red-avoiding path, while the path 6-2-7-10 serves as a green-avoiding path for internal vertices – this last condition is only required for strong color-avoiding connectivity. Vertex 3 and vertex 10 are weakly color-avoiding connected since the path 3-5-10 serves as both a red-avoiding and blue-avoiding path, while they are not strongly color-avoiding connected since no path from vertex 3 can avoid green vertices as internal nodes. Vertex 7 and vertex 11 are not strongly/weakly color-avoiding connected since no green-avoiding path exists between them. (Color figure online)

Fig. 2. The vertices v_2 and v_3 are color-avoiding edge-connected. But v_1 and v_2 are not: the removal of the blue edges disconnects them. (Color figure online)

Multiple Edges. A natural modification of the previous definition is if we allow for multiple edges, meaning that there can be more types of connection between two nodes (see Fig. 3). Multiple edges make the network less vulnerable: if there are at least two edges of different colors between two nodes, then they are color-avoiding edge-connected since their connection cannot be destroyed by attacking only one color at a time.

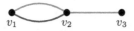

Fig. 3. The vertices v_1 and v_2 are color-avoiding edge-connected. But v_2 and v_3 are not: the removal of the red edges disconnects them. (Color figure online)

Every Edge Has a List of Colors. Another possible generalization of the framework is to modify Definition 1 in such a way that we make the edges more sensitive to attack by assigning a list of colors to the vertices representing all the vulnerabilities that an edge has and an edge is destroyed whenever one of its colors is attacked (see Fig. 4). Formally, the function c is modified: $c : E(G) \to 2^C$, where 2^C is the power set of the color set $C = \{c_1, \ldots, c_k\}$. Furthermore, in this scenario we say that the vertices $u, v \in V(G)$ are c_i-avoiding edge-connected if after the removal of the edges that contain c_i on their lists of colors, u and v are in the same component in the remaining graph.

2.2   Coloring the Vertices

It is also interesting to consider the case when the colors are assigned to vertices that are exposed to attack or failure while the edges remain indistinguishable and unharmed. Possible real-world scenarios of having vulnerable classes of nodes include AS-level Internet with routers registered in different countries, telecommunication networks with transmission towers operated by different providers. The color of the nodes can also represent e.g. ownership, geographical location, dependence on a critical material [12]. The strong/weak vertex color-avoiding connectivity is illustrated in Fig. 1 on an Erdős-Rényi random graph.

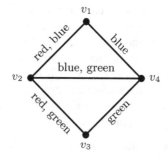

Fig. 4. The vertices v_2 and v_4 are color-avoiding edge-connected: a red-, blue- and green-avoiding path between them are v_2v_4, $v_2v_3v_4$ and $v_2v_1v_4$, respectively. But v_1 and v_2 are not: the removal of the edges containing the color blue on their lists disconnects them. (Color figure online)

Strong Color-Avoiding Connectivity

Definition 2 ([12]). *Let G be a graph, $C = \{c_1, \ldots, c_k\}$ a color set and $c : V(G) \to C$ a function that assigns colors to the vertices.*

*For any $i \in \{1, \ldots, k\}$ the vertices $u, v \in V(G)$ are called **strongly c_i-avoiding vertex-connected** (or **strongly c_i-avoiding connected**) if after the removal of the vertices of color c_i excluding these two vertices, they are in the same component in the remaining graph, i.e. there exists a path between u and v whose internal vertices are not of color c_i – such a path is called a c_i-avoiding path.*

*We say that the vertices $u, v \in V(G)$ are **strongly color-avoiding vertex-connected** (or **strongly color-avoiding connected**) if they are strongly c_i-avoiding connected for all $i \in \{1, \ldots, k\}$.*

In this case the relation of strong color-avoiding connectivity is not transitive, therefore it is not an equivalence relation (see Fig. 5). The *strongly color-avoiding connected components* are maximal sets of vertices such that any two of them are strongly color-avoiding connected.

More Colors on the Vertices. Similarly to the case when the edges were colored, here we can also assign multiple colors to vertices. One can think of the multiple colors (lists of colors) as multiple vulnerabilities on the nodes making the network less robust. Another approach is to consider the scenario analogously to multiple edges. Here it is important to note that if a node has at least two different colors, it makes the vertex immortal under the previously mentioned color attacks analogously to the "multiple edges" approach.

Weak Color-Avoiding Connectivity. Contrary to color-avoiding edge-connectivity, if we color the vertices, it is dubious how to handle the source and target nodes in the definition of color-avoiding vertex-connectivity. In Definition 2 a possible approach was presented that can capture several realistic scenarios. Considering eavesdropping, it is reasonable that the sender and receiver guarantee the security of the message but vulnerability may affect the nodes as transmitters. However, when the attack of the vertices rather means destroying the entities, it is more natural to consider another approach to define color-avoiding vertex-connectivity: attacking red vertices makes it pointless which nodes a red vertex can reach on a path without other red vertices. This scenario is captured by the concept of weak color-avoiding connectivity (see Fig. 5).

Definition 3. *Let G be a connected graph, $C = \{c_1, \ldots, c_k\}$ a color set and $c : V(G) \to C$ a function that assigns colors to the vertices.*

*For any $i \in \{1, \ldots, k\}$ the vertices $u, v \in V(G)$ are called **weakly c_i-avoiding vertex-connected** (or **weakly c_i-avoiding connected**) if after the removal of the vertices of color c_i, either at least one of u or v is deleted or they are in the same component in the remaining graph, i.e. if neither u nor v are of color c_i, there exists a path between them whose vertices are not of color c_i – such a path is called a c_i-**avoiding path**.*

*We say that the vertices $u, v \in V(G)$ are **weakly color-avoiding vertex-connected** (or **weakly color-avoiding connected**) if they are weakly c_i-avoiding connected for all $i \in \{1, \ldots, k\}$.*

Similarly to the strong case, weak color-avoiding connectivity is also not an equivalence relation (see Fig. 5). The *weakly color-avoiding connected components* are maximal sets of vertices such that any two of them are weakly color-avoiding connected.

We can extend the definition to non-connected graphs with the extra condition that two vertices can be weakly color-avoiding connected only if they are in the same component in the original graph.

Remark 1. The notion of weakly color-avoiding connectivity is indeed a weaker concept than the one defined in Definition 2. It is easy to see that if two vertices are strongly color-avoiding connected then it implies that they are weakly color-avoiding connected as well.

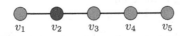

Fig. 5. The vertices v_1 and v_2 are weakly/strongly color-avoiding connected, and so are the vertices v_2 and v_3. But v_1 and v_3 are not: the removal of the blue vertices (i.e. the removal of v_2) disconnects them. The vertices v_3 and v_5 are weakly but not strongly color-avoiding connected. The above observation also shows that neither the strong nor the weak color-avoiding connectivity is a transitive relation. (Color figure online)

More Colors on the Vertices. Weak color-avoiding connectivity can be extended to multiple colors as well, in the exact same manner as strong color-avoiding connectivity.

Other Generalizations. It is worth mentioning that other generalizations have been also proposed. Krause et al. [13] consider nodes with differentiated functions, either as senders/receivers or transmitters. They introduce a flexible trust scenario where vertices can be trusted or avoided in both functions. Trusting colors for transmission naturally increases color-avoiding connectivity [13].

3   Computational Complexity of Finding the Color-Avoiding Components

After presenting the problem of color-avoiding percolation and various modeling approaches, in this section we analyze the computational complexity of finding the robustly (color-avoiding) connected components considering the different problem definitions. In the following we assume the reader's acquaintance with standard concepts of computational complexity theory that may be found e.g. in [21]. We will use in this section that the following well-known decision problem is NP-complete [21].

CLIQUE
Instance: a graph G and a positive integer l.
Question: does G have a clique of size at least l?

Now we list the decision problems for which the computational complexity will be presented in this section. Although the problems are seemingly very similar, they differ considerably concerning their complexity.

COLORAVOIDINGEDGECONNECTEDCOMPONENT
Instance: a graph G, a color set $C = \{c_1, \ldots, c_k\}$, a function $c : E(G) \to C$ and a positive integer l.
Question: is it true that G has a color-avoiding edge-connected component of size at least l?

STRONGLYCOLORAVOIDINGCONNECTEDCOMPONENT
Instance: a graph G, a color set $C = \{c_1, \ldots, c_k\}$, a function $c : V(G) \to C$ and a positive integer l.
Question: is it true that G has a strongly color-avoiding connected component of size at least l?

WEAKLYCOLORAVOIDINGCONNECTEDCOMPONENT
Instance: a graph G, a color set $C = \{c_1, \ldots, c_k\}$, a function $c : V(G) \to C$ and a positive integer l.
Question: is it true that G has a weakly color-avoiding connected component of size at least l?

WEAKLYCOLORAVOIDINGCONNECTEDCOMPONENT-LISTOFCOLORS
Instance: a graph G, a color set $C = \{c_1, \ldots, c_k\}$, a function $c : V(G) \to 2^C$ and a positive integer l.
Question: is it true that G has a weakly color-avoiding connected component of size at least l?

First we prove that the color-avoiding edge-connected components can be found in polynomial time.

Theorem 1. *The problem* COLORAVOIDINGEDGECONNECTEDCOMPONENT *is in P. More precisely, the color-avoiding edge-connected components of G can be found in polynomial time.*

Proof. Let G' be a graph on the vertex set of G where two vertices are connected if and only if they are color-avoiding edge-connected (for an example see Fig. 6). Obviously, G' can be constructed in polynomial time: we need to check for every pair of vertices whether they remain in the same component after erasing the edges of each color separately.

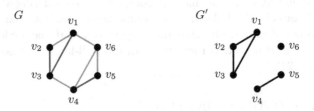

Fig. 6. Two vertices are adjacent in G' if and only if they are color-avoiding edge-connected in G.

Since the color-avoiding edge-connectivity is an equivalence relation, the graph G' is P_3-free, i.e. it cannot contain a path on 3 vertices as an induced subgraph. Obviously, the color-avoiding edge-connected components of G are exactly the maximal cliques of G'.

It is easy to see that the components of a P_3-free graph are cliques, therefore the maximal cliques of G' are its components. Hence, the color-avoiding edge-connected components of G can be found in polynomial time.

The above theorem obviously can be applied when there are multiple edges or when lists of colors are associated with the edges. Clearly, the same proof works in both cases.

Now, we move on to the analysis of color-avoiding vertex percolation. First, we prove that the stronger definition (Definition 2) leads to an NP-complete problem.

Theorem 2. *The problem* STRONGLYCOLORAVOIDINGCONNECTEDCOMPO-NENT *is NP-complete.*

Proof. Obviously, this problem is in NP: a witness is a strongly color-avoiding connected component of size at least l. To show that this problem is NP-hard we reduce CLIQUE to it.

If we use only one color, then by definition the strongly color-avoiding connected components of G are exactly its maximal cliques, therefore our problem is indeed NP-complete.

Next, we present that using the weak definition of color-avoiding connectivity (Definition 3) the connected components can be found in polynomial time. The proof consists of two main parts. First, we show that finding the weakly color-avoiding connected components in any graph is equivalent to finding the cliques of an associated locally chordal graph. This together with the fact that cliques can be found in polynomial time in a locally chordal graph gives us the desired result.

Theorem 3. *Let G be a graph, $C = \{c_1, \ldots, c_k\}$ a set of colors and $c : V(G) \to C$ a function that assigns colors to the vertices. Let G' be a graph on the vertex set of G where two vertices are connected if and only if they are weakly color-avoiding connected. Then the graph G' is locally chordal, i.e. the neighborhood of any vertex cannot contain an induced cycle of length at least 4.*

For an example on the construction of graph G' from Theorem 3 see Fig. 7.

Fig. 7. Two vertices are adjacent in G' if and only if they are weakly color-avoiding connected in G.

Proof. We note that throughout this proof the notion "color-avoiding" always stands for "weakly color-avoiding".

It is easy to see that a graph is locally chordal if and only if it does not contain a wheel on at least five vertices as an induced subgraph: if the graph contains an induced wheel on at least five vertices, then the outer cycle of this wheel is an induced cycle of length at least four in the neighborhood of the center vertex, therefore the graph is not locally chordal. To prove the reverse direction, suppose that the graph is not locally chordal, i.e., there exists a vertex whose neighborhood contains an induced cycle of length at least four. Then this vertex and this cycle together form an induced wheel on at least five vertices.

Suppose to the contrary that G' contains a wheel on $l + 1 \geq 5$ vertices as an induced subgraph. Let u be the center vertex of this wheel, and $w_1, \ldots, w_l$ be the vertices of the outer cycle (in this order), see Fig. 8.

We can assume that the color of the vertex w_2 is c_1. Now consider the vertices w_1 and w_3. Since they are not connected in G', there exists at least one color such that the removal of the vertices of that color disconnects them. On the other hand, the c_i-avoiding paths from w_1 to w_2 and from w_2 to w_3 (which exist since $w_1 w_2, w_2 w_3 \in E(G')$) can be combined into c_i-avoiding paths from w_1 to w_3 for every color $c_i \in C \setminus \{c_1\}$. (Obviously, this procedure does not work with color c_1 since the vertex w_2 is of color c_1.) Thus, only the removal of the vertices of color c_1 can disconnect w_1 and w_3. Therefore, u must have also color c_1 (otherwise the c_1-avoiding paths from w_1 to w_2 and from w_2 to w_3 could be combined into a c_1-avoiding path from w_1 to w_3).

Now consider the vertices w_2 and w_4. Since they are not connected in G', there exists at least one color such that the removal of the vertices of that color disconnects them. However, the c_i-avoiding paths from w_2 to u and from u to w_4 (which exist since $uw_2, uw_4 \in E(G')$) can be combined into c_i-avoiding paths from w_2 to w_4 for every color $c_i \in C \setminus \{c_1\}$. Again, this procedure does not work with color c_1 since the vertex u is of color c_1. But since w_2 is also of color c_1, w_2 and w_4 are weakly c_1-avoiding connected by definition. Hence, they are weakly color-avoiding connected, which is a contradiction.

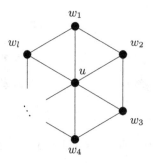

Fig. 8. The wheel on $l + 1$ vertices.

Theorem 4 ([6])**.** *The maximal cliques of any locally chordal graph can be found in polynomial time.*

Corollary 1. *The problem* WEAKLYCOLORAVOIDINGCONNECTEDCOMPONENT *is in P. More precisely, the weakly color-avoiding connected components of G can be found in polynomial time.*

The above theorem obviously can be applied in the more robust case when there may be multiple colors on the vertices resulting in indestructible nodes.

On the other hand, in the other case – when the vertices have multiple colors (lists of colors) and a vertex is destroyed whenever one of its colors is attacked – seemingly paradoxically – leads to a much harder, NP-complete problem.

Theorem 5. *The* WEAKLYCOLORAVOIDINGCONNECTEDCOMPONENT-LIST-OFCOLORS *problem is NP-complete.*

Proof. Obviously, this problem is in NP. To show that this problem is NP-hard we reduce SHAPE CLIQUE to it.

Assign a color to any two vertices, and add this color to the list of every other vertex (so altogether we use $\binom{n}{2}$ colors and every vertex has $\binom{n}{2} - (n-1)$ colors on its list). For an example on the construction of lists of colors see Fig. 9. Now, two vertices are weakly color-avoiding connected if and only if they are adjacent in G. Hence, the weakly color-avoiding connected components of G are exactly its maximal cliques, therefore our problem is indeed NP-complete.

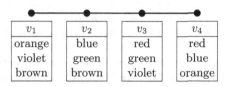

Fig. 9. Constructing the lists of colors: we assign red to v_1 and v_2 (and add the color red to the list of v_3 and v_4), blue to v_1 and v_3, green to v_1 and v_4, orange to v_2 and v_3, violet to v_2 and v_4 and brown to v_3 and v_4. (Color figure online)

Remark 2. In the above proof we can reduce the number of used colors by assigning colors only to nonadjacent pair of vertices; we can also reduce the lengths of the lists by adding this color only to a minimum vertex cut for these two nodes.

4   Conclusion

In this paper, we presented different notions to model various scenarios of shared vulnerabilities in complex networks by assigning colors to the edges or vertices using the framework of color-avoiding percolation developed by Krause et al. [13]. We also analyzed the complexity of finding the color-avoiding connected components. Despite the similarity of the presented concepts, the associated percolation problems – seemingly paradoxically – differ significantly regarding computational complexity. We showed that the color-avoiding edge-connected components can be found in polynomial time. However, the complexity of finding the color-avoiding vertex-connected components highly depends on the exact definition, using a strong version the problem is NP-hard, while using a weaker notion makes it possible to find the components in polynomial time.

Acknowledgment. We thank Michael Danziger, Panna Fekete and Balázs Ráth for useful conversations. The research reported in this paper was supported by the BME-Artificial Intelligence FIKP grant of EMMI (BME FIKP-MI/SC). The publication is also supported by the EFOP-3.6.2-16-2017-00015 project entitled "Deepening the activities of HU-MATHS-IN, the Hungarian Service Network for Mathematics in Industry and Innovations" through University of Debrecen. The work of both authors is partially supported by the NKFI FK 123962 grant. R. M. is supported by NKFIH K123782 grant and by MTA-BME Stochastics Research Group.

References

1. Albert, R., Jeong, H., Barabási, A.-L.: Error and attack tolerance of complex networks. Nature **406**(6794), 378 (2000)
2. Barabási, A.-L., et al.: Network Science. Cambridge University Press, Cambridge (2016)
3. Callaway, D.S., Newman, M.E.J., Strogatz, S.H., Watts, D.J.: Network robustness and fragility: percolation on random graphs. Phys. Rev. Lett. **85**(25), 5468 (2000)
4. Dondi, R., Sikora, F.: Finding disjoint paths on edge-colored graphs: more tractability results. J. Comb. Optim. **36**(4), 1315–1332 (2018)
5. Dorogovtsev, S.N., Goltsev, A.V., Mendes, J.F.F.: Kcore organization of complex networks. Phys. Rev. Lett. **96**(4), 040601 (2006)
6. Gavril, F.: Intersection graphs of Helly families of subtrees. Discrete Appl. Math. **66**(1), 45–56 (1996)
7. Gourves, L., Lyra, A., Martinhon, C.A., Monnot, J.: On paths, trails and closed trails in edge-colored graphs. Discrete Math. Theor. Comput. Sci. **14**(2), 57–74 (2012)
8. Granata, D., Behdani, B., Pardalos, P.M.: On the complexity of path problems in properly colored directed graphs. J. Combin. Optim. **24**(4), 459–467 (2012)
9. Hackett, A., Cellai, D., Gómez, S., Arenas, A., Gleeson, J.P.: Bond percolation on multiplex networks. Phys. Rev. X **6**(2), 021002 (2016)
10. Kadović, A., Krause, S.M., Caldarelli, G., Zlatic, V.: Bond and site color-avoiding percolation in scale free networks. arXiv preprint arXiv:1807.08553 (2018)
11. Kadović, A., Zlatić, V.: Color-avoiding edge percolation on edge-colored network. In: Complenet (2017)
12. Krause, S.M., Danziger, M.M., Zlatić, V.: Hidden connectivity in networks with vulnerable classes of nodes. Phys. Rev. X **6**(4), 041022 (2016)
13. Krause, S.M., Danziger, M.M., Zlatić, V.: Color-avoiding percolation. Phys. Rev. E **96**(2), 022313 (2017)
14. Malaguti, E., Toth, P.: A survey on vertex coloring problems. Int. Trans. Oper. Res. **17**(1), 1–34 (2010)
15. Newman, M.: Networks. Oxford University Press, Oxford (2018)
16. Penrose, M.D.: On k-connectivity for a geometric random graph. Random Struct. Algorithms **15**(2), 145–164 (1999)
17. Santos, R.F., Andrioni, A., Drummond, A.C., Xavier, E.C.: Multicolour paths in graphs: NP-hardness, algorithms, and applications on routing in WDM networks. J. Combin. Optim. **33**(2), 742–778 (2017)
18. Shao, S., Huang, X., Stanley, H.E., Havlin, S.: Percolation of localized attack on complex networks. New J. Phys. **17**(2), 023049 (2015)
19. Shekhtman, L.M., et al.: Critical field-exponents for secure message-passing in modular networks. New J. Phys. **20**(5), 053001 (2018)
20. Siganos, G., Faloutsos, M., Faloutsos, P., Faloutsos, C.: Power laws and the AS-level internet topology. IEEE/ACM Trans. Netw. (TON) **11**(4), 514–524 (2003)
21. Sipser, M.: Introduction to the Theory of Computation. Cengage Learning, Boston (2012)
22. Son, S.-W., Bizhani, G., Christensen, C., Grassberger, P., Paczuski, M.: Percolation theory on interdependent networks based on epidemic spreading. EPL (Europhys. Lett.) **97**(1), 16006 (2012)
23. Stauffer, D., Aharony, A.: Introduction to Percolation Theory: Revised Second Edition. CRC Press, Boca Raton (2014)

24. Wu, B.Y.: On the maximum disjoint paths problem on edge-colored graphs. Discrete Optim. **9**(1), 50–57 (2012)
25. Yuan, X., Dai, Y., Stanley, H.E., Havlin, S.: k-core percolation on complex networks: comparing random, localized, and targeted attacks. Phys. Rev. E **93**(6), 062302 (2016)
26. Zhao, L., Park, K., Lai, Y.-C., Ye, N.: Tolerance of scale-free networks against attack-induced cascades. Phys. Rev. E **72**(2), 025104 (2005)

Lackadaisical Quantum Walks
with Multiple Marked Vertices

Nikolajs Nahimovs[✉]

Center for Quantum Computer Science, Faculty of Computing, University of Latvia,
Raina bulv. 19, Riga 1586, Latvia
nikolajs.nahimovs@lu.lv

Abstract. The concept of lackadaisical quantum walk – quantum walk
with self loops – was first introduced for discrete-time quantum walk on
one-dimensional line [8]. Later it was successfully applied to improve the
running time of the spacial search on two-dimensional grid [16].

In this paper we study search by lackadaisical quantum walk on
the two-dimensional grid with multiple marked vertices. First, we show
that the lackadaisical quantum walk, similarly to the regular (non-
lackadaisical) quantum walk, has exceptional configuration, i.e. placements of marked vertices for which the walk has no speed-up over the
classical exhaustive search. Next, we demonstrate that the weight of the
self-loop suggested in [16] is not optimal for multiple marked vertices.
And, last, we show how to adjust the weight of the self-loop to overcome
the aforementioned problem.

1 Introduction

Quantum walks are quantum counterparts of classical random walks [9]. Similarly to classical random walks, there are two types of quantum walks:
discrete-time quantum walks (DTQW), introduced by Aharonov *et al.* [1], and
continuous-time quantum walks (CTQW), introduced by Farhi *et al.* [4]. For
the discrete-time version, the step of the quantum walk is usually given by two
operators – coin and shift – which are applied repeatedly. The coin operator
acts on the internal state of the walker and rearranges the amplitudes of going
to adjacent vertices. The shift operator moves the walker between the adjacent
vertices.

Quantum walks have been useful for designing algorithms for a variety of
search problems [10]. To solve a search problem using quantum walks, we introduce the notion of marked elements (vertices), corresponding to elements of the
search space that we want to find. We perform a quantum walk on the search
space with one transition rule at the unmarked vertices, and another transition rule at the marked vertices. If this process is set up properly, it leads to
a quantum state in which the marked vertices have higher probability than the
unmarked ones. This method of search using quantum walks was first introduced
in [12] and has been used many times since then.

© Springer Nature Switzerland AG 2019
B. Catania et al. (Eds.): SOFSEM 2019, LNCS 11376, pp. 368–378, 2019.
https://doi.org/10.1007/978-3-030-10801-4_29

Most of the papers studying quantum walks consider a search space containing a single marked element only. However, in contrary of classical random walks, the behavior of the quantum walk can drastically change if the search space contains more that one marked element. Ambainis and Rivosh [3] have studied DTQW on two-dimensional grid and showed that if the diagonal of the grid is fully marked then the probability of finding a marked element does not grow over time. Wong [15] analyzed the spatial search problem by CTQW on the simplex of complete graphs and showed that the placement of marked vertices can dramatically influence the required jumping rate of the quantum walk. Wong and Ambainis [17] analysed DTQW on the simplex of complete graphs and showed that if one of the complete graphs is fully marked then there is no speed-up over classical exhaustive search. Nahimovs and Rivosh [5,6] studied DTQW on two-dimensional grid for various placements of multiple marked vertices and proved several gaps in the running time of the walk (depending on the placement of marked vertices). Additionally the authors have demonstrated placements of a constant number of marked vertices for which the walk have no speed-up over classical exhaustive search. They named such placements *exceptional configurations*. Nahimovs and Santos [7] have extended their work to general graphs.

The concept of lackadaisical quantum walk (quantum walk with self loops) was first studied for DTQW on one-dimensional line [8,13]. Later on, Wong showed an example of how to apply the self-loops to improve the DTQW based search on the complete graph [14] and two-dimensional grid [16]. The running time of the lackadaisical walk heavily depends on a weight of the self-loop. Saha *et al.* [11] showed that the weight $l = \frac{4}{N}$ suggested by Wong for two-dimensional grid of N vertices with a single marked vertex may be not optimal for multiple marked vertices (i.e. result in larger number of steps and lower probability). They have demonstrated that for a block of $\sqrt{m} \times \sqrt{m}$ marked vertices one should use the weight $l = \frac{4}{N(m+\sqrt{m}/2)}$.

In this paper, we study search by discrete-time lackadaisical quantum walk on two-dimensional grid with multiple marked vertices. First, we show that the lackadaisical quantum walk, similarly to the regular (non-lackadaisical) quantum walk, has exceptional configurations, i.e. placements of marked vertices for which the walk have no speed-up over the classical exhaustive search. Next, we study an arbitrary placement of m marked vertices and demonstrate that the weight l suggested by Wong is not optimal for multiple marked vertices. The same holds for the weight suggested by Saha *et al.*, which seems to work only for a block of $\sqrt{m} \times \sqrt{m}$ marked vertices. Last, we analyze how to adjust the weight to overcome the aforementioned problem. We propose two better constructions – $l = \frac{4m}{N}$ and $l = \frac{4(m-\sqrt{m})}{N}$ – and discuss their boundaries of application.

2   Quantum Walk on the Two-Dimensional Grid

2.1   Regular (Non-lackadaisical) Quantum Walk

Consider a two-dimensional grid of size $\sqrt{N} \times \sqrt{N}$ with periodic (torus-like) boundary conditions. The locations of the grid are labeled by the coordinates (x, y) for $x, y \in \{0, \ldots, \sqrt{N} - 1\}$. The coordinates define a set of state vectors, $|x, y\rangle$, which span the Hilbert space $\mathcal{H}_P$ associated with the position. Additionally, we define a 4-dimensional Hilbert space $\mathcal{H}_C$, spanned by the set of states $\{|c\rangle : c \in \{\uparrow, \downarrow, \leftarrow, \rightarrow\}\}$, associated with the direction. We refer to it as the coin subspace. The Hilbert space of the quantum walk is $\mathcal{H}_P \otimes \mathcal{H}_C$.

The evolution of a state of the walk (without searching) is driven by the unitary operator $U = S \cdot (I \otimes C)$, where S is the flip-flop shift operator

$$S|x, y, \uparrow\rangle = |x, y + 1, \downarrow\rangle \tag{1}$$

$$S|x, y, \downarrow\rangle = |x, y - 1, \uparrow\rangle \tag{2}$$

$$S|x, y, \leftarrow\rangle = |x - 1, y, \rightarrow\rangle \tag{3}$$

$$S|x, y, \rightarrow\rangle = |x + 1, y, \leftarrow\rangle, \tag{4}$$

and C is the coin operator, given by the Grover's diffusion transformation

$$C = 2|s_c\rangle\langle s_c| - I_4 \tag{5}$$

with

$$|s_c\rangle = \frac{1}{\sqrt{4}}(|\uparrow\rangle + |\downarrow\rangle + |\leftarrow\rangle + |\rightarrow\rangle).$$

The system starts in

$$|\psi(0)\rangle = \frac{1}{\sqrt{N}} \sum_{x,y=0}^{\sqrt{N}-1} |x, y\rangle \otimes |s_c\rangle, \tag{6}$$

which is uniform distribution over vertices and directions. Note, that this is a unique eigenvector of U with eigenvalue 1.

To use quantum walk for search, we extend the step of the algorithm with a query to an oracle, making the step

$$U' = U \cdot (Q \otimes I_4).$$

Here Q is the query transformation which flips the sign at a marked vertex, irrespective of the coin state. Note that $|\psi(0)\rangle$ is a 1-eigenvector of U but not of U'. If there are marked vertices, the state of the algorithm starts to deviate from $|\psi(0)\rangle$. In case of a single marked vertex, after $O(\sqrt{N \log N})$ steps the inner product $\langle \psi(t)|\psi(0)\rangle$ becomes close to 0. If the state is measured at this moment, the probability of finding a marked vertex is $O(1/\log N)$ [2]. With amplitude amplification this gives the total running time of $O(\sqrt{N} \log N)$ steps.

2.2 Lackadaisical Quantum Walk

In case of lackadaisical quantum walk the coin subspace of the walk is 5-dimensional Hilbert space spanned by the set of states $\{|c\rangle \; : \; c \in \{\uparrow, \downarrow, \leftarrow, \rightarrow, \circlearrowleft\}\}$. The Hilbert space of the quantum walk is $\mathbb{C}^N \otimes \mathbb{C}^5$.

The shift operator acts on a self loop as

$$S|x, y, \circlearrowleft\rangle = |x, y, \circlearrowleft\rangle. \tag{7}$$

The coin operator is

$$C = 2|s_c\rangle\langle s_c| - I_5 \tag{8}$$

with

$$|s_c\rangle = \frac{1}{\sqrt{4+l}}(|\uparrow\rangle + |\downarrow\rangle + |\leftarrow\rangle + |\rightarrow\rangle + \sqrt{l}|\circlearrowleft\rangle).$$

The system starts in

$$|\psi(0)\rangle = \frac{1}{\sqrt{N}} \sum_{x,y=0}^{\sqrt{N}-1} |x, y\rangle \otimes |s_c\rangle, \tag{9}$$

which is uniform distribution over vertices, but not directions. As before $|\psi(0)\rangle$ is a unique 1-eigenvector of U.

The step of the search algorithm is $U' = U \cdot (Q \otimes I_5)$. As it is shown in [16], in case of a single marked vertex, for the weight $l = \frac{4}{N}$, after $O(\sqrt{N \log N})$ steps the inner product $\langle\psi(t)|\psi(0)\rangle$ becomes close to 0. If one measures the state at this moment, he will find the marked vertex with $O(1)$ probability, which gives $O(\log N)$ improvement over the loopless algorithm.

3 Stationary States of the Lackadaisical Quantum Walk

In this section we will show that the lackadaisical quantum walk, similarly to the regular (non-lackadaisical) quantum walk, has exceptional configurations, i.e. placements of marked vertices for which the walk have no speed-up over the classical exhaustive search.

Consider a state $|s_\uparrow^a\rangle = a(-(3+l)|\uparrow\rangle + |\downarrow\rangle + |\leftarrow\rangle + |\rightarrow\rangle + \sqrt{l}|\circlearrowleft\rangle)^T$. Similarly one can define states $|s_\downarrow^a\rangle$, $|s_\leftarrow^a\rangle$ and $|s_\rightarrow^a\rangle$. The defined states are orthogonal to $|s_c\rangle$. Consider an effect of the coin transformation on $|s_\uparrow^a\rangle$:

$$C|s_\uparrow^a\rangle = (2|s_c\rangle\langle s_c| - I_5)|s_\uparrow^a\rangle = -|s_\uparrow^a\rangle.$$

As one can see, the coin transformation inverts a sign of the state.

Now, consider a two-dimensional grid with two marked vertices (i, j) and $(i + 1, j)$. Let $|\phi_{stat}^a\rangle$ be a state where the coin part of all unmarked vertices is $|s_c^a\rangle = a(|\uparrow\rangle + |\downarrow\rangle + |\leftarrow\rangle + |\rightarrow\rangle + \sqrt{l}|\circlearrowleft\rangle)^T$, the coin part of (i, j) is $|s_\rightarrow^a\rangle$ and the coin part of $(i + 1, j)$ is $|s_\leftarrow^a\rangle$ (see Fig. 1), that is,

$$|\phi_{stat}^a\rangle = \sum_{x,y=0}^{\sqrt{N}-1} |x, y\rangle|s_c^a\rangle - (4+l)a\left(|i, j, \rightarrow\rangle + |i+1, j, \leftarrow\rangle\right). \tag{10}$$

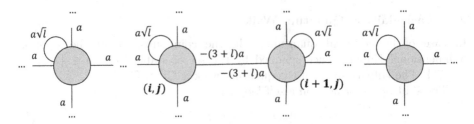

Fig. 1. Stationary state of two marked vertices (i, j) and $(i + 1, j)$.

We claim that this state is not changed by a step of the algorithm.

Lemma 1. *Consider a grid of size $\sqrt{N} \times \sqrt{N}$ with two adjacent marked vertices (i, j) and $(i + 1, j)$. Then the state $|\phi_{stat}^a\rangle$, given by Eq. (10), is not changed by the step of the algorithm, that is, $U'|\phi_{stat}^a\rangle = |\phi_{stat}^a\rangle$.*

Proof. Consider the effect of a step of the algorithm on $|\phi_{stat}^a\rangle$. The query transformation flips the sign of marked vertices. The coin transformation has no effect on $|s_c^a\rangle$ but flips the signs of $|s_\leftarrow^a\rangle$ and $|s_\rightarrow^a\rangle$. Thus, $(I \otimes C)(Q \otimes I)$ does not change the amplitudes of unmarked vertices and twice flips the signs of amplitudes of marked vertices. Therefore, we have $(I \otimes C)(Q \otimes I)|\phi_{stat}^a\rangle = |\phi_{stat}^a\rangle$. The shift transformation swaps the amplitudes of near-by vertices. For $|\phi_{stat}^a\rangle$, it swaps a with a and $-(3 + l)a$ with $-(3 + l)a$. Thus, we have $S(I \otimes C)(Q \otimes I)|\phi_{stat}^a\rangle = |\phi_{stat}^a\rangle$. □

The initial state of the algorithm, given by Eq. (9), can be written as

$$|\psi_0\rangle = |\phi_{stat}^a\rangle + (4 + l)a(|i, j, \rightarrow\rangle + |i + 1, j, \leftarrow\rangle), \tag{11}$$

for $a = 1/\sqrt{(4 + l)N}$. The only part of the initial state which is changed by the step of the algorithm is

$$\frac{\sqrt{4 + l}}{\sqrt{N}}(|i, j, \rightarrow\rangle + |i + 1, j, \leftarrow\rangle). \tag{12}$$

Let us establish an upper bound on the probability of finding a marked vertex.

Lemma 2. *Consider a grid of size $\sqrt{N} \times \sqrt{N}$ with two adjacent marked vertices (i, j) and $(i + 1, j)$. Then for any number of steps, the probability of finding a marked vertex p_M is $O\left(\frac{1}{N}\right)$.*

Proof. We have $M = \{(i, j), (i + 1, j)\}$. The only part of the initial state $|\psi(0)\rangle$ changed by the step of the algorithm is $|\phi\rangle = (4+l)a(|i, j, \rightarrow\rangle + |i+1, j, \leftarrow\rangle)$. The basis states $|i, j, \rightarrow\rangle$ and $|i + 1, j, \leftarrow\rangle$ have the biggest amplitudes of $-(3 + l)a$ in the stationary state. Therefore, the maximum probability of finding a marked vertex is reached if the state $|\phi\rangle$ becomes

$$|\phi'\rangle = -\alpha|i, j, \rightarrow\rangle - \beta|i + 1, j, \leftarrow\rangle, \tag{13}$$

for $\alpha, \beta \geq 0$. Thus, p_M is at most

$$p_M \leq 6a^2 + 2(a\sqrt{l})^2 + (-(3+l)a - \alpha)^2 + (-(3+l)a - \beta)^2. \qquad (14)$$

Since the evolution is unitary, we have $\alpha^2 + \beta^2 = |||\phi\rangle||^2 = 2\left((4+l)a\right)^2$. Due to symmetry α and β should be equal, so the expression (14) reaches the maximum when $\alpha = \beta = (4+l)a$.

We have $l = \frac{4}{N}$ and $a = \frac{1}{\sqrt{(4+l)N}} = \frac{1}{\sqrt{4(N+l)}}$. Each of summands in the expression (14) is $O\left(\frac{1}{N}\right)$ and, therefore, we have $p_M = O\left(\frac{1}{N}\right)$.   □

That is the probability of finding a marked vertex is of the same order as for the classical exhaustive search.

Note that if we have a block of marked vertices we can construct a stationary state as long as we can tile the block by the sub-blocks of size 1×2 and 2×1. For example, consider $M = \{(0,0), (1,0), (1,1), (1,2)\}$ for $n \geq 3$. Then the stationary state is given by

$$|\phi^a_{stat}\rangle = \sum_{x,y=0}^{n-1} |x,y\rangle|s^a_c\rangle - (4+l)a \left(|0,0,\rightarrow\rangle + |1,0,\leftarrow\rangle + |1,1,\uparrow\rangle + |1,2,\downarrow\rangle\right).$$

For more details on constructions of stationary states for blocks of marked vertices on two-dimensional grid see [7]. The paper focuses on the non-lackadaisical quantum walk, nevertheless, the results can be easily extended to the lackadaisical quantum walk.

4   Optimality of l for Multiple Marked Vertices

In [16] Wong showed that in case of a single marked vertex, for the weight $l = \frac{4}{N}$, after $O(\sqrt{N \log N})$ steps the inner product $\langle \psi(t)|\psi(0)\rangle$ becomes close to 0. If one measures the state at this moment, one will find the marked vertex with $O(1)$ probability[1]. The suggested value of l, however, is optimal for a single marked vertex only. Saha et al. [11] studied search for a block of $\sqrt{m} \times \sqrt{m}$ marked vertices and showed that optimal weight in this setting is $l = \frac{4}{N(m+\sqrt{m}/2)}$.

In this section we study search for an arbitrary placement of multiple marked vertices. The presented data is obtained from numerical simulations. The values listed in the tables are calculated in the following way. The number of steps of the algorithm T is the smallest t for which $|\langle \psi(t)|\psi(0)\rangle|$ reaches its minimum (that is the current and the initial states are maximally orthogonal). By the probability we mean the probability of finding a marked vertex when $|\psi(T)\rangle$ is measured.

Tables 1 and 2 give the number of steps and the probability of finding a marked vertex for random placements of 2 and 3 marked vertices on 100×100 grid for $l = \frac{4}{N}$. As one can see the probability of finding a marked vertex is no more close to 1 as it is for a single marked vertex.

---

[1] The numerical results in [16] show that probability of finding a marked vertex is close to 1 and approaches 1 as N goes to infinity.

Table 1. The number of steps and the probability of finding a marked vertex for different placements of two marked vertices for 100×100 grid for $l = \frac{4}{N}$.

Marked vertices	T	Pr
(0, 0), (23, 27)	153	0.586377681077719
(0, 0), (35, 68)	150	0.591030741055657
(0, 0), (30, 69)	151	0.588384716869901
(0, 0), (42, 4)	152	0.590451037529614
(0, 0), (84, 60)	151	0.584982804352049

Table 2. The number of steps and the probability of finding a marked vertex for different placements of three marked vertices for 100×100 grid for $l = \frac{4}{N}$.

Marked vertices	T	Pr
(0, 0), (34, 52), (93, 53)	117	0.440756928151790
(0, 0), (26, 12), (22, 32)	126	0.434581723157292
(0, 0), (40, 94), (13, 62)	119	0.430688837061525
(0, 0), (7, 44), (7, 98)	131	0.430029225026132
(0, 0), (80, 78), (28, 31)	118	0.454915029501263

Table 3 shows the number of steps and the probability for 200×200 grid with the set of marked vertices

$$M_m = \{(0, 10i) \mid i \in [0, m - 1]\} \tag{15}$$

for weights of a self-loop suggested by Wong (the 2nd and the 3rd columns) and by Saha *et al.* (the last two columns). As one can see for both weights the probability goes down with the number of marked vertices.

Table 3. The number of steps and the probability of finding a marked vertex for 200×200 grid with the set of marked vertices M_m for different l.

m	$l = \frac{4}{N}$		$l = \frac{4}{N(m+\sqrt{m/2})}$	
	T	Pr	T	Pr
1	602	0.987103466750771	602	0.9871034667507710
2	374	0.556471227830710	355	0.3290596740364150
3	320	0.393873564782729	307	0.1901285270921410
4	288	0.318205769345174	278	0.1362737798676850
5	266	0.269653054659757	258	0.1120867687513450
6	250	0.234725633256426	243	0.0963898188447711
7	235	0.205158185237765	229	0.0847091122232096
8	223	0.184324335272977	218	0.0764074018340319
9	213	0.168420810292804	208	0.0694735116911546
10	203	0.153267792359668	198	0.0634301283171891

We tried to adjust the value of l to increase the probability of finding a marked vertex. We searched for a better value of l in the form $l = \frac{4}{N}a$. The Figs. 2 and 3 show the probability of finding a marked vertex for 100×100 grid with the sets of marked vertices M_2 and M_3, respectively, for different values of a.

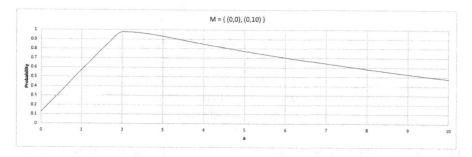

Fig. 2. Probability of finding a marked vertex for 100×100 grid with the set of marked vertices M_2 for different values of a.

As one can see the optimal value of a for M_2 is close to 2 and for M_3 is close to 3. The similar results were obtained for bigger grids with larger sets of marked vertices. Table 4 gives the optimal value of a and the corresponding number of steps and the probability for 200×200 grid with the set of marked vertices M_m.

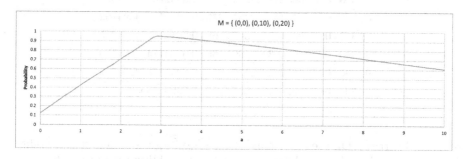

Fig. 3. Probability of finding a marked vertex for 100×100 grid with the set of marked vertices M_3 for different values of a.

This raises a conjecture that the optimal weight of a self loop is $l = \frac{4(m-O(m))}{N}$. The Table 5 shows the number of steps and the probability for 200×200 grid with the set of marked vertices M_m for the weight $l = \frac{4m}{N}$ (the 2nd and the 3rd columns) and $l = \frac{4m-\sqrt{m}}{N}$ (the last two columns). As one can see $l = \frac{4m}{N}$ results in a high probability of finding a marked vertices for a small number of marked vertices, however, the probability goes down with the number

Table 4. The number of steps and the probability of finding a marked vertex for 200×200 grid with the set of marked vertices M_m for optimal a.

m	a_{opt}	T	Pr
2	1.94	470	0.970784853767743
3	2.90	419	0.968156591210997
4	3.82	394	0.957428109231279
5	4.66	374	0.93524432034913
6	5.44	358	0.910278544128265
7	6.17	329	0.884824083920976
8	7.06	301	0.884650346189075
9	8.00	295	0.891195819702051
10	8.86	292	0.889060897077511

Table 5. The number of steps and the probability of finding a marked vertex for 200×200 grid with the set of marked vertices M_m for different l.

m	$l = \frac{4m}{N}$		$l = \frac{4m - \sqrt{m}}{N}$	
	T	Pr	T	Pr
1	602	0.987103466750771	421	0.138489015636136
2	480	0.973610115577208	358	0.368553562270952
3	426	0.970897595293325	326	0.474753065755793
4	400	0.957956584718826	305	0.541044821578945
5	376	0.933005243569973	288	0.593276362658860
6	352	0.904811189309431	277	0.633876384394702
7	312	0.885901799105365	268	0.661120674334215
8	300	0.891698403206386	260	0.678417412900138
9	296	0.892165251874117	254	0.694145864271432
10	293	0.884599315314024	250	0.709033853082403

of marked vertices. On the other hand, $l = \frac{4(m - \sqrt{m})}{N}$ gives a modest probability for a small number of marked vertices, but the probability grows with the number of marked vertices (and, moreover, seems to tend to a constant). Therefore, we would suggest to use the last of the proposed value of l, especially for bigger grids and large number of marked vertices.

The found values of l result in high probability not only for M_i sets of marked vertices, but work equivalently well for other placements of m marked vertices, including a random placement. For example, Table 6 gives the mean and the standard deviation over 100 runs for the number of steps and the probability of finding a marked vertex for 200×200 grid with m randomly placed marked vertices for $l = \frac{4m}{N}$.

Table 6. The mean and standard deviation for the number of steps and the probability of finding a marked vertex for 200×200 grid with m randomly placed marked vertices.

m	$\mu(T)$	$\sigma(T)$	$\mu(Pr)$	$\sigma(Pr)$
2	426.53	18.19115947	0.986380083	0.005025848
3	351.24	14.97683059	0.970760080	0.006463305
4	305.01	11.52379620	0.952193422	0.011523802
5	276.77	15.78636095	0.922327254	0.023677911
6	250.28	13.54385320	0.892480153	0.035932293
7	231.25	16.57573680	0.871024499	0.043598408
8	213.52	13.51428724	0.857905654	0.049283674
9	199.78	13.08525271	0.841684604	0.049829490
10	186.70	10.11199907	0.843118514	0.058874363

5   Conclusions

In this paper, we have demonstrated the existence of exceptional configurations of marked vertices for search by lackadaisical quantum walk on two-dimensional grid. We also numerically showed that weights of the self-loop l, suggested by previous papers [11,16], are not optimal for multiple marked vertices (both weight seems to work in specific cases only). We proposed two values of l resulting in a much higher probability of finding a marked vertex than previously suggested weights. Moreover, for the found values, the probability of finding a marked vertex does not decrease with number of marked vertices.

References

1. Aharonov, Y., Davidovich, L., Zagury, N.: Quantum random walks. Phys. Rev. A **48**(2), 1687–1690 (1993)
2. Ambainis, A., Kempe, J., Rivosh, A.: Coins make quantum walks faster. In: Proceedings of the 16th ACM-SIAM Symposium on Discrete Algorithms, pp. 1099–1108 (2005)
3. Ambainis, A., Rivosh, A.: Quantum walks with multiple or moving marked locations. In: Geffert, V., Karhumäki, J., Bertoni, A., Preneel, B., Návrat, P., Bieliková, M. (eds.) SOFSEM 2008. LNCS, vol. 4910, pp. 485–496. Springer, Heidelberg (2008). https://doi.org/10.1007/978-3-540-77566-9_42
4. Farhi, E., Gutmann, S.: Quantum computation and decision trees. Phys. Rev. A **58**, 915–928 (1998)
5. Nahimovs, N., Rivosh, A.: Exceptional configurations of quantum walks with Grover's coin. In: Kofroň, J., Vojnar, T. (eds.) MEMICS 2015. LNCS, vol. 9548, pp. 79–92. Springer, Cham (2016). https://doi.org/10.1007/978-3-319-29817-7_8
6. Nahimovs, N., Rivosh, A.: Quantum walks on two-dimensional grids with multiple marked locations. In: Proceedings of SOFSEM 2016, vol. 9587, pp. 381–391 (2016). arXiv:quant-ph/150703788

7. Nahimovs, N., Santos, R.A.M.: Adjacent vertices can be hard to find by quantum walks. In: Steffen, B., Baier, C., van den Brand, M., Eder, J., Hinchey, M., Margaria, T. (eds.) SOFSEM 2017. LNCS, vol. 10139, pp. 256–267. Springer, Cham (2017). https://doi.org/10.1007/978-3-319-51963-0_20

8. Inui, N., Konno, N., Segawa, E.: One-dimensional three-state quantum walk. Phys. Rev. E Stat. Nonlin. Soft Matter Phys. **72**(5), 168–191 (2005)

9. Portugal, R.: Quantum Walks and Search Algorithms. Springer, New York (2013). https://doi.org/10.1007/978-1-4614-6336-8

10. Reitzner, D., Nagaj, D., Buzek, V.: Quantum walks. Acta Physica Slovaca **61**(6), 603–725 (2011). arxiv.org/abs/1207.7283

11. Saha, A., Majumdar, R., Saha, D., Chakrabarti, A., Sur-Kolay, S.: Search of clustered marked states with lackadaisical quantum walks (2018). arXiv:1804.01446

12. Shenvi, N., Kempe, J., Whaley, K.B.: A quantum random walk search algorithm. Phys. Rev. A **67**, 052307 (2003)

13. Stefanak, M., Bezdekova, I., Jex, I.: Limit distributions of three-state quantum walks: the role of coin eigenstates. Phys. Rev. A **90**(1), 124–129 (2014)

14. Wong, T.G.: Grover search with lackadaisical quantum walks. J. Phys. A Math. Gen. **48** (2015)

15. Wong, T.G.: Spatial search by continuous-time quantum walk with multiple marked vertices. Quantum Inf. Process. **15**(4), 1411–1443 (2016)

16. Wong, T.G.: Faster search by lackadaisical quantum walk. Quantum Inf. Process. **17**, 68 (2018)

17. Wong, T.G., Ambainis, A.: Quantum search with multiple walk steps per oracle query. Phys. Rev. A **92**, 0022338 (2015)

A 116/13-Approximation Algorithm for $L(2,1)$-Labeling of Unit Disk Graphs

Hirotaka Ono[✉] and Hisato Yamanaka[✉]

Department of Mathematical Informatics, Graduate School of Informatics,
Nagoya University, Nagoya 464-8601, Japan
ono@i.nagoya-u.ac.jp, ymnk0114@nagoya-u.ac.jp

Abstract. Given a graph, an $L(2,1)$-labeling of the graph is an assignment ℓ from the vertex set to the set of nonnegative integers such that for any pair of vertices (u,v), $|\ell(u) - \ell(v)| \geq 2$ if u and v are adjacent, and $\ell(u) \neq \ell(v)$ if u and v are at distance 2. The $L(2,1)$-labeling problem is to minimize the span of ℓ (i.e., $\max_{u \in V}(\ell(u)) - \min_{u \in V}(\ell(u)) + 1)$. In this paper, we propose a new polynomial-time 116/13-approximation algorithm for $L(2,1)$-labeling of unit disk graphs. This improves the previously best known ratio 12.

Keywords: Frequency/channel assignment · Graph algorithm $L(2, 1)$-labeling · Approximation algorithm · Unit disk graphs

1 Introduction

Let G be an undirected graph. An $L(2,1)$-labeling of a graph G is an assignment ℓ from the vertex set $V(G)$ to the nonnegative integers such that $|\ell(x) - \ell(y)| \geq 2$ if x and y are adjacent and $|\ell(x) - \ell(y)| \geq 1$ if x and y are at distance 2 for all x and y in $V(G)$. A k-$L(2,1)$-labeling is an $L(2,1)$-labeling $\ell : V(G) \to \{0, \ldots, k\}$, and the $L(2,1)$-labeling problem asks the minimum k among all possible $L(2,1)$-labelings. We call this invariant, the minimum value k, the $L(2,1)$-labeling number, and denote it by $\lambda(G)$. Notice that we use at most $\lambda(G) + 1$ different labels in an optimal $L(2,1)$-labeling, and we call this $\lambda(G) + 1$ the *optimal span* of $L(2,1)$-labeling, and denote it by $\sigma(G)$. As the objective function of $L(2,1)$-labeling problem, we adopt not $\lambda(G)$ but $\sigma(G)$, which is more natural for the approximation guarantee.

The original notion of $L(2,1)$-labeling can be seen in Hale [11] and Roberts [16] in the context of frequency/channel assignment, where 'close' transmitters must receive different frequencies and 'very close' transmitters must receive frequencies that are at least two apart so that they can avoid interference. Due to its practical importance, the $L(2,1)$-labeling problem has been intensively and extensively studied. Furthermore, this problem is attractive from the graph theoretical point of view since it is a kind of vertex coloring problem.

This work is partially supported by KAKENHI 17K19960, 17H01698.

B. Catania et al. (Eds.): SOFSEM 2019, LNCS 11376, pp. 379–391, 2019.
https://doi.org/10.1007/978-3-030-10801-4_30

Related Work

In general, $L(h,k)$-labelings of a graph G are defined for arbitrary nonnegative integers h and k, as an assignment of nonnegative integers to $V(G)$ such that adjacent vertices receive labels at least h apart and vertices connected by a path of length 2 receive labels at least k apart. This problem is one of the generalizations of the vertex coloring problem since the $L(h,0)$-labeling problem is equivalent to it. Therefore, we can hardly expect that the $L(h,k)$-labeling problem is tractable, and in fact, $L(0,1)$- and $L(1,1)$-labeling problems are known to be NP-hard, for example. We can find a lot of related results on $L(h,k)$-labelings in comprehensive surveys by Calamoneri [3] and Yeh [18].

There are also a number of studies about the $L(2,1)$-labeling problem from the algorithmic point of view [13]. It is known to be NP-hard for general graphs [10], and it still remains NP-hard for some restricted classes of graphs, such as planar, bipartite, chordal graphs [1], and graphs of treewidth 2 [7]. In contrast, only a few graph classes are known to have polynomial time algorithms for this problem. Outerplanar graphs and its sub-classes of graphs (e.g., trees) are such rare classes of graphs [12,15].

From the viewpoint of frequency/channel assignment in wireless networks, it is natural and interesting to restrict instances of graphs to disk graphs. Unfortunately, even for unit disk graphs, the $L(2,1)$-labeling problem is known to be NP-hard [8]. As a positive result, a 12-approximation algorithm is proposed by Fiala et al. [6].

Our Contributions

The contribution of this paper is to provide a new 116/13-approximation polynomial-time algorithm for $L(2,1)$-labeling of unit disk graphs, which improves the previously best known ratio 12. The previous 12-approximation algorithm first divides the plane into wide strips, and after that, it labels vertices in each strip by 18ω labels, where ω is the size of a maximum clique in a given graph. Since $2\omega - 1$ is a lower bound on the optimal span of $L(2,1)$-labeling, this labeling achieves approximation ratio 12, which comes from $18\omega/(2\omega - 1) \leq 12$.

Our algorithm divides the plane into big squares with a certain size instead. We can label vertices in a big square by $16\omega + 7$ labels, but this does not immediately yield approximation ratio 9. In fact, if we adopt $2\omega - 1$ as a lower bound of σ, we just have $(16\omega + 7)/(2\omega - 1) \leq 13$, which is even worse. As an extra idea, we focus on the lower bound $2\omega - 1$, which means that $\sigma = 2\omega - 1$ or $\sigma \geq 2\omega$. If we can distinguish these two cases and label vertices by a smaller number of labels than $16\omega + 7$ for the former case, we can expect a better approximation ratio. Although the task itself might be difficult, a necessary condition of $\sigma = 2\omega - 1$ is helpful to achieve a smaller upper bound on labels. This is the key idea to achieve the approximation ratio 116/13.

Organization of This Paper

The rest of this paper is organized as follows. Section 2 introduces the terminology and gives basic definitions. In Sect. 3, we introduce basic results on $L(2,1)$-labeling number of unit disk graphs and review the 12-approximation algorithm [6]. Section 4 is the main part of this paper and we present a new polynomial-time 116/13-approximation algorithm.

2 Preliminaries

2.1 Definitions and Notations

A graph G is an ordered set of its vertex set $V(G)$ and edge set $E(G)$ and denoted by $G = (V(G), E(G))$. We often denote $G = (V(G), E(G))$ simply by $G = (V, E)$, if no confusion arises. We assume throughout this paper that all graphs are undirected, simple and connected, unless otherwise stated. An edge $e \in E$ is an unordered pair of vertices u and v, which are end vertices of e, though we often denote it conventionally by $e = (u, v)$. Two vertices u and v are adjacent if $(u, v) \in E$, and two edges are adjacent if they share one of their end vertices. A graph G is called *complete* if every two vertices in V are adjacent.

A subset $V' \subseteq V$ is a clique if a subgraph $G[V']$ of G induced by V' is complete. A maximum clique of G is, naturally, a clique whose cardinality is largest, and the number is called the *clique number* of G, which is denoted by $\omega(G)$. Also we use $\Delta(G)$ to denote the maximum degree of graph G. These are just denoted by Δ and ω, if no confusion arises.

For the definitions of an $L(2,1)$-labeling of a graph G and related concepts, see Introduction. We denote the $L(2,1)$-labeling number and the optimal span of $L(2,1)$-labeling of G by $\lambda(G)$ and $\sigma(G)$, respectively. These are also simply denoted by λ and σ, if no confusion arises. In $L(2,1)$-labeling ℓ, the condition that $|\ell(x) - \ell(y)| \geq 2$ if x and y are adjacent is called the *condition of distance one* and the condition that $|\ell(x) - \ell(y)| \geq 1$ if x and y are at distance 2 for all x and y is called the *condition of distance two*. These two conditions together are referred as just $L(2,1)$-*condition*.

2.2 Unit Disk Graphs (UDG)

Let D be a set of disks in the Euclidean plane. Any disk in D is defined by its center and the value of its diameter. Then, the intersection graph G of the disks in D is called a *disk graph*, and D is called its *disk representation*. Let $d_{\min}$ and $d_{\max}$ be the minimum and maximum diameter values of the disks in D. Then, the value of $d_{\max}/d_{\min}$ is called the diameter ratio of D, denoted also by $\rho(D)$. Let ρ be some constant. A disk graph G is called a ρ-disk graph if there exists its representation D whose diameter ratio $\rho(D) \in (1, \rho]$. If $\rho(D) = 1$, then G is called a *unit disk graph*. Figure 1 is an example of a collection of unit disks, and Fig. 2 shows a unit disk graph constructed from the intersection set of Fig. 1. In this paper, we assume that all the disks in D have unit diameter since we deal with UDG.

By definition, we can easily construct a graph from its disk representation, but the opposite is not trivial. In fact, the recognition problem of an unit disk graph is known to be NP-hard [2]. Whether it is natural to be given graph representation or disk representation as the input depends on the original problem.

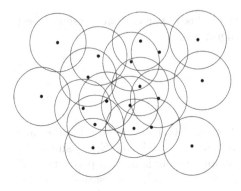

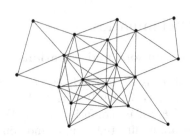

Fig. 2. Graph representation of Fig. 1

Fig. 1. Intersections of unit disks

As mentioned above, the original notion of $L(2,1)$-labeling can be seen in the context of frequency/channel assignment. Since the vertices in G correspond to the wireless sites and the diameter values of the disks in D correspond to communication coverage, we assume that our algorithm is given a disk representation as the input. It should be noted that there is a polynomial time algorithm to find a maximum clique for disk representation [5].

In this paper, we use word "distance" in two contexts. One is Euclidean distance between two vertices in D and the other is the shortest path length between two vertices on G. We distinguish them by D and G.

3   Basic Results and an Existing Algorithms

In this section, we introduce basic results on $L(2,1)$-labeling number of unit disk graphs and review the 12-approximation algorithm [6].

3.1   Upper and Lower Bounds on $L(2,1)$-Labeling Numbers

The following are general lower bounds on $\lambda(G)$ (or $\sigma(G)$): $\lambda(G) \geq \Delta(G) + 1$ and $\lambda(G) \geq 2(\omega(G) - 1)$. These are easily shown by focusing on a subgraph that forms $K_{1,\Delta}$ or ω-clique.

Concerning upper bounds, several results are known. Although it is conjectured that $\lambda(G) \leq \Delta^2$, there is still a gap between Δ^2 and the current best one. For general graphs, Goncalves gives $\lambda(G) \leq \Delta^2 + \Delta - 2$, for example [9]. It is a constructive proof, and the labeling itself is generated by a polynomial-time algorithm based on Chang and Kuo (CK algorithm) [4].

For restricted graph classes, Shao et al. [17] showed that $\lambda(G) \leq (k - 2)\Delta^2/(k-1) + 2\Delta$ holds for $K_{1,k}$-free graphs. Since unit disk graphs are $K_{1,6}$-free, this bound implies that $\lambda(G) \leq 4\Delta^2/5 + 2\Delta$ for unit disk graphs. Note that this bound is also obtained by CK algorithm. Furthermore, the algorithm by Fiala et al. [6] proposes an approximation algorithm for $L(2,1)$-labeling of unit disk graphs, which is reviewed in the next subsection. Since the algorithm constructs an $L(2,1)$-labeling of a unit disk graph with the maximum label $18\omega - 1$, it implies that $\lambda(G) \leq 18\omega - 1$. Also Junosza-Szaniawski et al. [14] show $\lambda(G) \leq 3\Delta^2/4 + 3(\Delta - 1)$ for unit disk graphs with $\Delta \geq 7$.

3.2 Existing Algorithms and Its Improvement

In this subsection, we roughly review two existing algorithms for $L(2,1)$-labeling. One is CK algorithm for general graphs by Chang and Kuo [4] and the other is a 12-approximation algorithm for unit disk graphs, which is proposed by Fiala et al. [6].

CK algorithm is a greedy algorithm based on the concept of 2-*stable set* of G, which is a subset S of $V(G)$ such that every two distinct vertices in S are of distance greater than 2. Let X_0 be a maximal 2-stable set of G. We then define Y_i and X_i for each $i \geq 1$ by $Y_i = \{x \in V \setminus (X_0 \cup \cdots \cup X_{i-1}) \mid \forall y \in X_{i-1} : d(x,y) \geq 2\}$, where $d(x,y)$ is the distance between x and y on G, and X_i is a maximal 2-stable set of $G[Y_i]$. This yields $X_0, X_1, \ldots, X_k$ such that $V = X_0 \cup X_1 \cup \cdots \cup X_k$. For $X_1, X_2, \ldots, X_k$, CK algorithm labels vertices in X_i by label i.

CK algorithm would be originally proposed not for approximating the $L(2,1)$-labeling problem but for giving an upper bound on optimal $L(2,1)$-labelings. As mentioned in the previous subsection, (slightly modified) CK algorithm finds an $L(2,1)$-labeling with at most $\min\{\Delta^2 + \Delta - 1, 4\Delta^2/5 + 2\Delta + 1\}$ labels [9,17]. Since $\Delta + 2 \leq \sigma$, the approximation ratio of CK is bounded by

$$\frac{\min\{\Delta^2 + \Delta - 1, 4\Delta^2/5 + 2\Delta + 1\}}{\Delta + 2}. \tag{1}$$

Unfortunately, since this value is monotonically increasing with respect to Δ, the approximation ratio of CK algorithm is not bounded by a constant, but the ratio is not large for small Δ. For example, it is at most $(4 \cdot 11^2 + 10 \cdot 11 + 5)/5(11+2) = 599/65 = 9.215\cdots$ if $\Delta \leq 11$.

Fiala et al.'s algorithm (FFF algorithm) works as follows: It first divides the plane into wide strips with a certain width, each of which contains 6 sub-strips. Then it periodically labels vertices in each sub-strip from left to right with a set of 3ω labels. Adjacent sub-strips are labeled by different sets of 3ω labels, and 18ω labels are used for one wide strip in total; the next wide strip is labeled by the same 18ω labels and so on in a periodic manner. Combining these 18ω labels and $\sigma(G) = \lambda(G) + 1 \geq 2\omega - 1$, Fiala et al. analyzed the approximation ratio as follows:

$$\frac{\text{ALG}(G)}{\text{OPT}(G)} \leq \frac{18\omega}{2\omega - 1} \leq 12,$$

where $\text{ALG}(G)$ is the span output by the algorithm for G. This is the argument that the approximation ratio is bounded by 12.

Actually, this bound 12 can be improved to 9.6 by combining with CK algorithm as follows: It is easy to see that $18\omega/(2\omega - 1)$ is monotonically decreasing and maximum; FFF algorithm achieves approximation ratio $18 \cdot 8/(2 \cdot 8 - 1) = 9.6$ for unit disk graph with $\omega \geq 8$. On the other hand, if $\Delta \leq 11$, the approximation ratio of CK algorithm is at most $9.215 \cdots$ as seen above. The remaining case is $\omega \leq 7$ and $\Delta \geq 12$. In this case, the approximation ratio of FFF algorithm can be bounded by $\mathrm{ALG}(G)/\mathrm{OPT}(G) \leq 18\omega/(\Delta + 2) \leq 18 \cdot 7/14 = 9$. In total, the approximation ratio is bounded by 9.6.

However, it would be difficult to achieve approximation ratio 9 by using FFF algorithm as a main routine, because $(18\omega)/(2\omega - 1) > 9$ holds for positive ω.

4   116/13-Approximation Algorithm

In this section, we present a 116/13-approximation algorithm. Similar to the 12-approximation algorithm, our algorithm also utilizes the division of the plane, though the division is different. The 12-approximation algorithm divides the plane into wide strips with a certain width, each of which contains 6 sub-strips. Then it periodically labels vertices in each sub-strip from left to right with a set of 3ω labels, where ω is the size of a maximum clique in a given graph. Adjacent sub-strips are labeled by different sets of 3ω labels, and 18ω labels are used for one wide strip in total; the next wide strip is labeled by the same 18ω labels and so on in a periodic manner. The approximation ratio is analyzed by the comparison with a lower bound $2\omega - 1$ on the optimal span σ of $L(2,1)$-labelings of a given graph.

Our algorithm divides the plane into big squares with a certain size instead, each of which contains 16 sub-squares. As we see later, we can label vertices in a big square by $16\omega + 7$ labels, but this does not immediately yield approximation ratio 116/13. In fact, if we adopt $2\omega - 1$ as a lower bound of σ, we just have $(16\omega + 7)/(2\omega - 1) \leq 13$, which is even worse. Even though we exclude small ω by a similar analysis of in the end of Sect. 3.2, we obtain $(16 \cdot 7 + 7)/(14 - 1) = 119/13 = 9.1538 \cdots \leq 9.215 \cdots$, which is still worse than 116/13. As an extra idea, we focus on the lower bound $2\omega - 1$, which means that $\sigma = 2\omega - 1$ or $\sigma \geq 2\omega$. If we can distinguish these two cases and label vertices by a smaller number of labels than $16\omega + 7$ for the former case, we can expect a better approximation ratio. Although the task itself might be difficult, a necessary condition of $\sigma = 2\omega - 1$ is helpful to achieve a smaller upper bound on labels. This is the idea to achieve the approximation ratio 116/13.

In the following subsections, we explain the details of the above ideas. In Sect. 4.1, we introduce *Square division* and consider its properties. Based on the division, we design a *basic labeling*, which can always label vertices in a big square by $16\omega + 7$ labels. We then give a necessary condition for $\sigma = 2\omega - 1$ in Sect. 4.2. Since vertices in every big square are sparsely located in such a situation, we can design another labeling called *2-phase labeling* that can label vertices in each big square by $16\omega + 4$ labels. This is explained in Sect. 4.3. Finally, we give an analysis that the approximation ratio is bounded by 116/13 in total in Sect. 4.4.

4.1   Square Division and Basic Labeling

In this subsection, we introduce the notion of *square division*, which plays a key role of our algorithm.

Suppose that a unit disk graph is given as a collection of points in a 2-dimensional Euclidean plane. We then divide the area where the points are placed into small squares with side length $1/\sqrt{2}$ as Fig. 3. Each small square is referred by its row and column location (i,j), say $S_{i,j}(i,j = 0,1,\ldots)$. Here, we assume without loss of generality that no vertex is located at boundaries of two or four small squares in the division; otherwise, we can avoid such a case by perturbing the division. By this assumption, every vertex belongs to exactly one small square. For convenience of explanation, for $p = 0,1,2,\ldots$ and $q = 0,1,2\ldots,$ we define a *big square* $B_{p,q}$ as a collection of consecutive 16 small squares $S_{i,j}$ of $i = 4p, 4p+1, 4p+2, 4p+3$, $j = 4q, 4q+1, 4q+2, 4q+3$. We denote $S_{i,j} \in B_{p,q}$ if $S_{i,j}$ is a sub-square of $B_{p,q}$. We also denote $v \in B_{p,q}$ if some $S_{i,j}$ contains v and $S_{i,j} \in B_{p,q}$.

We next consider a unit disk graph induced by a small square, called *square-UDG*. Let us denote a square-UDG induced by $S_{i,j}$ by $G_{i,j}$. See figures in Fig. 4 as square-UDG's of the instance of Fig. 3. Note that $G_{i,j}$ forms a complete graph, since the diagonal length of each $S_{i,j}$ is 1. Furthermore, $|V(G_{i,j})| \le \omega$. Thus we can label vertices in every $G_{i,j}$ by $2\omega - 1$ labels, e.g., $0, 2, \ldots, 2\omega - 2$ if we ignore labels used in surrounding square-UDG's.

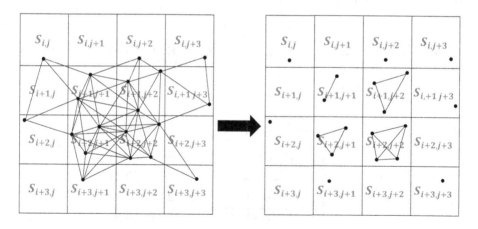

Fig. 3. Square division

We next define sets of labels. Let $L(i) = \{(2\omega+1)i, (2\omega+1)i+2, (2\omega+1)i+4, \ldots, (2\omega+1)i+2\omega-2\}$ and $L(\bar{i}) = \{(2\omega+1)i+1, (2\omega+1)i+3, (2\omega+1)i+5, \ldots, (2\omega+1)i+2\omega-1\}$ for $i = 0,1,\ldots,7$. We call these $L(i)$ and $L(\bar{i})$'s *label sets*. The concrete elements of label sets are listed in Table 1.

We are now ready to explain the basic labeling. We define a mapping f from the set of small squares to the set of indices of label sets, that is, f :

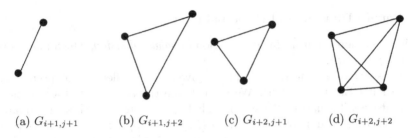

(a) $G_{i+1,j+1}$ (b) $G_{i+1,j+2}$ (c) $G_{i+2,j+1}$ (d) $G_{i+2,j+2}$

Fig. 4. Square-UDGs in Fig. 3

$\bar{0}$	$\bar{1}$	$\bar{2}$	$\bar{3}$	$\bar{0}$	$\bar{1}$	$\bar{2}$	$\bar{3}$
$\bar{4}$	$\bar{5}$	$\bar{6}$	$\bar{7}$	$\bar{4}$	$\bar{5}$	$\bar{6}$	$\bar{7}$
2	3	0	1	2	3	0	1
6	7	4	5	6	7	4	5
$\bar{0}$	$\bar{1}$	$\bar{2}$	$\bar{3}$	$\bar{0}$	$\bar{1}$	$\bar{2}$	$\bar{3}$
$\bar{4}$	$\bar{5}$	$\bar{6}$	$\bar{7}$	$\bar{4}$	$\bar{5}$	$\bar{6}$	$\bar{7}$
2	3	0	1	2	3	0	1
6	7	4	5	6	7	4	5

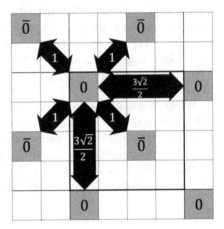

Fig. 5. Positions of label sets $L(0)$, $L(\bar{0}), L(1), \ldots, L(7), L(\bar{7})$

Fig. 6. Euclidean distances between $L(0)$ and $L(\bar{0})$ squares

$\{0, 1, \ldots, \} \times \{0, 1, \ldots, \} \to \{0, 1, \ldots, 7\} \cup \{\bar{0}, \bar{1}, \ldots, \bar{7}\}$ as follows:

$$f(i,j) = \begin{cases} j \pmod 4 & i \equiv 0 \pmod 4, \\ j \pmod 4 + 4 & i \equiv 1 \pmod 4, \\ j + 2 \pmod 4 & i \equiv 2 \pmod 4, \\ j + 2 \pmod 4 + 4 & i \equiv 3 \pmod 4. \end{cases} \tag{2}$$

The mapping (or assignment) is illustrated in Fig. 5. The part indicated by the color in Fig. 5 represents a big square, and the assignment is done in a periodic manner. The basic labeling is just to label vertices in $S_{i,j}$ by labels in $L(f(i,j))$. The following lemma is about the correctness of the basic labeling.

Lemma 1. *A labeling output by the basic labeling satisfies the $L(2,1)$-condition, and its span is $16\omega + 7$.*

Table 1. Label set for basic labeling

Label set	Labels	Label set	Labels
$L(0)$	$0, 2, 4, \ldots, 2\omega - 2$	$L(\bar{0})$	$1, 3, 5, \ldots, 2\omega - 1$
$L(1)$	$2\omega + 1, 2\omega + 3, \ldots, 4\omega - 1$	$L(\bar{1})$	$2\omega + 2, 2\omega + 4, \ldots, 4\omega$
$L(2)$	$4\omega + 2, 4\omega + 4, \ldots, 6\omega$	$L(\bar{2})$	$4\omega + 3, 4\omega + 5, \ldots, 6\omega + 1$
$L(3)$	$6\omega + 3, 6\omega + 5, \ldots, 8\omega + 1$	$L(\bar{3})$	$6\omega + 4, 6\omega + 6, \ldots, 8\omega + 2$
$L(4)$	$8\omega + 4, 8\omega + 6, \ldots, 10\omega + 2$	$L(\bar{4})$	$8\omega + 5, 8\omega + 7, \ldots, 10\omega + 3$
$L(5)$	$10\omega + 5, 10\omega + 7, \ldots, 12\omega + 3$	$L(\bar{5})$	$10\omega + 6, 10\omega + 8, \ldots, 12\omega + 4$
$L(6)$	$12\omega + 6, 12\omega + 8, \ldots, 14\omega + 4$	$L(\bar{6})$	$12\omega + 7, 12\omega + 9, \ldots, 14\omega + 5$
$L(7)$	$14\omega + 7, 14\omega + 9, \ldots, 16\omega + 5$	$L(\bar{7})$	$14\omega + 8, 14\omega + 10, \ldots, 16\omega + 6$

Proof. Since the cardinality of $L(i)$ or $L(\bar{i})$ is $2\omega - 1$ and $G_{i,j}$ forms a complete graph with at most ω vertices, each $G_{i,j}$ is always appropriately labeled. Therefore, what we need to consider is the relationship between $L(f(i,j))$ and the assigned label set for its neighbor small squares.

First notice that the difference between label values in two different label sets except the case $L(i)$ and $L(\bar{i})$ are at least 2. The labels in $L(i)$ and $L(\bar{i})$ are different but the difference could be just 1. Keeping this in mind, we will see that the mapping (label assignment) does not violate both the conditions of distance 2 and distance 1.

First we consider the condition of distance 2. See Fig. 5 and focus on a square assigned with 1, without loss of generality. As we can see in Figs. 5 and 6, the Euclidean distance between a square assigned with 1 and another square assigned with 1 is greater than $3/\sqrt{2} > 2$. This implies that vertices labeled by a same label in $L(0)$ are at distance more than 2 in G. Thus this does not violate the condition of distance 2. We next consider the condition of distance 1; we focus on square assigned with 0 and $\bar{0}$ without loss of generality. Again see Fig. 6. The Euclidean distance between a square assigned with 0 and a square with $\bar{0}$ is greater than 1. This implies that vertices labeled by $L(0)$ and $L(\bar{0})$ are at distance at least 2 in G, which satisfies the condition of distance 1.

Obviously, the span of this labeling is $16\omega + 7$ (See Table 1).   □

By a similar analysis of FFF algorithm, the basic labeling achieves approximation ratio 13, but the performance is not good enough.

4.2 Necessary Condition for $\sigma = 2\omega - 1$

The lower bounds used in the analyses of FFF and CK algorithms are $2\omega - 1$ and $\Delta + 2$. For more detailed analyses, we focus on the lower bound $2\omega - 1$. If we can distinguish instances with $\sigma = 2\omega - 1$ from $\sigma \geq 2\omega$ and label vertices by a smaller number of labels than $16\omega + 7$ for the former case, we can expect a better approximation ratio. The goal of this subsection is to give a good characterization of instances with $\sigma = 2\omega - 1$.

We first focus on how $G_{i,j}$'s of K_ω are located.

Lemma 2. *If there are two adjacent $G_{i,j}$ and $G_{i',j'}$ such that they are both K_ω and directly connected in G, $\sigma \geq 2\omega$ holds.*

Proof. Prove by contradiction. Assume that $\sigma = 2\omega - 1$ and the usable labels are $\{0, 1, \ldots, 2\omega - 2\}$. Let $G_{i,j}$ and $G_{i',j'}$ be such K_ω's, and thus there is an edge (u, v) such that $u \in V(G_{i,j})$ and $v \in V(G_{i',j'})$. Since $G_{i,j}$ is K_ω, the ω vertices in $G_{i,j}$ are labeled by ω labels, say $0, 2, \ldots, 2\omega - 2$, and same for $G_{i',j'}$. However, vertices in $G_{i,j}$ and v are at distance two via u; the labels for v and one vertex in $G_{i,j}$ must conflict, which violates the condition of distance two.   □

Lemma 2 implies that if $\sigma = 2\omega - 1$, any two $G_{i,j}$'s forming K_ω are not directly connected. This property enables a smaller span of labeling as we see in the next subsection.

4.3   2-Phase Labeling

In this subsection, we assume that there are no two adjacent $G_{i,j}$ and $G_{i',j'}$ such that they are both K_ω and directly connected in G. We design 2-phase labeling based on the property.

We pick up one vertex from every $G_{i,j}$ forming K_ω, and name it $v_{i,j}$. Let T be the set of such $v_{i,j}$'s. The idea of 2-phase labeling is as follows: we first give a labeling with $\{0, 1, \ldots, 16(\omega - 1) + 6\}$ based on the basic labeling by ignoring vertices in T. This is possible because every $G_{i,j}$ becomes a clique with size at most $\omega - 1$ by ignoring T. After that, we label vertices in T with extra 12 labels. In the resulting labeling, the span is at most $16\omega + 4$.

To explain 2-phase labeling, let us define $L'(i) = \{(2\omega-1)i, (2\omega-1)i+2, (2\omega-1)i+4, \ldots, (2\omega-1)i+2\omega-4\}$ and $L'(\bar{i}) = \{(2\omega-1)i+1, (2\omega-1)i+3, (2\omega-1)i+5, \ldots, (2\omega-1)i+2\omega-5\}$ for $i = 0, 1, \ldots, 7$. The concrete elements of label sets are listed in Table 2. Additionally, we define $L''(0) = \{16\omega - 8, 16\omega - 7, 16\omega - 6\}$, $L''(1) = \{16\omega - 5, 16\omega - 4, 16\omega - 3\}$, $L''(2) = \{16\omega - 2, 16\omega - 1, 16\omega\}$, $L''(3) = \{16\omega + 1, 16\omega + 2, 16\omega + 3\}$. These are the label set for T.

Now we are ready to explain the 2-phase labeling.

Lemma 3. *Assume that there are no two adjacent $G_{i,j}$ and $G_{i',j'}$ such that they are both K_ω and directly connected in G. A labeling generated by the 2-phase labeling satisfies the $L(2, 1)$-condition, and its span is $16\omega + 4$.*

Proof. Labeling in line 2. of Algorithm 1 is essentially same as the basic labeling with maximum clique size $\omega - 1$. Hence except for the excluded vertices in T, the correctness of this step follows from the proof of Lemma 1. Used labels a line 3., that is, in L'', are 2 different from ones in $L'(i)$ and $L'(\bar{i})$ used at line 2; we just need to be careful about the relations among T. Notice that T is an independent set of G by the construction and that the width of $S_i (= S_{i,1} \cup S_{i,2} \cup \cdots)$ is $1/\sqrt{2}$. If u_1, u_2, u_3 and u_4 are aligned on S_i in order from the left, $d(u_1, u_4) > 2$ since the width of S_i is $/\sqrt{2}$ and $d(u_1, u_2), d(u_2, u_3), d(u_3, u_4) > 1$. Thus, it is always possible to label $v_{i,j} \in T$ by $L''(k)$ by the left-to-right manner, periodically. Since vertices in T are at distance at least 2, what we need to be careful about is the condition of distance two, which is clearly satisfied.   □

Table 2. Label set for 2-phase labeling

Label set	Labels	Label set	Labels
$L'(0)$	$0, 2, 4, \ldots, 2\omega - 4$	$L'(\bar{0})$	$1, 3, 5, \ldots, 2\omega - 3$
$L'(1)$	$2\omega - 1, 2\omega + 1, \ldots, 4\omega - 5$	$L'(\bar{1})$	$2\omega, 2\omega + 2, \ldots, 4\omega - 4$
$L'(2)$	$4\omega - 2, 4\omega + 2, \ldots, 6\omega - 6$	$L'(\bar{2})$	$4\omega - 1, 4\omega + 1, \ldots, 6\omega - 5$
$L'(3)$	$6\omega - 3, 6\omega - 1, \ldots, 8\omega - 7$	$L'(\bar{3})$	$6\omega - 2, 6\omega, \ldots, 8\omega - 6$
$L'(4)$	$8\omega - 4, 8\omega - 2, \ldots, 10\omega - 8$	$L'(\bar{4})$	$8\omega - 3, 8\omega - 1, \ldots, 10\omega - 7$
$L'(5)$	$10\omega - 5, 10\omega - 3, \ldots, 12\omega - 9$	$L'(\bar{5})$	$10\omega - 4, 10\omega - 2, \ldots, 12\omega - 8$
$L'(6)$	$12\omega - 6, 12\omega - 4, \ldots, 14\omega - 10$	$L'(\bar{6})$	$12\omega - 5, 12\omega - 3, \ldots, 14\omega - 9$
$L'(7)$	$14\omega - 7, 14\omega - 5, \ldots, 16\omega - 11$	$L'(\bar{7})$	$14\omega - 6, 14\omega - 4, \ldots, 16\omega - 10$

Algorithm 1. 2-phase labeling

1: Pick up $v_{i,j}$ from every $G_{i,j}$ with ω vertices, and let T be the set of such $v_{i,j}$'s.
2: (first phase) Label vertices in $V[G_{i,j}] \setminus \{v_{i,j}\}$ by labels in $L'(f(i,j))$.
3: (second phase)
4: **for all** i **do**
5:     Let $i \equiv k \pmod 4$. Periodically label $v_{i,j} \in T$ by $L''(k)$ by the left-to-right manner.
6: **end for**

4.4 Overall Algorithm and Approximation Ratio

By combining all the ideas, our algorithm forms as stated in Algorithm 2.

Algorithm 2. $L(2,1)$-labeling of UDG

1: Compute the maximum degree Δ and the maximum clique number ω.
2: **if** $\Delta \leq 10$ **then**
3:     Apply CK algorithm, output the result, and halt.
4: **else**
5:     Define a Square Division.
6:     **if** there exist two $G_{i,j}$ and $G_{i',j'}$ such that they are both K_ω and directly connected in G. **then**
7:         Apply basic labeling, output the result, and halt.
8:     **else**
9:         Apply 2-phase labeling, output the result, and halt. .
10:     **end if**
11: **end if**

Now we analyze the approximation ratio. Our algorithm outputs a labeling in lines 3, 7 and 9. We see each of these cases.

Line 3. The output is based on CK algorithm under $\Delta \leq 10$. By Eq. (1), the approximation ratio is bounded by

$$\frac{4\Delta^2/5 + 2\Delta + 1}{\Delta + 2} \leq \frac{101}{12} = 8.416 \cdots < \frac{116}{13}.$$

Line 7. By Lemma 2, $\sigma \geq 2\omega$ holds. Since the span output by the basic labeling is $16\omega + 7$ and $\Delta \geq 11$, the approximation ratio is bounded by

$$\frac{16\omega + 7}{\max\{2\omega, \Delta + 2\}} \leq 8 + \frac{7}{\Delta + 2} \leq 8.5384 \cdots < \frac{116}{13}.$$

Line 9. By Lemma 3, the 2-phase labeling outputs a labeling with span $16\omega + 4$. Since $\Delta \geq 11$ for this case, the approximation ratio is bounded by

$$\frac{16\omega + 4}{\max\{2\omega - 1, \Delta + 2\}} \leq 8 + \frac{12}{\Delta + 2} \leq \frac{116}{13}.$$

Thus in all the cases, the approximation ratio is bounded by $116/13$.

Theorem 1. *The approximation ratio of Algorithm 2 is bounded by 116/13.*

References

1. Bodlaender, H.L., Kloks, T., Tan, R.B., Van Leeuwen, J.: Approximations for λ-colorings of graphs. Comput. J. **47**(2), 193–204 (2004)
2. Breu, H., Kirkpatrick, D.G.: Unit disk graph recognition is NP-hard. Comput. Geom. **9**(1), 3–24 (1998)
3. Calamoneri, T.: The $L(h,k)$-labelling problem: an updated survey and annotated bibliography. Comput. J. **54**(8), 1344–1371 (2011)
4. Chang, G.J., Kuo, D.: The $L(2,1)$-labeling problem on graphs. SIAM J. Discrete Math. **9**(2), 309–316 (1996)
5. Clark, B.N., Colbourn, C.J., Johnson, D.S.: Unit disk graphs. Discrete Math. **86**(1), 165–177 (1990)
6. Fiala, J., Fishkin, A.V., Fomin, F.: On distance constrained labeling of disk graphs. Theor. Comput. Sci. **326**(1), 261–292 (2004)
7. Fiala, J., Golovach, P.A., Kratochvíl, J.: Distance constrained labelings of graphs of bounded treewidth. In: Caires, L., Italiano, G.F., Monteiro, L., Palamidessi, C., Yung, M. (eds.) ICALP 2005. LNCS, vol. 3580, pp. 360–372. Springer, Heidelberg (2005). https://doi.org/10.1007/11523468_30
8. Fiala, J., Kloks, T., Kratochvíl, J.: Fixed-parameter complexity of λ-labelings. Discrete Appl. Math. **113**(1), 59–72 (2001)
9. Gonçalves, D.: On the $L(p,1)$-labelling of graphs. Discrete Math. **308**(8), 1405–1414 (2008). Third European Conference on Combinatorics
10. Griggs, J.R., Yeh, R.K.: Labelling graphs with a condition at distance 2. SIAM J. Discrete Math. **5**(4), 586–595 (1992)
11. Hale, W.K.: Frequency assignment: theory and applications. Proc. IEEE **68**(12), 1497–1514 (1980)

12. Hasunuma, T., Ishii, T., Ono, H., Uno, Y.: A linear time algorithm for $L(2,1)$-labeling of trees. Algorithmica **66**(3), 654–681 (2013)
13. Hasunuma, T., Ishii, T., Ono, H., Uno, Y.: Algorithmic aspects of distance constrained labeling: a survey. Int. J. Netw. Comput. **4**(2), 251–259 (2014)
14. Junosza-Szaniawski, K., Rzążewski, P., Sokół, J., Wesek, K.: Coloring and $L(2,1)$-labeling of unit disk intersection graphs. In: European Workshop on Computational Geometry (EuroCG), pp. 83–86 (2016)
15. Koller, A.E.: The frequency assignment problem. Ph.D. thesis. Brunel University, School of Information Systems, Computing and Mathematics (2005)
16. Roberts, F.S.: T-colorings of graphs: recent results and open problems. Discrete Math. **93**(2), 229–245 (1991)
17. Shao, Z., Yeh, R.K., Poon, K.K., Shiu, W.C.: The $L(2,1)$-labeling of $K_{1,n}$-free graphs and its applications. Appl. Math. Lett. **21**(11), 1188–1193 (2008)
18. Yeh, R.K.: A survey on labeling graphs with a condition at distance two. Discrete Math. **306**(12), 1217–1231 (2006)

Minimizing the Cost of Team Exploration

Dorota Osula[(⊠)]

Faculty of Electronics, Telecommunications and Informatics,
Gdańsk University of Technology, Gdańsk, Poland
dorurban@student.pg.edu.pl

Abstract. A group of *mobile agents* is given a task to *explore* an edge-weighted graph G, i.e., every vertex of G has to be visited by at least one agent. There is no centralized unit to coordinate their actions, but they can freely communicate with each other. The goal is to construct a deterministic *strategy* which allows agents to complete their task optimally. In this paper we are interested in a *cost-optimal* strategy, where the cost is understood as the total distance traversed by agents coupled with the cost of invoking them. Two graph classes are analyzed, rings and trees, in the *off-line* and *on-line* setting, i.e., when a structure of a graph is known and not known to agents in advance. We present algorithms that compute the optimal solutions for a given ring and tree of order n, in $O(n)$ time units. For rings in the on-line setting, we give the 2-competitive algorithm and prove the lower bound of 3/2 for the competitive ratio for any on-line strategy. For every strategy for trees in the on-line setting, we prove the competitive ratio to be no less than 2, which can be achieved by the DFS algorithm.

Keywords: Graph exploration · Distributed searching
Cost minimization · Mobile agents · On-line searching

1 Introduction

A group of *mobile agents* is given a task to *explore* the edge-weighted graph G, i.e., every vertex of G has to be visited by at least one agent. Initially agents are placed on one vertex, called *homebase*[1], they are distinguishable (each entity has its unique *id*) and they can communicate freely during the whole exploration process. The goal is to find a deterministic *strategy* (*protocol* or *algorithm*), which is a sequence of *steps*, where each step consists of parallel *moves*. Each move is one of the two following types: (1) traversing an edge by an agent or (2) invoking a new agent in the homebase. The strategy should be optimal in specified sense; in the literature we discuss the following approaches: *exploration time*, *number of entities*, *energy* and *total distance* optimization. Exploration

---

Research partially supported by National Science Centre (Poland) grant number 2015/17/B/ST6/01887.

[1] After finishing the exploration, agents do not have to come back to the homebase.

B. Catania et al. (Eds.): SOFSEM 2019, LNCS 11376, pp. 392–405, 2019.
https://doi.org/10.1007/978-3-030-10801-4_31

time is the number of *time units* required to complete the exploration, with the assumption that a walk along an edge e takes $w(e)$ time units (where $w(e)$ is the weight of the edge e). As one agent is sufficient to explore the whole graph, in the problem of minimizing the number of entities additional restrictions of the size of the *battery* of searchers (i.e., the maximum distance each agent can travel) or the maximum exploration time are added. Energy is understood as the maximum value taken over all agents traversed distances. Lastly, the total distance is the sum of distances traversed by all agents. In this work we introduce a new approach: we are looking for the *cost-optimal* strategy, where cost is the sum of the distances traversed by agents and a cost of invoking them. We consider the problem in the *off-line* setting, where a graph is known in advance for searchers and the *on-line* one, where agents have no *a priori* knowledge about the graph. We assume, for simplicity, that in one step only one agent can perform a move.[2] As the measure for an on-line algorithm the *competitive ratio* is used (formally defined later), which is the maximum taken over all networks of the results of the on-line strategy divided by the optimal one in the off-line setting.

Related Work. For exploration in the off-line setting (referred often as *searching*) many different models were extensively studied and numerous deep results have been obtained. Interestingly, there is a strong connection between graph exploration and many different graph parameters, e.g., pathwidth, treewidth, vertex separation number; see e.g., [10] for a survey and further references. In [4] edge-weighted trees in the off-line setting were studied, where a group of k mobile agents has a goal to explore the tree minimizing the total distance. Agents (as in the model presented in this paper) do not have to return to the homebase after the exploration. Thus, for k big enough, it is a special case of our model, for which the invoking cost is equal to zero. For k greater or equal to the number of leaves authors present the $O(n)$ time algorithm solving the problem. In the on-line setting, the most results were established in minimizing the time of the exploration. Algorithms and bounds of competitive ratio were investigated, mostly for trees [9,11,13] in different communication settings. As for the edge-exploration of general graphs (where apart from vertices also every edge has to be explored) see [3,11]. The competitive ratio of exploration arbitrary graphs for teams of size bigger than $\sqrt{n}$ was studied in [6,7]. As for different graph classes, grids [15] and rings [13] were investigated. Several studies have been also undertaken minimizing the energy [8,9] and the number of agents [5] for trees. As one can observe trees are very important graph class and in this paper we give the algorithms in off-line and on-line setting and prove the upper bound for the competitive ratio. We note that our results may be of particular interest not only by providing theoretical insight into searching dynamics in agent computations, but may also find applications in the field of robotics. This model describes well the real life problems, where every traveled unit costs (e.g., used fuel or energy) and entities costs itself (e.g., equipping new machines or software license cost). It can be viewed also as a special case of traveling salesmen, vehicle routing or

---

[2] One may notice, that in order to reduce the number of time units of the algorithm, agents moves, when possible, should be perform simultaneously.

pickup and delivery problem. Many deep results were established in these fields, see e.g., [1,2,12,14,16] for the further references.

This work is constructed as follows: in the next Section we introduce the necessary notation and formally define the problem. The further two Sections present results for rings. In Sect. 3 the cost-optimal algorithm for the off-line setting is presented, whereas in Sect. 4 the 2-competitive algorithm in the on-line setting is described. It is also proved, that for a positive invoking cost and any on-line strategy there exist a ring, which force the strategy to produce at least 3/2 times higher cost than the optimal, off-line one. Section 5 contains the algorithm and its analysis for trees in the off-line setting, while Sect. 6 provides the proof that no algorithm can perform better on trees in the on-line setting than DFS. In other words, the competitive ratio for every on-line algorithm is no less than 2. We finish this work with the summary and a future outlook, and suggest areas of further research.

The missing proofs will appear in the full version of the paper.

2 Notation

Let $\mathcal{G}$ be a class of non-directed, edge-weighted graphs. For every $G \in \mathcal{G}$ we denote the sets of vertices and edges of G as $V(G)$ and $E(G)$, respectively, $n = |V(G)|$ and a weight function $w : E(G) \to \mathbb{R}^+$. The sum of all weights of a subgraph H of G is denoted by $w(H) = \sum_{e \in E(H)} w(e)$. For every tree T and the pair of vertices $v, u \in V(T)$ we denote a path between them as $P_T(v, u)$ (as a path is understood an open walk with no repeated vertices). We refer to the sum of all weights between two vertices $v, u \in V(T)$ as to a *distance* and denote it by $d_T(v, u)$, i.e., $d_T(v, u) = w(P_T(v, u))$. We omit the bottom index, when a graph is clear from the context.

We define a strategy $\mathcal{S}$ as a sequence of moves of the following two types: (1) traversing an edge by an agent and (2) invoking a new agent in the homebase. We say that a strategy *explores* a vertex, when it is reached for the first time. Let $k \in \mathbb{N}^+$ be the number of agents used by $\mathcal{S}$ (notice that k is not fixed) and $d_i \in \mathbb{R}^+ \cup \{0\}$ the distance that i-th agent traversed, $i = 1, 2, \ldots, k$. The *invoking* cost $q \in \mathbb{R}^+ \cup \{0\}$ is the cost connected to the agents: every time the strategy uses a new agent it has to 'pay' for it q. In other words, before exploring any vertex, the strategy needs to decide what is more profitable: invoke a new agent (and pay for it q) or use an agent already present in the graph. The number of agents, that can be invoked, is unbounded. The cost c is understood as the sum of invoking costs and the total distance traversed by entities, i.e., $c = kq + \sum_{i=1}^{k} d_i$. The goal is to find a cost-optimal strategy, which explores every graph $G \in \mathcal{G}$.

Let $\mathcal{S}$ be an on-line strategy and $\mathcal{S}^{opt}$ be the cost-optimal, off-line strategy for every graph in $\mathcal{G}$. We denote as $\mathcal{S}(G)$ and $\mathcal{S}^{opt}(G)$ the cost of proceeding the strategy $\mathcal{S}$ and $\mathcal{S}^{opt}$, respectively, on $G \in \mathcal{G}$. As a measure for an on-line algorithm $\mathcal{S}$ the *competitive ratio* is used, which is the maximum taken over all networks of the results of the on-line strategy divided by the optimal one in the off-line setting, i.e., $r(\mathcal{S}) = \max_{G \in \mathcal{G}} \{\mathcal{S}(G) \backslash \mathcal{S}^{opt}(G)\}$. In the on-line setting it is assumed that an agent, which occupies the vertex v, knows the length of edges incident to v and the *status* of vertices adjacent to v, i.e., if they have been already explored.

3   Rings in the Off-Line Setting

Let $C = (v_1, v_2, \ldots, v_n) \in \mathcal{G}$ be a ring with a homebase in v_1. We enumerate edges in E as $e_i = (v_i, v_{i+1})$, $i \in \{1, 2, \ldots, n-1\}$ and $e_n = (v_n, v_1)$. We define the problem in the off-line setting as follows: *Given the ring C, the invoking cost q and the homebase h, find a strategy of the minimum cost.* Due to space limitations we give only an informal description of procedure RingOffline.

In the cost-optimal solution exactly one of the edges does not have to be traversed. Procedure RingOffline finds in $O(n)$ steps, which edge is optimal to omit. If this edge is incident to the homebase, then only one agent is used, which simply traverses the whole ring without it. Otherwise, depending on the cost q, there might be one or two agents in use. Let e be an omitted edge and let $C' = C \backslash e$, i.e., C' is a tree rooted in v_1 with two leaves. We denote as v^{min} and v^{max} the closer and further, respectively, leaf in C'. If the invoking cost q is lower than $d_{C'}(v_1, v^{min})$, then it is more efficient to invoke two agents, which traverse two paths $P_{C'}(v_1, v^{min})$ and $P_{C'}(v_1, v^{max})$. On the other hand, if $q \geq d_{C'}(v_1, v^{min})$, then only one agent is used, which traverses the path $P_{C'}(v_1, v^{min})$ twice. For any invoking cost q and ring C, the strategy $\mathcal{S}$, returned by the procedure RingOffline is cost-optimal.

4   Rings in the On-Line Setting

In this section we present the procedure RingOnline, which produces in $O(n)$ steps an 2-competitive strategy, which explores any unknown ring C. We also prove the lower bound of $3/2$ for the competitive ratio for any $q > 0$.

We start by giving the informal description of the procedure RingOnline. Let $\mathcal{S}$ be the on-line strategy returned by the procedure RingOnline for a given homebase v_0 and invoking cost q. Firstly $\mathcal{S}$ invokes an agent a_1 in v_0 and denotes as e_1 and e_{-1} edges incident to v_0, with the lower and higher weight respectively (lines 4–5). Searcher a_1 traverses first e_1 and then continues the exploration process as long as it is profitable, i.e., the cost of traversing the next edge is less or equal to the invoking cost plus $w(e_{-1})$ (lines 7–10). If at some point a new agent is invoked, then it traverses the edge e_{-1} (lines 11–16). We notice here that the lines 11–16 are executed at most once, as these are initial steps for the second agent. Later, the greedy approach is performed: an edge with lesser weight is traversed either by a_1 (lines 7–10) or by a_2 (lines 17–21). Below we give a formal statement of the procedure RingOnline.

The next lemma says that for any invoking cost and any ring procedure RingOnline returns the solution at most twice worse than the optimum, which is tight. The following theorem shows that for any positive invoking cost and any on-line strategy there exist a ring for which the strategy achieves at least $3/2$ times higher cost than the optimal one.

Lemma 1. *The strategy returned by* RingOnline *is 2-competitive.*

Theorem 1. *For any invoking cost $q > 0$, every on-line strategy S is at least $\frac{3}{2}$-competitive.*

At the end we observe, that for $q = 0$ the strategy returned by the procedure RingOnline is cost-optimal for every ring.

Procedure. RingOnline

Input: Homebase v_0, invoking cost q
Result: Strategy S
1: $i_r \leftarrow 1$
2: $i_l \leftarrow -1$
3: $s \leftarrow 1$
4: Add a move to S: invoke an agent a_1 in v_0
5: Denote as e_1 and e_{-1} edges adjacent to v_0, with the lower and higher weight respectively
6: **while** Graph is not explored **do**
7:     **while** $(w(e_{i_l}) + q \cdot s) \geq w(e_{i_r})$ and graph is not explored **do**
8:         Add a move to S: traverse e_{i_r} by a_1
9:         Denote the unexplored edge incident to the vertex occupied by a_1 as e_{i_r+1}
10:         $i_r \leftarrow i_r + 1$
11:     **if** $s == 1$ and $(w(e_{-1}) + q) < w(e_{i_r})$ **then**
12:         Add a move to S: invoke an agent a_2 in v_0
13:         Add a move to S: traverse e_{-1} by a_2
14:         Denote the unexplored edge incident to the vertex occupied by a_2 as e_{-2}
15:         $i_l \leftarrow -2$
16:         $s \leftarrow 0$
17:     **if** $s == 0$ **then**
18:         **while** $w(e_{i_l}) < w(e_{i_r})$ and graph is not explored **do**
19:             Add move to S: traverse e_{i_l} by a_2
20:             Denote the unexplored edge incident to the vertex occupied by a_2 as e_{i_l-1}
21:             $i_l \leftarrow i_l - 1$
22: **return** S

---

5   Tree in the Off-Line Setting

Let $T \in \mathcal{G}$ be a tree rooted in a homebase r and $\mathcal{L}(T)$ be the set of all leaves in T. For every $v \in V(T)$ we denote by T_v a subtree of T rooted in v, $c(v)$ list of its children and $p(v)$ its parent vertex.

Vertex $v \in V(T)$ is called a *decision vertex* if $|c(v)| \geq 2$ and an *internal vertex* if $|c(v)| = 1$ and v is different from the root. We say that an agent *terminates* in $v \in V(T)$, if v is its last visited vertex. We state the problem in the off-line setting formally: *Given the tree T, the invoking cost q and the homebase in the root of T, find a strategy of the minimum cost.*

5.1   The Algorithm

In order to simplify our algorithm, a *compressing* operation on a tree T is proceeded. Let $v \in V(T)$ be a decision vertex and $u \in V(T)$ be a decision vertex, a leaf or the root. The new tree T' is obtained by substituting every path $P_T(v, u)$, which apart from u and v consists only internal vertices, with a single edge $e = (v, u)$. The weight of e is set as the weight of the whole path, i.e., $w(e) = w(P_T(v, u))$. See Fig. 1 for an example of the compressing operation.

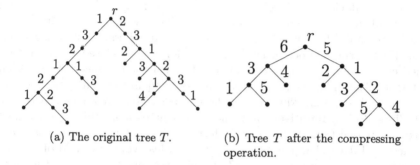

(a) The original tree T.   (b) Tree T after the compressing operation.

Fig. 1. The compressing operation on an exemplary tree T. The new tree T' has no internal vertices.

Observation 1. *In every cost-optimal strategy if an agent enters a subtree T_v, it has to explore at least one leaf in it.*

Observation 2. *In every cost-optimal strategy once an agent leaves any subtree, it never comes back to it.*

Remark 1. Let v be any internal vertex. It is never optimal for an agent, which occupies v, to return to the previously occupied vertex in its next move.

In other words, it is always optimal for agents to continue movement along the path once entered. Thus, if we find the optimal strategy for compressed tree T', then we can easily obtain the optimal strategy for T. The only difference is that instead of walking along one edge (v, u) in T', the agent has to traverse the whole path $P_T(v, u)$ in T. From now on, till the end of this Section, whenever we talk about trees, we refer to its compressed version.

For all vertices $v \in V(T)$ we consider a labeling Λ_v, which is a triple (k, u_l, u_c), where k stands for the minimum number of agents needed to explore the whole subtree T_v by any cost-optimal strategy. The second one, u_l, is the furthest leaf from v in T_v (if there is more than one, then v is chosen arbitrary) and u_c is the child of v, such that $u_l \in T_{u_c}$. We will refer to this values using the dot notation, e.g., the number of agents needed to explore tree rooted in v is denoted by $\Lambda_v.k$. The set of labels for all vertices is denoted by $\Lambda = \{\Lambda_v, \ v \in V(T)\}$.

Procedures. The algorithm is built on the principle of dynamic programming: first the strategy is set for leaves, then gradually for all subtrees and finally for the root. We present three procedures: firstly, labeling Λ is calculated by SetLabeling, which is the main core of our algorithm. Once labels for all the vertices are set, the procedure SetStrategy builds a strategy based on them. The main procedure CostExpl describes the whole algorithm.

Procedure SetLabeling for every subtree T_v, calculates and returns labeling Λ_v. We give a formal statement of the procedure and its informal description followed by an example. Firstly, for every leaf v label $\Lambda_v = (1, v, \text{null})$ is set, as one agent is sufficient to explore v. Then, by recursion, labels for the ancestors are set until the root r is reached. Let us describe now how the labeling for the vertex v is established based on the labeling of its children (main loop, lines 9–16). Firstly, the number of needed agents for v is increased by the number of needed agents for its child u (line 10). Then, if the distance between v and the furthest leaf in T_u (i.e., $d(v, \Lambda_u.u_l)$) is less or equal to the distance from the root r to v plus the invoking cost q, the number of required agents is reduced by 1 (lines 12–13). Intuitively, it is more efficient to reuse this agent, than to invoke a new one from r. As we show formally later at most one agent can be returned, and it can happen only if $\Lambda_u.k = 1$. Meanwhile the child of v, which is an ancestor of the furthest leaf in T_v is being set (lines 14–16). See the formal statement of the procedure and an example on the Fig. 2.

Procedure. SetLabeling

Input: Tree T, vertex v, invoking cost q, labeling Λ
Result: Updated Λ
 1: **if** $v \in \mathcal{L}(T)$ **then**
 2:     $\Lambda_v \leftarrow (1, v, \text{null})$
 3:     **return** Λ
 4: **for each** $u \in c(v)$ **do**
 5:     Invoke Procedure SetLabeling for T, u, q and Λ
 6: $k, d^{max} \leftarrow 0$
 7: $u_c^{max} \leftarrow \text{null}$
 8: $d_r \leftarrow d(r, v) + q$
 9: **for each** $u \in c(v)$ **do**
10:     $k \leftarrow k + \Lambda_u.k$
11:     $d \leftarrow d(v, \Lambda_u.u_l)$
12:     **if** $\Lambda_u.k == 1$ and $d \leq d_r$ **then**
13:         $k \leftarrow k - 1$
14:     **if** $d > d^{max}$ **then**
15:         $d^{max} \leftarrow d$
16:         $u_c^{max} \leftarrow u$
17: $k \leftarrow \max\{1, k\}$
18: $\Lambda_v \leftarrow (k, \Lambda_{u_c^{max}}.u_l, u_c^{max})$
19: **return** Λ

Procedure `SetStrategy` builds a strategy for a given subtree T_v based on the labeling Λ. If $v \in V(T) \backslash \mathcal{L}(T)$, then for each of its child u, firstly, the required number of agents is sent to u (line 7) and then the strategy is set for u (line 8). Lastly, for all children u of v (apart from the one, which has to be visited as the last one) if it is efficient for the searcher, which finished exploration of T_u in $\Lambda_u.u_l$, to come back to v, then the 'return' sequence of moves is added (lines 9–10). It is crucial that for every v the subtree $T_{\Lambda_v.u_c}$ is explored as the last one, but the order of the remaining subtrees is not important (line 5). To summarize, we give a formal statement of the procedure.

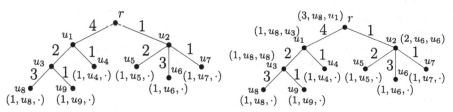

(a) Firstly, labels for leaves are set.   (b) Then, gradually labels are being set for the ancestors, until the root is reached.

Fig. 2. Example of the performing of the procedure `SetLabeling` for $q = 0$. Three agents are required to explore this tree in the cost-optimal way.

Procedure. SetStrategy

Input: Tree T, vertex v, invoking cost q, labeling Λ, strategy $\mathcal{S}$
Result: Strategy $\mathcal{S}$

1: **if** $v \notin \mathcal{L}(T)$ **then**
2:     **if** $v == r$ **then**
3:         Add a move to $\mathcal{S}$: invoke $\Lambda_r.k$ agents in r
4:     $d_r \leftarrow d(r, v) + q$
5:     Let $c_1, \ldots, c_l$ be children of v, where $c_l = \Lambda_v.u_c$
6:     **for** $i = 1, \ldots, l$ **do**
7:         Add a sequence of moves to $\mathcal{S}$: traverse (v, c_i) by $\Lambda_{c_i}.k$ agents
8:         Invoke Procedure `SetStrategy` for T, c_i, q, Λ and $\mathcal{S}$
9:         **if** $d(v, \Lambda_{c_i}.u_l) \leq d_r$ and $c_i \neq \Lambda_v.u_c$ **then**
10:         Add a sequence of moves to $\mathcal{S}$: send an agent back from $\Lambda_{c_i}.u_l$ to v

---

Procedure `CostExpl` consists of two procedures presented in the previous subsections. Firstly, `SetLabeling` is being invoked for the whole tree T. And then the strategy $\mathcal{S}$ is being calculated from the labeling Λ by the procedure `SetStrategy`. We observe that `CostExpl` finds a strategy in $O(n)$ time. To summarize, we give a formal statement of the procedure.

5.2   Analysis of the Algorithm

In this Section, we analyze the algorithm by providing the necessary observations and lemmas and give the lower and upper bounds. Firstly, let us make a simple observation about the behavior of agents in the cost-optimal strategies.

Observation 3. *In every cost-optimal strategy all agents terminates in leaves and every leaf is visited exactly once.*

Procedure. CostExpl

Input: Tree T, invoking cost q
Result: Strategy S
    Invoke Procedure `SetLabeling` for T, r, q and $\emptyset$; set Λ as an output
    Invoke Procedure `SetStrategy` for T, r, q, Λ and $\emptyset$; set S as an output
    Return S

In our strategies, subtrees T_v of the maximum height $d(r, v) + q$ are always explored by one agent. The next observation says that in the cost-optimal solution this agent finishes in the furthest leaf of T_v.

Observation 4. *If one agent is cost-optimal to search a tree T, then it terminates in one of the furthest leaves.*

Let $v \in V(T)$ different then root. Lemma 2 guarantees us that after the exploration of T_v at most one agents returns to $p(v)$. Lemma 3 and Theorem 2 present our main results.

Lemma 2. *In every cost-optimal strategy if an agent leaves any subtree, it has explored it on its own.*

Lemma 3. *Let Λ be a labeling returned by the procedure* `SetLabeling` *for an arbitrary tree T. Every cost-optimal strategy uses at least $\Lambda_v.k$ agents to explore T_v, $v \in V(T)$.*[3]

Theorem 2. *Procedure* `CostExpl` *for every tree T returns a strategy, which explores T in the cost-optimal way.*

Lower and Upper Bounds. For any tree T the value of the optimal cost c is bounded by $q + w(T) \leq c \leq q + 2w(T) - H$. A trivial lower bound is achieved on the path graph, where one agent traverses the total distance of $w(T)$. The upper bound can be obtained by performing DFS algorithm by one entity, which set it on $q + 2w(T)$. Let DFS' be the modified version of DFS, such that the agent does not return to the homebase (i.e., terminates in one of the leaves). Then

---

[3] There exist cost-optimal strategies that can use more than $\Lambda_v.k$ agents. Indeed, if $d(v, \Lambda_v.u_l) = d(r, v) + q$ reusing the agent and calling a new one generates equal cost.

we get an improved upper bound of $q + 2w(T) - H$, where H is the height of T, which is tight (e.g., for paths). It is worth to mention that although DFS' performs well on some graphs, it can be twice worse than CostExpl. Let $q \geq 0$ be any invoking cost and $K_{1,n}$ be a star rooted in the internal vertex with edges of the weight $l > q$. While DFS' produces the cost of $c' = q + 2ln - l$, the optimal solution is $c = qn + ln$. The ratio c'/c grows to 2 with the growth of l and n.

6   Tree in the On-Line Setting

In this Section we take a closer look at the algorithms for trees in the on-line setting. Because the height of tree T is not known, the upper bound of the cost, set by DFS', is $q + 2w(T) - \epsilon$, where ϵ is some small positive constant. This leads to the upper bound of 2 for the competitive ratio. We are going to prove that it is impossible to construct an algorithm that achieves better competitive ratio than 2.

For any tree T and vertex $v \in V(T)$ we define *branch* as a subtree rooted in a child c of v enlarged by the vertex v and edge (v, c). Denote as $\mathcal{G}$ an infinite class of rooted in v_0 trees, where every edge has weight equal to 1. For every integer $l \in \mathbb{N}^+$, $i \in \{1, 2, \ldots, l\}$ and $l_i \in \{1, 2, \ldots, l\}$, we add to the class $\mathcal{G}$ a tree constructed in the following way: (1) construct $l + 1$ paths $P(v_i, v_{i+1})$, $i \in \{0, 1, \ldots, l\}$ of the length l; (2) for every $i \in \{1, 2, \ldots, l\}$ construct a path $P(u'_i, u_i)$ of the length $l_i - 1$ (if $l_i = 1$, then $u_i = u'_i$) and add edge (v_i, u'_i). In other words, every graph in $\mathcal{G}$ has a set of decision vertices $\{v_1, v_2, \ldots, v_l\}$ and set of leaves $\{v_{l+1}, u_1, u_2, \ldots, u_l\}$. Every decision vertex has exactly two children, v_i is an ancestor of v_j and $d(v_i, v_{i+1}) = l$ for every $1 \leq i < j \leq l + 1$. See Fig. 3.

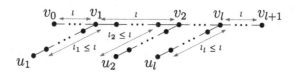

Fig. 3. Illustration of graphs from the class $\mathcal{G}$, where $l \in \mathbb{N}^+$ and $l_i \in \{1, 2, \ldots, l\}$, $i \in \{1, 2, \ldots, l\}$.

Theorem 3. *Any on-line cost-optimal solution for trees is 2-competitive.*

Proof. DFS' is an example of an algorithm at most twice worse than the best solution, which sets the upper bound. We are going now to show that for any invoking cost $q \geq 0$ and strategy $\mathcal{S}$ there exists a tree $T \in \mathcal{G}$ and a strategy $\mathcal{S}'$, such that $\mathcal{S}(T) \geq 2\mathcal{S}'(T)$. Let $l \in \mathbb{N}^+$ be any integer. Values of l_i, $i \in \{1, 2, \ldots, l\}$ are set during the execution of $\mathcal{S}$. For every v_i, $i \in \{1, 2, \ldots, l\}$ three cases can occur.

A1: More than one agent reaches v_i before any child of v_i is explored. The value of l_i is set as 1.

A2: One of the agents explores one of the branches of v_i at the depth $0 \le h < l$ and the second branch at the depth l, before any other agent reaches v_i for the first time. In this situation we choose a set of graphs from $\mathcal{G}$ for which the explored vertex at the depth l is v_{i+1} and $l_i = h + 1$. The value of h might be 0, as it takes place e.g., for DFS.

A3: One of the agents explores two branches of v_i at the depth $0 \le h_1 < l$, $1 \le h_2 < l$, before any other agent reaches v_i for the first time. Without loss of generality, we assume that the branch explored to the level h_2 is visited as the last one. In this situation we choose a set of graphs from $\mathcal{G}$ for which vertex v_{i+1} belongs to the branch of v_i explored till the level h_2 and $l_i = h_1 + 1$. Once again, $h_1 = 0$ means that the branch was not explored at all.

When $\mathcal{S}$ explores v_{l+1}, all l_i are defined and the set of graphs is narrowed to the exactly one graph, which we denote as T. We claim first that the distance d_0 traversed along the path $P(v_1, v_{l+1})$ is at least $2l^2 - l$ in any $\mathcal{S}$. Let agent a_1 be the one, which explores v_{l+1} and let $k \in \{0, 1, \ldots, l\}$ be the number of decision vertices visited by more than one agent.

B1: $k = 0$, i.e., T is explored by one agent. In other words, for all v_i holds the case *A2*. We notice that, whenever a strategy $\mathcal{S}$ explores v_{i+1}, $i \in \{1, 2, \ldots, l\}$, exactly one vertex (i.e., leaf u_i) on the path $P(v_i, u_i)$ is unexplored. Thus, $P(v_1, v_{l+1})$ has to be traversed at least twice and $d_0 \ge 2l^2$.

B2: $k = l$. Path $P(v_1, v_l)$ has to be obviously traversed at least twice and $P(v_l, v_{l+1})$ once, i.e., $d_0 \ge 2l(l-1) + l = 2l^2 - l$.

B3: $0 < k < l$. In other words, $T_{v_{k+1}}$ is explored by one agent. Paths $P(v_1, v_k)$ and $P(v_{k+1}, v_{l+1})$ are traversed at least twice and $P(v_k, v_{k+1})$ at least once. Thus, $d_0 \ge 2l(k-1) + 2l(l-k) + l = 2l^2 - l$.

Now, we have to analyze paths $P(v_i, u_i)$, $i \in \{1, 2, \ldots, l\}$. We divide decision vertices into the four groups based on the performance of $\mathcal{S}$:

- $V_1 = \{v_i | l_i = 1, \text{no agent terminates in } u_i, \ i \in \{1, 2, \ldots, l\}\}$;
- $V_2 = \{v_i | l_i = 1, \text{at least one agent terminates in } u_i, \ i \in \{1, 2, \ldots, l\}\}$;
- $V_3 = \{v_i | l_i > 1, \text{no agent terminates in any vertices of the path}$
  $P(v_i, u_i), \ i \in \{1, 2, \ldots, l\}\}$;
- $V_4 = \{v_i | l_i > 1, \text{at least one agent terminates in a vertex from the path}$
  $P(v_i, u_i), \ i \in \{1, 2, \ldots, l\}\}$.

Notice that V_1, V_2, V_3 and V_4 form a partition of decision vertices. Let us denote as d_i the total distance traversed by all the agents along $P(v_i, u_i)$ in $\mathcal{S}$. For any $v_i \in V_1$ we have $d_i \ge 2$ and $v_i \in V_2$ we have $d_i \ge 1$. From the way how T is constructed follows, that if $l_i > 1$, then either holds *A2* and $h > 0$ or *A3* and $h_1 > 0$. In both situations path $P(v_i, p(u_i))$ is first traversed at least twice, leaving u_i unexplored. If now, no agent terminates in any vertex of $P(v_i, u_i)$, then $P(v_i, p(u_i))$ has to be traversed at least twice more. Thus, for any $v_i \in V_3$ stays $d_i \ge 4(l_i - 1) + 2 \ge 4l_i - 2$. On the other hand, if at least one agent terminates in any vertex of $P(v_i, u_i)$, then $P(v_i, p(u_i))$ can be traversed only one extra time. Which leads to, $d_i \ge 3(l_i - 1) + 1 \ge 3l_i - 2$.

Lastly, we have to consider the extra cost d' generated by searchers. Every invoked agent, which terminates on some path $P(v_i, u_i)$ has to traverse the edge (v_0, v_1), thus $d' \geq (q + l)(|V_2| + |V_4|) \geq |V_2| + l|V_4|$.

The total cost of exploring T by S can be lower bounded by

$$\mathcal{S}(T) \geq 2l^2 - l + 2|V_1| + |V_2| + \sum_{v_i \in V_3}(4l_i - 2) + \sum_{v_i \in V_4}(3l_i - 2) + |V_2| \tag{1}$$

$$+ l|V_4| \geq 2l^2 - l + 2|V_1| + 2|V_2| + 4\sum_{v_i \in V_3} l_i - 2|V_3| + 4\sum_{v_i \in V_4} l_i - 2|V_4| \tag{2}$$

$$= 2l^2 - l + 4\sum_{i=1}^{l} l_i - 2(|V_1| + |V_2| + |V_3| + |V_4|) = 2l^2 + 4\sum_{i=1}^{l} l_i - 3l. \tag{3}$$

Consider now the following off-line strategy $\mathcal{S}'$, which explores the same graph T by using one agent, which after reaching the decision vertex v_i, $i \in \{1, 2, \ldots, l\}$, firstly traverses the path $P(v_i, u_i)$, then returns to v_i and explores further the tree. The agent finally terminates in v_{l+1}. Thus, the path $P(v_0, v_{l+1})$ of the length $(l + 1)l$ is traversed only once and paths $P(v_i, u_i)$, $i \in \{1, 2, \ldots, l\}$ twice. The optimal strategy can be then upper bounded by

$$\mathcal{S}^{opt}(T) \leq \mathcal{S}'(T) = q + l^2 + 2\sum_{i=1}^{l} l_i + l. \tag{4}$$

This leads to the following competitive ratio

$$r(\mathcal{S}) = \lim_{l \to \infty} \frac{\mathcal{S}(T)}{\mathcal{S}^{opt}(T)} \geq \lim_{l \to \infty} \frac{2l^2 + 4\sum_{i=1}^{l} l_i - 3l}{q + l^2 + 2\sum_{i=1}^{l} l_i + l} \tag{5}$$

$$= 2 - \lim_{l \to \infty} \frac{5l + 2q}{q + l^2 + 2\sum_{i=1}^{l} l_i + l} = 2, \tag{6}$$

which finishes the proof.

7   Conclusion

In this work we propose a new cost of the team exploration, which is the sum of total traversed distances by agents and the invoking cost which has to be paid for every searcher. This model describes well the real life problems, where every traveled unit costs (e.g., used fuel or energy) and entities costs itself (e.g., equipping new machines or software license cost). The algorithms, which construct the cost-optimal strategies for any given edge-weighted ring and tree in $O(n)$ time are presented. As for the on-line setting the 2-competitive algorithm for rings is given and the lower bounds of $3/2$ and 2 for the competitive ratio for rings and trees, respectively, are proved. While there is very little done in this area, a lot of new questions have been pondered. Firstly, it would be interesting to consider other classes of graphs, also for the edge-exploration (where not only every vertex has to be visited, by also every edge). Intuitively, for some of them, the problem would be easy and for some might be NP-hard (e.g., cliques). Another direction is to look more into the problem in the on-line setting, which is currently rapidly expanding due to its various application in many areas. It would be highly interesting to close the gap between lower and upper bounds of the competitive ratio for rings. Another idea is to bound communication for agents, which will make this model truly distributed. One may notice, that a simple solution of choosing one leader agent to pass messages between the other entities might not be cost-optimal, as it significantly rises the traveling cost. Lastly, different variation of this model might be proposed, e.g., the invoking cost might increase/decrease with the number of agents in use or time might be taken under consideration as the third minimization parameter.

References

1. Bellmore, M., Nemhauser, G.L.: The traveling salesman problem: a survey. Oper. Res. **16**(3), 538–558 (1968)
2. Berbeglia, G., Cordeau, J.F., Gribkovskaia, I., Laporte, G.: Static pickup and delivery problems: a classification scheme and survey. Top **15**(1), 1–31 (2007)
3. Brass, P., Vigan, I., Xu, N.: Improved analysis of a multirobot graph exploration strategy. In: 2014 13th International Conference on Control Automation Robotics & Vision (ICARCV), pp. 1906–1910. IEEE (2014)
4. Czyzowicz, J., Diks, K., Moussi, J., Rytter, W.: Energy-optimal broadcast in a tree with mobile agents. In: Fernández Anta, A., Jurdzinski, T., Mosteiro, M.A., Zhang, Y. (eds.) ALGOSENSORS 2017. LNCS, vol. 10718, pp. 98–113. Springer, Cham (2017). https://doi.org/10.1007/978-3-319-72751-6_8
5. Das, S., Dereniowski, D., Karousatou, C.: Collaborative exploration by energy-constrained mobile robots. In: Scheideler, C. (ed.) Structural Information and Communication Complexity. LNCS, vol. 9439, pp. 357–369. Springer, Cham (2015). https://doi.org/10.1007/978-3-319-25258-2_25
6. Dereniowski, D., Disser, Y., Kosowski, A., Pajak, D., Uznanski, P.: Fast collaborative graph exploration. Inf. Comput. **243**, 37–49 (2015)
7. Disser, Y., Mousset, F., Noever, A., Škorić, N., Steger, A.: A general lower bound for collaborative tree exploration. In: Das, S., Tixeuil, S. (eds.) SIROCCO 2017. LNCS, vol. 10641, pp. 125–139. Springer, Cham (2017). https://doi.org/10.1007/978-3-319-72050-0_8
8. Dynia, M., Korzeniowski, M., Schindelhauer, C.: Power-aware collective tree exploration. In: Grass, W., Sick, B., Waldschmidt, K. (eds.) ARCS 2006. LNCS, vol. 3894, pp. 341–351. Springer, Heidelberg (2006). https://doi.org/10.1007/11682127_24
9. Dynia, M., Łopuszański, J., Schindelhauer, C.: Why robots need maps. In: Prencipe, G., Zaks, S. (eds.) SIROCCO 2007. LNCS, vol. 4474, pp. 41–50. Springer, Heidelberg (2007). https://doi.org/10.1007/978-3-540-72951-8_5
10. Fomin, F., Thilikos, D.: An annotated bibliography on guaranteed graph searching. Theor. Comput. Sci. **399**(3), 236–245 (2008)

11. Fraigniaud, P., Gasieniec, L., Kowalski, D.R., Pelc, A.: Collective tree exploration. Networks **48**(3), 166–177 (2006)
12. Golden, B.L., Raghavan, S., Wasil, E.A.: The Vehicle Routing Problem: Latest Advances and New Challenges, vol. 43. Springer, Heidelberg (2008). https://doi.org/10.1007/978-0-387-77778-8
13. Higashikawa, Y., Katoh, N., Langerman, S., Tanigawa, S.: Online graph exploration algorithms for cycles and trees by multiple searchers. J. Comb. Optim. **28**(2), 480–495 (2014)
14. Kumar, S.N., Panneerselvam, R.: A survey on the vehicle routing problem and its variants. Intell. Inf. Manag. **4**(03), 66 (2012)
15. Ortolf, C., Schindelhauer, C.: Online multi-robot exploration of grid graphs with rectangular obstacles. In: Proceedings of the Twenty-Fourth Annual ACM Symposium on Parallelism in Algorithms and Architectures, pp. 27–36. ACM (2012)
16. Vaishnav, P., Choudhary, N., Jain, K.: Traveling salesman problem using genetic algorithm: a survey. Int. J. Sci. Res. Comput. Sci. Eng. Inf. Technol. **2**(3), 105–108 (2017)

Two-Head Finite-State Acceptors with Translucent Letters

Benedek Nagy[1] and Friedrich Otto[2(✉)]

[1] Department of Mathematics, Faculty of Arts and Sciences, Eastern Mediterranean University, Famagusta, North Cyprus, via Mersin 10, Turkey
nbenedek.inf@gmail.com
[2] Fachbereich Elektrotechnik/Informatik, Universität Kassel, 34109 Kassel, Germany
f.otto@uni-kassel.de

Abstract. Finite-state acceptors are studied that have two heads that read the input from opposite sides. In addition, a set of translucent letters is associated with each state. It is shown that these two-head automata are strictly more expressive than the model with a single head, but that they still only accept languages that have a semi-linear Parikh image. In fact, we obtain a characterization for the class of linear context-free trace languages in terms of a specific class of two-head finite-state acceptors with translucent letters.

Keywords: Two-head finite-state acceptor · Translucent letter
Linear context-free language · Semi-linear Parikh set · Trace language

1 Introduction

The finite-state acceptor is one of the most fundamental computing devices for accepting languages, and it is being used in many areas like compiler construction, text editors, hardware design, etc. A finite-state acceptor reads its input strictly sequentially from left to right. However, in the literature one finds many extensions of this model that process their inputs in different ways like, e.g.,

- the *multi-head finite-state acceptor*, which has a finite number of heads that all read the input from left to right [17], the *Watson-Crick* (WK for short) *automaton* [3], which has two heads that read the input from left to right, but which works on double stranded words where letters on corresponding positions are connected by a complementarity relation,
- the $5' \rightarrow 3'$ *sensing Watson-Crick automaton* [7,9], which has two heads that start from the two ends of an input word, one reading the word from left to right and the other reading it from right to left, halting when the two heads meet,
- the nondeterministic linear automata of [1] (see also [8,18]), and
- the finite-state acceptor with translucent letters [10,14], which has a single head starting at the left end of an input word, but depending on the actual

© Springer Nature Switzerland AG 2019
B. Catania et al. (Eds.): SOFSEM 2019, LNCS 11376, pp. 406–418, 2019.
https://doi.org/10.1007/978-3-030-10801-4_32

state, it skips across a prefix of letters that are translucent, in this way reading (and deleting) a letter from the input. This type of automaton is equivalent to the cooperating distributed systems of stateless deterministic restarting automata with window size 1 that were introduced and studied in [11,15].
– Finally, there is the *jumping finite automaton* [6], which has a single head starting at the left end of the input, but that jumps in each step to an arbitrary position reading (and deleting) the letter at that position.

In this paper we propose a new type of two-head finite-state acceptor, the *two-head finite-state acceptor with translucent letters* (2hNFAwtl, for short). It is obtained by combining the concept of the $5' \to 3'$ sensing WK automaton with the idea of translucent letters, in this way giving a new more powerful finite state model. Such a device is given an input word surrounded by sentinels, and it has two heads that start at the two ends of the given input, being positioned on the sentinels, one scanning the input from left to right, the other scanning it from right to left. However, depending on the actual state, certain letters are translucent. This means that the left head, that is, the one scanning the input from left to right, skips across a prefix of translucent letters and reads (and deletes) the first letter that is not translucent for the current state. Analogously, the right head, that is, the head that scans the input from right to left, skips across a suffix of translucent letters and reads (and deletes) the first letter from the right that is not translucent for the current state. If no such letter is found, then the automaton halts, accepting if the current state is final.

As the 2hNFAwtl extends the $5' \to 3'$ sensing WK automaton, it accepts all linear context-free languages. Actually, the 2hNFAwtl even accepts some languages that are not context-free, but we will see that each language L accepted by a 2hNFAwtl contains a linear context-free sublanguage L' that is letter-equivalent to L, that is, L and L' have the same Parikh image. This implies in particular that all these languages have semi-linear Parikh images. Further, we will see that all linear context-free trace languages are accepted by 2hNFAwtls, and we can even characterize this class of trace languages by a restricted type of 2hNFAwtls. In addition, we will see that the 2hDFAwtl, the deterministic variant of the 2hNFAwtl, is less expressive than the 2hNFAwtl, and we establish some closure and some non-closure properties for the classes of languages accepted by 2hNFAwtls and by 2hDFAwtls. Also we consider a number of decision problems for 2hNFAwtls.

This paper is structured as follows. In Sect. 2, we recall basic notions and notation of formal language theory. In Sect. 3, we introduce the two-head finite-state acceptor with translucent letters, we present an example illustrating its expressive power, and we establish a normal form for 2hNFAwtls. In Sect. 4, we derive the aforementioned results on the expressive power of the 2hNFAwtl. Finally, we study closure and decidability results in Sect. 5.

2   Preliminaries

For a finite alphabet Σ, we use Σ^+ to denote the set of non-empty words over Σ and Σ^* to denote the set of all words over Σ including the empty word ε. For a word $w \in \Sigma^*$, $|w|$ denotes the length of w, and $|w|_a$ is the a-length of w, that is, the number of occurrences of the letter a in w. For any automaton A, we will use the notation $L(A)$ to denote the language that consists of all words that are accepted by A, and for any type of automaton $\mathfrak{A}$, $\mathcal{L}(\mathfrak{A})$ is the class of languages that are accepted by automata of type $\mathfrak{A}$. Here we assume that the reader is familiar with the basics of formal language and automata theory for which we refer to the textbook [5]. By REG, LIN, and CFL we denote the classes of regular, linear context-free, and context-free languages, respectively. Finally, two languages over the same alphabet $\Sigma = \{a_1, a_2, \ldots, a_n\}$ are called *letter-equivalent* if they have the same image under the Parikh mapping $\psi : \Sigma^* \to \mathbb{N}^n$.

A *two-head finite automaton* (2hNFA) is described by a 5-tuple $A = (Q, \Sigma, I, F, \delta)$, where Q is a finite set of states, Σ is a finite input alphabet, $I \subseteq Q$ is the set of initial states, $F \subseteq Q$ is the set of final (or accepting) states, and $\delta : Q \times (\Sigma \cup \{\varepsilon\}) \times (\Sigma \cup \{\varepsilon\}) \to 2^Q$ is a transition relation.

A 2hNFA A works as follows. On an input word $w \in \Sigma^*$, it starts in an initial state $q_0 \in I$ with its first (or left) head on the first letter of w and its second (or right) head on the last letter of w. This configuration is encoded as $q_0 w$. Now, depending on the allowed transitions, it reads the first and/or the last letter (or nothing) of w, and changes its state. Formally, the configuration $qavb$ can be transformed into the configuration pv if $p \in \delta(q, a, b)$, where $p, q \in Q$ and $a, b \in \Sigma \cup \{\varepsilon\}$. If no transition can be applied, then A halts without accepting if $w = avb$ is nonempty. Otherwise, it continues reading (and deleting) letters until w has been consumed completely. We say that A accepts w if A has a computation that is in a final state $q_f \in F$ after reading w completely. It is known that the language class $\mathcal{L}(\text{2hNFA})$ coincides with the class LIN of linear context-free languages [7,9].

In [14] the finite-state acceptor with translucent letters (NFAwtl) was introduced. Essentially it is a nondeterministic finite-state acceptor A for which a subset $\tau(q)$ of translucent letters is associated with each state q. Thus, when in state q, then A does not read the first letter on its input tape, but it looks for the first occurrence from the left of a letter that is not translucent for state q. Thus, if $w = uav$ such that $u \in (\tau(q))^*$ and $a \notin \tau(q)$, then A nondeterministically chooses a state $q_1 \in \delta(q, a)$, erases the letter a from the tape thus producing the tape contents $uv \triangleleft$, and enters state q_1. In case $\delta(q, a) = \emptyset$, A halts without accepting. Finally, if $w \in (\tau(q))^*$, then A reaches the endmarker $\triangleleft$ and the computation halts, accepting iff q is a final state. NFAwtls accept a class of languages that properly contains all rational trace languages [14].

3   Two-Head Finite-State Acceptors with Translucent Letters

Definition 1. *A two-head finite-state acceptor with translucent letters (2hNFAwtl) consists of a finite-state control, a single flexible tape with endmarkers, and two heads that are positioned on these endmarkers. It is defined by an 8-tuple $A = (Q, \Sigma, \triangleright, \triangleleft, \tau, I, F, \delta)$, where Q is a finite set of states that* *is partitioned into two subsets* $Q = Q_L \cup Q_R$ *of* left *and* right *states,* Σ *is a finite input alphabet,* $\triangleright, \triangleleft \notin \Sigma$ *are special symbols that are used as* endmarkers, $\tau : Q \to 2^\Sigma$ *is a* translucency mapping, $I \subseteq Q$ *is a set of* initial *states,* $F \subseteq Q$ *is a set of* final *states, and* $\delta : Q \times \Sigma \to 2^Q$ *is a* transition relation. *For each state* $q \in Q$, *the letters from the set* $\tau(q)$ *are translucent for* q, *that is, in state* q *the automaton* A *cannot see them.*

A *is called a* deterministic two-head finite-state acceptor with translucent letters, *abbreviated as* 2hDFAwtl, *if* $|I| = 1$ *and if* $|\delta(q,a)| \le 1$ *for all* $q \in Q$ *and all* $a \in \Sigma$.

A 2hNFAwtl $A = (Q, \Sigma, \triangleright, \triangleleft, \tau, I, F, \delta)$ works as follows. For an input word $w \in \Sigma^*$, an initial configuration consists of the automaton being in an initial state q_0 chosen nondeterministically from the set I with the word $\triangleright w \triangleleft$ on its tape. On the set $Q \cdot \triangleright \Sigma^* \triangleleft$ of configurations, A executes the following computation relation:

$$q \triangleright w \triangleleft \vdash_A \begin{cases} q' \triangleright uz\triangleleft, \text{ if } q \in Q_L \text{ and } w = uaz \text{ for } u \in (\tau(q))^*, a \in \Sigma \smallsetminus \tau(q), \\ \qquad \text{ and } q' \in \delta(q,a), \\ \text{Reject,} \quad \text{ if } q \in Q_L \text{ and } w = uaz \text{ for } u \in (\tau(q))^*, a \in \Sigma \smallsetminus \tau(q), \\ \qquad \text{ and } \delta(q,a) = \emptyset, \\ q' \triangleright vy\triangleleft, \text{ if } q \in Q_R \text{ and } w = vby \text{ for } y \in (\tau(q))^*, b \in \Sigma \smallsetminus \tau(q), \\ \qquad \text{ and } q' \in \delta(q,b), \\ \text{Reject,} \quad \text{ if } q \in Q_R \text{ and } w = vby \text{ for } y \in (\tau(q))^*, b \in \Sigma \smallsetminus \tau(q), \\ \qquad \text{ and } \delta(q,b) = \emptyset, \\ \text{Accept,} \quad \text{ if } w \in (\tau(q))^* \text{ and } q \in F, \\ \text{Reject,} \quad \text{ if } w \in (\tau(q))^* \text{ and } q \notin F. \end{cases}$$

A word $w \in \Sigma^*$ is *accepted* by A if there exist an initial state $q_0 \in I$ and a computation $q_0 \triangleright w \triangleleft \vdash_A^*$ Accept, where $\vdash_A^*$ denotes the reflexive transitive closure of the single-step computation relation $\vdash_A$.

If A is a 2hNFAwtl such that $\tau(q) = \emptyset$ for all states q of A, then A can actually be seen as a 2hNFA. Conversely, if B is a 2hNFA, then by splitting each state q of B into a left state q_L and a right state q_R and by rearranging the transitions accordingly, we obtain a 2hNFAwtl A with empty transparency sets such that $L(A) = L(B)$. Hence, the 2hNFAwtl is an extension of the 2hNFA, which shows that $\mathsf{LIN} \subseteq \mathcal{L}(\text{2hNFAwtl})$.

Obviously, the 2hNFAwtl is also an extension of the NFAwtl. The linear language $\{ a^n b^n \mid n \ge 0 \}$ is not accepted by any NFAwtl [14], while $\mathcal{L}(\text{NFAwtl})$ does even contain some non-context-free languages, for example, the language $\{ w \in \{a,b,c\}^* \mid |w|_a = |w|_b = |w|_c \}$. However, the 2hNFAwtl is more expressive than the NFAwtl.

Example 2. Let L be the non-context-free language

$$L = \{ w_1 \# u \# w_2 \mid w_1, w_2 \in \{a,b\}^*, |w_1|_a = |w_2|_a, |w_1|_b = |w_2|_b \\ \text{and } u \in \{c,d\}^* \text{ is a palindrome} \}.$$

As L does not contain any regular subset that is letter-equivalent to L itself, L is not accepted by any NFAwtl [14]. However, L is accepted by the 2hNFAwtl $A = (Q, \Sigma, \triangleright, \triangleleft, \tau, I, F, \delta)$ that is defined as follows:

- $Q = Q_L \cup Q_R$, where $Q_L = \{q_a, q_b, q_\#, q_c, q_d\}$ and $Q_R = \{p_a, p_b, p_\#, p_c, p_d\}$,
- $\Sigma = \{a, b, c, d, \#\}$, $I = \{q_a, q_b, q_\#\}$, and $F = \{q_c, q_d, p_c, p_d\}$,
- $\tau(q_a) = \tau(p_a) = \{b\}$, $\tau(q_b) = \tau(p_b) = \{a\}$, and $\tau(q) = \emptyset$ for all other states,
- and the transition relation is given through the following table:

	q_a	q_b	$q_\#$	q_c	q_d	p_a	p_b	$p_\#$	p_c	p_d
a	p_a	–	–	–	–	$q_a, q_b, q_\#$	–	–	–	–
b	–	p_b	–	–	–	–	$q_a, q_b, q_\#$	–	–	–
$\#$	–	–	$p_\#$	–	–	–	–	q_c, q_d	–	–
c	–	–	–	p_c	–	–	–	–	q_c, q_d	–
d	–	–	–	–	p_d	–	–	–	–	q_c, q_d

Thus, we have the following proper inclusions.

Proposition 3. $\mathcal{L}(\text{2hNFA}) = \text{LIN} \subsetneq \mathcal{L}(\text{2hNFAwtl})$ *and* $\mathcal{L}(\text{NFAwtl}) \subsetneq \mathcal{L}(\text{2hNFAwtl})$.

Next we introduce a restricted type of 2hNFAwtls that yields very transparent computations, in this way simplifying constructions and proofs.

Definition 4. *Let* $A = (Q, \Sigma, \triangleright, \triangleleft, \tau, I, F, \delta)$ *be a 2hNFAwtl. For each state* $q \in Q$ *we define* $\mu(q) = \{a \in \Sigma \mid \delta(q, a) \neq \emptyset\}$, *that is,* $\mu(q)$ *is the set of letters which A can read in state q. Observe that we can assume w.l.o.g. that* $\mu(q) \cap \tau(q) = \emptyset$ *for all states* $q \in Q$. *Now the 2hNFAwtl A is said to be in* normal form *if*

1. *it always accepts with empty tape, that is, each word from $L(A)$ is read (and deleted) completely before A accepts,*
2. $|\mu(q)| \leq 1$ *for each state* $q \in Q$, *that is, for each state* $q \in Q$, *there exists at most one letter* $a \in \Sigma$ *such that* $\delta(q, a)$ *is defined.*

Concerning this normal form we have the following result.

Proposition 5. *From a given 2hNFAwtl* $A = (Q, \Sigma, \triangleright, \triangleleft, \tau, I, F, \delta)$ *one can effectively construct a 2hNFAwtl* $B = (Q_B, \Sigma, \triangleright, \triangleleft, \tau_B, I_B, F_B, \delta_B)$ *in normal form such that* $L(B) = L(A)$.

If A is a 2hNFAwtl that is in normal form, then by removing the translucency relation from A, we obtain a 2hNFA A'. In fact, the following result holds.

Proposition 6. *By removing the translucency relation from a 2hNFAwtl A that is in normal form, we obtain a 2hNFA A' such that $L(A')$ is a sublanguage of $L(A)$ that is letter-equivalent to $L(A)$.*

Proof. Let $A = (Q, \Sigma, \triangleright, \triangleleft, \tau, I, F, \delta)$ be a 2hNFAwtl that is in normal form, and let $B = (Q, \Sigma, I, F, \delta_B)$ be the 2hNFA that is obtained from A by removing the translucency relation τ and the endmarkers $\triangleright$ and $\triangleleft$. In addition, if $q \in Q_L$ and $a \in \Sigma \smallsetminus \tau(q)$, then $\delta_B(q, a, \varepsilon) = \delta(q, a)$ and $\delta_B(q, b, \varepsilon) = \emptyset$ for all $b \in \tau(q)$, and if $q \in Q_R$ and $a \in \Sigma \smallsetminus \tau(q)$, then $\delta_B(q, \varepsilon, a) = \delta(q, a)$ and $\delta_B(q, \varepsilon, b) = \emptyset$ for all $b \in \tau(q)$. Then each accepting computation of B corresponds to an accepting computation of A in which no translucent letter is ever skipped over. Thus, $L(B) \subseteq L(A)$.

Conversely, assume that $w \in L(A)$, where $w = a_1 a_2 \cdots a_n$. From the first condition in Definition 4, it follows that an accepting computation of A on input w must read and erase the word w completely, that is, it has the form

$$q_0 \triangleright w \triangleleft \vdash_A q_{i_1} \triangleright w_1 \triangleleft \vdash_A q_{i_2} \triangleright w_2 \triangleleft \vdash_A \cdots \vdash_A q_{i_{n-1}} \triangleright w_{n-1} \triangleleft \vdash_A q_{i_n} \triangleright \triangleleft,$$

where $q_0 \in I$ and $q_{i_n} \in F$. We claim that there exists a word $z \in \Sigma^*$ such that $q_0 z \vdash_B^n q_{i_n}$ and $\psi(z) = \psi(w)$ hold. We proceed by induction on the length n of w.

If $n = 0$, then $w = \varepsilon$, which means that $q_0 = q_{i_n}$. Hence, we can take $z = \varepsilon = w$.

Now assume that $q_0 \in Q_L$ and that $w = xay$ for some $x \in (\tau(q_0))^*$ and $a \in \Sigma \smallsetminus \tau(q_0)$ such that $|\delta(q_0, a)| > 0$. Then $w_1 = xy$ and $q_{i_1} \in \delta(q_0, a)$. From the induction hypothesis we see that there exists a word $z_1 \in \Sigma^*$ such that $q_{i_1} z_1 \vdash_B^{n-1} q_{i_n}$ and $\psi(z_1) = \psi(w_1)$. Let $z = az_1$. Then $q_0 z = q_0 a z_1 \vdash_B q_{i_1} z_1 \vdash_B^{n-1} q_{i_n}$, that is, B accepts on input z, and $\psi(z) = \psi(az_1) = \psi(aw_1) = \psi(axy) = \psi(xay) = \psi(w)$.

Finally, if $q_0 \in Q_R$ and $w = ubv$ for some $v \in (\tau(q_0))^*$ and $b \in \Sigma \smallsetminus \tau(q_0)$ such that $|\delta(q_0, b)| > 0$, then $w_1 = uv$ and $q_{i_1} \in \delta(q_0, b)$. Again from the induction hypothesis we see that there exists a word $z_1 \in \Sigma^*$ such that $q_{i_1} z_1 \vdash_B^{n-1} q_{i_n}$ and $\psi(z_1) = \psi(w_1)$. Let $z = z_1 b$. Then $q_0 z = q_0 z_1 b \vdash_B q_{i_1} z_1 \vdash_B^{n-1} q_{i_n}$, that is, B accepts on input z, and $\psi(z) = \psi(z_1 b) = \psi(w_1 b) = \psi(uvb) = \psi(ubv) = \psi(w)$. Thus, we see that, for each word $w \in L(A)$, there exists a word $z \in L(B)$ such that w and z are letter-equivalent. □

This yields the following consequence.

Corollary 7. *Each language $L \in \mathcal{L}(\text{2hNFAwtl})$ contains a sublanguage that is linear context-free and letter-equivalent to L. In particular, this implies that L is semi-linear, that is, $\psi(L)$ is a semi-linear subset of $\mathbb{N}^n$.*

As the language $L' = \{ a^m b^n \# u \# a^m b^n \mid m, n \geq 0, u \in \{c, d\}^* \text{ is a palindrome} \}$ does not contain a linear sublanguage that is letter-equivalent to L' itself, it follows from Corollary 7 that L' is not accepted by any 2hNFAwtl. As

$$L' = L \cap a^* \cdot b^* \cdot \# \cdot \{c, d\}^* \cdot \# \cdot a^* \cdot b^*,$$

where L is the language from Example 2, this yields the following non-closure property.

Corollary 8. *$\mathcal{L}(\text{2hNFAwtl})$ is not closed under intersection with regular languages.*

As all regular languages are accepted by 2hNFAwtls, we see that the class $\mathcal{L}(\text{2hNFAwtl})$ is not closed under intersection. However, it is easily seen that this class is closed under union. Thus, we obtain the following result.

Corollary 9. *The language class $\mathcal{L}(\text{2hNFAwtl})$ is closed under union, but it is neither closed under intersection nor under complementation.*

The languages $L_1 = \{\, a^m b^m \mid m \geq 0 \,\}$ and $L_2 = \{\, c^n d^n \mid n \geq 0 \,\}$ are both linear context-free and they are accepted by 2hDFAwtls. However, their product $L_1 \cdot L_2 = \{\, a^m b^m c^n d^n \mid m, n \geq 0 \,\}$ does not contain a sublanguage that is linear context-free and letter-equivalent to the language itself. Hence, by Corollary 7, $L_1 \cdot L_2$ is not accepted by any 2hNFAwtl. This yields the following non-closure property.

Corollary 10. $\mathcal{L}(\text{2hNFAwtl})$ *and* $\mathcal{L}(\text{2hDFAwtl})$ *are not closed under product.*

Finally, we separate the 2hDFAwtl from the 2hNFAwtl.

Proposition 11. *The language class $\mathcal{L}(\text{2hDFAwtl})$ is closed under complementation.*

Proof. Let $A = (Q, \Sigma, \rhd, \lhd, \tau, q_0, F, \delta)$ be a 2hDFAwtl. We define a 2hDFAwtl $A^c = (Q \cup \{q_f\}, \Sigma, \rhd, \lhd, \tau_c, q_0, F_c, \delta_c)$, where q_f is a new left state, by taking $\tau_c(q_f) = \Sigma$ and $\tau_c(q) = \tau(q)$ for all $q \in Q$, $F_c = (Q \setminus F) \cup \{q_f\}$, and by defining δ_c as follows:

$$\delta_c(q, a) = \delta(q, a) \text{ for all } q \in Q \text{ and all } a \in \mu(q),$$
$$\delta_c(q, b) = q_f \quad \text{ for all } q \in Q \text{ and all } b \in \Sigma \setminus (\tau(q) \cup \mu(q)).$$

Thus, given a word $w \in \Sigma^*$ as input, A^c performs the same steps as A until A halts. Let $q \in Q$ and $u \in \Sigma^*$ be the current state and the current tape contents at that point.

- If $q \in F$ and $u \in (\tau(q))^*$, then A accepts, which means that $w \in L(A)$. However, as $q \notin F_c$, we see that A^c does not accept on input w.
- If $q \notin F$, but $u \in (\tau(q))^*$, then A does not accept, which means that $w \notin L(A)$. However, as $q \in F_c$, we see that A^c accepts on input w.
- Finally, if q is a left state and $u = xay$ for some word $x \in (\tau(q))^*$ and a letter $a \notin (\tau(q) \cup \mu(q))$, then A just gets stuck without accepting, which means that $w \notin L(A)$. However, A^c continues the current computation by $q \rhd xay \lhd \vdash_{A^c} q_f \rhd xy \lhd$, and then A^c accepts as $q_f \in F_c$ and $xy \in (\tau(q_f))^*$, that is, $w \in L(A^c)$.
- If q is a right state and $u = xay$ for some word $y \in (\tau(q))^*$ and a letter $a \notin (\tau(q) \cup \mu(q))$, then it follows analogously that A^c accepts on input w.

Thus, we see that $L(A^c) = \Sigma^* \setminus L(A)$.   □

From the definition it follows immediately that $\mathcal{L}(\text{2hDFAwtl}) \subseteq \mathcal{L}(\text{2hNFAwtl})$. By Proposition 11, $\mathcal{L}(\text{2hDFAwtl})$ is closed under complementation, while by Corollary 9, $\mathcal{L}(\text{2hNFAwtl})$ is not. Thus, we obtain the following separation result.

Corollary 12. $\mathcal{L}(\text{2hDFAwtl}) \subsetneq \mathcal{L}(\text{2hNFAwtl})$.

4   Linear Context-Free Trace Languages

Let Σ be a finite alphabet, and let D be a binary relation on Σ that is reflexive and symmetric. Then D is called a *dependency relation* on Σ, and $I_D = (\Sigma \times \Sigma) \setminus D$ is called the corresponding *independence relation*. Obviously, the relation I_D is irreflexive and symmetric. The independence relation I_D induces a binary relation $\equiv_D$ on Σ^* that is defined as the smallest congruence relation containing the set of pairs $\{(ab, ba) \mid (a, b) \in I_D\}$. For $w \in \Sigma^*$, the congruence class of $w \bmod \equiv_D$ is denoted by $[w]_D$, that is, $[w]_D = \{z \in \Sigma^* \mid w \equiv_D z\}$. These congruence classes are called *traces*, and the factor monoid $M(D) = \Sigma^*/\equiv_D$ is a *trace monoid*. In fact, $M(D)$ is the *free partially commutative monoid* presented by (Σ, D) (see, e.g., [2]).

To simplify the notation, we introduce the following notions. For $w \in \Sigma^*$, we use $\mathrm{Alph}(w)$ to denote the set of all letters that occur in w, and we extend the independence relation from letters to words by defining, for all words $u, v \in \Sigma^*$, $(u, v) \in I_D$ if and only if $\mathrm{Alph}(u) \times \mathrm{Alph}(v) \subseteq I_D$.

A language $L \subseteq \Sigma^*$ is called a *rational trace language*, if there exist a trace monoid $M(D)$ and a regular language $R \subseteq \Sigma^*$ such that $L = \varphi_D^{-1}(\varphi_D(R)) = \bigcup_{w \in R}[w]_D$. Here φ_D denotes the morphism that maps a word w to the corresponding trace $[w]_D$. By $\mathcal{LRAT}(D)$ we denote the set of rational trace languages $\varphi_D^{-1}(\mathrm{RAT}(M(D)))$ over $M(D)$. In [11] (see also [14]) it is shown that $\mathcal{LRAT}(D) \subsetneq \mathcal{L}(\mathsf{NFAwtl})$.

Here we are interested in more general trace languages. A language $L \subseteq \Sigma^*$ is called a *(linear) context-free trace language*, if there exist a dependency relation D on Σ and a (linear) context-free language $R \subseteq \Sigma^*$ such that $L = \varphi_D^{-1}(\varphi_D(R)) = \bigcup_{w \in R}[w]_D$. By $\mathcal{LLCF}(D)$ we denote the set of linear context-free trace languages obtained from (Σ, D). In [12,13] it is shown that the context-free trace languages are accepted by certain cooperating distributed systems of a very restricted type of restarting automata, which can be interpreted as nondeterministic pushdown automata with translucent letters. Here we derive the following result.

Theorem 13. *For each trace monoid $M(D)$ presented by (Σ, D), where D is a dependency relation on the alphabet Σ, $\mathcal{LLCF}(D) \subseteq \mathcal{L}(\mathsf{2hNFAwtl})$.*

Proof. Let R be a linear context-free language over Σ, let $S = \varphi_D(R) \subseteq M(D)$, and let $L = \varphi_D^{-1}(S) \subseteq \Sigma^*$ be the linear context-free trace language defined by R and D. As $R \subseteq \Sigma^*$ is a linear context-free language, there exists a 2hNFA $A = (Q, \Sigma, I, F, \delta)$ such that $L(A) = R$. Applying well-known techniques (see, e.g., [1,7]), we may assume without loss of generality that the set Q of states of A is partitioned into two disjoint subsets Q_L and Q_R such that, in a state $q \in Q_L$, A reads (and deletes) the first letter of the current tape contents, and in a state $p \in Q_R$, A reads (and deletes) the last letter of the current tape contents. Further, we may assume that, for each left state $q \in Q_L \setminus F$, there is at most a single letter $a_q \in \Sigma$ such that $\delta(q, a_q, \varepsilon) \neq \emptyset$, for each right state $q \in Q_R \setminus F$, there is at most a single letter $a_q \in \Sigma$ such that $\delta(q, \varepsilon, a_q) \neq \emptyset$, that $F = \{q_+\}$ consists of a single left state only, and that $\delta(q_+, a, \varepsilon) = \emptyset$ for all letters $a \in \Sigma$ (that is, no transition is allowed from the accepting state). From A we now construct a 2hNFAwtl $B = (Q_B, \Sigma, \triangleright, \triangleleft, \tau, I_B, F_B, \delta_B)$ as follows:

- $Q_B = Q$, $I_B = I$, and $F_B = F = \{q_+\}$,
- $\tau(q) = \{\, b \in \Sigma \mid (b, a_q) \in I_D \,\}$ for all $q \in Q \smallsetminus F$ and $\tau(q_+) = \emptyset$, and
- the transition relation δ_B is defined as follows:

$$\begin{aligned}
\delta_B(q, a_q) &= \delta(q, a_q, \varepsilon) \text{ for all } q \in Q_L \smallsetminus F, \\
\delta_B(q, a_q) &= \delta(q, \varepsilon, a_q) \text{ for all } q \in Q_R \smallsetminus F, \\
\delta_B(q, b) &= \emptyset \qquad \text{for all } q \in Q \smallsetminus F \text{ and all } b \neq a_q, \\
\delta_B(q_+, a) &= \emptyset \qquad \text{for all } a \in \Sigma.
\end{aligned}$$

It remains to show that $L(B) = L = \varphi_D^{-1}(S) = \bigcup_{u \in R}[u]_D$. Notice that in order to accept a word w, B must read (and delete) it completely.

Claim 1. $\bigcup_{u \in R}[u]_D \subseteq L(B)$.

Proof. Assume that $w \in \bigcup_{u \in R}[u]_D$. Then there exists a word $u \in R$ such that $w \equiv_D u$, and so there exists a sequence of words $u = w_0, w_1, \ldots, w_n = w$ such that, for each $i = 1, 2, \ldots, n$, w_i is obtained from w_{i-1} by replacing a factor ab by ba for some pair of letters $(a, b) \in I_D$. By induction on i, $w_i \in L(B)$ can now be shown for all i. $\qquad\square$

Claim 2. $L(B) \subseteq \bigcup_{u \in R}[u]_D$.

Proof. Let $w \in L(B)$, and let

$$\begin{aligned}
q_n \triangleright w \triangleleft = q_n \triangleright w_n \triangleleft &\vdash_B q_{n-1} \triangleright w_{n-1} \triangleleft \vdash_B q_{n-2} \triangleright w_{n-2} \triangleleft \vdash_B \\
\cdots &\vdash_B \quad q_1 \triangleright w_1 \triangleleft \quad \vdash_B \quad q_+ \triangleright \triangleleft
\end{aligned}$$

be an accepting computation of B on input w, where $q_n \in I_B$. We claim that, for each $i = 1, 2, \ldots, n$, there exists a word $u_i \in \Sigma^*$ such that $u_i \equiv_D w_i$ and $q_i u_i \vdash_A^* q_+$, that is, the 2hNFA A accepts the word u_i when starting from state q_i.

We prove this claim by induction on i. For $i = 1$ we have $w_i = a_1$, and $q_+ \in \delta_B(q_1, a_1)$. From the definition of B, either $q_+ \in \delta(q_1, a_1, \varepsilon)$ or $q_+ \in \delta(q_1, \varepsilon, a_1)$ depending on whether q_1 is a left or a right state. According to this, we can simply take $u_1 = a_1 = w_1$. Now assume that, for some $i \geq 1$, $u_i \equiv_D w_i$ and $q_i u_i \vdash_A^* q_+$ hold. The above computation of B contains the step $q_{i+1} \triangleright w_{i+1} \triangleleft \vdash_B q_i \triangleright w_i \triangleleft$. Again from the definition of B, we see that either q_{i+1} is a left state and $q_i \in \delta(q_{i+1}, a_{i+1}, \varepsilon)$, where $w_{i+1} = x a_{i+1} y$ and $w_i = xy$ for some words $x, y \in \Sigma^*$ such that $(x, a_{i+1}) \in I_D$, or q_{i+1} is a right state and $q_i \in \delta(q_{i+1}, \varepsilon, a_{i+1})$ with $w_{i+1} = x a_{i+1} y$ and $w_i = xy$ for some words $x, y \in \Sigma^*$ such that $(y, a_{i+1}) \in I_D$. In the former case let u_{i+1} be the word $u_{i+1} = a_{i+1} u_i$, and in the latter case, let u_{i+1} be the word $u_{i+1} = u_i a_{i+1}$. Then

$$u_{i+1} = a_{i+1} u_i \equiv_D a_{i+1} w_i = a_{i+1} xy \equiv_D x a_{i+1} y = w_{i+1}$$

and $q_{i+1} u_{i+1} = q_{i+1} a_{i+1} u_i \vdash_A q_i u_i$, or

$$u_{i+1} = u_i a_{i+1} \equiv_D w_i a_{i+1} = xy a_{i+1} \equiv_D x a_{i+1} y = w_{i+1}$$

and $q_{i+1}u_{i+1} = q_{i+1}u_i a_{i+1} \vdash_A q_i u_i$. As by the induction hypothesis $q_i u_i \vdash_A^* q_+$, we see that in either case $q_{i+1}u_{i+1} \vdash_A^* q_+$ follows.

For $i = n$ we obtain a word $u \in \Sigma^*$ such that $u \equiv_D w$ and A accepts u starting from state $q_n \in I$. Hence, $u \in R$, and it follows that $L(B) \subseteq \bigcup_{u \in R}[u]_D$ holds.   □

Now Claims 1 and 2 together show that $L(B) = \bigcup_{u \in R}[u]_D = \varphi_D^{-1}(S) = L$, which completes the proof of Theorem 13.   □

Next we present a restricted class of 2hNFAwtls that accept exactly the linear context-free trace languages. Here the second condition of Definition 4 is essential.

Definition 14. *Let $B = (Q, \Sigma, \triangleright, \triangleleft, \tau, I, F, \delta)$ be a 2hNFAwtl in normal form that satisfies the following additional condition:*

$$(*) \ \forall p, q \in Q : \mu(p) = \mu(q) \ \text{implies that} \ \tau(p) = \tau(q).$$

With B we associate the binary relation $I_B = \bigcup_{q \in Q} (\tau(q) \times \mu(q))$, that is, $(a, b) \in I_B$ if and only if there exists a state $q \in Q$ such that $a \in \tau(q)$ and $\delta(q, b)$ is defined. Further, by D_B we denote the relation $D_B = (\Sigma \times \Sigma) \setminus I_B$.

Observe that the relation I_B defined above is necessarily irreflexive, but that it will in general not be symmetric. On the other hand, consider the 2hNFAwtl B that is constructed in the proof of Theorem 13. It satisfies the condition $(*)$ and the relation I_B coincides with the relation I_D restricted to the subset of letters of Σ that actually occur in the language L. Hence, this relation is in fact symmetric. The following result shows that also the converse of this observation holds.

Theorem 15. *Let B be a 2hNFAwtl in normal form over Σ satisfying condition $(*)$ above. If the associated relation I_B is symmetric, then $L(B)$ is a linear context-free trace language over Σ. In fact, from B one can construct a 2hNFA A over Σ such that $L(B) = \varphi_{D_B}^{-1}(\varphi_{D_B}(L(A)))$.*

Theorems 13 and 15 together yield the following characterization.

Corollary 16. *A language $L \subseteq \Sigma^*$ is a linear context-free trace language if and only if there exists a 2hNFAwtl B in normal form satisfying condition $(*)$ such that the relation I_B is symmetric and $L = L(B)$.*

Let D_1 be the Dyck language over $\Sigma = \{a, b\}$. This language is accepted by a 2hNFAwtl A that satisfies the above condition $(*)$, but the corresponding relation I_A is not symmetric. And indeed, the language D_1 is not a linear context-free trace language.

5   Further Closure Properties and Decidability Results

We have seen above that the language class $\mathcal{L}(\text{2hNFAwtl})$ is closed under union, but it is not closed under intersection (with regular sets), complementation, and product. Further, the language class $\mathcal{L}$(2hDFAwtl) is closed under complementation, but not under product. Obviously, both classes are closed under reversal. In addition, the former class is closed under a weaker product operation.

Proposition 17. $\mathcal{L}$(2hNFAwtl) *is closed under product with regular sets, that is, if* $L \in \mathcal{L}$(2hNFAwtl) *and* $R \in$ REG, *then* $L \cdot R$ *and* $R \cdot L$ *are accepted by 2hNFAwtls.*

The language L of Example 2 is accepted by a 2hDFAwtl, but as

$$L' = L \cap a^* \cdot b^* \cdot \# \cdot \{c, d\}^* \cdot \# \cdot a^* \cdot b^*$$

is not accepted by any 2hNFAwtl (see Sect. 3), the following non-closure results follow.

Corollary 18. *The language class* $\mathcal{L}$(2hDFAwtl) *is not closed under intersection with regular sets, and hence, it is not closed under union and intersection.*

In [16] it has been noted that the rational trace language

$$L_\vee = \{\, w \in \{a, b\}^* \mid \exists n \geq 0 : |w|_a = n \text{ and } |w|_b \in \{n, 2n\} \,\}$$

is not accepted by any stl-det-global-CD-R(1)-system, that is, it is not accepted by any DFAwtl. Actually, the proof of that result (Proposition 4.7 of [16]) generalizes to 2hDFAwtls.

Proposition 19. *The language* $L_\vee$ *is not accepted by any 2hDFAwtl.*

This observation yields the following result.

Corollary 20. $\mathcal{L}$(2hDFAwtl) *does not even contain all rational trace languages.*

Let $L'_\vee$ denote the following variant of the language $L_\vee$:

$$L'_\vee = \{\, wc \mid w \in \{a, b\}^+, |w|_a = |w|_b - 1 \,\} \cup \{\, wd \mid w \in \{a, b\}^+, 2 \cdot |w|_a = |w|_b - 1 \,\}.$$

It is easy to design a 2hDFAwtl A for this language. Now let $\varphi : \{a, b, c, d\}^* \rightarrow \{a, b\}^*$ be the alphabetic morphism that is defined by $a \mapsto a$ and $b, c, d \mapsto b$. Then

$$\varphi(L'_\vee) = \{\, wb \mid w \in \{a, b\}^+, \exists n \geq 0 : |w|_a = n \text{ and } |w|_b + 1 \in \{n, 2n\} \,\},$$

which is not accepted by any 2hDFAwtl (see the proof of Proposition 19). Thus, we obtain the following non-closure property.

Corollary 21. $\mathcal{L}$(2hDFAwtl) *is not closed under alphabetic morphisms.*

However, it remains open whether any of the classes $\mathcal{L}$(2hDFAwtl) and $\mathcal{L}$(2hNFAwtl) is closed under Kleene star or inverse morphisms, and whether $\mathcal{L}$(2hNFAwtl) is closed under ε-free morphisms.

Finally, we turn to decision problems. A 2hNFAwtl can easily be simulated by a nondeterministic one-tape Turing machine that is simultaneously linearly space-bounded and quadratically time-bounded. Hence, we have the following complexity result.

Proposition 22. $\mathcal{L}(\text{2hNFAwtl}) \subseteq \text{NSpaceTime}(n, n^2)$.

By Proposition 6, a 2hNFA A' can be obtained from a 2hNFAwtl A such that $E = L(A')$ is a subset of $L = L(A)$ that is letter-equivalent to L. Hence, E is non-empty if and only if L is non-empty, and E is infinite if and only if L is infinite. This shows that emptiness and finiteness are decidable for 2NFAwtls. Since universality is undecidable for linear languages (see, e.g., [5]), and as one can easily design a 2hNFAwtl for the language Σ^*, universality, containment of a regular set, inclusion and equivalence are undecidable for 2hNFAwtls. Finally, using the effective closure under product with a regular set (Proposition 17), Greibach's general undecidability result from [4] implies that it is undecidable in general whether the language accepted by a given 2hNFAwtl is linear context-free.

6 Conclusion

We have presented two-head finite-state acceptors with translucent letters, and we have seen that they are quite expressive. In fact, they accept a subclass of the class of all languages with semi-linear Parikh image that properly contains all linear context-free trace languages. In addition, we have stated a number of closure and non-closure results for the language classes $\mathcal{L}(\text{2hNFAwtl})$ and $\mathcal{L}(\text{2hDFAwtl})$, but some operations still remain to be considered.

References

1. Bedregal, B.R.C.: Some subclasses of linear languages based on nondeterministic linear automata. arXiv:10276v1 (2016)
2. Diekert, V., Rozenberg, G.: The Book of Traces. World Scientific, Singapore (1995)
3. Freund, R., Păun, Gh., Rozenberg, G., Salomaa, A.: Watson-Crick finite automata. In: Rubin, H., Wood, D.H. (eds.) DNA Based Computers, Proceedings of DIMACS Series in Discrete Mathematics and Theoretical Computer Science, pp. 297–328. DIMACS/AMS (1999)
4. Greibach, S.: A note on undecidable properties of formal languages. Math. Syst. Theory **2**, 1–6 (1968)
5. Hopcroft, J.E., Ullman, J.D.: Introduction to Automata Theory, Languages, and Computation. Addison-Wesley, Reading (1979)
6. Meduna, A., Zemek, P.: Jumping finite automata. Int. J. Found. Comput. Sci. **23**, 1555–1578 (2012)
7. Nagy, B.: On $5' \to 3'$ sensing Watson-Crick finite automata. In: Garzon, M.H., Yan, H. (eds.) DNA 2007. LNCS, vol. 4848, pp. 256–262. Springer, Heidelberg (2008). https://doi.org/10.1007/978-3-540-77962-9_27
8. Nagy, B.: A class of 2-head finite automata for linear languages. Triangle: Lang. Math. Approaches **8**, 89–99 (2012)
9. Nagy, B.: On a hierarchy of $5' \to 3'$ sensing Watson-Crick finite automata languages. J. Log. Comput. **23**, 855–872 (2013)

10. Nagy, B., Kovács, L.: Finite automata with translucent letters applied in natural and formal language theory. In: Nguyen, N.T., Kowalczyk, R., Fred, A., Joaquim, F. (eds.) Transactions on Computational Collective Intelligence XVII. LNCS, vol. 8790, pp. 107–127. Springer, Heidelberg (2014). https://doi.org/10.1007/978-3-662-44994-3_6

11. Nagy, B., Otto, F.: CD-systems of stateless deterministic R(1)-automata accept all rational trace languages. In: Dediu, A.-H., Fernau, H., Martín-Vide, C. (eds.) LATA 2010. LNCS, vol. 6031, pp. 463–474. Springer, Heidelberg (2010). https://doi.org/10.1007/978-3-642-13089-2_39

12. Nagy, B., Otto, F.: An automata-theoretical characterization of context-free trace languages. In: Černá, I., et al. (eds.) SOFSEM 2011. LNCS, vol. 6543, pp. 406–417. Springer, Heidelberg (2011). https://doi.org/10.1007/978-3-642-18381-2_34

13. Nagy, B., Otto, F.: CD-systems of stateless deterministic R(1)-automata governed by an external pushdown store. RAIRO Theor. Inform. Appl. **45**, 413–448 (2011)

14. Nagy, B., Otto, F.: Finite-state acceptors with translucent letters. In: Bel-Enguix, G., Dahl, V., De La Puente, A.O. (eds.) BILC 2011: AI Methods for Interdisciplinary Research in Language and Biology, Proceedings, pp. 3–13. SciTePress, Portugal (2011)

15. Nagy, B., Otto, F.: On CD-systems of stateless deterministic R-automata with window size one. J. Comput. Syst. Sci. **78**, 780–806 (2012)

16. Nagy, B., Otto, F.: Globally deterministic CD-systems of stateless R-automata with window size 1. Int. J. Comput. Math. **90**, 1254–1277 (2013)

17. Rosenberg, A.L.: On multi-head finite automata. IBM J. Res. Dev. **10**, 388–394 (1966)

18. Rosenberg, A.L.: A machine realization of the linear context-free languages. Inf. Control **10**, 175–188 (1967)

Do Null-Type Mutation Operators Help Prevent Null-Type Faults?

Ali Parsai[1]($^{(\boxtimes)}$) (iD) and Serge Demeyer[2] (iD)

[1] University of Antwerp, Antwerp, Belgium
ali.parsai@uantwerpen.be
[2] University of Antwerp and Flanders Make, Antwerp, Belgium
serge.demeyer@uantwerpen.be

Abstract. The null-type is a major source of faults in Java programs, and its overuse has a severe impact on software maintenance. Unfortunately traditional mutation testing operators do not cover null-type faults by default, hence cannot be used as a preventive measure. We address this problem by designing four new mutation operators which model null-type faults explicitly. We show how these mutation operators are capable of revealing the missing tests, and we demonstrate that these mutation operators are useful in practice. For the latter, we analyze the test suites of 15 open-source projects to describe the trade-offs related to the adoption of these operators to strengthen the test suite.

Keywords: Software maintenance · Software testing
Mutation testing · Null-type · Test quality

1   Introduction

The null-type is a special type in Java that has no name, cannot be casted, and practically equates to a literal that can be of any reference type [11]. The null-type is commonly misused, and frequently reported and discussed as an issue by developers [24]. The null-type is the source of the majority of faults in Java programs [25], and its overuse has a severe impact on software maintenance [15]. On the one hand, this scenario should push developers to build test suites capable of identifying null-type faults. On the other hand, developers without specific test requirements may struggle to identify all code elements or properties that the test must satisfy. To address this problem, we propose mutation testing as a way for improving the test suite to handle potential null-type faults.

Mutation testing is a technique to measure the quality of a test suite by assessing its fault detection capabilities [5]. Mutation testing is a two-step process. First, a small syntactic change is introduced in the production code. This change is obtained by applying a "mutation operator", and the resulting changed code is called a "mutant". Then, the test suite is executed for that mutant; if any of the tests fail, the mutant is "killed", otherwise, the mutant has "survived". Herein lies the aspect of mutation testing that we want to exploit: the identification of survived mutants that need to be killed.

© Springer Nature Switzerland AG 2019
B. Catania et al. (Eds.): SOFSEM 2019, LNCS 11376, pp. 419–434, 2019.
https://doi.org/10.1007/978-3-030-10801-4_33

Mutation operators are modeled after the common developer mistakes [14]. Over the years, multiple sets of mutation operators have been created to fit in different domains. By far the most commonly used mutation operators are the ones introduced in Mothra by Offutt et al. [21]. They use 10 programs written in Fortran to demonstrate that their reduced-set mutation operators is enough to produce a mutation-adequate test suite that can kill almost all of the mutants generated by the mutation operators of the complete-set. Later on, several attempts have been made to extend Offutt's mutation operators, for instance, to cope with the specificities of object-oriented programming [19]. Yet, none of the proposed mutation operators explicitly model null-type faults. As a result, mature general-purpose mutation testing tools currently used in literature, such as PITest [4] and Javalanche [35], do not cope explicitly with this type of faults by default. Therefore, the created mutants *risk* not being adequate to derive test requirements that handle null-type faults. Whether this risk is concrete or not depends on the ability of the available mutation operators to account for these faults. Yet, no study has explored this aspect.

This paper investigates the usefulness of mutation operators able to model null-type faults in order to strengthen the test suite against these faults. For this reason, we introduce four new mutation operators related to null-type faults. These mutation operators are modeled to cover the typical null-type faults introduced by developers [24]. We incorporate these mutation operators in LittleDarwin, an extensible open-source tool for mutation testing [32], creating a new version called LittleDarwin-Null. We organize our research in two steps: we show that (i) the current general-purpose mutation testing tools do not account for null-type faults by default, and modeling operators for null-type faults can drive the improvement of the test suite in practice, and (ii) the test suites of real open-source projects cannot properly catch null-type faults. The paper is driven by the following research questions:

- **RQ1:** *Are traditional mutation operators enough to prevent null-type faults?*
- **RQ2:** *To what extent is the addition of null-type mutation operators useful in practice?*

The rest of the paper is organized as follows: In Sect. 2, background information and related work is provided. In Sect. 3, the details of the experiment are discussed. In Sect. 4, the results are analyzed. In Sect. 5, we discuss the threats that affect the results. Finally, we present the conclusion in Sect. 6.

2   Background and Related Work

Mutation testing is the process of injecting faults into a software system and then verifying whether the test suite indeed fails, and thus detects the injected fault. First, a faulty version of the software is created by introducing faults into the system *(Mutation)*. This is done by applying a transformation *(Mutation Operator)* on a certain part of the code. After generating the faulty version of the software *(Mutant)*, it is passed onto the test suite. If a test fails, the mutant is marked as killed *(Killed Mutant)*. If all tests pass, the mutant is marked as survived *(Survived Mutant)*.

Mutation Operators. A mutation operator is a transformation which introduces a single syntactic change into its input. The first set of mutation operators were reported in King et al. [16]. These mutation operators work on essential syntactic entities of programming languages such as arithmetic, logical, and relational operators. For object-oriented languages, new mutation operators were proposed [19]. The mature mutation testing tools of today still mostly use the *traditional* (i.e. method-level) mutation operators [27].

Equivalent Mutants. An *equivalent mutant* is a mutant that does not change the semantics of the program, i.e. its output is the same as the original program for any possible input. Therefore, no test case can differentiate between an equivalent mutant and the original program. The detection of equivalent mutants is undecidable due to the halting problem [22].

Mutation Coverage. Mutation testing allows software engineers to monitor the fault detection capability of a test suite by means of mutation coverage [13]. A test suite is said to achieve *full mutation test adequacy* whenever it can kill all the non-equivalent mutants, thus reaching a mutation coverage of 100%. Such test suite is called a *mutation-adequate test suite*.

Mutant Subsumption. Mutant subsumption is defined as the relationship between two mutants A and B in which A subsumes B if and only if the set of inputs that kill A is guaranteed to kill B [18]. The subsumption relationship for faults has been defined by Kuhn in 1999 [17]. Later on, Ammann et al. tackled the theoretical side of mutant subsumption [2] where they define *dynamic* mutant subsumption as follows: Mutant A dynamically subsumes Mutant B if and only if (i) A is killed, and (ii) every test that kills A also kills B. The main purpose behind the use of mutant subsumption is to detect redundant mutants. These mutants create multiple threats to the validity of mutation analysis [26]. This is done by determining the dynamic subsumption relationship among a set of mutants, and keep only those that are not subsumed by any other mutant.

Mutation Testing Tools. In this study, we use three different mutation testing tools: Javalanche, PITest, and LittleDarwin. Javalanche is a mutation testing framework for Java programs that attempts to be efficient, and not produce equivalent mutants [35]. It uses byte code manipulation in order to speed up the process of mutation testing. Javalanche has been used in numerous studies in the past (e.g. [9,10]). PITest is a state-of-the-art mutation testing system for Java, designed to be fast and scalable [4]. PITest is the de facto standard for mutation testing within Java, and it is used as a baseline in mutation testing research (e.g. [12,34]). LittleDarwin is a mutation testing tool designed to work out of the box with complicated industrial build systems. For this, it has a loose coupling with the test infrastructure, instead relying on the build system to run the test suite. LittleDarwin has been used in several studies, and is capable of performing mutation testing on complicated software systems [30,31,33]. For more information about LittleDarwin please refer to Parsai et al. [32]. We implemented the new null-type mutation operators in a special version of LittleDarwin called LittleDarwin-Null. LittleDarwin and LittleDarwin-Null only differ in the set of mutation operators used, and are identical otherwise.

Related Work. Creating new mutation operators to deal with the evolution of software languages is a trend in mutation testing research. For example, mutation operators have been designed to account for concurrent code [3], aspect-oriented programming [7], graphical user interfaces [23], modern C++ constructs [29], and Android applications [6]. Nanavati et al. have previously studied mutation operators targeting memory-related faults [20]. However, the difference in the semantics of null object of Java and NULL macro of C is sufficient to grant the need for a separate investigation.

3   Experimental Setup

In this section, we first introduce our proposed mutation operators, and then we discuss the experimental setup we used to address our research questions.

3.1   Null-Type Mutation Operators

We derived four null-type mutation operators to model the typical null-type faults often encountered by developers [25]. These mutation operators are presented in Table 1.

Table 1. Null-type faults and their corresponding mutation operators

Mutation operator	Description
NullifyReturnValue	If a method returns an object, it is replaced by null
NullifyInputVariable	If a method receives an object reference, it is replaced by null
NullifyObjectInitialization	Wherever there is a new statement, it is replaced with null
NegateNullCheck	Any binary relational statement containing null at one side is negated

3.2   Case Study

For RQ1, we use a didactic project. For RQ2, we use 15 open-source projects.

RQ1. In order to address RQ1, we chose a modified version of VideoStore as a small experimental project [8]. Choosing a small project allows us to (i) create a mutation-adequate test suite ourselves, (ii) find out which mutants are equivalent, and (iii) avoid complexities when using multiple mutation testing tools. The source code for VideoStore is available in the replication package.

Table 2. Projects sorted by mutation coverage

Project	Ver.	Size (LoC)		#C	TS	SC	BC	MC
		Prod.	Test					
Apache Commons CLI	1.3.1	2,665	3,768	816	15	96%	93%	94%
JSQLParser	0.9.4	7,342	5,909	576	19	81%	73%	94%
jOpt Simple	4.8	1,982	6,084	297	14	99%	97%	92%
Apache Commons Lang	3.4	24,289	41,758	4,398	30	94%	90%	91%
Joda Time	2.8.1	28,479	54,645	1,909	42	90%	81%	82%
Apache Commons Codec	1.10	6,485	10,782	1,461	10	96%	92%	82%
Apache Commons Collections	4.1	27,914	32,932	2,882	26	85%	78%	81%
VRaptor	3.5.5	14,111	15,496	3,417	65	87%	81%	81%
HTTP Request	6.0	1,391	2,721	446	15	94%	75%	78%
Apache Commons FileUpload	1.3.1	2,408	1,892	846	19	76%	74%	77%
jsoup	1.8.3	10,295	4,538	888	43	82%	72%	76%
JGraphT	0.9.1	13,822	8,180	1,150	31	79%	73%	69%
PITest	1.1.7	17,244	19,005	1,044	19	79%	73%	63%
JFreeChart	1.0.17	95,354	41,238	3,394	4	53%	45%	35%
PMD	r7706	70,767	43,449	7,706	20	62%	54%	34%

Acronyms: Version (Ver.), Line of code (LoC), Production code (Prod.), Number of commits (#C), Team size (TS), Statement coverage (SC), Branch coverage (BC), Mutation coverage (MC)

RQ2. We selected 15 open-source projects for our empirical study (Table 2). The selected projects differ in size of their production code and test code, number of commits, and team size to provide a wide range of possible scenarios. Moreover, they also differ in the adequacy of their test suite based on statement, branch, and mutation coverage (Table 2). We used JaCoCo and Clover for statement and branch coverage, and LittleDarwin for mutation coverage.

4   Results and Discussion

RQ1: Are Traditional Mutation Operators Enough to Prevent Null-type Faults?
We are interested to compute the number of killed, survived and equivalent mutants along with three versions of VideoStore. The first version we analyze

is the original one (VideoStore Orig). This version has only 4 tests. Then, we create a mutation-adequate test suite that kills all mutants generated by the general-purpose tools (Javalanche, PITest, and LittleDarwin). In this version (VideoStore TAdq) we added 15 tests. Finally, we create a mutation-adequate test suite that kills all mutants, included the ones generate by LittleDarwin-Null. In this version (VideoStore NAdq) we added 3 more tests.

Table 3. Mutation testing results for VideoStore

Program	LittleDarwin			PITest			Javalanche			LittleDarwin-Null		
	K	S	E	K	S	E	K	S	E	K	S	E
VideoStore Orig	24	18	2	25	43	5	87	69	11	11	14	1
VideoStore TAdq	42	0	2	68	0	5	202	0	11	22	3	1
VideoStore NAdq	42	0	2	68	0	5	202	0	11	25	0	1

K: Killed, S: Survived, E: Equivalent

Table 3 shows the number of remaining mutants after each phase of test development: VideoStore Orig, VideoStore TAdq, and VideoStore NAdq. The discrepancy in total number of generated mutants for the three versions of the program in case of Javalanche is due to its particular optimizations. In VideoStore Orig, there are several survived mutants according to all the tools. This is because the test suite accompanying the VideoStore program was not adequate.

In VideoStore TAdq, we create a mutation-adequate version of the test suite with respect to the results of PITest, Javalanche, and LittleDarwin. In the process of creating this test suite, we noticed that all of these tools produce equivalent mutants. Two of such mutants are shown in Fig. 2. Mutant A is equivalent because the method `super.determineAmount` always returns 0, so it does not matter whether it is added to or subtracted from `thisAmount`.

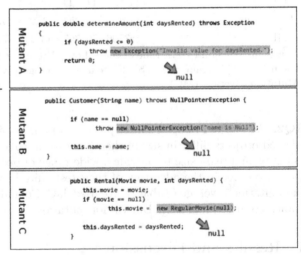

Fig. 1. The surviving non-equivalent null-type mutants

Mutant B is also equivalent, because if `daysRented` is 2, the value added to `thisAmount` is 0. We analyzed VideoStore TAdq with LittleDarwin-Null in order to find out whether the mutation-adequate test suite according to three general-purpose tools is able to kill all the null-type mutants. By analyzing the 26 generated mutants, we noticed that 22 mutants were killed and 4 survived. The manual review of these mutants show that one of them is an equivalent mutant.

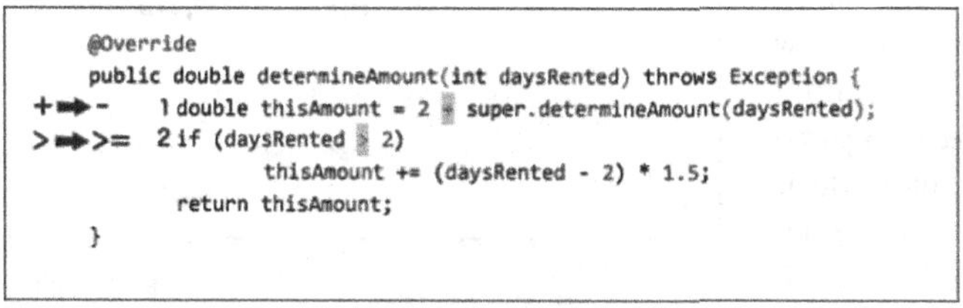

```
    @Override
    public double determineAmount(int daysRented) throws Exception {
+ ➡ -     1 double thisAmount = 2 * super.determineAmount(daysRented);
> ➡ >=    2 if (daysRented > 2)
                thisAmount += (daysRented - 2) * 1.5;
            return thisAmount;
    }
```

Fig. 2. Two of the equivalent mutants generated by traditional mutation operators

```
    public void addRental(Rental rental) throws NullPointerException {
        if (rental == null)
                rentals.addElement(new Rental(new RegularMovie(null), 0));
        else
                rentals.addElement(rental);                    ➘
    }                                                         null
```

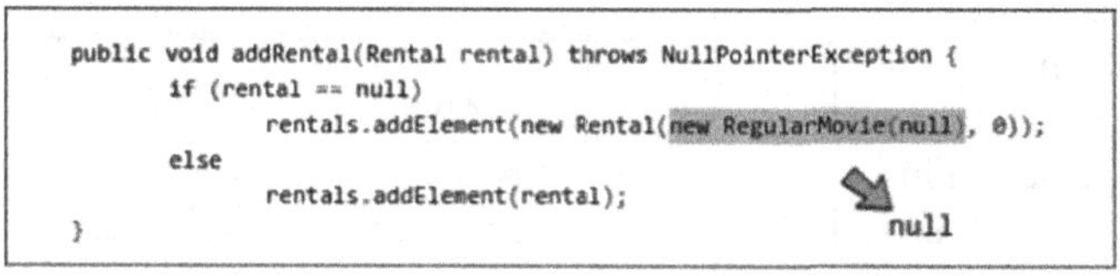

Fig. 3. One of the equivalent mutants generated by null-type mutation operators

Considering that 3 mutants generated by null-type mutation operators are not equivalent, and yet the mutation-adequate test suite we created according to the general-purpose tools cannot kill them, we conclude that **using traditional mutation operators to strengthen the test suite does not necessarily prevent null-type faults**.

The four mutants survived in VideoStore TAdq are all of type *NullifyObjectInitialization*. Figure 3 shows the equivalent null-type mutant. Here the default behavior of `Rental` object is to create a new `RegularMovie` object when it receives `null` as its input. So, replacing `new RegularMovie(null)` with null does not change the behavior of the program.

The three remaining surviving mutants are described in Fig. 1. Here, mutants A and B replace the exception with `null`. Consequently, as opposed to the program throwing a detailed exception, the mutant always throws an empty `NullPointerException`. Such a mutant is desirable to kill, since the program would be able to throw an unexpected exception due to a fault that the test suite cannot recognize. In the case of Mutant C, it replaces the initialization of a `RegularMovie` object with `null`. This means that as opposed to the program that guarantees the private attribute `movie` is always instantiated, the same attribute contains

```
@Test                                          Test Killing Mutant A
public void testMutantA() {
    movieInstance1 = new ChildrensMovie("null");
    movieInstance2 = new RegularMovie("null");
    movieInstance3 = new NewReleaseMovie("null");
    try {
            movieInstance1.determineAmount(-8);
    } catch (Exception e) {
            assertTrue(e.getMessage().equals("Invalid value for daysRented."));
    }
    try {
            movieInstance2.determineAmount(-48);
    } catch (Exception e) {
            assertTrue(e.getMessage().equals("Invalid value for daysRented."));
    }
    try {
            movieInstance3.determineAmount(-248);
    } catch (Exception e) {
            assertTrue(e.getMessage().equals("Invalid value for daysRented."));
    }
}
```

```
@Test                                          Test Killing Mutant B
public void testMutantB()
{
        try {
                new Customer(null);
        } catch (NullPointerException e) {
                assertTrue(e.getMessage().equals("name is Null"));
        }
}
```

```
@Test                                          Test Killing Mutant C
public void testMutantC()
{
        Rental rental = new Rental(null, 42);
        Movie movie = rental.getMovie();
        assertTrue(movie instanceof RegularMovie);
}
```

Fig. 4. The tests written to kill the surviving null-type mutants

a null literal in the mutant. If not detected, a `NullPointerException` might be thrown when another object tries to access the `movie` attribute of this object.

We created three new tests to kill each of the survived mutants. These tests are shown in Fig. 4. Here, `testMutantA` and `testMutantB` verify whether the unit under test throws the correct exception if called with an invalid input value. `testMutantC` verifies whether the unit under test is able to handle a null input correctly. These three tests are not "happy path tests", namely a well-defined test case using known input, which executes without exception and produces an expected output. Consequently, they might not be intuitive for a test developer to consider, even though they are known as good testing practice [1]. If not for the three survived null-type mutants, these tests would not have been written. This leads us to conclude that **traditional mutation operators are not enough to prevent null-type faults.**

RQ2: To What Extent is the Addition of Null-type Mutation Operators Useful in Practice?

RQ1 shows for the VideoStore project that mutation testing tools need to introduce explicit mutation operators for modeling null-type faults. Yet, such a project is not representative of real projects. In this RQ, we want to verify to what extent null-type mutation operators are useful *in practice*. For this reason, we perform an experiment that involves real open-source projects. After introducing null-type mutants, two groups of mutants are affected: (i) survived mutants are the targets the developer needs during test development, (ii) killed mutants show the types of faults the test suite can already catch.

Considering this, we can justify the effort needed for extending mutation testing by incorporating null-type mutants only if: (i) the real test suites do not already kill most of the null-type mutants, (ii) the null-type mutants are not increasing redundancy by a large margin. Otherwise, the current mutation testing tools are already "good enough" for preventing null-type faults.

To verify to what extent the null-type mutants "do matter" when testing for null-type faults we analyze both killed and survived mutants:

In case of survived mutants, we analyze the number of survived mutants that each mutation operator generates for each project. We divide this analysis into two parts. First, we analyze survived mutants for null-type and traditional mutation operators. Second, we analyze each mutation operator individually to find out which one produces the most surviving mutants. This analysis shows whether the survived mutants produced by the null-type mutation operators are "enough" to drive the test development process.

In case of killed mutants, we take all projects as a whole, and we analyze whether the killed null-type mutants are redundant when used together with traditional mutation operators. We measure redundancy using dynamic mutant subsumption: we analyze the distributions of subsuming, killed, and all null-type mutants. This way we can tell whether or not the null-type mutation operators are producing "valuable" mutants to strengthen the test suite.

Survived Mutants. Table 4 shows for each project the number of survived, killed, and total generated mutants for both groups of mutation operators. The first noticeable trend is a strong correlation ($R^2 = 0.81$) between survived to killed ratio (SKR) of the traditional mutants and SKR of the null-type mutants. One exception to this trend is JSQLParser, in which there are significantly more survived null-type mutants than survived traditional mutants. Investigating further, we find that this happens because 50 small classes lack statements that can be mutated by the traditional mutation operators. However, null-type mutation operators are able to generate mutants for these classes. This uncovers many of the weaknesses of the test suite. On the other side of the fence, there is PITest, in which a single class (`sun.pitest.CodeCoverageStore`) contains many arithmetic operations while poorly tested, so it produces 129 out of 398 survived traditional mutants. This shows that **the usefulness of the null-type mutation operators is program-dependent.**

Table 4. Mutants generated by LittleDarwin and LittleDarwin-Null

Project	Traditional mutation operators			Null-type mutation operators		
	Survived	Killed	Total	Survived	Killed	Total
Apache Commons CLI	24	318	342	71	415	486
JSQLParser	31	457	488	358	1,062	1,420
jOpt Simple	17	189	206	37	494	531
Apache Commons Lang	559	5,455	6,014	564	5,469	6,033
Joda Time	892	3,978	4,870	836	5,371	6,207
Apache Commons Codec	364	1,612	1,976	147	927	1,074
Apache Commons Collections	638	2,705	3,343	1,179	5,851	7,030
VRaptor	111	478	589	795	2,111	2,906
HTTP Request	49	178	227	69	383	452
Apache Commons FileUpload	81	273	354	137	211	348
jsoup	291	928	1,219	553	1,455	2,008
JGraphT	416	940	1,356	834	1,457	2,291
PITest	398	672	1,070	551	2,964	3,515
JFreeChart	10,558	5,603	16,161	8,563	6,248	14,811
PMD	5,205	2,734	7,939	5,099	4,613	9,712
Total	**19,634**	**26,520**	**46,154**	**19,793**	**39,031**	**58,824**

Figure 5 shows the number of killed and survived mutants for each mutation operator. We see that among the traditional mutation operators, *Arithmetic-OperatorReplacementBinary*, *LogicalOperatorReplacement*, and *ArithmeticOperatorReplacementUnary* have the highest ratio of survived to killed mutants. This means that these mutation operators are generating mutants that are harder to kill than the rest. The same can be observed among the null-type mutation operators, where *NullifyObjectInitialization* produces harder to kill mutants than the others. This is as we expected, since *NullifyInputVariable* applies a major change to the method (removal of an input), and *NegateNullCheck* negates a check that the developer deemed necessary. However, the unexpected part of the result is that so many of the mutants generated by *NullifyReturnValue* have survived. This means that lots of methods are not tested on their output correctly. This can be due to the fact that many of such methods are not tested directly, and when tested indirectly, their results only affect a small part of the program state of the method under test.

In general, the number of survived null-type mutants has a strong correlation with the number of survived traditional mutants for most projects. This implies that not all parts of the code are tested well. However, the exceptions to this rule are caused by classes that produce many more mutants of a particular type. Here, our results show that **the null-type mutation operators complement the traditional mutation operators and vice versa by each providing a large portion of survived mutants.**

Killed Mutants. Considering all projects as a whole, the number of generated mutants is 104,978. Out of this total, the number of killed and subsuming

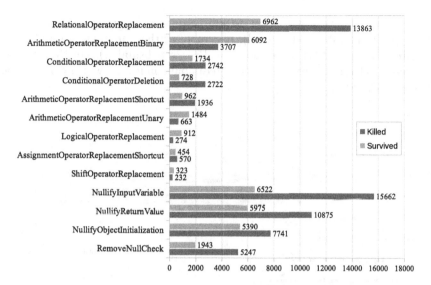

Fig. 5. Number of killed and survived mutants for each mutation operator

mutants are 65,551 and 16,205 respectively. This means that at least 50,029 were subsumed, and thus redundant. To put null-type and traditional mutants in perspective, Fig. 6 shows the percentages for all, killed, and subsuming mutants for both groups. Here, we notice that the percentage of the null-type mutants remains similar in these three categories. The null-type mutants have a higher impact on the semantics of the program due to being applied at the entry and exit points of a method, the branching statements, and the declaration of an object. Therefore, the fact that they comprise a higher percentage of the killed mutants is not surprising. However, it is important to note that the distribution of null-type mutants differs only 4% in all and killed mutants. While 60% of the killed mutants are null-type, they still account for almost 55% of subsuming mutants. This indicates that **the inclusion of the null-type mutants increases the mutant redundancy only marginally**.

To delve deeper, Fig. 7 shows for each mutation operator the percentage of killed and subsuming mutants. Among the traditional mutation operators, *RelationalOperatorReplacement* and *ConditionalOperatorReplacement* produce the most subsuming mutants. The rest of the mutation operators create mutants that have the same distribution among subsuming and killed mutants. As this figure shows, the

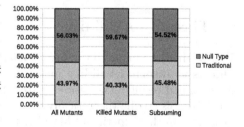

Fig. 6. Ratio of null-type and traditional mutants in all, killed, and subsuming

marginal increase in redundancy by the null-type mutation operators can be blamed on *NullifyInputVariable* mutation operator. This mutation operator

produces mutants that are easier to kill compared to other mutation operators (21% of all, 24% of the killed), and more of these mutants are redundant
compared to others (24% of killed, only 15% of subsuming). On the contrary,
NullifyReturnValue is producing fewer redundant mutants, which confirms our
previous observation.

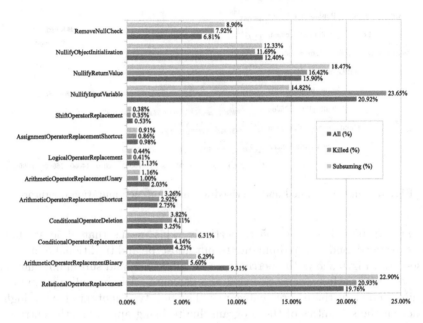

Fig. 7. Ratio of mutants by each mutation operator in all, killed, and subsuming

Given the results of RQ2, we can conclude that **while the inclusion of the
null-type mutation operators increases the redundancy marginally,
they complement the traditional mutation operators in their role of
strengthening the test suite against null-type faults.**

5   Threats to Validity

To describe the threats to validity we refer to the guidelines reported by Yin [36].
Threats to **internal validity** focus on confounding factors that can influence
the obtained results. These threats stem from potential faults hidden inside our
analysis tools. While theoretically possible, we consider this chance limited. The
tools used in this experiment have been used previously in several other studies,
and their results went through many iterations of manual validation. In addition,
the code of LittleDarwin and LittleDarwin-Null along with all the raw data of
the study is publicly available for download in the replication package [28].

Threats to **external validity** refer to the generalizability of the results. In
RQ1 we advocate for the adoption of null-type mutation operators by using

a didactic project. We alleviate the non-representativeness of this project, by analyzing 15 real open-source projects in RQ2. Although our results are based on projects with various levels of test adequacy in terms of traditional and null-type mutation coverage, we cannot assume that this sample is representative of all Java projects. We use PITest, LittleDarwin, and Javalanche as mutation testing tools. We cannot assume that these tools are representative of all mutation tools available in literature. For this reason, we refer to these tools as general-purpose since they can work with little effort on many open-source projects. We modeled null-types mutation operators upon the typical null-type faults described by Osman et al. [24]. However, there may be other types of null-type faults that we did not consider. Even if this was the case, our results should still hold since we already demonstrate with four mutation operators that they are in need of explicit modeling.

Threats to **construct validity** are concerned with how accurately the observations describe the phenomena of interest. The problem of equivalent mutants affects the analysis of surviving mutants on the test suites of the 15 open-source projects. Due to the large number of created mutants, it is impractical to filter equivalent mutants in the final results. Still, we believe this threat is minimal, because we analyze two different aspects of mutation testing, which lead to converging results. The total number of generated mutants can be different based on the set of mutation operators that are used in each tool. However, this difference has been taken into account when discussing the results of the experiments. To measure redundancy among the mutants, we use dynamic subsumption relationship. However, the accuracy of the dynamic subsumption relationship depends on the test suite itself. This is a compromise, as the only way to increase the accuracy is to have several tests that kill each mutant, which is not practical.

6   Conclusion

Developers are prone to introduce null-type faults in Java programs. Yet, there is no specific approach devoted to helping developers strengthen the test suite against these faults. On the one hand, mutation testing provides a systematic method to create tests able to prevent common faults. On the other hand, the general-purpose mutation testing tools available today do not model null-type faults explicitly by default.

In this paper, we advocate for the introduction of null-type mutation operators for preventing null-type faults. As a first step, we show that traditional mutation operators are not enough to cope with null-type faults as they cannot lead to the creation of a mutation-adequate test suite that can kill all of them. Then we demonstrate, by means of code examples, how the null-type mutants can drive the extension of the test suite. Finally, we highlight that null-type mutation operators are helpful in practice by showing on 15 open-source projects that real test suites are inadequate in detecting null-type faults. In this context, we explore the trade-offs of having null-type mutants. On the downside, we show that the inclusion of null-type mutants increases the mutant redundancy. Yet, this increment is only marginal. On the upside, we show that null-type mutants complement traditional mutants in two ways. First, they provide a large number of survived mutants to the developer to strengthen the test suite. Second, they comprise a large part of subsuming mutants.

As a consequence, developers can increase their confidence in the test suite regarding to the null-type faults by (i) prioritizing the classes that have a large difference in traditional and null-type mutation coverage, (ii) creating tests to kill the survived null-type mutants in these classes, and (iii) repeating the process until all classes have similar levels of traditional and null-type mutation coverage.

References

1. Alexander, I.: Misuse cases: use cases with hostile intent. IEEE Softw. **20**(1), 58–66 (2003). https://doi.org/10.1109/ms.2003.1159030
2. Ammann, P., Delamaro, M.E., Offutt, J.: Establishing theoretical minimal sets of mutants. In: 2014 IEEE Seventh International Conference on Software Testing, Verification and Validation, pp. 21–30. IEEE, March 2014. https://doi.org/10.1109/icst.2014.13
3. Bradbury, J.S., Cordy, J.R., Dingel, J.: Mutation operators for concurrent Java (J2SE 5.0). In: MUTATION 2006, Proceedings of the Second Workshop on Mutation Analysis (Mutation 2006 - ISSRE Workshops 2006), School of Computing, Queen's University, Kingston, p. 11. IEEE, Washington, DC, November 2006. https://doi.org/10.1109/mutation.2006.10
4. Coles, H., Laurent, T., Henard, C., Papadakis, M., Ventresque, A.: PIT: a practical mutation testing tool for Java (Demo). In: Proceedings of the 25th International Symposium on Software Testing and Analysis - ISSTA 2016, pp. 449–452. ACM Press, New York (2016). https://doi.org/10.1145/2931037.2948707
5. DeMillo, R.A., Lipton, R.J., Sayward, F.G.: Hints on test data selection: help for the practicing programmer. Computer **11**(4), 34–41 (1978). https://doi.org/10.1109/c-m.1978.218136
6. Deng, L., Offutt, J., Ammann, P., Mirzaei, N.: Mutation operators for testing Android apps. Inf. Softw. Technol. **81**, 154–168 (2017). https://doi.org/10.1016/j.infsof.2016.04.012
7. Ferrari, F.C., Maldonado, J., Rashid, A.: Mutation testing for aspect-oriented programs. In: 2008 International Conference on Software Testing, Verification, and Validation. ICST 2008, Department of Computer System, Sao Paulo University, Sao Carlos, pp. 52–61. IEEE, Washington, DC, April 2008. https://doi.org/10.1109/icst.2008.37
8. Fowler, M.: Refactoring: Improving the Design of Existing Code. Addison-Wesley, Boston (1999)
9. Fraser, G., Zeller, A.: Mutation-driven generation of unit tests and oracles. IEEE Trans. Softw. Eng. **38**(2), 278–292 (2012). https://doi.org/10.1109/tse.2011.93
10. Gligoric, M., Groce, A., Zhang, C., Sharma, R., Alipour, M.A., Marinov, D.: Comparing non-adequate test suites using coverage criteria. In: Proceedings of the 2013 International Symposium on Software Testing and Analysis - ISSTA 2013, pp. 302–313. ACM Press, New York (2013). https://doi.org/10.1145/2483760.2483769
11. Gosling, J., Joy, B., Steele, G., Bracha, G., Buckley, A.: The Java Language Specification (Java SE). Oracle, 8th edn. Addison-Wesley, Boston (2014)

12. Inozemtseva, L., Holmes, R.: Coverage is not strongly correlated with test suite effectiveness. In: Proceedings of the 36th International Conference on Software Engineering - ICSE 2014, pp. 435–445. ACM Press, New York (2014). https://doi.org/10.1145/2568225.2568271

13. Jia, Y., Harman, M.: An analysis and survey of the development of mutation testing. IEEE Trans. Softw. Eng. **37**(5), 649–678 (2011). https://doi.org/10.1109/tse.2010.62

14. Just, R., Jalali, D., Inozemtseva, L., Ernst, M.D., Holmes, R., Fraser, G.: Are mutants a valid substitute for real faults in software testing? In: Proceedings of the 22nd ACM SIGSOFT International Symposium on Foundations of Software Engineering - FSE 2014, pp. 654–665. ACM Press, New York (2014). https://doi.org/10.1145/2635868.2635929

15. Kimura, S., Hotta, K., Higo, Y., Igaki, H., Kusumoto, S.: Does return null matter? In: 2014 Software Evolution Week - IEEE Conference on Software Maintenance, Reengineering, and Reverse Engineering (CSMR-WCRE), pp. 244–253. IEEE, February 2014. https://doi.org/10.1109/csmr-wcre.2014.6747176

16. King, K.N., Offutt, A.J.: A fortran language system for mutation-based software testing. Softw.: Pract. Exp. **21**(7), 685–718 (1991). https://doi.org/10.1002/spe.4380210704

17. Kuhn, D.R.: Fault classes and error detection capability of specification-based testing. ACM Trans. Softw. Eng. Methodol. **8**(4), 411–424 (1999). https://doi.org/10.1145/322993.322996

18. Kurtz, B., Ammann, P., Offutt, J.: Static analysis of mutant subsumption. In: 2015 IEEE Eighth International Conference on Software Testing, Verification and Validation Workshops (ICSTW), pp. 1–10. IEEE, April 2015. https://doi.org/10.1109/icstw.2015.7107454

19. Ma, Y.S., Kwon, Y.R., Offutt, J.: Inter-class mutation operators for Java. In: 13th International Symposium on Software Reliability Engineering, 2002. Proceedings, pp. 352–363. IEEE Computer Society (2002). https://doi.org/10.1109/issre.2002.1173287

20. Nanavati, J., Wu, F., Harman, M., Jia, Y., Krinke, J.: Mutation testing of memory-related operators. In: 2015 IEEE Eighth International Conference on Software Testing, Verification and Validation Workshops (ICSTW), pp. 1–10. IEEE, April 2015. https://doi.org/10.1109/icstw.2015.7107449

21. Offutt, A.J., Lee, A., Rothermel, G., Untch, R.H., Zapf, C.: An experimental determination of sufficient mutant operators. ACM Trans. Softw. Eng. Methodol. **5**(2), 99–118 (1996). https://doi.org/10.1145/227607.227610

22. Offutt, A.J., Pan, J.: Automatically detecting equivalent mutants and infeasible paths. Softw. Test. Verif. Reliab. **7**(3), 165–192 (1997). https://doi.org/10.1002/(SICI)1099-1689(199709)7:3⟨165::AID-STVR143⟩3.0.CO;2-U

23. Oliveira, R.A., Alegroth, E., Gao, Z., Memon, A.: Definition and evaluation of mutation operators for GUI-level mutation analysis. In: 2015 IEEE Eighth International Conference on Software Testing, Verification and Validation Workshops (ICSTW), pp. 1–10. IEEE, April 2015. https://doi.org/10.1109/icstw.2015.7107457

24. Osman, H., Leuenberger, M., Lungu, M., Nierstrasz, O.: Tracking null checks in open-source java systems. In: 2016 IEEE 23rd International Conference on Software Analysis, Evolution, and Reengineering (SANER), vol. 1, pp. 304–313. IEEE, March 2016. https://doi.org/10.1109/saner.2016.57

25. Osman, H., Lungu, M., Nierstrasz, O.: Mining frequent bug-fix code changes. In: 2014 Software Evolution Week - IEEE Conference on Software Maintenance, Reengineering, and Reverse Engineering (CSMR-WCRE), pp. 343–347. IEEE, February 2014. https://doi.org/10.1109/csmr-wcre.2014.6747191

26. Papadakis, M., Henard, C., Harman, M., Jia, Y., Traon, Y.L.: Threats to the validity of mutation-based test assessment. In: Proceedings of the 25th International Symposium on Software Testing and Analysis - ISSTA 2016, pp. 354–365. ACM Press, New York (2016). https://doi.org/10.1145/2931037.2931040

27. Papadakis, M., Kintis, M., Zhang, J., Jia, Y., Traon, Y.L., Harman, M.: Mutation testing advances: an analysis and survey. In: Advances in Computers (2018). https://doi.org/10.1016/bs.adcom.2018.03.015

28. Parsai, A.: Replication package. http://parsai.net/files/research/SofSemReplicationPackage.7z

29. Parsai, A., Demeyer, S., De Busser, S.: C++11/14 mutation operators based on common fault patterns. In: Medina-Bulo, I., Merayo, M.G., Hierons, R. (eds.) ICTSS 2018. LNCS, vol. 11146, pp. 102–118. Springer, Cham (2018). https://doi.org/10.1007/978-3-319-99927-2_9

30. Parsai, A., Murgia, A., Demeyer, S.: Evaluating random mutant selection at class-level in projects with non-adequate test suites. In: Proceedings of the 20th International Conference on Evaluation and Assessment in Software Engineering - EASE 2016, pp. 11:1–11:10. ACM Press, New York (2016). https://doi.org/10.1145/2915970.2915992

31. Parsai, A., Murgia, A., Demeyer, S.: A model to estimate first-order mutation coverage from higher-order mutation coverage. In: 2016 IEEE International Conference on Software Quality, Reliability and Security (QRS), pp. 365–373. IEEE, August 2016. https://doi.org/10.1109/qrs.2016.48

32. Parsai, A., Murgia, A., Demeyer, S.: LittleDarwin: a feature-rich and extensible mutation testing framework for large and complex Java systems. In: Dastani, M., Sirjani, M. (eds.) FSEN 2017. LNCS, vol. 10522, pp. 148–163. Springer, Cham (2017). https://doi.org/10.1007/978-3-319-68972-2_10

33. Parsai, A., Murgia, A., Soetens, Q.D., Demeyer, S.: Mutation testing as a safety net for test code refactoring. In: Scientific Workshop Proceedings of the XP2015 on - XP 2015 workshops, pp. 8:1–8:7. ACM Press, New York (2015). https://doi.org/10.1145/2764979.2764987

34. Parsai, A., Soetens, Q.D., Murgia, A., Demeyer, S.: Considering polymorphism in change-based test suite reduction. In: Dingsøyr, T., Moe, N.B., Tonelli, R., Counsell, S., Gencel, C., Petersen, K. (eds.) XP 2014. LNBIP, vol. 199, pp. 166–181. Springer, Cham (2014). https://doi.org/10.1007/978-3-319-14358-3_14

35. Schuler, D., Zeller, A.: Javalanche: efficient mutation testing for Java. In: Proceedings of the 7th Joint Meeting of the European Software Engineering Conference and the ACM SIGSOFT Symposium on the Foundations of Software Engineering on European Software Engineering Conference and Foundations of Software Engineering Symposium - ESEC/FSE 2009, pp. 297–298. ACM Press, New York (2009). https://doi.org/10.1145/1595696.1595750

36. Yin, R.K.: Case Study Research: Design and Methods. Applied Social Research Methods. SAGE Publications, Thousand Oaks (2003)

Towards Combining Multitask and Multilingual Learning

Matus Pikuliak[✉], Marian Simko, and Maria Bielikova

Faculty of Informatics and Information Technologies,
Slovak University of Technology in Bratislava, Ilkovicova 2, Bratislava, Slovakia
matus.pikuliak@stuba.sk

Abstract. Machine learning is an increasingly important approach to Natural Language Processing. Most languages however do not possess enough data to fully utilize it. When dealing with such languages it is important to use as much auxiliary data as possible. In this work we propose a combination of multitask and multilingual learning. When learning a new task we use data from other tasks and other languages at the same time. We evaluate our approach with a neural network based model that can solve two tasks – part-of-speech tagging and named entity recognition – with four different languages at the same time. Parameters of this model are partially shared across all data and partially they are specific for individual tasks and/or languages. Preliminary experiments show that this approach has its merits as we were able to beat baseline solutions that do not combine data from all the available sources.

Keywords: Transfer learning · Multilingual learning
Deep natural language processing

1 Introduction

Modern machine learning approaches to natural language processing (NLP) are notoriously data hungry. Currently there is a significant disparity in volume of available datasets between various languages. While English, Chinese and a handful of other major languages have the most data, other languages are seriously lacking. This is naturally slowing down the research and development of crucial NLP algorithms, models and services for these low-resource languages.

Collecting new data is laborious and expensive. Transfer learning is sometimes proposed as a possible remedy. Instead of creating new datasets we might try to utilize existing ones, even though they are not completely related to our problem. In NLP this means using data from other tasks, languages or domains. Research so far predominately focused on only one of these options at the time. The novelty of our work lies in the fact that we are combining multitask and multilingual learning.

We combine them to utilize as much available data during learning as possible. We believe that combining data from multiple sources might be crucial

© Springer Nature Switzerland AG 2019
B. Catania et al. (Eds.): SOFSEM 2019, LNCS 11376, pp. 435–446, 2019.
https://doi.org/10.1007/978-3-030-10801-4_34

when trying to create robust and efficient solutions. This is especially important for low-resource languages as it might significantly reduce data requirements.

We propose a method of training a model with multiple tasks from multiple languages at the same time. In theory the task-specific part can be adapted to solve any task that can use contextualized word representations. However so far we experimented only with two: part-of-speech tagging (POS) and named entity recognition (NER). Both these tasks were solved for four languages (English, German, Spanish, Czech). We evaluated the performance of this model when trained on train data from target task only versus when trained on train data from all tasks. In some cases we noted significant score improvements.

2   Related Work

Parameter Sharing. Parameter sharing is a popular technique of multitask learning. Multiple models trained on different tasks share the values of subset of parameters. Such sharing can boost the results for any of the relevant tasks by the virtue of having additional data influencing the training process. Various combinations of sequence tagging tasks were already considered [14,17,20,24]. These approaches usually share certain layers (e.g. LSTM layer processing word embeddings or character embeddings) between tasks or languages. These layers are then supervised from multiple tasks and these task regularize each other in a sense. Subramanian et al. use parameter sharing with multiple tasks to create robust sentence representations [21].

Multilingual Learning. Multilingual learning can be perceived as a special type of multitask learning, when the samples come from different languages. The goal is to transfer knowledge from one language to others. Various techniques to multilingual learning exist, using annotation projection [1,5], language independent representations [2,25] or parameter sharing [7,24]. Parameter sharing techniques are the most relevant to us. We chose to use this technique because in contrast with other techniques it can be used for both multilingual and multitask learning at the same time. They share certain layers between more languages and these layers are therefore becoming multilingual. We extend this idea and combine multilingual learning with multitask learning. Recently concurrently with our research Lin et al. experimented with a setup similar to ours [13]. They have a hierarchical LSTM model that is solving sequential tagging tasks. Compared to our task some details of model implementations differ. Our unique contribution is that we really use multiple auxiliary tasks, as opposed to their approach when they always use just two - one with the same task but a different language and one with the same language but with a different task. Other works use different techniques for combining multilingual and multitask learning, e.g., [1,22] use annotation projection to solve POS tagging and dependency parsing with multiple languages at the same time.

3   Proposed Method

We propose a multitask multilingual learning method based on parameter sharing. In our approach, the training is done with multiple datasets at the same time. In this work we work with two tasks – NER and POS – and four natural languages. In effect we have 8 unique training datasets and for each of these datasets we have a separate model that is being trained. All these models have the same neural network based architecture. In various experiments selected parameters are shared between certain models to achieve transfer of information between them. Sharing parameters in this case means having identical weight matrices on selected layers of a neural network.

3.1   Architecture

The architecture of our model needs to be general enough to allow us to effectively solve multiple tasks at once. In our work we use sentence level tasks, i.e., we expect a sentence as an input to our model. The form of the output depends on the task. Our model is not suited to process higher level units, such as paragraphs or documents. We propose a model with three consecutive parts:

Part 1: Word Embeddings. Word embeddings are fixed-length vector representation of words that became very popular in previous years [8,15] as they are able to significantly outperform other types of word representations, namely one-hot encodings. They are based on an idea of using language modeling as auxiliary task when learning about words. The first step of our architecture is to project the words into vector space. For each language L we have a dictionary of words D_L and a function d_L that takes a given word and returns an integer identifier of this word in given dictionary. The id of Czech word *on* is $d_{cs}(on)$, while the id of English word *on* is $d_{en}(on)$. Even though the form (the characters they are composed of) of these two words is the same, they have different ids in their respective dictionaries.

For each language we also have an embedding matrix E_L whose i-th row is a word representation for a word W for which $d_L(W) = i$. The matrix E_L has dimensions $a \times b$, where a is the number of words in dictionary D_L and b is the arbitrarily set length of word representation that is set before the word embeddings are created. For words that are not in the dictionary of its language we use a zero vector.

The input of our model is a sentence of N words from language L:

$$I = \langle W_1, W_2, W_3, \ldots, W_N \rangle \tag{1}$$

For clarity, we define a function d' that return a vector for a given word as:

$$d'_L(W) = E_L(d_L(W)) \tag{2}$$

With this we can then define the output of this model layer as a sequence of embeddings:

$$e = \langle d'_L(W_1), d'_L(W_2), d'_L(W_3), \ldots, d'_L(W_K), \rangle \tag{3}$$

In our case we use so called multilingual word embeddings as pre-trained word representations that are stored in the E matrices. Multilingual word embeddings are an extension of standard word embeddings technique where words from multiple languages share one semantic vector space [18]. Semantically similar words have similar representations even when they come from different languages. This is the only information that explicitly tells our model what is the relation between various languages and their words. Sometimes researchers let the word embeddings be trainable parameters. In our case we fix them so we do not lose this link between languages.

Part 2: LSTM. Word embeddings are processed in a bi-directional LSTM recurrent layer [9]. The weights of this layer are shared across both tasks and languages – the same LSTM layer is used during each pass in the network. This is the part that contains the majority of trainable parameters and is therefore also responsible for most of the computation that is performed. This is also where most of the information sharing happens. The output of this layer is a sequence of contextualized word representations. While in the previous layer the same word always has the same representation, here the same word will have different representation if it is used in different contexts. The context in our case is the whole sentence.

We used LSTM recurrent layer as they are able to partially tackle the forgetting problem of basic recurrent networks, which tend to forget what they saw in previous steps if the input sequence length is too big. LSTMs use several gates that let the model learn what parts of the signal should it keep between the steps and which parts should be forgotten. The LSTM is traditionally defined as follows:

$$f_t = \sigma(W_f x_t + U_f h_{t-1} + b_f) \tag{4}$$

$$i_t = \sigma(W_i x_t + U_i h_{t-1} + b_i) \tag{5}$$

$$o_t = \sigma(W_o x_t + U_o h_{t-1} + b_o) \tag{6}$$

$$c_t = f_t \circ c_{t-1} + i_t \circ \sigma(W_c x_t + U_x h_{t-1} + b_c) \tag{7}$$

$$h_t = o_t \circ \sigma(c_t) \tag{8}$$

where x_t, h_t and c_t is the LSTM input, output and cell state at the time t. The size of h and c can be set arbitrarily. f_t, i_t and o_t are forgetting, input and output activation gates that govern how much signal is kept during the computation. W's, U's and b's are trainable weights and biases. Finally $\circ$ is Hadamard product and σ is a non-linear activation function.

This defines an $LSTM$ function that takes a sequence of inputs and returns a sequence of outputs that encode the state of the LSTM at individual timestamps:

$$LSTM(\langle x_1, x_2, x_3, \ldots, x_K \rangle) = \langle h_1, h_2, h_3, \ldots, h_K \rangle \tag{9}$$

The bi-directional LSTM layer of our model is then defined with two LSTM networks. The first one processes the word embeddings from the start, while the second one processes it from the end:

$$h_1, h_2, h_3, \ldots, h_K = LSTM(e) \tag{10}$$

$$\langle h'_K, h'_{K-1}, h'_{K-2}, \dots h'_1 \rangle = LSTM(reverse(e)) \tag{11}$$

The output q of this layer is finally defined as:

$$q = \langle (h_1; h'_1), (h_2; h'_2), (h_3; h'_3), \dots, (h_K; h'_K) \rangle \tag{12}$$

with a semicolon marking a concatenation of the two vectors and e being an output of previous layer.

Part 3: Output Layers. Finally the output of the LSTM is processed by task-specific layers (architectures might differ depending on the tasks). The parameters of this part might or might not be shared across languages. So far we experimented with two tasks, part-of-speech tagging and named entity recognition. As both of them are sequence tagging tasks we use the same architecture for this part.

Each contextualized word representation from the bi-LSTM layer is used to predict the appropriate tag for a given word with a linear projection:

$$p = Wh + b \tag{13}$$

where p is the prediction vector containing probabilities for each possible tag within given task, W and b are weights and biases and h is a contextualized vector for one particular word from previous layer. We use the same parameters (W and b) for each word.

All these predictions for one sentence are then processed by a CRF sequence modeling algorithm [12] to calculate final results. Using this algorithm means that instead of simply optimizing for the p to predict the correct tag as much as possible we also take into account the order of individual tags. To this end a transition matrix counting how many times one particular tag followed all the other tags is used. During training this step is differentiable but during inference we need to use dynamic programming to calculate the final tags from the predictions p our model generates while also taking the transition matrix into account. A detailed description of this part of the network is outside of the scope of this article, cf. Lample et al. [12].

The complete architecture is depicted in Fig. 1. From bottom up we can see the word being transformed into their respective embedding representations and processed by a bi-directional layer. The output then flows into a task-specific part. In our case we always use a CRF component that predicts the tag for each word. If we were to extend our experiments with additional tasks that might not be sequential tagging tasks, this final part would differ.

3.2   Training

We consider several training modes, based on what kind of data the model is exposed to:

1. **Single task, single language (ST-SL).** This is the standard machine learning setting when we have data for one task from one language only.

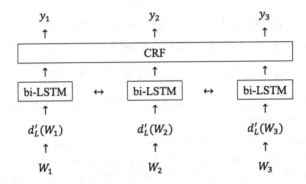

Fig. 1. Architecture of our solution. The final depicted layer is CRF, however in theory an arbitrary architecture can follow the bi-LSTM layer.

2. **Multitask (MT).** More tasks from single language are solved at once, e.g., we train both NER and POS for English.
3. **Multilingual (ML).** Data from multiple languages are used to train one shared task, e.g. we train POS on all languages at the same time.
4. **Multitask, multilingual (MT-ML).** A combination of multitask and multilingual learning. Multiple tasks are solved for multiple languages all at the same time.

We use epochs with fixed amount of training steps. When training with more languages and/or tasks, each training step consists of several minibatches – one minibatch for each relevant dataset. For example, during multitask learning we might have two relevant datasets, English POS and English NER. This means that we first run one English POS minibatch followed by one English NER minibatch as one training step. Minibatches always contain the same amount of samples.

Each model processes *epochs* × *steps* × *datasets* minibatches during training. In effect this means that the model gets exposed to more data with an increasing number of datasets used for training. Naturally during each pass only the parameters that are relevant for a given task and language gets updated. Rest of the parameters lie unchanged.

4   Experiments and Results

4.1   Datasets

In our experiments we used four languages (English, German, Spanish and Czech) and two tasks (part-of-speech tagging, named entity recognition). This means that in overall we had 8 datasets, each with training, development and testing part. The amount of annotated sentences for each dataset can be found in Table 1.

Table 1. Number of sentences in datasets (in thousands).

	en	es	de	cs
NER train	38.4	7.1	24.0	7.2
NER dev	4.8	1.6	2.2	0.9
NER test	4.8	1.4	5.1	0.9
POS train	12.5	14.1	13.8	68.5
POS dev	2.0	1.4	0.8	9.3
POS test	2.0	0.4	1.0	10.1

Part-of-Speech Tagging. For POS we used Universal Dependencies [16] datasets for each language. These are annotated using the universal POS tagging schema containing 17 common tags.

Named Entity Recognition. We used Groningen Meaning Bank [4] for English, GermEval 2014 NER dataset [3] for German, CoNLL 2002 [19] for Spanish and Czech Named Entity Corpus [11] for Czech. The tagging schemata differ between these datasets so we had to unify them ourselves. We converted them to standard BIO schema used for NER. We used 4 types of named entities (persons, locations, organizations and miscellaneous). English dataset was the only one that did not have separated training, development and testing data so we split it with 8:1:1 ratio.

Word Embeddings. For multilingual word embeddings we use publicly available MUSE embeddings [6]. These can be trained for any two languages if we have a monolingual corpus for each language and a bilingual dictionary. They have word vectors of size 300 for 200,000 words in each language. Vectors for other words were set to zero.

4.2 Experiment

We trained our model in all modes as mentioned in Sect. 3.2. Every time we have 8 models for each task-language combination. In various settings they share different parameter subspaces. When using multilingual learning they share all the parameters (in effect this means they are identical so it is one model being trained with more data). When using multitask learning the two models share an LSTM layer, but the task-specific weights used to make the final tag predictions are naturally not shared across tasks. When using multitask multilearning models, they still all share the LSTM layer, while the output layer is shared only between the models with the same task. To explain more clearly what models are connected through parameter sharing we illustrate our settings in Fig. 2.

Hyperparameters. We used RMSProp [23] optimization algorithm with learning rate $1e-4$ with gradient clipping set to 1. Dropout was used before and after LSTM layer and it was set to 50%. For each run we had 60 epochs, each with 512 training steps. Batch size was 32. LSTM hidden cell size was 300.

(a) SL-ST

	en	es	cs	de
POS	1	2	3	4
NER	5	6	7	8

(b) MT

	en	es	cs	de
POS				
	1	2	3	4
NER				

(c) ML

	en	es	cs	de
POS		1		
NER		2		

(d) ML-MT

	en	es	cs	de
POS				
		1		
NER				

Fig. 2. Illustration of how different training modes use all the datasets. E.g. we can see that in MT we have 4 model pairs that share parameters.

Results from these experiments are presented in Table 2 for NER and Table 3 for POS. We use tag F1 score for NER and tag accuracy for POS. The precision for NER is calculated as a number of correctly predicted NER tags (excluding O tag for words without named entity tag) divided by a number of all NER tags in labels. Recall on the other hand is the number of correctly predicted NER tags divided by a overal number of predicted NER tags. In all cases by NER tags we still mean only non-O tags. F1 is then traditionally defined as a harmonic average of precision and recall.

Table 2. NER results for various learning modes for individual languages. Results are per tag F1 scores.

	en	es	de	cs
ST-SL	77.3	73.0	73.3	66.2
MT	77.4	74.3	**75.3**	67.8
ML	**78.1**	75.6	74.6	68.1
MT-ML	77.5	**77.1**	74.3	**69.8**

Combination of multilingual and multitask learning managed to beat other learning modes in 4 out of 8 cases. The most significant was 4.1% improvement for Spanish NER. In all but two cases it beat the single task single language baseline. The result was a slight decline of 0.11% was measured for Czech POS. When reviewing these results we noted that there seems to be a negative correlation between the amount of training samples for the task and the improvement we achieved with MT-ML training. This is depicted in Fig. 3. The two datasets with highest and lowest number of samples are in fact those with the lowest and

Table 3. POS results for various learning modes for individual languages. Results are per tag accuracy scores.

	en	es	de	cs
ST-SL	90.66	94.16	91.19	**94.06**
MT	90.90	94.19	91.27	94.05
ML	91.17	94.30	**91.42**	93.95
MT-ML	**91.21**	**94.41**	91.16	93.95

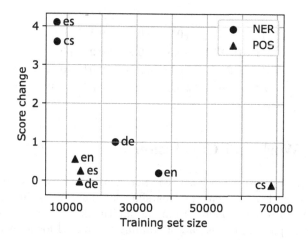

Fig. 3. Relation between training set size and the change in score when using MT-ML instead of ST-SL for each dataset. The score is F1 for NER and accuracy for POS.

highest improvement in score. This indicates that our method is well suited for low-resource scenarios but it loses its effectiveness when we have enough training data.

4.3   Sharing the Output Layer

In previous experiments when performing multilingual learning (both ML and MT-ML) the output layers with CRF were not shared across languages. Each language had its own private subset of parameters. Our goal was to let the model learn specifics of each language this way. To confirm our hypothesis that it is beneficial to have private output layers we run the same experiments as before for these two learning modes but this time with only one set of output parameters shared across all four languages. We compare the absolute change in score (F1 for NER, accuracy for POS) in Table 4.

We can see slight improvement in NER (on average +0.17) and slight fall in POS (on average −0.02). The way parts of speech are used in various languages differ more than the way named entities behave. Instinctively this difference in results makes sense. During analysis we noticed that with shared output layers it took longer for the model to converge to a near optimal solution in all cases. We think that private output layers make the work easier for the rest of the model as they are able to correct the mistakes of model. When the output layer is shared the LSTM is forced to predict correct tags as there is no fallback mechanism. It is ultimately able to overcome this challenge but it takes longer because the task solved is harder.

Table 4. The absolute change in score when output layer parameters are shared between languages.

	en	es	de	cs
ML NER	+0.2	+0.8	−0.3	−0.6
MT-ML NER	+0.5	+0.2	+0.2	+0.3
ML POS	−0.06	+0.06	−0.02	−0.04
MT-ML POS	+0.16	−0.06	+0.05	−0.25

5 Future Work and Conclusion

The most important future work is the experimenting with additional languages and also tasks, such as dependency parsing, language modeling and machine translation. We also plan to shift from the multitask learning paradigm to the transfer learning paradigm. Instead of training the model for all available tasks at once, we are interested if we could specialize it only for one specifically selected task (perhaps an extremely low-resource one). To do this we will need an agent capable of dynamically changing the selection policy. Using partially private models [14], adversarial learning [10] or sub-word level representations [24] are several other ideas we plan to experiment with.

So far our model proved itself to be capable of multitask multilingual learning. We were able to beat reasonable single task baselines that use less auxiliary data, especially in small training set cases. Combining data from various heterogeneous sources might be crucial for developing effective solutions especially for low-resource languages. We perceive this work as one step towards this goal.

The proposed model work on the sentence level which is compatible with many NLP tasks the community is solving so far. It can be easily extended by modifying the output layer to solve other tasks. By aggregation we could even combine representations from several sentences to form representations for paragraphs or documents. From these we could gather additional signal for learning as some tasks are traditionally solved in a document level fashion, e.g. document classification.

Acknowledgements. This work was partially supported by the Slovak Research and Development Agency under the contract No. APVV-15-0508, and by the Scientific Grant Agency of the Slovak Republic, grants No. VG 1/0667/18 and No. VG 1/0646/15.

References

1. Agić, Ž., Johannsen, A., Plank, B., Alonso, H.M., Schluter, N., Søgaard, A.: Multilingual projection for parsing truly low-resource languages. Trans. Assoc. Comput. Linguist. **4**, 301–312 (2016)
2. Ammar, W., Mulcaire, G., Ballesteros, M., Dyer, C., Smith, N.: Many languages, one parser. Trans. Assoc. Comput. Linguist. **4**, 431–444 (2016)
3. Benikova, D., Biemann, C., Reznicek, M.: NoSta-D named entity annotation for German: guidelines and dataset. In: Proceedings of the Ninth International Conference on Language Resources and Evaluation. LREC 2014, 26–31 May 2014, Reykjavik, Iceland, pp. 2524–2531 (2014)
4. Bos, J., Basile, V., Evang, K., Venhuizen, N.J., Bjerva, J.: The Groningen meaning bank. In: Ide, N., Pustejovsky, J. (eds.) Handbook of Linguistic Annotation, pp. 463–496. Springer, Dordrecht (2017). https://doi.org/10.1007/978-94-024-0881-2_18
5. Buys, J., Botha, J.A.: Cross-lingual morphological tagging for low-resource languages. In: Proceedings of the 54th Annual Meeting of the Association for Computational Linguistics (Volume 1: Long Papers), pp. 1954–1964. Association for Computational Linguistics (2016)
6. Conneau, A., Lample, G., Ranzato, M., Denoyer, L., Jégou, H.: Word translation without parallel data. In: 6th International Conference on Learning Representations, Vancouver, Canada, May 2018
7. Cotterell, R., Heigold, G.: Cross-lingual character-level neural morphological tagging. In: Proceedings of the 2017 Conference on Empirical Methods in Natural Language Processing, pp. 759–770. Association for Computational Linguistics (2017)
8. Gallay, L., Šimko, M.: Utilizing vector models for automatic text lemmatization. In: Freivalds, R., Engels, G., Catania, B. (eds.) SOFSEM 2016. LNCS, vol. 9587, pp. 532–543. Springer, Heidelberg (2016). https://doi.org/10.1007/978-3-662-49192-8_43
9. Hochreiter, S., Schmidhuber, J.: Long short-term memory. Neural Comput. **9**(8), 1735–1780 (1997)
10. Joty, S., Nakov, P., Màrquez, L., Jaradat, I.: Cross-language learning with adversarial neural networks. In: Proceedings of the 21st Conference on Computational Natural Language Learning (CoNLL 2017), pp. 226–237. Association for Computational Linguistics (2017)
11. Kravalova, J., Zabokrtsky, Z.: Czech named entity corpus and SVM-based recognizer. In: Proceedings of the 2009 Named Entities Workshop: Shared Task on Transliteration (NEWS 2009), pp. 194–201. Association for Computational Linguistics (2009)
12. Lample, G., Ballesteros, M., Subramanian, S., Kawakami, K., Dyer, C.: Neural architectures for named entity recognition. In: Proceedings of the 2016 Conference of the North American Chapter of the Association for Computational Linguistics: Human Language Technologies, pp. 260–270. Association for Computational Linguistics (2016)
13. Lin, Y., Yang, S., Stoyanov, V., Ji, H.: A multi-lingual multi-task architecture for low-resource sequence labeling. In: Proceedings of the 56th Annual Meeting of the Association for Computational Linguistics (Volume 1: Long Papers), pp. 799–809. Association for Computational Linguistics (2018). http://aclweb.org/anthology/P18-1074

14. Liu, P., Qiu, X., Huang, X.: Adversarial multi-task learning for text classification. In: Proceedings of the 55th Annual Meeting of the Association for Computational Linguistics (Volume 1: Long Papers), pp. 1–10. Association for Computational Linguistics (2017)
15. Mikolov, T., Yih, W., Zweig, G.: Linguistic regularities in continuous space word representations. In: Human Language Technologies: Proceedings and Conference of the North American Chapter of the Association of Computational Linguistics, 9–14 June 2013, Westin Peachtree Plaza Hotel, Atlanta, Georgia, USA, pp. 746–751 (2013)
16. Nivre, J., et al.: Universal dependencies v1: a multilingual treebank collection. In: Proceedings of the Tenth International Conference on Language Resources and Evaluation. LREC 2016, 23–28 May 2016, Portorož, Slovenia (2016)
17. Peng, N., Dredze, M.: Multi-task domain adaptation for sequence tagging. In: Proceedings of the 2nd Workshop on Representation Learning for NLP, pp. 91–100. Association for Computational Linguistics (2017)
18. Ruder, S.: A survey of cross-lingual embedding models. CoRR abs/1706.04902 (2017)
19. Sang, E.F.T.K., Meulder, F.D.: Introduction to the CoNLL-2003 shared task: language-independent named entity recognition. In: Proceedings of the Seventh Conference on Natural Language Learning. CoNLL 2003, 31 May–1 June 2003, Held in Cooperation with HLT-NAACL 2003, Edmonton, Canada, pp. 142–147 (2003)
20. Søgaard, A., Goldberg, Y.: Deep multi-task learning with low level tasks supervised at lower layers. In: Proceedings of the 54th Annual Meeting of the Association for Computational Linguistics (Volume 2: Short Papers), pp. 231–235. Association for Computational Linguistics (2016)
21. Subramanian, S., Trischler, A., Bengio, Y., Pal, C.J.: Learning general purpose distributed sentence representations via large scale multi-task learning. In: 6th International Conference on Learning Representations, Vancouver, Canada, May 2018
22. Tiedemann, J.: Rediscovering annotation projection for cross-lingual parser induction. In: Proceedings of COLING 2014, The 25th International Conference on Computational Linguistics: Technical Papers, pp. 1854–1864. Dublin City University and Association for Computational Linguistics (2014)
23. Tieleman, T., Hinton, G.: Lecture 6.5-rmsprop: divide the gradient by a running average of its recent magnitude. COURSERA: Neural Netw. Mach. Learn. 4(2), 26–31 (2012)
24. Yang, Z., Salakhutdinov, R., Cohen, W.W.: Transfer learning for sequence tagging with hierarchical recurrent networks. In: 5th International Conference on Learning Representations, Toulon, France, April 2017
25. Zirikly, A., Hagiwara, M.: Cross-lingual transfer of named entity recognizers without parallel corpora. In: Proceedings of the 53rd Annual Meeting of the Association for Computational Linguistics and the 7th International Joint Conference on Natural Language Processing (Volume 2: Short Papers), pp. 390–396. Association for Computational Linguistics (2015)

On the Size of Logical Automata

Martin Raszyk[(✉)]

Department of Computer Science, ETH Zürich, Universitätstrasse 6, 8092 Zürich,
Switzerland
m.raszyk@gmail.com

Abstract. The state complexity of simulating 1NFA by 2DFA is a long-standing open question, which is of particular interest also due to its connection to the DLOG vs. NLOG problem for Turing machines.

What makes proving lower bounds on the size of deterministic two-way automata particularly hard is the fact that one has to consider *any* automaton, and unlike the designer, one does not have any meaning of the states at hand. This motivates the notion of *logical automata* whose states are annotated by formulas representing the meaning of a state.

In the paper at hand, we first introduce the notion of logical automata and present a *general approach* to proving lower bounds on the number of states of logical automata. We then apply this approach to derive an *exponential* lower bound on the size of logical automata over formulas with a restricted set of atomic predicates. Finally, we complement the lower bound with an (also exponential) upper bound.

1 Introduction

A two-way finite-state automaton is a natural generalization of a (one-way) finite-state automaton. It is known for a long time that the two models (and both their deterministic and non-deterministic versions) accept the same set of languages—the *regular* languages. Nevertheless, a long-standing open question is the state complexity of simulating 1NFA by 2DFA, i.e., the minimum number of states of 2DFA accepting a language for which there exists a 1NFA with a given number of states. The question is of particular interest also due to its connection to the DLOG vs. NLOG problem for Turing machines. As shown in [1], if the simulation of 1NFA by an equivalent 2DFA requires a superpolynomial number of states when restricted to words of polynomial length, then DLOG $\neq$ NLOG.

Sakoda and Sipser [5] conjecture that the minimum number of states of 2DFA simulating a 1NFA with a given number of states is not polynomial. Nevertheless, up to now, exponential lower bounds have only been proved for certain restrictions of 2DFA, such as sweeping automata [6] (reversing the direction of the reading head movement only at the left and right delimiter of the input word) and oblivious automata [4] (exhibiting only a sublinear number of trajectories of the reading head on input words of a given length).

Refer to [2] for a more comprehensive survey of known results.

© Springer Nature Switzerland AG 2019
B. Catania et al. (Eds.): SOFSEM 2019, LNCS 11376, pp. 447–460, 2019.
https://doi.org/10.1007/978-3-030-10801-4_35

1.1   Logical Automata

What makes proving lower bounds on the size of deterministic two-way automata particularly hard is the fact that one has to consider *any* automaton, and unlike the designer, one does not have any meaning of the states at hand. This motivates a line of research which has been initiated by introducing the notion of reasonable automata in [3] and further followed in [2]. Based on this concept, we introduce the notion of *logical automata* in this paper.

Every state of a logical automaton is annotated by a formula. The formula assigned to a state of a logical automaton is over a finite or countably infinite set of atomic predicates. We represent formulas as functions, mapping any binary vector of truth assignments to the corresponding truth value. The unbounded number of atomic predicates allows a logical automaton to operate on words of an unbounded length.

Unlike in reasonable automata, a state of a logical automaton does not have any focus (i.e., an assigned position on the input word). This way, the transitions of a logical automaton behave more like that of a standard two-way deterministic finite-state automaton. Note that information about the current position on the input word can still be encoded in the formula assigned to a state.

One more difference with respect to reasonable automata is that the computation of a logical automaton started from any valid configuration must be finite (a valid configuration is such that the input word and current position satisfy the formula assigned to the current state). This property prevents the automaton from storing some information only into the state and not into the formula associated with it. Otherwise, a valid logical automaton could, for instance, start the computation on states which are assigned the tautology as their formula, and thus store all the information about the input word transparently into the state. Then, after reaching a state at which the automaton knows whether to accept or reject, but still annotated by the tautology, the automaton could properly finish the computation on the input words which reach this state and loop indefinitely on those input words which do not reach this state (which would not violate the condition that the computation on any input word is finite).

It is not hard to see that, with a suitable choice of the set of formulas, any two-way deterministic finite-state automaton is also a logical automaton (one just has to choose a set of formulas expressive enough, e.g., being able to uniquely describe the current word on the input tape and setting the formula of a state so that it is only satisfied by configurations which do occur in a computation on a word). Hence, we want to study the state complexity of logical automata with some restricted classes of formulas, in order to get more insight into the general (hard) problem.

1.2   A Complete Family of Languages

As shown in [5], there exists a family of languages $\{B_n\}_{n\in\mathbb{N}}$ which is *complete* with respect to the simulation of 1NFA by 2DFA. This means that, among all languages accepted by a 1NFA with n states, the language B_n, which is accepted by a 1NFA with n states, requires a 2DFA with the maximum number of states. The family $\{B_n\}_{n\in\mathbb{N}}$ is also referred to as *one-way liveness*.

Let Σ_n denote the alphabet of the language B_n. A symbol $s \in \Sigma_n$ corresponds to a directed bipartite graph consisting of two columns of n nodes each. Directed edges then connect nodes from the left column to nodes in the right column.

A non-empty word $w \in \Sigma_n^*$ is to be interpreted as a directed graph obtained by concatenating the graphs corresponding to the individual symbols and merging the right column of a symbol with the left column of the next symbol. It is in the language B_n if there exist some vertices in the leftmost column (left side) and rightmost column (right side) of the corresponding directed graph which are connected by a directed path.

Example 1. Fig. 1 contains two words over Σ_5, each of length 3. The corresponding graph for a word from Fig. 1 thus consists of four columns. The word $w^{(1)} \notin B_5$, since there exists no directed path from any vertex $a_1, \ldots, a_5$ in the leftmost column to any vertex $d_1, \ldots, d_5$ in the rightmost column. On the other hand, the word $w^{(2)} \in B_5$, since there exists, e.g., a directed path a_3, b_5, c_1, d_4 from the left side to the right side.

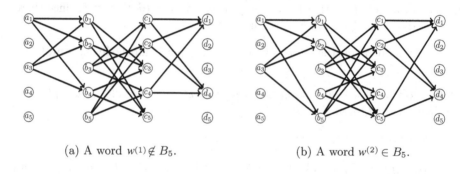

(a) A word $w^{(1)} \notin B_5$.   (b) A word $w^{(2)} \in B_5$.

Fig. 1. Words of length 3.

1.3   Definitions

Let $\mathbb{N} := \{0, 1, 2, \ldots\}$, and $[n] := \{1, 2, \ldots, n\}$, for any $n \in \mathbb{N}$.

For a word $w \in \Sigma^*$, let $|w|$ denote the length of the word w, and for any $k \in [|w|]$, let w_k denote the k-th symbol of w. We further define $w_0 := \vdash$, $w_{|w|+1} := \dashv$.

Definition 1. *We call a pair $(w, k) \in \Sigma^* \times \mathbb{N}$ of a word over an alphabet Σ and a position on it a* positioned word.

Definition 2. *An* atomic predicate h *on a positioned word is a function* $h : \Sigma^* \times \mathbb{N} \to \{0, 1\}$. *A positioned word (w, k) satisfies h if and only if $h(w, k) = 1$.*

Definition 3. *For a (finite or countably infinite) set $\mathcal{U}$ of atomic predicates, let $\mathcal{F}(\mathcal{U})$ denote the set of functions defined as:* $\mathcal{F}(\mathcal{U}) := \{f : \{0,1\}^{|\mathcal{U}|} \to \{0,1\}\}$. *The set $\mathcal{F}(\mathcal{U})$ is the set of all* formulas *over $\mathcal{U}$.*

Definition 4. *A positioned word (w,k) satisfies a formula $f \in \mathcal{F}(\mathcal{U})$ over a set $\mathcal{U} = \{h_1, h_2, \ldots, h_j, \ldots\}$ of atomic predicates if and only if*

$$f(h_1(w,k), h_2(w,k), \ldots, h_j(w,k), \ldots) = 1.$$

We write $(w,k) \vDash f$ if and only if (w,k) satisfies f.

Example 2. Let us look at words over Σ_5. Let $h_i : \Sigma_5^* \times \mathbb{N} \to \{0,1\}$, for $i \in [5]$, be an atomic predicate which is satisfied for a positioned word if and only if there exists a (directed) edge from the i-th vertex at the left part of the current symbol (in particular, the current symbol must not be $\vdash$, $\dashv$). Let the set $\mathcal{U}$ of atomic predicates be $\mathcal{U} = \{h_1, \ldots, h_5\}$.

Let a function $f : \{0,1\}^5 \to \{0,1\}$ be defined as follows: $f(b) = 1 \iff b_2 = 1 \vee b_5 = 1$, for any $b \in \{0,1\}^5$. The function $f \in \mathcal{F}(\mathcal{U})$ is a formula which can also be written as: $f \equiv h_2 \vee h_5$. The positioned word $(w^{(1)}, 1) \nvDash f$, with $w^{(1)}$ from Fig. 1a, since $h_2(w^{(1)}, 1) = h_5(w^{(1)}, 1) = 0$, i.e., there is no directed edge from a_2 or a_5. On the other hand, the positioned word $(w^{(1)}, 2) \vDash f$, since there exists a directed edge from b_2 (and also b_5).

We formally define the notion of a logical automaton for the liveness problem.

Definition 5. *A logical* automaton $\mathcal{A}$ *over a set $\mathcal{F}$ of formulas is a tuple $\mathcal{A} = (Q, \Sigma, \delta, q_s, Q_F, Q_R, \kappa)$ such that*

- *Q is a (finite) set of states,*
- *Σ is the input alphabet,*
- *$\delta : (Q \setminus (Q_F \cup Q_R)) \times (\Sigma \cup \{\vdash, \dashv\}) \to Q \times \{-1, 1\}$ is the transition function which has the following properties:*
  - *$\delta(q, \vdash) \in Q \times \{1\}$ for any $q \in Q \setminus (Q_F \cup Q_R)$,*
  - *$\delta(q, \dashv) \in Q \times \{-1\}$ for any $q \in Q \setminus (Q_F \cup Q_R)$,*
- *$q_s \in Q \setminus (Q_F \cup Q_R)$ is the initial state,*
- *$Q_F \subseteq Q$ is the set of accepting states,*
- *$Q_R \subseteq Q$ is the set of rejecting states such that $Q_F \cap Q_R = \emptyset$,*
- *$\kappa : Q \to \mathcal{F}$ is a function assigning formulas to the states.*

A configuration of $\mathcal{A}$ is a tuple $(q, w, k) \in Q \times \Sigma^* \times \mathbb{N}$, where q is the current state of $\mathcal{A}$, w is the word on the input tape, and $k \in \mathbb{N}$ is the current position of the reading head on the input tape.

The set $\mathcal{C}$ of valid configurations of $\mathcal{A}$ is the set $\mathcal{C} = \{(q, w, k) \in Q \times \Sigma^* \times \mathbb{N} \mid (w,k) \vDash \kappa(q)\}$, i.e., the set of all configurations (q, w, k) for which the positioned word (w,k) satisfies the formula $\kappa(q)$ assigned to the state q.

A step $\vdash_{\mathcal{A}}$ of $\mathcal{A}$ is a relation on the configurations of $\mathcal{A}$ such that $(q, w, k) \vdash_{\mathcal{A}} (q', w', k')$ if and only if $w = w'$ and $\delta(q, w_k) = (q', k' - k)$.

A computation of $\mathcal{A}$ is a sequence of configurations in which all pairs of consecutive configurations are contained in the relation $\vdash_{\mathcal{A}}$, i.e., the latter configuration can be reached in one step of $\mathcal{A}$ from the former one. The computation of $\mathcal{A}$ on an input word $w \in \Sigma^*$ is a computation that starts in the configuration $(q_s, w, 0)$, i.e., on the symbol $\vdash$, and finishes in a configuration (r, w, k), for some $r \in Q_F \cup Q_R$ and $k \in \mathbb{N}$.

A logical automaton must satisfy the following properties:

(i) the initial state q_s is assigned the formula $\kappa(q_s) \equiv \top$ (i.e., the tautology),
(ii) for any $q \in Q_F$, $(w, k) \vDash \kappa(q)$ implies $w \in B_n$,
(iii) for any $q \in Q_R$, $(w, k) \vDash \kappa(q)$ implies $w \notin B_n$,
(iv) for any $c_1 \in \mathcal{C}$, if $c_1 \vdash_{\mathcal{A}} c_2$, then $c_2 \in \mathcal{C}$,
(v) the computation started from any valid configuration $c \in \mathcal{C}$ is finite.

The condition (i) reflects the fact that initially the automaton does not know anything about the input word. The condition (ii) ensures that the formula assigned to an accepting state is only satisfied by words in the language B_n, and the condition (iii) ensures that the formula assigned to a rejecting state is only satisfied by words not in the language B_n, so that B_n is the language accepted by $\mathcal{A}$. The condition (iv) makes the set $\mathcal{C}$ of valid configurations closed under $\vdash_{\mathcal{A}}$, i.e., no invalid configurations can be reached from a valid configuration. In particular, this ensures that all configurations reached in a computation of $\mathcal{A}$ on any word are valid. The condition (iv) can also be phrased as follows: whenever $\delta(q, a) = (q', d)$ for any $q \in Q \setminus (Q_F \cup Q_R), q' \in Q, a \in (\Sigma \cup \{\vdash, \dashv\}), d \in \{-1, 1\}$, then for all words $w \in \Sigma^*$ and positions $k \in \mathbb{N}$, $(w, k) \vDash \kappa(q)$ and $w_k = a$ imply $(w, k + d) \vDash \kappa(q')$. Finally, the condition (v) ensures that the computation on any input word is finite.

In the next definition, we also restrict the set $\mathcal{U}$ of available atomic predicates.

Definition 6. A logical automaton $\mathcal{A} = (Q, \Sigma_n, \delta, q_s, Q_F, Q_R, \kappa)$ over a set $\mathcal{F}$ of all formulas over atomic predicates in a set $\mathcal{U}$ is called reachability-logical if the set $\mathcal{U}$ consists only of the following types of atomic predicates:

- $p_t : \Sigma_n^* \times \mathbb{N} \to \{0, 1\}$, where $t \in \mathbb{N}$ is a fixed position on the input word. The atomic predicate p_t is defined as follows: $p_t(w, k) = 1$ if and only if $k = t$ (i.e., if the current position equals t),
- $d_t : \Sigma_n^* \times \mathbb{N} \to \{0, 1\}$, where $t \in \mathbb{N}$. The atomic predicate d_t is defined as follows: $d_t(w, k) = 1$ if and only if $k = |w| + 1 - t$, i.e., the current symbol is at offset t from the end of the input word,
- $r_{i,j} : \Sigma_n^* \times \mathbb{N} \to \{0, 1\}$, where $u_{i,j}$ denotes the i-th vertex in the j-th column. The atomic predicate $r_{i,j}$ is defined as follows: $r_{i,j}(w, k) = 1$ if and only if w contains the vertex $u_{i,j}$, i.e., $i \in [n], j \in [|w| + 1]$, and $u_{i,j}$ is reachable from a vertex in the leftmost column via a directed path in w,
- $s_{i,j} : \Sigma_n^* \times \mathbb{N} \to \{0, 1\}$, where $u_{i,j}$ denotes the i-th vertex in the j-th column. The atomic predicate $s_{i,j}$ is defined as follows: $s_{i,j}(w, k) = 1$ if and only if w contains the vertex $u_{i,j}$, i.e., $i \in [n], j \in [|w| + 1]$, and a vertex in the rightmost column is reachable from the vertex $u_{i,j}$ via a directed path in w.

The following example shows that the atomic predicates from Definition 6 allow to express the reachability of some vertex at the left part of the *current* symbol from the left side.

Example 3. Let n be an arbitrary fixed positive integer. Then the formula

$$f \equiv \forall j \geq 1. \left(p_j \Rightarrow \bigvee_{i=1}^{n} r_{i,j} \right)$$

is satisfied for a positioned word (w, k), $k \in [\|w\|]$, if and only if some vertex in the k-th column is reachable from the left side, i.e., if and only if some vertex at the left part of the current symbol is reachable from the left side.

The positioned word $(w^{(1)}, 3)$, with $w^{(1)}$ from Fig. 1a, satisfies $(w^{(1)}, 3) \vDash f$, since $p_j(w^{(1)}, 3) = 1 \iff j = 3$ and $r_{i,3}(w^{(1)}, 3) = 1 \iff i = 3 \vee i = 5$. The positioned word $(w^{(1)}, 4) \nvDash f$, since $p_j(w^{(1)}, 4) = 1 \iff j = 4$, but $r_{i,4}(w^{(1)}, 4) = 0$, for all $i \in [5]$.

1.4   Related Results

Up to now, the only exponential lower bound for reasonable automata over all propositional formulas on a restricted set of atomic predicates is contained in Theorem 7 in [2]. It attempts to show an exponential lower bound on the size of reasonable automata over all propositional formulas on the set of available atomic predicates restricted to $r_{i,j}$. We noticed that the proof of the theorem in [2] is wrong, and provide an explanation with a counterexample in the appendix. Nevertheless, the lower bound claimed by the theorem holds, and can be proved by adapting the technique from our paper to the model of reasonable automata.

2   Main Results

2.1   General Approach

We prove a theorem which provides lower bounds on the number of states of logical automata. To apply the theorem, one has to find two sequences $w^{(i)}$, $z^{(i)}$ of t words each, and a fixed position k which form a *witness* for a logical automaton.

Definition 7. *Let* $\mathcal{A} = (Q, \Sigma_n, \delta, q_s, Q_F, Q_R, \kappa)$ *be a logical automaton over a set* $\mathcal{F}$ *of formulas. Let* $\mathcal{W} = \{w^{(1)}, \ldots, w^{(t)}\} \subseteq \Sigma_n^*$ *and* $\mathcal{Z} = \{z^{(1)}, \ldots, z^{(t)}\} \subseteq \Sigma_n^*$ *be two sequences of* t *words each. Let a position* k *be fixed. Suppose that*

(a) $w^{(i)} \notin B_n$ *holds for any word* $w^{(i)} \in \mathcal{W}$,
(b) whenever

$$(p, w^{(i)}, k') \vdash_{\mathcal{A}} (q, w^{(i)}, k''),$$

for some states $p, q \in Q$, *some word* $w^{(i)} \in \mathcal{W}$, *and some positions* $k', k'' \in \mathbb{N}$ *such that* $(z^{(i)}, k') \vDash \kappa(p)$ *and* $(z^{(i)}, k'') \nvDash \kappa(q)$, *then* $k' = k$, *and for all* $i < j \leq t$, $(z^{(j)}, k') \vDash \kappa(p)$ *implies* $(z^{(j)}, k'') \vDash \kappa(q)$, *(c)* $(z^{(i)}, k') \not\models \kappa(q_r)$ *holds for all* $q_r \in Q_R, z^{(i)} \in \mathcal{Z}, k' \in \mathbb{N}$.

Then $(\mathcal{W}, \mathcal{Z}, k)$ *is a* witness *for* $\mathcal{A}$.

Informally, $(\mathcal{W}, \mathcal{Z}, k)$ is a witness for a logical automaton $\mathcal{A}$ if none of the words $w^{(i)} \in \mathcal{W}$ is in the one-way liveness (a) and no formula assigned to a rejecting state of $\mathcal{A}$ is satisfied under any word $z^{(i)} \in \mathcal{Z}$ and any position $k' \in \mathbb{N}$ (c). Furthermore, it must hold that, whenever $\mathcal{A}$ learns in one step on the i-th word $w^{(i)}$ that $z^{(i)}$ and the current position do not satisfy the formula assigned to the current state, then the current position is k and $\mathcal{A}$ may not learn that $z^{(j)}$ and the current position do not satisfy the formula assigned to the current state, for any $j > i$, in the same step of computation (b).

Theorem 1. *Let* $\mathcal{A} = (Q, \Sigma_n, \delta, q_s, Q_F, Q_R, \kappa)$ *be any logical automaton over a set* $\mathcal{F}$ *of formulas. Let* $\mathcal{W} = \{w^{(1)}, w^{(2)}, \ldots, w^{(t)}\} \subseteq \Sigma_n^*$ *be a sequence of* t *words and let* $\mathcal{Z} = \{z^{(1)}, z^{(2)}, \ldots, z^{(t)}\} \subseteq \Sigma_n^*$ *be a sequence of* t *words. Let* $k \in \mathbb{N}$ *be a fixed position. Suppose that* $(\mathcal{W}, \mathcal{Z}, k)$ *is a witness for* $\mathcal{A}$. *Then* $\mathcal{A}$ *must have at least* $\sqrt{t}/2$ *states.*

Proof. Note that any word $w^{(i)} \in \mathcal{W}$ should be rejected by (a) of Definition 7. Since for all $q_r \in Q_R, z^{(i)} \in \mathcal{Z}, k' \in \mathbb{N}$, $(z^{(i)}, k') \models \kappa(q_s)$ holds by (i) of Definition 5, and $(z^{(i)}, k') \not\models \kappa(q_r)$ holds by (c) of Definition 7, there exists a step in the computation of $\mathcal{A}$ on any $w^{(i)} \in \mathcal{W}$ such that

$$(q_s, w^{(i)}, 0) \vdash_{\mathcal{A}}^* (P(w^{(i)}), w^{(i)}, k_1^{(i)}) \vdash_{\mathcal{A}} (S(w^{(i)}), w^{(i)}, k_2^{(i)}), \tag{1}$$

for some positions $k_1^{(i)}, k_2^{(i)} \in \mathbb{N}$, where P, S are functions mapping the words $w^{(i)} \in \mathcal{W}$ to the corresponding states, $(z^{(i)}, k_1^{(i)}) \models \kappa(P(w^{(i)}))$ and $(z^{(i)}, k_2^{(i)}) \not\models \kappa(S(w^{(i)}))$.

Let us now distinguish the following two cases with respect to the size of the image of P:

(a) the size of the image $|P(\mathcal{W})| \geq \sqrt{t}$, i.e., there are at least $\sqrt{t}$ distinct states which are assigned to the words in $\mathcal{W}$ by the function P,

(b) the size of the image $|P(\mathcal{W})| < \sqrt{t}$, i.e., in total less than $\sqrt{t}$ distinct states are assigned to the words in $\mathcal{W}$ by the function P.

Case (a)
Since $P(\mathcal{W}) \subseteq Q$, we conclude

$$|Q| \geq |P(\mathcal{W})| \geq \sqrt{t} \geq \frac{\sqrt{t}}{2}.$$

Case (b)
Refer to the diagram in Fig. 2 for an overview of this case. Since $|\mathcal{W}| = t$, it is easy to see that there exists some state $p \in P(\mathcal{W})$ in the image of P whose preimage under P is of size at least $\sqrt{t}$ (otherwise the size of the domain of $P : \mathcal{W} \to Q$ would be strictly less than $|P(\mathcal{W})| \cdot \sqrt{t} < \sqrt{t} \cdot \sqrt{t} = t = |\mathcal{W}|$). Let $\mathcal{W}' \subseteq \mathcal{W}$ be the largest subset of all words $w^{(i)} \in \mathcal{W}$ such that $P(w^{(i)}) = p$ and $k_2^{(i)}$ is constant in (1). Since $k_1^{(i)} = k$ is constant by (b) of Definition 7 and $|k_2^{(i)} - k_1^{(i)}| \in \{-1, 1\}$, it is easy to see that $|\mathcal{W}'| \geq \frac{\sqrt{t}}{2}$.

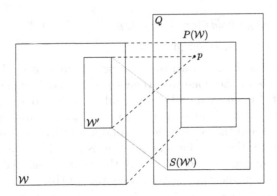

Fig. 2. The diagram of case (b) in the proof of Theorem 1. The dashed lines represent the mapping by the function P, whereas the dotted lines represent the mapping by the function S.

Lemma 1. *The function $S|_{\mathcal{W}'} : \mathcal{W}' \to Q$ restricted to the set $\mathcal{W}'$ is injective.*

Proof. For the sake of contradiction, assume that $S(w^{(i)}) = S(w^{(j)})$ for some $w^{(i)}, w^{(j)} \in \mathcal{W}'$, $i \neq j$. Without loss of generality, assume that $i < j$.

By (1), we have $(z^{(i)}, k_1^{(i)}) \models \kappa(P(w^{(i)}))$ and $(z^{(i)}, k_2^{(i)}) \not\models \kappa(S(w^{(i)}))$. Since $P(\mathcal{W}') = \{p\}$, we have $P(w^{(i)}) = P(w^{(j)})$. By (1), $(z^{(j)}, k_1^{(j)}) \models \kappa(P(w^{(j)}))$ with $k_1^{(j)} = k = k_1^{(i)}$, and thus $(z^{(j)}, k_1^{(i)}) \models \kappa(P(w^{(i)}))$.

By (b) of Definition 7, $(z^{(j)}, k_1^{(i)}) \models \kappa(P(w^{(i)}))$ and $i < j$ imply $(z^{(j)}, k_2^{(i)}) \models \kappa(S(w^{(i)}))$.

Since $S(w^{(i)}) = S(w^{(j)})$, we get $(z^{(j)}, k_2^{(i)}) \models \kappa(S(w^{(j)}))$, and with $k_2^{(i)} = k_2^{(j)}$ $(z^{(j)}, k_2^{(j)}) \models \kappa(S(w^{(j)}))$, which is a contradiction to $(z^{(j)}, k_2^{(j)}) \not\models \kappa(S(w^{(j)}))$ by (1).

Hence, the function $S|_{\mathcal{W}'} : \mathcal{W}' \to Q$ restricted to the set $\mathcal{W}'$ is injective. $\square$

From Lemma 1, we immediately get

$$|Q| \geq |\mathcal{W}'| \geq \frac{\sqrt{t}}{2},$$

which concludes the case (b), and thus also the proof of Theorem 1. $\square$

2.2   Lower Bounds

We are going to use Theorem 1 to prove a lower bound on the number of states of reachability-logical automata. To this end, we define a witness $(\mathcal{W}, \mathcal{Z}, k)$ for an arbitrary reachability-logical automaton. Let us point out that the witness consists solely of words of length 3.

In the following, let A, B, C, D denote the sets of vertices in the first, second, third, and fourth column. For any $\emptyset \subsetneq R \subsetneq [n]$, let $w^{(R)} \in \Sigma_n^3$ be a graph which contains exactly the edges (a_1, b_i) for all $i \in R$, $(b_i, c_j), (b_j, c_i)$ for all

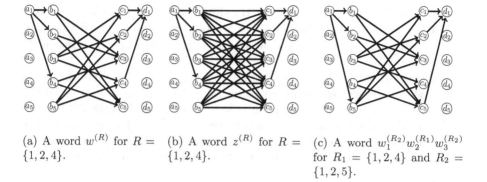

(a) A word $w^{(R)}$ for $R = \{1, 2, 4\}$.

(b) A word $z^{(R)}$ for $R = \{1, 2, 4\}$.

(c) A word $w_1^{(R_2)} w_2^{(R_1)} w_3^{(R_2)}$ for $R_1 = \{1, 2, 4\}$ and $R_2 = \{1, 2, 5\}$.

Fig. 3. Sample words from the witness $(\mathcal{W}, \mathcal{Z}, k)$.

$i \in R, j \in [n] \setminus R$, and (c_i, d_1) for all $i \in R$. An example of a word $w^{(R)}$ is shown in Fig. 3a.

For any $\emptyset \subsetneq R \subsetneq [n]$, let $z^{(R)} \in \Sigma_n^3$ be a graph which contains exactly the edges (a_1, b_i) for all $i \in R$, (b_i, c_j) for all $i, j \in [n]$, and (c_i, d_1) for all $i \in R$. An example of a word $z^{(R)}$ is shown in Fig. 3b.

Let $R_1, R_2, \ldots, R_m$ be the sequence of all subsets $\emptyset \subsetneq R \subsetneq [n], |R| > \frac{n}{2}$, ordered such that $|R_i| \leq |R_j|$ for all $1 \leq i \leq j \leq m$. Here $m \in \mathbb{N}$ denotes the number of all such subsets.

Let $\mathcal{W}, \mathcal{Z}$ be two sequences of m words each which are defined as follows:

$$\mathcal{W} := \{w^{(R_1)}, w^{(R_2)}, \ldots, w^{(R_m)}\},$$
$$\mathcal{Z} := \{z^{(R_1)}, z^{(R_2)}, \ldots, z^{(R_m)}\}.$$

Let the position $k = 2$ be fixed. We prove some observations we will use to establish that $(\mathcal{W}, \mathcal{Z}, k)$ as defined above is a witness for an arbitrary reachability-logical automaton.

Let us consider words of the form $w_1^{(R_2)} w_2^{(R_1)} w_3^{(R_2)}$ with a sample word shown in Fig. 3c.

Lemma 2. *Let $w = w_1^{(R_2)} w_2^{(R_1)} w_3^{(R_2)}$ for some $\emptyset \subsetneq R_1, R_2 \subsetneq [n]$ with $R_1 \neq R_2$, $\frac{n}{2} < |R_1| \leq |R_2|$. Then $(w, k'), (z^{(R_2)}, k')$ evaluate to the same truth values of all atomic predicates in Definition 6, for any $k' \in \mathbb{N}$.*

Proof. Since $w_1 = w_1^{(R_2)}$, exactly the vertices with indices in R_2 are reachable from the left side among the vertices in the second column. Since $R_1 \neq R_2$, $|R_1| \leq |R_2|$, it is easy to see that $R_2 \not\subseteq R_1$, and thus there exists some index $i \in R_2, i \notin R_1$ so that all the vertices with indices in R_1 are reachable from the left side among the vertices in the third column (through the vertex b_i). Since $\frac{n}{2} < |R_1|, |R_2|$, it is easy to see that $R_2 \cap R_1 \neq \emptyset$, and thus there exists some index $i' \in R_2, i' \in R_1$ so that all the vertices with indices in $[n] \setminus R_1$ are reachable from the left side among the vertices in the third column (through $b_{i'}$). It follows that all the vertices in the third column are reachable from the left side, and since $R_2 \neq \emptyset$ and $w_3 = w_3^{(R_2)}$, the right side (exactly the first vertex) is reachable from the left side.

Analogously, we can show that the right side is reachable exactly from the vertices with indices in R_2 among the vertices in the third column, from all the vertices in the second column, and from the first vertex in the first column.

Finally, the current position is k' in both (w, k'), $(z^{(R_2)}, k')$, which means that p_t and d_t evaluate to the same truth values. Hence, (w, k'), $(z^{(R_2)}, k')$ evaluate to the same truth values of all atomic predicates in Definition 6, for any $k' \in \mathbb{N}$.

$\square$

Lemma 3. Let $\mathcal{A} = (Q, \Sigma_n, \delta, q_s, Q_F, Q_R, \kappa)$ be any reachability-logical automaton. Let $\emptyset \subsetneq R \subsetneq [n]$ with $\frac{n}{2} < |R|$. Suppose that

$$(p, w^{(R)}, k') \vdash_{\mathcal{A}} (q, w^{(R)}, k''),$$

for some states $p, q \in Q$, and positions $k', k'' \in \{0, 1, 2, 3, 4\}$, where $(z^{(R)}, k'') \vDash \kappa(p)$ and $(z^{(R)}, k'') \nvDash \kappa(q)$. Then $k' = k$, and for all $\emptyset \subsetneq R' \subsetneq [n]$ with $R \neq R'$, $|R| \leq |R'|$, $(z^{(R')}, k') \vDash \kappa(p)$ implies $(z^{(R')}, k'') \vDash \kappa(q)$.

Proof. We first prove that $k' = 2 = k$. For the sake of contradiction, assume that $k' \neq 2$. Since $w^{(R)}, z^{(R)}$ only differ at the second symbol, we get $(p, z^{(R)}, k') \vdash_{\mathcal{A}} (q, z^{(R)}, k'')$ which is a contradiction to (iv) of Definition 5.

Let $\emptyset \subsetneq R' \subsetneq [n]$ be such that $R \neq R'$, $|R| \leq |R'|$, and $(z^{(R')}, k') \vDash \kappa(p)$. We show that $(z^{(R')}, k'') \vDash \kappa(q)$.

Consider the word $w' := w_1^{(R')} w_2^{(R)} w_3^{(R')}$. Since $(z^{(R')}, k') \vDash \kappa(p)$, and both (w', k'), $(z^{(R')}, k')$ evaluate to the same truth values of all atomic predicates in Definition 6 (see Lemma 2), we get that $(w', k') \vDash \kappa(p)$. Since $k' = 2$, it holds that $w_{k'}' = w_2' = w_2^{(R)} = w_{k'}^{(R)}$, and thus $(p, w', k') \vdash_{\mathcal{A}} (q, w', k'')$, from which we get that $(w', k'') \vDash \kappa(q)$. Finally, since both (w', k''), $(z^{(R')}, k'')$ evaluate to the same truth values of all atomic predicates in Definition 6 (see Lemma 2), we get that $(z^{(R')}, k'') \vDash \kappa(q)$.

$\square$

Now, we can state that $(\mathcal{W}, \mathcal{Z}, k)$ is a witness for an arbitrary reachability-logical automaton.

Lemma 4. $(\mathcal{W}, \mathcal{Z}, k)$ as defined above form a witness for an arbitrary reachability-logical automaton.

Proof. It is easy to see that any $w^{(R_i)} \notin B_n$, i.e., (a) of Definition 7 holds. Lemma 3 then implies (b) of Definition 7. Finally, $z^{(R_i)} \in B_n$ and (iii) of Definition 5 imply (c) of Definition 7.

$\square$

The lower bound using the witness $(\mathcal{W}, \mathcal{Z}, k)$ is contained in the next theorem.

Theorem 2. Let $\mathcal{A} = (Q, \Sigma_n, \delta, q_s, Q_F, Q_R, \kappa)$ be any reachability-logical automaton. Then $\mathcal{A}$ must have at least $\Omega(2^{\frac{n}{2}})$ states.

Proof. Recall that m denotes the number of all subsets $\emptyset \subsetneq R \subsetneq [n], |R| > \frac{n}{2}$. It is easy to show that $m \in \Omega(2^n)$. By Lemma 4, $(\mathcal{W}, \mathcal{Z}, k)$ is a witness for $\mathcal{A}$. Hence, Theorem 1 allows us to conclude that $\mathcal{A}$ must have at least $\frac{\sqrt{m}}{2} \in \Omega(2^{\frac{n}{2}})$ states. $\square$

2.3 Upper Bounds

We complement the exponential lower bound with an (also exponential) upper bound. In fact, the following theorem shows that the straightforward 1DFA for one-way liveness, which stores the subset of vertices of the current symbol which are reachable from the left side into the state, can be represented as a reachability-logical automaton.

Theorem 3. *There exists a reachability-logical automaton $\mathcal{A}$ with $2^n + 3$ states.*

Proof. Informally, the reachability-logical automaton $\mathcal{A}$ keeps track of the set of vertices at the left part of the current symbol which are reachable from the left side. Then upon reaching the right delimiter $\dashv$ after a single pass through the input word, $\mathcal{A}$ checks whether the set of reachable vertices is empty, or not.

More formally, let $\mathcal{A} = (Q, \Sigma_n, \delta, q_s, Q_F, Q_R, \kappa)$ be defined as follows. The set of states is $Q = \{q_s, q_f, q_r\} \cup \bigcup_{\emptyset \subseteq R \subseteq [n]} q_R$, the set of accepting states is $Q_F = \{q_f\}$, and the set of rejecting states is $Q_R = \{q_r\}$.

The formula assigned to a state q_R, $\emptyset \subseteq R \subseteq [n]$, is

$$\kappa(q_R) \equiv \forall j \geq 1. \left(p_j \Rightarrow \bigwedge_{i=1}^{n} l_{i,j} \right),$$

where $l_{i,j} \equiv r_{i,j}$ if $i \in R$, and $l_{i,j} \equiv \neg r_{i,j}$ if $i \notin R$. It expresses that exactly the vertices with indices in R are reachable among the vertices at the left part of the current symbol. This (relative) property is expressed by quantifying over the (absolute) position $j \in \{1, 2, \dots\}$.

The formula assigned to the starting state q_s is $\kappa(q_s) \equiv \top$, i.e., the tautology. The formulas assigned to the accepting and rejecting states are $\kappa(q_f) \equiv d_1 \wedge \forall j \geq 1. (p_j \Rightarrow \bigvee_{i=1}^{n} r_{i,j+1})$ and $\kappa(q_r) \equiv d_1 \wedge \forall j \geq 1. (p_j \Rightarrow \bigwedge_{i=1}^{n} \neg r_{i,j+1})$. They express that the current position is the last symbol of the word (i.e., the position just before the right delimiter $\dashv$), and that some (none, respectively) vertex at the right part of the current (last) symbol is reachable from the left side. Again, the (relative) property is expressed by quantifying over the (absolute) position $j \in \{1, 2, \dots\}$.

The transition function δ from the starting state q_s is defined as follows: $\delta(q_s, \vdash) = (q_{[n]}, 1)$, and $\delta(q_s, a) = (q_s, -1)$, for all $a \neq \vdash$.

The transition function δ from a state q_R, $\emptyset \subseteq R \subseteq [n]$, is defined as follows: $\delta(q_R, \vdash) = (q_{[n]}, 1)$. For an arbitrary symbol $a \in \Sigma_n$, let $R' \subseteq [n]$ be the set of indices of vertices at the right part of a which are reachable from the vertices at the left part of a with indices in R via directed arcs on a. Then $\delta(q_R, a) = (q_{R'}, 1)$. Finally, $\delta(q_\emptyset, \dashv) = (q_r, -1)$, and $\delta(q_R, \dashv) = (q_f, -1)$, for any $R \neq \emptyset$.

Let us remark that we define the (complete) transition function δ so that the property (v) of Definition 5 holds for $\mathcal{A}$. In particular, valid configurations (q_s, w, k), for $k \geq 1$, which do not occur in any computation of $\mathcal{A}$ on a word, lead to the initial configuration $(q_s, w, 0)$, from which a finite computation follows. $\square$

Finally, let us show the actual computation of the reachability-logical automaton $\mathcal{A}$ from Theorem 3 on the words $w^{(R)}$, $z^{(R)}$, as in Fig. 3:

$$(q_s, w^{(R)}, 0) \vdash_{\mathcal{A}} (q_{[n]}, w^{(R)}, 1) \vdash_{\mathcal{A}} (q_R, w^{(R)}, 2)$$
$$\vdash_{\mathcal{A}} (q_{[n] \setminus R}, w^{(R)}, 3) \vdash_{\mathcal{A}} (q_{\emptyset}, w^{(R)}, 4) \vdash_{\mathcal{A}} (q_r, w^{(R)}, 3),$$
$$(q_s, z^{(R)}, 0) \vdash_{\mathcal{A}} (q_{[n]}, z^{(R)}, 1) \vdash_{\mathcal{A}} (q_R, z^{(R)}, 2)$$
$$\vdash_{\mathcal{A}} (q_{[n]}, z^{(R)}, 3) \vdash_{\mathcal{A}} (q_{\{1\}}, z^{(R)}, 4) \vdash_{\mathcal{A}} (q_f, z^{(R)}, 3).$$

Acknowledgements. The author would like to thank Hans-Joachim Böckenhauer, Juraj Hromkovič, and the referees for their valuable comments and suggestions.

A   Wrong Proof in Previous Work

In this section, we show that the proof of Theorem 7 in [2] is wrong. Following [2], we define 1-LIV$_2$ to denote the family of languages B_n restricted to words of length 2. Let us first restate the theorem.

Theorem 7 (Bianchi et al. [2]). *Consider propositional variables $r(a)$ ($\neg r(a)$, resp.) with the interpretation that the vertex a is reachable (non reachable, resp.) from the left side, and let $\mathcal{F}$ be the set of all propositional formulæ on such variables. Then, any reasonable automaton over $\mathcal{F}$ solving the n-th language from 1-LIV$_2$ must have $\Omega(2^{\frac{n}{2}})$ states.*

Analogously to Sect. 2.2, let $A = \{a_1, \ldots, a_n\}$, $B = \{b_1, \ldots, b_n\}$, and $C = \{c_1, \ldots, c_n\}$ be the vertices in the first, second, and third column, respectively.

The proof in [2] proceeds as follows. For any $\emptyset \subsetneq R \subsetneq [n]$, let $w^{(R)} \in \Sigma_n^2$ be a graph which contains exactly the edges (a_1, b_i) for all $i \in R$, and (b_j, c_1) for all $j \in [n] \setminus R$. Hence, any word $w^{(R)}$ should be rejected by a valid reasonable automaton. For any set R, let f_R be the formula which is defined as follows: $f_R \equiv \bigwedge_{i \in R} r(b_i) \wedge \bigwedge_{j \in [n] \setminus R} \neg r(b_j)$. Let L be the set of all $2^n - 2$ words $w^{(R)}$.

To arrive at a contradiction, let $\mathcal{A}$ be a reasonable automaton over $\mathcal{F}$ solving the n-th language from 1-LIV$_2$, which is in the normal form defined in [2], with $m < 2^{\frac{n}{2}}$ states. Then there are at most $m^2 < 2^n - 2$ pairs of states, and thus there exists a pair of words $w^{(R_1)}, w^{(R_2)} \in L$, such that the pair of states preceding the (rejecting) final state in the computation of $\mathcal{A}$ on $w^{(R_1)}, w^{(R_2)}$ is identical on both of them. Let q_{pp}, q_p be these two states in the order in which they appear in the computation of $\mathcal{A}$ on $w^{(R_1)}, w^{(R_2)}$. Since $\mathcal{A}$ is in the normal form, the formula $\kappa(q_{pp})$ together with the information carried by either $w^{(R_1)}_{\tau(q_{pp})}$ or $w^{(R_2)}_{\tau(q_{pp})}$ cannot imply either $r(c_1)$ or $\neg r(c_1)$.

The next step in the proof is the conclusion that the formula $\kappa(q_{pp})$ must hold for all words satisfying any of the following four formulas: $f_{R_1} \wedge r(c_1)$, $f_{R_1} \wedge \neg r(c_1)$, $f_{R_2} \wedge r(c_1)$, and $f_{R_2} \wedge \neg r(c_1)$. This conclusion is wrong as demonstrated by the formula $\kappa(q_{pp}) \equiv (f_{R_1} \Rightarrow \bigwedge_{c \in C} \neg r(c)) \wedge (f_{R_2} \Rightarrow \bigwedge_{c \in C} \neg r(c))$. If $\tau(q_{pp}) = 2$, then the formula $\kappa(q_{pp})$ together with the information carried by either $w_2^{(R_1)}$ or $w_2^{(R_2)}$ does not imply either $r(c_1)$ or $\neg r(c_1)$ (observe that the first symbol of the current word could potentially contain no edges, or contain edges to all vertices in the second column, and still satisfy the formula $\kappa(q_{pp})$ which is a conjunction of two *implications*). However, the formula $\kappa(q_{pp})$ contradicts the formula $f_{R_1} \wedge r(c_1)$, and thus it does not hold for any word satisfying the formula $f_{R_1} \wedge r(c_1)$.

Another minor issue in the reasoning of the proof is that the word $w_1^{(R_1)} w_2^{(R_2)}$ is in the n-th language from 1-LIV$_2$ only if $R_1 \not\subseteq R_2$, which can be nevertheless assumed without loss of generality.

In the following, we construct a concrete counterexample to the (wrong) step in the proof in [2]. Let $\mathcal{A}' = (\Sigma_n, Q', 2, q'_s, Q'_F, Q'_R, \delta', \tau', \kappa')$ be an arbitrary reasonable automaton over $\mathcal{F}$ solving the n-th language from 1-LIV$_2$, which is in the normal form. We construct a *valid* reasonable automaton $\mathcal{A}$ over $\mathcal{F}$ solving the n-th language from 1-LIV$_2$, which is in the normal form, and still contradicts the reasoning in the proof of Theorem 7 in [2].

Let us fix two words $w^{(R_1)}, w^{(R_2)} \in L$, $R_1 \neq R_2$, such that $R_1 \not\subseteq R_2$. In the following, let us abbreviate $w^{(R_1)} \equiv x$, $w^{(R_2)} \equiv y$.

Let us define the reasonable automaton $\mathcal{A} = (\Sigma_n, Q, 2, q_s, Q_F, Q_R, \delta, \tau, \kappa)$ as follows. The set of states $Q = \{q_s, q_x, q_y, q_{pp}, q_p, q_f, q_r\} \cup Q'$. The starting state is $q_s \in Q$. The set of accepting states is $Q_F = \{q_f\} \cup Q'_F$, and the set of rejecting states is $Q_R = \{q_r\} \cup Q'_R$. The focus of a state $q \in Q$ is as follows: $\tau(q_s) = \tau(q_p) = 1$, $\tau(q_x) = \tau(q_y) = \tau(q_{pp}) = 2$, and $\tau(q') = \tau'(q')$ for any $q' \in Q' \setminus (Q'_F \cup Q'_R)$.

Let us define $f_x \equiv f_{R_1}$, and $f_y \equiv f_{R_2}$. The formula assigned to a state $q \in Q$ is as follows: $\kappa(q_s) = \top$, $\kappa(q_x) = f_x \vee \bigwedge_{b \in B} r(b)$, $\kappa(q_y) = f_y \vee \bigwedge_{b \in B} r(b)$, $\kappa(q_{pp}) = \kappa(q_p) = (f_x \Rightarrow \bigwedge_{c \in C} \neg r(c)) \wedge (f_y \Rightarrow \bigwedge_{c \in C} \neg r(c))$, $\kappa(q_f) = \bigvee_{c \in C} r(c)$, $\kappa(q_r) = \bigwedge_{c \in C} \neg r(c)$, and $\kappa(q') = \kappa'(q')$ for any $q' \in Q'$.

The transition function is as follows: $\delta(q_s, x_1) = q_x$, $\delta(q_s, y_1) = q_y$, $\delta(q_x, x_2) = \delta(q_y, y_2) = q_{pp}$, $\delta(q_{pp}, x_2) = \delta(q_{pp}, y_2) = q_p$, $\delta(q_p, x_1) = \delta(q_p, y_1) = q_r$, and $\delta(q, a) \in \{q_f, q_r, q'_s\}$ for any $q \in \{q_s, q_x, q_y, q_{pp}, q_p\}$ and $a \in \Sigma_n$ such that $\delta(q, a)$ has not been defined previously. Note that $\delta(q, a) \in \{q_f, q_r\}$ whenever possible, so that $\mathcal{A}$ is in the normal form. Finally, $\delta(q', a) = \delta'(q', a)$ for any $q' \in Q' \setminus (Q'_F \cup Q'_R)$ and $a \in \Sigma_n$.

The computation of $\mathcal{A}$ on the words x, y looks as follows:

$$(x, q_s) \vdash_{\mathcal{A}} (x, q_x) \vdash_{\mathcal{A}} (x, q_{pp}) \vdash_{\mathcal{A}} (x, q_p) \vdash_{\mathcal{A}} (x, q_r),$$
$$(y, q_s) \vdash_{\mathcal{A}} (y, q_y) \vdash_{\mathcal{A}} (y, q_{pp}) \vdash_{\mathcal{A}} (y, q_p) \vdash_{\mathcal{A}} (y, q_r).$$

Hence, the computation of $\mathcal{A}$ on x, y has the same pair of states preceding the (rejecting) final state q_r. According to the proof of Theorem 7 in [2], the formula $\kappa(q_{pp})$ must hold for all words satisfying $f_x \wedge r(c_1)$, which is not the case as we see from the definition of $\kappa(q_{pp})$.

It only remains to show that $\mathcal{A}$ is a valid reasonable automaton in the normal form. Since $\mathcal{A}'$ is assumed to be a valid reasonable automaton in the normal form, it suffices to only check the transitions from the states $q \in \{q_s, q_x, q_y, q_{pp}, q_p\}$. We omit the details.

References

1. Berman, P., Lingas, A.: On complexity of regular languages in terms of finite automata. Technical report 304, Institute of Computer Science, Polish Academy of Sciences (1977)
2. Bianchi, M.P., Hromkovič, J., Kováč, I.: On the size of two-way reasonable automata for the liveness problem. In: Potapov, I. (ed.) DLT 2015. LNCS, vol. 9168, pp. 120–131. Springer, Cham (2015). https://doi.org/10.1007/978-3-319-21500-6_9
3. Hromkovič, J., Královič, R., Královič, R., Štefanec, R.: Determinism vs. nondeterminism for two-way automata. In: Yen, H.-C., Ibarra, O.H. (eds.) DLT 2012. LNCS, vol. 7410, pp. 24–39. Springer, Heidelberg (2012). https://doi.org/10.1007/978-3-642-31653-1_4
4. Hromkovič, J., Schnitger, G.: Nondeterminism versus determinism for two-way finite automata: generalizations of Sipser's separation. In: Baeten, J.C.M., Lenstra, J.K., Parrow, J., Woeginger, G.J. (eds.) ICALP 2003. LNCS, vol. 2719, pp. 439–451. Springer, Heidelberg (2003). https://doi.org/10.1007/3-540-45061-0_36
5. Sakoda, W.J., Sipser, M.: Nondeterminism and the size of two way finite automata. In: Proceedings of the Tenth Annual ACM Symposium on Theory of Computing. ACM (1978)
6. Sipser, M.: Lower bounds on the size of sweeping automata. In: Proceedings of the Eleventh Annual ACM Symposium on Theory of Computing. ACM (1979)

Bayesian Root Cause Analysis
by Separable Likelihoods

Maciej Skorski[✉]

DELL, Klosterneuburg, Austria
maciej.skorski@gmail.com

Abstract. Root Cause Analysis for anomalies is challenging because of the trade-off between the accuracy and its explanatory friendliness, required for industrial applications. In this paper we propose a framework for simple and friendly RCA within the Bayesian regime under certain restrictions (namely that Hessian at the mode is diagonal, in this work referred to as *separability*) imposed on the predictive posterior. Within this framework anomalies can be decomposed into independent dimensions which greatly simplifies readability and interpretability.

We show that the separability assumption is satisfied for important base models, including Multinomial, Dirichlet-Multinomial and Naive Bayes. To demonstrate the usefulness of the framework, we embed it into the Bayesian Net and validate on web server error logs (real world data set).

Keywords: Bayesian modeling · Anomaly detection
Root Cause Analysis

1 Introduction

1.1 Anomaly Detection and Root Cause Analysis

In the likelihood-based approaches to anomaly detection, a generative probabilistic model for data is learned and used to evaluate new data records. Records with unusually low likelihoods are marked as anomalies. A classical example is the Z-score measure, which fits a Gaussian distribution to 1-dimensional data (estimating the mean and variance) and scores observations in the decreasing order with respect to the log-likelihood; for its simplicity it is widely used in explanatory data analysis, quality controls and other industrial applications.

The challenge with real data sets, however, is that they usually contain both continuous and categorical features, as well as inter-dependencies (in particular anomaly scores cannot be applied independently across features). Interactions and dependencies can be effectively modeled by the modern framework of probabilistic graphical models [KF09]. Further, simplicity can be traded for accuracy

Full version available on arXiv https://arxiv.org/pdf/1808.04302.pdf.

© Springer Nature Switzerland AG 2019
B. Catania et al. (Eds.): SOFSEM 2019, LNCS 11376, pp. 461–472, 2019.
https://doi.org/10.1007/978-3-030-10801-4_36

by using more sophisticated models as building blocks (for example more exotic base distributions or mixtures); only for multivariate counts several models have been proposed [ZZZS17].

This paper concerns the constrained scenario of Root Cause Analysis (RCA) where in addition to identifying anomalies, a readable explanation (in terms of predefined features) is required. Because the purpose of RCA is to support business decision making, complexity and fit accuracy are often traded for explanatory abilities. This makes some powerful models (such as neural set) not adequate for this task [SMRE17]. In this paper we show how to build, out of simple building blocks, an anomaly detection system for error logs. While our model is a fairly simple variant of Bayes Network, the main added value is the proposed paradigm of *determining anomaly contributions*, which is used to estimate how different features contribute to the likelihood of the anomaly data record. These scores can be used directly to perform efficient RCA which is illustrated by a case study on real data.

1.2   Contribution

Root-Cause Analysis for Separable Posteriors. For the task of anomaly detection the main quantity of interest is the likelihood of the new data record $x = (x_1, \ldots, x_d)$ given the training data $\mathcal{D}$, called the *predictive posterior*. Assuming a generative process $\Pr(\cdot|\theta)$ for the data, with some parameter vector θ, the likelihood of a new point can be computed by one of the following formulas

$$L(x) = \Pr(x|\mathcal{D}) = \int \Pr(x|\theta) \Pr(\theta|\mathcal{D}) \mathrm{d}\theta$$

$$L(x) \approx \Pr(x|\theta^*), \quad \theta^* = \mathrm{argmax}_\theta \Pr[\mathcal{D}|\theta]$$

where the first formula is the fully Bayesian predictive posterior (parameters integrated) and the second one is the *maximum a posteriori* approximation (MAP), preferred for analytical tractability and justified when the parameter posterior $\Pr(\theta|\mathcal{D})$ is sharply peaked around its mode θ^*.

For successful RCA it would be helpful to answer the following question

Q: How individual components x_i impact the likelihood L?

Obtaining analytical estimates on these impacts is not possible in general, because posteriors are often not analytically tractable and can be only approximated by sampling. However in certain cases the predictive posterior, after subtracting its mode, factorizes into terms depending on individual terms θ_i. More precisely, suppose that the predictive posterior log-likelihood can be written as

$$\log L(x) \approx \log L(x^*) + \sum_i I(x_i). \tag{1}$$

where $x^* = \mathrm{argmax}_x \log L(x)$ is the mode. When the posterior obeys Eq. (1) we say it is *separable*. The term $I(x_i)$ can be then thought as *influence* of the i-th coordinate of the data point x. Moreover, similarly to the notion of the averaged log-likelihood, these influences can be aggregated over several independent observations x (e.g. daily). We note conceptual similarity of this decomposition technique to independent component analysis (ICA) which decompose signals into (almost) independent components [HO00].

This formula has the following intuitive meaning: we decompose *the deficiency w.r.t. the mode* per individual dimensions; the deficiency is understood as the difference in the log-likelihood with respect to the mode and can be seen as a natural anomaly measure (note that $\sum_i I(x_i) \leqslant 0$ by the definition of x^*). We stress that it is important to subtract the mode in Eq. (1), otherwise we explain the likelihood of the whole point, rather than its abnormal part.

It is worth mentioning that Eq. (1) can be characterized alternatively, by noticing that the hessian matrix H satisfies (assuming x_i are unconstrained[1])

$$H_{i,j}(x^*) = \frac{\partial^2 L}{\partial x_i \partial x_j}(x^*) = [i = j] \cdot \frac{\partial I(x_i)}{\partial x_i} \cdot \frac{\partial I(x_j)}{\partial x_j} \tag{2}$$

hence is *diagonal at the mode*. In fact, from the diagonal hessian an approximation of form Eq. (1) can be obtained by means of *Laplace approximation*.

We will show theoretical results on separability for two popular building blocks: the posterior of Dirichlet-Multinomial distribution and the posterior of categorical variable given category-dependent multivariate Bernoulli or Multinomial observations (for example, naive bayes text classification on the bag-of-words representation). They will be discussed in Sect. 2; now we sketch a simpler example. Consider the multinomial model with total counts of k and probability $p = (p_1, \ldots, p_d)$. The likelihood of counts $x = (k_1, \ldots, k_d)$ is

$$L(x) = \binom{k}{k_1, \ldots, k_d} \prod_{i=1}^{d} p_i^{k_i}.$$

Denote by $q_i = \frac{k_i}{k}$ the observed frequencies. The log-likelihood normalized by the number of observations can be approximated by Stirling formulas [Shl14] establishing the connection to the *Kullback-Leibler divergence of observed and real frequencies*, respectively $q = (q_1, \ldots, q_d)$ and $p = (p_1, \ldots, p_d)$.

$$\frac{1}{k} \log L(x) \approx O(k^{-1} \log k) + D_{\mathrm{KL}}(q \| p)$$

$$= O(k^{-1} \log k) - \sum_{i=1}^{d} q_i \log \frac{q_i}{p_i}$$

It is not hard to see that the mode is at $k \cdot p$ and the log-likelihood at this point equals $O(\log k)$. Thus we obtain Eq. (1) with $I(x_i) = q_i \log \frac{q_i}{p_i}$.

---

[1] For example in the multinomial model $\sum_i x_i$ is fixed.

Case Study on Real Data. We apply our framework to the real data set of error logs from company servers. Each record contains the number of errors for a given zone, project, procedure and the error message. The data was collected for more than 120 consecutive days. Every day consists of a full log of errors and has size of several GB. For efficient training the data was preaggregated on Hive (by considering multiplicities) which decreased the size to few MB. A sample of the data set is shown in Table 1. The model we build is analytically tractable, we avoid complicated posterior sampling. The results will be discussed in Sect. 3.

Table 1. Dataset for log errors.

row_id	date	region	project_name	procedure_name	error_detail	err_cnt
15362	2018-04-01	EMEA	GLOBAL_ONLINE_SERVICE	EXPLODE_BUNDLE	Object reference not set to an instance of an ...	3
29308	2018-04-01	EMEA	Y_API	Y.Controllers.Configurator.Global.Glo...	VerifyError:Invalid option selected	1
29222	2018-04-01	EMEA	G Services: CustomerService	NaN	Operation: GetSalesPerson	26
3157	2018-04-01	EMEA	G Services: CustomerService	NaN	Operation: GetCustomer Exception: G.Ex...	77
7801	2018-04-01	EMEA	Y_API	Y.Controllers.Configurator.Global.Glo...	BuildError:InvalidOrderCodeOrCustomerSet	5

1.3 Organization

In Sect. 2 we derive theoretical results for separable posteriors. In Sect. 3 we apply our framework to real-world data. The paper is concluded in Sect. 4.

2 Separable Posteriors

2.1 Dirichlet-Multinomial Model

The Dirichlet-Multinomial Model (DM) is popular for modeling multivariate counts. As opposed to the plain multinomial model, it captures uncertainty in the probability parameter, which helps avoiding over-dispersion.

$$
\begin{aligned}
(p_1, \ldots, p_d) &\sim \mathsf{Dir}(\alpha_1, \ldots, \alpha_d) \\
(k_1, \ldots, k_d) &\sim \mathsf{Mult}(p_1, \ldots, p_d | k)
\end{aligned}
\tag{3}
$$

This model is analytically tractable, we utilize formulas derived in [Tu15].

$$
P((k_i)_i | \mathcal{D}) = \frac{\Gamma(k+1)}{\prod_i \Gamma(k_i + 1)} \cdot \frac{\Gamma(\alpha')}{\prod_i \Gamma(\alpha'_i)} \cdot \frac{\prod_i \Gamma(k_i + \alpha'_i)}{\Gamma(k + \alpha')}
\tag{4}
$$

where $\alpha'_i = \alpha_i + \sum_{x \in \mathcal{D}} \sum x^i$ and $\alpha' = \sum_i \alpha'_i$ or the sake of concise notation. By using the Stirling approximation we obtain

$$
\log L \approx
\tag{5}
$$

$$
- k \sum_i \frac{k_i}{k} \log \frac{k_i}{k} - \alpha' \sum_i \frac{\alpha'_i}{\alpha'} \log \frac{\alpha'_i}{\alpha'} + (k + \alpha') \sum_i \frac{k_i + \alpha'_i}{k + \alpha'} \log \frac{k_i + \alpha'_i}{k + \alpha'}
$$

In order to see separability we will apply the well known trick called *Laplace approximation*, which is merely a multivariate Gaussian approximation to the predictive posterior (see for example [Deh17] for theoretical justifications). Technically, we expand the log-likelihood in a Talyor series around its mode, so that linear term disappear (by the first-derivative test, as the mode maximizes the likelihood!) and quadratic terms correspond to the Gaussian terms. In our case, the second-order terms turn out to be diagonal hence we obtain separability.

In order to find the mode we need to use the Lagrangian because of the implicit constraint $k = \sum_i k_i$. For some constant C, the mode satisfies[2]

$$- \log k_i^* + \log(k_i^* + \alpha_i') + C = 0 \tag{6}$$

which implies

$$k_i^* = \frac{k}{\alpha'} \cdot \alpha_i'. \tag{7}$$

By the Taylor expansion around the mode we obtain (note that the linear part disappears and the coefficients of the quadratic part are determined from the first order conditions Eq. (6))

$$\log L((k_i)) \approx \log L((k_i^*)) - \frac{1}{2} \sum_i \frac{\alpha_i'}{(k_i^* + \alpha_i')k_i^*} \left(k_i - \frac{k}{\alpha'} \cdot \alpha_i'\right)^2$$

$$= \log L((k_i^*)) - \frac{1}{2} \sum_i \frac{1}{1 + \frac{k}{\alpha'}} \cdot \frac{\left(k_i - \frac{k}{\alpha'} \cdot \alpha_i'\right)^2}{\frac{k}{\alpha'} \cdot \alpha_i'} \tag{8}$$

denoting $q_i = \frac{k_i}{k}$ (observed frequency) and $p^* = \frac{\alpha_i'}{\alpha'}$ (mode frequency) we have

Lemma 1 (Predictive Posterior vs Mode for DM)

$$\log L(q) \approx \log L(p^*) - \frac{1}{2} \cdot \frac{\alpha'}{\alpha' + k} \cdot k \sum_i \frac{(q_i - p_i^*)^2}{p_i^*} \tag{9}$$

Since usually $k \ll \alpha'$ (α' collects all occurrences over the training data) we have $\frac{\alpha'}{k + \alpha'} \approx 1$ and we conclude

Corollary 1 (DM Posterior Predictive Impacts). *For the DM-model the impact for the i-th component in Eq. (1) equals*

$$I(k_i) \approx \frac{1}{2} \cdot k \sum_i \frac{(q_i - p_i^*)^2}{p_i^*}. \tag{10}$$

Remark 1 (Intuition). The major reason for impacts being large negative is a significant relative increase in frequencies (observed vs posterior), under large volume. Indeed, let $q_i = (1 + r_i)p_i^*$ then the i-th impact equals $I(k_i) = r_i^2 p_i^*$.

---

[2] We extend the likelihood over non-integer frequencies as the gamma function is well-defined and the Stirling approximation works.

2.2 BNB Model

We prove separability only for Bernoulli Naive Bayes (BNB) as we will be using this model in our case study. However, separability is not limited to the Bernoulli variant and can be also proved for Multinomial Naive Bayes.

The BNB model is popular for classification of short text messages. Texts are represented as as the $|V|$-dimensional boolean vectors where V is the vocabulary. Each entry is a boolean number indicating occurrence of the word w in a given text $\mathbf{w}$; we will use the notation $I(w \in \mathbf{w})$. The model with Beta prior (which smooths zero-frequencies adding extra "pseudocounts" of one for each class-word) can be written as

$$\forall c \in C \forall w \in V \quad p_{w|c} \sim \mathsf{Beta}(1,1)$$
$$\forall c \in C \forall w \in V \quad I(w \in \mathbf{w}|c) \sim \mathsf{Ber}(p_{w|c})$$

where C is the set of classes (categories). Let $p_{w|c}$ and p_c be posterior probabilities for word given class and for class (bayesian estimation is discussed in the appendix in the full version).

Proposition 1 (Predicitve Posterior for BNB). *Probability of the class c given the vector of words $\mathbf{w} \in \mathbb{R}^V$ is given by*

$$L(c|\mathbf{w}) \propto p_c \cdot \prod_{w \in \mathbf{w}} p_{w|c}^{I(w \in \mathbf{w})} (1 - p_{w|c})^{I(w \notin \mathbf{w})} \tag{11}$$

where the proportionality constant is independent on c (but depends on $\mathbf{w}$).

By taking the logarithm of Eq. (11) evaluated at c and c^*, and subtracting (the unknown constant cancels) we obtain

Lemma 2 (Predictive Posterior vs Mode for BNB). *For the BNB model, let c^* be the most likely class given the sequence of words $\mathbf{w} \subset V$. We have*

$$\log L(c|\mathbf{w}) - \log L(c^*|\mathbf{w}) = \tag{12}$$
$$\log \frac{p_c}{p_{c^*}} + \sum_{w \in \mathbf{w}} \left[I(w \in \mathbf{w}) \log \frac{p_{w|c}}{p_{w|c^*}} + I(w \notin \mathbf{w}) \log \frac{1 - p_{w|c}}{1 - p_{w|c^*}} \right]$$

From we immediately obtain the word impact.

Corollary 2 (BNB Posterior Predictive Impact). *For the BNB-model the impact for the w-th word in Eq. (1) equals*

$$I(w) = \sum_{w \in \mathbf{w}} \left[I(w \in \mathbf{w}) \log \frac{p_{w|c}}{p_{w|c^*}} + I(w \notin \mathbf{w}) \log \frac{1 - p_{w|c}}{1 - p_{w|c^*}} \right] \tag{13}$$

where c is the actual class.

Remark 2 (Intuition). The major reason for impacts $I(w)$ to be large negative is the presence of class-untypical words (so that $p_{w|c} \ll p_{w|c^*}$). The effect is stronger with large volume when evaluating averaged likelihoods.

3 Root Cause Analysis of Anomalies

3.1 Generative Model

Before we apply the results of the previous section, we need to construct the joint model for all features in our data set. We model the Data by a Bayes Net illustrated in Fig. 1. Every feature is dependent on zone (justification: different zones use servers in different location) and at most one other feature (in the natural hierarchical way). Thus, the model is actually a Tree-Augmented Network (TAN). These models generally allow for a feature-root relation and one more level of interaction. While TANs can capture non-trivial dependencies, they are computationally attractive since every node has at most two parents which reduces the size of internal conditional probability tables [Pad14].

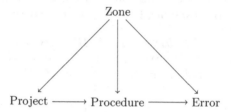

Fig. 1. TAN model for occurrences of a *single error*.

More precisely, we assume

$$\mathsf{Proj}|\mathsf{Zone} \sim \mathsf{Cat}(p = p(\mathsf{Zone}))$$
$$\mathsf{Proc}|\mathsf{Proj}, \mathsf{Zone} \sim \mathsf{Cat}(p = p(\mathsf{Proj}, \mathsf{Zone})) \qquad (14)$$
$$\mathsf{Err}|\mathsf{Proc}, \mathsf{Proj}, \mathsf{Zone} \sim \mathsf{Ber}(p = p(\mathsf{Proc}, \mathsf{Zone}))$$

with empirical Dirichlet priors (estimated from data) for Proj, Proc and non-informative Beta prior for Err). Bernoulli distributions are over the (binarized) bag-of-word text representation of Err.

Given the graph, the likelihood factorizes into likelihoods of individual features given parents; these models can be fit separately [Pad14]. In our case

$$\Pr[\mathsf{Proj}, \mathsf{Proc}, \mathsf{Err}|\mathsf{Zone}] = \Pr[\mathsf{Err}|\mathsf{Proc}, \mathsf{Zone}] \cdot \Pr[\mathsf{Proc}|\mathsf{Proj}, \mathsf{Err}, \mathsf{Zone}] \cdot \Pr[\mathsf{Proj}|\mathsf{Zone}]$$

We also use this fact to structure our anomaly detection: we will analyze separately anomalies in Proj, Zone and separately in tuples Err, Proc, Zone. Since we are interested in discovering and explaining anomalies on the daily bases, we perform the inference day by day, training the algorithm on the past data. The model was implemented under Python package PyMC3 [SWF16].

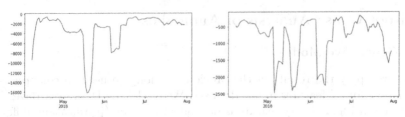

(a) Likelihood for Proj, Zone = EMEA   (b) Likelihood for Proj, Zone = APJ

Fig. 2. Project counts likelihoods by zone.

3.2 RCA for Projects

The posterior for Proj given observed projects counts is Dirichlet-Multinomial. The daily-averaged likelihood is illustrated in Fig. 2. Anomalies are detected as unusually low values (for example, lowest values over recent 14 days).

Anomalies 2018/05/17 and 2018/06/11, EMEA. By applying Corollary 1 we obtain most impacting projects. We see that the anomalies corresponds to peaks in project hits as illustrated in Fig. 3.

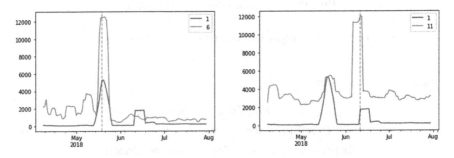

Fig. 3. Daily hits by project (Zone = EMEA).

Anomalies 2018/05/07 and 2018/07/28, APJ. By applying Corollary 1 we obtain most impacting projects (we pick two). The anomalies again corresponds to peaks in project hits as illustrated in Fig. 4.

3.3 RCA for Procedures and Error Messages

According to our model, the distribution of procedures given error descriptions follows the classification Bernoulli Naive Bayes (BNB) model (where Proc is the class and Err is text; class priors are determined by fitting Proc[Proc|Proj, Zone]). To detect anomalies in errors, we evaluate the likelihood of *procedures given observed error messages* rather than investigating for individual errors.

To see daily anomalies, we use daily-averaged likelihoods (see Fig. 5).

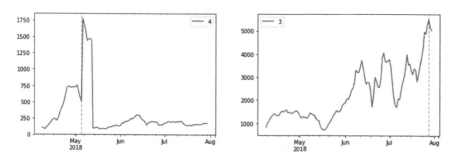

Fig. 4. Daily hits by project (Zone = APJ).

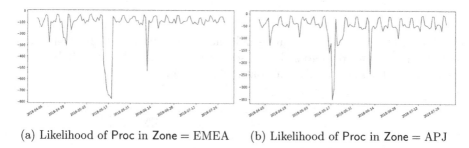

(a) Likelihood of Proc in Zone = EMEA   (b) Likelihood of Proc in Zone = APJ

Fig. 5. Likelihood of Proc split by Zone.

Anomaly 2018/05/17 in EMEA. By Corollary 2 we identify the set

$$S = \{\text{'object', 'set', 'reference', 'instance', 'connection'}\}$$

of 3 keywords with biggest negative influence on the likelihood. By inspecting hits on these keywords (by hit we understand every message matching at least one word in S) across the classes we notice a huge difference between the anomaly day and the reference data set (see Fig. 6).

By inspecting message texts we also recognize the specific messages related to the keywords S. The result is summarized in Table 2.

Table 2. RCA for anomaly 2018/05/17 EMEA.

Procedure	Error message
$Proc_1$	Object reference not set to an instance of an object
$Proc_4$	Object reference not set to an instance of an object

Anomaly 2018/06/11 in EMEA. By Corollary 2 we identify the set

$$S = \{\text{'channel', 'timed', 'remote', 'returned', 'request'}\}$$

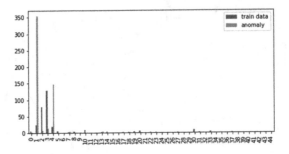

Fig. 6. Average daily hits of the keywords 'object', 'set', 'reference', 'instance', 'connection' split by class (Proc), for EMEA zone.

of 5 keywords with biggest negative influence on the likelihood. By inspecting hits on these keywords across the classes we notice a significant shift between the anomaly day and the reference data set (see Fig. 7).

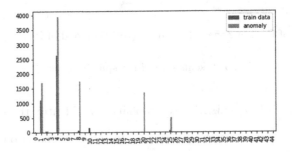

Fig. 7. Average daily hits of the keywords 'channel', 'timed', 'remote', 'returned', 'request' split by class (Proc) for EMEA zone.

Having localized the keywords, we easily find procedures with biggest shifts and also the messages. The explanation is summarized in Table 3.

Anomaly 2018/05/18 APJ. By Cororllary 2 we identify the set

$$S = \{\text{'null', 'reference', 'set'}\}$$

of 3 keywords with biggest negative influence on the likelihood. By inspecting hits on these keywords across the classes we notice a significant shift between the anomaly day and the reference data set (see Fig. 8). The explanation by procedures and error messages is shown in Table 4 below.

Anomaly 2018/06/11 APJ. By Corollary 2 we identify the set

$$S = \{\text{'contract', 'G', 'contracts', 'target', 'invocation'}\}$$

Table 3. RCA for anomaly 2018/06/11 EMEA.

Procedure	Error message
$Proc_8$	The operation has timed out
$Proc_{25}$	The request channel timed out
$Proc_{20}$	The request failed with HTTP status 404

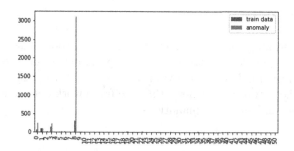

Fig. 8. Daily average hits for the keywords 'null', 'reference', 'set' split by Proc, APJ zone.

Table 4. RCA for anomaly 2018/05/18 APJ.

Procedure	Error message
$Proc_8$	Argument is null

of 5 keywords with biggest negative influence on the likelihood. By inspecting hits on these keywords across the classes we notice a significant shift between the anomaly day and the reference data set (see Fig. 9). The explanation by procedures and error messages is shown in Table 5 below.

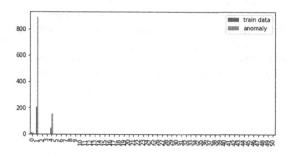

Fig. 9. Average daily hits of the keywords 'contract', 'G', 'contracts', 'target', 'invocation' split by class (Proc), for APJ zone.

Table 5. RCA for anomaly 2018/06/11 APJ.

Procedure	Error message
$Proc_1$	Operation: G.Exceptions.CustomerNotFoundException
$Proc_4$	Exception has been thrown by the target of an invocation

4   Conclusion

We proposed a framework for anomaly detection and root cause analysis based on *separable posterior approximation*. This approximation has been proved for the case of Multinomial, Dirichlet-Multinomial and Naive Bayes Models. The validation on the real data set shows that the framework detects anomalies and offers reasonable and simple explanations.

References

[Deh17]   Dehaene, G.P.: Computing the quality of the Laplace approximation, November 2017. http://adsabs.harvard.edu/abs/2017arXiv171108911D

[HO00]   Hyvrinen, A., Oja, E.: Independent component analysis: algorithms and applications. Neural Netw. **13**, 411–430 (2000)

[KF09]   Koller, D., Friedman, N.: Probabilistic Graphical Models: Principles and Techniques - Adaptive Computation and Machine Learning. The MIT Press, Cambridge (2009)

[Pad14]   Padmanaban, H.: Comparative analysis of naive Bayes and tree augmented naive Bayes models (2014). http://scholarworks.sjsu.edu/etd_projects/356

[Shl14]   Shlens, J.: Notes on Kullback-Leibler divergence and likelihood. http://arxiv.org/abs/1404.2000 (2014)

[SMRE17]   Solé, M., Muntés-Mulero, V., Rana, A.I., Estrada, G.: Survey on models and techniques for root-cause analysis. http://arxiv.org/abs/1701.08546 (2017)

[SWF16]   Salvatier, J., Wiecki, T.V., Fonnesbeck, C.: Probabilistic programming in Python using PYMC3. Peer J. Comput. Sci. **2**, e55 (2016)

[Tu15]   Stephen, T.: The Dirichlet-multinomial and Dirichlet-categorical models for Bayesian inference (2015). https://people.eecs.berkeley.edu/~stephentu/writeups/dirichlet-conjugate-prior.pdf

[ZZZS17]   Zhang, Y., Zhou, H., Zhou, J., Sun, W.: Regression models for multivariate count data. J. Comput. Graph. Stat. **26**(1), 1–13 (2017)

Algorithms and Complexity Results for the Capacitated Vertex Cover Problem

Sebastiaan B. van Rooij and Johan M. M. van Rooij[✉]

Department of Information and Computing Sciences, Utrecht University,
PO Box 80.089, 3508 TB Utrecht, The Netherlands
J.M.M.vanRooij@uu.nl

Abstract. We study the capacitated vertex cover problem (CVC). In this natural extension to the vertex cover problem, each vertex has a pre-defined capacity which indicates the total amount of edges that it can cover. In this paper, we study the complexity of the CVC problem. We give NP-completeness proofs for the problem on modular graphs, tree-convex graphs, and planar bipartite graphs of maximum degree three. For the first two graph classes, we prove that no subexponential-time algorithm exist for CVC unless the ETH fails.

Furthermore, we introduce a series of exact exponential-time algorithms which solve the CVC problem on several graph classes in $\mathcal{O}((2 - \epsilon)^n)$ time, for some $\epsilon > 0$. Amongst these graph classes are, graphs of maximum degree three, other degree-bounded graphs, regular graphs, graphs with large matchings, c-sparse graphs, and c-dense graphs. To obtain these results, we introduce an FPT treewidth algorithm which runs in $\mathcal{O}^*((k + 1)^{tw})$ or $\mathcal{O}^*(k^k)$ time, where k is the solution size and tw the treewidth, improving an earlier algorithm from Dom et al.

Keywords: Capacitated vertex cover
Exact exponential-time algorithms · Treewidth
Fixed parameter tractability

1   Introduction

In the last couple of decades, exact exponential-time algorithms for NP-hard problems gained more attention. It is expected that there are no algorithms which solve NP-complete problems in polynomial time. Therefore, we must try different approaches, such as approximation algorithms, (meta) heuristics, parameterized algorithms, and moderately exponential-time algorithms. Many NP-hard problems have a trivial exponential-time algorithm running in $\mathcal{O}^*(2^n)$ time, where the $\mathcal{O}^*$-notation suppresses all polynomial factors. An active field of research is to find algorithms that improve these trivial bounds and solve these problems in $\mathcal{O}^*(c^n)$ time for some constant $c < 2$.

© Springer Nature Switzerland AG 2019
B. Catania et al. (Eds.): SOFSEM 2019, LNCS 11376, pp. 473–489, 2019.
https://doi.org/10.1007/978-3-030-10801-4_37

Many vertex subset problems are NP-hard problems. If there exists a polynomial-time check to test whether a given vertex subset is a solution to the given problem, then these problems can be solved trivially in $\mathcal{O}^*(2^n)$ time by checking every possible subset of vertices. Many of these vertex subset problems, such as VERTEX COVER[1] (e.g., see [3,9,23,25,28,29,33,34]) and DOMINATING SET (e.g., see [9,12,16,22,27,31,32]), can be solved in $\mathcal{O}^*(c^n)$ time for some constant $c < 2$.

A natural extension to the these two problems is to add a capacity constraint. Each vertex v in the graph gets a predefined capacity $c(v)$, which indicates the maximum amount of edges/vertices it can cover/dominate. The resulting CAPACITATED VERTEX COVER and CAPACITATED DOMINATING SET problems are in many ways more difficult that their counterparts without capacities. For both these problems, the questions whether there exist $\mathcal{O}((2 - \epsilon)^n)$-time algorithms, for some $\epsilon > 0$, were stated at IWPEC 2008 by van Rooij [2]. For the CAPACITATED DOMINATING SET problem, this question was resolved by Cygan et al. [7], who gave an $\mathcal{O}(1.89^n)$-time algorithm. This runtime was later improved by Liedloff et al. [26] to $\mathcal{O}(1.8463^n)$. An $\mathcal{O}((2 - \epsilon)^n)$-time algorithm for the CAPACITATED VERTEX COVER problem is still unknown.

Definition 1 (Capacitated Vertex Cover Problem (CVC)). *Let $G = (V,E)$ be an undirected graph, $c : V \rightarrow \mathbb{N}$ a capacity function on the vertices V, and k a positive integer. A subset of vertices $V' \subseteq V$ is a capacitated vertex cover if there exists a function $f : E \rightarrow V'$ such that for every edge $e = \{v,w\} \in E : f(e) \in \{v,w\}$ and for every vertex $v \in V' : |f^{-1}(v)| \leq c(v)$. The CAPACITATED VERTEX COVER problem asks whether there exists a capacitated vertex cover of size $\leq k$.*

In prior work, Guo et al. [18] proved that the capacitated vertex cover problem is FPT with respect to the solution size k and can be solved in $\mathcal{O}^*(1.2^{k^2})$ time. Dom et al. [8] improved this bound to $\mathcal{O}^*(k^{3k})$ using a treewidth algorithm, and they proved that the CVC problem is $W[1]$-hard when parameterized by treewidth. Guha et al. [17] considered the more general weighted CVC problem and gave a primal-dual based approximation algorithm with an approximation ratio of 2. They also proved that the problem restricted to trees can be solved in $\mathcal{O}(n \log n)$ time.

This Work. We show that the CAPACITATED VERTEX COVER problem is NP-complete on tree-convex and modular graphs and that it cannot be solved in subexponential time, unless the ETH fails. We also show that CVC is NP-complete on planar bipartite graphs of maximum degree three.

Next, we introduce a treewidth algorithm that runs in $\mathcal{O}^*((k+1)^{tw})$ or $\mathcal{O}^*(k^k)$ time, where k is the solution size, improving the bound from Dom et al. [8]. Thereafter, we introduce a series of algorithms that solve CVC on some specific graph classes in $\mathcal{O}((2 - \epsilon)^n)$ time, for some $\epsilon > 0$. We will give an algorithm on

---

[1] Note that VERTEX COVER equals INDEPENDET SET and CLIQUE when considered from the viewpoint of exact exponential-time algorithms.

graphs which contain a degree-bounded spanning tree, which includes the graphs of bounded degree. We generalize this algorithm to solve the CVC problem on graphs with a significantly sized matching. Using a combination of these approaches, we show that there are $\mathcal{O}((2-\epsilon)^n)$-time algorithms for CVC on graphs of maximum degree three, other degree-bounded graphs, regular graphs, graphs with large matchings, sparse graphs, and c-dense graphs.

2   Difficulty of the Problem

We start by giving several NP-completeness proofs for CAPACITATED VERTEX COVER. There is a trivial reduction from VERTEX COVER, which shows that CAPACITATED VERTEX COVER is NP-complete on any graph class on which the vertex cover problem is NP-complete. There are, however, many graph classes on which VERTEX COVER is in P, while CVC is NP-complete.

Below, we will give reductions from NP-complete problems to CVC on planar bipartite graphs with maximum degree 3, modular graphs and tree-convex graphs. Note that the problem is in NP as we can check in polynomial time whether a given subset of vertices V' is a valid capacitated vertex cover using a maximum flow algorithm.

Theorem 1. *Capacitated vertex cover is NP-complete on planar bipartite graphs of maximum degree three.*

Proof. We use that VERTEX COVER is NP-complete on planar graphs with maximum degree 3 [15]. Given a VERTEX COVER instance that is a planar graph $G = (V, E)$ of maximum degree 3 and an integer k, we will create a new planar bipartite graph G' of maximum degree 3 which has a capacitated vertex cover of size at most $|E| + k$ if and only if G has a vertex cover of size at most k.

Let $G' = (V', E')$ be a copy of G, and let all copied vertices have a capacity equal to their degree. Next, subdivide every edge $e \in E'$, which introduces a new node v_e per edge e. We give these vertices v_e a capacity of 2. Finally, for every v_e, add a new capacity zero vertex of degree one connected only to the vertex v_e. Note that the new graph G' remains planar, that every vertex in G' has degree at most 3, and that G' is a bipartite graph.

Now, we show that G has a vertex cover of size at most k if and only if G' has a capacitated vertex cover of size at most $|E| + k$. First, we note that any capacity zero vertex is never in an optimal capacitated vertex cover. This forces the vertices v_e in the cover giving us at least $|E|$ vertices in any capacitated vertex cover in G'. Notice that the vertices v_e have capacity 2 and can cover at most one edge besides the edge to its capacity zero neighbour, thus the third edge of v_e must be covered by one of the original endpoints of e. As such, any vertex cover of size k in G corresponds to a capacitated vertex cover of size $|E| + k$ in G' and vice versa.   □

This theorem implies that CAPACITATED VERTEX COVER on bipartite graphs in general is also NP-complete. We know that CVC on planar graphs can be solved in subexponential time (see, Corollary 2). However, assuming the ETH, this does not seem to be the case for bipartite graphs.

Definition 2 (Exponential Time Hypothesis (ETH) *[20]). There exists an $\epsilon > 0$ such that $3 - SAT$ cannot be solved in $\mathcal{O}^*(2^{\epsilon n})$ time.*

Theorem 2. *Assuming the ETH, capacitated vertex cover on bipartite graphs cannot be solved in subexponential time.*

Proof. We will give a reduction from SAT to CAPACITATED VERTEX COVER on bipartite graphs that transforms a formula with n variables and m clauses into a graph with $\mathcal{O}(n + m)$ vertices. Because of the Sparsification Lemma [21], this implies that, assuming the ETH, CVC on bipartite graphs cannot be solved in subexponential time. See [11] for a discussion on this standard approach.

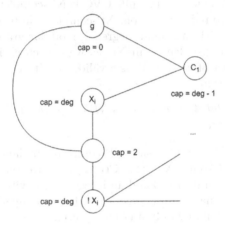

Fig. 1. SAT variable gadget of three vertices and the special vertex g and a clause vertex for clause C_1. The vertex g and the variable assignment vertices (X_i and $!X_i$) are in vertex partition X, the clause vertices (C_1) and the centre vertex of the gadget are in vertex partition Y.

Let $x_1, \ldots, x_n$ be the variables of the SAT instance and let $C_1, \ldots, C_m$ be the clauses. We will construct a bipartite graph $G = (X, Y, E)$ with capacities $c(v)$ which has a capacitated vertex cover of size $2n + m$ if and only if the SAT instance is satisfiable.

We construct G in the following way. First add a single vertex g with capacity 0 to vertex partition X. For every variable x_i, create a gadget (see Fig. 1) that consists of a capacity two vertex and connected to two more new vertices representing the true and false assignments for the variable x_i. Next add, for every clause C_i, a vertex to partition Y connected to g, and connect the clause vertices by new edge to the corresponding variable assignment vertices. Set the capacity of the clause vertices equal to their degree minus one. Finally, connect the vertex g to the centre vertex of all variable gadgets.

We will now show that the SAT instance is satisfiable if and only if the minimum capacitated vertex cover on G has size $2n + m$. First note that the capacity zero vertex g is connected to all vertices in Y, and therefore all vertices in Y must be included in any minimum (size) capacitated vertex cover. This results in a cover of size at least $n + m$ because there are m clause vertices and n variable gadgets. Also, the centre vertex of the variable gadget has three edges and a capacity of two, since g had capacity zero, we conclude that either the true or the false assignment vertex must also be in any minimum capacitated vertex cover. Therefore, any minimum capacitated vertex cover is of size at least $m + 2n$. Lastly, observe that every clause vertex requires that at least one of its incident edges is covered by a neighbouring vertex.

It follows that any satisfying assignment to the SAT problem corresponds to a capacitated vertex cover of size $m + 2n$ by taking all vertices in Y plus the variable assignment vertices corresponding to the assignment. Vice versa, any minimum capacitated vertex cover of size $m + 2n$ must include the $m + n$ vertices in Y plus at most one variable assignment vertex per variable, and thus corresponds to a satisfying assignment of the SAT formula.   □

The reduction in Theorem 2 is also valid for tree-convex graphs, a subclass of bipartite graphs.

Definition 3. *A bipartite graph $G = (X, Y, E)$ is called tree convex if there exists a tree $T = (X, F)$ such that for any $v \in Y$, $N(v)$ is a connected subtree in T.*

The tree $T = (X, F)$ in this definition is not a subtree of the graph G but has its own set of edges F.

It is not hard to see that the graph G constructed in the proof of Theorem 2 is a tree-convex graph. To construct the tree $T = (X, A)$, select g as the root of T and every other vertex in X as a leaf in T. Because all the vertices in Y are connected to the vertex g, no matter what other neighbours they might have, they will form a connected subtree in T.

Corollary 1. *The capacitated vertex cover problem is NP-complete on tree-convex graphs, and it cannot be solved in subexponential time on tree-convex graphs, unless the ETH fails.*

Next, we consider another subclass of bipartite graphs, namely modular graphs (e.g., see [1]).

Definition 4. *A bipartite graph G is modular if for every vertex triplet x, y, z there exist three shortest paths, one from x to y, one y to z, and one from x to z, that share a common vertex.*

We will modify the reduction of Theorem 2 to construct a modular graph.

Theorem 3. *The capacitated vertex cover problem is NP-complete on modular bipartite graphs, and it cannot be solved in subexponential time on modular graphs, unless the ETH fails.*

Proof. Given an instance of SAT, let $G = (X, Y, E)$ be the bipartite graph constructed with the SAT reduction described in Theorem 2. We will transform G into a new graph $G' = (X', Y', E')$ such that G' is a modular graph. Start with $G' = G$. Then, add a new vertex h to Y' connected to all vertices in X', and let the capacity of h equal its degree. Because h is connected to g (see Theorem 2), it has to be included in any capacitated vertex cover, and because it has capacity equal to its degree, it can cover all its edges. It is easy to see that G' is equivalent to G in the sense that any capacitated vertex cover on G can be transformed into a capacitated vertex cover on G' and vice versa, by adding or removing h from the cover, respectively. Hence by the proof of Theorem 2, G' has a capacitated vertex cover of size $2n + m + 1$ if and only if the SAT instance is satisfiable.

We will now show that the new graph G' is a modular graph by proving that, for any vertex triplet x, y, $z \in V'$, three shortest paths between the three vertices share a common vertex. Without loss of generality, we will look at four cases.

- $x, y, z \in X'$. The three vertices can reach each other directly via h, with h as common vertex.
- $x, y \in X', z \in Y'$ *with none of the three vertices adjacent.* A shortest path from x to y goes directly through h. Because z is not adjacent to x and y, a shortest path from z to x has to pass through a third vertex $w \in X'$, which is a neighbour of z. From w we can go through h and reach both x and y. Thus h is the common vertex on three shortest paths.
- $x, y \in X', z \in Y'$ *and z is adjacent to both x and y.* Now, z is on the shortest paths from z to x and from z to y. And, as stated by this subcase, z also lies on a direct path from x to y.
- $x, y \in X', z \in Y'$ *and, without loss of generality, z adjacent to x and z not adjacent to y.* The shortest path from z to x and from y to x all go through x. A shortest path from z to y goes through x followed by h to y. Thus x lies on three shortest paths.

All other combination of x, y, and z, where two or more vertices are in Y' can be reduced to these four cases by replacing the vertex h with the vertex g in the arguments.

This reduction proves the NP-completeness. And, analogous to Theorem 2, we have an $\mathcal{O}(n+m)$ sized reduction from SAT to the CVC problem on modular graphs, implying that, assuming the ETH, we cannot solve the capacitated vertex cover on modular graphs in subexponential time. $\qquad\square$

3   Treewidth

In this section, we introduce a dynamic programming algorithm on a tree decomposition of a graph G. The algorithm improves a result by Dom et al. [8] and has some interesting corollaries.

Definition 5. *A tree decomposition of a graph $G = (V, E)$ is pair (T, X) with $T = (I, F)$ a tree and $X = \{X_i | i \in I\}$ a family of subsets of V, called bags, one for each node of T such that:*

- For every $v \in V$, there exists a bag X_i such that $v \in X_i$.
- For every edge $\{v, w\} \in E$, there exists a bag X_i such that $v, w \in X_i$.
- For every $v \in V$, all the bags X_i which contain v form a connected subtree of T.

The width of a tree decomposition is the size of the largest bag minus one. The treewidth of a graph G, $tw(G)$, is the minimum width of a tree decomposition over all tree decompositions of G.

A dynamic programming algorithm on a tree decomposition works as follows. First, set one arbitrary node of the tree decomposition as the root, and define for every node i, $G_i = G[\bigcup_{j \in \Delta(i)} X_j]$ where $\Delta(i)$ is the set of descendants of i in T. Then, for each node i in a bottom-up fashion on the decomposition tree T, compute all partial solutions to the problem on G_i that can be extended to a global optimal solution. As a result, the solution to the problem is found at the root node.

To simplify the formulation of a dynamic programming algorithm, one uses *nice tree decompositions* [24]. We will use a *nice tree decomposition with edge introduce bags* [5], where a vertex set X_i and an edge set Y_i are used to define the subgraph G_i for which partial solutions are computed.

Definition 6. *A nice tree decomposition with edge introduce bags of a graph G is a rooted tree decomposition (T, X) with tree $T = (I, F)$, $X = \{X_i | i \in I\}$ a family of subsets of V, and $Y = \{Y_i | i \in I\}$ a family of subsets of E, such that every node i of T is of one the following types:*

- Leaf node: *the node i has no child nodes, $X_i = \{v\}$, and $Y_i = \emptyset$.*
- Vertex introduce node: *the node i has one child node j, $X_i = X_j \cup \{v\}$ for a $v \in V$, and $Y_i = Y_j$.*
- Vertex forget node: *the node i has one child node j, $X_i = X_j \setminus \{v\}$ for a $v \in X_j$, and $Y_i = Y_j$.*
- Edge introduce node: *the node i has one child node j, $X_i = X_j$, and $Y_i = Y_j \cup \{e\}$ for some edge e with both endpoints in X_i. Such a node exists exactly once for every edge $e \in E$.*
- Join node: *the node i has two children j and k, $X_i = X_j = X_k$, and $Y_i = Y_j \cup Y_k$.*

The root node r of a nice tree decomposition with edge introduce bags is a special vertex forget node with $X_r = \emptyset$. Finally, we define for every node $i \in I$, the subgraph G_i for which partial solutions are computed at each node as $G_i = (\bigcup_{j \in \Delta(i)} X_j, Y_i)$.

A nice tree decomposition with edge introduce bags of the same width can be calculated from a given tree decomposition in polynomial time [5,24].

Theorem 4. *Given a graph $G = (V, E)$ with a capacity function $c(v)$ with a maximum capacity of $M = \max\{c(v) : v \in V\}$, and a tree decomposition of G of width tw, capacitated vertex cover can be solved in $\mathcal{O}^*((M+1)^{tw})$ time.*

Proof. First, we compute a nice tree decomposition with edge introduce bags T of G of width k.

In the algorithm, we will decide which incident vertex should cover the edge considered in every edge introduce node. The algorithm bottom-up computes values $c_i(f)$, for every node i of T, indicating the smallest partial solution to CAPACITATED VERTEX COVER on G_i corresponding to a capacity function f on X_i. These capacity functions $f : X_i \to \mathbb{N}$ define $f(v)$ as the total amount of capacity vertex v has used to cover edges on G_i. The optimal solution to the problem is found in the single entry of the root node of the tree decomposition T.

We will now give the recursive formulas to calculate the table $c_i(f)$ for a given node i for every type of node in a nice tree decomposition with edge introduce bags.

- **Leaf node:** Let i be a leaf node with $X_i = \{v\}$ for some vertex v and $Y_i = \emptyset$. If the capacity function assigns more than 0 capacity, we have an invalid solution, as there are no edges in G_i on which the capacity could be spent. Otherwise, we have a solution of size 0.

$$c_i(f) = \begin{cases} \infty & \text{if } f(v) > 0 \\ 0 & \text{if } f(v) = 0 \end{cases}$$

- **Vertex introduce node:** Let i be a vertex introduce node with child node j, $X_i = X_j \cup \{v\}$, for a vertex v, and $Y_i = Y_j$. $X_i = X_{i-1} \cup \{v\}$ with v a vertex. Because the new vertex has no incident edges in G_i, the capacity function cannot assign more than 0 spent capacity.

$$c_i(f) = \begin{cases} \infty & \text{if } f(v) > 0 \\ c_j(f') & \text{if } f(v) = 0 \end{cases} \text{ where } \forall w \in X_{i-1} : f'(w) = f(w)$$

- **Vertex forget node:** Let i be a vertex forget node with child node j, $X_i = X_j \setminus \{v\}$, for a vertex $v \in X_j$, $Y_i = Y_j$. Because in a vertex forget node a vertex is removed from the bag X_i compared to X_j, it will no longer receive new edges in edge introduce nodes. Now, we check whether the vertex is used in partial solutions represented by $c_j(f)$ by checking whether v has spent any of its capacity. That is, if $f(v) = 0$ it has not spent any capacity and thus can be left out of the cover, while if $f(v) > 0$ we must add one to the size of the partial solution.

$$c_i(f) = \min \begin{cases} \min_{f'(v)=1,\dots,c(v)} \{1 + c_{i-1}(f')\} & \text{where } f' \text{ equals } f \text{ except on } v \\ c_j(f') & \text{where } f' \text{ equals } f \text{ except } f'(v) = 0 \end{cases}$$

- **Edge introduce node:** Let i be an edge introduce node with child node j, $X_i = X_j$ and $Y_i = Y_j \cup \{e\}$ for some edge e with both endpoints in X_i. The introduced edge has to be covered by one of its end points. Let $e = \{v, w\}$, we then get:

$$c_i(f) = \min\{c_j(f'), c_j(f'')\}$$

where f' and f'' equal f with the exceptions $f'(v) = f(v) - 1$ and $f''(w) = f(w) - 1$. Here, we use the value ∞ for $c_j(f')$ and $c_j(f'')$ if $f(v) = 0$ or $f(w) = 0$, respectively. This results in ∞ for $c_i(f)$ if there is no valid partial cover (when $f(v) = f(w) = 0$).

- **Join node:** Let i be a join node with children j and k, $X_i = X_j = X_k$ and $Y_i = Y_j \cup Y_k$. For a join node, we need to add the spent capacity from both subtrees rooted at the child nodes.

$$c_i(f) = \min_{g,h}\{c_j(g) + c_k(h)\} \tag{1}$$

where we take the minimum over all pairs of capacity functions g, h such that $\forall v \in X_i : g(v) + l(v) = f(v)$.

For the running time of the algorithm, notice that for any vertex $v \in X_i$ the capacity function can have $c(v) + 1$ different values. This results in an upper bound of $(M+1)^{tw}$ different capacity functions per node i. For all types of nodes except the join node, it is easy to see that all values $c_i(f)$ can be computed in $\mathcal{O}^*(M + 1)^{tw}$ time. For the join node, we can use the fast fourier transform based join algorithm from [6] to obtain the same time bound. Since the number of nodes in the tree decomposition is linear in the size of the graph, the total running time is $\mathcal{O}^*((M + 1)^{tw})$.   □

We emphasise the following corollary, that also follows from the results by Dom et al. [8],

Corollary 2. *Capacitated vertex cover can be solved in $\mathcal{O}(2^{\mathcal{O}(\sqrt{n}\log n)})$ time on planar graphs.*

Proof. For planar graphs, we can compute a tree decomposition of width $\mathcal{O}(\sqrt{n})$ in polynomial time [13]. Using $M \leq n$, the treewidth algorithm runs in $\mathcal{O}^*((n + 1)^{\mathcal{O}(\sqrt{n})}) = \mathcal{O}^*(2^{\mathcal{O}(\sqrt{n}\log n)})$ time.   □

We can use the treewidth algorithm to obtain a fast exponential-time algorithm for CAPACITATED VERTEX COVER on graphs of maximum degree three. For this, we need the following result.

Lemma 1 ([10]). *Graphs of maximum degree three have treewidth at most $(\frac{1}{6} + \epsilon)n$, for any $\epsilon > 0$.*

Clearly, any vertex in a graph with maximum degree three has a capacity of at most 3. Therefore, we can apply the treewidth algorithm and solve CAPACITATED VERTEX COVER in $\mathcal{O}^*(4^{\frac{1}{6}n})$ time. The worst case in this algorithm is achieved when we have a bag whose vertices all have a capacity 3. Then, the capacity function has four possible values for each vertex. However, if a vertex has capacity 3 in a graph with maximum degree three, then we could either take the vertex and cover all the edges (no longer needing to keep track which one exactly), or we do not take the vertex and cover no edges. In particular, we can reduce the amount of values we store by only allowing $f(v) = 0$ or $f(v) = 3$, resulting in at most 3 different values for f per vertex. We obtain a run time of $\mathcal{O}^*(3^{\frac{1}{6}n})$.

Corollary 3. *Capacitated vertex cover can be solved in $\mathcal{O}^*(3^{\frac{1}{6}n})$ time on graphs of maximum degree three.*

We can alter the treewidth algorithm to obtain an FPT algorithm with respect to the solution size k, improving the $\mathcal{O}^*(k^{3k})$ time result by Dom et al. [8]. To do this, we will modify what the capacity functions f in the formulation of the algorithm keep track of: $f(v)$ no longer defines the amount of edges covered by v in G_i, but the amount of edges not covered by v in G_i (this is similar to [8]). These new capacity functions f effectively define the amount of edges in G_i incident to v covered by the neighbours of v. In an FPT algorithm computing a solution of size at most k, $f(v)$ can thus have most $k + 1$ values (including zero), leading to $\mathcal{O}^*((k + 1)^{tw})$ values $c_i(f)$ per node i.

The recursions given in the proof of Theorem 4 stay mostly the same, which we leave to the reader to verify. The only differences are in the vertex forget nodes and edge introduce nodes. In a vertex forget node, the algorithm has to check whether a vertex v is included in the cover, which is now done by checking whether the amount of edges covered by neighbours is less than the degree of v. In an edge introduce node, the algorithm decides which endpoint of an edge e covers e and checks whether a vertex does not spend more capacity than it has available (then we have an invalid solution). This can be checked by comparing difference between the amount of edges covered by neighbours and the degree of the vertex to the capacity of v.

Theorem 5. *Given a graph $G = (V, E)$ with a capacity function $c(v)$, a tree decomposition of G of width tw, and an integer k, we can check whether there exists a capacitated vertex cover of size at most k in G in $\mathcal{O}^*((k + 1)^{tw})$ time.*

Corollary 4. *Given an instance of capacitated vertex cover and an integer k. We can check whether there exists a solution of size at most k in $\mathcal{O}^*(k^k)$ time.*

Proof. The treewidth of any graph is at most the size of its minimum vertex cover, and a minimum vertex cover has at most the size of a minimum capacitated vertex cover. Thus, we can use a fast FPT-algorithm (e.g. [4]) to find a vertex cover of size at most k in G. If it does not exist we are done as there also cannot be a capacitated vertex cover of size k. and, if it does exists, we use the vertex cover to obtain an tree decomposition of width at most k and apply Theorem 5. $\square$

4   Exact Exponential Time Algorithms

In this section, we look at algorithms solving CAPACITATED VERTEX COVER on several graph classes in $\mathcal{O}((2 - \epsilon)^n)$ time, for some $\epsilon > 0$.

4.1   Graphs of Bounded Degree

We start with a branching algorithm solving CVC on graphs containing a bounded degree spanning tree, which includes graph of bounded degree. Our algorithm is based on the observation that any capacitated vertex cover is a vertex cover on a spanning tree of the same graph. The algorithm enumerates all vertex covers on a spanning tree T of G, and for every enumerated vertex cover on T, it checks whether it is a capacitated vertex cover on G. The algorithm then returns the smallest solution found. To prove the running time of the algorithm, we need the following lemma.

Lemma 2. *All vertex covers on a tree with maximum degree $k \geq 3$ can be enumerated in $\mathcal{O}^*((1 + 2^{k-1})^{\frac{n}{k}})$ time. All vertex covers on a tree with maximum degree 2 (a path) can be enumerated in $\mathcal{O}(1.619^n)$ time.*

Proof. Let T be a tree with maximum degree k. Choose a node of the tree as the root of T such that each node has at most $k - 1$ children. We propose a branching algorithm. See e.g. [11], for an introduction to the fundamentals of branching algorithms.

On the current instance, let v be a leaf with maximum depth in T. The algorithm considers the set of vertices S consisting of v, the parent node of v, and all the sibling nodes of v. Given this set S, we let E_S be the edges of T between nodes in S. The algorithm branches into a subproblem for each subset $C \subseteq S$ that contains at least one endpoint of every edge in E_S: it puts the chosen vertices in C in the vertex cover and removes all vertices in S. Notice, that if the parent node is not in C, then all other nodes in S must be in C. Thus, the algorithm branches into at most $1 + 2^{|S|-1}$ subproblems. Note that T remains a tree after removing S because v is of maximum depth. It is not hard to see that this branching algorithm enumerates all vertex covers as it forces all edges incident to vertices in S to be covered by vertices assigned to the vertex cover. The algorithm can, however, output a vertex set that is not a vertex cover, not covering the edge between the parent of v and the parent of the parent of v, but we can discard such sets afterwards.

We will measure the running time of this branching rule with respect to the number of vertices removed in each branch. By standard methods (see, e.g. [11]), the branching rule has a branching factor of $\tau(|S|, |S|, \dots, |S|) = (1 + 2^{|S|-1})^{\frac{1}{|S|}}$, where the τ-function has $1 + 2^{|S|-1}$ arguments. The running time of the algorithm is $\mathcal{O}^*(c^n)$ where c is the largest branching factor. We first note that $|S| \leq k$. Second, note that we can ignore the case $|S| = 1$ here, as this only happens when the algorithm ends with a single root node. Because this occurs at most once in each branch of the search tree, this has only a constant effect on the running time. As such, we restrict our analysis of the worst-case running time to the cases where $2 \leq |S| \leq k$.

For all $k \geq 4$, the running time of the algorithm is $\mathcal{O}^*((1 + 2^{k-1})^{\frac{n}{k}})$ because it is dominated by the case $|S| = k$ since $(1 + 2^{k-1})^{\frac{n}{k}} < (1 + 2^k)^{\frac{n}{k+1}}$ for any $k \geq 4$. For the case $k = 3$, the running time is dominated by the case where a parent node has one child ($|S| = 2$), instead of two children since $\tau(2, 2, 2) > \tau(3, 3, 3, 3, 3)$. However, in this case, we can branch differently: we either take v in the vertex cover, or we discard it and take the parent of v in the vertex cover. This results in a run time of at most $\tau(1, 2) < 1.619$. This improves the running

time for the cases where $k = 2$ and $k = 3$. For $k = 3$, we get a runtime of $\mathcal{O}^*((1 + 2^{k-1})^{\frac{n}{k}}) = \mathcal{O}(1.733^n)$. For $k = 2$, we get a runtime of $\mathcal{O}(1.619^n)$ time.
□

For any given constant k, the runtime is of the form $\mathcal{O}((2 - \epsilon)^n)$ for some $\epsilon > 0$.

Theorem 6. *Capacitated vertex cover on graphs of bounded degree k, for any $k \geq 3$, can be solved in $\mathcal{O}^*((1 + 2^{k-1})^{\frac{n}{k}})$ time.*

Proof. Let T be any spanning tree in G. Use Lemma 2 to enumerate all vertex covers on T and notice that this enumeration must include all vertex covers on G as well. For each enumerated vertex cover, check in polynomial time whether it is a valid capacitated vertex cover.
□

We note that CVC on graphs of bounded degree $k \leq 2$ can be solved in polynomial time.

Corollary 5. *Capacitated vertex cover on graphs given with a spanning tree with maximum degree $k \geq 3$ can be solved in $\mathcal{O}^*((1 + 2^{k-1})^{\frac{n}{k}})$ time. If the graph is given with a spanning tree with maximum degree 2 (a Hamiltonian path), then CVC can be solved in $\mathcal{O}(1.619^n)$ time. If the spanning tree is not given in advance, then CVC on graphs containing a spanning tree of maximum degree $k \geq 2$ can be solved in $\mathcal{O}^*((1 + 2^k)^{\frac{n}{k+1}})$ time.*

Proof. We can use the construction in Theorem 6 on any graph with a given bounded degree spanning tree as long as we have such a spanning tree. If the spanning tree is not given, we cannot compute the required spanning tree as finding a minimum maximum degree spanning tree is NP-hard. In that case, we use a polynomial time approximation which results in a spanning tree of maximum degree at most $k + 1$ [14], and apply Lemma 2 with slightly worse running times.
□

4.2 Graphs with Large Matchings

The main observation which allowed us to break the $\mathcal{O}^*(2^n)$ time barrier on graphs with a bounded degree spanning tree, is that for every edge at least one endpoint needs to be in the cover. This gave us three possibilities per set of edge endpoints, instead of four. We have a similar branching rule when we are given a single edge for which we have not yet decided whether both endpoints should be in the capacitated vertex cover. This branching rules branches in the three possibilities, as we cannot have that both endpoints are not in the capacitated vertex cover, in $\tau(2, 2, 2)$ time.

This rule requires that both endpoints are undecided. Therefore, to use it effectively, we need a significantly large set of disjoint edges, i.e. a matching. The next theorem formalizes this idea.

Theorem 7. *Let $G = (V, E)$ be a graph. If G contains a matching M with $|M| \geq \epsilon n$ for some $\epsilon > 0$, then we can solve capacitated vertex cover on G in $\mathcal{O}^*(3^{\epsilon n} 2^{(1-2\epsilon)n})$ time.*

Proof. Let M be a matching in G with $|M| \geq \epsilon n$ for some $\epsilon > 0$. For every edge $e \in M$ at least one of the endpoints needs to be in the cover. Hence, we can branch into the three relevant subproblems, for every edge $e \in M$, with a branching factor of $\tau(2,2,2) = \sqrt{3}$. Together, these branches remove exactly 2ϵ vertices and take $\mathcal{O}^*(\sqrt{3}^{2\epsilon n})$ time. For every branch, we will brute force the remaining vertices in $\mathcal{O}^*(2^{(1-2\epsilon)n})$ time, which results in the total running time of $\mathcal{O}^*(3^{\epsilon n}2^{(1-2\epsilon)n})$. $\qquad\square$

This matching-based algorithm allows us to solve CVC on any graph which contains a significantly sized matching. In particular, we can apply this to regular graphs.

Definition 7. *A graph $G = (V, E)$ is regular if all vertices have the same degree. A graph G is called k-regular if all vertices have a degree k.*

For k-regular graphs ($k \geq 3$), we can apply Theorem 6 to solve CVC in $\mathcal{O}^*((1 + 2^{k-1})^{\frac{n}{k}})$ time.

To solve CVC on regular graphs in general, we need the following Theorem by Henning and Yeo [19] which states that k-regular graphs have a significant large matching.

Lemma 3. *([19]). Let G be a connected k-regular graph with $k \geq 3$. If k is even, G has a maximum matching of size at least $\min\{(\frac{k^2+4}{k^2+k+2})\frac{n}{2}, \frac{n-1}{2}\}$. If k is uneven, G has a maximum matching of size at least $\frac{(k^3-k^2-2)n-2k+2}{2(k^3-3k)}$.*

We use this lemma to give a lower bound on the size of the maximum matching in any regular graph, and then solve CVC on regular graphs in general by applying Theorem 7.

Corollary 6. *The capacitated vertex cover problem on regular graphs can be solved in $\mathcal{O}^*(3^{\frac{4}{9}n}2^{\frac{1}{9}n}) = \mathcal{O}(1.760^n)$ time.*

Proof. From Lemma 3, we can conclude that for even k, the minimum lower bound of the size of the maximum matching over all connected k-regular graphs is achieved for both $k = 4$ and $k = 6$, which gives a lower bound of $\frac{10n}{22}$. For odd $k \geq 3$, the minimum lower bound on the size of the maximum matching is achieved for $k = 3$, which attains a value of $\frac{4n}{9}$. We conclude that for any k-regular graph with $k \geq 3$, we have a lower bound on the size of a maximum matching of $\frac{4n}{9}$. Applying Theorem 7 completes the proof if $k \geq 3$, while we note that CVC can be solved in polynomial time on 1 and 2-regular graphs. $\qquad\square$

4.3 Instances with Capacity and Edge Restrictions

Thus far we have seen, amongst others, a treewidth-based and a matching-based algorithm to solve CAPACITATED VERTEX COVER on some graph classes. Next, we will combine these two algorithms to solve CVC in other cases. In particular, we will look at instances of CVC where the capacities are bounded by some constant k, followed by CVC on graphs with restrictions on the number of edges. For this we will make frequent use of the following lemma.

Lemma 4. *For any graph G with a maximal matching M, the treewidth of G is at most $2|M|$.*

Proof. Let I be the set of vertices without endpoint in M: this is an independent set. The constructed tree decomposition is a path with a node for every $i \in I$ such that the associated bag equals $X_i = (V \setminus I) \cup \{i\}$. It is easy to verify that this satisfies all properties of a tree decomposition. $\square$

Theorem 8. *Capacitated vertex cover restricted to graphs with maximum vertex capacity $k \geq 2$ can be solved in $\mathcal{O}^*(2^{\frac{1}{1 - \log_{k+1}(\frac{1}{2}\sqrt{3})}n})$ time.*

Proof. Compute a maximum matching M in G. If $|M| \geq \delta n$, for a yet to determine δ, use Theorem 7 to solve the problem in the problem in $\mathcal{O}^*(\sqrt{3}^{2\delta n}2^{(1-2\delta)n})$ time. Otherwise, use Lemma 4 to obtain a tree decomposition of G of width of at most $2\delta n$. Because every vertex has capacity of at most k, we can use Theorem 4 to solve the problem in $\mathcal{O}^*((k+1)^{2\delta n})$ time. If we choose δ such that both running times are equal, this results in a running time of $\mathcal{O}^*(2^{\frac{1}{1 - \log_{k+1}(\frac{1}{2}\sqrt{3})}n})$. $\square$

Note that, for any constant k, the proven running time is $\mathcal{O}^*((2 - \epsilon)^n)$, for some $\epsilon > 0$.

Next, we look at c-sparse graphs: graphs with at most cn edges, for a fixed $c \geq 1$. The restriction on the number of edges allows us to improve the running time of the treewidth algorithm.

Theorem 9. *Capacitated vertex cover on c-sparse graphs can be solved in $\mathcal{O}((2 - \epsilon)^n)$ time, for some $\epsilon > 0$.*

Proof. Compute a maximum matching M in G. If $|M| \geq \frac{1}{2}\delta n$, for a yet to determine δ, use Theorem 7 to solve the problem in $\mathcal{O}((2 - \epsilon)^n)$ time. Otherwise, use Lemma 4 to obtain a tree decomposition T of G of width of at most δn and apply the algorithm of Theorem 4.

Observe that sum of the capacities over all vertices in the graph is at most $2cn$. Consider a bag X_i associated node i of T. Now, we will bound total number of capacity functions f used to index the values $c_i(f)$ in the algorithm in Theorem 4 differently. There are at most $\prod_{v \in X_i}(c(v) + 1)$ such capacity functions $f : X_i \to \mathbb{N}$, which should also satisfy $\sum_{v \in X_i} c(v) \leq 2cn$ due to the capacity restriction. If we maximise $\prod_{v \in X_i}(c(v) + 1)$ under the constraint $\sum_{v \in X_i} c(v) \leq 2cn$, we find that the worst-case running time occurs when $c(v) = \frac{2cn}{|X_i|}$ for all $v \in X_i$, resulting in $(\frac{2cn}{|X_i|} + 1)^{|X_i|}$ different capacity functions f. This is an increasing function with respect to $|X_i|$. Therefore, it takes a maximum value when $|X_i|$ is it's maximum: $|X_i| = \delta n$. The runtime of the treewidth algorithm then becomes $\mathcal{O}^*((\frac{2cn}{\delta n} + 1)^{\delta n}) = \mathcal{O}^*((\frac{2c}{\delta} + 1)^{\delta n})$ time. For any constant c we can find a small enough constant δ such that $(\frac{2c}{\delta} + 1)^{\delta} < 2$, which results in a run time of $\mathcal{O}^*((2 - \epsilon)^n)$ for some $\epsilon > 0$. $\square$

Finally, we will look at c-dense graphs: graphs having at least cn^2 edges, for a fixed $c > 0$.

Theorem 10. *Capacitated vertex cover on c-dense graphs can be solved in* $\mathcal{O}^*(\sqrt{3}^{cn}2^{(1-c)n})$ *time.*

Proof. We will prove that c-dense graphs have a matching M of size at least $\frac{1}{2}cn$. The result then follows from Theorem 7.

Assume a c-dense graph G does not have a matching M of size at least $\frac{1}{2}cn$. Then, G has an independent set I with $|I| > (1-c)n$. Because of this independent set, the graph has $|E| < cn \cdot (1 - c)n + (cn)^2 = cn^2$. This is a contradiction to G being a c-dense graph. $\qquad\square$

5   Conclusion

We have shown several NP-completeness proofs, showing that CAPACITATED VERTEX COVER (CVC) is NP-complete on planar bipartite graphs with maximum degree 3, modular graphs and tree-convex graphs. We showed that, assuming the ETH, CVC on two subclasses of bipartite graphs, namely modular graphs and tree-convex graphs, cannot be solved in subexponential time.

We have also given a treewidth algorithm, which runs in $\mathcal{O}^*((k+1)^{tw})$ or $\mathcal{O}^*(k^k)$ time, improving the results of Dom et al. [8]. Finally, we have introduced a series of algorithms which break the $\mathcal{O}^*(2^n)$ time barrier for CAPACITATED VERTEX COVER on some graph classes.

We have looked for, but have not found, an algorithm which solves the general CVC problem in $\mathcal{O}((2-\epsilon)^n)$ time, which remains an open problem for further research. In this search [30], we have considered an interesting variation of CVC, namely PARTIAL CAPACITATED VERTEX COVER (PCVC). In this problem, we are given a graph with capacities and a subset $V' \subseteq V$ such that $I = V \setminus V'$ forms an independent set. For the vertices in V' it is decided whether they are included in or excluded from the capacitated vertex cover. The problem asks to determine if there exists a capacitated vertex cover smaller than some integer k respecting the choices already made on V'.

For the PCVC problem, one can show [30] that it cannot be solved in $\mathcal{O}((2-\epsilon)^{|I|})$ time, for any $\epsilon > 0$, unless the Strong Exponential Time Hypothesis fails. This shows the difficulty of the CVC problem with respect to branching algorithms, as a branching algorithm can hit instances of the PCVC as subproblems. In order to get an $\mathcal{O}((2-\epsilon)^n)$-time algorithm for CVC, we might need to use more involved analyzing techniques like measure and conquer to overcome the difficulty of the PCVC subproblems. We have not tried this because in all of our attempts we already found PCVC instance very early in the branching tree, for example on split graphs with a very large independent set and a small clique. It might be the case that no $\mathcal{O}((2-\epsilon)^n)$ time algorithm exists for CVC.

We would like to conclude this paper by a comment on the $\mathcal{O}(n \log n)$ time algorithm for the *weighted* capacitated vertex cover problem on trees by Guha et al. [17]. When one considers this algorithm for the *unweighted* CVC problem, one directly sees that their sorting step that costs $\mathcal{O}(n \log n)$ time can be replaced by a counting sort in $\mathcal{O}(n)$ time. One then obtains the following.

Proposition 1. *Capacitated vertex cover on trees can be solved in $\mathcal{O}(n)$ time.*

References

1. Bandelt, H.J.: Hereditary modular graphs. Combinatorica **8**(2), 149–157 (1988)
2. Bodlaender, H.L., et al.: Open problems in parameterized and exact computation - IWPEC 2008. Technical report UU-CS-2008-017, Department of Information and Computing Sciences, Utrecht University (2008)
3. Bourgeois, N., Escoffier, B., Paschos, V.Th., van Rooij, J.M.M.: Fast algorithms for max independent set. Algorithmica **62**(1), 382–415 (2012)
4. Chen, J., Kanj, I.A., Xia, G.: Improved upper bounds for vertex cover. Theoret. Comput. Sci. **411**(40), 3736–3756 (2010)
5. Cygan, M., Nederlof, J., Pilipczuk, M., Pilipczuk, M., van Rooij, J.M.M., Wojtaszczyk, J.O.: Solving connectivity problems parameterized by treewidth in single exponential time. arXiv.org. The Computing Research Repository abs/1103.0534 (2011)
6. Cygan, M., Pilipczuk, M.: Exact and approximate bandwidth. Theoret. Comput. Sci. **411**(40–42), 3701–3713 (2010)
7. Cygan, M., Pilipczuk, M., Wojtaszczyk, J.O.: Capacitated domination faster than $O(2^n)$. Inf. Process. Lett. **111**(23), 1099–1103 (2011)
8. Dom, M., Lokshtanov, D., Saurabh, S., Villanger, Y.: Capacitated domination and covering: a parameterized perspective. In: Grohe, M., Niedermeier, R. (eds.) IWPEC 2008. LNCS, vol. 5018, pp. 78–90. Springer, Heidelberg (2008). https://doi.org/10.1007/978-3-540-79723-4_9
9. Fomin, F.V., Grandoni, F., Kratsch, D.: A measure & conquer approach for the analysis of exact algorithms. J. ACM **56**(5), 25:1–25:32 (2009)
10. Fomin, F.V., Høie, K.: Pathwidth of cubic graphs and exact algorithms. Inf. Process. Lett. **97**(5), 191–196 (2006)
11. Fomin, F.V., Kratsch, D.: Exact Exponential Algorithms. TTCSAES. Springer, Heidelberg (2010). https://doi.org/10.1007/978-3-642-16533-7
12. Fomin, F.V., Kratsch, D., Woeginger, G.J.: Exact (exponential) algorithms for the dominating set problem. In: Hromkovič, J., Nagl, M., Westfechtel, B. (eds.) WG 2004. LNCS, vol. 3353, pp. 245–256. Springer, Heidelberg (2004). https://doi.org/10.1007/978-3-540-30559-0_21
13. Fomin, F.V., Thilikos, D.M.: A simple and fast approach for solving problems on planar graphs. In: Diekert, V., Habib, M. (eds.) STACS 2004. LNCS, vol. 2996, pp. 56–67. Springer, Heidelberg (2004). https://doi.org/10.1007/978-3-540-24749-4_6
14. Fürer, M., Raghavachari, B.: Approximating the minimum degree spanning tree to within one from the optimal degree. In: 3rd Annual ACM-SIAM Symposium on Discrete Algorithms, SODA 1992. pp. 317–324. Society for Industrial and Applied Mathematics (1992)
15. Garey, M.R., Johnson, D.S.: The rectilinear steiner tree problem is NP-complete. SIAM J. Appl. Math. **32**(4), 826–834 (1977)
16. Grandoni, F.: A note on the complexity of minimum dominating set. J. Discret. Algorithms **4**(2), 209–214 (2006)
17. Guha, S., Hassin, R., Khuller, S., Or, E.: Capacitated vertex covering with applications. In: 13th Annual ACM-SIAM Symposium on Discrete Algorithms, SODA 2002, pp. 858–865. Society for Industrial and Applied Mathematics (2002)
18. Guo, J., Niedermeier, R., Wernicke, S.: Parameterized complexity of vertex cover variants. Theory Comput. Syst. **41**(3), 501–520 (2007)
19. Henning, M.A., Yeo, A.: Tight lower bounds on the size of a maximum matching in a regular graph. Graphs Comb. **23**(6), 647–657 (2007)

20. Impagliazzo, R., Paturi, R.: On the complexity of k-SAT. J. Comput. Syst. Sci. **62**(2), 367–375 (2001)

21. Impagliazzo, R., Paturi, R., Zane, F.: Which problems have strongly exponential complexity? J. Comput. Syst. Sci. **63**(4), 512–530 (2001)

22. Iwata, Y.: A faster algorithm for dominating set analyzed by the potential method. In: Marx, D., Rossmanith, P. (eds.) IPEC 2011. LNCS, vol. 7112, pp. 41–54. Springer, Heidelberg (2012). https://doi.org/10.1007/978-3-642-28050-4_4

23. Jian, T.: An $O(2^{0.304n})$ algorithm for solving maximum independent set problem. IEEE Trans. Comput. **35**(9), 847–851 (1986)

24. Kloks, T. (ed.): Treewidth. LNCS, vol. 842. Springer, Heidelberg (1994). https://doi.org/10.1007/BFb0045375

25. Kneis, J., Langer, A., Rossmanith, P.: A fine-grained analysis of a simple independent set algorithm. In: 29th IARCS Annual Conference on Foundations of Software Technology and Theoretical Computer Science, FSTTCS 2009, pp. 287–298. Schloss Dagstuhl - Leibniz-Zentrum fuer Informatik (2009)

26. Liedloff, M., Todinca, I., Villanger, Y.: Solving capacitated dominating set by using covering by subsets and maximum matching. Discret. Appl. Math. **168**, 60–68 (2014)

27. Nederlof, J., van Rooij, J.M.M., van Dijk, T.C.: Inclusion/exclusion meets measure and conquer. Algorithmica **69**, 685–740 (2014)

28. Robson, J.M.: Algorithms for maximum independent sets. J. Algorithms **7**(3), 425–440 (1986)

29. Robson, J.M.: Finding a maximum independent set in time $O(2^{n/4})$. Technical report, Laboratoire Bordelais de Recherche en Informatique, Université Bordeaux I, 1251-01, Bordeaux, France (2001)

30. van Rooij, S.B.: A search for faster algorithms for the capacitated vertex cover problem. Master's thesis. Department of Information and Computing Sciences, Utrecht University (2018)

31. van Rooij, J.M.M., Bodlaender, H.L.: Exact algorithms for dominating set. Discret. Appl. Math. **159**(17), 2147–2164 (2011)

32. Schiermeyer, I.: Efficiency in exponential time for domination-type problems. Discret. Appl. Math. **156**(17), 3291–3297 (2008)

33. Tarjan, R.E., Trojanowski, A.E.: Finding a maximum independent set. SIAM J. Comput. **6**(3), 537–546 (1977)

34. Xiao, M., Nagamochi, H.: Exact algorithms for maximum independent set. Inf. Comput. **255**, 126–146 (2017)

Comparative Expressiveness of Product Line Calculus of Communicating Systems and 1-Selecting Modal Transition Systems

Mahsa Varshosaz[1]([⊠]) and Mohammad Reza Mousavi[2]

[1] Centre for Research on Embedded Systems, Halmstad University,
Halmstad, Sweden
mahsa.varshosaz@hh.se

[2] Department of Informatics, University of Leicester, Leicester, UK
mm789@le.ac.uk

Abstract. Product line calculus of communicating systems (PL-CCSs) is a process calculus proposed to model the behavior of software product lines. Modal transition systems (MTSs) are also used to model variability in behavioral models. MTSs are known to be strictly less expressive than PL-CCS. In this paper, we show that the extension of MTSs with hyper transitions by Fecher and Schmidt, called 1-selecting modal transition systems (1MTSs), closes this expressiveness gap. To this end, we propose a novel notion of refinement for 1MTSs that makes them more suitable for specifying variability for software product lines and prove its various essential properties.

Keywords: Product line calculus of communicating systems (PL-CCS)
Modal transition system (MTSs)
1-selecting modal transition system (1MTS)
Comparative expressiveness

1 Introduction

Variability modeling is a cornerstone of software product line (SPL) engineering, through which an inventory of commonalities and differences among different products are specified in a structured manner. Efficient analysis of variability-intensive systems is a major challenge due to the potentially large number of valid products. To this end, many techniques have been adapted, which exploit variability in different types of analysis. A basic building block of many of these techniques is a model for capturing variability at the structural or behavioral level. In this paper, we focus on formal behavioral models that can be used to capture variability; examples of such models include modal transition systems (MTSs) [18], product line calculus of communicating systems (PL-CCS) [14], and featured transition systems (FTSs) [10].

In a previous paper [9], we studied the comparative expressiveness of these formalisms with respect to the set of products (labeled transition systems

© Springer Nature Switzerland AG 2019
B. Catania et al. (Eds.): SOFSEM 2019, LNCS 11376, pp. 490–503, 2019.
https://doi.org/10.1007/978-3-030-10801-4_38

(LTSs)) they can specify. There, we proved that MTSs are strictly less expressive than PL-CCS (and its underlying semantic model, product line labeled transition systems (PL-LTSs)). A formalism that was not studied in our previous paper [9] is 1-Selecting Modal Transition System (1MTS) [12], which extends modal transition systems with (must/may) hyper transitions. Such hyper transitions bundle a number of possible behavior, of which exactly one will be included in each valid product. Using 1MTSs it is possible to model alternative behaviour (choices with XOR relation) in products, which cannot be modeled using MTSs. Intuitively, this seems the very missing modeling feature in order to fill the expressiveness gap between MTSs and PL-LTSs.

In this paper, we show that this extension is indeed sufficient to close the expressiveness gap between MTS and PL-LTS (see Sect. 5). Furthermore, we observed that by considering the current refinement relation provided for 1MTSs, some aspects of behavioral variability, such as persistent choices in recursive specifications, cannot be modeled satisfactorily (see Sect. 3). Hence, we propose a new refinement relation for 1MTSs which addresses these concerns (see Sect. 4) and also leads to more succinct models, and we show that the new refinement relation enjoys the same intuitive properties as the original one [12]. The other direction of comparison (from 1MTSs to PL-LTSs) is left as a future work. However, we conjecture that encoding 1MTSs into PL-LTSs is also possible.

2   Preliminaries

In this section, we explain some basic concepts regarding software product lines, 1-selecting modal transition systems, and product-line labeled transition systems that are used throughout the rest of the paper.

2.1   Software Product Lines

The products in a software product line are developed from a common core. The commonalities and variabilities among products are usually described in terms of features. A feature is a distinctive user-visible aspect or characteristic of the system [15]. The products in a product line can be described as sets of features. There are different types of relations between features in a product line. We explain some of these relations using an example of a vending machine product line. The product line includes three mandatory features, namely, *Coin*, *Drink*, and *Coffee*, which means all the products in this product line should include these three features. There are two types of coin, namely, *Dollar* and *Euro* which have *alternative* relation. This means that a product in this product line can either accept dollar or euro coins but not both. The *Drink* feature has two sub-features as well, namely, *Tea* and *Coffee*. The *Tea* is an optional feature. This means that a product in this product line can offer both tea and coffee or only coffee as drinks (since, *Coffee* is a mandatory feature).

2.2   1-Selecting Modal Transition Systems

Fecher and Schmidt [12] introduced the following definition of 1MTSs.

Definition 1 (1MTS). *A 1-selecting modal transition system, is a tuple* $(\mathbb{S}, A, \rightarrow, -\rightarrow, s_{init})$, *where:*

- $\mathbb{S}$ *is a set of states or processes,*
- *A is a set of actions,*
- $\rightarrow \subseteq \mathbb{S} \times (2^{A \times \mathbb{S}} \setminus \emptyset)$, *is the must hyper transition relation,*
- $-\rightarrow \subseteq \mathbb{S} \times (2^{A \times \mathbb{S}} \setminus \emptyset)$ *is the may hyper transition relation,*
- $s_{init} \subseteq \mathbb{S}$, *is a non-empty set of initial states.*

In each 1MTS, the relation $\rightarrow \subseteq -\rightarrow$ *holds between the sets of may- and must hyper transitions. This means that must hyper transitions also implicitly represent may hyper transitions.*

We use 1MTS, to denote the class of all 1MTSs.

Based on the above definition, there are two types of hyper transitions in a 1MTS, called may- and must hyper transitions. A may hyper transition represents a set of alternative choices which are optional (at most one of the choices can be selected). On the other hand, a must hyper transition represents a set of alternative choices where selecting one of the choices is obligatory. Furthermore, we assume that for each state s, $(s-\rightarrow) = \{\gamma \mid (s, \gamma) \in -\rightarrow\}$. A simple example of a 1MTS is provided in Fig. 1. This 1MTS represents the behavior of products in the vending machine product line.

In order to define how a transition among those in a hyper transition is chosen, the following notion of choice function is used.

Definition 2 (Choice Function). *Let A be a set, and $B \subseteq 2^A$ and $\gamma : B \to A$. Then γ is a choice function if $\forall_{b \in B} : \gamma(b) \in b$. The set of all choice functions on B is defined by choice(B).*

As 1MTSs are abstract models, one can associate with each 1MTS a set of 1MTSs that refine it by allowing for fewer optional choices. The refinement relation on 1MTSs is defined as follows [12].

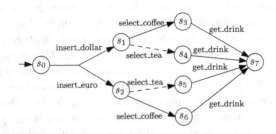

Fig. 1. 1MTS for vending machine product line.

Definition 3 (Refinement for 1MTSs). *A refinement relation between two 1MTSs such as $M = (\mathbb{S}, L, \rightarrow, \dashrightarrow, s_{init})$ and $\bar{M} = (\bar{\mathbb{S}}, L, \bar{\rightarrow}, \bar{\dashrightarrow}, \bar{s}_{init})$, is defined as a relation $\mathcal{R}_{1MTS} \subseteq \mathbb{S} \times \bar{\mathbb{S}}$ such that $\forall_{s \in s_{init}} \cdot \exists_{\bar{s} \in \bar{s}_{init}} \cdot s \; \mathcal{R}_{1MTS} \; \bar{s}$ and $\forall (s, \bar{s}) \in \mathcal{R}_{1MTS} \cdot \forall \gamma \in choice(s\text{-}\dashrightarrow) \cdot \exists \bar{\gamma} \in choice(\bar{s}\text{-}\dashrightarrow)$, such that:*

1. $\forall \, \omega \in (s\text{-}\dashrightarrow) \cdot \exists \, \bar{\omega} \in (\bar{s}\text{-}\dashrightarrow) \cdot \exists \, a \in L, s' \in \mathbb{S}, \bar{s}' \in \bar{\mathbb{S}} \; \cdot \; \gamma(\omega) = (a, s') \wedge \bar{\gamma}(\bar{\omega})$
   $= (a, \bar{s}') \wedge s' \; \mathcal{R}_{1MTS} \; \bar{s}'$,
2. $\forall \, \bar{\omega} \in (\bar{s}\bar{\rightarrow}) \cdot \exists \, \omega \in (s \rightarrow) \cdot \; \exists \, a \in L, s' \in \mathbb{S}, \bar{s}' \in \bar{\mathbb{S}} \; \cdot \; \gamma(\omega) = (a, s') \wedge \bar{\gamma}(\bar{\omega})$
   $= (a, \bar{s}') \wedge s' \; \mathcal{R}_{1MTS} \; \bar{s}'$.

M is said to refine $\bar{M}$, written as $M \triangleright \bar{M}$, when there exists a refinement relation $\mathcal{R}_{1MTS}$ relating each of the initial states of M to one of the initial states of $\bar{M}$.

As a simple example in Fig. 2(1), a 1MTS is shown which refines the 1MTS in Fig. 1. In this 1MTS, the may hyper transitions are not present.

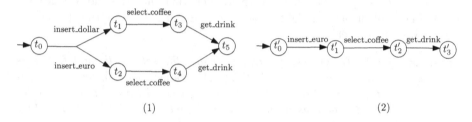

$$(1) \qquad\qquad\qquad\qquad\qquad\qquad (2)$$

Fig. 2. (1) 1MTS and (2) LTS refining the model in Figs. 1 and 2(1).

We define the concrete implementations of a 1MTS as labeled transition systems, defined below.

Definition 4 (LTS). *An LTS is a tuple $(S, A, \rightarrow, s_{init})$, where S is a set of states, A is a set of actions, $\rightarrow: S \times A \times S$ is the transition relation, and s_{init} is the initial state. We denote the class of LTSs by $\mathbb{LTS}$. (We follow the definition given for LTSs as implementations of 1MTSs with single initial states in [12]).*

As a simple example in Fig. 2(2), an LTS is shown which refines the 1MTSs in Figs. 1 and 2(1). In this LTS, the may hyper transitions are not present and the alternative choice among the *insert_euro* and *insert_dollar* is resolved by choosing the former.

2.3 Product Line Process Algebras

Milner's Calculus of Communicating Systems (CCS) [20] is extended by Gruler et al. [14] into PL-CCS by introducing a new operator, called *binary variant*, to represent the alternative behavior. The introduced binary variant operator $\oplus_i$ is different from the ordinary alternative composition operator $+$ in CCS in that the binary variant choice is made once and for all. As an example, consider the process terms $s = a.(b.s + c.s)$ and $t = a.(b.t \oplus_1 c.t)$; recursive process s keeps making choices between b and c in each recursion; while process t makes a choice between b and c in the first recursion after performing a, and the choice is recorded and respected in all the following iterations. This means that process t behaves deterministically after the first iteration with respect to the choice between b and c. To simplify the formal development of the theory, Gruler et al. assume that in every PL-CCS term, there is at most one appearance of the operator $\oplus_i$ for each and every index i. We use the same assumption throughout the rest of the paper, as well.

The semantics of a PL-CCS term is defined based on PL-LTSs [14], using a structural operational semantics, which is explained informally next. The states of a product line labeled transition system are pairs of ordinary states, i.e., process terms, and *configuration vectors*. The transitions of a PL-LTS are also labeled with configuration vectors. These vectors are of type $\{L, R, ?\}^I$ with I being an index set, L and R, respectively, denoting that the choice has been made in favor of the left- or right-hand-side term and ? denoting that the choice has not been made yet.

Definition 5 (PL-LTS). *Let $\{L, R, ?\}^I$ denote the set of all total functions from an index set I to the set $\{L, R, ?\}$. A product line labeled transition system is a 5-tuple $(\mathbb{P} \times \{L, R, ?\}^I, A, I, \rightarrow, p_{init})$ consisting of a set of states $\mathbb{P} \times \{L, R, ?\}^I$, a set of actions A, and a transition relation $\rightarrow \subseteq (\mathbb{P} \times \{L, R, ?\}^I) \times (A \times \{L, R, ?\}^I) \times (\mathbb{P} \times \{L, R, ?\}^I)$, and an initial state $p_{init} \in \mathbb{P} \times \{L, R, ?\}^I$, satisfying the following restrictions:*

1. $\forall_{P,\nu,a,Q,\nu',\nu''} \; (P, \nu) \xrightarrow{a,\nu'} (Q, \nu'') \implies \nu' = \nu''.$

2. $\forall_{P,\nu,a,Q,\nu',i} \; (P, \nu) \xrightarrow{a,\nu'} (Q, \nu') \wedge \nu(i) \neq ? \implies \nu'(i) = \nu(i).$

3. $\forall_{P_0,\nu_0,a,Q_0,\nu_0',i,P_1,\nu_1,b,Q_1,\nu_1',i} \; (P_0, \nu_0) \xrightarrow{a,\nu_0'} (Q_0, \nu_0') \wedge (P_1, \nu_1) \xrightarrow{b,\nu_1'} (Q_1, \nu_1') \wedge$
   $\nu_0(i) = \nu_1(i) = ? \wedge \nu_0'(i) \neq ? \neq \nu_1'(i) \implies (P_0, \nu_0) = (P_1, \nu_1).$

In Definition 5, the conditions follow from the operational rules given by Gruler et al. [14]. The first condition indicates that the change in the configuration is identically reflected in the label and the target. The second condition indicates that a decision made on a choice is recorded as L or R in the configuration vector and would not change in the future. The third condition reflects that the configuration at index i can be resolved in at most one state; this follows immediately from the uniqueness of indices in PL-CCS terms.

In order to define the valid implementations of a PL-LTS, we start with the following relation between the configuration vectors [9].

Definition 6 (Configuration Ordering). *The ordering relation $\sqsubseteq$ on the set $\{L, R, ?\}$ is defined as $\sqsubseteq = \{(?, ?), (L, L), (R, R), (?, L), (?, R)\}$. We lift this ordering relation to the level of configuration vectors by defining $\nu \sqsubseteq \nu' \iff \forall_{i \in I} \; \nu(i) \sqsubseteq \nu'(i)$, for any $\nu, \nu' \in \{L, R, ?\}^I$.*

Considering the above definition, for each $\nu, \nu' \in \{R, L, ?\}^I$, we say $\nu(i) \bowtie \nu(j) \Leftrightarrow \nu(i) \sqsubseteq \nu(j) \vee \nu(j) \sqsubseteq \nu(i)$, for each $i, j \in I$. We lift this ordering relation to the level of configuration vectors by defining $\nu \bowtie \nu' \iff \forall_{i \in I} \nu(i) \bowtie \nu'(i)$, for any $\nu, \nu' \in \{L, R, ?\}^I$.

In order to compare the expressiveness of 1MTS with PL-LTS, we define product derivation relation for a PL-LTS as follows [9].

Definition 7 (Refinement for PL-LTSs). *Let* $(\mathbb{P} \times \{L, R, ?\}^I, A, \rightarrow, p_{init})$ *be a PL-LTS and let* $(\mathbb{S}, A, \rightarrow, s_{init})$ *be an LTS. A binary relation* $\mathcal{R}_\theta \subseteq \mathbb{S} \times (\mathbb{P} \times \{L, R, ?\}^I)$ *(parameterized by every product configuration* $\theta \in \{L, R\}^I$*) is a product-derivation relation if and only if the following transfer properties are satisfied:*

(a) $\forall_{P,Q,a,\nu,\nu',s}\ s\ \mathcal{R}_\theta\ (P, \nu) \wedge (P, \nu) \xrightarrow{a,\nu'} (Q, \nu') \wedge \nu' \sqsubseteq \theta \Rightarrow \exists_t\ s \xrightarrow{a} t \wedge t\ \mathcal{R}_\theta\ (Q, \nu')$,

(b) $\forall_{P,a,\nu,s,t}\ s\ \mathcal{R}_\theta\ (P, \nu) \wedge s \xrightarrow{a} t \Rightarrow \exists_{Q,\nu'}\ (P, \nu) \xrightarrow{a,\nu'} (Q, \nu') \wedge \nu' \sqsubseteq \theta \wedge t\ \mathcal{R}_\theta\ (Q, \nu')$.

A state $s \in \mathbb{S}$ *in an LTS is (the initial state of) a product of a PL-LTS* (P, ν) *with respect to a configuration vector* θ, *denoted by* $(P, \nu) \vdash_\theta s$, *if* $\nu \sqsubseteq \theta$ *and there exists an* $\mathcal{R}_\theta$ *product-derivation relation such that* $s\ \mathcal{R}_\theta\ (P, \nu)$.

We say an LTS $T = (\mathbb{S}, A, \rightarrow, s_{init})$ is a valid implementation of a PL-LTS $P = (\mathbb{P} \times \{L, R, ?\}^I, A, I, \rightarrow, p_{init})$, denoted by $T \prec P$ if and only if there exists a configuration $\theta \in \{L, R, ?\}^I$ such that $p_{init} \vdash_\theta s_{init}$.

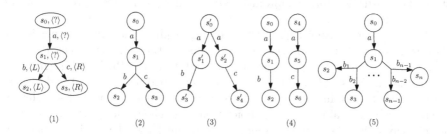

(1)   (2)   (3)   (4)   (5)

Fig. 3. (1) A PL-LTS example (2) A 1MTS with an LTS implementation (3). (4) A 1MTS modeling the same behavior as the PL-LTS in (1). (5) An example of a 1MTS to demonstrate conciseness problem.

3 Design Decisions

In this section, we study the refinement relation provided for 1MTSs by Fecher and Schmidt [12] (see Definition 3) and use some examples to point out a few issues in using this notion of refinement for product derivation. These issues lead us to design decisions for a new notion of refinement, introduced in the next section, that is more suitable for the setting of software product lines.

The first example concerns alternative behavior. Consider the PL-CCS terms $s_0 = a.s_1$ and $s_1 = b.s_2 \oplus_1 c.s_3$. The corresponding underlying PL-LTS is represented in Fig. 3(1). Then, consider the 1MTS shown in Fig. 3(2). Intuitively, this model may be considered as a solution to represent the same set of products using 1MTSs: it bundles the choice between the b- and c-labeled transitions into a must hyper transition. (Recall from Definition 1 that must hyper transitions intuitively represent mandatory choices.) However, in Fig. 3(3), a valid implementation of this 1MTS based on the refinement relation in Definition 3 is depicted. (The dashed arrows show how the states of the LTS and 1MTS are related using the refinement relation.) In the LTS implementation, both the b- and c-labeled transitions are included. A 1MTS that has the same implementations as the PL-LTS in Fig. 3(1), is given in Fig. 3(4); namely, the choice has been lifted to the initial states. This way, the exclusive behavior can be separated among the two parts of the model initiated in these two states.

The process of lifting choices to the initial states can lead to an exponential blow up in 1MTS representation of product lines. This is already hinted at by the 1MTS given in Fig. 3(4) and can be generalized as follows. Consider the 1MTS shown in Fig. 3(5). This model is similar to the 1MTS given in Fig. 3(2) with $k = n/2$ independent exclusive choices (modeled by k must hyper transitions). There are 2^k possible combinations of all choices. This model suffers from the same problem as described above, namely, the alternative transitions can be included simultaneously in some LTS implementations. As mentioned above, in order to model alternative behavior the solution is to use a model with several initial states where each part of the model includes one of the possible combinations. Hence, the model should include 2^k separate parts each with a different initial state. This issue severely compromises succinctness in 1MTS representation of product lines.

Another issue in using 1MTSs for modeling product lines concerns persistent choices. Assume that we add the term $s_3 = d.s_1$ to the aforementioned PL-CCS process term. This will lead to having a new state in the PL-LTS $(s_1, \langle R \rangle)$ and a transition from $(s_3, \langle R \rangle)$ to this state. As mentioned in Sect. 2.3, the decisions made about the exclusive choices are stored in configuration vectors. Hence, when going back again to s_1, the choice that was made before, which is R, will not change. Using the current notion of refinement for 1MTSs, it is not possible to keep track of the choices that are made in the past. Assume that we want to model the same behavior (as in Fig. 3(1)) using 1MTSs. Assume a transition from state s_3 to state s_1 with label d is added to the 1MTS represented in Fig. 3(2). One of the valid implementations of such 1MTS is an LTS where b is chosen the first time reaching state s_1 and then c is chosen the next time that this state is reached. The solution to solve this problem, is the same as above (using several initial states) in addition to unrolling loops.

To address these 3 issues, namely, alternative behavior, succinct representation of choice, and persistence choice, we introduce a new notion of refinement for 1MTSs in the next section.

4   Revisiting the Refinement Relation

In this section, we propose a new refinement relation for 1MTSs to address the issues pointed out in the previous section regarding the original refinement relation [12]. Then, we show that our new refinement relation preserves the intuitive properties posed for the original one [12].

4.1   New Refinement Relation

We revisit the refinement relation in Definition 3, and provide a new refinement relation for 1MTSs as follows. First, we define an auxiliary function, namely, the choice resolution function.

Definition 8 (Choice Resolution Function). *Consider a 1MTS $M = (\mathbb{S}, L, \rightarrow, \dashrightarrow, s_{init})$. A choice resolution function is a total function $\Gamma : \mathbb{S} \rightarrow \bigcup_{s \in \mathbb{S}} choice(s\text{-}\twoheadrightarrow)$. We denote the set of all choice resolution functions of the 1MTS M by Γ_M.*

The purpose of defining the choice resolution function is to assign a choice function to each state of the 1MTS once and for all. Next, we give the refinement relation for 1MTSs as follows.

Definition 9 (New Refinement for 1MTS). *Consider two arbitrary 1MTSs $M = (\mathbb{S}, L, \rightarrow, \dashrightarrow, s_{init})$ and $M' = (\mathbb{S}', L, \rightarrow', \dashrightarrow', s'_{init})$, we say M refines M', denoted by $M \triangleright M'$, iff there exists a refinement relation $\mathcal{R}_{1MTS} \subseteq S \times S' \times \Gamma_M \times \Gamma_{M'}$ such that $\forall f \in \Gamma_M \exists f' \in \Gamma_{M'} \forall s_0 \in s_{init} \exists s'_0 \in s'_{init} \cdot (s_0, s'_0, f, f') \in \mathcal{R}_{1MTS}$ and $\forall (s, s', f, f') \in \mathcal{R}_{1MTS}$, the following conditions hold:*

(i) $\forall \omega \in (s\text{-}\twoheadrightarrow) \cdot \exists \omega' \in (s'\text{-}\twoheadrightarrow') \cdot \exists a \in L, s'' \in \mathbb{S}, s''' \in \mathbb{S}' \cdot f(s)(\omega) = (a, s'')$
    $\wedge f'(s')(\omega') = (a, s''') \wedge (s'', s''', f, f') \in \mathcal{R}_{1MTS}$, and
(ii) $\forall \omega' \in (s' \rightarrow') \cdot \exists \omega \in (s \rightarrow) \cdot \exists a \in L, s'' \in \mathbb{S}, s''' \in \mathbb{S}' \cdot f(s)(\omega) = (a, s'')$
    $\wedge f'(s')(\omega') = (a, s''') \wedge (s'', s''', f, f') \in \mathcal{R}_{1MTS}$.
(iii) Additionally, $\forall s_1 \in S, f'' \in \Gamma_{M'} \cdot (s_1, s', f, f'') \in \mathcal{R}_{1MTS} \Rightarrow f' = f''$.

In the rest of the paper, we use $\mathcal{R}_{1MTS}^{f, f'}$ to denote a 1MTS refinement relation that follows the above definition (that uses choice resolution functions f and f'). In Fig. 4(1), an example of a 1MTS is given. Based on the Definition 3, the 1MTS in Fig. 4(2) is refining this 1MTS. However, based on the Definition 9, this is not a valid refinement for the 1MTS in Fig. 4(1). Hence, the problem with modeling alternative behavior that was mentioned in Sect. 3 is solved in the new definition. Similarly the problems with modeling the conciseness and the persistent behavior are solved.

4.2   Refinement Relation Properties

We prove a set of properties for the new refinement relation as follows. This is the same set of properties proven for the original 1MTS refinement relation by Fecher and Schmidt in [12]. (Due to space limitation, the proofs are omitted and we will include them in an extended version of the paper.) First, we show that the new refinement relation is a preorder.

Proposition 1. *The refinement relation given in Definition 9, is a preorder.*

Next, we show that all the LTS implementations of a 1MTS also implement the 1MTSs that are refined by this 1MTS.

Proposition 2. *Consider two 1MTSs M and M' such that $M \triangleright M'$. Then $\forall lts \in$ $\mathbb{LTS} \cdot lts \triangleright M \Rightarrow lts \triangleright M'$.*

Next, we prove that the bisimulation relation satisfies the properties of the refinement relation in Definition 9.

Proposition 3. *Consider two arbitrary LTSs lts_1 and lts_2 such that $lts_1 \sim lts_2$, where $\sim$ denotes strong bisimilarity; it follows that $lts_1 \triangleright lts_2$.*

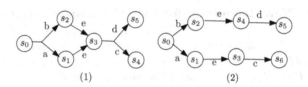

Fig. 4. (1) A 1MTS example. (2) A 1MTS refining (1).

5   Encoding PL-LTSs into 1MTSs

In order to compare the expressiveness of PL-LTSs with 1MTSs, following the approach provided by Beohar et al. in [9], we define an encoding from PL-LTSs into 1MTSs. The main idea of giving an encoding is to define a transformation from one class of models into the other class of models that is semantic preserving. First, we give the following auxiliary definitions taken from [9].

Definition 10 (Product Line Structure). *A product line structure is a tuple $\mathbf{M} = (\mathbb{M}, [\![\;]\!])$, where $\mathbb{M}$ is the class of the intended product line models (in this paper 1MTSs and PL-LTSs) and $[\![\;]\!] : \mathbb{M} \to \mathbb{LTS}$ is the semantic function mapping a product formalism to a set of product LTSs that can be derived from each product line model.*

Next, we give the formal definition of an encoding.

Definition 11 (Encoding). *An encoding from a product line structure* $\mathbf{M} = (\mathbb{M}, [\![\;]\!])$ *into* $\mathbf{M}' = (\mathbb{M}', [\![\;]\!]')$, *is defined as a function* $E : \mathbb{M} \to \mathbb{M}'$ *satisfying the following correctness criterion:* $[\![\;]\!] = [\![\;]\!]' \circ E$. *We say a product line structure* $\mathbf{M}'$ *is at least as expressive as* $\mathbf{M}$ *if and only if there exists an encoding* $E : \mathbf{M} \to \mathbf{M}'$.

Before elaborating on the proposed encoding, we give two auxiliary definitions which are used for encoding the transitions of a PL-LTS into must/may hyper transitions of a 1MTS. As (hyper) transitions in a 1MTS are transitions with multiple targets (see Definition 1), we need to group some of the transitions in a PL-LTS, which correspond to resolving the same alternative choice, and encode them as a (may/must) hyper transition. To this end, we consider the type of changes that is made by a transition to the configuration vector of a PL-LTS. A transition for which the configuration vectors in the source and target states are not identical, is corresponding to resolving a choice (making a decision about one of the variant operators). We formally define the hyper must closed set and hyper may closed set as follows.

Definition 12 (Hyper Must Closed Set). *Consider a state* (P, ν) *of a PL-LTS such as* $(\mathbb{P} \times \{L, R, ?\}^I, A, I, \to, p_{init})$; *we assume that* $Out^{(P,\nu)}$ *denotes the set of all outgoing transitions from state* (P, ν) *and* $Out_{\delta}^{(P,\nu)}$ *denotes the set of outgoing transitions form* (P, ν) *that make a change in at least one of the elements of the configuration vector of the source state, i.e., for each* $(P, \nu) \xrightarrow{a, \nu'} (P', \nu') \in Out_{\delta}^{(P,\nu)}$, *there exists an* $i \in I$ *s.t.* $\nu(i) = ? \wedge \nu'(i) \neq ?$. *A set* $T \subseteq \to$ *of transitions is hyper must-closed for* (P, ν) *when it is a maximal subset of* $Out_{\delta}^{(P,\nu)}$ *such that:*

- *For each* $(P, \nu) \xrightarrow{a_0, \nu_0} (Q_0, \nu_0) \in T$, *and each* $i \in I$ *s.t.* $\nu(i) \neq \nu_0(i)$ *there exists a* $(P, \nu) \xrightarrow{a_1, \nu_1} (Q_1, \nu_1) \in T$ *s.t.* $\neg(\nu_0(i) \bowtie \nu_1(i))$ *and for all* $j \neq i$, $\nu_0(j) \bowtie \nu_1(j)$.
- *For each two different transitions* $(P, \nu) \xrightarrow{a_0, \nu_0} (Q_0, \nu_0) \in T$ *and* $(P, \nu) \xrightarrow{a_1, \nu_1} (Q_1, \nu_1) \in T$, *exists* $i \in I$ *s.t.* $\neg(\nu_0(i) \bowtie \nu_1(i))$.

We denote the set of all such maximal subsets for a state (P, ν), *by* $\mathcal{T}_{\to}^{(P,\nu)}$.

Definition 13 (Hyper May Closed Set). *The hyper may closed set for a state* (P, ν), *denoted by* $\mathcal{T}_{\dashrightarrow}^{(P,\nu)}$, *is defined the same as the hyper must closed set as given in Definition 12, with the only difference that the first condition is replaced with the following condition.*

- *For each* $(P, \nu) \xrightarrow{a_0, \nu_0} (Q_0, \nu_0) \in T$, *for some* $i \in I$ *s.t.* $\nu(i) \neq \nu_0(i)$ *there exists a* $(P, \nu) \xrightarrow{a_1, \nu_1} (Q_1, \nu_1) \in T$ *s.t.* $\neg(\nu_0(i) \bowtie \nu_1(i))$ *and for all* $j \neq i$, $\nu_0(j) \bowtie \nu_1(j)$.

Next, we formalise the encoding of a PL-LTS into a 1MTS.

Definition 14 (PL-LTS to 1MTS Encoding). *Let* $(\mathbb{P}, A, I, \to, p_{init})$ *be a PL-LTS. We construct a 1MTS* $M = (\mathbb{S}, A, \to, \dashrightarrow, s_{init})$ *as an encoding of such a PL-LTS as follows.*

- *The set $\mathbb{S}$ of states is defined as $\mathbb{P}$, i.e., the set of states in the PL-LTS, $p_{init} = s_{init}$, A is the same set of actions,*
- *We construct the $\rightarrow$ and $\dashrightarrow$, which, respectively, denote the must and may hyper transition relations for each state of the the the encoding 1MTSs as follows. Given Definitions 12 and 13, we define the following transition rules:*

$$((P,\nu) \rightarrow) = \bigcup_{\substack{\Lambda \in \mathcal{T}_{\rightarrow}^{(P,\nu)} \\ |\Lambda| > 1}} \{ \bigcup_{1 \leq i \leq |\Lambda|} \{(a_i, (P_i, \nu_i))\} | (P,\nu) \xrightarrow{a_i, \nu_i} (P_i, \nu_i) \in \Lambda\} \cup$$

$$\bigcup_{1 \leq i \leq |Out^{(P,\nu)} \setminus Out_\delta^{(P,\nu)}|} \{\{(a_i, (P_i, \nu))\} | (P,\nu) \xrightarrow{a_i, \nu} (P_i, \nu) \in Out^{(P,\nu)} \setminus Out_\delta^{(P,\nu)}\}$$

$$((P,\nu) \dashrightarrow) = \bigcup_{\Lambda \in \mathcal{T}_{\dashrightarrow}^{(P,\nu)}} \{ \bigcup_{1 \leq i \leq |\Lambda|} \{(a_i, (P_i, \nu_i))\} | (P,\nu) \xrightarrow{a_i, \nu_i} (P_i, \nu_i) \in \Lambda\} \cup$$

$$\{(a, (P', \nu'))| (P,\nu) \xrightarrow{a, \nu'} (P', \nu') \in \big(Out_\delta^{(P,\nu)} \setminus (\mathcal{T}_{\rightarrow}^{(P,\nu)} \cup \mathcal{T}_{\dashrightarrow}^{(P,\nu)})\big)\}$$

Given the above encoding, we prove that the class of 1MTSs is at least as expressive as the class of PL-LTSs. (Due to space limitation, the proofs are omitted and we will include them in an extended version of the paper.)

Theorem 1. *The class of 1MTSs is at least as expressive as the class of PL-LTSs.*

6   Related Work

In this section, we discuss related work regarding formalisms used for modeling product lines and the comparison of their expressiveness. We limit our consideration to the models which have LTSs as the semantic domain.

Considering the comparison of the expressiveness of the formalisms used for modeling variability, Beohar et al. in [9] provide a comparison between the expressiveness of three fundamental models, namely, MTSs, PL-CCSs, and Feature Transition Systems (FTSs). (FTSs [11] are extensions of LTSs with propositional formulas called feature expressions.) A novel notion of encoding, based on the set of implementing LTSs, from one class of models to the other is provided. The existence of mutual encodings between two classes of models is described as having the same expressiveness. As a result a hierarchy of formalisms based on their expressiveness is provided. Furthermore, Benduhn et al. in [7], provide a survey on formalisms focusing on the suitability of these models in applying different analysis techniques.

Considering the formalisms proposed for modeling product lines; In [13], Fischbein et al. for the first time argued that MTSs are adequate for modeling variability. In several works, MTSs have been used for modeling variability in the behavior of product lines [1–3,16,19]. As shown in [9], MTSs are the least expressive in the provided hierarchy. In order to tackle the limited expressiveness of MTSs, several extensions of such models have been proposed. In a set of works, MTSs are used with variability constraints [6], which are constraints expressed in Modal-Hennessy-Milner-Logic (MHML) [1–3]. In [17], an extension of MTSs, namely, Disjunctive Modal Transition Systems (DTMSs) are introduced which provides the possibility to model an *or* relation between choices in the behavior using hyper transitions. Fecher and Schmidt in [12], introduce 1MTSs, which (as mentioned in Sect. 2) can be used for modeling alternative choices. Furthermore, in this work, a comparison between the expressiveness of these two models is provided, which shows that the two classes of models have the same expressiveness concerning the sets of implementing LTSs. Benes et al. in [8], introduce an extension of MTSs, namely, parametric modal transition systems in which the concept of obligation functions is used. The obligation functions are defined upon atomic propositions of states, the transitions, and a set of parameters, which can be used for representing features. By setting the valuation of parameters the presence or absence of states and transitions in a specific product model can be specified. Moreover, an extension of contract automata with modality [5] is introduced by Basile et al. in [4]. In this extension of the model, permitted and necessary requests are distinguished using feature constraints. There have been other approaches introduced that use some interface theories principles to indicate the set of derivable variants from an MTS as the ones that are compatible under parallel composition with regards to a given environmental specification [16,19].

As mentioned in Sect. 2, PL-CCS [14], introduced by Gruler et al. [14], is an extension of Milner's CCS [20] by means of an alternative choice operator called "binary variant". This operator provides the possibility of modeling persistent choices in the behavior. The validity of variants can be further restricted using the multi-valued modal *mu*-calculus [21].

To the best of our knowledge, the provided encoding from PL-LTSs into 1MTSs, the results regarding the expressiveness, and the provided refinement relation for 1MTSs that addresses the limitations of such models in modeling variability in the behavior in this paper are novel.

7 Conclusion

In this paper, we compared the expressiveness of PL-LTSs and 1MTSs. To this end, we defined the set of products for specifications in both formalisms, of which the behaviors are commonly specified in the domain of LTSs. We then showed that 1MTSs can capture all products that can be specified by the product line calculus of communicating systems. Furthermore, we provided a set of observations regarding the limitations in modeling variability in the behavior which are enforced by the refinement relation given for 1MTSs. We proposed a new refinement relation for 1MTSs to tackle these limitations and proved a set of properties for the new refinement relation.

An immediate question to ask is whether the two formalism have the same expressive power or not. We conjecture that the answer is positive and leave this for immediate future work. We also would like to combine the results of this paper with our earlier results in [9] and present a comprehensive lattice of expressive power among all fundamental behavioral models for software product lines. As another part of our future work, we plan to provide a stronger relation between PL-LTSs and PL-CCS terms by introducing a set of conditions (on the configuration vectors of states) in a PL-LTS which guarantee that the PL-LTS is induced from a PL-CCS term.

References

1. Asirelli, P., ter Beek, M.H., Fantechi, A., Gnesi, S.: A model-checking tool for families of services. In: Bruni, R., Dingel, J. (eds.) FMOODS/FORTE -2011. LNCS, vol. 6722, pp. 44–58. Springer, Heidelberg (2011). https://doi.org/10.1007/978-3-642-21461-5_3

2. Asirelli, P., ter Beek, M.H., Gnesi, S., Fantechi, A.: Formal description of variability in product families. In: Proceedings of the 15th International Software Product Line Conference, SPLC 2011, pp. 130–139. IEEE (2011)

3. Asirelli, P., ter Beek, M.H., Fantechi, A., Gnesi, S.: A compositional framework to derive product line behavioural descriptions. In: Margaria, T., Steffen, B. (eds.) ISoLA 2012. LNCS, vol. 7609, pp. 146–161. Springer, Heidelberg (2012). https://doi.org/10.1007/978-3-642-34026-0_12

4. Basile, D., ter Beek, M.H., Di Giandomenico, F., Gnesi, S.: Orchestration of dynamic service product lines with featured modal contract automata. In: Proceedings of the 21st International Systems and Software Product Line Conference, SPLC 2017, vol. B, pp. 117–122. ACM, New York (2017). https://doi.org/10.1145/3109729.3109741

5. Basile, D., Di Giandomenico, F., Gnesi, S., Degano, P., Ferrari, G.L.: Specifying variability in service contracts. In: Proceedings of the Eleventh International Workshop on Variability Modelling of Software-Intensive Systems, VAMOS 2017, pp. 20–27. ACM, New York (2017). https://doi.org/10.1145/3023956.3023965

6. ter Beek, M.H., Fantechi, A., Gnesi, S., Mazzanti, F.: Modelling and analysing variability in product families: model checking of modal transition systems with variability constraints. J. Logic. Algebraic Methods Program. 85(2), 287–315 (2016)

7. Benduhn, F., Thüm, T., Lochau, M., Leich, T., Saake, G.: A survey on modeling techniques for formal behavioral verification of software product lines. In: Proceedings of the Ninth International Workshop on Variability Modelling of Software-Intensive Systems, VaMoS 2015, New York, NY, USA, pp. 80:80–80:87 (2015)

8. Beneš, N., Křetínský, J., Larsen, K.G., Møller, M.H., Srba, J.: Parametric modal transition systems. In: Bultan, T., Hsiung, P.-A. (eds.) ATVA 2011. LNCS, vol. 6996, pp. 275–289. Springer, Heidelberg (2011). https://doi.org/10.1007/978-3-642-24372-1_20

9. Beohar, H., Varshosaz, M., Mousavi, M.R.: Basic behavioral models for software product lines: expressiveness and testing pre-orders. Sci. Comput. Program. (2015, in press)

10. Classen, A., Cordy, M., Schobbens, P.Y., Heymans, P., Legay, A., Raskin, J.F.: Featured transition systems: foundations for verifying variability-intensive systems and their application to LTL model checking. IEEE Trans. Softw. Eng. 39(8), 1069–1089 (2013)

11. Classen, A., Heymans, P., Schobbens, P.Y., Legay, A., Raskin, J.F.: Model checking lots of systems: efficient verification of temporal properties in software product lines. In: Proceedings of the 32nd International Conference on Software Engineering, ICSE 2010, vol. 1, pp. 335–344. ACM (2010)
12. Fecher, H., Schmidt, H.: Comparing disjunctive modal transition systems with an one-selecting variant. J. Logic Algebraic Program. **77**(1–2), 20–39 (2008)
13. Fischbein, D., Uchitel, S., Braberman, V.: A foundation for behavioural conformance in software product line architectures. In: Proceedings of the ISSTA Workshop on Role of Software Architecture for Testing and Analysis, pp. 39–48. ACM (2006)
14. Gruler, A., Leucker, M., Scheidemann, K.: Modeling and model checking software product lines. In: Barthe, G., de Boer, F.S. (eds.) FMOODS 2008. LNCS, vol. 5051, pp. 113–131. Springer, Heidelberg (2008). https://doi.org/10.1007/978-3-540-68863-1_8
15. Kang, K., Cohen, S., Hess, J., Novak, W., Peterson, S.: Feature-oriented domain analysis (FODA) feasibility study. Technical report CMU/SEI-90-TR-21. Software Engineering Institute, Carnegie Mellon University (1990)
16. Larsen, K.G., Nyman, U., Wąsowski, A.: Modal I/O automata for interface and product line theories. In: De Nicola, R. (ed.) ESOP 2007. LNCS, vol. 4421, pp. 64–79. Springer, Heidelberg (2007). https://doi.org/10.1007/978-3-540-71316-6_6
17. Larsen, K.G., Xinxin, L.: Equation solving using modal transition systems. In: Proceedings of the Fifth Annual Symposium on Logic in Computer Science, LICS 1990, pp. 108–117. IEEE Computer Society (1990). https://doi.org/10.1109/LICS.1990.113738
18. Larsen, K., Thomsen, B.: A modal process logic. In: Proceedings of the 3rd Annual Symposium on Logic in Computer Science, LICS 1988, pp. 203–210. IEEE (1988)
19. Lochau, M., Kamischke, J.: Parameterized preorder relations for model-based testing of software product lines. In: Margaria, T., Steffen, B. (eds.) ISoLA 2012. LNCS, vol. 7609, pp. 223–237. Springer, Heidelberg (2012). https://doi.org/10.1007/978-3-642-34026-0_17
20. Milner, R. (ed.): A Calculus of Communicating Systems. LNCS, vol. 92. Springer, Heidelberg (1980). https://doi.org/10.1007/3-540-10235-3
21. Shoham, S., Grumberg, O.: Multi-valued model checking games. J. Comput. Syst. Sci. **78**(2), 414–429 (2012)

A Hierarchy of Polynomial Kernels

Jouke Witteveen[1]([✉])(iD), Ralph Bottesch[2], and Leen Torenvliet[1]

[1] Institute for Logic, Language, and Computation, Universiteit van Amsterdam,
Amsterdam, The Netherlands
j.e.witteveen@uva.nl
[2] Department of Computer Science, Universität Innsbruck, Innsbruck, Austria

Abstract. In parameterized algorithmics the process of *kernelization* is defined as a polynomial time algorithm that transforms the instance of a given problem to an equivalent instance of a size that is limited by a function of the parameter. As, afterwards, this smaller instance can then be solved to find an answer to the original question, kernelization is often presented as a form of *preprocessing*. A natural generalization of kernelization is the process that allows for a number of smaller instances to be produced to provide an answer to the original problem, possibly also using negation. This generalization is called *Turing* kernelization. Immediately, questions of equivalence occur or, when is one form possible and not the other. These have been long standing open problems in parameterized complexity. In the present paper, we answer many of these. In particular we show that Turing kernelizations differ not only from regular kernelization, but also from intermediate forms as truth-table kernelizations. We achieve absolute results by diagonalizations and also results on natural problems depending on widely accepted complexity theoretic assumptions. In particular, we improve on known lower bounds for the kernel size of compositional problems using these assumptions.

Keywords: Kernelization · Parameterized complexity
Turing reductions · Truth-table reductions · Kernel lower bounds

1 Introduction

Fixed-Parameter Tractability. For many important computational problems, the best known algorithms have a worst-case running time that scales exponentially or worse with the size of the input. Generally however, the size of an input instance is a poor indicator of whether the instance is indeed difficult to solve. This is because for most natural problems, a good fraction of all instances of a given size can be solved much more efficiently than the worst-case instance of that size. To gain a better understanding of the *complexity of individual instances*, we might define a function $\kappa : \{0,1\}^* \to \mathbb{N}$ that assigns to each instance x a numeric *parameter* $\kappa(x)$. This parameter then indicates the extent to which certain features that we have identified as a potential cause of computational

© Springer Nature Switzerland AG 2019
B. Catania et al. (Eds.): SOFSEM 2019, LNCS 11376, pp. 504–518, 2019.
https://doi.org/10.1007/978-3-030-10801-4_39

hardness are present in the given instance. If the function κ is itself polynomial-time computable, we call it a *parameterization*. We shall assume that $\kappa(x) \leq |x|$ holds for all $x \in \{0,1\}^*$.

Consider a problem for which the fastest known algorithm has a worst-case running time in $2^{O(|x|)}$. If, for some parameterization κ, we can give an algorithm of which the worst-case running time on any instance x is in $2^{O(\kappa(x))}\mathrm{poly}(|x|)$ and, furthermore, we have that $\kappa(x) \ll |x|$ holds for at least *some* arbitrarily large instances, then we can argue that κ is a more accurate measure of the complexity of instances than is their size, since the running time of the second algorithm is exponential *only* in the parameter value. Note that this implies that interesting parameterizations cannot be monotonic functions. More generally, for $X \subseteq \{0,1\}^*$ and a parameterization κ, a *parameterized problem* (X, κ) is said to be *fixed-parameter tractable (fpt)* if, for some computable function f and constant $c \geq 0$, there is an algorithm solving any instance x of X in time $f(\kappa(x))|x|^c$.[1] The essential feature of such running times is that the parameter value and instance size appear only in separate factors.

Kernelization. An important notion in the study of fixed-parameter tractability is that of *kernelization*. Informally, a kernelization (or kernel) for a parameterized problem is a polynomial-time algorithm that, for any input instance, outputs an equivalent instance of which the size is upper-bounded by a function of the parameter. This type of algorithm is usually presented as a formalization of preprocessing in the parameterized setting. It reduces any instance with large size but small parameter value to an equivalent smaller instance, after which some other algorithm (possibly one with large complexity) is used to solve the reduced instance. Another explanation, which fits well with the idea of studying the complexity of individual instances, is that a kernelization extracts the *hard core* of an instance.

Of particular interest is the case where the upper bound on the size of the output instance of a kernelization is itself a polynomial function in the parameter. Such *polynomial kernelizations* are important because they offer a quick way to obtain efficient fpt-algorithms for a problem. If X is solvable in exponential-time, then the existence of a polynomial kernelization for (X, κ) means that the problem can be solved in time $2^{\mathrm{poly}(k)}\mathrm{poly}(n)$, which roughly corresponds to what we might reasonably consider to be useful in practice. Conversely, for many parameterized problems that can be solved by algorithms with such running times (for example, k-VERTEX-COVER), it is also possible to show the existence of polynomial kernelizations. However, there are also exceptions, such as the k-PATH problem, where an algorithm with time complexity $2^{O(k)}\mathrm{poly}(n)$, but no polynomial kernelization, is known. It was a long-standing open question whether the existence of polynomial kernelizations is equivalent to having fpt-algorithms with a particular kind of running time. Eventually, Bodlaender et al. [2] showed that for many fixed-parameter tractable problems (including

---

[1] From here onward, we may write k for $\kappa(x)$ when there is no risk of confusion. Also, n stands for $|x|$ when specifying the complexity of an algorithm.

k-PATH), the existence of polynomial kernels would imply the unlikely complexity-theoretic inclusion $\mathbf{NP} \subseteq \mathbf{coNP}_{/\mathbf{poly}}$. This framework for proving conditional lower bounds against polynomial kernels was subsequently considerably extended and strengthened [3,6] (see also the survey of Kratsch [12]). In the same paper, Bodlaender et al. also unconditionally prove the existence of a parameterized problem that is solvable in time $O(2^k n)$, but has no polynomial kernels, thus ruling out the possibility of an equivalence between polynomial kernels and fpt-algorithms with running times of the form $2^{\mathrm{poly}(k)}\mathrm{poly}(n)$.

Generalized Kernelization. A *Turing kernelization* is an algorithm that can solve any instance of a parameterized problem in polynomial-time, provided it can query an oracle for the same problem with instances of which the size is upper-bounded by a function of the parameter value of the input. The idea here is that if we are willing to run an inefficient algorithm on an instance of size bounded in terms of the parameter alone (as was the case with *regular* kernelizations), then we might as well run this algorithm on more than one such instance. A regular kernelization can be regarded as a particular, restricted type of Turing kernelization that (a) runs the polynomial kernelization algorithm on the input, (b) queries the oracle for the resulting output instance, and (c) outputs the oracle's answer. As in the case of regular kernelizations, a *polynomial Turing kernelization* is such that the bound on the size of the query instances is itself a polynomial function.

Polynomial Turing kernelizations are not as well-understood as regular kernels. The methods for proving lower bounds against the size of regular kernels do not seem to apply to them. Indeed, there are problems that most likely have no polynomial kernels, but which *do* admit a polynomial Turing kernelization. An example being k-LEAF-SUBTREE (called MAX-LEAF-SUBTREE in [4]). Furthermore, there are only a few examples of non-trivial polynomial Turing kernelizations for problems that are not believed to admit polynomial regular kernelizations, such as restricted versions of k-PATH [9,10] and of k-INDEPENDENT SET [15]. Whether the general versions of these problems also have polynomial Turing kernels are major open questions in this field.

Compared to the regular kind, polynomial Turing kernelizations have a number of computational advantages, such as the ability to output the opposite of the oracle's answer to a query (non-monotonicity), the ability to make polynomially (in the size of the input) many queries, and the ability to adapt query instances based on answers to previous queries (adaptiveness). Rather than focus on specific computational problems to determine the difference in strength between Turing and regular kernelizations, we instead look into the possibility of unconditionally separating the computational strengths of these two types of algorithms in general. We investigate and answer a number of questions that, to our knowledge, were all open until now:

– Without relying on any complexity-theoretic assumptions, can we prove the existence of parameterized problems that admit polynomial Turing but not polynomial regular kernelizations? If so, which of the computational advantages of Turing kernelizations are sufficient for an unconditional separation?

Note that for k-LEAF-SUBTREE, only a larger number of queries is used, the
known polynomial Turing kernel being both monotone and non-adaptive (see
[4], Sect. 9.4). On the other hand, the kernels in [9] and [15] are adaptive.
- Does every parameterized problem that is decidable in time $2^{poly(k)}poly(n)$,
  also admit a polynomial Turing kernelization?
- To what extent can we relax the restrictions on regular kernelizations (viewed
  as Turing kernelizations), while still being able to apply known lower bound
  techniques? For example, can we rule out, for some natural problems, the
  existence of non-monotone kernels that make a few adaptive oracle queries?

1.1 Overview of Our Results

We show that each of the advantages of polynomial Turing kernelizations over polynomial
regular kernelizations is, by itself, enough
to unconditionally separate the two notions.
This produces a hierarchy of kernelizability
within the class of problems that admit polynomial Turing kernelizations, Fig. 1. Specifically, we show that:

- there are problems that are not polynomially kernelizable, but do admit
  a polynomial Turing kernelization
  that makes a single oracle query (Theorem 1);
- there are problems that admit non-
  adaptive polynomial Turing kernelizations
  (also known as polynomial *truth-table* kernelizations), but cannot be solved by polynomial Turing kernelizations making a
  constant number of queries, even adaptively (Theorems 2 and 3);
- there are problems that admit adaptive polynomial Turing kernelizations but
  not polynomial truth-table kernelizations
  (Theorem 4).

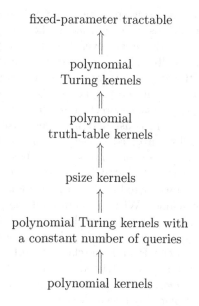

Fig. 1. A hierarchy of polynomial
kernels. Arrows signify a strict
increase in computational power.

Next, we show (Theorem 5) that it is not enough for a problem to be decidable
in time $2^{poly(k)}poly(n)$ in order for it to have a polynomial Turing kernelization.
In fact, the problem we construct can be solved in time $O(2^k n)$. Our theorem
is stronger than a comparable result of Bodlaender et al., who only exclude
regular kernelizations. We obtain a considerably simpler proof, harnessing the
Time Hierarchy Theorem in favor of a direct diagonalization.

Finally, we ask how far up the hierarchy the known methods for proving
lower bounds against polynomial kernelization can be applied. The example of

k-LEAF-SUBTREE shows that they should already fail somewhere below polynomial truth-table kernelizations. Indeed, we identify what we call *psize kernelizations* as the apparently strongest type of polynomial Turing kernel that can be ruled out by current lower bound techniques (Sect. 4). A psize kernelization makes $\operatorname{poly}(k)$ non-adaptive oracle queries (of size $\operatorname{poly}(k)$), and then feeds the oracle's answers into a poly-sized circuit to compute its own final answer. In terms of computational power, this type of kernelization stands between polynomial Turing kernelizations that make only a constant number of queries and polynomial truth-table kernelizations (Sect. 3, Theorems 2 and 3).

1.2   Proof Techniques

The price we pay for being able to prove unconditional separations is that the problems we construct in the proofs are artificial rather than natural. This is unavoidable, however, because computational problems that arise naturally will typically belong to classes that are hard to separate from **P** (such as **NP**, **PH**, **PP**, etc.). Thus, any claim that some parameterized version of a natural problem admits no polynomial kernelization, would currently have to rely on some complexity-theoretic assumptions.

In the construction of every problem witnessing a separation, diagonalization will be involved, in one way or another. However, the application of diagonalization arguments in this context has some subtle issues. An intuitive reason for this is the fact that it is very difficult to control the complexity of a problem that is constructed via an argument using diagonalization against polynomial-time machines. Without additional complexity-theoretic assumptions, such problems can be forced to reside in powerful classes such as **EXP**. Positioning them in any interesting smaller classes is not straightforward. By contrast, the difference between **P** and the class of problems that can be decided in polynomial-time with a very restricted form of access to an oracle, seems rather thin, and it is by no means clear whether a problem that is constructed via diagonalization can be placed between these two classes. In Sect. 3 we discuss these issues, as well as how to overcome them, in detail. Here, let us mention that the overall structure of our artificial problems resembles that of examples of natural problems which, subject to complexity-theoretic assumptions, admit polynomial Turing but not regular kernelizations. Because of this, even the artificial examples we construct provide new insights into the power of Turing kernelization.

2   Preliminaries

We assume familiarity with standard notations and the basics of parameterized complexity theory, and refer the reader to [7] for the necessary background. Here we review only the definitions of the notions most important for our work.

Definition 1. *A* kernelization *(or* kernel*) for a parameterized problem* (X, κ), *where* $X \subseteq \{0, 1\}^*$ *and* κ *is a parameterization, is a polynomial-time algorithm* that, on a given input $x \in \{0, 1\}^*$, outputs an instance $x' \in \{0, 1\}^*$ such that $x \in X \Leftrightarrow x' \in X$ holds, and, for some fixed computable function f, we have $|x'| \leq f(\kappa(x))$. The function f is referred to as the size of the kernel. The kernel is said to be polynomial if f is a polynomial.

Definition 2. A Turing kernelization for a parameterized problem (X, κ) is a polynomial-time algorithm that decides any instance x of X using oracle queries to X of restricted size. For some fixed computable function f that is independent of the input, the size of the queries must be upper bounded by $f(\kappa(x))$. A Turing kernelization is polynomial if f is a polynomial.

A Turing kernelization is a truth-table kernelization if, on every input, all of its oracle queries are independent of the oracle's answers. Thus, as an oracle machine, a truth-table kernelization is non-adaptive.

A parameterized problem that exemplifies the relevance of our results is k-LEAF-SUBTREE, where a graph G and integer k are given, and the question is whether G has a subtree with at least k leaves. This problem admits a polynomial Turing kernelization but no polynomial regular kernelization, unless **NP** $\subseteq$ **coNP**$_{/\mathbf{poly}}$. See Sect. 9.4 of [4] for a proof of the former, and Chap. 15 of the same reference for a proof of the latter fact.

3 Separations

To prove an unconditional separation between polynomial Turing kernelizability and polynomial regular kernelizability (or between two intermediate kinds of kernelizability), we construct a problem of which the instances can be solved in polynomial-time with oracle queries for small instances of the same problem. We shall make sure that the instances cannot be solved in polynomial-time without such queries (remember, polynomial kernelizations are also poly-time decision procedures). These requirements prevent us from constructing the classical part of our parameterized problem via simple diagonalization against polynomial-time machines. The instances of the resulting language would not depend on each other in a way that would allow oracle queries to be useful, nor would all instances be solvable in time $p(n)$ for some *fixed* polynomial p. Solving an instance of such a language requires simulating Turing machines (*TM*s) for a polynomial number of steps, but the degree of these polynomials increases with n. Thus, a hypothetical polynomial Turing kernelization would neither be able to solve the instances of such a language directly within the allowed time, nor use its oracle access to speed up the computation. An additional difficulty arises due to the bound on the size of the oracle queries (polynomial in k). If the parameter value of an instance x is too small relative to $|x|$, then the restricted oracle access of a polynomial Turing kernelization may offer no computational advantage, since the instances for which the oracle can be queried will be small enough to be solved directly within the required time bound.

These issues can be overcome by designing a problem that shares what seems to be the essential feature of natural problems that, under complexity-theoretic

assumptions, admit polynomial Turing but not polynomial (regular) kernelizations, such as the k-LEAF-SUBTREE problem. Recall that for this problem, a
quadratic kernelization exists for the case when the input graph is connected,
but that a polynomial kernelization for general graphs is unlikely to exist. The
known polynomial Turing kernelization for this problem works on general graphs
by computing the kernel for each connected component of the input graph, and
then querying the oracle for each of the $O(n)$ resulting instances of size $O(k^2)$
(see [4], Sect. 9.4). The crucial aspect here is that although the general problem may not admit polynomial kernelizations, it has a subproblem that does.
Furthermore, the polynomial Turing kernelization only queries instances of this
subproblem.

The problems we construct will also have a polynomially kernelizable "core,"
as well as a "shell" of instances that can be solved efficiently with small queries
to the core. Taking V to be some decidable language, we can define

$$X(V) = \{0x \mid x \in V\} \cup \{1x \mid \ldots\},$$

where the ellipsis stands for a suitable condition that can be verified with small
queries to V. With the parameterization κ such that $\kappa(0x) = |x|$ and $\kappa(1x) =
\log|x|$ for all $x \in \{0,1\}^*$, the first set in the above disjoint union plays the role of
the polynomially kernelizable core (it admits the trivial kernelization), while the
second set plays the role of the shell. The crucial observation now is that we can
choose the condition that determines membership of an element of the form $1x$
in $X(V)$ in such a way that a polynomial-time algorithm can decide the instance
using small queries of the form $0w$, *regardless of the choice of V*. Having thus
secured the existence of a polynomial Turing kernelization (perhaps one that is
further restricted), we are now free to construct V via diagonalization against
some weaker type of kernelization, so as to get the desired separation.

Using this approach, we prove that each of the computational advantages
a polynomial Turing kernelization has over polynomial (regular) kernelizations,
results in a strictly stronger type of kernelization, as shown in Fig. 1.

Theorem 1. *There is a parameterized problem that has a polynomial Turing kernelization using only a single oracle query, but admits no polynomial
kernelizations.*

Proof. Given any decidable set V, we can define

$$X(V) = \{0x \mid x \in V\} \cup \left\{1x \;\middle|\; \log|x| \in \mathbb{N} \text{ and } 0^{\log|x|} \notin V\right\},$$

parameterized so that for all $x \in \{0,1\}^*$, $\kappa(0x) = |x|$ and $\kappa(1x) = \log|x|$.

Clearly, the problem $(X(V), \kappa)$ has a polynomial Turing kernelization making
a single query, regardless of the decidable set V. For instances of the form $0x$,
the answer can be obtained by querying the oracle directly for the input, and if
the input is $1x$, one can query $0^{\log|x|+1}$ and output the opposite answer.

We shall construct the set V by diagonalization, ensuring that $X(V)$ does
not admit a polynomial (regular) kernelization. Note that the kernelization procedures we diagonalize against can query $X(V)$, whereas we only decide the

elements of V. Because every problem that admits a polynomial kernelization can also be decided by a polynomial-time TM that makes a single query of size poly(k) and then outputs the oracle's answer, we only need to diagonalize against this type of TM. As in a standard diagonalization argument, we run every such machine for an increasing number of steps, using as input the string 10^{2^n} (the parameter value of which is n), where n is chosen large enough for decisions made at previous stages to not interfere with the current simulation. Each machine is simulated until it runs out of time or makes an oracle query. Whenever the machine makes an oracle query different from 10^{2^n}, we answer it according to the current state of the set V. To complete the diagonalization, we either add 0^n to V or not, so as to ensure the machine's answer is incorrect.

Note that for sufficiently large values of n, the string 10^{2^n} cannot be queried, because 2^n outgrows any fixed polynomial in n ($\in$ poly(k)). Additionally, a query to 00^n is of no concern as the machine is incapable of negating the answer of the oracle. □

Next, we show that polynomial truth-table kernelizations, which can make poly(n) oracle queries of size poly(k) but cannot change their queries based on the oracle's previous answers, are more powerful than a restricted version of the same type of kernelization that makes at most poly(k) queries. This restricted form of polynomial truth-table kernelization is of further interest because it can be ruled out by the current lower bounds techniques (see Sect. 4). We give the definition here.

Definition 3. *A polynomial truth-table kernelization is a psize kernelization if, on any input instance with parameter value k, it makes at most poly(k) oracle queries and its output can be expressed as the output of a poly(k)-sized circuit that takes the answers of the oracle queries as input.*

The proof of the next theorem follows the same pattern as that of Theorem 1, except that in the diagonalization part of the proof we now use the restriction on the number of queries the machines can make. Recall that in Theorem 1 we made use of the machine's monotonicity, that is, the fact that its output must be equivalent to the outcome of its single oracle query.

Theorem 2. *There is a parameterized problem that has a polynomial truth-table kernelization but no psize kernelization.*

A proof is available in the appendix. The condition used for the shell is that V contains a string of length $\log |x|$. The conclusion of the proof is actually that there exists a parameterized problem with a polynomial truth-table kernelization making $n - 1$ oracle queries, that admits no polynomial (possibly adaptive!) Turing kernelization making fewer than $n - 2$ queries on certain inputs of length n. A psize kernel fits this condition, but is much more restricted (in particular, the number of allowed queries is polynomial in the parameter value).

Via a very similar proof, with a diagonalization argument relying on the number of oracle queries a machine can make, we can show that psize kernelizations are stronger than polynomial Turing kernelizations making any fixed finite number of queries, even adaptively.

Theorem 3. *There is a parameterized problem that has a psize kernelization but no polynomial Turing kernelization making only a constant number of (possibly adaptive) queries.*

We can also show that adaptive queries provide a concrete computational advantage. The proof of the separation between general polynomial Turing and truth-table kernelizations also follows the pattern of the previous three theorems, but with a more involved diagonalization argument, due to the need to distinguish between adaptive and non-adaptive oracle TMs.

Theorem 4. *There is a parameterized problem that has a polynomial Turing kernelization but no polynomial truth-table kernelization.*

A proof is included in the appendix and hinges on a series of $(\log |x|)^2$ queries to V, each query depending on the outcome of the one before it.

Finally, we show that decidability in time $2^{\text{poly}(k)}\text{poly}(n)$ does not guarantee the existence polynomial Turing kernelizations for the same problem. This strengthens a theorem of Bodlaender et al. [2], who construct a problem with the above complexity but rule out only polynomial regular kernelizations.

Theorem 5. *For every time-constructible function $g(k) \in 2^{o(k)}$, there is a problem that is solvable in time $O(2^k n)$ but admits no Turing kernelization of size $g(k)$. In particular, there is a problem that is solvable in time $O(2^k n)$ but admits no polynomial Turing kernelization.*

Proof. Let $g(k)$ be a time-constructible function in $2^{o(k)}$. Without loss of generality, we may assume that $g(k)$ is also in $\Omega\left(2^{(\log k)^2}\right)$. Let $\kappa : \mathbb{N} \to \mathbb{N}$ be a time-constructible function such that we have $\kappa(n) \in \omega(\log n) \cap o(n)$ as well as $\kappa(g(k)) \in o(k)$ (for example, $\kappa(n) = \log n \log\left(\frac{g^{-1}(n)}{\log n}\right)$ is suitable). Let $t(n) = 2^{\kappa(n)}n$ and let L be a language in **DTIME**$(t(n)) \setminus$ **DTIME**$(o(t(n)/\log(t(n)))$. Such a language exists by the Time Hierarchy Theorem. Assigning each instance x of L the parameter value $k = \kappa(|x|)$, we find that L can be solved in time $O(2^k n)$.

Furthermore, we have

$$\frac{t(n)}{\log t(n)} = \frac{2^{\kappa(n)}n}{\kappa(n) + \log n} \in \Omega\left(2^{\kappa(n)}\right),$$

so we may conclude $2^{o(\kappa(n))} \subseteq o(t(n)/\log(t(n)))$.

Assume now that for some polynomial p, there exists a Turing kernelization for L that runs in time $p(n)$ and queries the oracle with instances of size bounded by $g(k)$, where we set $k = \kappa(n)$. We show that such a Turing kernelization can be used to solve L in time $o(t(n)/\log(t(n))$, contradicting the choice of the language. Our new algorithm will solve any instance x with parameter value $k = \kappa(|x|)$ by running the Turing kernelization on it, except that the instances for which the oracle is supposed to be queried are solved directly using the $O(2^{\kappa(n)}n)$-time algorithm whose existence is guaranteed by the choice of L. The total running time of this new algorithm is then upper-bounded by:

$$p(n) + p(n)2^{\kappa(g(k))}g(k) = 2^{o(k)} = 2^{o(\kappa(n))},$$

which contradicts the lower bound on the deterministic time complexity of L. $\square$

4   Lower Bounds

An immediate consequence of the separations arrived at in the previous section is that not all fixed-parameter tractable problems have polynomial kernelizations. However, for any particular parameterized problem the (non-)existence of a polynomial kernelization may not be easy to establish. The most fruitful program for deriving superpolynomial lower bounds on the size of regular kernelizations was started by Bodlaender et al. [2]. While a straightforward application of their technique to Turing kernelizations is not possible, an extension to the psize level in our hierarchy, Fig. 1, is feasible.

In order to keep our presentation focussed, we shall include only a limited exposition of the lower bound technique. For a more complete overview, refer to [5,12], or turn to [3] for an in-depth treatment. Central to the lower bounds engine are two similar looking classifications of instance aggregation. The first of these does not involve a parameterization.

Definition 4. *A* weak **and**-distillation *(*weak **or**-distillation*) of a set* X *into a set* Y *is an algorithm that*

- *receives as input a finite sequence of strings* $x_1, x_2, \ldots, x_t$,
- *uses time polynomial in* $\sum_{i=1}^{t} |x_i|$,
- *outputs a string* y *such that*
  - *we have* $y \in Y$ *if and only if for all (any)* i *we have* $x_i \in X$,
  - $|y|$ *is bounded by a polynomial in* $\max_{1 \leq i \leq t} |x_i|$.

Note how the size of the output of a distillation is bounded by a polynomial in the *maximum* size of its inputs and not by the sum of the input sizes. Originally, distillations where considered where the target set Y was equal to X, hence the *weak* designator in this more general definition. The parameterized counterpart to distillations is, as we shall soon see, more lenient than the non-parameterized one.

Definition 5. *An* **and**-compositional *(*or-compositional*) parameterized problem* (X, κ) *is on for which there is an algorithm that*

- *receives as input a finite sequence of strings* $x_1, x_2, \ldots, x_t$ *sharing a parameter value* $k = \kappa(x_1) = \kappa(x_2) = \ldots = \kappa(x_t)$,
- *uses time polynomial in* $\sum_{i=1}^{t} |x_i|$,
- *outputs a string* y *such that*
  - *we have* $y \in X$ *if and only if for all (any)* i *we have* $x_i \in X$,
  - $\kappa(y)$ *is bounded by a polynomial in* k.

Here, a bound is placed on the *parameter value* of the output of the algorithm, instead of on the *length* of the output. Additionally, this bound is a function of the *unique* parameter value shared by all input strings. Conceptually, a bound of this kind makes sense as parameter values serve as a proxy of the computational hardness of instances. Thus, a parameterized problem is compositional, when instances can be combined efficiently, without an increase in computational hardness.

Generalizing the results of Bodlaender et al. [2,3], we find that not just regular polynomial kernelizations, but also psize kernelizations tie the two ways of aggregating instances together. For our proof to work, two aspects of the definition of psize kernelizations on page 8 that were not made explicit are crucial. Firstly, because a psize kernelization is a *polynomial* truth-table kernelization, the size of the queries can be bounded by a polynomial of the parameter value. Secondly, it is important to note that the circuits involved must be uniformly computable from the input instances.

Theorem 6. *If (X, κ) is an **and**-compositional (**or**-compositional) parameterized problem that has a psize kernelization, then X has a weak **and**-distillation (weak **or**-distillation).*

Proof. Given a set X, consider the following set based on circuits and inputs derived from membership in X,

$$C(X) = \{\langle \phi, (x_1, x_2, \dots, x_t)\rangle \mid \phi \text{ is a circuit with } t \text{ inputs},$$
$$\text{accepting } (x_1 \in X, x_2 \in X, \dots, x_t \in X)\}.$$

Note that a pairing of the specification of a circuit ϕ and t strings $(x_1, x_2, \dots, x_t)$ can be done so that $|\langle \phi, (x_1, x_2, \dots, x_t)\rangle|$ is bounded by a polynomial in $|\phi| + |x_1| + |x_2| + \dots + |x_t|$.

We sketch the proceedings of a distillation that is given $x_1, x_2, \dots, x_t$ as input. This procedure is adapted from [2].

First, the inputs are grouped by their parameter value $k_i = \kappa(x_i)$ and the composition algorithm is applied to each group, obtaining $(y_1, k_1'), (y_2, k_2'), \dots, (y_s, k_s')$. Taking $k_{\max} = \max_{1 \le i \le t} k_i$, we have $s \le k_{\max}$ and, for some polynomial p, all k_i' are bounded by $p(k_{\max})$.

Next, the psize kernelization is applied to each (y_i, k_i'), obtaining s polynomial sized circuits and s sequences of strings to query in order to get the inputs of the circuits. These circuits and strings can be amalgamated (dependent on the type of composition) into a single circuit ϕ and sequence of strings $(z_1, z_2, \dots, z_r)$.

We claim that the mapping of $(x_1, x_2, \dots, x_t)$ to $\langle \phi, (z_1, z_2, \dots, z_r)\rangle$ constitutes a weak distillation of X into $C(X)$. Both s and $k_{\max}$ are bounded by $\max_{1 \le i \le t} |x_i|$, since, for all i, we have $k_i \le |x_i|$. Therefore, the proposed weak distillation procedure produces an output of which the size is bounded by a polynomial in $\max_{1 \le i \le t} |x_i|$ and its running time is indeed polynomial in $\sum_{i=1}^{t} |x_i|$. Moreover, by definition of a psize kernelization the required preservation of membership is satisfied, hence the procedure is truly a weak distillation of X into $C(X)$. □

Assuming we have $\mathbf{NP} \not\subseteq \mathbf{coNP}_{/\mathbf{poly}}$, it has been shown that $\mathbf{NP}$-hard problems admit neither weak **or**-distillations [8], nor weak **and**-distillations [6]. Thus we can further our generalization of the results of Bodlaender et al. [3].

Corollary 1. *If* (X, κ) *is an* **and**-*compositional (***or***-compositional) parameterized problem and* X *is* $\mathbf{NP}$-*hard, then* (X, κ) *does not have a psize kernelization unless we have* $\mathbf{NP} \subseteq \mathbf{coNP}_{/\mathbf{poly}}$.

Accordingly, our hierarchy of polynomial kernels is not merely synthetic and the place of many natural problems in the hierarchy is lower bounded. In light of the more general setting of Bodlaender et al. [3], we remark that a generalization of our results to cross-composition (generalizing compositionality) and psize compression (generalizing psize kernelization) is immediate.

5   Classical Connections

Algorithms for fixed-parameter tractable problems are not easily diagonalized against. Such algorithms have a running time of the form $f(\kappa(x))|x|^c$, where f is a computable function and c a constant. The challenge in diagonalizing is caused by the absence of a computable sequence of computable functions such that every computable function is outgrown by a member of the sequence. However, as witnessed by this document, diagonalization can be used to uncover structure *inside* **FPT**. Key to this possibility is that a problem is fixed-parameter tractable precisely when it is kernelizable, and the running time bound for kernelizations does not include arbitrary computable functions.

While, to our knowledge, not done before in a parameterized context, separating many–one, truth-table, and Turing reductions is an old endeavour, dating back to Ladner et al. [13]. Indeed, kernelizations are in essence reductions, more specifically, they are *autoreductions* in the spirit of Trakhtenbrot [16]. Since kernelizations come with a time bound, a Turing kernelization could more accurately be described as a *bounded Turing* kernelization, or *weak truth-table* kernelization (see [14], Sect. 3.8). However, the adaptiveness of a Turing kernelization entails that the number of different queries it *could* make (unaware of the answers of the Oracle) is much higher than that of a truth-table kernelization, given the same time bound. In that sense, our separation based on adaptiveness, Theorem 4, is also a separation based on the number of queries made.

An important feature of kernelizations is not covered by an interpretation of kernelizations as autoreductions. Where the definition of an autoreduction excludes querying the input string, the definition of a kernelization imposes a stronger condition on the queries, namely a size bound as a function of the parameter value. In this light, it may be worthwhile comparing kernelizations to a more restrictive type of autoreduction, the *self-reduction* (see [1], Sect. 4.5). Self-reducibility is defined in [1] as autoreducibility where all queries are shorter than the input. However, many of the results around self-reducibility extend to more general orders than the "shorter than"-order and the definition can be generalized [11]. While the size bound on the queries that is required of kernelizations does not fit the self-reducibility scheme perfectly, the similarities in the definitions urge the consideration of other forms of self-reducibility in a parameterized context. In particular, reducibility with a decreasing parameter value may be of interest.

Appendix (Deferred Proofs)

Theorem 2. *There is a parameterized problem that has a polynomial truth-table kernelization but no psize kernelization.*

Proof. Given any decidable set V, we can define

$$X(V) = \{0v \mid v \in V\} \cup \left\{ 1x \mid \log|x| \in \mathbb{N} \text{ and } \{0,1\}^{\log|x|} \cap V \neq \emptyset \right\},$$

parameterized so that for all $x \in \{0,1\}^*$, $\kappa(0x) = |x|$ and $\kappa(1x) = \log|x|$. Clearly, $(X(V), \kappa)$ has a polynomial truth-table kernelization regardless of V: on input $0x$ it queries the oracle for the input, and on input $1x$, with $\log|x| \in \mathbb{N}$, it queries the oracle with each string $0y$, for all $y \in \{0,1\}^{\log|x|}$, and accepts if one of the queries has a positive answer (otherwise it rejects). This procedure runs in polynomial time and makes at most n oracle queries on any input of length $n+1$.

We construct V by diagonalizing against psize kernelization algorithms. To do this, we consider a computable list of TMs such that every machine appears infinitely often. At stage i of the construction we choose n, a power of 2, so that membership in V has not been decided at a previous stage for any strings of length at least $\log n$. We then run the i-th machine on input 10^n for n^i steps. All new oracle queries are answered with 'no', all other queries are answered so as to be consistent with previous answers. If the machine at stage i terminates without having queried the oracle for all strings of the form $0y$ with $y \in \{0,1\}^{\log n}$, we add an unqueried string of this length to V if and only if the machine rejects.

If P is a psize kernelization, then the number of oracle queries it makes on an input $1x$ is upper-bounded by $q(\log|x|)$, for some fixed polynomial q. This is clearly $o(|x|)$, so for some sufficiently large i and n, P will terminate without having queried all n strings which can determine the correct answer. Thus, our diagonalization procedure will ensure that it terminates with the incorrect answer. On the other hand, the above-mentioned polynomial truth-table kernelization will always query all necessary strings in order to output the correct answer. □

Theorem 4. *There is a parameterized problem that has a polynomial Turing kernelization but no polynomial truth-table kernelization.*

Proof. For any decidable set V we can define the function: $s^V : \{0,1\}^* \to \{0,1\}^*$ by

$$s^V(q) = \begin{cases} 0q & \text{if } q \notin V, \\ 1q & \text{if } q \in V. \end{cases}$$

Also for a decidable set V, we define the following parameterized problem:

$$X(V) = \{0x \mid x \in V\} \cup \left\{ 1x \ \middle| \ \log|x| \in \mathbb{N} \text{ and } \underbrace{(s^V \circ s^V \circ \cdots \circ s^V)}_{(\log|x|)^2 \text{ times}}(0^{\log|x|}) \in V \right\},$$

where the parameterization is defined so that for all $x \in \{0,1\}^*$, $\kappa(0x) = |x|$ and $\kappa(1x) = \log|x|$. The problem $X(V)$ has a polynomial Turing kernelization regardless of the set V: On inputs of the form $0x$, the machine queries the oracle with its input (whose size is linear in the parameter value), and outputs the answer. On inputs of the form $1x$ the machine makes the following $(\log|x|)^2$ queries: $0^{\log|x|+1}$, $0b_1 0^{\log|x|}$, $0b_2 b_1 0^{\log|x|}, \ldots, 0b_{(\log|x|)^2} \ldots b_1 0^{\log|x|}$, where b_i is the outcome of the i-th query, for each $i \leq (\log|x|)^2$. The output is the answer of the last oracle query. Since each of the queries in the second case is of size at most quadratic in $\kappa(1x) = \log|x|$, this procedure is a polynomial Turing kernelization.

We now construct the set V so that no polynomial truth-table kernelization can solve $X(V)$. Consider a variant of oracle TMs where the oracle can be queried for an arbitrary number of queries at once. Let $P_1, P_2, \ldots$ be a computable list of all such TMs in which each machine appears infinitely often.

At each stage $i \in \mathbb{N}$, we set n to be the smallest positive integer so that no oracle queries to $X(V)$ at any previous stage of the simulation depend on instances of V of size at least n, and so that $n > i$ and $2^n > n^i$. At stage i of the construction, we run P_i on input 10^{2^n} for $(2^n)^i$ steps (note that this is a polynomial of degree i in $2^n + 1$, the size of the input). In case the machine queries the oracle, let S be the set of strings it queries. If S includes strings of length at least 2^n, we move on to the next stage. In particular, when no query of length $2^n + 1$ is made, P_i is not making a query with prefix 1 that is equivalent to the input. By the time bound, we have $|S| \leq 2^{ni} < 2^{n^2}$, so there must be a string $y = b_{n^2} \ldots b_2 b_1 0^n$, $b_j \in \{0,1\}$, such that $0y$ is not in S. The queries in S are answered as follows: all queries also made at previous stages are answered so as to be consistent with previous answers; all queries of the form $0b_j \ldots b_2 b_1 0^n$, with $j \leq n^2 - 1$, are answered with b_{j+1}; all other queries are answered with 0 ('no'). For all $j \leq n^2 - 1$ such that $b_{j+1} = 1$, we place $b_j \ldots b_2 b_1 0^n$ into V. After thus answering the queries in S, we resume the simulation of P_i for the remainder of its allotted 2^{ni} steps and treat every subsequent invocation of the query instruction as a crash. Finally, we place y into V if and only if P_i terminated within the time bound and rejected, making 10^{2^n} a 'yes'-instance if and only the P_i rejects it.

Assume now that there is a polynomial truth-table kernelization for $X(V)$. Such a procedure will eventually be targeted in the above construction. Indeed, a problem has a truth-table kernelization precisely when it is decided by a machine that runs in polynomial time and can make all its queries at once. Let i be such that P_i is a polynomial truth-table kernelization for $X(V)$, running in time $p(|x|)$ on any input of the form $1x$, and non-adaptively making oracle queries of size at most $q(\log|x|)$, where p and q are fixed polynomials. As this machine occurs infinitely often in the list $P_1, P_2, \ldots$, we may assume that i and its corresponding n are large enough for P_i to terminate on input 10^{2^n}, because we have $p(2^n+1) < 2^{ni}$. Moreover, we may assume that i and n are large enough for $q(n) < n^i < 2^n$ to hold. As P_i will not be able to query all strings of the form $0y0^n$ with $|y| = n^2$, it will, by our construction of V, incorrectly decide some instance of $X(V)$. $\square$

References

1. Balcázar, J.L., Díaz, J., Gabarró, J.: Structural Complexity I. Springer, Heidelberg (1995). https://doi.org/10.1007/978-3-642-79235-9
2. Bodlaender, H.L., Downey, R.G., Fellows, M.R., Hermelin, D.: On problems without polynomial kernels. J. Comput. Syst. Sci. **75**(8), 423–434 (2009)
3. Bodlaender, H.L., Jansen, B.M., Kratsch, S.: Kernelization lower bounds by cross-composition. SIAM J. Discret. Math. **28**(1), 277–305 (2014)
4. Cygan, M., et al.: Parameterized Algorithms. Springer, Heidelberg (2016). https://doi.org/10.1007/978-3-319-21275-3
5. Downey, R.G., Fellows, M.R.: Fundamentals of Parameterized Complexity. Springer, Heidelberg (2016). https://doi.org/10.1007/978-1-4471-5559-1
6. Drucker, A.: New limits to classical and quantum instance compression. SIAM J. Comput. **44**(5), 1443–1479 (2015)
7. Flum, J., Grohe, M.: Parameterized Complexity Theory. TTCSAES. Springer, Heidelberg (2006). https://doi.org/10.1007/3-540-29953-X
8. Fortnow, L., Santhanam, R.: Infeasibility of instance compression and succinct PCPs for NP. J. Comput. Syst. Sci. **77**(1), 91–106 (2011)
9. Jansen, B.M.: Turing kernelization for finding long paths and cycles in restricted graph classes. J. Comput. Syst. Sci. **85**, 18–37 (2017)
10. Jansen, B.M., Pilipczuk, M., Wrochna, M.: Turing kernelization for finding long paths in graphs excluding a topological minor. In: 12th International Symposium on Parameterized and Exact Computation (IPEC 2017), vol. 89, pp. 23:1–23:13. Schloss Dagstuhl-Leibniz Zentrum fuer Informatik (2018)
11. Ko, K.I.: On self-reducibility and weak P-selectivity. J. Comput. Syst. Sci. **26**(2), 209–221 (1983)
12. Kratsch, S.: Recent developments in kernelization: a survey. Bull. EATCS **2**(113), 57–97 (2014)
13. Ladner, R.E., Lynch, N.A., Selman, A.L.: A comparison of polynomial time reducibilities. Theor. Comput. Sci. **1**(2), 103–123 (1975)
14. Soare, R.I.: Turing Computability. Springer, Heidlberg (2016). https://doi.org/10.1007/978-3-642-31933-4
15. Thomassé, S., Trotignon, N., Vušković, K.: A polynomial Turing-kernel for weighted independent set in bull-free graphs. Algorithmica **77**(3), 619–641 (2017)
16. Trakhtenbrot, B.A.: On autoreducibility. Doklady Akademii Nauk SSSR **192**(6), 1224–1227 (1970)

Behavioral Strengths and Weaknesses of Various Models of Limited Automata

Tomoyuki Yamakami$^{(\boxtimes)}$

Faculty of Engineering, University of Fukui, 3-9-1 Bunkyo, Fukui 910-8507, Japan
TomoyukiYamakami@gmail.com

Abstract. We examine the behaviors of various models of k-limited automata, which naturally extend Hibbard's [Inf. Control, vol. 11, pp. 196–238, 1967] scan limited automata, each of which is a linear-bounded automaton satisfying the k-limitedness requirement that the content of each tape cell should be modified only during the first k visits of a tape head. One central model is k-limited probabilistic automaton (k-lpa), which accepts an input exactly when its accepting states are reachable from its initial state with probability more than $1/2$. We further study the behaviors of one-sided-error and bounded-error variants of such k-lpa's as well as deterministic and nondeterministic models. We discuss fundamental properties of those machine models and obtain inclusions and separations among language families induced by these machine models. In due course, we study special features—the blank skipping property and the closure under reversal—which are keys to the robustness of k-lpa's.

Keywords: Limited automata · Pushdown automata
Probabilistic computation · Bounded-error probability
One-sided error · Blank skipping property · Reversal

1 Background and Main Contributions

1.1 Limited Automata and Probabilistic Computation

In 1967, Hibbard [3] studied a novel computational model of so-called *scan limited automata* to characterize context-free languages by conducting direct simulations between one-way nondeterministic pushdown automata (or 1npda's) and his model. Hibbard's model seems to have been paid little attention until Pighizzini and Pisoni [10] reformulated the model from a modern-machinery perspective and reproved a characterization theorem of Hibbard in a more sophisticated manner. A *k-limited automaton*,[1] for each fixed index $k \geq 0$, is in general a one-tape (or a single-tape) Turing machine whose tape head is allowed

---

[1] Hibbard's original formulation of "k-limited automaton" is equipped with a semi-infinite tape that stretches only to the right with no endmarker but is filled with the blank symbols outside of an input string. Our definition in this paper is different from Hibbard's but it is rather similar to Pighizzini and Pisoni's [10].

© Springer Nature Switzerland AG 2019
B. Catania et al. (Eds.): SOFSEM 2019, LNCS 11376, pp. 519–530, 2019.
https://doi.org/10.1007/978-3-030-10801-4_40

to rewrite each tape cell between two endmarkers only during the first k scans or visits (except that, whenever a tape head makes a "turn," we count this move as double visits). Although these automata can be viewed as a special case of linear-bounded finite automata, the restriction on the number of times that they rewrite tape symbols brings in quite distinctive effects on the computational power of the underlying automata, different from other restrictions, such as upper bounds on the numbers of nondeterministic choices or the number of tape-head turns. Hibbard actually proved that k-*limited nondeterministic automata* (or k-lna's) for $k \geq 2$ are exactly as powerful as 1npda's, whereas 1-lna's are equivalent in power to 2-way deterministic finite automata (or 2dfa's) [12].

In a subsequent paper [11], Pighizzini and Pisoni discussed a close relationship between k-*limited deterministic automata* (or k-lda's) and one-way deterministic pushdown automata (or 1dpda's). In fact, they proved that 2-lda's embody exactly the power of 1dpda's; in contrast, Hibbard observed that, when $k \geq 3$, k-lda's do not, in general, coincide in computational power with 1dpda's. This observation gives a clear structural difference between determinism and nondeterminism on the machine model of "limited automata" and this difference naturally raises a question of whether other variants of limited automata matches their corresponding models of one-way pushdown automata.

Lately, a computation model of *one-way probabilistic pushdown automata* (or 1ppda's) has been discussed extensively to demonstrate computational strengths as well as weaknesses in [5,7,9,17]. Hromkovič and Schnitger [5] as well as Yamakami [17], in particular, demonstrated clear differences in computational power between two pushdown models, 1npda's and 1ppda's.

While nondeterministic computation is purely a theoretical notion, probabilistic computation could be implemented in real life by installing a mechanism of generating (or sampling) random bits (e.g., by flipping fair or biased coins). A *bounded-error* probabilistic machine makes error probability bounded away from $1/2$, whereas an *unbounded-error* probabilistic machine allows error to take arbitrarily close to probability $1/2$. In most cases, a probabilistic approach helps us solve a target mathematical problem algorithmically faster, and probabilistic (or randomized) computation often exhibits its superiority over its deterministic counterpart. For example, 2-way probabilistic finite automata (or 2pfa's) running in expected exponential time can recognize non-regular languages with bounded-error probability [2]. By contrast, when restricted to expected subexponential runtime, bounded-error 2pfa's recognize only regular languages [1,6]. As this example shows, the expected runtime bounds of probabilistic machines largely affect the computational power of the machines, and thus its probabilistic behaviors significantly differ from deterministic behaviors.

The usefulness of probabilistic algorithms motivates us to take a probabilistic approach toward an extension of Hibbard's original model of k-limited automata. This paper in fact introduces k-*limited probabilistic automata* (or k-lpa's) and their variants, including one-sided-error and bounded-error variants, and to explore their fundamental properties to obtain strengths and weaknesses of families of languages recognized by those machine models.

1.2   Main Contributions

Our first goal is to provide in the field of probabilistic computation a complete characterization of finite and pushdown automata in terms of limited automata. All probabilistic machines are assumed to run in expected polynomial time.

For any error bound $\varepsilon \in [0, 1/2)$, the notations 1PPDA_ε and 2PFA_ε refer to the families of all languages recognized by ε-error 1ppda's and ε-error 2pfa's, respectively. As a restriction of 2PFA_ε, 2RFA_ε denotes the family of all languages recognized by 2pfa's with one-sided error probability at most ε. Similarly, we define 1RPDA_ε as the one-sided-error variant of 1PPDA_ε. In addition, we often use more familiar notation of PCFL, BPCFL, and RCFL respectively for 1PPDA_{ub}, $\bigcup_{0 \leq \varepsilon < 1/2} 1\text{PPDA}_\varepsilon$, and $\bigcup_{0 \leq \varepsilon < 1} 1\text{PPDA}_\varepsilon$, while CFL denotes the family of context-free languages. Since $\{a^n b^n c^n \mid n \geq 0\}$ is in BPCFL [5], BPCFL $\nsubseteq$ CFL follows, further leading to PCFL $\neq$ CFL.

For limited automata, $k\text{-LPA}_\varepsilon$ with an index $k \geq 1$ refers to the family of all languages recognized by k-lpa's with error probability at most ε. Using bounded-error k-lpa's, we denote by $k\text{-LBPA}$ the union $\bigcup_{\varepsilon \in [0,1/2)} k\text{-LPA}_\varepsilon$. In the unbounded-error case, we write $k\text{-LPA}$ (or $k\text{-LPA}_{ub}$ for clarity). Similarly, $k\text{-LRA}_\varepsilon$ is characterized by one-sided ε-error k-lpa's. Let $k\text{-LRA} = \bigcup_{\varepsilon \in [0,1)} k\text{-LRA}$.

Using k-lda's and k-lna's, we define $k\text{-LDA}$ and $k\text{-LNA}$, respectively. Pighizzini and Pisoni [11] demonstrated that 2-LDA coincides with DCFL, which is the deterministic variant of CFL. Hibbard [3] proved that $k\text{-LNA} = \text{CFL}$ for any $k \geq 2$. It is also possible to show that PCFL $\subseteq$ 2-LPA and BPCFL $\subseteq$ 2-LBPA; however, the opposite inclusions are not known to hold. Therefore, our purpose of exact characterizations of PCFL and BPCFL requires a specific property of k-lpa's, called *blank skipping*, for which a k-lpa writes only a unique blank symbol, say, B during the kth visit and it makes only the same deterministic moves while reading B in such a way that it neither changes its inner state nor changes the head direction (either to the right or to the left); in other words, it behaves exactly in the same way while reading consecutive blank symbols. This property plays an essential role in simulating various pushdown automata by limited automata. To emphasize the use of the *blank skipping property*, we append the prefix "bs-", as in bs-2-LPA_ε. We then obtain the following characterizations.

Theorem 1. *Let $\varepsilon \in [0, 1/2)$ be any error bound.*

1. $2\text{PFA}_\varepsilon = 1\text{-LPA}_\varepsilon$ *and* $2\text{PFA}_{ub} = 1\text{-LPA}_{ub}$.
2. $1\text{PPDA}_\varepsilon = \text{bs-2-LPA}_\varepsilon$, $1\text{RPDA}_\varepsilon = \text{bs-2-LRA}_\varepsilon$, *and* PCFL = bs-2-$\text{LPA}_{ub}$.

Theorem 1(2), in particular, follows from the fact shown in Sect. 3.2 that 1ppda's can be converted into their "ideal shapes."

In the case of k-lda's, as shown in Proposition 2, we can transform limited automata into their blank skipping form and this is, in fact, a main reason that 2-LDA equals DCFL (due to Theorem 1(2) with setting $\varepsilon = 0$).

Proposition 2. *For each index $k \geq 2$, $k\text{-LDA} = \text{bs-}k\text{-LDA}$.*

For other limited automata, it is not yet clear that, for example, k-LPA = bs-k-LPA.

The second goal of this paper is to argue on various separations of the aforementioned language families. Earlier, Hibbard [3] devised an example language that can separate $(k+1)$-LDA from k-LDA for each index $k \geq 2$. In the case of $k = 2$, a much simpler example language was given in [11]: $L = \{a^n b^n c, a^n b^{2n} d \mid n \geq 0\}$, which is in 3-LDA but not in 2-LDA.

Proposition 3. *For any $k \geq 2$, k-LDA $\subsetneq k$-LRA $\subsetneq k$-LBPA $\subseteq k$-LPA.*

Unfortunately, it is unknown whether k-LBPA $\neq (k+1)$-LBPA for each index $k \geq 2$. Proposition 3 will be shown by exploring basic closure properties of target language families. In Sect. 4.4, we will explore these properties in depth.

The language family 2-LRA turns out to be relatively large since it contains languages not recognized by any k-lda for every fixed index $k \geq 2$.

Theorem 4. *For any index $k \geq 2$, 2-LRA $\not\subseteq k$-LDA.*

Let ω-LDA stand for $\bigcup_{k \geq 1} k$-LDA. Notice that Theorem 4 is not strong enough to yield the separation of 2-LRA $\not\subseteq \omega$-LDA. We also do not know whether or not 3-LDA $\not\subseteq$ 2-LRA and 3-LRA $\not\subseteq$ 2-LBPA.

We seek a refinement of CFL using *unambiguous computation* (i.e., nondeterministic computation with at most one accepting path). Let us define UCFL, from CFL, by restricting 1npda's to have unambiguous computation (see [15]).

Theorem 5. *ω-LDA $\subseteq$ UCFL $\subsetneq$ CFL.*

To show Theorem 5, we need to (1) introduce a new model of k-*limited unambiguous automata* (or k-lua's, for short) and its corresponding language family k-LUA, (2) show that k-LUA = bs-k-LUA by a similar argument used for k-LDA, and (3) prove that k-LUA = $(k+1)$-LUA for each index $k \geq 1$ by employing a similar argument for k-LNA. Item (3) then yields a conclusion that ω-LUA $(= \bigcup_{k \geq 1} k$-LUA) equals 2-LUA. Since k-LDA $\subseteq k$-LUA, we immediately obtain ω-LDA $\subseteq$ 2-LUA = UCFL. This obviously implies Theorem 5. Due to page limit, we omit the details of the above proof.

Wang [13] showed that DCFL contains all languages recognized with bounded-error probability by 2pfa's having rational transition probabilities. Let k-LBPA$^{(rat)}$ denote the subclass of k-LBPA defined by k-lpa's using only rational transition probabilities. Let k-LRA$_{<1/2} = \bigcup_{\varepsilon \in [0,1/2)} k$-LRA$_\varepsilon$. Theorem 1(1) thus implies the following.

Corollary 6. *1-LBPA$^{(rat)} \subseteq$ DCFL $\subseteq$ 2-LRA$_{<1/2}$.*

2   Limited Automata

Let us formally introduce various computational models of limited automata, in which we can rewrite the content of each tape cell only during the first k scans or visits of the cell.

Let $\mathbb{N}$ be the set of all non-negative integers and set $\mathbb{N}^+ = \mathbb{N} - \{0\}$. We denote by $[m, n]_{\mathbb{Z}}$ the set $\{m, m + 1, m + 2, \ldots, n\}$ for any two integers m and n with $m \leq n$. In addition, we abbreviate as $[m]$ the integer interval $[1, m]_{\mathbb{Z}}$ for any integer $m \geq 1$.

2.1 Definitions of k-lpa's with the k-Limitedness Requirement

A k-limited probabilistic automaton (or a k-lpa, for short) M is formally defined as a tuple $(Q, \Sigma, \{\math00a2, \$\}, \{\Gamma_i\}_{i \in [k]}, \delta, q_0, Q_{acc}, Q_{rej})$, which accesses only tape area in between two endmarkers (those endmarkers can be accessible but not changeable), where Q is a finite set of (inner) states, Q_{acc} ($\subseteq Q$) is a set of accepting states, Q_{rej} ($\subseteq Q$) is a set of rejecting states, Σ is an input alphabet, $\{\Gamma_i\}_{i \in [k]}$ is a collection of mutually disjoint finite sets of tape symbols, q_0 is an initial state in Q, and δ is a probabilistic transition function from $(Q - Q_{halt}) \times \Gamma \times Q \times \Gamma \times D$ to the real unit interval $[0, 1]$ with $D = \{-1, +1\}$, $Q_{halt} = Q_{acc} \cup Q_{rej}$, and $\Gamma = \bigcup_{i=0}^{k} \Gamma_i$ for $\Gamma_0 = \Sigma$ and $\math00a2, \$ \in \Gamma_k$. We implicitly assume that $Q_{acc} \cap Q_{rej} = \emptyset$. The k-lpa has a rewritable tape, on which an input string is initially placed, surrounded by two endmarkers $\math00a2$ (left endmarker) and $\$$ (right endmarker). In our formulation of k-lpa, the tape head always moves either to the right or to the left without stopping still. Along each computation path, M probabilistically chooses one of all possible transitions given by δ.

Purely for clarity reason, we express $\delta(q, \sigma, p, \tau, d)$ as $\delta(q, \sigma \mid p, \tau, d)$. Each value $\delta(q, \sigma \mid p, \tau, d)$ indicates the probability that, when M scans σ on the tape in inner state q, M changes its inner state to p, overwrites τ onto σ, and moves its tape head in direction d. We set $\delta[q, \sigma] = \sum_{(p,\tau,d) \in Q \times \Gamma \times D} \delta(q, \sigma \mid p, \tau, d)$. The function δ must satisfy $\delta[q, \sigma] = 1$ for every pair $(q, \sigma) \in Q \times \Gamma$.

The k-lpa M must satisfy the following k-limitedness requirement: during the first k scans of each tape cell, at the ith scan with $0 \leq i < k$, if M reads the content of the cell containing a symbol in Γ_i, then M rewrites it to another symbol in Γ_{i+1}. After the the kth scan, the cell becomes unchangeable (or frozen); that is, M still reads a symbol in the cell but M no longer alters the symbol. For the above rule, there is one exception: whenever the tape head makes a turn (either from the left to the right or from the right to the left) at any tape cell, we count this move as "double scans" or "double visits." To make the endmarkers special, we assume that no symbol in $\bigcup_{i=}^{k-1} \Gamma_i$ can be replaced by any endmarker. This k-limitedness requirement is formally stated as follows: for any transition $\delta(q, \sigma \mid p, \tau, d) \neq 0$ with $p, q \in Q$, $\sigma \in \Gamma_i$, $\tau \in \Gamma_j$, and $d \in \{+1, -1\}$, (1) if $i = k$, then $\sigma = \tau$ and $j = i$, (2) if $i < k$ and i is even, then $j = i + 2^{(1-d)/2}$, and (3) if $i < k$ and i is odd, then $j = i + 2^{(1+d)/2}$.

The probability of each computation path is determined by the multiplication of all chosen transition probabilities along the path. The acceptance probability of M on input x is the sum of all probabilities of accepting computation paths of M starting with the input x. We express by $p_{M,acc}(x)$ the total acceptance probability of M on x. Similarly, we define $p_{M,rej}(x)$ to be the rejection probability of M on x. Given a k-lpa M, we say that M accepts x if $p_{M,acc}(x) > 1/2$ and rejects x if $p_{M,rej}(x) \geq 1/2$. The notation $L(M)$ stands for the set of all strings x accepted by M. Given a language L, we say that M *recognizes* L exactly when $L = L(M)$. We further say that M makes *bounded* error if there exists a constant $\varepsilon \in [0, 1/2)$ (called an *error bound*) such that, for every input x, either $p_{M,acc}(x) \geq 1 - \varepsilon$ or $p_{M,rej}(x) \geq 1 - \varepsilon$. With or without this condition, M is said to make *unbounded* error. For a language L, the *error probability of M on x for L* is the probability that M's outcome is different from L.

Generally, a k-lpa may produce an extremely long computation path or even an infinite computation path. Following an early discussion in Sect. 1.1 on the expected runtime of probabilistic machines, it is desirable to restrict our attention to k-lpa's whose *computation paths have a polynomial length on average*; that is, there is a polynomial p for which the expected length of all terminating computation paths on input x is bounded from above by $p(|x|)$. In what follows, we implicitly assume that all k-lpa's should satisfy this expected polynomial termination requirement. Given an input x, we say that M *accepts (resp., rejects)* x *with probability* p if the total probability of accepting (resp., rejecting) computation paths is exactly p.

Let us recall the language families introduced in Sect. 1.2, associated with limited automata. Among these language families, for each index $k \geq 2$, it follows from the above definitions and by [3] that k-LDA $\subseteq k$-LRA$_\varepsilon \subseteq 2$-LNA $=$ CFL and k-LBPA$_{\varepsilon'} \subseteq k$-LPA$_{\varepsilon'}$ for any constants $\varepsilon \in [0,1)$ and $\varepsilon' \in [0, 1/2)$. Moreover, by amplifying the success probability of k-lra's, we easily obtain the inclusion: k-LRA $\subseteq k$-LBPA for every index $k \geq 1$.

3   One-Way Pushdown Automata

We will formally describe various one-way pushdown automata.

3.1   One-Way Probabilistic Pushdown Automata

One-way deterministic and nondeterministic pushdown automata (abbreviated as 1dpda's and 1npda's, respectively) can be viewed as special cases of the following *one-way probabilistic pushdown automata* (or *1ppda's*, for short).

Formally, a 1ppda M is a tuple $(Q, \Sigma, \{\cent, \$\}, \Gamma, \Theta_\Gamma, \delta, q_0, Z_0, Q_{acc}, Q_{rej})$, in which Q is a finite set of (inner) states, Σ is an input alphabet, Γ is a stack alphabet, Θ_Γ is a finite subset of Γ^* with $\lambda \in \Theta_\Gamma$, δ is a *probabilistic transition function* (with $\check{\Sigma} = \Sigma \cup \{\lambda, \cent, \$\}$) from $(Q - Q_{halt}) \times \check{\Sigma} \times \Gamma \times Q \times \Theta_\Gamma$ to $[0,1]$, q_0 $(\in Q)$ is an initial state, Z_0 $(\in \Gamma)$ is a bottom marker, Q_{acc} $(\subseteq Q)$ is a set of accepting states, and Q_{rej} $(\subseteq Q)$ is a set of rejecting states, where λ is the empty string and $Q_{halt} = Q_{acc} \cup Q_{rej}$.

For clarity reason, we express $\delta(q, \sigma, a, p, u)$ as $\delta(q, \sigma, a \mid p, u)$. Let $\delta[q, \sigma, a] = \sum_{(p,u) \in Q \times \Theta_\Gamma} \delta(q, \sigma, a \mid p, u)$ with $\sigma \in \check{\Sigma}$. When $\sigma = \lambda$, we call its transition a λ-*move* (or a λ-*transition*) and the tape head must stay still. At any point, M can probabilistically select either a λ-move or a non-λ-move. This is formally stated as $\delta[q, \sigma, a] + \delta[q, \lambda, a] = 1$ for any given tuple $(q, \sigma, a) \in (Q - Q_{halt}) \times (\Sigma \cup \{\cent, \$\}) \times$

Γ. In a way similar to k-lpa's, we can define the notions of unbounded-error, bounded-error, acceptance/rejection probability, etc. We require every 1ppda to run *in expected polynomial time*. Two 1ppda's M_1 and M_2 are *(recognition) equivalent* if $L(M_1) = L(M_2)$. Let us recall the language families described in Sect. 1.2. It is well-known that DCFL $\subseteq$ BPCFL $\subseteq$ PCFL.

For two language families $\mathcal{F}$ and $\mathcal{G}$, the notation $\mathcal{F} \vee \mathcal{G}$ (resp., $\mathcal{F} \wedge \mathcal{G}$) denotes the *2-disjunctive closure* $\{A \cup B \mid A \in \mathcal{F}, B \in \mathcal{G}\}$ (resp., the *2-conjunctive closure* $\{A \cap B \mid A \in \mathcal{F}, B \in \mathcal{G}\}$). For any index $d \in \mathbb{N}^+$, define $\mathcal{F}(1) = \mathcal{F}$ and $\mathcal{F}(d+1) = \mathcal{F} \wedge \mathcal{F}(d)$. Notice that CFL($k$) $\neq$ CFL($k+1$) for any $k \in \mathbb{N}^+$ [8].

3.2 An Ideal Shape of 1ppda's

We want to show how to convert any 1ppda to a "pop-controlled form" (called an *ideal shape*), in which the pop operations always take place by first reading an input symbol σ and then making a series (one or more) of the pop operations without reading any further input symbol. In other words, a 1ppda *in an ideal shape* is restricted to take only the following transitions. Let $\Gamma^{(-)} = \Gamma - \{Z_0\}$. (1) Scanning $\sigma \in \Sigma$, preserve the topmost stack symbol (called a *stationary operation*). (2) Scanning $\sigma \in \Sigma$, push a new symbol u ($\in \Gamma^{(-)}$) without changing any other symbol in the stack. (3) Scanning $\sigma \in \Sigma$, pop the topmost stack symbol. (4) Without scanning an input symbol (i.e., λ-move), pop the topmost stack symbol. (5) The stack operations (4) comes only after either (3) or (4).

Lemma 7 states that any 1ppda can be converted into its "equivalent" 1ppda in an ideal shape. We say that two 1ppda's are *error-equivalent* if they are recognition equivalent and their error probabilities coincide on all inputs. The *push size* of a 1ppda is the maximum length of any string pushed into a stack by any single move.

Lemma 7 (Ideal Shape Lemma for 1ppda's). *Let $n \in \mathbb{N}^+$. Any n-state 1ppda M with stack alphabet size m and push size e can be converted into another error-equivalent 1ppda N in an ideal shape with $O(en^2m^2(2m)^{2enm})$ states and stack alphabet size $O(enm(2m)^{2enm})$. This is true for the model with no end-marker.*

The proof of this lemma is lengthy, consisting of a series of transformations of automata, and is proven by utilizing, in part, ideas of Hopcroft and Ullman [4, Chap. 10] and of Pighizzini and Pisoni [11, Sect. 5].

The ideal shape lemma is useful for simplifying certain proofs. In what follows, we give one such example. Given a language A, the notation A^R denotes the *reverse language* $\{x^R \mid x \in A\}$. For a family $\mathcal{F}$ of languages, $\mathcal{F}^R$ expresses the collection of A^R for any language A in $\mathcal{F}$.

Lemma 8. PCFL *is closed under reversal; that is,* PCFL^R = PCFL.

4   Behaviors of Limited Automata

In the subsequent subsections, we intend to verify our main results stated in Sect. 1.2 by making structural analyses on the behaviors of k-lpa's.

4.1   Blank Skipping Property, Theorem 1, and Proposition 2

We will show the proofs of Theorem 1 and Proposition 2. For the former proof, we want to restrict the behaviors of k-lpa's so that we can control their computation. Firstly, we give the formal description of the notion of blank skipping property. A k-lpa is *blank skipping* if (1) $\Gamma_k = \{\text{¢}, \$, B\}$, where B is a unique blank symbol, and (2) there are two disjoint subsets Q_{+1}, Q_{-1} of Q for which $\delta(q, B \mid q, B, d) = 1$ for any direction $d \in \{\pm 1\}$ and any state $q \in Q_d$. In other words, when a k-lpa passes a cell for the kth time, it must make the cell *blank* (i.e., the cell has B) and the cell becomes frozen afterward.

Let us begin with the proof of Theorem 1.

Proof Sketch of Theorem 1. (1) It is rather easy to simulate a 2pfa by a 1-lpa that behaves like the 2pfa but changes each input symbol σ to its corresponding symbol σ'. On the contrary, we want to simulate a 1-lpa M by the following 2pfa N. A key idea is that it is possible to maintain and update a list of state pairs, each (p, q) of which indicates that, if M's tape head enters the tape area left of the currently scanning cell from the right in state p, then M eventually leaves the area to the right in state q with positive probability.

(2) This directly comes from Lemmas 9 and 10.   □

In what follows, we describe Lemmas 9 and 10 and present their proofs.

Lemma 9. *Let $n \geq 2$ and $l \geq 1$. Every n-state blank-skipping 2-lpa working on an l-letter alphabet can be converted into a recognition-equivalent 1ppda with the same error probability and of states at most $2n$.*

Proof Sketch. Given a blank-skipping 2-lpa M, we simulate it as follows. On input x, when M reads a new input symbol σ by changing it to τ, we read σ and push τ into a stack. In contrast, when M moves its tape head leftwards by skipping B to the first non-blank symbol τ and changes it to B, we simply pop a topmost stack symbol. This simulation can be done by a certain 1ppda.   □

The ideal shape form of 1ppda's is a key to the next lemma.

Lemma 10. *Let $n, l \in \mathbb{N}^+$. Let L be a language over an alphabet Σ of size l recognized by an n-state 1ppda M in an ideal shape with error probability at most ε. There is a blank-skipping 2-lpa N that has $O(nl)$ states and recognizes L with the same error probability.*

Proof Sketch. Let M be a 1ppda and assume that M is in an ideal shape. We simulate M by an appropriate 2-lpa in the following way. Let x be any input string. Assume that M reads a new input symbol σ. If M pushes τ into a stack, then we read σ and change it into τ. If M pops a topmost stack symbol, then we move a tape head leftwards to read the first non-blank symbol τ and then replace it with B. On the contrary, assume that M makes a λ-move. Since M's move must be a pop operation, we move the tape head leftwards and replace the first non-blank symbol by B.   □

It is possible to convert any k-lda into its equivalent blank-skipping k-lda. The following is a key lemma, from which Proposition 2 follows immediately. Our proof partly takes an idea from [11].

Lemma 11. *Let k be any integer with $k \geq 2$. Given any k-lda M, there exists another k-lda N such that (1) N is blank-skipping and (2) N agrees with M on all inputs.*

4.2   Properties of ω-LPFA

As done in [14–16], we equip each 1nfa with a *write-only output tape*.[2] Let 1NFAMV denote the class of all multi-valued partial functions from Σ_1^* to Σ_2^* whose output values are produced on write-only tapes along only accepting computation paths of 1nfa's, where Σ_1 and Σ_2 are arbitrary alphabets. We further write 1NFAMV_t for the collection of all *total* functions in 1NFAMV. Let $k \geq 2$. For any $f : \Sigma_1^* \to \Sigma_2^*$ in 1NFAMV_t witnessed by a 1nfa, say, M_f with an output tape and for any k-lpa M over Σ_2, let $L_{f,M} = \{x \in \Sigma_1^* \mid \sum_{y \in \Sigma_2^*} |AP_f(x|y)| \text{Prob}_M[M(y) = 1]/|AP_f(x)| > 1/2\}$, where $AP_f(x|y)$ is the set of all accepting computation paths of M_f producing y on input x and $AP_f(x) = \bigcup_{y \in \Sigma_2^*} AP_f(x|y)$. By abusing the notation, we write $k\text{-LPA} \circ 1\text{NFAMV}_t$ to denote the set of all such $L_{f,M}$'s.

We argue that k-LPA is "invariant" with an application of 1NFAMV_t-functions in the following sense.

Lemma 12. *For any index $k \geq 2$, $k\text{-LPA} \circ 1\text{NFAMV}_t = k\text{-LPA}$.*

Proof Sketch. Let $k \geq 2$. Since it is obvious that $k\text{-LPA} \subseteq k\text{-LPA} \circ 1\text{NFAMV}_t$, we want to show the opposite inclusion. Take a function $f \in 1\text{NFAMV}_t$ and a k-lpa M, and consider $L_{f,M}$. There is a 1nfa M_f computing f. Consider the following machine N. On input x, run M_f and, whenever M_f produces one output symbol σ, run M to process σ. Along each computation path of M_f, if M_f enters an accepting state, then N does the same, otherwise, N enters both accepting and rejecting states with equiprobability. Clearly, N is also a k-lpa and it recognizes L with unbounded-error probability.   □

Consider any k-lpa M used in the definition of $k\text{-LPA} \circ 1\text{NFAMV}_t$. If we feed such an M with the reverses x^R of inputs x, then we obtain $k\text{-LPA}^R \circ 1\text{NFAMV}_t$. We show the following relationship between $(k+1)$-LPA and $k\text{-LPA}^R$.

Lemma 13. *For any $k \geq 2$, $k\text{-LPA}^R \circ 1\text{NFAMV}_t \subseteq (k+1)\text{-LPA}$.*

Proof Sketch. Fix $k \geq 2$.

We show the inclusion of $k\text{-LPA}^R \circ 1\text{NFAMV}_t \subseteq (k+1)\text{-LPA}$. Let M be a k-lpa and let f be a function in 1NFAMV_t witnessed by a certain 1nfa, say, M_f. Define $L_{f,M^R} = \{x \in \Sigma^* \mid \sum_y |AP_f(x|y)| \text{Prob}_M[M(y^R) = 1]/|AP_f(x)| > 1/2\}$.

---

[2] An output tape is *write only* if its cells are initially blank and its tape head moves to the right whenever it writes a non-blank symbol.

Our goal is to verify that $L_{f,M^R} \in (k+1)$-LPA. Consider the following machine N. On input x, run M_f on x, change x into y in $f(x)$ along all computation paths in $AC_f(x|y)$, and run M on $\textcent y\$$ starting at $\$$ and ending at $\textcent$. Since M is k-limited, N must be $(k+1)$-limited. We thus conclude that $L_{f,M^R} \in (k+1)$-LPA. $\square$

4.3   Power of Probabilistic Computation and Theorem 4

We will give the proof of Theorem 4. The proof requires, for each index $k \geq 2$, a certain language, which is in $(k+1)$-LDA but outside of k-LDA. The example languages shown below are slight modifications of the ones given by Hibbard [3].

(1) When $k = 3$, the language L_3 is composed of all strings of the forms $a^{n_2} b^{n_2} c^{p_2} \# a$ and $a^{n_2} b^{m_2} c^{m_2} \# b$ for all integers $n_2, m_2, p_2 \geq 0$.

(2) Let $k \geq 4$ be any index and assume that L_{k-1} is already defined. For each index $i \in [2, k-1]_{\mathbb{Z}}$, we succinctly write w_i in place of $a^{n_i} b^{m_i} c^{p_i}$ for certain numbers $n_i, m_i, p_i \in \mathbb{N}$. The desired language L_k is composed of all strings w of the form $w_2 \# w_4 \# \cdots \# w_{k-1} \# \cdots \# w_5 \# w_3 \# x$ with $x \in \{a, b\}$ satisfying Conditions (i)–(iv) given below. For each index $i \in [3, k-2]_{\mathbb{Z}}$, we define $\tilde{w}_i^{(k)} = w_{i-1}$ if i is even, and w_{i+1} otherwise. Moreover, let $\tilde{w}_{k-1}^{(k)} = w_{k-1}$ if k is even, and w_{k-2} otherwise. Finally, let $\tilde{w}^{(-)}$ express the string $\tilde{w}_4^{(k)} \# \tilde{w}_6^{(k)} \# \cdots \# \tilde{w}_{k-1}^{(k)} \# \cdots \# \tilde{w}_5^{(k)} \# \tilde{w}_3^{(k)}$. (i) If $x = a$ and $n_2 = m_2$, then $\tilde{w}^{(-)} \# a \in L_{k-1}$. (ii) If $x = a$ and $n_2 < m_2$, then $\tilde{w}^{(-)} \# b \in L_{k-1}$. (iii) If $x = b$ and $m_2 = p_2$, then $\tilde{w}^{(-)} \# a \in L_{k-1}$. (iv) If $x = b$ and $m_2 < p_2$, then $\tilde{w}^{(-)} \# b \in L_{k-1}$.

An argument similar to [3, Sect. 4] verifies that, for each index $k \geq 2$, the language L_{k+1} is included in $(k+1)$-LDA but excluded from k-LDA.

Fix $k \geq 2$. For each symbol $x \in \{a, b\}$, let $L_x = \{w\#x \in L_{k+1} \mid w = w_2 \# w_4 \# \cdots \# w_k \# \cdots \# w_5 \# w_3\}$. Note that $L_{k+1} = L_a \cup L_b$. Since $L_a, L_b \in k$-LDA, it follows that $L_{k+1} \in k$-LDA $\vee$ k-LDA. Therefore, we obtain the following corollary.

Corollary 14. *For every $k \geq 2$, k-LDA $\vee$ k-LDA $\nsubseteq k$-LDA.*

Lemma 15. *For any $k \geq 3$, $L_k \in 2$-LRA$_{(1-2^{-2k+5})}$.*

Proof. We first show that $L_3 \in 2$-LRA$_{1/2}$. Let $L'_a = \{a^n b^n c^p \# a \mid n, p \geq 0\}$ and $L'_b = \{a^n b^m c^m \# b \mid n, m \geq 0\}$. Clearly, $L_3 = L'_a \cup L'_b$ holds. Since $L'_a, L'_b \in$ DCFL, for each symbol $x \in \{a, b\}$, we take a 2-lda M_x that recognizes L'_x. Consider the following 2-lra N. Let w be any input. Initially, choose an index $x \in \{a, b\}$ with equiprobability and then run M_x. If $w \in L_3$, then N accepts w with probability $1/2$; otherwise, N rejects w with probability 1.

By induction hypothesis, we assume that $L_k \in 2$-LRA$_{(1-2^{-2k+5})}$. Let us consider L_{k+1}. Using the aforementioned notation, define $L_{a=} = \{w\#a \mid n_2 = m_2, \tilde{w}^{(-)} \# a \in L_k\}$ and $L_{a<} = \{w\#a \mid n_2 < m_2, \tilde{w}^{(-)} \# a \in L_k\}$, where w is of the form $w_2 \# w_4 \# \cdots \# w_k \# \cdots \# w_5 \# w_3 \# x$ for a certain symbol $x \in \{a, b\}$. Similarly, we define $L_{b=}$ and $L_{b<}$. Note that $L_{k+1} = L_{a=} \cup L_{a<} \cup L_{b=} \cup L_{b<}$. It is not difficult to show that $L_{a=}, L_{a<}, L_{b=}, L_{b<}$ all belong to 2-LRA$_{(1-2^{-2k+5})}$ since

$L_k \in 2\text{-LRA}_{(1-2^{-2k+5})}$. Consider the following machine N. On the input of the form $w\#x$, choose one of the pairs $\{a =, a <, b =, b <\}$ with equal probability. Suppose that we have chosen $a =$. As an example, let $M_{a=}$ be a 2-lra recognizing the language $L_{a=}$. In this case, run $M_{a=}$ on $w\#x$. When $w\#a \in L_{k+1}$, N accepts the input with probability $\frac{1}{4} \times 2^{-2k+5}$, which equals $2^{-2(k+1)+5}$. The other cases are similarly treated. □

Lemma 15 implies that $L_{k+1} \in 2\text{-LRA}$. Since $L_{k+1} \notin k\text{-LDA}$, we instantly conclude that $2\text{-LRA} \nsubseteq k\text{-LDA}$. This completes the proof of Theorem 4.

4.4 Closure Properties of Probabilistic Classes and Proposition 3

We will explore basic closure properties of k-LRA, k-LBPA, and k-LDA. By utilizing some of those properties, we will prove Proposition 3 in the end.

Lemma 16. *For any $k \geq 2$, k-LRA is closed under finite union but not under finite intersection.*

Lemma 17. *For any $k \geq 2$, k-LBPA is closed under complementation but k-LRA is not.*

Proof. It is not difficult to show that $k\text{-LBPA} = \text{co-}k\text{-LBPA}$ for all indices $k \geq 2$. By Lemma 16, k-LRA is closed under finite union. If $k\text{-LRA} = \text{co-}k\text{-LRA}$, then k-LRA must be closed under finite intersection. This contradicts the second part of Lemma 16. □

Recall that $2\text{-LDA} = \text{DCFL}$ [11]. Although $k\text{-LDA} \neq \text{DCFL}$ for all $k \geq 3$, k-LDA still satisfies many non-closure properties as DCFL does.

Lemma 18. *For any $k \geq 2$, k-LDA is not closed under reversal, concatenation, λ-free homomorphism, or Kleene star.*

Next, we look at the closure properties of k-LBPA.

Lemma 19. *For each operator $\diamond \in \{\wedge, \vee\}$, $k\text{-LDA} \diamond k\text{-LDA} \subseteq k\text{-LBPA}$; thus, $k\text{-LDA}(2) \subseteq k\text{-LBPA}$.*

Proof. It suffices to consider the case of $\diamond = \vee$ because k-LBPA is closed under complementation. Let M_1, M_2 be two k-lpa's working over the same alphabet Σ. We design a new k-lpa N to work as follows. On input x, choose an index $i \in \{1, 2\}$ uniformly at random, run M_i on x. If M_i enters an accepting state, accept x with probability 1; otherwise, accept x with probability $1/3$ and reject with probability $2/3$. If $x \in L(M_1) \cup L(M_2)$, then the acceptance probability of N is at least $2/3$. In contrast, if $x \notin L(M_1) \cup L(M_2)$, then the rejection probability is at least $2/3$. Therefore, $L(M_1) \cup L(M_2)$ is in k-LBPA. □

It is, however, unknown that $k\text{-LDA}(d) \subseteq k\text{-LBPA}$ for every index $d \geq 3$.

Proof of Proposition 3. All inclusions obviously hold. We want to show the remaining two separations. Note that $k\text{-LDA} = \text{co-}k\text{-LDA}$ for any $k \geq 1$. By Lemma 16, it follows that $k\text{-LDA} \neq k\text{-LRA}$. Similarly, from $k\text{-LBPA} = \text{co-}k\text{-LBPA}$, we obtain $k\text{-LRA} \neq k\text{-LBPA}$. □

References

1. Dwork, C., Stockmeyer, L.: Finite verifiers I: the power of interaction. J. ACM **39**, 800–828 (1992)
2. Freivalds, R.: Probabilistic two-way machines. In: Gruska, J., Chytil, M. (eds.) MFCS 1981. LNCS, vol. 118, pp. 33–45. Springer, Heidelberg (1981). https://doi.org/10.1007/3-540-10856-4_72
3. Hibbard, T.N.: A generalization of context-free determinism. Inf. Control **11**, 196–238 (1967)
4. Hopcroft, J.E., Ullman, J.D.: Introduction to Automata Theory, Languages, and Computation. Addison-Wesley, Boston (1979)
5. Hromkovič, J., Schnitger, G.: On probabilistic pushdown automata. Inf. Comput. **208**, 982–995 (2010)
6. Kaņeps, J., Freivalds, R.: Minimal nontrivial space complexity of probabilistic one-way turing machines. In: Rovan, B. (ed.) MFCS 1990. LNCS, vol. 452, pp. 355–361. Springer, Heidelberg (1990). https://doi.org/10.1007/BFb0029629
7. Kaņeps, J., Geidmanis, D., Freivalds, R.: Tally languages accepted by Monte Carlo pushdown automata. In: Rolim, J. (ed.) RANDOM 1997. LNCS, vol. 1269, pp. 187–195. Springer, Heidelberg (1997). https://doi.org/10.1007/3-540-63248-4_16
8. Liu, L.Y., Weiner, P.: An infinite hierarchy of intersections of context-free languages. Math. Syst. Theory **7**, 185–192 (1973)
9. Macarie, I.I., Ogihara, M.: Properties of probabilistic pushdown automata. Theor. Comput. Sci. **207**, 117–130 (1998)
10. Pighizzini, G., Pisoni, A.: Limited automata and regular languages. Int. J. Found. Comput. Sci. **25**, 897–916 (2014)
11. Pighizzini, G., Pisoni, A.: Limited automata and context-free languages. Fund. Inform. **136**, 157–176 (2015)
12. Wagner, K., Wechsung, G.: Computational Complexity. D. Reidel Publishing, Dordrecht (1986)
13. Wang, J.: A note on two-way probabilistic automata. Inf. Process. Lett. **43**, 321–326 (1992)
14. Yamakami, T.: Oracle pushdown automata, nondeterministic reducibilities, and the hierarchy over the family of context-free languages. In: Geffert, V., Preneel, B., Rovan, B., Štuller, J., Tjoa, A.M. (eds.) SOFSEM 2014. LNCS, vol. 8327, pp. 514–525. Springer, Cham (2014). https://doi.org/10.1007/978-3-319-04298-5_45
15. Yamakami, T.: Structural complexity of multi-valued partial functions computed by nondeterministic pushdown automata. In: Proceedings of ICTCS 2014, CEUR Workshop Proceedings, vol. 1231, pp. 225–236 (2014)
16. Yamakami, T.: Not all multi-valued partial CFL functions are refined by single-valued functions (extended abstract). In: Diaz, J., Lanese, I., Sangiorgi, D. (eds.) TCS 2014. LNCS, vol. 8705, pp. 136–150. Springer, Heidelberg (2014). https://doi.org/10.1007/978-3-662-44602-7_12
17. Yamakami, T.: One-way bounded-error probabilistic pushdown automata and kolmogorov complexity (preliminary report). In: Charlier, É., Leroy, J., Rigo, M. (eds.) DLT 2017. LNCS, vol. 10396, pp. 353–364. Springer, Cham (2017). https://doi.org/10.1007/978-3-319-62809-7_27

Locality Sensitive Hashing Schemes, Similarities, and Distortion (Invited Talk)

Flavio Chierichetti[(✉)]

Dipartimento di Informatica, Sapienza University of Rome, Rome, Italy
flavio@di.uniroma1.it

Abstract. Locality sensitive hashing (LSH) is a key algorithmic tool that lies at the heart of many information retrieval and machine learning systems [1, 2, 8]. LSH schemes are used to sketch large objects (e.g., Web pages, fields of flowers, or – more generally – sets and vectors) into fingerprints of few bits each; the fingerprints are then used to quickly, and approximately, reconstruct some similarity relation between the objects.

A LSH scheme for a similarity (or, analogously, for a distance) can significantly improve the computational cost of many algorithmic primitives (e.g., nearest neighbor search, and clustering). For this reason, in the last two decades, researchers have tried to understand which similarities admit efficient LSH schemes: such schemes were obtained for many similarities [1–3, 7–9], while the non-existence of LSH schemes was proved for a number of other similarities [3].

In our talk, we will introduce the class of LSH-preserving transformations [4] (functions that, when applied to a similarity that admits a LSH scheme, return a similarity that also admits such a scheme). We will give a characterization of this class of functions: they are precisely the probability generating functions, up to scaling. We will then show how this characterization can be used to construct LSH schemes for a number of well-known similarities.

We will then discuss a notion of similarity distortion [6], in order to deal with similarities which are known to not admit LSH schemes — this notion aims to determine the minimum distortions that these similarities have to be subject of, before starting to admit a LSH scheme. We will introduce a number of general theoretical tools that can be used to determine the optimal distortions of some important classes of similarities.

Finally, we will consider the computational problem of checking whether a similarity admits a LSH scheme [5], showing that, unfortunately, this problem is computationally hard in a very strong sense.

Supported in part by the ERC Starting Grant DMAP 680153 and by a Google Focused Research Award.

B. Catania et al. (Eds.): SOFSEM 2019, LNCS 11376, pp. 531–532, 2019.
https://doi.org/10.1007/978-3-030-10801-4

References

1. Andoni, A., Indyk, P.: Near-optimal hashing algorithms for approximate nearest neighbor in high dimensions. In: FOCS, pp. 459–468 (2006)
2. Broder, A.Z.: On the resemblance and containment of documents. In: Proceedings of the SEQUENCES, pp. 21–29 (1997)
3. Charikar, M.: Similarity estimation techniques from rounding algorithms. In: Proceedings of the STOC, pp. 380–388 (2002)
4. Chierichetti, F., Kumar, R. LSH-preserving functions and their applications. J. ACM **62**(5), 33:1–33:25 (2015)
5. Chierichetti, F., Kumar, R., Mahdian, M.: The complexity of LSH feasibility. Theor. Comput. Sci. **530**, 89–101 (2014)
6. Chierichetti, F., Kumar, R., Panconesi, A., Terolli, E.: The distortion of locality sensitive hashing. In: ITCS (2017)
7. Christiani, T., Pagh, R.: Set similarity search beyond minhash. In: Proceedings of the 49th Annual ACM SIGACT Symposium on Theory of Computing, STOC 2017, pp. 1094–1107, New York, NY, USA. ACM (2017)
8. Gionis, A., Indyk, P., Motwani, R., et al.: Similarity search in high dimensions via hashing. In: VLDB, pp. 518–529 (1999)
9. Indyk, P., Motwani, R.: Approximate nearest neighbors: Towards removing the curse of dimensionality. In: STOC, pp. 604–613 (1998)

Author Index

Printed in the United States
By Bookmasters